Solar and Stellar Granulation

NATO ASI Series

Advanced Science Institutes Series

A Series presenting the results of activities sponsored by the NATO Science Committee, which aims at the dissemination of advanced scientific and technological knowledge, with a view to strengthening links between scientific communities.

The Series is published by an international board of publishers in conjunction with the NATO Scientific Affairs Division

A Life Sciences	Plenum Publishing Corporation
B Physics	London and New York
C Mathematical	Kluwer Academic Publishers
and Physical Sciences	Dordrecht, Boston and London
D Behavioural and Social Sciences	
E Applied Sciences	
F Computer and Systems Sciences	Springer-Verlag
G Ecological Sciences	Berlin, Heidelberg, New York, London,
H Cell Biology	Paris and Tokyo

Series C: Mathematical and Physical Sciences - Vol. 263

Solar and Stellar Granulation

edited by

Robert J. Rutten

Sterrekundig Instituut Utrecht, The Netherlands

and

Giuseppe Severino

Osservatorio Astronomico di Capodimonte,
Naples, Italy

Kluwer Academic Publishers

Dordrecht / Boston / London

Published in cooperation with NATO Scientific Affairs Division

Proceedings of the Third International Workshop of the OAC and
the NATO Advanced Research Workshop on
Solar and Stellar Granulation
Capri, Italy
21–25 June 1988

Library of Congress Cataloging in Publication Data

```
Osservatorio astronomico di Capodimonte.  International Workshop (3rd
  : 1988 : Capri, Italy)
   Solar and stellar granulation : Third International Workshop of
the OAC and NATO advanced research workshop, Capri, Italy, 21-25
June 1988 / edited by Robert J. Rutten and Giuseppe Severino.
     p.   cm. -- (NATO ASI series. Series C, Mathematical and
physical sciences ; vol. 263)
   "Published in cooperation with NATO Scientific Affairs Division."
   Includes bibliographies and index.
```

ISBN-13: 978-94-010-6899-4   e-ISBN-13: 978-94-009-0911-3

DOI: 10.1007/978-94-009-0911-3

```
   1. Solar granulation--Congresses.  2. Stellar granulation-
-Congresses.   I. Rutten, Robert J.  II. Severino, Giuseppe.
III. North Atlantic Treaty Organization.  IV. North Atlantic Treaty
Organization.  Scientific Affairs Division.  V. Title.  VI. Series:
NATO ASI series.  Series C, Mathematical and physical sciences ; no.
263.
QB539.G7O87  1988
523.7'4--dc19                                          88-35031
```

Published by Kluwer Academic Publishers,
P.O. Box 17, 3300 AA Dordrecht, The Netherlands.

Kluwer Academic Publishers incorporates the publishing programmes of
D. Reidel, Martinus Nijhoff, Dr W. Junk and MTP Press.

Sold and distributed in the U.S.A. and Canada
by Kluwer Academic Publishers,
101 Philip Drive, Norwell, MA 02061, U.S.A.

In all other countries, sold and distributed
by Kluwer Academic Publishers Group,
P.O. Box 322, 3300 AH Dordrecht, The Netherlands.

printed on acid free paper

TABLE OF CONTENTS

PART 1: OBSERVATIONAL TECHNIQUES

PART 2: OBSERVATIONS

PART 3: MODELLING

x

Participant List

Solar and Stellar Granulation

Capri, 21–25 June 1988

Dr. Claude Aime
Département d'Astrophysique
Université de Nice
Parc Valrose
F-06034 Nice Cedex, France

Dr. Jacques M. Beckers
European Southern Observatory
Karl Schwarzschildstrasse 2
D-8046 Garching bei München, FRG

Dr. Luca Bertello
Osservatorio Astrofisico di Arcetri
Largo E. Fermi 5
I-50125 Firenze, Italy

Dr. Thomas J. Bogdan
High Altitude Observatory/NCAR
P.O. Box 3000
Boulder, Colorado 80307-3000, USA

Dr. Jose A. Bonet
Instituto de Astrofisica de Canarias
E-38200 La Laguna, Tenerife, España

Mr. Jo H.M.J. Bruls
Sterrekundig Instituut
Postbus 80000
NL-3508 TA Utrecht, The Netherlands

Dr. David H. Bruning
Department of Physics
University of Louisville
Louisville, Kentucky 40292, USA

Dr. Fausto Cattaneo
Joint Institute for Laboratory Astrophysics
University of Colorado
Boulder, Colorado 80309, USA

Dr. Fabio Cavallini
Osservatorio Astrofisico di Arcetri
Largo E. Fermi 5
I-50125 Firenze, Italy

Dr. Guido Ceppatelli
Osservatorio Astrofisico di Arcetri
Largo E. Fermi 5
I-50125 Firenze, Italy

Mr. Manfred Cuntz
Institut für Theoretische Astrophysik
Im Neuenheimer Feld 561
D-6900 Heidelberg 1, FRG

Dr. Luc Damé
Laboratoire de Physique,
Stellaire et Planétaire
BP 10, Verrières-le-Buisson
F-91371 Cedex, France

Mr. Tron-André Darvann
Institute of Theoretical Astrophysics
University of Oslo
P.O. Box 1029, Blindern
N-0315 Oslo 3, Norway

Dr. Maria Teresa Gomes
Osservatorio Astronomico di Capodimonte
Via Moiariello 16
I-80131 Napoli, Italy

Prof. Frans-Ludwig Deubner
Institut für Astronomie und Astrophysik
Universität Würzburg
D-8700 Würzburg - Am Hubland, FRG

Prof. David F. Gray
Astronomy Department,
University of Western Ontario
London, Ontario, N6A 3K7 Canada

Dr. Dimitris Dialetis
National Observatory of Athens
P.O. Box 20048
GR-11810 Athens, Greece

Dr. Deborah A. Haber
National Solar Observatory
P.O. Box 62
Sunspot, New Mexico 88349-0062, USA

Prof. Dainis Dravins
Lund Observatory
Box 43
S-22100 Lund, Sweden

Mr. Göran Hammarbäck
Astronomiska Observatoriet
Box 515
S-751 20 Uppsala, Sweden

Mr. Bernhard Fleck
Institut für Astronomie und Astrophysik
Universität Würzburg
D-8700 Würzburg - Am Hubland, FRG

Dr. Jan Högbom
Stockholms Observatoriet
S-133 00 Saltsjöbaden, Sweden

Dr. Peter A. Fox
Center for Solar and Space Research
Yale University
P.O. Box 6666
New Haven, Connecticut 06511, USA

Prof. Hartmut Holweger
Institut für Theoretische Physik
und Sternwarte der Universität
Olshausenstrasse 40
D-2300 Kiel 1, FRG

Dr. Alexei S. Gadun
The Main Astronomical Observatory
Academy of Sciences of the Ukr. SSR
252127 Kiev Goloseevo, USSR

Dr. Neal Hurlburt
ACS University of Colorado
Boulder, Colorado 80309-0455, USA

Mr. Detlef Gigas
Institut für Theoretische Physik
und Sternwarte der Universität
Olshausenstrasse 40
D-2300 Kiel 1, FRG

Prof. Wolfgang Kalkofen
Center for Astrophysics
60 Garden Street
Cambridge, MA 02138, USA

Dr. Stephen L. Keil
Air Force Geophysics Laboratory
Sacramento Peak Observatory
P.O. Box 62
Sunspot, New Mexico 88349–0062, USA

Mr. Dan Kiselman
Astronomiska Observatoriet
Box 515
S-751 20 Uppsala, Sweden

Prof. Franz Kneer
Universitäts-Sternwarte
Geismarlandstrasse 11
D-3400 Göttingen, FRG

Dr. Serge Koutchmy
Air Force Geophysics Laboratory
National Solar Observatory
P.O. Box 62
Sunspot, New Mexico 88349-0062, USA

Dr. Yu-Qing Lou
High Altitude Observatory/NCAR
P.O. Box 3000
Boulder, Colorado 80307-3000, USA

Dr. Günter Lustig
Institut für Astronomie
Universitätsplatz 5
A-8010 Graz, Austria

Prof. Constantin I. Macris
National Observatory of Athens
P.O. Box 20048,
GR-11810 Athens, Greece

Dr. Jean-Pierre Maillard
Institut d'Astrophysique
98bis Boulevard Arago
F-75014 Paris, France

Dr. Ciro Marmolino
Dipartimento di Fisica Teorica
Università di Napoli
Mostra d'Oltremare Pad. 19
I-80100 Napoli, Italy

Dr. Ines Marquez
Instituto de Astrofisica de Canarias
E-38200 La Laguna, Tenerife, España

Mrs. Milena Martić
Institut d'Astrophysique
98bis, Boulevard Arago
F-75014 Paris, France

Prof. Wolfgang Mattig
Kiepenheuer-Institut für Sonnenphysik
Schöneckstrasse 6,
D-7800 Freiburg, FRG

Dr. Richard Müller
Observatoire du Pic du Midi
F-65200 Bagnères de Bigorre, France

Prof. Edith A. Müller
Rennweg 15,
CH-4052 Basel, Switzerland

Dr. Daniel Nadeau
Département de physique,
Université de Montréal,
C.P. 6128, Succ. "A." Montréal,
Québec, Canada H3C 3J7

Dr. Anastasios Nesis
Kiepenheuer-Institut für Sonnenphysik
Schöneckstrasse 6,
D-7800 Freiburg, FRG

xiv

Dr. Åke Nordlund
Astronomiske Observatorium
Øster Volgade 3
DK-11350 København K, Danmark

Dr. Göran Scharmer
Stockholms Observatorium
S-133 00 Saltsjöbaden, Sweden

Mrs. Michal L. Peri
California Institute of Technology
c/o Solar Astronomy 264-33
Pasadena CA 91125, USA

Dr. Hans-Jozef Schober
Institut für Astronomie
Universitätsplatz 5
A-8010 Graz, Austria

Dr. Theodore Prokakis
National Observatory of Athens
P.O. Box 20048
GR-11810 Athens, Greece

Prof. Egon H. Schröter
Kiepenheuer-Institut für Sonnenphysik
Schöneckstrasse 6
D-7800 Freiburg, FRG

Prof. Alberto Righini
Istituto di Astronomia
Università di Firenze
Largo E. Fermi 5,
I-50125 Firenze, Italy

Dr. Manfred Schüssler
Kiepenheuer-Institut für Sonnenphysik
Schöneckstrasse 6
D-7800, Freiburg, FRG

Prof. Mario Rigutti
Osservatorio Astronomico di Capodimonte
Via Moiariello 16
I-80131 Napoli, Italy

Dr. Giuseppe Severino
Osservatorio Astronomico di Capodimonte
Via Moiariello 16
I-80131 Napoli, Italy

Prof. Jean Rösch
Observatoire du Pic du Midi
F-65200 Bagnères de Bigorre, France

Dr. George W. Simon
Air Force Geophysics Laboratory
National Solar Observatory
P.O. Box 62
Sunspot, New Mexico 88349-0062, USA

Dr. Robert J. Rutten
Sterrekundig Instituut
Postbus 80000
NL-3508 TA Utrecht, The Netherlands

Dr. Luigi A. Smaldone
Dipartimento di Scienze Fisiche
Università di Napoli
Mostra d'Oltremare Pad. 19
I-80125 Napoli, Italy

Dr. Steven Saar
Center for Astrophysics
60 Garden Street
Cambridge MA 02138, USA

Mr. Dirk Soltau
Kiepenheuer-Institut für Sonnenphysik
Schöneckstrasse 6
D-7800 Freiburg, FRG

Dr. Matthias Steffen
Institut für Theoretische Physik
und Sternwarte der Universität
Olshausenstrasse 40
D-2300 Kiel 1, FRG

Prof. Robert F. Stein
Physics and Astronomy Department
Michigan State University
East Lansing, MI 48824, USA

Dr. Alan M. Title
S2 – 10 Solar Physics
Lockheed Research Laboratories
3251 Hanover Street
Palo Alto, CA 94304, USA

Dr. Javier Trujillo-Bueno
Universitäts-Sternwarte Göttingen
Geismarlandstrasse 11
D-3400 Göttingen, FRG

Prof. Peter Ulmschneider
Institut für Theoretische Astrophysik
Im Neuenheimer Feld 561
D-6900 Heidelberg 1, FRG

Dr. Aad A. van Ballegooijen
Center for Astrophysics
Harvard-Smithsonian
60 Garden Street
Cambridge, MA 02138, USA

Dr. Istvan Vince
Astronomical Observatory
Volgina 7
YU-11050 Belgrade, Yugoslavia

Dr. Alberto A. Vittone
Osservatorio Astronomico di Capodimonte
Via Moiariello 16
I-80131 Napoli, Italy

Prof. Nigel O. Weiss
D.A.M.T.P.
University of Cambridge
Silver Street
Cambridge CB3 9EW, United Kingdom

Dr. Hubertus Wöhl
Kiepenheuer-Institut für Sonnenphysik
Schöneckstrasse 6
D-7800 Freiburg, FRG

Dr. Yusef D. Zhugzhda
Institute of Terrestrial Magnetism,
Ionosphere and Wave Propagation
SU-142092 Troitsk, USSR

Acknowledgments

This workshop has been the third one in a series organised by the Osservatorio Astronomico di Capodimonte in Naples, Italy. It was co-sponsored by the NATO and the OAC. On behalf of all participants, we thank the NATO Science Committee and the OAC Directorate gratefully for their generous sponsorship.

The *Scientific Organising Committee* consisted of D.F. Gray, H. Holweger, S.L. Keil, R.I. Kostik, R. Müller, Å. Nordlund, M. Rigutti, R.J. Rutten (Chair), G.B. Scharmer, G. Severino and A.M. Title.

The *Local Organising Committee* consisted of Mrs. M.T. Gomez, C. Marmolino and G. Severino (Chair).

We are very grateful to all of the above individuals for their help. Many of them have earned special thanks by acting as Session Chairman (respectively, in session order: M. Rigutti, H. Holweger, S.L. Keil, G.B. Scharmer, Å. Nordlund, A.M. Title and D.F. Gray) and as Discussion Recorder (in session order: S.L. Keil, S.L. Keil again, D.F. Gray, H. Holweger, A.M. Title, G.B. Scharmer, and S.L. Keil yet another time).

In addition, we thank the support staffs of the OAC and the Sterrekundig Instituut Utrecht for their ever-willing helpfulness, in particular the OAC Workshop Team (Mrs. G. Iaccarino, Mrs. T. Ievolella, R. Trentarosa and G. Cuccaro) and the OAC LaTeXnicians (Mrs. E. Acampa and Mrs. A. d'Orsi).

We are also indebted to Marjolein Dietzel for much editorial assistance rendered at the Sterrekundig Instituut Utrecht, to H.M.G. Burm, G.T. Geertsema, H. Uitenbroek and E.B.J. van der Zalm for solving TeXnical problems, and to Karen L. Harvey of the Solar Physics Research Corporation for compiling a granulation bibliography on very short notice.

Most of all, we are grateful to the Workshop participants for making this a lively and interesting workshop by actively attending rather than sampling the delights of Capri outside, and to the Workshop contributors for giving excellent presentations and for supplying carefully prepared manuscripts.

Finally, we wish to thank those authorities, unknown to us, who have enabled us to communicate efficiently between Naples and Utrecht, and with most of the participants as well, via CERN, Stanford and other EARN/Bitnet nodes.

The workshop organisers,

Robert J. Rutten

Giuseppe Severino

Editorial Note

The Workshop was divided into three parts:
- Part 1: Observational Techniques;
- Part 2: Observations;
- Part 3: Modelling.

These proceedings follow this order. Each part contains the oral presentations with the subsequent discussions in the order in which they were given, followed by the corresponding poster and video presentations in alphabetical order (by first author). Part 3 ends with the *Workshop Impressions*, given by J.M. Beckers.

Most of the *manuscripts* have been supplied as camera-ready texts, with the authors responsible for the layout and the style. However, a third of the manuscripts were mailed electronically in the form of a LaTeX input file, and were printed at Utrecht. Some of these electronscripts have been edited to some extent; I apologize for any editorial errors so made.

The *discussions* following the oral presentations were recorded using the sheet system. Questions and answers were written down on sheets that were distributed and collected by the session's Discussion Recorder. The discussions printed here are not complete because it was a participant's choice whether or not to enter his or her question or remark into the proceedings; indeed, many short questions and remarks are missing. The discussions printed here are also not verbatim, firstly because they are transcriptions from written versions rather than from a tape recording, and secondly because they have sometimes been re] ased and reordered for the sake of clarity. I apologize for all errors introduced in this e ing.

In addition, there were three *Position and Discussion* sessions in which two 'anchorpersons' led a general discussion on the state and the future of the field, triggering response by presenting provocative item lists. These three sessions were:
- Stellar Data (D. Dravins and D.F. Gray);
- Solar Data (A. Righini and E.H. Schröter);
- Modelling (A.A. van Ballegooijen and N.E. Hurlburt).

These sessions were very lively—in fact so lively that record keeping proved impractical. They are not reproduced here. However, D. Dravins has kindly supplied a written version of a position taken in the first session (page 153*ff*), and A. Righini and E.H. Schröter have kindly supplied a report on the second session (page 295*ff*).

Finally, a *Granulation Bibliography* has been added to these proceedings (page 623*ff*). It has been prepared by K.L. Harvey.

Robert J. Rutten

October 28, 1988
Sterrekundig Instituut Utrecht

WELCOME ADDRESS

M. Rigutti

Director, Osservatorio Astronomico di Capodimonte, Napoli, Italy

Authorities, colleagues, dear friends.

It is a great honour to me to welcome you on behalf of the Osservatorio Astronomico di Capodimonte, and it is a real personal pleasure to have had the opportunity to meet you solar physicists here in the island of Capri, where the sun is at home.

As you know, this is the third international workshop of the Capodimonte Observatory, and we hope that many other meetings like the present one will be held in the future. As a matter of fact, the Capodimonte Observatory's Management Committee decided to organize at least one workshop every year, but this is not an easy task because it depends on too many variables, the majority of which are not under our control. In particular, the most important of them is to find sponsors. In any case, we are confident in the continuity of our will-power and, above all, in our patience, which is really necessary to overcome all those little difficulties you find when try to organize something at an international scale.

The NATO joined us in organizing the present workshop, which appears also as a NATO advanced research workshop. We are very happy for this and warmly thank the NATO officers who decided to accept to work with us and put at our, to be honest, also to your, disposal its power, which permitted us to solve a number of money problems and have the pleasure and opportunity to see here at the same time so many nice and important names of the international solar physics community.

Some of you know that I also come from solar physics. This is another reason to me to be glad for your being here, and I certainly will be definitely happy if new and deeper insight into the so difficult and intriguing granulation phenomenon will be got as a product of your discussions in this place.

Before starting working, I would like to thank some persons and public bodies whose assistence and help made possible this meeting.

First of all, I want to express once again all my appreciation to the NATO for having joined us in organizing the present workshop. Then I have to thank the Ministero della Pubblica Istruzione, the Ministero della Ricerca Scientifica, the Consiglio Nazionale delle Ricerche, the Università degli Studi di Napoli, the Regione Campania, the Società Italiana di Fisica, the Società Astronomica Italiana, whose auspicies in organizing this workshop encouraged us to overcome all the difficulties I mentioned before. I also want to thank the Comune di Anacapri and the Azienda Autonoma di Cura, Soggiorno e

Turismo for their kind hospitality which certainly will help you in remembering us and this beautiful place when you are back home.

I also want to warmly thank the Scientific Organizing Committee, in particular its chairman and old friend R.J. Rutten, the members of the local Organizing Committee for the good job they did, and the staff of the Capodimonte Observatory who all worked so nicely to give us the opportunity to attend a smooth meeting.

Let me also thank my daughter Adriana, who so well interpreted the subject of our scientific interest and prepared the nice poster you will take home as a souvenir of this meeting.

And last, but not least of course, let me thank you dear colleagues and friends for coming, and wish you a pleasant and fruitful stay at Capri.

NATO SCIENCE COMMITTEE'S PROGRAMME: PROMOTING SCIENTIFIC MOBILITY AND INTERNATIONAL COHESION

Craig Sinclair

Director, Advanced Research Workshop Programme, Scientific Affairs Division NATO, Brussels-1110, Belgium

The NORTH ATLANTIC TREATY ORGANISATION, which came into existence some forty years ago, is frequently perceived solely as a military and political alliance. It is, indeed, primarily this, and its goals and focus change with the world political climate.

Thirty years ago, however, with a lessening in East-West tensions, the Alliance began to look towards the wider implications and, in fact, the injunctions, of the Treaties bringing it into being. The clause invoking wider collaboration by its signatory nations[1] in economic and social matters was used to give a 'third dimension' to the Alliance.

In the light of the contemporaneously orbiting Sputnik, three senior political figures reported their conclusions on the form such a dimension should take. The 'Three Wise Men' envisaged the establishment of a Science Committee which would fund and strengthen civil non-military science within the Alliance. This Committee which assisted at the birth of the science programme was, and is, in form unique in NATO. Composed not of national delegates but of eminent scientists (professors in the main), directors of national agencies, and representing the scientific spectrum of disciplines, it conceived and assisted at the birth of several programmes whose long existence is testimony to their usefulness in the financially constrained field of international scientific co-operation.

The initial task was to find the appropriate niche in the field. National administrations and a few goal-orientated international organisations, with laboratories and programmes directed towards particular fields, already covered equipment costs, installations and scientific salaries. The Science Committee therefore chose to promote scientific mobility, long accepted (by scientists at least) as a sure way to exchange ideas and help scientific creativity.

Fundamental science was chosen for a number of motives: distance from military application, a needy area then as now, an area where nation could speak to nation - an international language.

---

[1]Belgium, Canada, Denmark, France, Germany, Greece, Iceland, Italy, Luxembourg, Netherlands, Norway, Portugal, Spain, Turkey, United Kingdom, United States.

Programmes

These motives have the flavour of their time but, perhaps remarkably, the programmes have shown a resilience and a flexibility which still matches many current scientific needs.

What then, are the programmes? The largest in terms of funds is the *fellowship programme*. Administered by the capitals, it enables over 1,000 Alliance scientists annually to work periods of up to one year abroad. The exact graduate level and the topics covered are nationally decided, tailored to national needs. About half the annual budget of over US\$ 20 million is thus directly in the hands of the member countries.

For shorter visits the *collaborative research grants programme* provides for around 500 projects each year, involving well over 1,000 visits. It enables, for example, a Greek theoretician to spend some weeks in a US laboratory working experimentally on the same topic, and a reciprocal visit to be made to Greece.

The *meetings programme*, perhaps the best known of all, was modelled on the classic physics summer schools. It brings together a dozen or so internationally eminent scientists, who teach and interact intensively over a two weeks' period with sixty to eighty postdoctoral students. These produce scientific exchanges and often life-long connections. A total of almost 900 volumes demonstrates the material output.

A total of sixty such *Advanced Study Institutes* take place each year, covering topics ranging from supergravity to child abuse, from halide glasses to heavy ion collisions. A similar number of shorter *workshops* cover a similar range of topics, with thirty or forty senior scientists joining together for four or five days to exchange results, to plan future collaboration, to review the state-of-the-art. The workshops programme was instituted five years ago in response to the changing needs of the international scientific community.

This responsiveness also marks the work of the general programme panels. The major programmes outlined above absorb over 90% of the budget. Applications for funding in any of these are judged by international panels drawn, like the applications, across all the sciences. Success rates for these typically average two-thirds. No quotas are operated in respect of topics, nationality or location within the Alliance. These unsolicited applications are open to all within the Alliance and in the meetings programme, non-Alliance scientists are welcome. Perhaps a fifth of all participation in the latter are non-NATO nationals.

The non-military aspects of the programme are further exhibited by the fact that all work coming from the scientific collaboration in the programmes is published in the open literature.

As a demonstration of the evolutionary nature of the programme, several *specialised group programmes* are also supported. Small panels of experts in a particularly topical area are brought together for five years, during which they actively solicit meetings, travel, and work from scientists working at the forefront of the particular field chosen. Currently, these special programme panels are working in robotics, cell-to-cell signalling in plants and animals, chaos, order and pattern amongst other things. Previous groups have worked on areas as diverse as human factors, marine science, systems analysis and ecology.

There is in this way a continuous development of funding objectives. The major, unsolicited application programmes provide a basic competitive area across the sciences, while the special programme panels, with only a few percent of the budget, are designed on a 'sunset' basis - carrying out their more narrowly focussed work only while the subject is still highly topical and rapidly advancing. Recent evolution of the Scientific Affairs Division's work is further exemplified by a move towards technology and applied science, and also from the traditional, though still important, physical sciences towards the biological and social.

The administration of these programmes, actively involving each year around 10,000 scientists, is based on a small secretariat in Brussels, who support individual scientists working in the 16 countries of the Alliance.

Expectations

Other major areas of activity undertaken by the division are the so-called Science for Stability programme, and the work of the Committee on the Challenges of Modern Society[2], which assists the environmental collaboration of national authorities-but has little funding.

The Science of Stability programme is separately funded. It provides technological management support for development projects in the three less scientifically advanced, and poorer, countries Turkey, Greece and Portugal.

Returning to the international scientific exchange programme, it might be remarked that other agencies, in particular the European Community, have now adopted similar (and certainly better funded) schemes. The EC, for instance, certainly funds the movement of scientists. It is claimed an EC funded scientist steps on board a plane at the average rate of one every two hours. A similar calculation shows that the NATO Science Committee will have sent on similar missions some three or four scientists in the same time. Such comparisons do not go very deep, of course, and NATO's exchange programme budget corresponds to about a quarter of an hour of the yearly world spending on military R&D. The real difference from the EC is, of course, the transatlantic nature of the exchanges; over 50% of travel within the Science Committee programmes is across the Atlantic and this is likely to increase, maintaining a vital link beneficial to both sides of the ocean. In any case, the exchange programme will continue to support good scientists in the whole of the Alliance - a *bon* rather than a *mauvais quart d'heure*.

Mobility is central to the health of science and technology and is ultimately for the good of the economies and dependent facets of national life. Such ideas on the necessity of exchanging ideas through exchange of people is not always self-evident to the nonscientist. To the scientist it is a basic tenet. Science, and political confrontation with its ultimate manifestation of war, rest in an uneasy alliance at the present time. This is unlikely to change.

The idealistic notion of 30 years ago, which saw science as an international link for lessening tensions, is clearly naïve. However, it should be remembered that it is an idea of long standing. For instance, after the Peace of Amiens in 1804, which brought a temporary halt to the Napoleonic Wars, it is claimed the first Briton to enter France was Sir Humphrey Davy on a scientific mission. The ideal should not be lightly abandoned.

A recent special activity mounted in Brussels by the Science Committee itself, was to examine the quality of Soviet civil science[3]. The scientists at that conference showed a guarded optimism about the quality of science in the USSR for non-military ends. The political analysists, though more sceptical, were sufficiently enlightened to see the usefulness of a similar look at Comecon science as a whole. With science and technology playing an increasingly in society and in government policy, plans are being considered to widen this review and forecasting aspect of the programme.

---

[2]C. Sinclair, *Nato and the Environment*, in: Project Appraisal, vol. 3 no.2, June 1988
[3]C. Sinclair (editor), *The Status of Soviet Civil Science*, Nijhoff, Dordrecht, July 1987

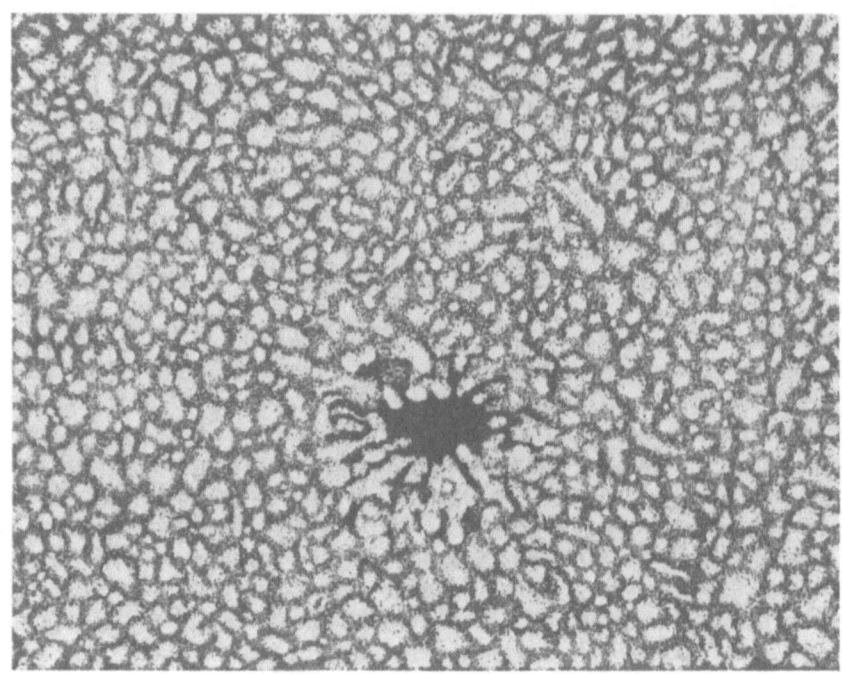

Granulations. [...] la surface est recouverte d'une multitude de petits grains, ayant presque tous les mêmes dimensions, mais des formes très-différentes, parmi lesquelles l'ovale semble dominer. Les interstices très-déliés qui séparent ces grains forment un réseau sombre sans être complétement noir. Dans la *fig.*, nous avons essayé de faire une esquisse qui représentât l'aspect caractéristique de la surface, car les détails sont impossibles à reproduire. [...] Cette structure est très-apparente dans les premiers moments de l'observation; mais elle ne tarde pas à devenir moins distincte, parce que l'œil se fatigue, en même temps que l'objectif s'échauffe, ainsi que l'air qui est contenu dans le tube.

P.A. Secchi *S.J*, 1875, *Le Soleil*, GAUTHIER-VILLARS, IMPRIMEUR-LIBRAIRE DU BUREAU DES LONGITUDE ET DE L'ÉCOLE POLYTECHNIQUE, Deuxième Édition, Paris, p. 50-51.

WORKSHOP INTRODUCTION

Robert J. Rutten

Sterrekundig Instituut Utrecht, The Netherlands

Why this workshop?

Why this workshop? Or rather, since the real question that arose about a year ago was not whether there should be another OAC workshop, but only what it should be about: *why a workshop on granulation?*

To answer this question I will play an unfair trick on you. I will simply present the scientific justification which I included last autumn in a grant application to NATO's Scientific Affairs Division. It lists the reasons why I thought a workshop on this particular topic and at this particular moment ought to be worthwhile. There must be something in its reasoning, because NATO has indeed agreed to co-sponsor this workshop, and because all of you have decided to spend time and effort on your contributions and to journey to this beautiful island in order to participate. But since the proof is in the pudding, I am eager to see whether indeed this workshop will be as outstanding as I have promised; in the meantime, you are entitled to know what we got you here for.

The justification went as follows:
    "The subject 'granulation' has recently become a hot topic, at the center of much new research, observational as well as interpretational and theoretical, and both in solar physics and in stellar physics.
    The reason for the renewed attention for this phenomenon (after all discovered on the sun already a century ago) is that recent advances have occurred, fortuitously simultaneous, which now enable a transition from morphological description to detailed physical interpretation: the subject is becoming of age.
    The solar granulation becomes the first example of a coming solar physics revolution: providing insights that can only be obtained by studying the sun with high spatial resolution. This era of studying basic physical processes rather than the phenomena they cause starts now. This is extremely important for all of stellar physics because the sun is the Rosetta stone of astronomy: it is the place where the processes can be identified and understood that are the machinery of astrophysics in general.

1

R. J. Rutten and G. Severino (eds.), Solar and Stellar Granulation, 1–5.
© *1989 by Kluwer Academic Publishers.*

Granulation is the signature of the convection that transports the stellar energy flux through the outer parts of all cool stars. Furthermore, it is now clear that it is also the agent which sweeps the concentrations of intense magnetic field which most cool stars possess, into the patterns of solar and stellar activity which underly the existence of the hot nonthermal chromospheres and coronae. The hydrodynamical, MHD and plasma processes that interact in the formation, heating and instabilities of these highly complex layers together represent a unique astrophysics laboratory for us to understand. The granulation is an important part of the engine and the main structuring agent in layers where the magnetic field is still frozen in. Knowledge of how granulation acts in stars that differ from the sun adds to our understanding of stellar structure.

The recent advances are:

- *Space Observation.*
  The last successful flight of *Challenger* carried Spacelab 2 with the SOUP instrument. It made the first granulation movie that is free from atmospheric image degradation, enabling granule astrometry. The reduction and analysis are in full swing; important results are in press.

- *New mountain-top solar telescopes.*
  The realisation that the future of solar physics lies in resolving the basic processes has led to the design of new telescopes specifically built to obtain maximum spatial resolution at the best sites worldwide. Granulation is their top priority subject. The Swedes have completed a small but superb solar telescope at the Roque de las Muchachos Observatory (La Palma). The first granulation studies are in and very exciting. The Germans are installing two major telescopes at the Izaña Observatory (Tenerife). The first one has started granulation studies and the second one will do so soon.

- *New observational techniques.*
  A forefront of new technology is exemplified by the development of active optics, enabling real-time correction of atmospheric image distortion. A prime target (outside SDI) is to enable correlation tracking of solar granulation (Sacramento Peak, Lockheed). The only solar telescope which reached high resolution before is the Pic du Midi refractor. It has now been equipped with the Meudon subtractive multi-slit spectrometer, enabling detailed two-dimensional mapping of intensity and velocity patterns strictly simultaneously.

- *Stellar observations.*
  Although there are typically a million granules on the visible hemisphere of a star, their averaged-together velocity signatures can now be measured using precision spectrometry. Presently, granulation is studied for a number of cool stars.

- *Supercomputer simulations.*
  An important breakthrough has been reached: ab-initio self-consistent modeling of the solar granulation has been achieved using supercomputers, which serves to interprete and diagnose the new observations in detail. This work is now being extended to numerical simulations of prototype stellar convection.

- *MHD theory.*
  The discovery that most magnetic field on the sun (and on other cool stars) exists in very strong concentrations has led to the 'fluxtube' concept and a large amount of theoretical study. At the moment, the interplay between granular buffeting and tube instabilities attracts much attention. On a larger scale, the role of the granular velocity fields in patterning the activity phenomena poses important theoretical questions.

- *Future projects.*
  Currently, two large-scale programs are being defined that both have solar granulation high on their list. The first is the LEST, the large solar telescope to be sited in the Canary or Hawaiian islands which is now being designed by a large international consortium. It will be the major ground-based solar physics facility of the late nineties. The second is NASA's Orbiting Solar Observatory, successor to the SOT and HRSO projects, which aims to put a major solar telescope in orbit in the mid-nineties. It will be the major space-based solar physics facility of the late nineties.

All these developments together make an expert workshop on solar and stellar granulation highly desirable, to review what is going on, to consolidate new results, and especially to provide a road map for future work."

This is what I promised to NATO, and hence, this is what you are supposed to do here!

Historical review

One talk that is missing on the program of this workshop is a historical review. This may seem strange for a field with such a long history of scientific endeavour, and indeed Peppe Severino and I originally planned to start the meeting with one. However, we canceled it when it became clear that the colleague whom we invited as reviewer would not be able to come, and also that we would be short on time in this morning's 'Telescope Caleidoscope' session. As compensation I will now give you a very brief historical review, consisting of two transparencies only.

The first one shows one of Father Secchi's beautiful drawings[1], to remind you that high-resolution imaging is not new under the sun.

The second shows a table taken from the monograph of Bray, Loughhead and Durrant. It summarizes the history of solar granulation studies from Herschel's first description until Project Stratoscope. I am happy to note that it mentions two participants of this workshop—Constantin Macris and Jean Rösch.

---

[1]Reproduced as frontispiece.

4

1.7   Chronological summary

1801   W. Herschel uses the term 'corrugations' to describe the mottled appearance of the solar disk.

1862   Announcement of Nasmyth's 'willow-leaf' pattern.

1864   Dawes introduces the term 'granule'.

1866   Publication of a paper by Huggins ends controversy over Nasmyth's 'willow-leaves'.

1877   Granulation successfully photographed by Janssen.

1896   Publication of Janssen's collected observations.

1908   Hansky estimates mean lifetime of granules to be about 5 min.

1914   Publication of Chevalier's collected observations of the granulation.

1930   Unsöld attributes the origin of the granulation to convection currents in the hydrogen ionization zone.

1933   Announcement of Strebel's discovery of the polygonal shapes of the granules.

1933   Siedentopf formulates a theory of the granulation based on Prandtl's mixing-length theory of turbulent convection.

1936   H. H. Plaskett identifies the granules as Bénard-type convection cells.

1949   Richardson obtains spectra showing the Doppler shifts of the granules.

1950   Richardson and Schwarzschild identify the granules as the eddies of a large-scale aerodynamic turbulence.

1953   First reliable determination of granule lifetimes made by Macris.

1955   Various workers claim that the solar surface shows random brightness fluctuations, not a cellular pattern.

1957   Rösch publishes granule observations made at the Pic-du-Midi.

1957   High-resolution photoheliograph brought into operation near Sydney by the CSIRO.

1957   Leighton re-asserts convective origin of the granulation.

1957   Granulation photographed from a manned balloon.

1957   Project Stratoscope I yields granulation photographs of unsurpassed definition.

Figure 1: Table 1.7 from R.J. Bray, R.E. Loughhead and C.J. Durrant, 1984, *The Solar Granulation*, Cambridge University Press, second edition. Permission granted by the publisher

The monograph furnishes an excellent survey of the developments since Stratoscope, in which many of you have taken part. There is no point in summarizing them here: we will re-address them in detail while we "review what is going on, consolidate new results, and provide a road map for future work". Let us begin!

Part 1
Observational Techniques

OBSERVATIONS OF THE SOLAR GRANULATION AT THE PIC
DU MIDI OBSERVATORY AND WITH THEMIS

R. MULLER
Pic du Midi Observatory
65200 Bagnères-de-Bigorre
France.

1. PIC DU MIDI 50 CM REFRACTOR.

Observations of the solar granulation are performed at the Pic du Midi
Observatory (elevation, 2860) with a 50 cm refractor (focal length,
6.5m) installed inside the entirely closed "Coupole Tourelle". The
àtmosphere can be continuously stable for a few hours in the morning, up
to 10 days a year (Figure 1). In some occasions the stability can last
as long as five hours. During those moments of superb seeing, the
resolution of the instrument (0''2 in the blue) is reached with
exposure times less than 1/10 sec.
      The observations consist either of filtergrams or of spectrograms
taken in the common slit mode or in the MSDP (Multichannel Soustractive
Double Pass) mode. The spectrograph moves with the equatorial mounting.

FILTERGRAMS. They are taken in white light at 5750 Å and in the CH band
at 4308 Å, where Network Bright Points and Faculae are visible even at
the disc center. Bursts of about 50 frames are taken at a speed of 24
frames per seconde. The best ones are then selected. Granulation, Bright
Points, Faculae, and Sunspot evolution movies can thus be obtained, up
to 3 hours long.

2-D MSDP SPECTROGRAMS. They are taken in the Na D_2 5696 Å photospheric
lines; the dispersion is 4.5 mm/Å; exposure time : 1/4 sec. , a 70 mm
movie camera is used. In this mode the spectral resolution is low. The
best spatial resolution reached so far is 0''5. Such kind of movies are
used to derive velocity power spectra of the granulation and to study
the evolution and penetration into the photosphere of the intensity and
velocity of granules.

SLIT-JAW SPECTROGRAPH. Only one wavelength range can be observed on
70 mm film (movie-camera). The dispersion is 5 mm/Å . A 0''5 slit-jaw is
commonly used. A polarization analyses will be placed in the near future
near the prime focus. It will allow   us to analyse polarized light free
of instrumental polarization.

2. THEMIS.

THEMIS is a magnetograph especially designed to minimize the instrumental
polarization, to be installed at Izana, Canary Islands.

R. J. Rutten and G. Severino (eds.), Solar and Stellar Granulation, 9–12.
© *1989 by Kluwer Academic Publishers.*

10

Figure 1. Exemples of evolution of the image qualité Q during the very best conditions at the Pic du Midi Observatory. The quality scale is that used during the JOSO site testing campaigns; images of Q = 0 are completely free of seeing on more than 3/4 of the field of view (2' x 1'5), only traces of seeing being visible on the rest of the image. Each plot represents the quality of the best image among 300 taken during 1 min.

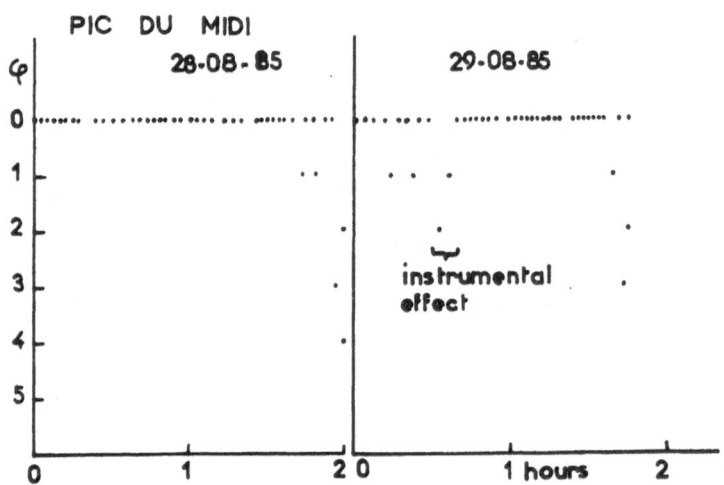

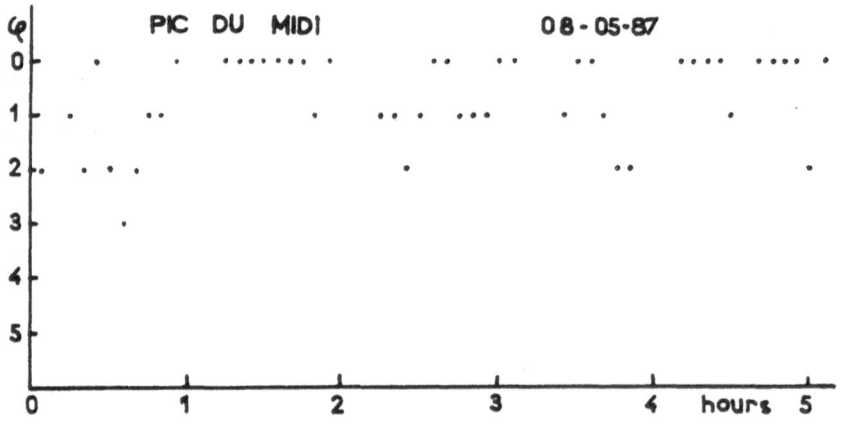

2.1. TECHNICAL CHARACTERISTICS.

- Ritchey-Chrétien telescope $\emptyset$ = 90 cm, f = 15 m
- guiding system
- polarization analyser
- echelle spectrograph (dispersion : 0 - 12 mm/Å)
                        7 lines simultaneously
- MSDP mode
- detectors : CCD or diode arrays.

2.2. SCIENTIFIC PROGRAMS.

Possibility, if permitted by the seeing, to observe the solar granulation with a resolution of 0".1 in the blue : structure, evolution, photometry. Spectrographic observations (slit-jaw and MSDP) with an expected resolution of less than 0".5.
    THEMIS will be particularly suitable to study the interaction between magnetic field and granulation, because its magnetographic performances

Discussion

RIGHINI — Can you please comment on the feasibility of installing LEST at Pic du Midi?

MÜLLER — Pic du Midi may yet be the best astronomical site as far as the stability of the atmosphere is concerned. However, we cannot expect more than 10 such stable days per year. You should also be aware that the number of clear days is only about 120 per year. Otherwise, Pic du Midi is an extremely good place for a large solar instrument. In any case, I think it should be worthwile to analyze the site carefully during a site testing campaign.

BECKERS — Is THEMIS a vacuum telescope and will it go to Pic du Midi?

MÜLLER — THEMIS is a vacuum telescope, but it will go to Tenerife.

THE GREGORY-COUDE-TELESCOPE AT THE OBSERVATORIO DEL TEIDE, TENERIFE

F. KNEER and E. WIEHR
Universitäts-Sternwarte
Geismarlandstr. 11
D-3400 Göttingen
Federal Republic of Germany

The Gregory-Coudé-telescope in Izaña, Tenerife, is in operation since July 1986. Since then, it has made possible high quality observations while the post-focal instrumentation is still improving. The instrument was described earlier (Kneer et al., 1987; Wiehr, 1987). Thus, only a summary of its features and its performance will be given here.

1. The Telescope. The focal length of the 45cm parabolic primary mirror is 2.40m. It is magnified by the Gregorian to an effective focal length of 25m. A 45^0 mirror deflects the sunlight into the δ-axis, the Coudé-mirror brings the light down into the hour axis. Focussing is done with the Gregorian. Under good seeing conditions we find the best focus within the Rayleigh tolerance of the primary. The light path is evacuated from the entrance window close to the dome aperture down to the exit at about 50cm distance from the spectrograph slit. The advantages of this type of telescope are: short optical device with large effective focal length, low scattered light level, low and daily constant instrumental polarisation.

2. The Spectrograph. The focal length of both the collimator and the camera mirror of the Czerny-Turner spectrograph is 10m. The echelle grating has a ruled area of $12.8 \times 25.6 cm^2$ with 300gr/mm and is blazed at $63^0\!.5$. Due to the symmetry the spectrograph is free of coma. Its astigmatism, which goes as $tan^2\omega$ (ω = angle between light rays and collimator axis) has been removed with the aid of the opposite astigmatism of the grating. Tests with a "tree"-pattern near the Sodium D lines show a spatial resolution of about 15l/mm (at 10% contrast) corresponding to 0.55 arcsec. This is close to the diffraction limited resolution of the configuration in which the size of the grating is the limiting element.

13

R. J. Rutten and G. Severino (eds.), Solar and Stellar Granulation, 13–16.
© 1989 by Kluwer Academic Publishers.

3. The Slit-Jaw Camera. This allows observation in white
light, (5800Å), Hα, and Ca K simultaneously with spectra.
The combination of the spectrograph and the slit-jaw camera
is the most relevant feature for high spatial resolution
work. The equipment of this telescope is not installed for
high resolution imaging but for high resolution spectros-
copy/polarimetry.

4. The Dome. The telescope does need a wind shield to prevent shaking. We chose a classical dome, but with highly
reflecting painting and with a small aperture to keep the
inside at low temperature, i.e. to avoid heating by the
sunlight. The dome drive is computer controlled such that
the aperture follows the telescope.

5. The Detectors. Up to now, photographic film is used
mainly. Exposure times for spectra from disc centre at
6000Å are of the order of 1 sec on Kodak Technical Pan 2415.
A RA 100 Reticon array is used also frequently, while a
RA 256 Reticon array is in its testing phase. CCD detectors
will come in use as common facilities with the Vacuum-Tower-
Telescope in Izaña.

6. The Computers. The control of instrument operation can
be performed by computers (Computer Automation and DEC).
Software has been developed by colleagues in Göttingen to
handle the telescope guiding, image scanning, grating drive,
data acquisition, and data handling.

7. The Future. We will install very soon a larger grating
to obtain higher spatial resolution in the spectra. A fast
spectrum scanner is almost completed. And finally, we expect a larger selection of CCD-detectors with forthcoming
experience.

Acknowledgements. We thank our colleagues of the Universitäts-Sternwarte in Göttingen, of the Kiepenheuer-Institut
in Freiburg, and the IAC in La Laguna, Tenerife, as well as
Prof. F.-L. Deubner for their collaboration. This observing
facility was financed in most parts by the Deutsche Forschungsgemeinschaft.

References

Kneer, F., Schmidt, W., Wiehr, E., Wittmann, A.D.: 1987,
     Mitt. Astron. Ges. 68, 181
Wiehr, E.: 1987, in Proc. Inaugural Workshop 'The Role of
     Finescale Magnetic Fields on the Structure of the
     Solar Atmosphere', eds. E.H. Schröter et al., Cambridge
     Universtiy Press, 354

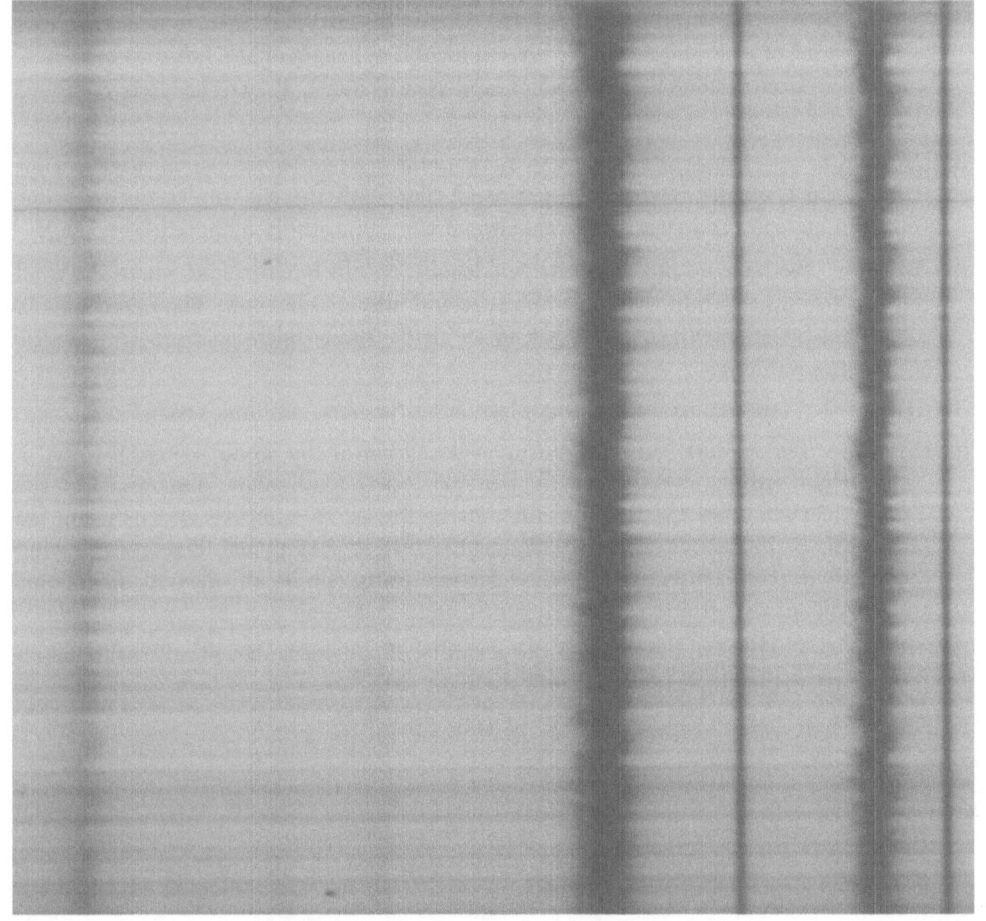

Spectrum around 6300Å covering 130 arcsec, observed with
the Gregory-Coudé-telescope from disc center.

Discussion

DRAVINS — The number of optical surfaces appears to be large (8 in the telescope, 11 in the spectrograph). Could you comment on how this affects the straylight and the contrast in the final spectrum image?

KNEER — The entrance vacuum window can be cleaned from dust from time to time. The mirrors inside have been recently recoated and have now very high reflectivity and low scattering. It should stay this way for a while. And the diaphragm at the prime focus takes out 99% of the light.

SCHRÖTER — Let me additionally comment on Dravins' question. The Tenerife Gregory-Coudé Telescope is essentially the one from the old Locarno station. I have once measured aureolae with this telescope and compared them with my previous aureolae measurements at Sac Peak with the Big Dome. They were almost identical when in the focus of the Big Dome Telescope an inverse occulting disc was placed of a size equal to the inverse focal diaphragm of the Locarno telescope. (The Big Dome telescope is a coronograph-type system). Hence the Gregory-Coudé Telescope is really a low scattering system.

KOUTCHMY — I understand that at Tenerife the seeing is better when there is some wind. So I wonder if you have problems then with the shaking of the telescope (image motion produced by the shaking of the equatorially mounted telescope)?

KNEER — We have no problem with windshake, except in situations where the wind can enter directly into the dome aperture and hit the telescope. The telescope and the post-focal instrumentation rest on an inside tower which is separated from the shielding building.

RIGHINI — Can you say something more about 'rumors' of dome seeing?

KNEER — The rumors came up during a discussion of the dome seeing observed in the neighbouring tower, which houses the 'Newtonian' solar telescope. But the dome of that telescope is quite different as far as its aperture and painting are concerned. My experience with Foucault's test is that under fair external wind conditions (sufficiently strong wind to take away the locally heated air), dome seeing is not detectable at the Gregory-Coudé Telescope.

DEUBNER — Have you estimated the stability of the spectrograph after the installation of a shielding mantle around the light path?

KNEER — No, we have not yet had the time to test it.

The Status of the Latest German Solar Facility on Tenerife

Dirk Soltau
Kiepenheuer-Institut fur Sonnenphysik
D-78 Freiburg, FRG

0. Abstract

First experiences and performance characteristics of the new German Vaccum Tower Telescope are presented.

1. Introduction

The German Vacuum Tower Telescope (VTT) is now very near to its completion and this workshop gives the opportunity to report for the first time some practical experiences and performance characteristics of the new instrument.

2. Some historical remarks

Very early in the history of this project decisions were made which placed principal constraints on the instrument. E. g. the coelostat mirrors were ordered as early as 1973. At that times it was not planned to bring this new telescope of the Kiepenheuer-Institut (in that times still named Fraunhofer Institut) to a site with outstanding seeing quality like the Canary Islands. This happended later as a result of the JOSO activities inspired by the late Prof. K.O. Kiepenheuer. Dr. J. P. Mehltretter undertook the task to design a solar telescope which should be

* of moderate size, because large scale telescope should be planned in an international effort (LEST)
* a versatile multi-purpose telescope for high resolution spectroscopy (mainly photographic work), because other JOSO-members already had plans for dedicated instruments (THEMIS)
* simple in design and construction to be well within the personal capacity of the institute
* cheap in order to be able to cover the expenses within the institute's yearly budget
* high in performance in order to beat the observational facilities so far available in Europe.

In the meantime the extensive site testing campaign of 1979 had compared two particular sites on La Palma and Tenerife and the later island was selected to be the site foe the new VTT. In April 1983 the Federal Republic of Germany signed the contracts which enabled it to start the construction work in May 1983. In April 1987 the building for the VTT was finished. To save time the assembly of the telescope went parallel to the building's construction and this allowed for having »first light« on May 18, 1987. Since then many further test and alignment procedures were performed.

R. J. Rutten and G. Severino (eds.), Solar and Stellar Granulation, 17–23.
© 1989 by Kluwer Academic Publishers.

3. Performance of the telescope

The Vacuum Tower Telescope has been described elsewhere(e.g. Mehltretter (1975), Schröter et al. (1985), Soltau (1986)) so only a very brief description of its principles shall be given here:
 A classical coelostat feeds the light into the evacuated telescope tube. As a result of this design there is no image rotation. The inclined spherical primary mirror forms the solar image, the beam beeing folded once by an additional flat.
For some characteristic data see Table I.

Table I:	Some characteristic data for the VTT
Diameter of the primary mirror:	70 cm
Focal length:	46 m
Image scale :	4.6 arcsec/mm
Field of view:	13.8 arcminutes

Because of the primary's inclination the dominating aberration is astigmatism. To check for it we photographed some focal series of a bright star and saw the preference orientation of the image beeing perpendicular inside and outside focus. Then by applying defined forces on four points at the limb of the mirror we deformed the surface of the sphere slightly until intra- and extrafocal images were both of circular symmetry. As a result the quality of the image is now seeing dominated. Under medium seeing conditions the diameter of the star image is now less then 1.2 arcsec.
 To have a more quantitative measure for the telescope's performance we plan to take advantage of the possibility to realize an autocollimation setup using the coelostat's secondary. Then an appropiate test target may be reimaged and its image quality may be investigated.

4. Performance of the Spectrograph

The spectrograph is mounted in a vertical steel tube reaching down to 16m below ground level. We therefore expect it to be thermal rather stable beeing on top warmer then at the bottom. There are two observational modes. On the one hand on can use a low order grating (e.g. 600 gr./mm) and interference filters for order selecting. In this case the light coming from the telescope enters directly the main spectrograph. On the other hand one may use a high order echelle grating. In this case it is possible to select the orders by means of a predisperser which is obligatory if one wants to observe in different orders simultanously.
A set of characteristic data for the spectrograph is given in Table II.

Table II:	Some characteristic data for the Echelle spectrograph

Focal length of predisperser:	4.2 m
Focal length of main spectrograph:	15 m
Echelle grating:	220 mm x 420 mm ruled area
	79 grooves per mm
	blaze angle: 63°.43

For 5000 Å the linear dispersion is 12.5 mm/Å, the separation of two neighbouring orders is about 110Å, the spectral field is about 40Å

Because of inclination of the collimating mirror in the main spectrograph also here as in the telescope the dominating aberration is astigmatism. Similar as in the telescope the fine tuning of the image quality was done by slightly deforming the collimator and the grating. For this purpose we observed the diffraction pattern of a laser illuminated pinhole having 30 μ in diameter. This is a very sensitive diagnostic. By comparing our diffraction patttern with those belonging to known aberrations we could estimate the overall performance in the average focus beeing better than $\lambda/3$.

There is a residual astigmatism which could not be eliminated by the above mentioned deforming procedures. This reduces the spectral resolution if one observes in the spatial focus. For the echelle mode the theoretical spectral resolution is in the order of 10^6 or better depending on wavelength.
 Because of the geometry of the spectrograph the spectral lines are inclined if the entrance slit is parallel to the grating rules. In order to rectify them one has to incline the slit with respect to the grating grooves which again reduces the spectral resolution. This is necessary if the spectrum scanner is used whose scanning direction is parallel to the direction of dispersion.
So we tested the system in this »practical case configuration« : spatial focus and inclined slit. For obtaining the instrumental profile the HeNe-laser line at 6329Å has been scanned photoelectrically after suppressing the second mode with a polaroid. It turned out that the FWHM is less then 8 mÅ (raw data, not deconvoluted for finite scanning slit of 5 mÅ), the width of the 10^{-2} level is 70 mÅ that of the 10^{-3} level is 290 mÅ. The shape is slightly asymmetric.

Even more interesting is the spatial resolution which should be expressed in terms of a MTF rather than as a single number giving »the resolution«. To obtain this information a test target with variable line frequency has been scanned. This can be compared with the calculated performance of an ideal system.

But before doing so the effect of the rectangular shape of the test target and the finite scanning slit width has to be taken into account. Fig. 1 shows how the theoretical MTF - defined for a sinoidal test pattern - transforms to the corresponding modulation function for a rectangular test target and a 50 μ slit. With that curve the measurements have to be compared.

Figure 1

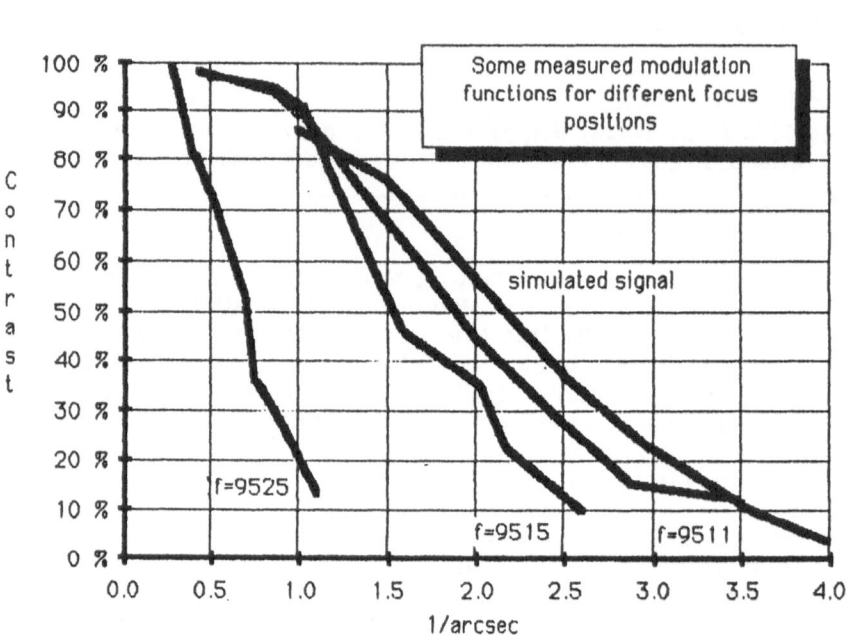

Fig. 1 shows also the results for three different focus positions. Note that already a small defocussing within the Rayleigh tolerance changes the measured contrast for higher wavenumbers significantly!

As a result we may state that our spectrograph has a MTF corresponding to about 70% of the theoretical one for a 0.3 arcsec feature. So we conclude that the optical performance of the spectrograph should allow for spectral observations of features with granular contrast of sizes less than say 0.25 arcsecs.

5. Present equipment and plans for the near future

At the present time the following equipment is either operational or in its final testing phase

• slit jaw camera, making it possible to obtain photographs of the vicinity of the entrance slit (field: 370 x 110 arcsec2)
• various camera housings for spectrum photography

• a high precision spectrum scanning device with continuum reference detector

Within the next few months we shall be able to perform computer controlled scanning with the telescope and the software for computer controlled data acquisition will be available. Moreover the tests of three CCD cameras will start this fall, one of them beeing a prototype for the OSL (Orbiting Solar Laboratory, former HRSO) project.

6. Concluding remarks

After many years of joint activities of a large number of people we are now near to the situation where a new telescope with promising performance on an excellent site can provide new observational possibilities.

References:

Mehltretter, J.P.: 1975, JOSO Annual report 1975
Schröter, E.H., Soltau D., Wiehr, E.: 1985, Vistas in Astronomy, Vol. 28, p. 519
Soltau, D.: 1986, Proceedings of the workshop »The Role of Fine-Scale Magnetic Fields on the Structure of the Solar Atmosphere«, La Laguna, Tenerife , October 1986

Discussion

KOUTCHMY — Are you prepared to cool the edges of your entrance window?

SOLTAU — Yes. The entrance window cell is surrounded by a copper pipe through which a cooling fluid can flow. However, we have not yet performed systematic investigations of the behaviour of the window, and so we have not used the cooling system up to now.

RIGHINI — Can you comment, please, on secondary mirror vibrations and guiding system eigenfrequencies?

SOLTAU — There are indeed mechanical vibrations stemming from the secondary's mounting which cause considerable image motion when the wind speed is above 10 m/s. This also affects the performance of the guiding system.
We are curing the problem in two ways.
1). The control for the dome will be changed so that it can be opened to an arbitrary position (rather than be opened completely), and so the dome will serve as a windshield.
2). The mechanical setup of the secondary's actuators has been changed. By introducing a counterforce using a spring, we stabilized the servo system which tracks the solar image.
We believe that the vibrational problems are now solved.

RUTTEN — I have two questions on the spectrograph.
1). I understand that you have copied the scheme of the Sacramento Peak échelle spectrograph, with slits after the predisperser to select segments of overlapping parallel-dispersed orders. How small can these segments be? Specifically, can you put two lines from different parts of the spectrum close enough to fit onto one CCD frame?
2). I noticed that your spectrograph resolution is very high compared to that of the Littrow spectrograph on the Swedish solar telescope. What have you sacrificed to get this high resolution, and what do you gain from it?

SOLTAU — Ad 1): the linear dispersion in the predisperser spectrum is in the order of 20 Å/mm. We use typical slit segment widths of 1 to 2 mm in order to discriminate between the different orders. In order to observe different orders on one CCD frame one should use either filters in front of the CCD—which is possible in our setup—or use relay mirrors in order to place the lines of interest together on the chip.
Ad 2): having high spectral resolution we sacrifice speed. But we gain—among other things—the possibility to work in the spatial focus (in the presence of residual astigmatism) and so obtain maximum spatial resolution, because the lower spectral resolution there is still high enough.

MÜLLER — From the German experience at Tenerife, what about the quality of the site: frequency of quality $Q = 0$ images, how many days a year, etc?

SOLTAU — As far as my work with the Vacuum Tower Telescope is concerned, I cannot answer your question since we were in the phase of assembly and alignment up to now, and have done no systematic measurements or guesses of the seeing quality solar. But up to now, there is no reason to question the results of the extended site testing campaign of 1979 which proved the excellent quality of the site.

Ground-Based Tunable Filter Observations

Alan M. Title, Theodore D. Tarbell, and C. Jacob Wolfson

Lockheed Palo Alto Research Laboratory, 3251 Hanover Street, Palo Alto, California 94304

1. Introduction
During the development phases of both the Solar Optical Universal Polarimeter (SOUP) and the Coordinated Instrument Package (CIP), we have assembled ground-based filtergraph systems to test components, validate procedures, and collect scientific data. These systems have produced some of the highest quality magnetic observations ever taken. Until the summer of 1987, the vast majority of the data was collected on film because we did not have the computer capacity before that time to handle gigabytes of digital image data.

In the summer of 1987, a 1024 x 1024 CCD camera was used on a five week observing run at Sacramento Peak Observatory. Built by the Jet Propulsion Laboratory CCD group, this camera is the brassboard model for the Orbiting Solar Laboratory (OSL, formerly HRSO). Nearly 12,000 digital images were recorded on high density computer tape. This data has all been transferred to digital magnetic and analog optical disk which allows easy access to the entire data set. Figure 1 shows an example of the processed images.

In the late summer and early fall of 1988, an upgraded version of the system will be used at the Swedish Solar Observatory on La Palma, Canary Islands. A second CCD camera and data acquistion system is currently under construction. We are planning a joint observing run with scientists from the Kiepenheuer Institut für Sonnenphysik at the Vacuum Tower Telescope (VTT) on Tenerife, Canary Islands. In this run, coordinated observations of the filtergraph and the VTT spectrograph will occur.

2. OSL Filtergraph Evaluation System
The purpose of our instrumentation is to collect very high resolution images in narrow wavelength bands and well-defined polarization states, so that magnetic fields, Doppler shifts, and intensities in various lines can be measured cospatially and nearly simultaneously. The OSL tunable filtergraph evaluation system in its ground observing configuration consists of various components of the SOUP instrument and engineering model and brassboard components from the OSL CIP instrument.

The baseline configuration consists of the following.

1. Reimaging optics to make an image of the telescope entrance pupil.

R. J. Rutten and G. Severino (eds.), Solar and Stellar Granulation, 25–28.
© *1989 by Kluwer Academic Publishers.*

2. A high-speed steering mirror near the image of the entrance pupil, for image motion compensation (IMC).

3. An optical system to reimage the sun onto both an IMC sensor image plane and the primary CCD image plane.

4. A quad photodetector as an IMC sensor (sunspot tracker) for generating 1 kHz displacement signals in two axes; or,

5. 32 x 32 element array IMC sensor and correlation tracking electronics for generating displacement signals once per millisecond. The correlation tracker also produces a real-time measurement of image sharpness suitable for use by a fast shutter.

6. An electronic servo system which converts the displacement signals from the IMC sensor into the drive signals for the high speed steering mirror.

7. An eight position polarization analyzer wheel which contains a pair of quarter wave plates at ±45 degrees, 4 half wave plates at 0, 45, 90, and 135 degrees, a blank, and a defocusing lens.

8. An eight position temperature-controlled blocking filter wheel for selecting lines for the tunable filter. Currently, this wheel contains blockers for Mg I 5173, Fe I 5247 and 5250, Fe I 5576, He D3 5876, Na D 5896, Fe I 6302, Hα 6563, and Ni I 6768.

9. A tunable birefringent filter with selectable bandpasses of 50 or 80 milliangstroms at Fe I 5250 (75 or 125 milliangstroms at H-alpha). The filter can be tuned to any wavelength within 4 Angstroms of one of the solar lines within a second. It is fully temperature-compensated and does not require a thermal control chamber.

10. A computer-controlled sector shutter.

11. A 1024 x 1024 Texas Instruments uniphase CCD detector and the OSL brassboard camera. This may be read out in 1.1 seconds as a 1024 x 1024 x 12 bit array, or in less than a second as a 512 x 512 x 12 bit summed array. The system noise is between 50 and 75 electrons, and the full well is about 200,000 electrons.

The observing sequence is executed by a Compaq computer (IBM PC clone), which selects the polarization analyser and blocking filter wheel position, controls the tunable filter and selects its wavelength, selects the instant and duration of exposures, and commands the readout of the CCD. The data generated by the CCD camera is collected on a DEC Vaxstation 3200. The workstation displays the filter images on a video monitor and stores the digital images on a 8mm high capacity tape system. The current tape recorder can store a 1024 x1024 image once every 9 seconds. We hope to be able to double the frame rate in 1989.

3. Seeing Improvement Techniques

Although observations from the ground will always be limited by the atmosphere, during the next few years we will use the tunable filter system with several new seeing "improvers". The first device is a fast, 2 millisecond open-close, liquid crystal shutter (LCS) which will be controlled by the image sharpness monitor of the correlation tracker.

The LCS will be mounted directly in front of the sector shutter. The assumption is that, in times of good seeing, the image is extremely sharp for a sizable fraction of the time on an instant-by-instant basis. This has been demonstrated by the success of the video selection method used by Scharmer at La Palma.

In the Scharmer system, incoming video frames are digitized and saved or rejected depending on their measured sharpness. In this way, the sharpest exposure from each preselected time interval (say 10 seconds) is saved. Each video exposure is 20 milliseconds. Seeing clearly fluctuates on the time-scale of a few frames. The correlation tracker generates an image sharpness number every millisecond. Using that signal we should be able to open and close the LCS to extract the 20 to 60 millisecond intervals of excellent seeing.

Since the typical exposure time for the tunable filter is 0.5 seconds, many such intervals will be integrated over perhaps a 3 second period. This technique will not remove the blurring caused by image distortions in the instantaneous images. However, we have measured the La Palma white light images and have found the rms distortion to be .05 arc seconds over 12 x 14 arcsecond fields-of-view. It is reasonable to expect that under the best conditions the LCS system might obtain time sequences of 0.4 arc second filtergrams.

The second improvement will be the use of effectively simultaneous filtergram pairs for making longitudinal magnetograms. This eliminates the major source of noise in our present magnetograms, the seeing differences between frames taken about 6 seconds apart. A liquid crystal half wave plate and a fixed quarter wave will replace the polarization analyzer. A second electric wave plate will be mounted in front of a polarizing beam switch. The beam switch makes a pair of images on the CCD. Operated together the system can cause either a RCP image to fall on the top half of the CCD or a LCP image on the bottom half. We will switch the circular polarization beam switch several hundred times a second to guarantee the same seeing on the filtergram pairs used for magnetograms.

The polarizing beam switch concept was tried in the early 70's by our group using KDP crystals. Although the technique worked to make the seeing the same in both images, problems with the KDP transmission degraded the image quality beyond acceptable levels. The transmission of the new liquid crystals is better than 95 percent and their scatter is very low, so that the technique should now be viable for high resolution imaging.

A third technique for image improvement would be to use the filtergraph behind the adaptive mirror system built by Bob Smithson at Lockheed. The adaptive mirror should allow capture of diffraction-limited images inside the isoplanatic patch (2-5 arcseconds), and should reduce the distortion over a much large area.

In 1988 we will collect the majority of our data at the Swedish Solar Observatory on La Palma using the primary system. Experiments will be carried out using a sharpness measure to control the LCS. The circular polarization beam switch will be used for the first time on our 1989 observing runs. The adaptive mirror had a very successful observing run at Sacramento Peak this summer. With some luck we should have a first observing run combining the adaptive mirror and the tunable filter systems in the fall of 1989.

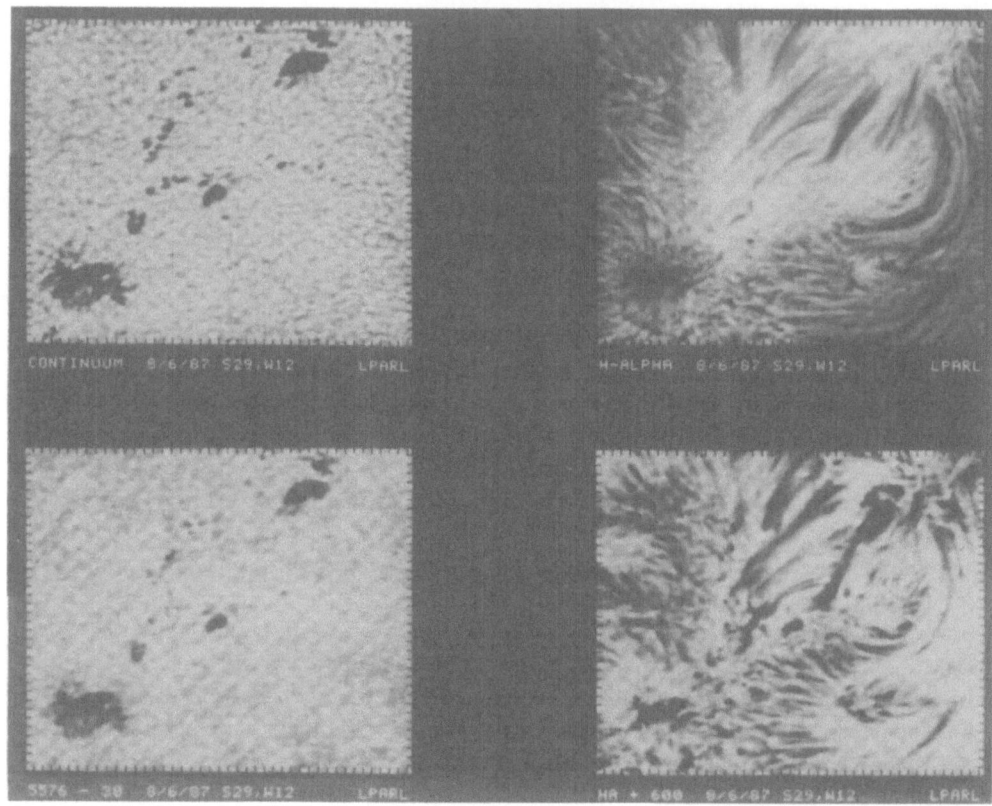

Figure 1: Snapshots of the active region in (a) continuum near 5576; (b) Fe I 5576, 30 mÅ blue; (c) H-α line center; (d) H-α wing, 600 mÅ red. All images in Figs. 1 and 2 were taken within 80 seconds of each other, in one cycle of the observing sequence. Field-of-view is 90 × 80 arcseconds, and ticks are at 2 arcsecond intervals.

An Overview of the Orbiting Solar Laboratory

Alan M. Title

Lockheed Palo Alto Research Laboratory, 3251 Hanover Street, Palo Alto, California 94304

1. Introduction

The Orbiting Solar Laboratory will be a free flying, polar orbiting complement of scientific instruments to observe the surface and upper atmosphere of the sun and to make precision measurements at high spatial and temporal resolution over a spectral range from the X-Ray to the near Infra-red. OSL contains a 1 meter telescope which feeds a narrowband tunable filter, a set of fixed broadband filters, and a spectrograph. These three instruments are mounted together in a common structure, share a unified science data and control system, and use a single image stabilization system for their common fields of view to form a Coordinated Instrument Package (CIP). An Ultraviolet Spectrograph and an XUV/X-ray imager are desired co-pointing instruments which have their own independent optical systems. The CIP and the copointing instruments operate with overlapping fields of view.

The 1 meter telescope, which is optimized for 2000 – 11,000 Å, can resolve 75 km on the solar surface which is about the mean free path for photons – the coupling length of radiatively dominated processes in the solar surface. This is crucial because the surface of the sun is the interface between the region of convective and radiative transport of energy. A detailed understanding of this interface is fundamental to solve a broad range of problems in solar and stellar physics. Not only is this the only astrophysical convection zone which can be studied in detail, moreover, solar observations also can complement laboratory convection experiments by extension to a much higher Rayleigh number.

The surface of the sun is also the region where the solar magnetic field first becomes visible. Convective motions in the atmosphere and interior interact with the magnetic fields to provide the energy which heats the outer solar atmosphere and drives the solar wind. These magneto-convective interactions range between those driven by the pervasive granulation and supergranulation flows of the quiet sun to large scale shear flows in active regions which cause the violent events of solar activity. The solar magnetic field, which expands outwards to fill the solar system, is responsible for the form of the magnetospheres of all the planets.

The Ultraviolet Spectrograph, 1175 – 1700 Å, will study the transition region between the 10,000 degree chromosphere and the 1,000,000 degree corona. The XUV/X-Ray Imager, 40 – 400 Å, will allow study of the corona and of multimillion degree solar transient events.

R. J. Rutten and G. Severino (eds.), Solar and Stellar Granulation, 29–42.
© *1989 by Kluwer Academic Publishers.*

The three year duration of the OSL mission implies that it will fly simultaneously with a variety of other spacecraft, which will significantly increase the joint scientific output. For example, the SoHO spacecraft has instruments that allow the study of the solar interior and the outer solar corona, while OSL is studying the critical interface region. In particular, the helioseismic instrument on SoHO will obtain data on the rotational gradients in the solar interior that drive the dynamo which is responsible for the solar magnetic cycle. Joint operation of OSL and SoHO offers the promise of actually observing the invisible processes in the subsurface layers by means of acoustic tomography, while magnetic flux tubes emerge at the surface. The corona instruments on SoHO and the fields and particle instruments on the rest of the ISTP complement of instruments will follow the effects of the magnetic field well into the environment of the earth, while OSL observes the physics of the processes which convert the convective energy to radiative and magnetohydrodynamic forms.

At present, OSL is the number one moderate mission of NASA. If funding becomes available now or very shortly, OSL could be launched in 1994-95. Unfortunately, because of the slow recovery from the shuttle accident and because of funding limitations, all NASA science programs have been delayed and may continue to be delayed.

The data from OSL should provide the basic physics for understanding the nature of the inputs to the Earth and its atmosphere to be studied by the Earth Observing System (EOS). Due to the significant advances in the last few years in data storage and data transmission, the OSL can act as the center piece in a network of groundbased observatories, which then can extend the total solar measurement capability with instruments in the near and far infrared and with ultra high spectral resolution. Once OSL has provided data on the basic physical processes in the solar atmosphere, a great body of information already collected as well as new ground data will be able to be correctly interpreted.

The interactions of convection, radiation, and the magnetic field–magnetohydrody-namics–together represent the key to understanding not only the sun and other stars, but also galaxies, pulsars, neutron stars, and black holes. While the sun does not have the range of conditions of all of these objects, it provides the only place where magneto-hydrodynamic process of astronomical scale can be studied in any detail.

The OSL instrument complement is summarized in section 2. Section 3 discusses some of the operational characteristics of the OSL mission.

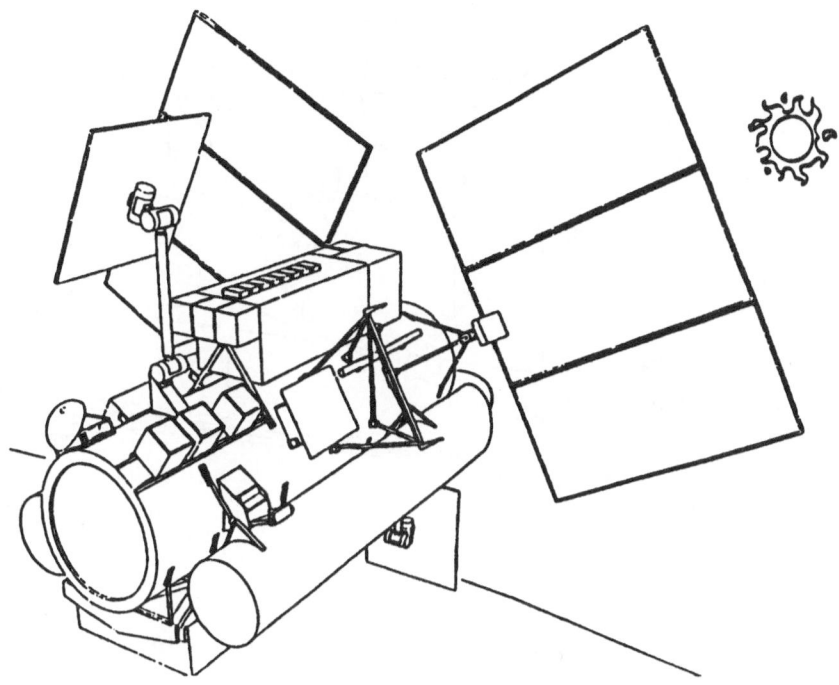

Figure 1: OSL pointing at the sun.

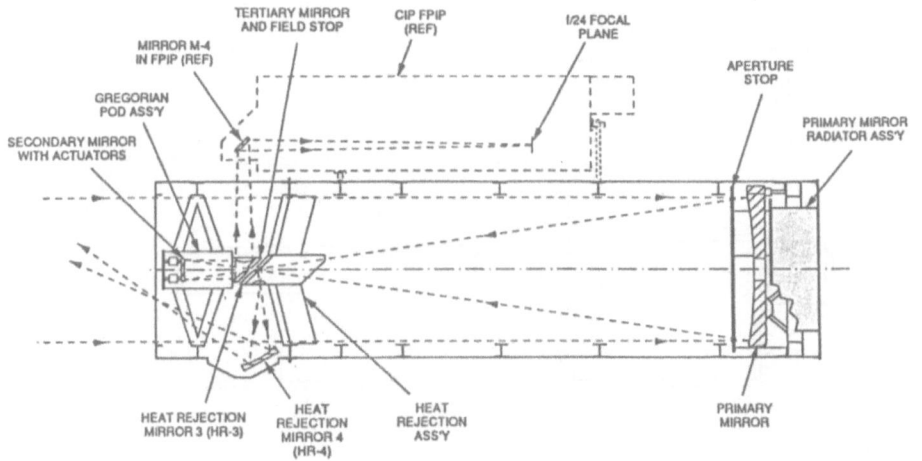

Figure 2: Optical Layout of the   meter telescope with CIP location indicated.

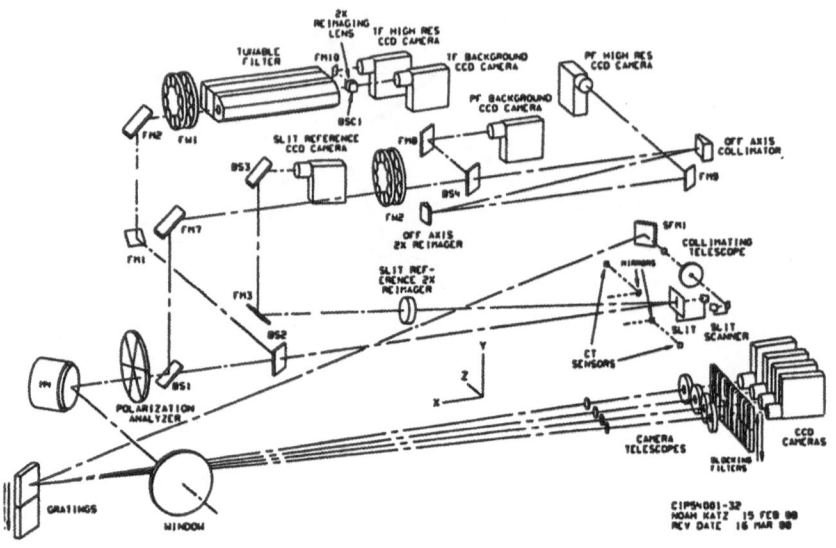

Figure 3: An optical cartoon of the CIP.

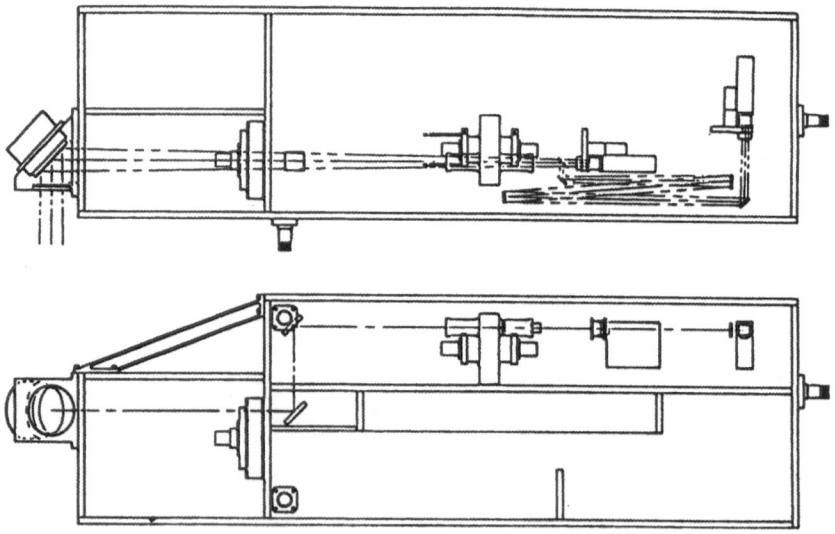

Figure 4: CIP with location of the Photometric Filtergraph components shown.

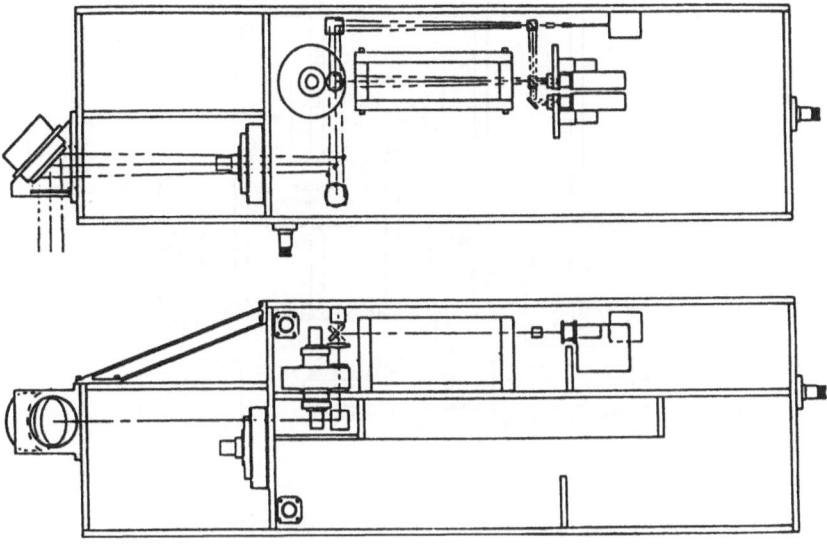

Figure 5: CIP with location of the Tunable Filtergraph components shown.

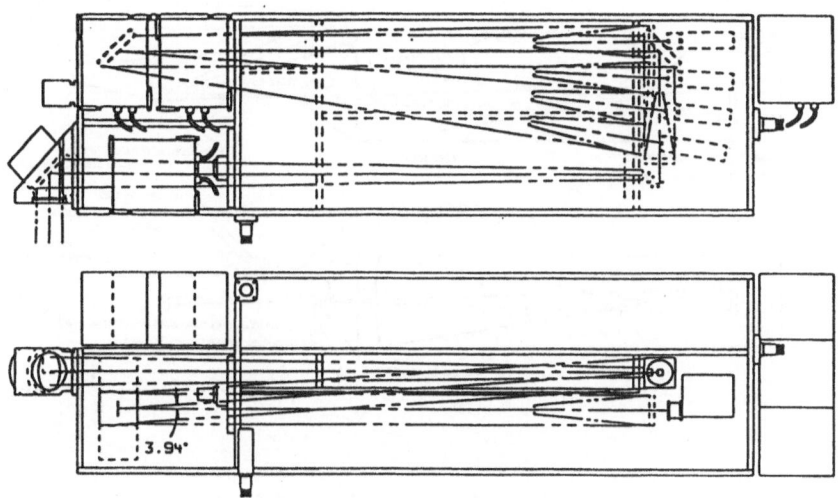

3.94°

Figure 6: CIP with location of the KIS spectograph components shown.

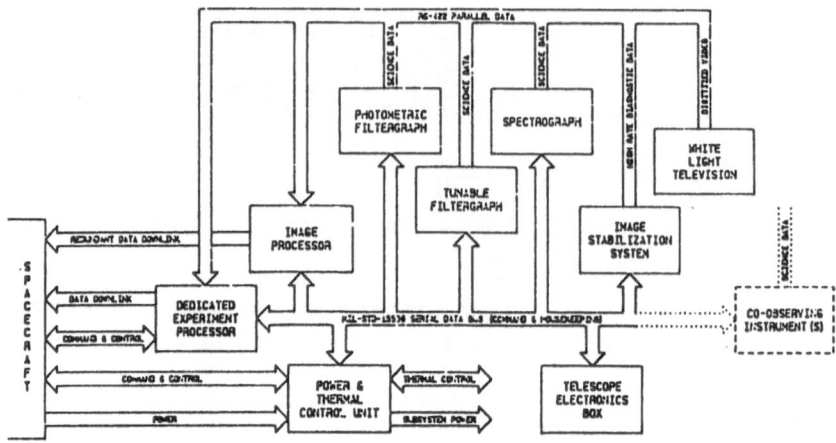

Figure 7: Top level electronic control diagram of CIP.

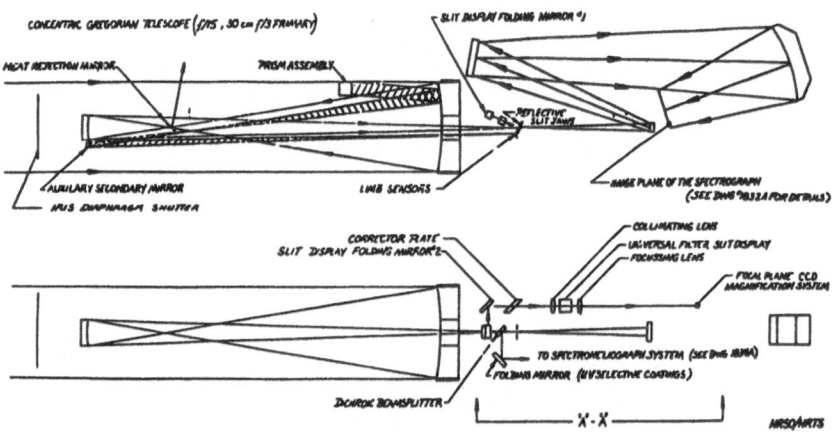

Figure 8: Optical layout of a modified HRTS for OSL.

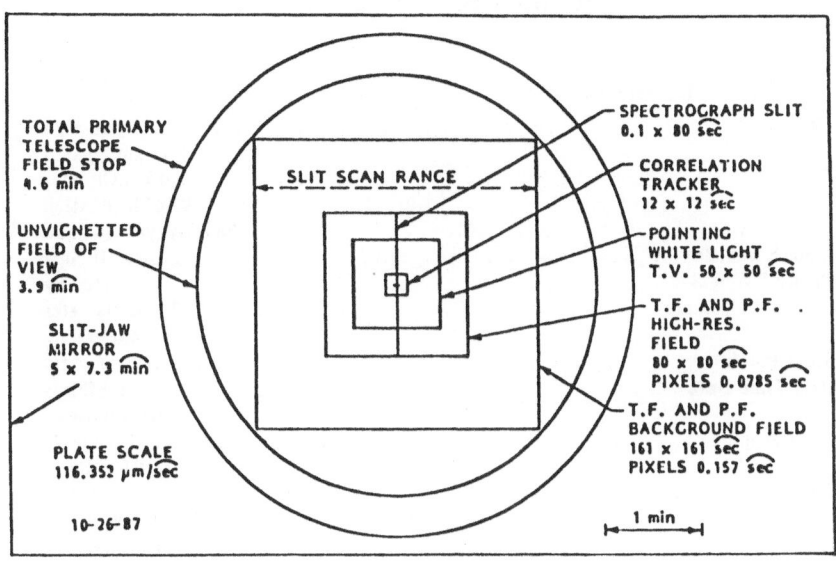

Figure 9: Fields of view of the CIP focal planes.

Orbiting Solar Laboratory

Mission Characteristics

Lifetime . . . . 3 yr design
8 yr anticipated
Orbit . . . . 510 km Circular
97.4 ° inclination
260 day/year full sun
Launch Vehicle . . . . Delta II class
Payload Weight . . . . 3400 kg
Maximum Volume . . . . 2.8 m diameter
4.6 m long
Total Power . . . . 7500 watts
Real Time Data . . . . 16 Mbit/s
TDRS K-Band
Housekeeping . . . . 1 kbit/s
TDRS S-Band
Data Coverage . . . . 8 hours/day
3 hours continuous
260 days/year
Operations Center . . . . 24 hours/day

Telescope for CIP

Design . . . . Gregorian with heat dump
Size . . . . 1 m clear aperture
Focal length . . . . 24 m
Resolution . . . . 0.13 arc second
Field of view . . . . 3.9 arc minute
Optical Alignment . . . . Secondary movable in six axes
Wavelength Range . . . . 2000 to 12,000 Å
Pointing: Offset Range . . . . ± 17 arc minutes
Stability . . . . ± 2 arc seconds

Table 1

Coordinated Instrument Package

System Characteristics

Focal length	24 m Low Res
	48 m High Res
Field of View	152 x 152 arc seconds (Low Res)
	(.15 arc second square pixels)
	76 x76 arc seconds (High Res)
	(.075 arc second square pixels)
Detectors	1024 x 1024 CCD
	18.3 micron square pixels
	200,000 electron full wheel
	S/N 316 (100,000 e average)
Polarization Analysis	RCP, LCP, Linear 0, 45, 90, 135
Data Rate	12 Mbits/s
Cycle Time	3.75 seconds
Internal Stablilization	0.02 arc seconds RMS
Drift WRT Solar Rotation	0.2 to 1.0 arc seconds/hr
Correlation Tracker Update	500 Hz
Stabilization 3 db point	65 Hz

Table 2

CIP Photometric Filtergraph

Wavelength Range . . . . 2000 to 11,000 Å
Detectors . . . . 1024 x 1024 CCD; 1 High Res, 1 Low Res

Wavelength (Å)	Band (Å)	Purpose
2100	100	UV continuum
2680	60	"Missing opacity" line blanketing
2795	12	MgII
3300	75	H–opacity minimum
3600	40	Balmer Continuum
3862	15	Fe, CN – local opac max
3915	10	Blue continuum between deep lines
3933	0.6	Calcium K narrow band; might also use broader
4305	4	G band – CH
4508	8	4503-4511 H – opacity minimum, clean
5670	10	Clean continuum
6563	0.4	H alpha line center
6687	10	Cleanest cont < 0.05 Å lines
8170	30	Paschen Continuum
9950	100	Furthest IR – for Planck baseline

CIP Tunable Filtergraph

Wavelength Range . . . . 4600 to 6600 Å
Positional Accuracy . . . . 1 mÅ Relative, 5 mÅ Absolute
Detectors . . . . 1024 x 1024 CCD; 1 High Res, 1 Low Res
Spectral Resolution . . . . 50 mÅ(narrow mode)
100 Å(wide mode)

W-Length	Species	Purpose
4571	Mg I	Temperature minimum diagnostic
4924	Fe II	Zeeman triplet
4607	Sr I	Weak magnetic fields via Hanle effect
5173	Mg I	Magnetic - velocities - low chromosphere
5247/5250	Fe I	Magnetic field strength pair
5324	Fe I	Strong g=0 line
5380	C I	Temperature diagnostic - photosphere
5394	Mn I	Temperature diagnostic - photosphere
5576	Fe I	Velocities - mid to upper photosphere
5714	Ti I	Sunspot umbral vel., g=0 (non-magnetic)
5876	He I	Prominences, spicules, activity
5896	Na I	Magnetic + vel. - low to mid chromosphere
6064	Ti I	Umbral magnetic fields
6302	Fe I	Magnetic line, less temperature sensitive
6563	H I	Chromospheric dynamics

Table 3

CIP Kiepenheuer Institut Solar Spectrograph

Wavelength Range . . . . 2700 to 10,000 A
Detectors . . . . 4 1024 x 1024 CCDs
Field of View . . . . 0.1 x 76 arc seconds
Scan Range . . . . 152 arc seconds
Resolution ($\lambda/\Delta\lambda$) 350,000
Configurations . . . . 2 grating positions, 3 sets of blocking filters
Dispersion . . . . 1.7 mm/Å (5000)

	Grating 1 (158 gr/mm)			Grating 2 (79 gr/mm)	
CCD-1:	3933.7	Ca K	CCD-1:	7771.9	IR O 1
	2795.5	Mg K		7774.2	IR O 2
				7775.4	IR O 3
CCD-2:	5582.0	Ca I		5895.9	Na D 1
	4468.5	Ti II		5172.7	Mg I
	5576.1	Fe g=0		8542.1	Ca IR
	5250.2	Fe g>0			
	5247.1	Fe g>0	CCD-2:	6302.5	Fe g>0
				5875.7	He I
CCD-3:	6562.8	H			
	5394.7	Mn I	CCD-3:	8248.8	Ca II
				8254.7	Ca II
CCD-4:	6302.5	Fe g>0		4657.0	Fe II
	5250.2	Fe g>0		4656.5	Ti II
	5247.1	Fe g>0		4657.2	Ti II
				4541.5	Fe II
				4128.7	Fe II
			CCD-4:	8498.0	Ca II
				5197.6	Fe II
				5193.0	Ti I
				4924.0	Fe II
				5052.2	C I
				4065.4	Fe I g=0
				7194.9	Fe I g=0

Table 4

Ultraviolet Spectrograph

Telescope

Design	Gregorian
Size	30 cm
Focal Ratio	F/15
Image Scale	20 microns/arc second
Resolution	0.5 arc second
Image Stabilization	0.2 arc seconds RMS
Data Rate	2 Mbits/second

Spectrograph

Type	Tandem Wadsworth
Spectral Range	1175 to 1700 Å
Field of View	0.5 x 256 arc seconds
Spectral Resolution	50 mÅ
Position	20 x 256 arc second band
Cycle time	3.75 seconds
Detectors	4 1024 x 1024 CCDs
Spectral Coverage	25.6 Å per CCD

Spectroheliograph

Field of View	256 x 256 arc seconds
Spatial Resolution	0.5 arc seconds (.25 arc second pixels)
Bandpass	15 Å
Wavelengths	1510 Å, 1550 Å, plus a visible band
Detector	1024 x 1024 CCD

Table 5

XUV/X-Ray Imager

Design	Normal Incidence Cassegrain Multilayer Mirrors
Wavelenght Range	40 – 400 Å
Spectral Resolution	2 – 4 Å
Spatial Resolution	0.5 arc seconds (0.25 arc second pixels)
Field of View	256 x 256 arc seconds
Detectors	4 1024 x 1024 CCDs
Temporal Resolution	30 s nominal
Data Rate	2 Mbits/s

Table 6

2. OSL Instruments

Figure 1 shows and artist's conception of the OSL pointing at the sun. The large central cylinder contains the 1 meter telescope, the two smaller cylinders attached to the sides indicate the volumes allocated for the Ultraviolet Spectrograph and the XUV/X-Ray imager, while the CIP is seen at the top of the telescope tube. The large square panels are solar arrays and the smaller ones are phased array TDRSS antennae. Figure 2 shows an optical layout of the 1 meter telescope. The characteristics of the entire instrument package are summarized in Table 1.

The CIP instrument has been under study since 1981. It consists of the Photometric Filtergraph (PI Professor Harold Zirin, California Institute of Technology), the Tunable Filtergraph (PI Alan Title, Lockheed Palo Alto Research Laboratory), and the Kiepenheuer Institute Solar Spectrograph (PI Professor Egon Schröter). An optical cartoon of the CIP and a series of optical mechanical drawings of the three instruments in the CIP are shown in figures 3 through 6. A top level electrical control diagram of the CIP is given in figure 7.

At present, two copointing instruments are in study as candidate for inclusion on the OSL mission. A version of the High Resolution Telescope and Spectrograph (HRTS) (PI Guenter Brueckner, Naval Research Laboratory) is being studied as the Ultraviolet Spectrograph. Figure 8 is an optical layout of the OSL/HRTS. The XUV/X-Ray Imager will probably be an array of normal incidence telescopes with multilayer coatings. These coatings have a high reflection coefficient in a spectral band of only a few angstroms and as such act as narrowband spectral filters. It is anticipated that the Air Force Geophysical Laboratory will provide the XUV/X-Ray Imager.

Figure 9 shows the fields of view of the CIP focal planes overlayed. The total science data stream of the OSL will be 16 Mbits/s proportioned 12:2:2 to the CIP:Ultraviolet Spectrograph:XUV/X-Ray Imager. The CIP data stream can be allocated in real-time, but normally the rates will be allocated internally as 4.5:4.5:3 to the PF:TF:KISS. The focal plane detectors of all instruments will be 1024 x 1024 CCD's. Those in the CIP will be front illuminated and have a full well of 200,000 electrons. The Ultraviolet instrument will probably use detectors that have been coated with a dye that absorbs at ultraviolet wavelengths and emits at visible wavelengths.

3. Operational Characteristics of the OSL Mission

The 510 km orbit will decay slowly, so that the OSL will have a useful life on the order of a decade. To conserve costs, the design life will be three years, but it is reasonable to expect that full scale operations should be possibe for 8 to 10 years. The OSL orbit will be fully illuminated about 260 days per year. During this time it can collect data constantly; however, its communication of data to the ground will be limited to TDRSS coverage. This has been guaranteed to be at least 3 hours of continuous coverage per day. The data handling capacity on the ground has been sized for 8 hours per day on average. However, the ground station has the capacity to capture 24 hours a day, and both the groundstation and the payload operations command and control center (POCC) will be manned 24 hours a day. The total duration of data collected on a given day will depend on the scheduling of the TDRSS, so that long data runs will occasionally be possible. An on-board tape recorder will be used to fill data gaps caused by switching TDRSS antennae. Its capacity, 6.5 gigabits, will store the full OSL data rate for 406 seconds (6.7

min). We expect that the recorder can be configured to collect data at a much lower rate as well. This would allow the recorder to store, for example, 540 1024 x 1024 frames, or 9 hours of images at one per minute. The OSL will not be operated during the 100 days per year when the orbit is in partial shadow.

The science programs of the OSL will normally be designed to use the total capacity of the instrument complement operating as a coordinated system. The CIP is designed to operate in this manner and the co-pointers will be designed to extend the concept of coordinated operations. In normal operation the CIP will collect images and spectra on a four second time base and run sequences which complete a cycle in 30 to 60 seconds. These cycles are designed to collect the basic data in one quarter or less of the 5 minute or 3 minute oscillation period. Examples of typical CIP observing sequences can be found in "Science Observing Programs for the Solar Optical Telescope".

The POCC will be at Goddard Spacecraft Center. The instruments will be run from science data and engineering stations designed and operated by PI team members. The CIP will normally capture all real time image and housekeeping data and analog and digital video data will be stored in real time. The image data will be formatted in near real time for distribution to observatories planning collaborative observing programs or for telescience usage. Within a few weeks after collection, the Goddard Data Center will provide data that have been time ordered, corrected for errors, and corrected for gain and dark current. The Science Working Group will be responsible for bringing the OSL into scientific operation. After full operation will have been established, the SWG will be responsible for selection of the science timeline for a significant fraction of the first observing campaign of 260 days. After the first observing campaign the observing time will be shared about 30:70 between the SWG and NASA selected Co-Investigators.

The OSL mission generates a great deal of data. Fortunately, new inexpensive mass storage media – 8 mm Digital Vidio cassette tape and digital and analog optical disks – allow a mechanism for distributing large data sets directly to the scientist at his work-station. Our experience is that the equipment required for serious OSL data analysis is not unusually costly compared to normal scientific analysis requirements, and that our scientific analysis would not be appreciably aided by the centralized production of other than the photometrically corrected data stream.

SOLAR IMAGE RESTORATION BY ADAPTIVE OPTICS

Jacques Maurice Beckers**
Advanced Development Program†
National Optical Astronomy Observatories*
Tucson, AZ   85726-6732

ABSTRACT. Adaptive Optics corrects the imaging by telescopes for
atmospheric wavefront distortions. I describe the concept and its
predicted performance. The concept of atmospheric tomography and
Multi Conjugate Adaptive Optics, or MCAO, is developed for solar
telescopes with a specific application to LEST. It increases the size
of the isoplanatic patch to make adaptive optics an attractive
alternative to space observations for observing extended (active)
regions on the solar surface.

1. INTRODUCTION:

Recently there have been a number of papers about the application of
the technique of adaptive optics to solar imaging (LEST Technical Report No. 28, Beckers 1987, 1988[b], Dunn 1987[a], Title et al. 1987). In
adaptive optics the imaging is corrected for atmospheric seeing in
real time by a complex servo loop which uses wavefront disturbance
measurements by a so-called Wavefront Sensor (WFS) and which applies
these error signals to a complex actuated mirror which corrects these
errors referred to frequently as an Adaptive Mirror (AM) or rubber
mirror.
    There are significant similarities as well as differences between
solar and nighttime adaptive optics and between visible and infrared
adaptive optics (Beckers 1988[b]). The similarities make it profitable
for both disciplines to work closely together. Differences result
mainly from the use of solar surface structures (pores, granulation)
for wavefront sensing, the use of smaller aperture telescopes in solar
astronomy, the shorter exposure times and from the insignificance of
the infrared sky and telescope background in solar research.

**Present address: European Southern Observatory, D-8046
Garching/Munchen, West Germany/FRG.

*Operated by the Association of Universities for Research in
Astronomy, Inc., under contract with the National Science Foundation.

R. J. Rutten and G. Severino (eds.), Solar and Stellar Granulation, 43–53.
© *1989 by Kluwer Academic Publishers.*

In Section 2, I will review the current status of solar adaptive optics. It's main predicted limitation is the small size of the so-called isoplanatic patch, the area on the sun over which the image is corrected (a few arcseconds). In Section 3, I will describe a proposed technique to increase this area. Such an increase is crucial for solar research where it is often necessary to study extended areas (sunspots, active regions, supergranule cells) to do the research necessary. In Section 4, I will outline future programs and prospects.

2. SOLAR ADAPTIVE OPTICS

2.1 Concept of Adaptive Optics

It is desirable to summarize the concept of adaptive optics before going into its application to solar physics. Figure 1 shows the

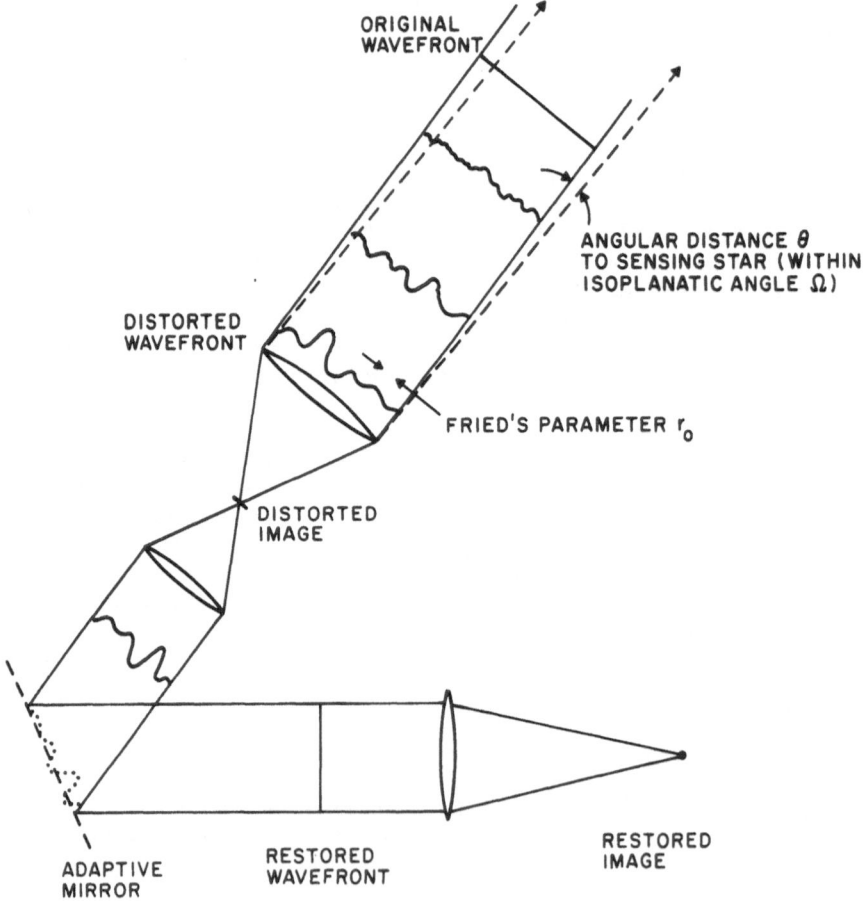

Figure 1: Concept of Adaptive Optics

essentials of an adaptive optics system for stellar observations. The
same considerations apply for solar observations. A wavefront sensor
using solar granulation or other solar structures (instead of a
"sensor star") measures the atmospheric disturbance for an area on the
solar disk within the so-called isoplanatic patch (diameter Ω). This
is done with a spatial resolution across the telescope aperture equal
to or better than the so-called Fried parameter r_0 and with a temporal
resolution τ better than r_0/v where v is the wind velocity of the
seeing layer. For a very good seeing site (Manua Kea, La Palma) where
the seeing is about 0.5 arcsecond r_0 equals 20 cm at $\lambda=500$ nm, $v=10$
m/sec, $\tau< 20$ msec and $\Omega\approx5$ arcseconds. The spatial and temporal
resolution of the adaptive mirror has to match that of the wavefront
sensor. Both r_0, τ and Ω are strongly wavelength dependent varying
proportional to $\lambda^{1.2}$ In the following discussion, I will assume these
values for these parameters and I will apply them to the LEST
Telescope (diameter D=240 cm) assumed to be located on Mauna Kea.

2.2 Solar Wavefront Sensors

Stellar wavefront sensors generally measure wavefront tilts using
stars located within the isoplanatic patch. This is done by shearing
interferometers but more commonly by Hartmann-Shack sensors. Since
the atmospheric wavefront disturbances at wavelengths >500 nm are
nearly achromatic when expressed in metric units (not in wavelengths!), the tilts are achromatic as well. Wavefront tilts therefore can be measured at any wavelength which is often advantageous as
e.g. in the solar case where granulation wavefront sensing is best
done at green wavelengths, where the granule contrast is high, even
when observing at longer wavelengths (e.g. 1.6μm). Dual wavelength
adaptive optics systems like this are referred to as "polychromatic
adaptive optics systems."
    Granule wavefront sensors have been described by von der Lühe
(1987), Beckers (1988[C]) and others. Figure 2 shows a sketch for the
granule Hartmann-Shack sensor proposed by Beckers (1988[C]). A mask M

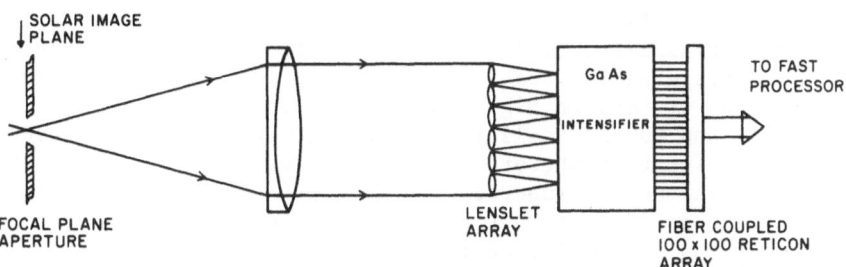

Figure 2: Solar Hartmann-Shack Sensor

less or equal to the isoplanatic patch size (5 arcsec) is placed in
the focus of the solar image. Behind it a lens L (focal lengths f)
images the telescope pupil onto a lenslet array LA (focal lengths

fa). Each lenslet images the solar area transmitted by M onto a 2D detector array A. In our case (D=240 cm, r_o=20 cm at 500 nm) the number of lenslets equal 127 or 13 across the pupil. This number decreases of course sharply with increasing wavelengths (to 7 at 1.6µm). For the present case, we assume the lenslets to be hexagonal and arranged in a close packed hexagonal pattern. Good achievable parameters for the lenslet array and detector arrays are: lenslet distance 600µm, pixel size 25µm equals 0.25 arcsec, reimaged mask M diameter 400µm, array size 350 x 350 or larger. With existing fast digitizers and correlators it should be possible to measure the relative displacements of the 127 granule images at a 100Hz rate using the edges of M as a zero wavefront distortion reference.

2.3 Solar Adaptive Mirrors

In stellar astronomy a major requirement that has to be placed on adaptive mirrors is their high reflectance, their low thermal emissivity and their low variation in thermal emissivity while they are being actuated. Adaptive mirrors with a continuous surface are therefore to be preferred. In solar astronomy, these requirements become less important (Beckers, 1988[b]) so that segmented mirrors are acceptable (Smithson et al. 1987, Dunn 1987). Even so the gaps between the segments have to be small to avoid a deterioration of the final modulation transfer function ("scattered light"). The Smithson et al. (1987) segmented mirror has the advantage over the Dunn (1987) mirror in that there is a one-to-one correlation between the segment actuation and the wavefront sensing since the spatial "influence function" for both are identical. That simplifies some control algorithms. That appears in fact to be the only advantage of segmented mirrors over continuous face plate mirrors. In general, I consider the latter to be preferred even in solar applications.

2.4 Expected Performance of Solar Adaptive Optics

No adaptive optics system works perfectly. Spatial and temporal limitations of the WFS and the AM cause the point spread function to be not a perfect Airy Disk. So do deviations from isoplanatism, failure to correct for amplitude variations (scintillation), deviations of achromatism for polychomatic adaptive optics etc. etc. Most of these imperfections show up more on high spatial frequency scales on the telescope pupil than on the small spatial frequency scale. The resulting point spread function therefore is a combination of an Airy Disk (A) and the original seeing function (S) as shown in figure 3. I have referred to such a profile as an A&S profile ("Airy and Seeing Profile"). It also has been referred to as a corrected profile surrounded by a halo. Smithson and Peri (1987) and Smithson et al. (1987) show some beautiful examples of the A&S profiles due to spatial WFS/AM limitations.

   I will refer to the fraction of the energy in the point spread function which ends up in the Airy part of the profile as the Strehl ratio (SR). For a good on-axis adaptive optics system, SR equals

≈75%. At the edge of the isoplanatic patch SR equals ≈35%. Even at this low value one ought to keep in mind that the width of the Airy disk equals ≈λ/D radians and that of the seeing disk ≈λ/r_0, which in the present case results in a difference in width of a factor of 12 and in area of 144. The amplitude of the Airy spike (figure 3) is therefore 77 x that of the S background even for the SR=35% case!

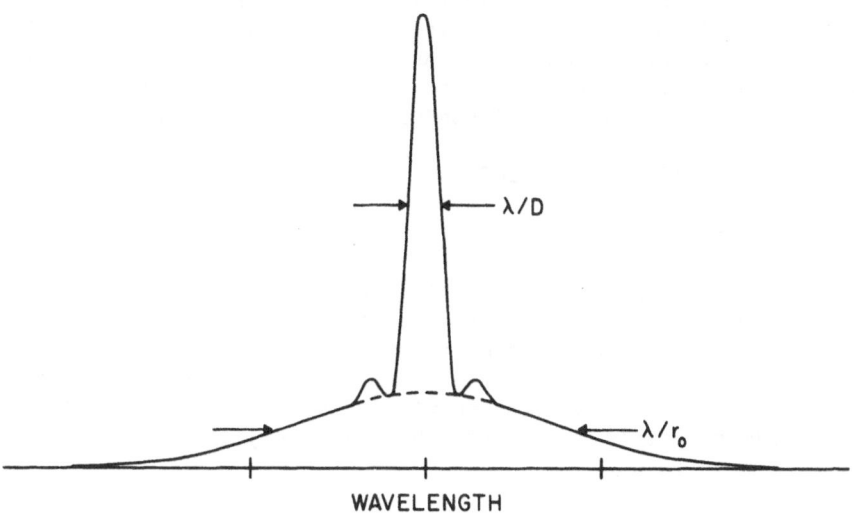

WAVELENGTH

Figure 3: Point Spread Function for an Adaptive Optics System

2.5 Promises and Limitations of Solar Adaptive Optics

The major promise of adaptive optics for solar physics is, of course, the high angular resolution which it will bring. For LEST this is 0.04 arcsecond at 500 nm and 0.14 arcsecond at the solar atmosphere opacity minimum near 1.6 μm. This will allow study of solar thermal, density, magnetic and velocity fields at scales at and below the pressure scale height of the solar atmosphere. This is an important threshold to overcome if we are to understand the physical processes which are acting in stellar atmospheres. Observations in the window at 1.6 μm are especially important since only at that wavelength the solar convection zone can be observed directly rather than the convective overshoot region at visible wavelengths. Velocity and magnetic fields will therefore be more intense. In addition it will be possible to measure vector magnetic fields better because of the larger Zeeman splittings.

There are however significant limitations to solar adaptive optics. They are:

(i) *The Limited Strehl Ratio*: which will cause an underestimate of the intensity, velocity and magnetic field contrast of ~25% on-axis.

By simultaneous measurement of r_o (e.g. from Hartmann-Shack test residuals) and by careful modeling this, and other SR effects, can be corrected by maybe a factor of 5 so that a residual uncertainity of ≈5% remains.

(ii) *The Small Size of the Isoplanatic Patch:* ≈5 arcsec at 500 nm and ≈20 arcsec at 1.6 μm is a severe limitation for the study of sunspots and active regions. It is, of course, possible to build multiple parallel adaptive optics systems which correct adjacent areas on the sun and then mosaic these areas. That is however very cumbersome, does not remove image distortions, and leaves large variations of the A&S function as a result of variations of deviation of isoplanatism. In the next section, I will describe another way to increase the size of the isoplanatic patch called Multi Conjugate Adaptive Optics (or MCAO).

3. MULTI CONJUGATE ADAPTIVE OPTICS

3.1 Description of Concept

The size of the isoplanatic patch is limited because the rays from different parts of the sun pass through different parts of the earth's atmosphere. They therefore experience a different integrated wavefront distortion by the time they reach the telescope aperture. This would even be the case if the atmospheric seeing were confined to a single layer (as sometimes approaches the real case) as shown in figure 4.

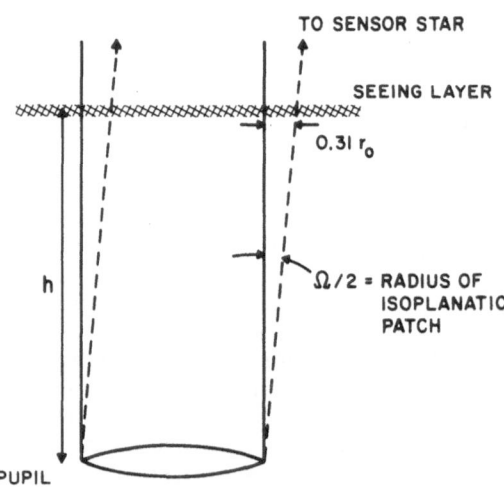

Figure 4: Single Layer Seeing Model

However, if the wavefront distortion at the single layer could be measured over an extended size, together with its height, one could place the AM at a conjugate of this layer rather than at a conjugate of the pupil. The resulting Single Conjugate Adaptive Optics (SCAO) system would have a much larger isoplanatic patch, in principle possibly up to the entire solar diameter.

In general the atmospheric seeing is, of course, not confined to a single layer. In that case, I have proposed (Beckers 1988[a]) compensating atmospheric seeing layer by layer. The increase of the isoplanatic patch in such a Multi Conjugate Adaptive Optics System (or MCAO) depends on the 3D structure of the wavefront disturbances, on our ability to measure and correct that, and of course on the number N and position of the layers in the assumed atmospheric model. For the following discussion, I have assumed an approximate model for the atmosphere above Mauna Kea derived from a model by Hufnagel (see Beckers 1988[a]) in which the C_n^2 parameter was taken as 2.2×10^{-17} m$^{-2/3}$ up to a local height of 10 km and zero above. This again gives r_0=20 cm at 500 nm.

3.2 Atmospheric Tomography Using Solar Wavefront Sensing

An essential ingredient for solar MCAO is the knowledge of the three dimensional structure of the wavefront above the telescope. Figure 5 explains the principle of atmospheric tomography applied to the measurement of the atmospheric wavefront disturbances at different heights. When observing the wavefront for different points on the solar disk the projection through the atmosphere varies. The

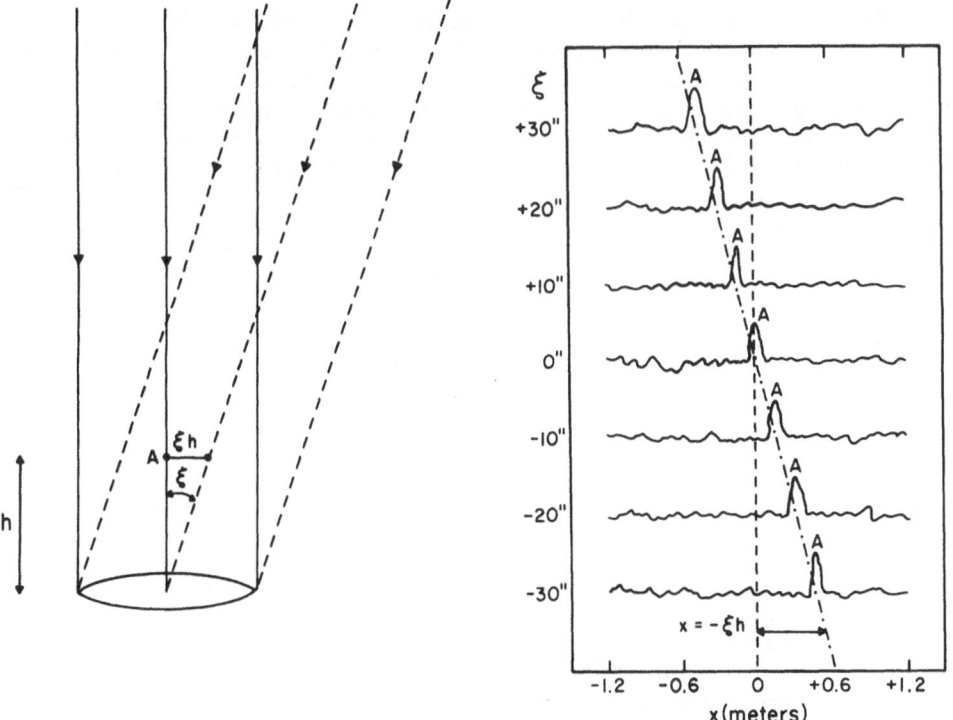

Figure 5: Principle of Atmospheric Tomography using the Sun.

refractive effects at a point A at height h located on the axis of the
telescope show up at the center of the pupil (x=o) in the wavefront
measurement of the part of the sun to which the telescope is pointed.
For a point on the sun ξ radians off-axis its effects show up displaced at x = $-\xi$h. In a two dimensional plane at right angles to the
telescope axis with (ξ,η) as angular and (x,y) as linear coordinates,
the displacement is ($-\xi$h, $-\eta$h). Since the phase disturbance processes
are linear in their behavior, tomography is directly applicable.

The simplest form of atmospheric tomography is therefore the
measurements of the wavefront for a grid of solar surface areas
(ξ_j, η_j) resulting in phases $\phi(x,y,\xi_j,\eta_j)$ in the pupil plane. By
shifting each of these by (ξ_jh, η_jh) and averaging the results, a
reasonable approximation is obtained for the phase disturbance
$\Delta\phi(x,y,h)$ at the height h. Better approximations can be obtained by
applying the extensive set of tools developed for other forms of
tomography. One might for example deconvolve the $\Delta\phi(x,y,h)$ by the
contribution function C(h,H) which can be derived for this simple
process describing the contribution of other layers H to the estimate
of $\Delta\phi$ at layer h.

In Beckers (1988[a]) I discussed the number of (ξ,η) values needed
to increase the isoplanatic patch. For a model of the atmosphere of N
layers divided so that each layer has the same Fried parameter r_i of
its own ($r_i^{-5/3}$ = $r_o^{-5/3}$/N) I found that the diameter of the isoplanatic patch is increased from Ω to $\omega=2N\Omega$. For 500nm that means an
increase from 5 arcseconds to 40 arcseconds for N=4, which is the number of layers which I will adopt for the present exercise. For 1.6μm
the increase is from 20 to 160 arcseconds. I assumed a rectangular grid
of (ξ,η) values 10 arcseconds (at 500nm) apart. Figure 6 shows this
grid superposed on the original (Ω) and enlarged (ω) isoplanatic patch.
A total of 21 positions on the solar disk is therefore indicated.

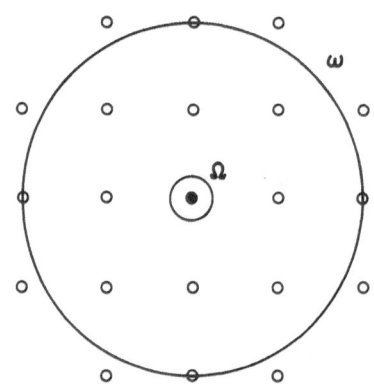

10" at 0.5 μm
40" at 1.6 μm
60" at 2.2 μm

Figure 6: Grid of Wavefront Sensing Points on Solar Disk

Probably the best way to accomplish the wavefront sensing at these 21 points is with a set of 21 separate but identical wavefront sensors of the type described in section 2.2. Figure 7 illustrates a possible configuration. Behind a set of 21 aperture masks each 4 arcsecond in size on 10 arcsecond centers (A) one may place a set of

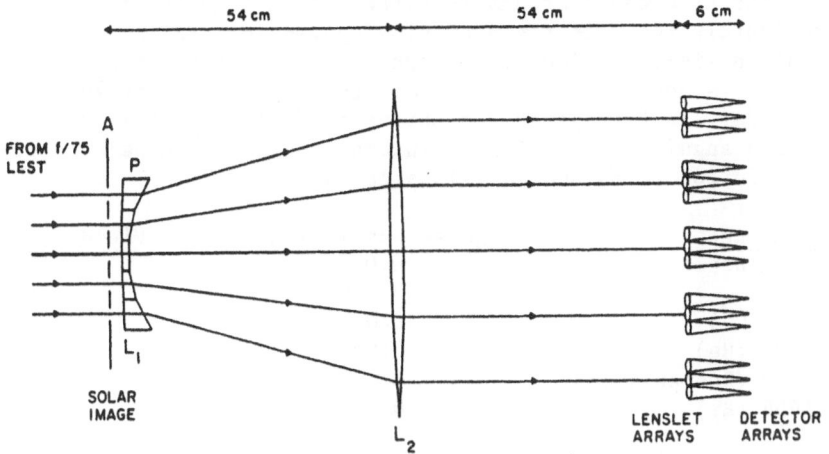

Figure 7: MCAO Wavefront Sensor for LEST. For Similicity Only Three Lenslets Are Shown For Each Solar Grid Point.

prisms P which deviate the principle rays for each aperture by an amount large enough to separate the pupils which have an angular extent of 1/75 radian for the f/75 LEST focus. For an angular deviation pattern of P similar to the aperture pattern (figure 6) and a deviation step of 0.02 radian the maximum deviation equals 0.045 radian. To keep chromatic effects small (corresponding to <1 cm on the 240 cm LEST aperture) the spectral bandwidth has to be limited to 100Å for BK7 glass used at green wavelengths. The prismlets array P is mounted on a field lens L_1 which places the LEST aperture at infinity. A lens L_2 collimates the aperture masks and images the 21 pupils on 21 lenslet arrays each of which has 127 lenslets and gives 127 images on each of the 21 detector arrays as described in section 2.2. This is, of course, a very extensive wavefront sensing system. One should however keep in mind that it uses 21 identical, parallel systems using parallel processing except for the final atmospheric tomography algorithms.

4. WHAT'S AHEAD?

The first order of business is, of course, to make the present prototype adaptive optics systems (Beckers 1988[b], Merkle 1987) into working systems including making them routinely user accessible. Solar and Stellar Adaptive Optics systems developers should work in close communication so that the optimum benefit is derived from their common experience. I expect adaptive optics to be a component of solar and stellar telescopes within the next 5 to 10 years.

Multi Conjugate Adaptive Optics will be the next step. Although of interest for nighttime astronomy, its promise for solar research far exceeds that. MCAO will give solar astronomy what it needs from adaptive optics systems: extended field of views so that the larger areas that one has to observe on the sun (e.g. active regions) fall within the extended isoplanatic patch. An N=4 system can deliver approximately the same field of view system for LEST as the high resolution field of view of the space solar optical telescope (or whatever its descendants are named these days) which is 76 x 76 arcseconds. An N=4 system for LEST should give the following field of views and angular resolution at interesting wavelengths (the present SOT resolution at Hα equals 0.13 arcsec).

Wavelength	Angular Resolution(")	Field of View (")
λ5250 (Zeeman line)	0.045	42
Hα	0.056	55
λ8542 (Ca)	0.074	76
λ10830 (He)	0.093	101
λ15648 (Zeeman line)	0.135	157
λ21656 (Bγ)	0.186	232

Solar Multi Conjugate Adaptive Optics systems have substantial advantages over stellar systems. They do not require arrays of artificial stars for wavefront sensing, they therefore do not suffer from the image distortion effects which are inherent in MCAO systems using artificial stars; the sun being an extended object can serve itself as the multiobject array needed for tomography, and being a bright object atmospheric tomography can be done at short wavelength and with high spatial resolution across the pupil.

Because of this the MCAO concept is an ideal one to be first explored by solar research. Technological developments are relatively minor since the same systems can be used as for the "conventional" systems under development now, one needs just more of them. Since $r_i > r_o$ the adaptive mirrors can in fact be simpler as discussed in section 3.4.

A sensible program for LEST MCAO optics is therefore to pursue the "conventional" prototypes now, to computer model atmospheric tomography using maybe actual observations used by experimental wavefront systems, to use the results for the design of the MCAO wavefront sensing and adaptive mirror system, and then to implement it as part of LEST.

5. REFERENCES:

Beckers, J.M. and Goad, L. 1987[a] in 'High Resolution Imaging in the Infrared using Adaptive Optics,' in Towards Understanding Galaxies at Large Redshifts (Eds. R. Kron and A. Renzini), Reidel Publishing Company.

Beckers, J.M. and Goad, L. 1987[b] 'Image Reconstruction using Adaptive Optics,' in Instrumentation for Ground Based Optical Astronomy (Ed. L. Robinson), Springer Verlag.

Beckers, J.M. 1987 'The NOAO/ADP Adaptive Optics Program and its Applica-ion to Solar Physics,' LEST Technical Report 28, 55.

Beckers, J.M. 1988[a] 'Increasing the Size of the Isoplanatic Patch with Multi conjugate Adaptive Optics,' Proceedings ESO Conference on "Very Large Telescopes and their Instrumentation," (Ed. M.H. Ulrich), in press.

Beckers, J.M. 1988[b] 'Adaptive Optics Work at NOAO,' Proceedings ESO Conference on "Very Large Telescopes and their Instrumentation", (Ed. M.H. Ulrich),in press.

Beckers, J.M. 1988[c] 'Hartmann-Shack Wavefront Sensors using Extended Structured Objects,' ADP R&D Note 88-01.

Dunn, R.B. 1987[a], 'Adaptive Optical System at NSO Sac Peak,' LEST Technical Report 28, 87.

Dunn, R.B. 1987[b], 'Specifications for the LEST Adaptive Optical System,' LEST Technical Report 28, 243.

Merkle, F. 1987, 'Adaptive Optics Program of ESO,' LEST Technical Report 28, 117.

Smithson, R.C., Peri, M.L., and Benson, R.S. 1987, 'A Quantitative Simulation of Image Correction for Astronomy with a Segmented Mirror,' LEST Technical Report 28, 179.
Smithson, R.C. and Peri, M.L. 1987, 'Partial Adaptive Correction of Astronomical Images,' LEST Technical Report 28, 193.

Title, A.M., Peri, M.L., Smithson, R.C. and Edwards, C.G. 1987, 'High Resolution Techniques at Lockheed Solar Observatory,' LEST Technical Report, 28 107.

von der Lühe, O. 1987, 'A Wavefront Sensor for Extended, Incoherent Targets,' LEST Technical Report 28, 155.

IMPROVING SOLAR IMAGE QUALITY BY IMAGE SELECTION

Jacques Maurice Beckers*
Advanced Development Program
National Optical Astronomy Observatories**
Tucson, AZ   85718   USA

ABSTRACT. Recent research has quantified the gain in angular resolution that can be achieved by image selection with high-quality, large-aperture solar and stellar telescopes. By applying the results to solar astronomy one might expect to improve solar image quality by a factor of 3-4 beyond average seeing quality.

1. INTRODUCTION

It has long been a practice in solar and planetary physics to take rapid sequences of images on photographic film or video tape and then select the best among these images for analysis. The best images were frequently significantly better than the average image quality, and substantial gains were achieved. Sometimes a seeing monitor has been used (e.g., Bray et al. 1959) to preselect the best images in order to avoid collecting useless data.

Recently Hecquet and Coupinot (1985) quantified the improvements that can be expected from image selection under the assumption of the now generally-accepted Kolmogorov-type atmospheric disturbances of the optical wavefront. Their results have been compared in a limited way to solar and stellar observations at the Pic du Midi Observatory (Hecquet and Coupinot 1984) and to the Canada-France-Hawaii Telescope (Lelievre et al. 1988). This paper will review these predictions and comment on their application to future telescopes.

*Now at European Southern Observatory, D-8046 Garching, West Germany.

**Operated by the Association of Universities for Research in Astronomy, Inc., under contract with the National Science Foundation.

R. J. Rutten and G. Severino (eds.), Solar and Stellar Granulation, 55-60.
© 1989 by Kluwer Academic Publishers.

2. EXPECTED IMAGE IMPROVEMENTS

In Table I, I summarize the results of Hecquet and Coupinot. The telescope diameter D has been expressed in units of the so-called Fried parameter r_o.

TABLE I

Expected Improvement of Image Size over the Expected Long-Exposure Image Size by Selection of the Listed Percentile of Best Images.

D/r_o	76%tile	10%tile	1%tile	0.1%tile
3	2.0	2.7	2.9	3.0
5	1.9	3.1	3.6	4.0
10	1.5	2.4	3.0	3.8
20	1.3	1.9	2.3	3.0

The long exposure image size equals $4\lambda/\pi r_o = 1.27\lambda/r_o$ radians and r_o changes as $\lambda^{1.2}$. It is therefore clear that even a modest amount of image selection can increase the image quality substantially.

For solar observations the selection percentile can be quite small. In order to "freeze" the wavefront distortion it is necessary to use exposure times less than the time of change of the wavefront. The latter is typically 0.02 second, so exposure times of 10 milliseconds or less are in order. Assuming a lifetime of the smallest observable structures of 20 seconds, that means a selection percentile of 0.1% for direct broadband imaging, giving an image improvement of a factor 3 to 4 over a long-exposure image.

Spectroscopy and narrow-band imaging require longer observing times (0.2-2.0 seconds) which exceed the wavefront change time. In that case multiple (10-100) exposures are needed on a rapidly guided image, with a seeing monitor to trigger the exposures. The selection percentile is reduced to 1 to 10%, bringing the image improvements to a factor 2-4.

3. WHY ARE THESE IMPROVEMENTS GENERALLY NOT ACHIEVED?

Although solar astronomers have used image improvement by image selection for a long time, they have not achieved anywhere as dramatic an improvement as a factor of 4. Why not? Since the Kolmogorov character of the wavefront disturbances has been well established one must look at the assumptions made for the telescope and its (dome) environment. To achieve the improvements one needs (according to Table I) a telescope diameter $D \approx 5\ r_o$, or 50 cm on a 1 arcsecond site and 100 cm on a 0.5 arcsecond site at 500 nm wavelength. In addition,

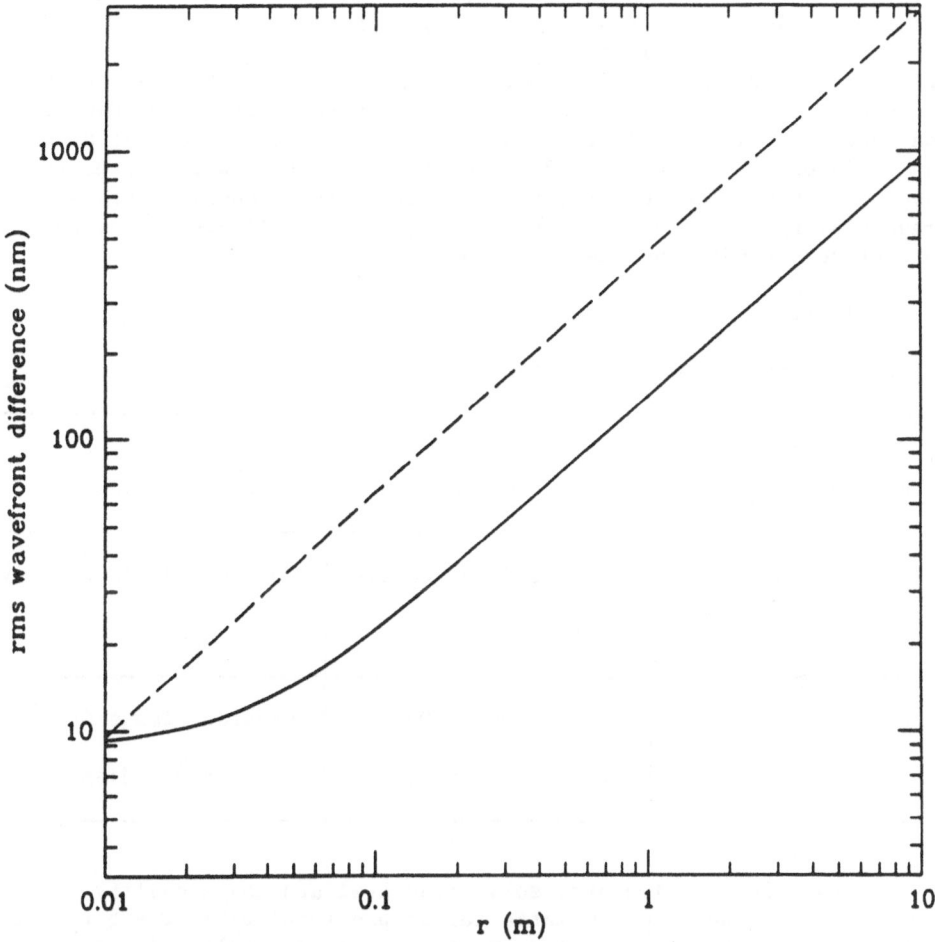

Figure 1. RMS wavefront difference between points on the primary mirror separated by r. Dashed curve: atmospheric differences for 0.25 arcsecond seeing (r_0=40 cm at 500 nm). Solid curve: specification for polishing and testing errors giving 0.06 arcsecond image size plus a small-scale RMS roughness of 6 nm. The same specifications apply for wavefront errors due to mirror support, mirror-generated seeing, thermal warping and coefficient of thermal expansion (CTE) warping (Angel and Beckers 1987).

one needs a telescope good enough to achieve the promised image quality. Modern nighttime telescopes have optical specifications that are coupled to the spectrum of wavefront disturbances. Figure 1 gives the specification for the polishing and testing errors for the mirror in the US National 8-meter telescope. The combined polishing, testing, support, mirror seeing and thermal and CTE warping errors will correspond to an equivalent atmospheric r_0 of 80 cm at 500 nm. I will refer to this as r_0^T, and to that of the atmosphere as r_0^A. As long as $r_0^T > r_0^A$ the calculations of Hecquet and Coupinet will be valid, the atmosphere occasionally canceling even the small errors in the optics. When $r_0^T \leq r_0^A$ this will not be the case and image selection will become ineffective, constant (poor) image quality being reached for small r_0^T. Table II summarizes the four extreme conditions one can encounter.

Table II

Image Variability as Function of Telescope Size and Quality

Telescope Quality:	$r_0^T > r_0^A$ ("Good Telescope")	$r_0^T \leq r_0^A$ ("Poor Telescope")
Telescope Size $D \leq r_0$ ("small")	Constant Image Quality (λ/D radians)	Constant Image Quality (λ/D radians)
$D > r_0$ ("large")	Variable Image Quality (λ/r_0^A radians)	Constant Image Quality (λ/r_0^T radians)

Image selection improves solar (and stellar) image quality therefore only when one has large telescopes (preferably $D \approx 5\ r_0^A$) of high quality ($r_0^T \gg r_0^A$). The spectrum of image quality is in fact a good measure of the merit of a solar telescope. An optimum solar telescope at a 0.5 arcsecond visible seeing site would have; at 500 nm wavelength, a diameter $D \approx 100$ cm and an r_0^T of 40 cm; and at 1.6 μm wavelength, $D = 400$ cm and $r_0^T = 160$ cm. LEST, with its 240 cm aperture, appears a good compromise.

4. ISOPLANATIC PATCH

Most solar image selection in the past shows the image quality to improve over a large area of the solar disk. In this sense, image selection may behave quite differently from adaptive optics (Beckers 1989). In adaptive optics the diameter of the isoplanatic patch is about $0.6\ r_0^A/h$ where h is the average height of the atmospheric

seeing layer. For image selection, the isoplanatic angle is more
likely determined by the telescope diameter D. Although it has not
been modeled yet, I guess it to be near $0.6D/\bar{h}$ or D/r_0^A, larger than
for adaptive optics. For the Sac Peak vacuum telescope, where D=75cm
and $r_0^A \approx 12.5$ cm, this amounts to a factor of 5 and isoplanatic
angles of 10 and 50 arcseconds respectively if $\bar{h}$ = 2500 meters. For
the case in which most seeing is near the telescope the angles will be
larger. For small and/or poor telescopes the angle will also be much
larger.

5. CONCLUSION

Image selection is likely to remain a powerful low-cost tool for
solar image improvements in the future. My experience with image
selection at the SPO vacuum telescope does not indicate the factor of
~4 improvements which would have been expected according to Table I--
which means that even with that powerful telescope there is room for
improvement. The predicted image selection improvements certainly
dictate size and quality for future solar telescopes like LEST.

One should keep in mind that the Fried parameter r_0 is a
statistical quantity characterizing the average atmospheric
behavior. It is therefore incorrect to refer to the 0.1 percentile
time of a factor 4 improvement (at D/r_0=5) as a momentary improvement
in r_0 to $4r_0$.

REFERENCES

Angel, J.R. and Beckers, J.M. 1987, Proposal to NSF.

Beckers, J.M. 1989 (this workshop).

Bray, R.J., Loughhead, R.E. and Norton, D.G. 1959, Observatory 79, 63.

Hecquet, J. and Coupinot, G. 1985, J. Optics (Paris) 16, 21.

Lelievre, G., Nieto, J.L., Salmon, D., Boulesteix, J. and Arnaud, J.
1988. Instrumentation in Astronomy (Ed. L. Robinson), Springer-
Verlag.

Discussion

KOUTCHMY — At the beginning of your talk, you used the work of Hecquet and Coupinot to infer the optimum size of a large solar telescope (presumably the LEST). Do you think it is really possible to extrapolate the results obtained at nighttime at Pic du Midi with a 2-meter telescope, to deduce the right size of a daytime telescope? And if so, why not employ the methods you described at the focus of the McMath telescope, which is a long focal length telescope with an aperture only a little bit smaller than what you think is the optimum?

BECKERS — I do not see why the theory of seeing should give different results for solar and stellar telescopes. The Hecquet and Coupinot calculation is valid for *good* solar telescopes, in which the optics and internal seeing cause no limitations to the image quality. I suspect that for the McMath and for many stellar telescopes this 'good telescope' assumption may not hold, at least at visible wavelengths.

RIGHINI — It may be considered that the existing best solar telescopes are not yet in the best sites, since they do not show for long periods of time, to my knowledge, the best attainable r_0, i.e. $r_0 \simeq 40$ cm, which is the value that may be obtained from the ground, as is indicated by several measurements.

I would therefore like to stress that, parallel to the development of adaptive optics, a lot of work remains to be done in searching for better sites in the 'optimal site belt' (Erasmus 1988)[1], and in designing the telescope structure as well.

BECKERS — I completely agree. For best solar imaging one needs good sites (r_0 large), very good large solar telescopes (r_0 for the telescope larger than for the atmosphere), and then adaptive optics—which become much simpler and more effective for good telescopes on good sites.

TITLE — Is it fair to argue that a multi-conjugate system is comparable to a space system? It seems to me that a multi system suffers from the 'scattered' light problem as a single system does.

BECKERS — That is not clear yet. The MCAO system may improve the 'scattered' light problem (= the seeing part of the profile) within the area of the increased isoplanatic path. It needs computer modelling. Anyway, I suspect that it will be possible to correct the image for the scattered light to some extent, since the Fried parameter can be measured from the residuals of the wavefront sensor signals.

DAMÉ — Could you comment on laser shooting techniques?

BECKERS — Laser stars in the Na line are faint. The best laser stars made so far are 14th magnitude, the best that can be done with existing lasers is 10th magnitude, and the best that can be done with a powerful CW laser is limited by stimulated emission in the sodium layer to 3rd magnitude. Artificial laser-generated stars are therefore useless to solar physics.

---

[1]D.A. Erasmus, 1988,*Meteorological Considerations in the Siting of Large Solar Telescopes*, LEST Technical Report Nr. 31, Institute of Theoretical Astrophysics, Oslo.

RECONSTRUCTION FROM FOCAL VOLUME INFORMATION

J. A. HÖGBOM
Stockholm Observatory
S-133 00 Saltsjöbaden
Sweden

ABSTRACT. The relevant information in the focal region is not restricted to the focal plane alone. Some correction for atmosphere disturbances is possible using a simultaneously recorded defocused image. An analysis of the full 3-dimensional interference pattern in the focal volume shows that this contains all the information that is needed for a determination of the object brightness distribution independent of atmospheric wavefront disturbances. Three dimensional detection over the focal volume offers an alternative for diffraction limited imaging from the ground.

1. On Information in Optical Images

Diffraction limited imaging from the ground is difficult because of instrumental and, in particular, atmospheric disturbances along the optical path. In this paper I discuss what information is available in the focal region, how this information is distributed and how it may be used to achieve diffraction limited imaging.

Hardware modifications can be made to the optical system in order to improve the image quality. A prime example of course is the use of adaptive optics which holds the promise of routine diffraction limited imaging from the ground. Future experience with operational systems will tell us what further processing may be needed. Optical phase closure has opened the way to phase coherent interferometry and aperture synthesis. Here, indeed, vital parts of the processing must be performed off line after detection. Speckle imaging has proven its worth also in solar imaging .

1.1. THE TRANSFER FUNCTION

Short exposures that 'freeze' the atmospheric disturbances are essential for diffraction limited work. The recorded image $I(\xi)$ can be formally described as wanted object brightness distribution $B(\xi)$ convolved with the point spread function $PSF(\xi)$:

$$I(\xi) = B(\xi) * PSF(\xi) \tag{1}$$

Convolution is a somewhat complex operation and the image itself will not be the best form of display on which to base a discussion about information, relevant or otherwise, in the data. This is better done in the transform domain, i.e. we do not primarily discuss the image itself but its Fourier transform. Since the image equals the convolution of the object brightness

61

distribution and the PSF, it follows that the image transform $T(u)$ is the product of the corresponding two individual transforms:

$$T(u) = \bar{B}(u) \cdot T(u) \qquad (2)$$

The transfer function $T(u)$ – the Fourier transform of the PSF – acts as a spatial frequency filter operating on the information, here represented by $\bar{B}(u)$, the transform of the object brightness distribution. $T(u)$ equals the autocorrelation function of the aperture (field) transmission $f(x)$:

$$T(u) = \int_{-\infty}^{+\infty} f(x+u/2) \cdot f^*(x-u/2) \cdot dx \qquad (3)$$

The aperture transmission for an ideal error free circular aperture is unity inside the aperture and zero outside. The corresponding (circularly symmetrical) transfer function has a nearly triangular profile. Information residing in spatial frequency components higher than those corresponding to the full aperture diameter are lost since the transfer function here disappears.

1.2. INVERSE FILTERING

The PSF is the Fourier transform of $T(u)$. The ideal error free circular aperture produces the well known Airy disc PSF. The transfer function simply weights the relevant information, and we may if so desired change this weighting and thereby also the shape of the PSF. This can be done simply by multiplying the image Fourier transform by $T_w(u)/T(u)$, where T_w is the wanted weight function. Enhancing the high spatial frequency information, we can increase the weight of the fine structure information. The PSF is narrowed, giving a sharper picture. There is a compromise here because in narrowing the PSF we also change its overall shape and, in particular, secondary maxima become more prominent. Still, the 'natural' strongly tapered transfer function is unnecessarily conservative for many applications.

It may seem a luxury to discuss the possible further sharpening of a diffraction limited image. However, such inverse filtering does improve images selected with the seeing monitor on the Swedish 50 cm La Palma telescope. Reweighted pictures become noticably sharper than the originals. This implies that the true transfer function, including the effects of atmospheric and telescope errors, cannot differ grossly from the ideal undisturbed form. In particular, if $T(u)$ had mean phase error variations larger than $\pi/2$, then the enhancement of the higher wave number information would have made the pictures worse, not better.

1.3. CORRECTION FOR ATMOSPHERIC DISTURBANCES BY INVERSE FILTERING

Even within the available portion of the transform domain, information about $\bar{B}(u)$ will be of doubtful value when $T(u)$ is badly known. This, of course, is the case when the incoming waveform has been disturbed by the atmosphere. Atmospheric phase and amplitude modulation of the wavefront, referred to the plane of the aperture, make $f(x)$ an incompletely known complex function. The same then holds for its autocorrelation function $T(u)$. However, if this can be measured, then one can attempt to correct the image by inverse filtering: the image transform is multiplied by T_w/T. There are two obvious problems: how to measure $f(x)$ and what to do for values of u at which the autocorrelation function T is close to zero. A way out of the latter problem would be to record two or more consecutive images; transfer functions corresponding to different wavefront perturbations will not in general have zeroes in the same places.

With the Fried parameter r_0 of the order of 10 cm at good sites, many solar telescopes have apertures of a few times r_0. Phase variations over distances $<r_0$ are tolerably small, and the main wavefront disturbances over such apertures may be adequately described in terms of a limited number of low wave number components, i.e. those whose wavelengths are $>r_0$. If these low wavenumber components can be measured separately, then one can calculate the corresponding transfer function (3) and attempt to restore the disturbed image by inverse filtering. I have made computer simulations in order to study a simple procedure that can easily be realized with existing solar telescopes, in particular the 50 cm Swedish La Palma instrument.

Two unknown functions combine to form the image (1) and one feels that two differently recorded images may allow a separation of the two, i.e. two images in which the two functions interact in different ways. This implies the simultaneous recording of two (or possible more) images, each taken with its own modification of the optical system. In practice, this can be achieved by introducing a beamsplitter at a suitable point in the optical system, making the required alterations to the optics in one or both beam paths, each leading to its own camera. The cameras must be synchronized to take simultaneous pictures. Modifications to the system such as apodization, the blocking of parts of the aperture or an additional phase modulation will change the aperture transmission in a way which is equivalent to a multiplication of $f(x)$ by a function $g(x)$, which only depends upon the (known) modifications.

2. Information in a Defocused Image

2.1. THE EQUIVALENT TRANSFER FUNCTION OF A DEFOCUSED IMAGE

A simple application of the approach discussed above would be to register, in addition to the normal image, a defocused image. The latter will be equivalent to a normal image taken with a square law phase modulation of the aperture. For simplicity I shall discuss the 1-dimensional case; computer simulations show that the 2-dimensional case is fully analogous. Let the (1-dimensional) aperture have a normalized full width of unity. The effective aperture transmission for defocused imaging (corresponding to a $k\pi$ radian phase shift at the aperture edges $x=\pm0.5$) including instrumental and atmospheric phase errors $\varphi(x)$ is :

$$f_k(x) = \exp\left[i\varphi(x) + i4\pi kx^2\right] \qquad \text{for } |x|<0.5 \qquad (4)$$

and zero outside the aperture. The corresponding transfer function is given by Equation (3). The transfer function of a defocused aperture with an additional weak sinusoidal phase modulation can be expressed in a fairly simple way. Let the phase error distribution be given by:

$$\varphi(x) = a \cdot \cos(2\pi sx) + b \cdot \sin(2\pi sx) \qquad \text{with } a^2+b^2 \ll 1 \qquad (5)$$

Then, after some algebra:

$$
\begin{aligned}
T_k(u) = &(1-|u|) \cdot \text{sinc}(A) - \\
&- a(1-|u|) \cdot \sin(\pi su) \cdot [\text{sinc}(A+B) - \text{sinc}(A-B)] \\
&- ib(1-|u|) \cdot \sin(\pi su) \cdot [\text{sinc}(A+B) + \text{sinc}(A-B)],
\end{aligned} \qquad (6)
$$

64

where $A = 4ku(1-|u|)$,
                 $B = s(1-|u|)$,
                 $\text{sinc}(q) \equiv \sin(\pi q)/(\pi q)$

The first term on the right hand side of (6) equals the transfer function of the error free defocused aperture (a=b=0). The remaining terms are due to the sinusoidal phase error, the real and imaginary parts proportional to a and b respectively.

2.2. THE FOCUSED-DEFOCUSED PAIR OF IMAGES

The transforms of simultaneously recorded focused and defocused images give the two corresponding transfer functions multiplied by $\bar{B}(u)$, the transform of the brightness distribution (2). Thus, the ratio $R_k = T_k/T_0$ of the two transfer functions is an observable function independent of the brightness distribution of the observed object. From (6):

$$R_k(u) = \text{sinc}(A) -$$
$$- a \cdot \sin(\pi su) \cdot [\text{sinc}(A+B) - \text{sinc}(A-B)] -$$
$$- ib \cdot \sin(\pi su) \cdot [\text{sinc}(A+B) + \text{sinc}(A-B) - 2\text{sinc}(A) \cdot \text{sinc}(B)] \qquad (7)$$

From the difference between the measured ratio R_k and the ratio we would have measured in the absence of the phase modulation, $\text{sinc}(A)$, we can find a and b as well as the modulation wave number s. If the defocus measure $k > s^2/4$, then the main features in this difference will be two symmetrically placed distinct maxima at

$$u = \pm s/4k, \qquad (8)$$

i.e. the values of u at which $\text{sinc}(A-B)$ or $\text{sinc}(A+B)=1$. Thus, with a defocusing corresponding to e.g. a 2 wavelengths path difference at the edge of the aperture (k=4) we can directly derive the amplitudes and phases of the three lowest aperture error wave numbers. Figure 1 shows a computer simulation illustrating that low wave number phase modulations can be recovered from simultaneous focused/defocused images. Knowing the main features of the effective aperture transmission f(x), we are in a position to correct the focused image by inverse filtering.

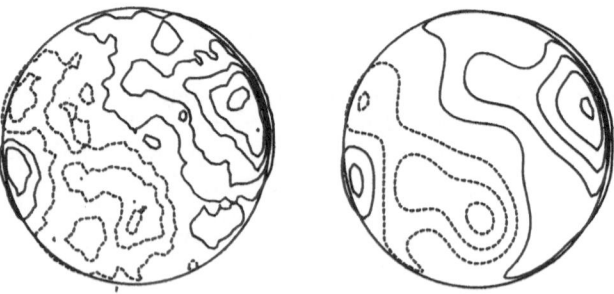

Figure 1. Left: model distribution of small phase errors (< 1 radian) over the aperture.
Right: result of a computer simulation where the main features of the phase error distribution (here the three lowest wavenumber components) was determined from a focused-defocused pair of images.

The Swedish La Palma telescope is being equipped for the registration of two simultaneous images and this facility will be used to test the procedure in practice. The frames selected by Scharmer's sharpness monitor are not far from being diffraction limited with phase errors not larger than a few radians. Still, an iterative procedure will be needed since the above equations are strictly valid only for phase modulations that are smaller than one radian.

3. Information in the Full Three Dimensional Focal Region

3.1 INFORMATION IN THE APERTURE PLANE

Information about the object brightness distribution and about the atmospheric modulation of incoming wavefronts is available in the aperture plane. One can show that the redundancy of aperture element spacings in a normal circular aperture allows these two functions to be separated, at least in theory. Thus, it should be possible to derive true diffraction limited images from measurements taken with ground based telescopes. This of course is the reason for the present interest in optical interferometry and optical aperture synthesis.

Information in the aperture plane exists mainly in the form of phase relations between waves passing through different parts of the aperture. The light is brought together at the focus where the phase relations generate amplitude (power) modulations which are registered by the detector. The question to be discussed now is what information is available in the focal region and how this information is distributed.

3.2. THE TRANSFER FUNCTION IN THREE DIMENSIONS

Let $f(x)$ be the (complex) aperture transmission function including the effects of instrumental as well as atmospheric disturbances in amplitude and phase. As before, I shall illustrate the main conclusions by considering a 1-dimensional aperture. The equivalent aperture transmission referred to a distance k from the nominal focal plane becomes:

$$f_k(x) = f(x) \cdot \exp(i4\pi kx^2), \tag{9}$$

and the corresponding transfer function (3) can be written

$$T_k(u) = \int_{-\infty}^{+\infty} \Delta(x) \cdot \exp(i8\pi kux) \cdot dx \tag{10}$$

where $\quad \Delta(x) \equiv f(x+u/2) \cdot f^*(x-u/2)$

If $f(x)$ contains no significant variations with periods shorter than u then, apart from a constant factor, $f(x)$ can be calculated from $\Delta(x)$. This is just a formal expression of the fact that the redundancy of any particular aperture element spacing u can be used to determine the true aperture transmission $f(x)$.

The shape of the equivalent transfer function for different distances k from the nominal focus is illustrated in Figure 3. Now, instead of discussing the shape of $T_k(u)$ at different distances k from the nominal focus, we shall consider T at a fixed value of u as a function of the distance k and rewrite (10) in the following form:

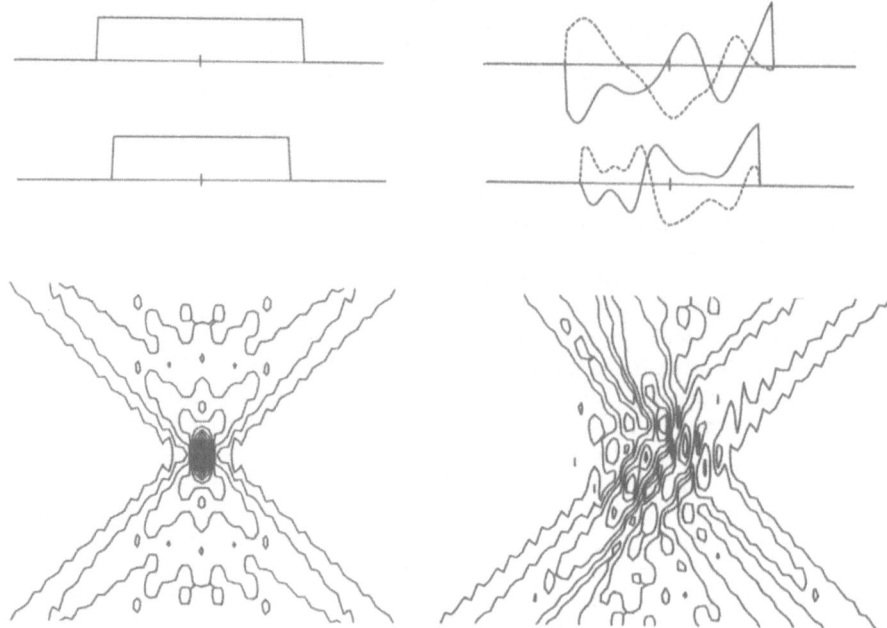

Figure 2. Computer simulations for a one-dimensional unit width aperture. Left: ideal error free aperture, right: aperture with amplitude and phase errors.
From top to bottom: The model aperture transmission f(x), real and imaginary (dashed) parts; The difference function $\Delta(x)$ with u=0.14; The PSF intensity distribution over a plane through the focal line (horizontal) and the optical axis (vertical). Note the in-depth speckle pattern of the disturbed aperture.

$$T_u(k) = \int_{-\infty}^{+\infty} \Delta(x) \cdot \exp\,(i8\pi kux) \cdot dx \qquad (11)$$

With the substitution t=4ux, this becomes a Fourier integral equation:

$$T_u(k) = \frac{1}{4u} \int_{-\infty}^{+\infty} \Delta\!\left(\frac{t}{4u}\right) \cdot \exp\,(i2\pi kt) \cdot dt$$

and $\qquad \bar{T}_u(t) = \frac{1}{4u} \cdot \Delta\!\left(\frac{t}{4u}\right)$ $\qquad\qquad\qquad (12)$

t is the conjugate variable to the distance k from the nominal focal plane. The transform of profiles parallel to the optical axis through the set of equivalent transfer functions equals the (scaled) difference function for this value of u. Observe that this result is not an approximation; it is <u>valid for large amplitude and phase modulations of the aperture.</u>

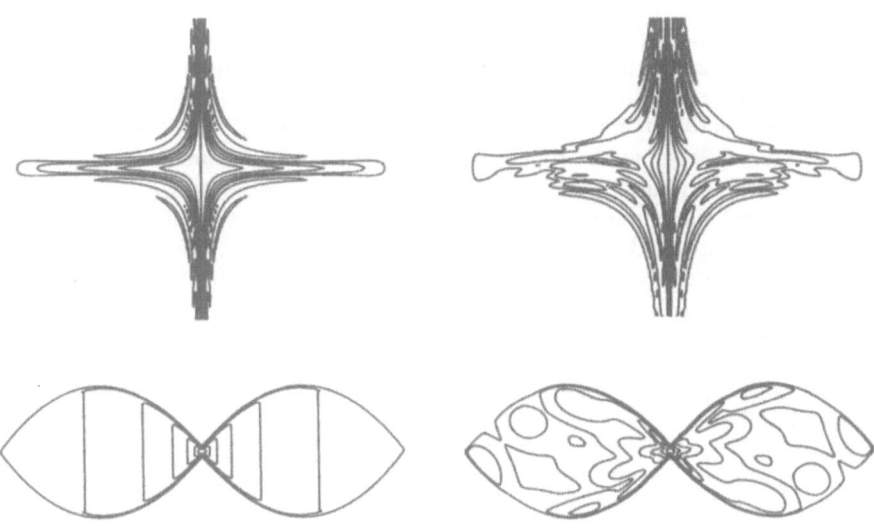

Figure 3. Computer simulations for a one-dimensional unit width aperture. Left: ideal error free aperture, right: aperture with amplitude and phase errors (see Figure 2).
Top: The (absolute value of the) transfer function $T_u(k)$; the u-axis is horizontal, the optical (k) axis vertical. Bottom: The (absolute value of the) multi dimensional transform $T_u(t)$; the u-axis is horizontal, the t-axis vertical. Contours are here drawn at increments of a factor of 2.

The transfer function $T_k(u)$ is the Fourier transform of the point spread function when measured at the distance k from the focal plane. For a two dimensional circular aperture, u becomes a vector coordinate in the two dimensional transfer function domain. The last transform in the axial dimension (12) completes the picture: we are now dealing with the three dimensional Fourier transform of the full three dimensional PSF, i.e. the three dimensional point source intensity distribution over the focal volume.

The function Δ is zero when the absolute value of the argument is $> (1-|u|)/2$, see (10). It follows that the transform $T_u(t)$ is restricted to a region defined by:

$$|t| < 2|u| (1-|u|) \qquad (13)$$

The boundary is clearly visible in the computer simulation, Figure 3. The full three dimensional transform of the three dimensional PSF is restricted to a doughnut shaped volume defined by the same equation (13). With an extended object, each column ($u=u_1$) of the transform is multiplied by $\bar{B}(u_1)$ (2). Relations between the functions $\Delta(x)$ for different u, equivalent to the closure relations, allow a reconstruction of $\bar{B}(u)$ and thereby also of the wanted brightness distribution $B(\xi)$.

4. Conclusions

The intensity distribution over the focal volume contains all the necessary information for a diffraction limited reconstruction of the object brightness distribution. Simultaneously recorded focused-defocused pairs of images should allow a reconstruction when atmospheric and instrumental phase disturbances are small (telescope aperture $< 4r_0$). With large phase and amplitude errors (telescopes apertures $> 4r_0$), more detailed measurements of the three dimensional intensity distribution will be needed. Three dimensional measurements over a suitable volume about the nominal focal plane offers an alternative to other methods for diffraction limited imaging. In particular, reconstructions can be performed locally for individual isoplanatic patch size regions covering the whole observed field of view.

Acknowledgement: This research has been supported by the Swedish Natural Science Research Council (NFR).

Discussion

KOUTCHMY — I wonder how your conclusions would be changed in case you take into account that measurements, especially in the case of 3-D observations of the "smearing function" (presumably the instantaneous one), are affected by noise?

HÖGBOM — I have not made a proper noise analysis so far. My ambition has been to clarify how relevant information is distributed over the focal volume. The trouble caused by the noise depends on how you choose to use this information. I expect to learn a great deal about the practical problems from the simultaneous focus/defocus image experiments which are to be performed with the Swedish La Palma telescope.

TITLE — Can you estimate the number of spatial steps perpendicular to the focal plane which one might require in a real system?

HÖGBOM — Not really, since it is not yet clear how this theoretical excercise is best translated into a working practical system. I hope to get some useful experience with the La Palma telescope.

VAN BALLEGOOIJEN — Are you not losing information from the fact that you can measure only the amplitude?

HÖGBOM — Not necessarily, because the amplitude structure is determined by the phase relations over the incoming wavefront.

BECKERS — Often we see that the seeing effects are different on both sides of the telescope focus. We believe this difference to be related to the height of the seeing layer. What can you tell us about that?

HÖGBOM — In the present version of the simulation, the atmospheric disturbances refer to the plane of the aperture, so you do not get such effects. This is of course an approximation, and the influence of the true height of the seeing layer should be included in future versions of the program.

GRANULATION IN THE PHOTOSPHERES OF STARS

David F. Gray
Department of Astronomy
University of Western Ontario

This is a short summary of some of the more important stellar results. The more complete discussion can be found in my monograph *Lectures on Spectral Lines Analysis: F, G, and K stars* (available from *The Publisher*, Box 141, Arva, Ontario, N0M 1C0, Canada), which is hot off the press.

I place the results under the following headings:

▢ The velocity span of line bisectors increases with effective temperature and luminosity (Vigor indicator #1).

▢ The lower "half" of the bisectors change systematically with velocity span, the core portion showing more blue shift when the span is larger (Vigor indicator #2).

▢ Classical macroturbulence shows the same temperature and luminosity dependences as the "vigor" of the granulation.

▢ There is a *granulation boundary*. Normal granulation exists on the cool side of the boundary. On the hot side, the line asymmetries are larger and in the opposite sense.

▢ Some line bisectors are variable because of *Starpatches* being carried across the disk of the star by rotation. No magnetic fields are detected in the starpatch even though the patch is nearly as bright as the surrounding photosphere. Instead, the distinguishing characteristic of a starpatch is the enhanced velocity field that exists in the patch area.

The first three results seem to indicate a "scaling" of the granulation phenomenon. The bisector velocity span, the height penetration, and the line broadening all vary in concert as we move around the HR diagram. Does the physics of granulation indicate this to be the case? Can we measure just one of these parameters and confidently predict the others?

Starpatches introduce two new aspects into stellar granulation studies: first, the

R. J. Rutten and G. Severino (eds.), Solar and Stellar Granulation, 71–73.
© *1989 by Kluwer Academic Publishers.*

obvious inconvenience of having to time-monitor the star to see if the bisectors are variable and identifying the phase when the patch is on the far side of the star (and this may not even occur) before we can attempt a granulation analysis. Second, the velocity fields within the starpatch will be of interest in their own right.

Simple two-stream numerical simulations of granulation have given us some numerical values for the main parameters affecting bisector shapes, but even more important, the simulations have brought to light some disk-integration effects:

□ the *Expanding-star Effect*, which shows that not only the difference between rise and fall velocities is important, but also the absolute values of these velocities, and

□ the *Rotation Effect* in which the Doppler shifts of rotation significantly redistribute those of the granulation, producing enhanced bisector displacements.

Both effects arise from the integration of Doppler shifts over the apparent disk of the star, and would not normally be relevant to solar studies. But for stellar work, some understanding of such phenomena is mandatory. Among other things, they offer the potential for establishing numerical values of parameters, such as the mean granule rise velocity, which otherwise are indeterminate for stars.

Some things that still need to be done on the stellar frontier:

□ separate the granule-to-lane areal ratio from the contrast

□ discriminate the ionization, excitation, and wavelength dependences and use them to map the depth dependence

□ see what connections exist between granulation and magnetic fields on short time scales such as those associated with bursts of activity (days-months) and on the longer stellar-cycle time scales (years-decades)

□ connect observed granulation parameters to convection-zone velocities and depth, i.e., the parameters of interest for dynamo theory.

These ideas lead into our "position and discussion" later today, so I shall not pursue them further here.

Discussion

BECKERS — How does the 6.43 day period for ξ Boo A compare with the binary period?

GRAY — There is no connection. The orbital period is 150 years.

TITLE — Dave, are you sure that just because the star patch is not caused directly by magnetic splitting, it is not caused by magnetic fields?

GRAY — Not at all. The physical mechanism causing the bisector variations seems not to depend on the magnetic sensitivity of the line, but whether or not magnetic fields play a role in causing or shaping starpatches is still unknown.

RIGHINI — Comparison of differential shifts in your observations as a function of time might suggest some information about the depth structure of your 'quasi-spot' pattern.

GRAY — Yes, differential line shifts might provide some additional information. Dravins and Nadeau will have more to say on this topic later in the meeting.

PERI — At the Kansas City AAS Conference Steven Vogt presented simulation work showing stars with 'starspots' covering up to 10–20% of the stellar disk[1]. These spots appeared dark and occurred largely at the stellar pole. Can you comment on the similarities of your work and his: whether the 'patches' you discussed are related to the 'spots' in his model?

GRAY — The starpatches seem to be distinctly different from starspots and also from sunspots. Some comparisons are made in an Ap.J. paper by Toner and Gray which will appear this November, and also in my book cited earlier.

VAN BALLEGOOIJEN — Could your starpatches correspond to sunspot penumbrae?

GRAY — They are certainly closer to solar penumbrae than to umbrae, but many of their other characteristics are different, e.g. patches can appear at higher latitudes, they may last years rather than weeks, and their areal coverage is an order of magnitude larger.

BECKERS — You did not describe results from stellar occultations. What do they tell us about surface structure and limb shift?

GRAY — There are not many such measurements, but some by Rodono[2] show fine structure during ingress and egress attributed to surface features.

[1] S.S.Vogt, G.D. Penrod and A.P. Hatzes, 1987, Ap. J. 321, 496.

[2] M. Rodono, 1986, Highlights of Astronomy 7, 429

SOLAR GRANULATION SPECKLE INTERFEROMETRY USING CROSS-SPECTRUM TECHNIQUES

C. AIME, J. BORGNINO, P. DRUESNE, F. MARTIN, G. RICORT
Département d'Astrophysique de l'I.M.S.P.
Parc Valrose, 06034 Nice Cedex
France

A study of the center-to-limb variation of the solar granulation spatial power spectrum, made at the McMath telescope of Kitt Peak, using speckle interferometric techniques is reported here. The experiment consists of one-dimensional photoelectric scans of the solar surface, with the telescope diaphragmed into a slit aperture of 152cm x 3.2 cm to overcome the problem of foreshortening effects. Power spectra up to angular frequencies u of about 2 arc second^{-1} are obtained free of photon noise bias by means of cross-spectrum data processing.

For angular frequencies higher than 0.4 arc second^{-1}, the power spectra are well approximated by exponentially decreasing curves of the form $\alpha \exp(-\beta u)$. The parameter α is almost independent of the heliocentric position μ, whilst β increases from 2.9 arc second at the disc centre to 4.5 arc second at $\mu = 0.25$, which indicates a fall of power in the highest frequencies towards the limb. The r.m.s. contrast of the solar granulation displays a monotonic centre-to-limb decrease. A morphological interpretation of these results is attempted. Besides that, by invoking the results of other authors, a variation of the power spectrum and of the value of β at the disc centre with the solar cycle is indicated.(Paper proposed to Astron. Astrophys.)

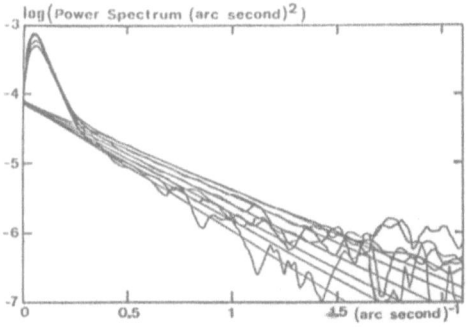

Radial slice in the unforshortened direction of the two dimensional power spectrum corrected for the forshortened effect. From top to bottom the measurements have been performed from the centre to the limb (at $\mu = 0.25$) of the solar disc. The spectra are fitted in the high frequency range by exponentially decreasing curves.

R. J. Rutten and G. Severino (eds.), Solar and Stellar Granulation, 75.

Active Optics, Anisoplanatism, and the Correction of Astronomical Images

M.L. Peri, R.C. Smithson, D.S. Acton, Z.A. Frank, A.M. Title

Lockheed Palo Alto Research Laboratory

Active optics technology has undergone great improvement in recent years. The new second generation Lockheed active mirror (shown in Figure 1) takes advantage of these innovations with the incorporation of new stable actuator technology, a more stable mechanical mount, neural network implementation of the phase reconstruction algorithm, and a computerized control system that provides automatic lock/unlock capability (to overcome beam interruptions), and automatic gain control (see Table 1). These features make the system more "user-friendly" than previous active optical systems and render it a functional scientific tool. A schematic of the system is shown in Figure 2.

The functionality of the new mirror will allow us to concentrate our efforts on using adaptive optics to make astronomical observations. Simulated images showing night-time (Figure 3) and solar observations (Figures 4-5) indicate qualitatively the image improvement that could be obtained under ideal conditions. Figures 6-7 show frames from movies that have been made with the Lockheed active mirror system: corrected images are compared with simultaneous uncorrected images to show the degree of correction obtained. The most pronounced effect of adaptive optical correction when used to make movies is to stabilize and increase the uniformity of resolution of the individual frames.

Studies of computer simulation of adaptive correction complement our observational program. We have simulated the wavefront correction produced by the Lockheed active mirror (Figure 8), and the resulting point spread functions (Figure 9). The simulations demonstrate that, as the seeing degrades, the images corrected with active mirrors retain diffraction limited resolution with decreased contrast. This suggests that adaptive optical correction will prove useful for morphological studies which require high resolution, but that it is of limited utility for photometric studies. We are currently working on a quantitative simulation of the effects of anisoplanatism (the deterioration that occurs as one looks away from the lock point of the active optics system, see Figure 10). Preliminary calculations agree with theoretical predictions by Roddier and Roddier

R. J. Rutten and G. Severino (eds.), Solar and Stellar Granulation, 77–89.
© *1989 by Kluwer Academic Publishers.*

that the point spread function is elongated asymmetrically—perpendicularly to the direction of the lock point—with higher spatial resolution (see Figure 11).

Adaptive optics can also be used as a diagnostic tool to increase our understanding of atmospheric behavior at observatory sites selected for their good seeing characteristics. The Lockheed active mirror system records the control signals sent to the mirror in real time, as observations are made. These control signals indicate the deviations introduced to the incident wavefronts by the atmosphere, and thus provide a probe to the atmospheric structure causing the distortion. We intend to use this technique to study the atmospheric seeing at Lockheed Solar Observatory in Palo Alto, California, at Sacramento Peak Observatory in New Mexico, and at the Swedish Solar Observatory at La Palma in the Canary Islands. Eventually, active optics may prove to be useful as a survey tool to analyze the micro-climateological seeing characteristics of proposed observatory sites.

Figure 1: The new Lockheed active mirror is the heart of the adaptive correction system. Each mirror segment is 2.8 cm across. The piezo-electric actuators that move the segments in piston and tilt can be seen behind the mirrors.

Table 1. Details of technological improvements incorporated in the second generation Lockheed active mirror system.

THE OLD SYSTEM	THE NEW SYSTEM
Creep and hysteresis in the piezoelectric stacks made frequent realignment of the mirror segments necessary.	Our new piezoelectric actuators have internal strain gauges which constantly monitor the length of each actuator and correct for discrepancies. There is essentially no creep or hysteresis in these actuators.
If the optical path of any closed loop in the system were interrupted, lock would be lost and would have to be reacquired. Often this meant realigning the mirror as well, since some of the actuators would be "railed" to their maximum length.	The new electronics employs analog switches and automatic gain controls. If the loop is ever broken, the actuators return gently to a neutral position. (The light beam is frequently interrupted by clouds, birds, moths, and people.) When the obstruction is removed, lock is automatically reacquired.
It is desirable to be able to lock onto low contrast objects such as dark lanes in solar granulation. Instabilities in the old system's gain and offset stability made this unlikely.	The new system contains improvements in the gain and offset stability which will allow tracking of dark lanes. An AC coupled tracking system with DC restoration is being considered which will allow tracking of low contrast objects very near the solar limb.
Stiction and "cross-talk" in the adjustable mounts for the actuators made it difficult to align the individual mirror segments to form a phase coherent mirror.	Virtually all of the problem with the old mounts is that they were designed to be too adjustable. We minimized this problem by simplifying the design considerably. The new mounting technique employs hardened steel instead of invar at pressure points to eliminate scratching and denting of the metal. Moly-disulfide is used as a lubricant on the threads of set screws to eliminate stiction. The new mount allows for easy adjustment and excellent stability of the actuators.
Since the active mirror has insufficient range to follow the image motion caused by guider error, wind shake, etc., it has always been necessary to use a separate tilt mirror to remove this overall image motion. Previously, the tilt mirror operated by stabilizing the image at a separate detector location from that of the wavefront sensor. Therefore, the system was very sensitive to small mechanical and thermal effects occurring after the tilt-stabilized image. An alignment accuracy of 1 micron over about 1 meter was required .	The present tilt mirror works as an auto-alignment system that keeps the active mirror in the center of its range, thus eliminating the need for precise alignment. Also, the separate tilt sensor is no longer needed.

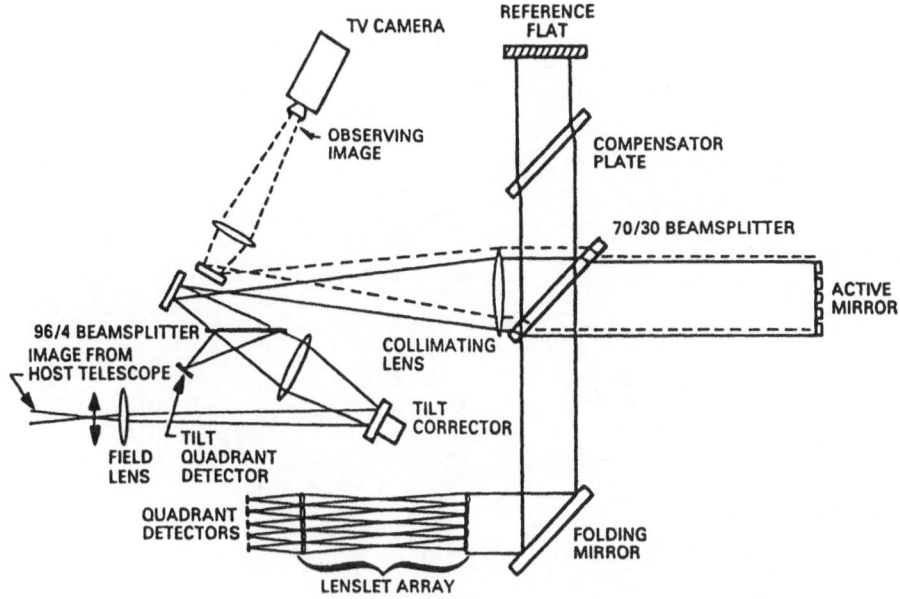

Figure 2: Schematic of the active mirror optical system.

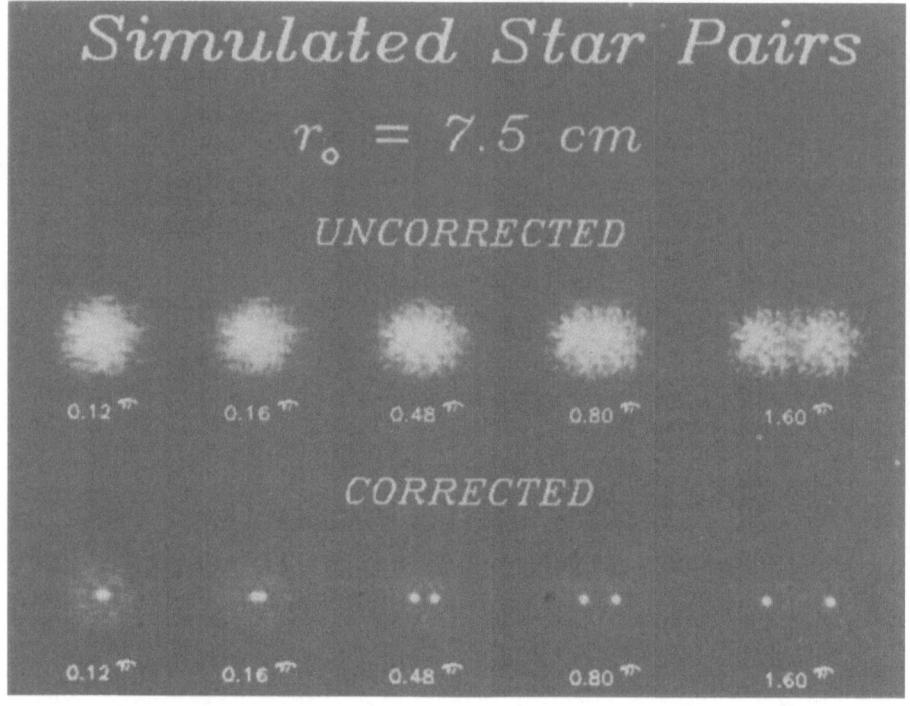

Figure 3: Log intensity images of simulated star pairs as they would appear after degradation by Kolmogorov turbulence with seeing parameter value $r_0 = 7.5$ cm, with and without active mirror correction. The resolution of the central peak corresponds to the diffraction limit of the telescope aperture (1.0 m for this simulation) and the diameter of the halo corresponds to the diffraction limit for an element of diameter r_0.

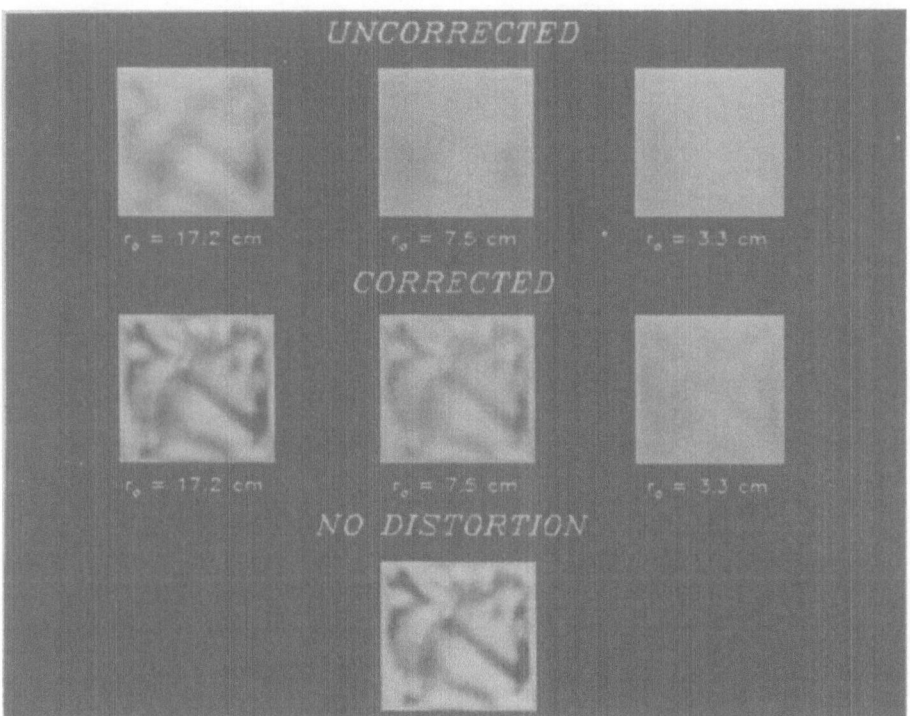

Figure 4: Simulation of the visual appearance of solar granulation as it would appear in varying seeing conditions, with and without active mirror correction. An undistorted image is provided for comparison. The field of view is 4.0 × 4.0 arcseconds.

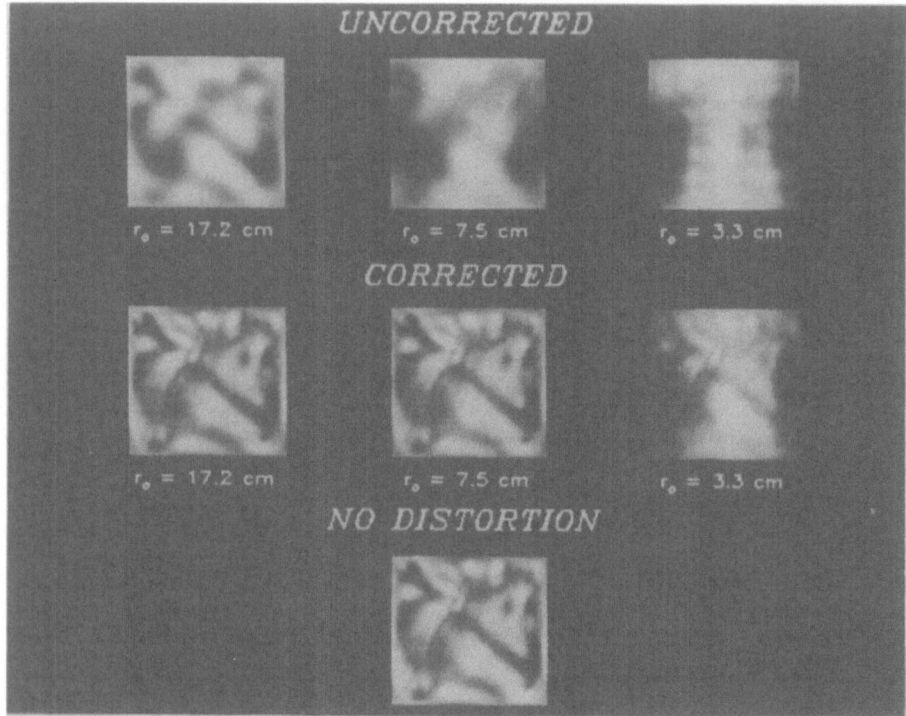

Figure 5: Simulation of the contrast enhanced images of solar granulation as it would appear in varying seeing conditions, with and without active mirror correction. An undistorted image is provided for comparison. The field of view is 4.0×4.0 arcseconds. Note the high spatial frequency content in the corrected image.

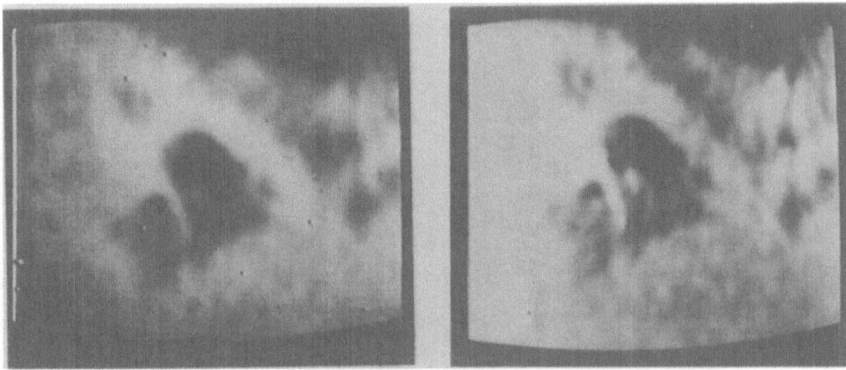

Figures 6 and 7: Improved resolution obtained with the prototype active mirror at Sacramento Peak Observatory in 1985. Figure 6 shows a picture taken with image motion compensation only. Figure 7 shows a picture taken with full wavefront compensation by the active mirror. The two pictures are simultaneous and represent 1/2 second time exposures of the video monitor screen. The resolution of the corrected image is believed to have been limited by residual alignment errors in the prototype mirrors. Such errors should not occur in the improved mirror as it is self-aligning.

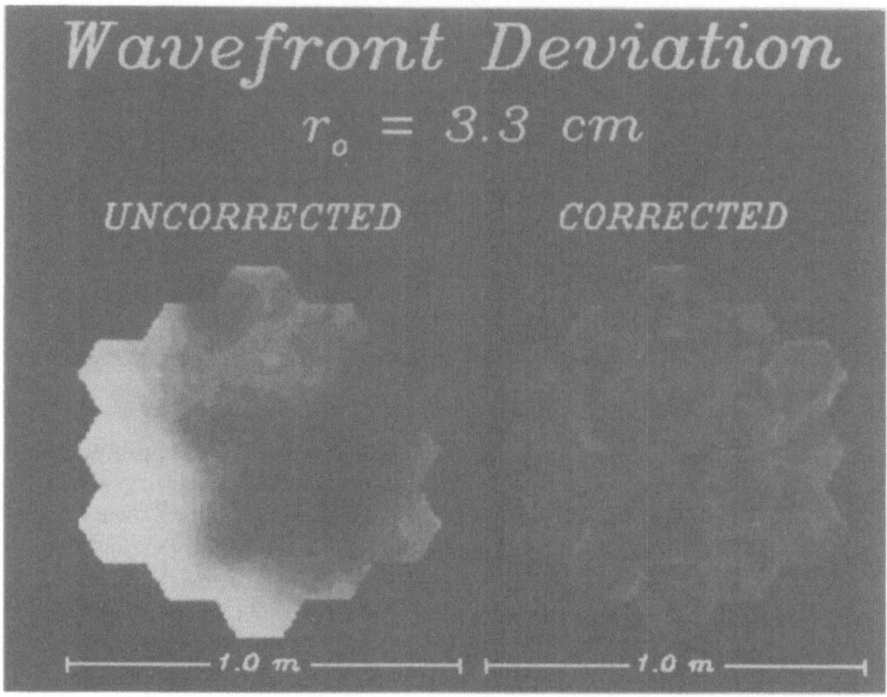

Figure 8: Wavefront deviations in a simulated random wavefront before and after active mirror correction. Contour interval is 0.5 waves. The uncorrected wavefront on the left has a Kolmogorov spatial frequency distribution of wavefront errors and shows a large amount of tilt. The corrected wavefront on the right exhibits phase coherence to within 1/2 wave across the face of the mirror, with superimposed deviations of which the dimension is on the order of r_o. The coherent phase level gives rise to a narrow peak of the point spread function, giving the corrected images diffraction-limited resolution. The residual errors scatter light incoherently, decreasing the contrast in corrected images.

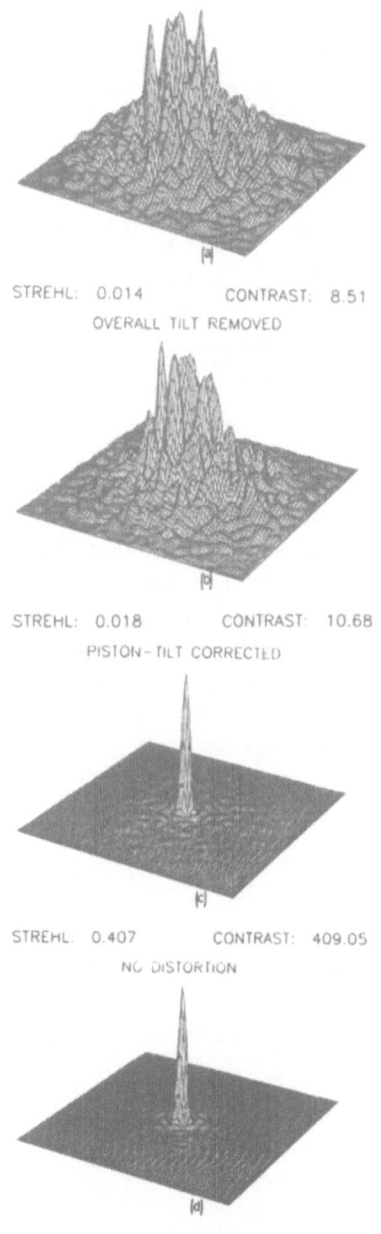

STREHL: 0.014   CONTRAST: 8.51
OVERALL TILT REMOVED

STREHL: 0.018   CONTRAST: 10.68
PISTON-TILT CORRECTED

STREHL: 0.407   CONTRAST: 409.05
NO DISTORTION

STREHL: 1.000   CONTRAST: 5218.26.

Figure 9: Simulated long-exposure point spread functions with $r_o = 7$. 5 cm. (a)-(c) Various corrections; (d) perfect diffraction pattern for comparison. Grid size is 2.75×2.75 sec of arc. Intensity is plotted on a linear axis and scaled to unit height.

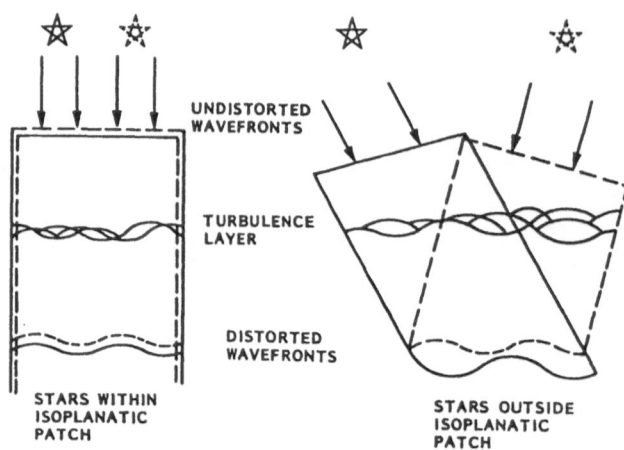

UNDISTORTED WAVEFRONTS

TURBULENCE LAYER

DISTORTED WAVEFRONTS

STARS WITHIN ISOPLANATIC PATCH

STARS OUTSIDE ISOPLANATIC PATCH

Figure 10: Anisoplanatic degradation occurs when one observes a feature that is not at the lock point of the active optical system. The wavefront incident from the object passes through atmospheric turbulence that differs from the turbulence in the direction of the lock point, and thus is not perfectly corrected. The angular separation between the lock point and the feature is called the "isoplanatic angle", and the field over which effective correction can be accomplished is called the "isoplanatic patch". The study of anisoplanatic effects is important to understand how adaptive correction will affect the imaging of extended objects such as the sun.

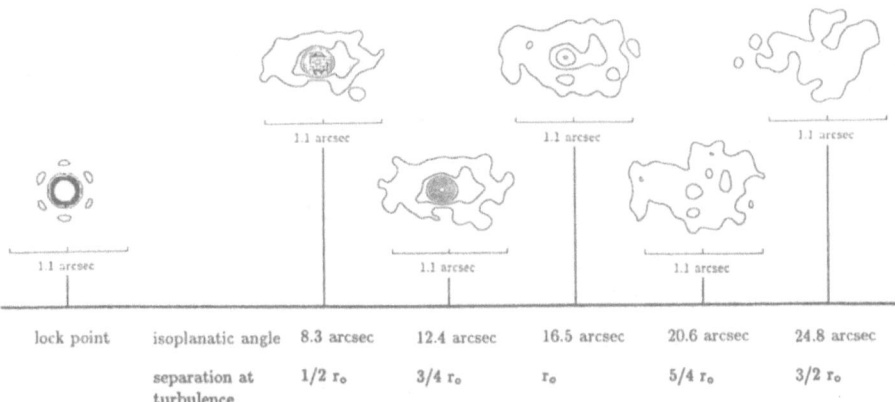

lock point	isoplanatic angle	8.3 arcsec	12.4 arcsec	16.5 arcsec	20.6 arcsec	24.8 arcsec
	separation at turbulence	$1/2\ r_0$	$3/4\ r_0$	r_0	$5/4\ r_0$	$3/2\ r_0$

Figure 11: Initial results from a quantitative simulation of anisoplanic degradation confirms theoretical predictions of Roddier and Roddier (1987). As one increases the isoplanic angle the seeing degrades in an asymmetric fashion, with resolution degrading more rapidly in the direction perpendicular to the lock point direction.

ON PRESSURE SHIFTS OF FeI LINES IN STELLAR ATMOSPHERES

Vladimir Kršljanin
Astronomical Observatory
Volgina 7
11050 Beograd
Yugoslavia

Recently, Dravins (1987) presented measured Fe line asymmetries of seven various A-K stars. Ussually, such data are interpreted as evidence of convective motion, and pressure broadening contribution is ussually neglected. Vince et al.(1985) showed that for detailed analysis of limb effect in solar NaI lines one should take into account the pressure shifts.

Here, we present calculated pressure shifts for FeI lines. Two main pressure shift contributions were taken into account: atomic hydrogen impact shift (calculated according to Hindmarsh et al.,1967) and electron impact shift (calculated according to Dimitrijević and Konjević, 1986). The results, as a function of T_{eff} and τ_{500} (based on Kurucz (1979) model atmospheres for A-G main sequence stars), for two representative cases (FeI λ536.487 and 541.52 nm lines), are presented in Table I. Hydrogen impact shift dominates the Stark shift for $T_{eff} \lesssim 6000$ K and Stark shift dominates for $T_{eff} \gtrsim 7000$ K. For the lines presented the hydrogen impact shifts are blue and the Stark shift is red (λ541.52 nm line) and blue (λ536.487 nm line). One can conclude that pressure shift contributes to the observed asymmetries in line wings, particularly in hot stellar atmospheres.

Table I. Pressure shifts (ms^{-1}) of two FeI lines in atmospheres of main sequence stars (log g=4.5), as a function of T_{eff} and τ_{500} (negative shifts are blue)

τ_{500}\ T_{eff}(K)		5500	5770[Δ]	6000	7000	8000	9000	10000
λ536.487nm	0.01	-14	-13	-14	-13	-16	-24	-39
(mult.1146)	0.1	-53	-50	-55	-42	-77	-110	-149
	1	-187	-189	-207	-606	-1010	-1030	-542
λ541.52nm	0.01	-10	-9	-9	-5	5	14	23
(mult.1165)	0.1	-37	-34	-36	-10	33	63	88
	1	-71	-31	-10	309	582	590	309

[Δ] log g=4.44

References

Dimitrijević,M.S. and Konjević,N.:1986,Astron.Ap.**163**,297
Dravins,D.:1987,Astron.Ap.**172**,211
Hindmarsh,W.R.,Petford,A.D. and Smith,G.:1967,Proc.Roy.Soc.A**297**,296
Kurucz,R.L.:1979,Ap.J.Suppl.**40**,1
Vince,I.,Dimitrijević,M.S. and Kršljanin,V.:1985,in Progress in Stellar Spectral Line Formation Theory, eds.J.E.Beckman and L.Crivellari, D.Reidel,Dordrecht,Boston,Lancaster,p.373

R. J. Rutten and G. Severino (eds.), Solar and Stellar Granulation, 91.
© 1989 by Kluwer Academic Publishers.

PRESSURE BROADENING AND SOLAR SPECTRAL LINE BISECTORS

I. VINCE and M.S. DIMITRIJEVIĆ
Astronomical observatory
Volgina 7
11050 Belgrade
Yugoslavia

ABSTRACT. In order to show that pressure broadening (atomic collisional processes) may have an important contribution to the spectral line asymmetries in some cases, what is of interest especially for the solar granulation studies, bisectors for NaI $3s^2$ S-np^2 P^0 solar lines have been calculated for different positions on the solar disk. Our numerical results clearly demonstrate that in the case of the examined spectral series the influence of pressure broadening has not a negligible role for the convective layer diagnostic.

1. INTRODUCTION

The study of line shapes and shifts is a powerful tool for the investigation of a number of problems in solar and stellar spectra (e.g. the line shift as function of the distance from the center of the apparent solar disk, commonly known as limb effect) may be studied using the line profiles analysis. Often (e.g. Dravins et al, 1981), line shape asymmetries are described by bisectors. The line bisector is formed by the loci of points midway between equal-intensity points on either side of the line profile dividing thus, the absorption line into two halves of equal equivalent widths.It showes the apparent radial velocity at each depth in the line what is of the special interest for solar granulation research. Such investigation does not require spatially resolved spectra and can be applied also in stellar studies.

In such research, the influence of collisions between the absorber and surrounding particles is usually neglected (e.g. Dravins et al, 1981). However, Hart (1974) and Vince et al. (1985, 1987) are demonstrated that the influence of collisions may be significant for some spectral line shifts.

The aim of our investigation is to show that pressure broadening (collision processes) may have an important

93

R. J. Rutten and G. Severino (eds.), Solar and Stellar Granulation, 93–97.
© *1989 by Kluwer Academic Publishers.*

contribution to the spectral line asymmetries in some cases and might not always be neglected in precise analysis. In order to demonstrate this, we performed an analysis of the influence of collisional processes on bisectors of some sodium atom spectral lines at different positions on solar disk.

2. THEORY

In order to investigate the influence of collision processes on the convective zone outer layer (observed as the granulation on solar surface) study, by using Fraunhofer lines bisectors, syntethic bisectors due to collisions with atomic hydrogen, electrons and protons have been calculated for NaI $3s^2S-np^2P^o$ lines (n=3,4,5,6).

The synthetic spectral line profile is determined from the equation of radiative transfer (e.g. Mihalas, 1972)

$$I_\lambda(0,\mu) = \int_0^\infty S(\tau_\lambda) \, e^{-\tau_\lambda/\mu} d\tau_\lambda/\mu$$

where $I_\lambda(0,\mu)$ is the emergent intensity, $S(\tau_\lambda)$ is the source function, τ_λ is the optical depth and μ is the cosine of heliocentric angle. The optical depth is a function of absorption coefficient, we assume that the absoption coefficient have Voigt profile which is defined by the following dimensionless parameters

$$a=2w/\Delta\lambda_D \quad \text{and} \quad v=(\lambda-\lambda_o+d)/\Delta\lambda_D ,$$

there, $\Delta\lambda_D$ is the Doppler width, λ_o is the wavelength of the unshifted line core. In the case of the solar atmosphere the impact approximation is valid. The profile for constant plasma conditions is then lorentzian and is defined with full width at half maximum (FWHM) 2w and shift d. Since a Fraunhofer line is formed in a wide layer of solar atmosphere where temperature, perturber density and other relevant parameters are functions of position and time, the emergent spectral line bisectors have been calculated using the VALC model (Vernazza et al, 1981) of the solar atmosphere.

The broadening and shift of spectral lines due to collisions with neutral perturbers has been calculated for NaI $3s^2S-np^2P^o$ (n=4,5,6) lines using the Smirnov-Roueff exchange potential, which takes into account the overlap at intermediate absorber-perturber distances of the electronic orbitals (Roueff, 1972, 1975, 1976).In the case of $3s^2S-3p^2P^o$ line, results of Monteiro et al. (1985) have been used.

Monteiro et al. (1985) used semiclassical impact-parameter method (see e.g. Lewis et al, 1971) which should be satisfactory at the temperatures of interest ($\sim$5000 K). Moreover, they used improved interatomic potentials.

For the electron- and proton-impact contribution, data by Dimitrijević and Sahal-Bréchot (1985), obtained with the semiclassical-perturbation formalism, have been used.

3. RESULTS AND DISCUSSION

Center to limb variation of NaI $3s^2S-np^2P^o$ (n=3-6) synthetic line asymmetries, i.e. bisectors, are shown in Fig.1 for five values of cosine of heliocentric angle. In laboratory conditions, it is expected that the variation of pressure broadening parameters within a spectral series is regular (see e.g. Dimitrijević and Sahal-Bréchot, 1985; Dimitrijević and Peach, 1987). However, we can see in Fig.1 that in the case of the examined bisectors the situation is not so simple.

The calculated bisectors depend on electron- proton- and neutral hydrogen atom-impacts. In spite of the fact that electron-impact broadening dominate in most cases, H- and p-impacts are also important and each contribution has a different trend whitin a spectral series. Moreover, the region of line formation as a function of the heliocentric angle changes within the spectral series examined becoming narrower when cosine of heliocentric angle tends to 1 for higher series number. Those changes are more important for the line wings than for the peaks which is clearly seen in Fig.1 for $3s^2S-6p^2P^o$ transition. In this case line wings are formed in the lower regions of solar atmosphera and are more sensitive to the particle density and temperature variations with heliocentric angle. On the other hand line peaks are formed in the higher part of solar atmosphere where the influence of collisional processes on the line shapes is smaller. Consequently, even the large variations in the height of line peak formation have not significant influence on examined bisectors.

We might conclude that our numerical results clearly demonstrate that in the case of choosen sodium spectral lines the influence of pressure broadening has not a negligible role for the convective layer diagnostic. Moreover, the influence of collisional processes become more significant for higher series members as clearly seen in Fig.1. In order to understand better the behaviour of Fraunhofer lines bisectors within a spectral series it is interesting to perform an analysis of the influence of various atomic processes contributions to line asymmetries (bisectors) within a spectral series.

96

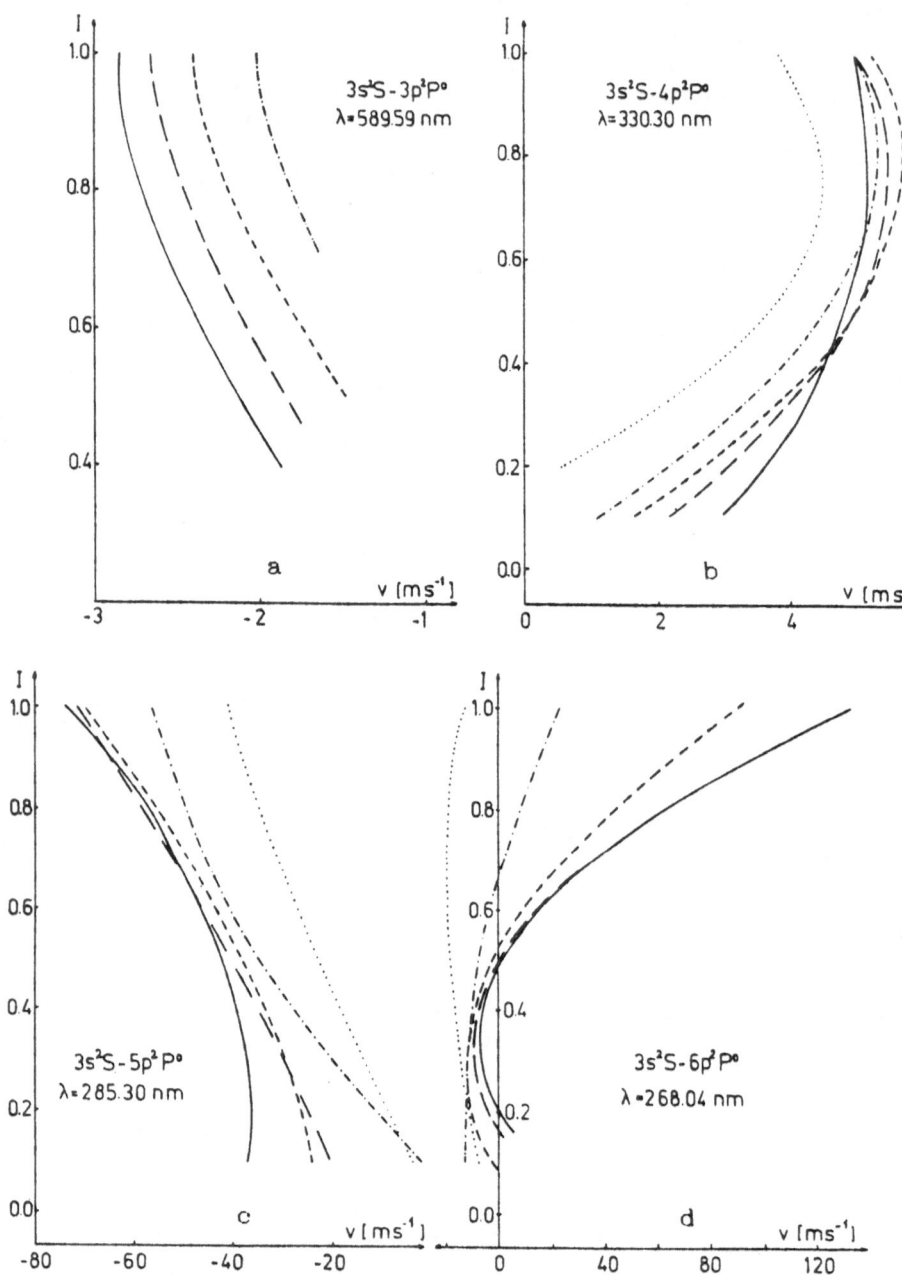

Fig.1. Bisectors of NaI $3s^2 S-3p^2 P^o$ (a), $3s^2 S-4p^2 P^o$ (b), $3s^2 S-5p^2 P^o$ (c) and $3s^2 S-6p^2 P^o$ (d) spectral lines for $\mu=1$ (———), $\mu=0.8$ (— —), $\mu=0.6$ (– – – –), $\mu=0.4$ (–·–·–·) and $\mu=0.2$ (·········)

4. REFERENCES

Dimitrijević, M.S. and Peach, G.: 1987, in Radiative
    Excitation and Ionization Processes, Zagreb, p. 19.
Dimitrijević, M.S. and Sahal-Bréchot, S.: 1985, JQSRT, $\underline{34}$,
    149.
Dravins, D., Lindgren, L., Nordlund, A.: 1981, Astron.
    Astrophys. $\underline{96}$, 345.
Hart, M.H.: 1974, Astrophys. J. $\underline{187}$, 393.
Lewis, E.L., McNamara, L.F., Michels, H.H.: 1971, Phys. Rev.
    $\underline{A3}$, 1939.
Mihalas, D.: 1978, Stellar atmospheres, W.H. Freeman and Co,
    San Francisco.
Monteiro, T.S., Dickinson, A.S., Lewis, E.L.: 1985, J. Phys.
    $\underline{B18}$, 3499.
Roueff, E.: 1972, J. Phys. $\underline{B5}$, 279.
Roueff, E.: 1975, Astron. Astrophys. $\underline{38}$, 41.
Roueff, E.: 1976, Astron. Astrophys. $\underline{46}$, 149.
Vernazza, J.E., Avrett, E.H., Loeser, R.: 1981, Astrophys.J.
    Suppl. Series, $\underline{45}$, 635.
Vince, I. and Dimitrijević, M.S.: 1987, in Spectral Line
    Shapes IV, ed. R. Exton, A. Deepak, Hampton (Virginia),
    631.
Vince, I. and Dimitrijević, M.S.: 1985, in Progress
    in Stellar Spectral Line Formation Theory,
    eds. J.E. Beckman, L. Crivellari, D. Reidel P.C., 373.

Part 2
Observations

SOLAR GRANULATION : OVERVIEW

R. Muller
Pic du Midi Observatory
65200 Bagnères de Bigorre
France

ABSTRACT. During the last few years the solar granulation has been extensively investigated thanks to new high quality material, and to the development of sophisticated and powerful image processing techniques. The material consists of distorsion-free granulation movies performed with the SOUP experiment on board of Spacelab-2 and of high resolution ground-based pictures and spectra. In this review paper emphazis is put on those properties which can help us to understand better the origin of the granulation. In particular several results favour a turbulent origin : the size histogram, the fractal dimension, the kinetic energy power spectrum, the fragmentation which appears to be the main process of disappearance. The mesogranulation is revealed by surface granular flows as well as by the spatial distribution of particular granules. The granular pattern is perturbed by the presence of magnetic fields, but in various ways depending on the situation. It undergoes variations over the solar cycle.

1. INTRODUCTION

When I was preparing the IAU Report on Astronomy (1981-1984) about Solar Granulation four years ago (Muller, 1985), my impression was that no fundamental progress was made during that period about our understanding of the granulation. Since then the situation has evolved, as many very important new results have been obtained; consequently our knowledge of the burbulent convection on the surface of the sun has been much improved. The main contributions are coming from the remarkable, distorsion free, movies collected by the SOUP experiment on board of Spacelab-2 and from the development of powerful image processing techniques applied both to the SOUP data and to high resolution ground based observations (filtergrams and spectrograms).

  In this review, the properties which can help us to better understand the turbulent convective origin of the solar granulation,will be emphasized. Morphological properties will first be described. Then the depth dependence of the vertical velocities of granules, as derived from

101

R. J. Rutten and G. Severino (eds.), Solar and Stellar Granulation, 101–123.
© *1989 by Kluwer Academic Publishers.*

various observation techniques, will be discussed. The energy power spectrum which appears to be very similar to the one expected in the case of turbulent eddies, opens the discussion on the turbulent properties of granules. The last sections are devoted to the perturbation of the granular pattern by magnetic fields and to the large scale distribution of granules, which surprisingly are not distributed at random on the surface of the sun; together with the organised granular flows, this granule distribution allows us to reveal the mesogranulation pattern. Finally results will be presented about the recently discovered variation of the granulation over the solar cycle, which is so important to help us understand the large scale convection-magnetic field interaction in the sun and stars. Owing to the reviving interest in solar granulation studies, new exciting results will surely be reported during this meeting, which I was not aware whilst preparing this review.

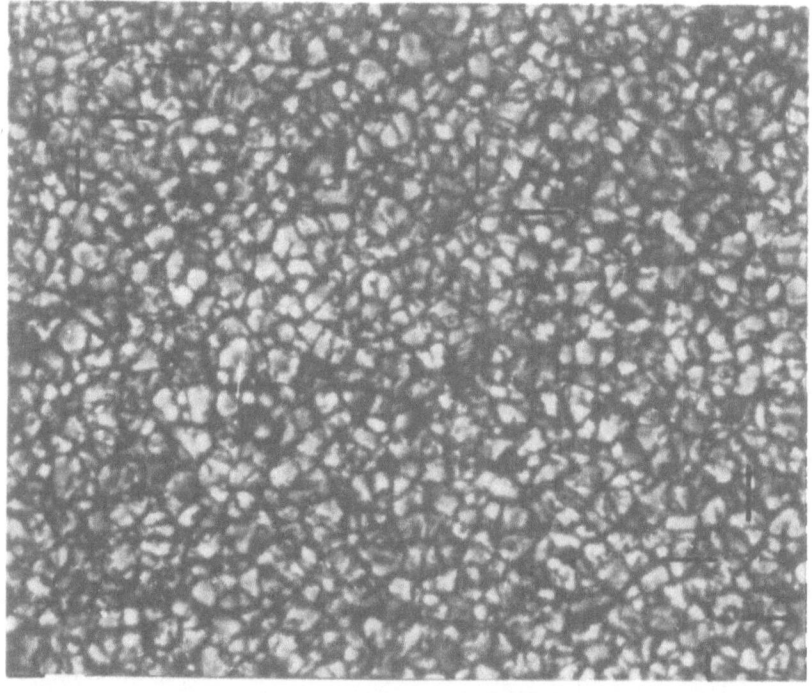

Figure 1. Solar granulation observed with a 0."25 resolution (Pic du Midi Observatory). Several examples of "daisy-like" arrangements of granules around Network Bright Points (NBPs) are shown (see section 7.1).

2. STRUCTURAL PROPERTIES : SIZE, SHAPE, EVOLUTION.

2.1. Size histogram

On the basis of visual impression or of the peaked shape of histograms, it has been accepted for a long time that the range of granule sizes is quite narrow, most granules having a size of 1"-2". However, most of the published histograms concern <u>intergranular distances</u>, not granule sizes, except the one published by Namba and Diemel (1969). The importance of a granule narrow range of sizes comes from the fact that it was used as one of the main arguments for regarding the granulation as a convection phenomenon. But recently, Roudier and Muller (1986) analysing a very high resolution (0".25) photograph (Figure 1) reported a size histogram showing that the number of granules increases steeply down to the limit of resolution (Figure 2). The important implication

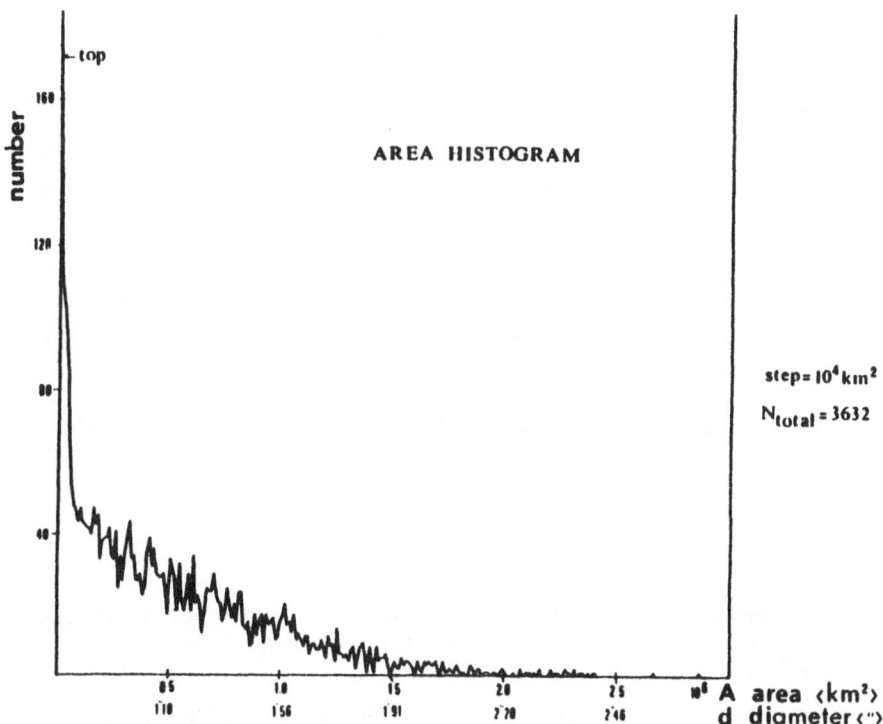

Figure 2. Histogram of granule areas. d is the effective diameter of granules with the area A, defined by $A = \pi d^2/4$ (Muller and Roudier, 1986).

of such a result is that the solar granulation has no characteristic scale and it cannot be inferred from this that we are in the presence of a convective phenomenon. Roudier and Muller also found that the histogram of intergranular distances has a peaked shape despite the fact that the size histogram has not, thus demonstrating that it cannot be used to discuss the nature of the granulation.

2.2. Evolution process and lifetime.

The first correct description of a typical granule evolution was given by Rösch and Hugon (1959). Detailed descriptions were later reported by Mehltretter (1978), Kawaguchi (1980) and Dialetis et al. (1986). Three processes of granule appearance and disappearance have been identified : fragmentation, merging, growth from a diffuse patch (in the case of appearance) or fading (in the case of disappearance). Statistically fragmentation is most frequent : 60 %, according to Dialetis et al.

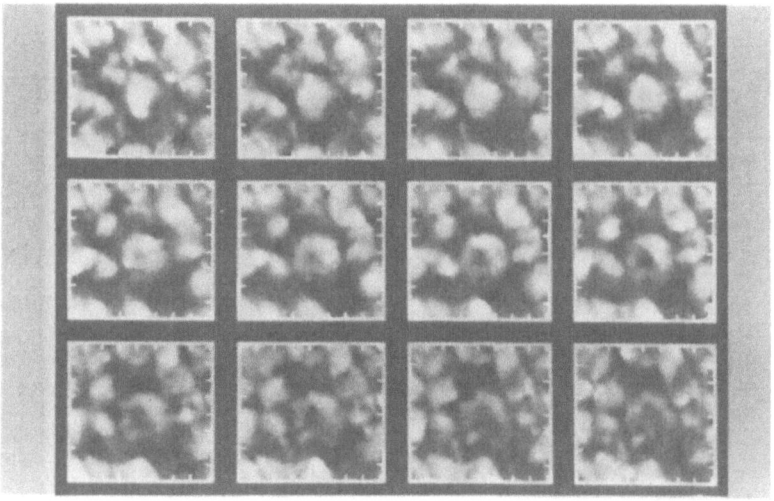

Figure 3. Prototypical exploding granule time history. Thick marks are at arc second intervals and the time between frames is one minute (Title et al. 1986).

Exploding granules (Carlier et al., 1969; Namba, 1986; Title et al. 1986) are spectacular and violent extreme cases of the process of fragmentation (Figure 3). From the SOUP movie, especially after it was filtered for the 5 min. oscillations (Title et al. 1986), it seems that exploding granules are much more frequent than reported before. In fact, when looking at this movie, we have the strong impression of large eddies breaking into smaller ones; this together with the increasing number of small granules, suggests that we are in the presence of a turbulent phenomenon.

The characteristic lifetime of granules depends on the quality of the observation and on the definition. Reported values range from 6 to more than 16 min. (Bray et al. 1984). Alissandrakis et al. (1987) demonstrated that the discrepencies are mainly due to the methods which were used. Applying the same technique to several previous measurements, based on the "survival function", they can get a unique value for the granule lifetime : 16 min (the survival function shows the decrease rate of the surviving granules from a set of granules present on a master frame). It has also been found that the lifetime increases with increasing granule size (Kawaguchi, 1980). If the granule lifetime is derived from time correlation functions, it is found to be as short as 4 to 7 min (Bahng and Schwarzschild, 1961; Title et al. 1986). However, after correction of the 5 min oscillations and of the granular flow, the lifetime rises to 12 min; this is still shorter than the value derived from direct measurements, probably because the autocorrelation lifetime is underestimated by the granule proper motions and their change of shape.

2.3. Shape and fractal dimension.

The solar granulation is visible on the surface of the Sun as a cellular pattern of bright elements (the granules) surrounded by dark intergranular lanes. Granules exhibit a wide variety of sizes and shapes. According to Groves and Pierce (1986) their shapes can be classified as follows : class I, single granules; class II : single granules embayed by a broad dark area or possessing a central darkening; class III : single granules split by very narrow rifts; class IV : complex exploding granules.
      The granule shapes can be characterized by the area (A) - perimeter (P) relation, $P \sim A^{D/2}$, where D is the fractal dimension. It is remarkable that D changes abruptly from a value of about 2 for the granules larger than $1\rlap{.}''37$ to a value of 1.25 for the smaller ones (Figure 4), which thus appear to be much more regular in shape (Roudier and Muller, 1986). The value 1.25 is close to 4/3, suggesting that the smaller granules are of turbulent origin, according to the Kolmogorov theory of isotropic, homogeneous three dimensional turbulence (Mandelbrot, 1977).

2.4. Granule critical size.

We have seen in section 2.1 that it means nothing to speak of a granule characteristic or mean size, because of the shape of the size histogram (Figure 2). Nevertheless, granules appear to have a <u>critical</u> scale of $1\rlap{.}''37$ (at the intensity level 1.01 I), at where drastic changes of their properties occur (Roudier and Muller, 1986) : change of the fractal dimension (this paper, Figure 4); change of the power law of the size histogram (Roudier and Muller. RM., Figure 7); change of the power law $\mathcal{F}$ (a) for a granule to be larger than an area (a) (RM, Figure 8); the typical size of the fragmentation of granules is about $1\rlap{.}''5$ (Kawaguchi, 1980); granules of critical size are the main contributors to the total granule area and radiation (RM, Figure 9). This point is important because it shows that the many small granules do not contribute much to

the total granule area or radiation; this means that the granulation
seems to have a well-defined scale because the small granules, although
very numerous do not catch the eye.

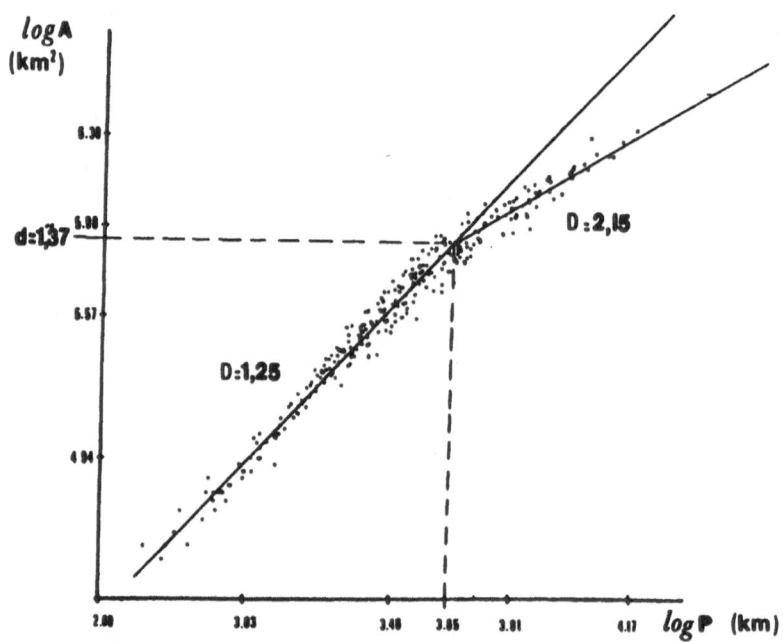

Figure 4. Area-Perimeter relation of the solar granulation. D is the
fractal dimension : $P \sim A^{D/2}$. Each point represents a granule (Roudier
and Muller, 1986).

3. INTENSITY FLUCTUATIONS AND TEMPERATURE STRUCTURE OF THE GRANULATION.

Intensity fluctuations, δI_{rms}, associated with the granulation, measured
at the disc center, near λ 5000 Å, are found in the range 8 - 18 %
when corrected for the seeing and instrumental scattered light fitted
by the sum of two Gaussians; the best estimate is probably close to 15 %
(Bray et al. 1984). However, the 3 - D numerical simulation of Nordlund
(1984) requires a much higher value of 25 - 30 %. Observations and
models can be, partially reconcile if one uses a Lorentzian fit for the
spread function, rather than a Gaussian one (Nordlund, 1984; Collados
and Vázquez, 1987) which makes the δI_{rms} rise up to about 20 %. I think
there is an alternative explanation for the discrepancy between observed and computed intensity fluctuations : in the Nordlund's simulation,

because of the lack of spatial resolution, there are no small granules, whereas in the real solar atmosphere many small granules are present in the dark spaces in between larger granules; they probably contribute to reducing the temperature difference as well as the intensity contrast between granules and intergranular lanes.

From intensity fluctuations measured in photospheric lines formed at various heights in the photosphere, it has been found that the temperature fluctuations associated with the solar granulation do not penetrate very high up into the photosphere (Figure 5), vanishing above 50 - 100 kms (Altrock, 1976; Keil and Canfield, 1978; Durrant et al. 1981). This has been confirmed by a coherence analysis of brightness fluctuations in the continuum and at various positions in the line Mgb$_2$, which sharply drops at $\Delta\lambda = 0.35$ Å from the line center (Kneer et al. 1980; Durrant and Nesis, 1981).

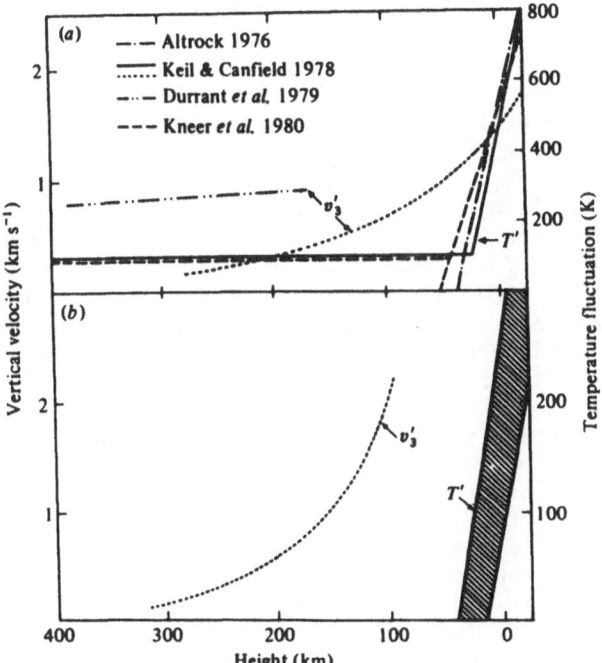

Figure 5. Empirical height dependence of the granular temperature fluctuations and vertical velocities in the solar atmosphere (from Bray et al., 1984).
(a) rms values. (b) Amplitude (half peak-to-peak). The band of temperature fluctuations is taken from Altrock and Musman (1976), the run of vertical velocity from Keil (1980b).

4. VERTICAL VELOCITY STRUCTURE OF THE GRANULATION.

The height-dependence of the granulation velocity field can be derived
through various kinds of spectral line measurements : Doppler shift
measurements; coherence and phase analysis; line asymmetry; line broadening. We compare here the results from all those various techniques,
except the last one.

a. Results from Doppler shift measurements.

In spatially resolved spectra, photospheric lines have a characteristic
wiggly shape caused by upflows and downflows correlated with continuum
bright (granules) and dark (intergranular spaces) features respectively.
The main difficulties in deriving granular velocities from observed line
shifts come from : a - smoothing of the shifts by seeing and instrumental stray light (the resolution of the best spectra is hardly better
than 1"); b - inaccuracy of spread function determination; c - mixing
of granular and oscillatory velocity fields which have to be separated;
insufficient knowledge of line heights of formation. Despite these
problems it is unanimously found that, while the temperature fluctuations vanish at 100 km or less above the photosphere, the velocity
fluctuations penetrate much higher into the photosphere, at least up to
300 km. But the effective height of penetration of the amplitude of the
velocity fluctuations $\delta_{rms}(h)$ is still a matter of debate. The Sac
Peak group (Canfield, 1976; Keil and Canfield, 1978; Keil, 1980, a,b)
reported empirical models where $\delta(h)$ decreases steeply (steep gradient
models) and no power is found above 300 km; inversely the Freiburg
group (Durrant et al., 1979; Bässgen and Deubner, 1982), reported flat
gradient models in which $\delta(h)$ decreases smoothly and penetrates well
above 300 km (Figure 5). The use of different procedures for separating
the granular and oscillatory components may explain this disagreement.

b. Coherence and phase analysis.

This kind of analysis consists of computing the degree of correlation
and the phase between intensity and velocity fluctuations at several
positions in photospheric lines (which means at various heights in the
photosphere). Coherence and phase are usually computed as a function of
horizontal wavenumber, in order to compare the depth dependence of
features of various sizes. The main conclusions can be summarized as
follows, from the results of Durrant and Nesis (1982); Pravdjuk (1982);
Nesis et al. (1984); Wiehr and Kneer (1988); Deubner (1988) :
    The coherence between line and continuum intensity fluctuations
is rapidly lost, which confirms that temperature fluctuations do not
penetrate high up into the photosphere.
    The coherence between line velocities and continuum intensity
fluctuations for scales larger than 1".5 is high and the phase close to
180°, even at heights larger than 300 km; this favours flat gradient
models (it has to be noted that a high degree of correlation between
upflows (downflows) and bright granules (dark intergranules) is the
main argument for regarding the granulation as a convective phenomenon).

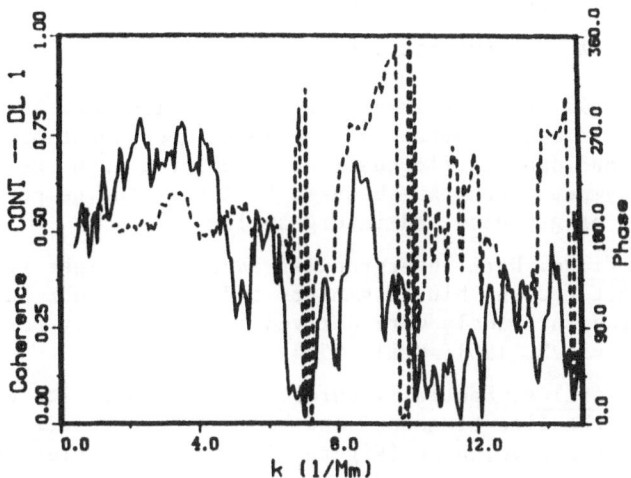

Figure 6. Coherence (full line) and phase (dashed, in degrees) between continuum intensity and line shift of Fe 6495. (Wiehr and Kneer, 1988)

As far as the smaller scales are concerned, the results are in disagreement : on the one hand, Wiehr and Kneer (1988) find that the coherence is lost for scales smaller than 1."5 (Figure 6) which according to these authors is due either to the rapid braking of the connective overshoot or to the onset of turbulence; on the other hand Deubner (1988) finds that the coherence is maintained down to about 1" (Figure 7)

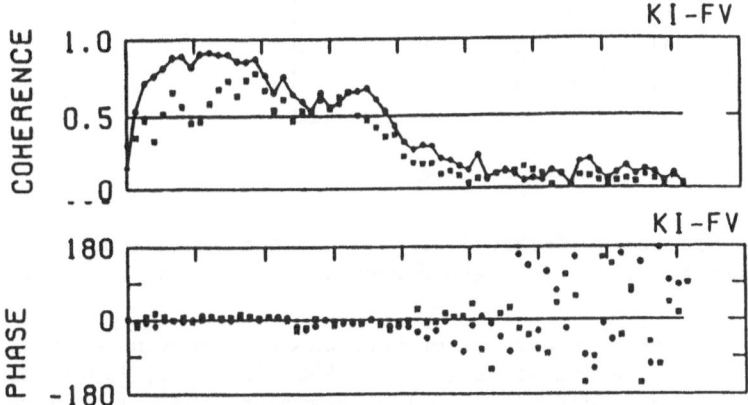

Figure 7. Coherence and phase of coherence continuum intensity (KI) and velocity measured in the FeI 5383 line (FV). (Deubner, 1988) Coherence and phase are maintained down to 8 Mm^{-1}.

These results will be discussed in section 6.

c. Asymmetry and shift of photospheric lines.

Spatially resolved as well as unresolved photospheric lines are asymmetric and blue-shifted, as a consequence of the depth dependence of velocity, temperature and opacity fluctuations associated with the granulation. Shift and asymmetry can thus be used to derive the temperature and velocity structure of granules and intergranules.

Spatially resolved profiles. Dark intergranular lanes and bright granules respectively exhibit red and blue asymmetry and shift (Wiehr and Kneer, 1988). Steep gradient models were derived from spatially resolved profiles (Keil and Yacovitch, 1981; Keil, 1984).

Spatially unresolved profiles. Spatially unresolved line profiles are asymmetric (C-shape of the line bisector) and blue shifted relative to their laboratory reference wavelength (Figure 8); such properties are

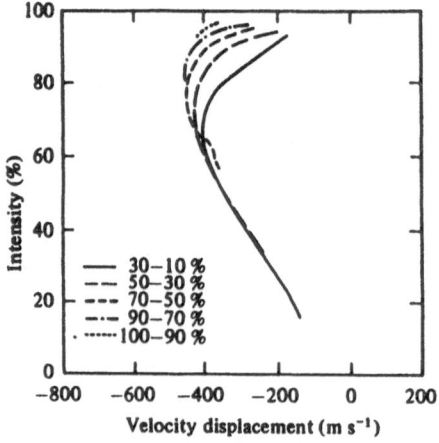

Figure 8. Bisectors (C-shape) of Fe I lines from unresolved observations at disc centers (Line strength depence, from Dravins et al. 1981).

signatures of the presence of convection or of buoyant motions on the surface of the sun and stars (Voigt, 1956; Schröter, 1957). Many granulation models were derived from line asymmetry and shift in recent years : Beckers and Nelson (1978); Dravins et al. (1981); Dravins (1982); Kaizig and Durrant (1982); Nordlund (1982); Kaisig and Schröter (1983); Kostik (1983); Marmolino et al. (1987). The main advantage of such kinds of observations is that they are independent of seeing; in couterpart the solution is not unique when deriving models, because the depth dependence of temperature, velocity and opacity fluctuations all contribute to the shaping and shifting of line profiles (Kaisig and Schröter, 1983; Nordlund, 1981). The granulation model of Kaisig and Durrant, which is the most elaborated one, is of the steep velocity gradient type, in agreement with models derived from resolved profiles.

Lines with cores formed at higher layers in the solar photosphere are less blue-shifted (Dravins et al. 1981; Balthasar, 1984). The blue-shift progressively decreases as one moves away from the disc center : this is known as the limb effect. This blue-shift reduction is a consequence of the decrease of granular velocities in higher layers. It must be noted that the limb effect can be properly explained only if horizontal granule motions are taken into account (Beckers and Nelson, 1977; Balthasar, 1985).

5. VELOCITY POWER SPECTRUM.

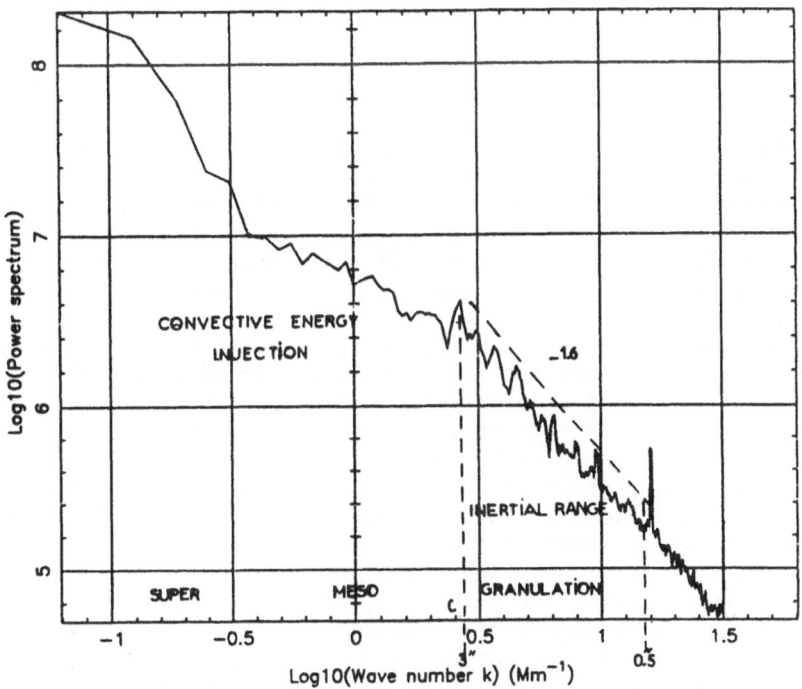

Figure 9. Power spectrum of the solar photospheric motions, derived from two superposed Doppler images taken two and half minutes apart (5 min oscillations filtering).

Analysing two dimensional, high resolution (0".5) NaD_2 5696 spectra

taken with the Multichannel Soustractive Double Pass spectrograph of the
Pic du Midi Observatory, Muller et al. (1987) have found that the energy
power decreases with a slope close to the value - 5/3, for the sizes
smaller than 3", (see also, Zahn, 1987). This trend becomes more prominent when the 5 min oscillations are (partially) filtered by the sum of
two spectra taken with a 2.5 min time interval (Figure 9). Compared to
previously published power spectra (Durrant et al. 1982), the statistics
(two-dimensional, 15"x240" field of view instead of one-dimensional slit
spectra) and the spatial resolution of our observations is better; in
addition we have filtered the 5 min oscillations which are carrying a
significant amount of power at scales larger than 3". It has to be
noted that the power spectrum shown in Figure 9 is not corrected for
blurring; however a spread function equivallent to a resolution of 0".5
should not significantly change the slope of the spectrum at scales
larger than 1".
    It is tempting to interpret the - 5/3 slope, which is characteris-
tic of the energy spectrum of homogeneous isotropic turbulence in the
inertial range, in terms of turbulent granules. In that case convective
energy is advected towards the surface by meso - and super - granular
cells; this energy is then cascading down to the smaller scales through
smaller and smaller granules.

6. THE SOLAR GRANULATION : TURBULENT CONVECTION ?

Solar granules appear to have several properties characteristic of
turbulent eddies, described in the previous sections :
- their number rapidly increases with the decreasing size;
- the most common process of disappearance is fragmentation;
- the fractal dimension is close to 4/3 (for those smaller than 1".4);
- the kinetic energy power spectrum decreases according to a - 5/3 law,
  for scales smaller than 3".
    However the degree of coherence between continuum intensity and
velocity is high, at least for scales larger than 1".5. (Durrant and
Nesis, 1982; Nesis et al., 1984; Wiehr and Kneer, 1988; Deubner, 1988).
For smaller scales the situation is not yet clear since, very recently,
contradictory results have been reported from spectra having apparently
the same quality : according to Deubner (1988) the correlation remains
at a high level down to 0".8, whereas for Wiehr and Kneer (1988) the
correlation is already lost at 1".4. The velocity-brightness correlation
has been used to regard the granulation as a convective phenomenon,
despite the high value of the Reynolds number in the top of the convection zone. Such a high degree of coherence does not necessarily imply
that granules are convective eddies; it may be explained, alternatively,
as follows : the convective energy is carried into the photosphere by
meso - and super - granules rather than by granules; the kinetic energy
is then transferred to smaller scales through the granules, which thus
appear to have a turbulent origin, however, because they are formed in
a superadiabatic atmosphere, they are forced to rise towards the surface,
explaining the observed intensity-velocity correlation. In this scenario
the scale where the turbulent cascade begins remains uncertain : 1".5
according to the fractal analysis, in which case only the smaller

ρranules would be turbulent; 3" according to the power spectrum, which
would mean that all granules are turbulent.

7. GRANULATION AND MAGNETIC FIELD.

7.1. Granules near isolated flux tubes.

In the quiet photosphere the magnetic flux mainly appears in the form of
concentrated magnetic flux tubes of kG field strength, (Stenflo, 1973)
of a few hundred kilometers in diameter and of a few 10^{17} Mx in magnetic
flux (Muller and Roudier, 1984). Flux tubes are located in the intergranular lanes (Title et al. 1987) where they are visible as Network
Bright Points (NBP; Muller et al 1988; Muller, 1983; Mehltretter, 1974).
The numerical simulation of Schmidt et al. (1985),which have shown that
weak magnetic fluxes are expelled by convective motions into the intergranular downflowing spaces, are in agreement with the observed flux
tube properties described above.
    The shape of the granules surrounding an NBP is modified : they
are smaller elongated, pointing in the direction of the NBP; they are
arranged so that they exhibit a characteristic "daisy-like" appearance
(Figure 1). Such a perturbation is very similar to the one which occurs
in laboratory experiments when a regular convective pattern is locally
disturbed by an obstacle (Muller et al., 1988).

7.2. Elongated granules.

They are visible in developping active regions, in areas between pores
or between sunspots and pores (Bray and Loughhead, 1964; Zwaan, 1985,
Figure 2a, Figure 10 in this review). The lanes separating these
elongated granules are darker than the average. They are attributed to
the action of the horizontal field at the top of a magnetic flux loop
rising through the photosphere.

7.3. Smaller granules near sunspots.

Near sunspots granules have been found to be 10 to 25% smaller than
normal by several authors (Macris, 1953; Schröter, 1962; Macris, 1979;
Kitai et al. 1988; Schmidt et al. 1988). In such regions the line
asymmetry is reduced (Ichimoto et al., 1989) indicating an inhibition
of convection (Livingston, 1982). Intensity fluctuations are reduced
(Title et al., 1986; 1987; Schmidt et al. 1988); whereas velocity
fluctuations in the low photosphere are enhanced (Mattig and Nesis,
1974). Granule lifetime is longer (Muller, 1973; Title et al., 1988).
Thus the velocity amplitude, the contrast, the size and the lifetime
of granules is found to be modified near sunspot. The cause of the
perturbation is not yet well established : is it the magnetic field of
flux tubes embedded in intergranular lanes, as described in section 7.1.?
the proximity of the sunspot which perturbs the heat flow ? a background
weak magnetic field ? It has to be noted, however, that the decrease of
the granule size near sunspots has not been confirmed by Collados et
al. (1966) using an image processing rather than a visual inspection

to identify granules. Also the contribution of the magnetic elements (facular points and facular granules) to the measured parameter is unknown.

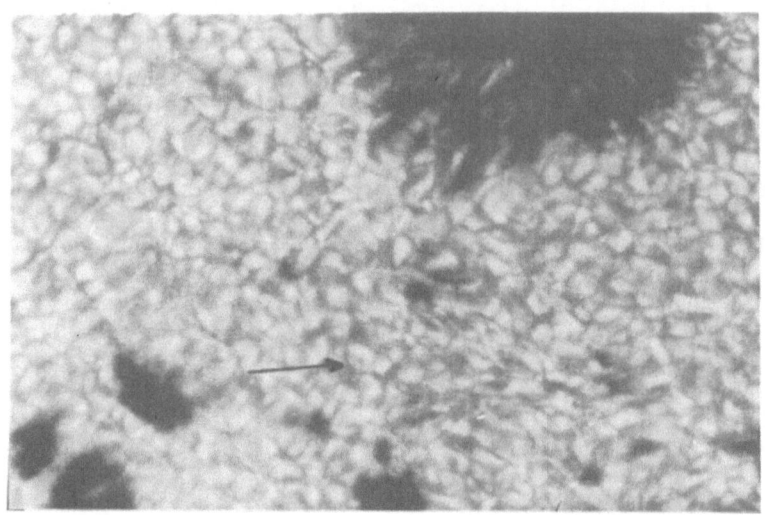

Figure 10. Elongated granules between pores.

7.4. Abnormal granulation.

In facular and filigree areas the granulation appears to be smeared and is called abnormal granulation (Dunn and Zirker, 1973; Muller, 1987; Title et al. 1987). The question is to know whether abnormal granulation is a kind of inhibited granulation or a granular pattern in which intergranular lanes are filled with facular points, smeared by seeing. Our high resolution observations from the Pic du Midi Observatory favours the second interpretation (Figure 11); however, granules in those areas are smaller than normal.

8. LARGE SCALE ORGANIZATION GRANULES.

Very recently a cellular pattern has been revealed by the position of particular granules, from high resolution Pic du Midi photographs and by horizontal granular flows observed on distorsion free SOUP movies. The cells are of mesogranulation scale and it is tempting to identify them with the mesogranules discovered by November et al. 1981, on dopplergrams; this is a confirmation of the existence of the mesogranulation.

The first, Oda (1984) reported that the active granules (those

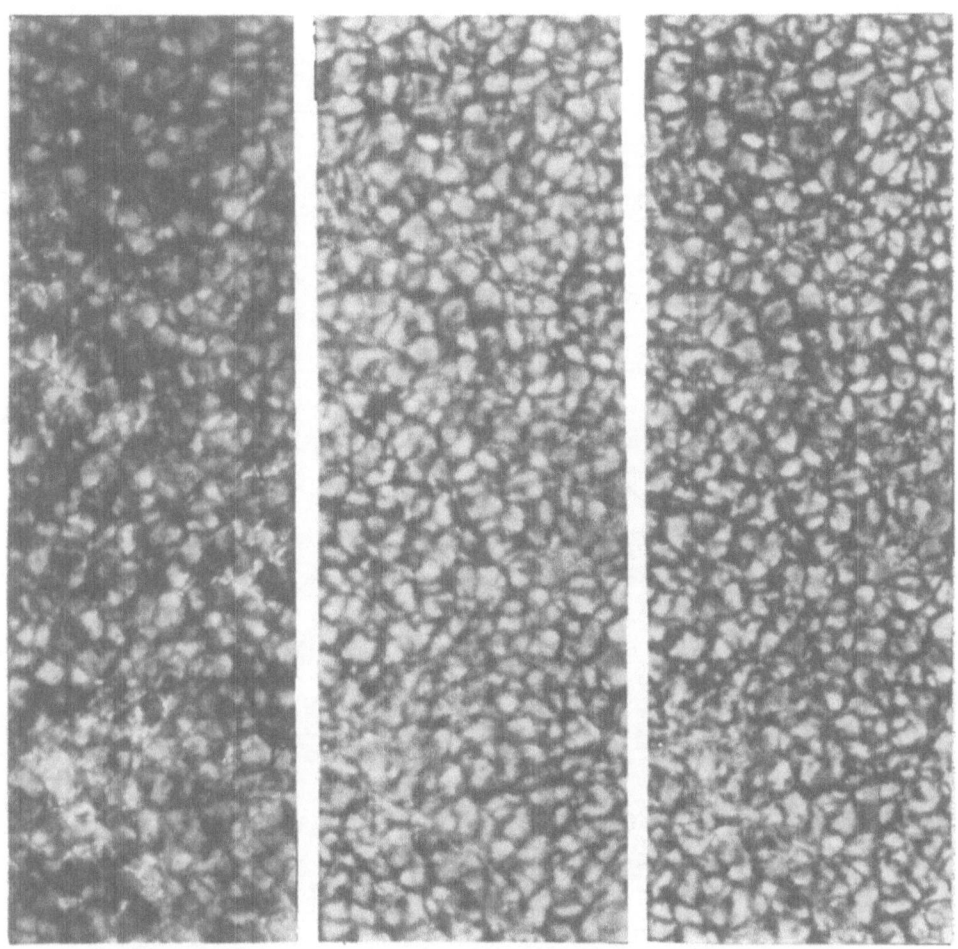

Figure 11. Abnormal granulation. Left : remnant of active region (filigree) observed at 4308 Å; taken at 7 h 24 min 20 s; middle : white light picture of the same region taken at 7.26.05, under good seeing conditions, granulation is smeared in the granulation area (abnormal granulation); right : the same region observed one second later under super good seeing conditions, abnormal granulation is resolved into tiny bright (magnetic) points and granules.

which fragment or merge) form a network of cells of average size 11".3, which coincides with the brightest areas of the photosphere (Figure 12). He tentatively identified this network with the mesogranulation pattern which has the same scale. (A large scale brightness pattern of mesogranulation scale has also been reported by Koutchmy and Lebecq, 1986). Then Dialetis et al. (1988) have found that long lived granules (lifetimes longer than the mean granule lifetime of 15 min) are not

116

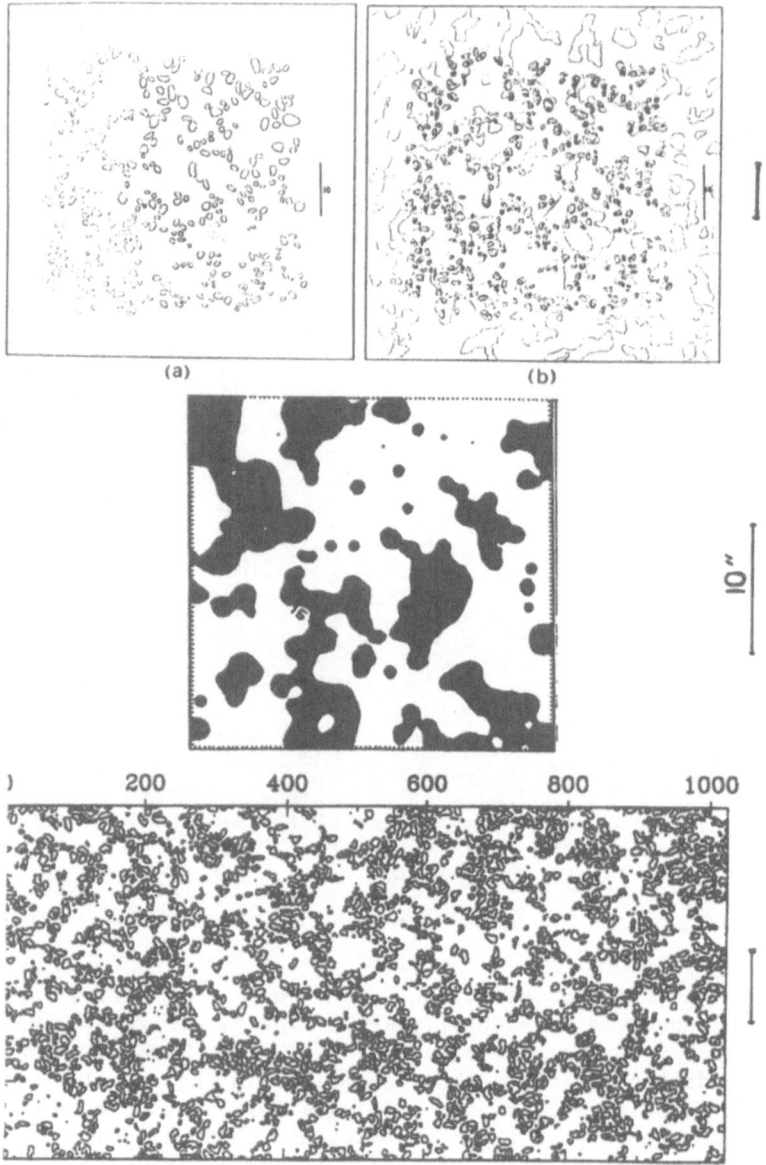

Figure 12. Large scale pattern formed by : (top a) active granules; (top b) brightest photospheric areas (Oda, 1984); (middle) long-lived granules (Dialetis et al. 1988); (bottom) granules smaller than 1".4 (Muller et al. 1988).

distributed at randon on the surface of the sun; they form a pattern of mean size 12". Finally, according to Muller et al. (1988), the granules larger than the critical scale 1".4 (section 2.4) also form a network; however the mean size of this network is found to be significantly smaller, 7", for a yet unknown reason. Of course the smaller granules form a similar and complementary pattern. Areas of smaller granules were mentioned earlier by Edmonds (1960).

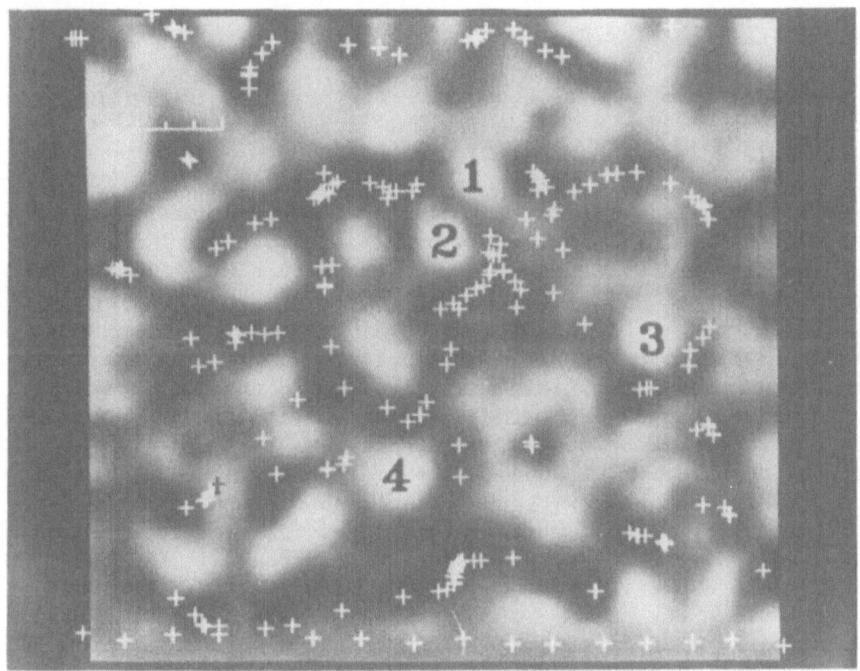

Figure 13. Divergence of the granular flow field. The crosses represent "corks" swept to the supergranular boundaries by the granular flow. Mesogranules appear embedded in the supergranular cells. Simon et al.

Summarizing those results, the spatial distribution of granules on the surface of the sun is not isotropic. The larger ones which also have a longer lifetime and are the brightest, form a network which is very likely to be associated with the mesogranulation. We may speculate that convective mesogranules advect a bunch of large granules up to the surface where they split into smaller granules, transferring the energy to smaller scales.

Thanks to a remarkable analysis of the distorsion-free SOUP movie as well as ground-based movies it has been possible to measure precisely the proper motion of granules. Their flow (Figure 13) is organized within cells of mesogranular and supergranular scales (Title et al., 1986; Simon et al. 1988; November and Simon, 1988). It has been demonstrated that this flow carries the magnetic flux to the supergranule boundaries, not to the mesogranule boundaries. The mesogranular and supergranular patterns appear to have some distinct properties : meso-granules do not retain magnetic flux at their boundaries but are asso-ciated with brightness and granule properties fluctuations; on the contrary supergranules retain magnetic flux at their boundaries, maybe because the flows are more vigorous. Moreover, according to Damé and Martic (1986), the 5 min oscillations may influence the white light intensity fluctuation of mesogranular scale.

9. VARIATION OVER THE SOLAR CYCLE.

The size of the granulation has been found to vary over the solar cycle, the granules being smaller near the maximum of activity. Either the mean intergranular distance (Macris et Rösch, 1983; Macris, 1988), or the number of granules per surface unity (Macris et al. 1984; Muller and Roudier, 1984b) was measured. The amplitude of the size variation is not yet well determined since it is of 5 % according to Muller and Roudier and as high as 20 % according to Macris. Such kinds of results are important because they, as well as the solar luminosity and diameter, probably reflect changes of the structure in deeper layers under the action of a variable magnetic field. The question is to know whether the interaction between magnetic field and convection occurs at the global scale of the sun as suggested by the correlation with the sunspot num-ber, or at a local scale between granules and flux tubes. It has to be noted that the number of flux tubes per surface unity, outside active regions, varies in antiphase with the granule number (Muller and Roudier, 1984; Muller, 1985), which rather suggests local interaction. Also related to the granule and photosphere variations, it has been found that the photospheric line equivalent width (Livingston, 1983, 1984) changes during the solar cycle. Latitude variations of the granu-lation can help us to understand the magnetic field influence. From the difference between the line shift variations toward the pole and along the equateur, Andersen (1984, 1988- inferred that granules are larger at high latitudes than near the equator. A latitude variation has not yet been detected from direct granule size measurements.

10. CONCLUSION AND PERSPECTIVES.

Although the solar granulation was discovered long ago and studied in depth since then, it remains a fascinating astrophysical phenomenon of very wide interest. In addition to their own interest (turbulent - convective eddies formed in compressible and stratified atmospheres of unusually high Reynolds number) granules can be used to investigate a very broad range of phenomena, from micro-phenomena of the smallest size permitted by observations (0".25 at present, but 0".1 in the future), to macro-phenomena like the supergranulation and even larger, of the scale of the global sun regarded as a star. Here I want to take up several important points which have been discussed in this review, and which need observational and theoretical improvements in order to be fully or at least better understood.

Concerning the nature of the granulation, several results were reported favouring a turbulent rather than convective origin of the granules (at least the smaller ones). This still has to be confirmed by improved observations : 0".1 in the case of filtergrams, in order to observe in more details the granule fragmentation, to extend the size histogram and the perimeter-area relation to smaller sizes; 0".3 or less in the case of spectrograms, with a high spectral resolution if possible, in order to extend the kinetic energy power spectrum and the coherence and phase analysis to the smallest possible sizes; such kinds of observations would also allow the study of the evolution and penetration of the temperature and velocity fluctuations of granules of various sizes up into the photosphere. The use of super-computers of increasing possibilities should help theoreticians to improve the spatial resolution of the 3-D numerical simulations, so that on one hand the turbulent energy can be simulated rather than parameterized and, on the other hand, the advection of convective energy can be extended to scales larger than the granulation scale. This concerns the sun and stars as well.

The asymmetry and shift of photospheric lines, integrated over a wide area of the sun or even over the global sun, can be a powerful tool for deriving granulation models (2 or multi-columns) of the sun and stars. However many efforts have yet to be made in order to disentagle the contributions of the velocity, temperature and opacity height dependance and obtain a unique solution.

We have seen that large scale patterns can be revealed by granule horizontal flows. We need to know from more observations and theoretical investigations how the supergranular, mesogranular and granular patterns are connected and how they interact with the magnetic field. One way to study the meridional circulation is to analyse the centre-to-limb variation of the asymmetry and shift of photospheric lines of various strengths (Brandt and Schröter, 1982; Andersen, 1984, 1988). But such kinds of observations require very precise Doppler shift measurements and the contribution of meridional motions and granulation variations have to be disentangled. May be meridional circulation and giant cells could be detected by granule horizontal flows, even under moderate resolution observations (November and Simon, 1988); for example, a 1000 x 1000 array detector covers one solar hemisphere if the pixel is 1"x 1"; it takes 10 hours of consecutive observations to detect a displacement of 1" for a motion of 20 ms^{-1}; such an observation could be performed from space.

The variation of the structure of the granulation, as observed through its size variation or the line asymmetry, is no doubt due to the convective-magnetic field interaction, and is connected to luminosity, diameter... convection zone variations. However it is not clear whether this interaction occurs at the global scale of the sun or very locally between granules and flux tubes. Long term meridional and equatorial observations are required, both of granulation and magnetic field. Improved image processing has to be developped in order to detect subtle granulation variations, especially latitude variations. Theories should also be developped in order to understand the origin of

the granule variation which is very probably related to the sun variability.

The solar magnetic field interacts not only at large scales, thus
inducing the solar variability, but also at very small scales as demonstrated by the perturbation of granules described in section 7. Those
perturbations should be observed in more details, including high
resolution magnetic field observations. Here also a lot of theoretical
work needs to be done.

I am sure that the perspectives discussed above are attractive
enough to incite us to play with solar granules for many extra years.

REFERENCES

Alissandrakis, E.C., Dialetis, D., and Tsiropoula, G. : 1987, Astron.
    Astrophys. 174, 275.
Altrock, R.C. : 1976, Solar Phys. 47, 517.
Andersen, B.N.: 1984, Solar Phys. 94, 49.
Andersen, B.N.: 1988, Solar Phys. 114, 2.
Bahng, J. and Schwarzschild, M. : 1961, Astrophys. J. 134, 337.
Balthasar, H. : 1984, Solar Phys. 93, 243.
Balthasar, H. : 1985, Solar Phys. 99, 31.
Bässgen, M. and Deubner, F.L. : 1982, Astron. Astrophys. 111, L 1.
Beckers, J.M. and Nelson, G.D. : 1978, Solar Phys. 58, 243.
Brandt, P.N. and Schröter, E.H. : 1982, Solar Phys. 79, 3.
Bray, R.J. and Loughhead, R.E. : 1964, Sunspots, Chapman and Hall,
    London.
Bray, R.J. , Loughhead, R.E., and Durrant, C.J. : 1984,'The Solar
    Granulation', 2nd edition, Cambridge University Press.
Canfield, R.C. : 1976, Solar Phys. 50, 239
Carlier, A., Chauveau, F., Hugon,M. and Rösch, J. : 1969, Comptes
    Rendus Acad. Sci. Paris, 266, 199.
Collados, M., Marco, E., Del Torro, J.C., and Vazquez, M. : 1986,
    Solar Phys. 105, 17.
Collados, M. and Vazquez, M. : 1987, Astron. Astrophys. 180, 223.
Damé, L. and Martic, M. : 1987, Astrophys. J. Letters, 314, L 15.
Deubner, F.L. : 1988, preprint.
Dialetis, D. Macris, C., Muller, R., and Prokakis, Th. : 1988, Astron.
    Astrophys. (in press).
Dialetis, D., Macris, C., Prokakis, Th., Sorris, E. : 1986, Astron.
    Astrophys. 168, 330.
Dravins, D. : 1982, Ann. Rev. Astron. Astrophys. 20, 61
Dravins, D., Lidegren, L. and Nordlund, Å. : 1981, Astron. Astrophys.
    96, 345.
Dunn, R.B. and Zirker, J.B. : 1973, Solar Phys. 33, 281.
Durrant, C.J., Kneer, F., and Maluck, G. : 1981, Astron. Astrophys. 104,
    211.
Durrant, C.J., Mattig, W., Nesis, A., Reiss, G., and Schmidt, W. : 1979,
    Solar Phys. 61, 251.
Durrant, C.J. and Nesis, A. : 1981, Astron. Astrophys. 95 , 272.

Durrant, C.J. and Nesis, A. : 1982, Astron. Astrophys. 111, 272.
Edmonds, F.N. / 1960, Astrophys. J. 131, 59.
Graves, J.E. and Pierce, A.K. : 1986, Solar Physics, 106, 249.
Ichimoto, K., Kitai, R., Muller, R. and Roudier, Th. : 1989, NATO
    Advanced Research Workshop, 'Solar and Stellar Granulation', Rutten,
    R. and Soverino, G. (eds)., Capri, Italy.
Kaisig, M. and Durrant, C.J. : 1982, Astron. Astrophys. 116, 332.
Kaizig, M. and Schröter, E.H. : 1983, Astron. Astrophys. 117, 305.
Kawaguchi, I. : 1980, Solar Phys. 65, 207.
Keil, S.L. : 1980a, Astrophys. J. 237, 1024
Keil, S.L. : 1980b, Astrophys. J. 237, 1035.
Keil, S.L. : 1984, in S.L. Keil (ed), 'Small-Scale Dynamical Processes
    in the Solar and Stellar Atmospheres', Sacramento Peak Observatory,
    Sunspot, New Mexico, p. 149.
Keil, S.L. and Canfield, R.C. : 1978, Astron. Astrophys. 70, 169.
Keil, S.L. and Yacovitch, F.H. : 1981, Solar Phys. 69, 213.
Kneer, F., Mattig, W., Nesis, A., and Werner, W. : 1980, Solar Phys
    60, 31.
Kostik, R. : 1983, Publ. of the Academy of Sci of Ukrain 83, 63.
Koutchmy, S. and Lebeck, Ch. : 1986, Astron. Astrophys. 169, 323.
Livingston, W.C. : 1982, Nature, 297, 208.
Livingston, W.C. : 1983, in J.O. Stenflo (ed.)'Solar and Stellar Magnetic
    Field' IAU Symp. 71, 149.
Livingston, W.C.: 1984, in S.L. Keil (ed.) 'Small-Scale Dynamical
    Processes in the Solar and Stellar Atmospheres' Sacramento Peak
    Observatory, Sunspot, New Mexico, p. 265.
Livingston, W. and Holweger, H. : 1982, Astrophys. J. 292, 375.
Livingston, W. and Holweger, H. : 1985,
Macris, C.J. : 1953, Ann. Astrophys. 16, 19.
Macris, C.J. : 1979, Astron. Astrophys. 78, 186.
Macris, C.J. : 1988, Compt. Rend. Acad. Sci. Paris,
Macris, C.J., Muller, R., Rösch, J., Roudier, Th. : 1984, in S.Keil (ed.),
    'Small-Scale Dynamical Processes in the Solar and Stellar Atmos-
    pheres' Sacramento Peak Observatory, Sunspot, New Mexico, p. 265.
Macris, C.J. and Rösch, J. : 1983, Compt. Rend. Acad. Sci. Paris, 296,
    265.
Mandelbrot, B. : 1977, Fractals, Freeman, San Francisco
Marmolino, C., Roberti, G., and Severino, G. : 1987, Solar Phys. 108, 21.
Mattig, W. and Nesis, A. : 1974, Solar Phys. 38, 337.
Mehltretter, J.P. : 1974, Solar Phys. 34, 33.
Mehltretter, J.P. : 1978, Astron. Astrophys. 62, 311.
Muller, R. : 1973, Solar Phys. 29, 55.
Muller, R. : 1983, Solar Phys. 85, 113.
Muller, R. : 1985a, Transactions of the IAU, 'Reports on Astronomy'
    vol. XIX A, p. 430 : Solar Granulation
Muller, R. : 1985b, Solar Phys. 100, 237.
Muller, R. : 1987, in 'Development of Active Regions', 10th. Regional
    European Assembly of the IAU, Praha; Publications of the Astrono-
    mical Institute of the Czeschoslovak Academy of Sciences.

Muller, R. and Roudier, Th. : 1984a, Solar Phys. 94, 33.

Muller, R. and Roudier, Th. : 1984b, '4th European Meeting on Solar Physics', Noordwijk, The Netherlands, p. 239.

Muller, R., Roudier, Th. and Hulot, J.C. : 1988, submitted to Solar Phys.

Muller, R., Roudier, Th., Malherbe, J.M., and Mein, P. : 1987, 5th European meeting on Solar and Stellar Physics. Titisee (poster).

Muller, R., Roudier, Th.,and Vigneau, J. : 1989, NATO Advanced Research Workshop and Stellar Granulation, Rutten R. and Severino, G. (eds), Capri, Italy.

Namba, O. : 1986, Astron. Astrophys. 161, 31.

Namba, O. and Diemel, W.E. : 1969, Solar Phys. 7, 167.

Nesis, A., Durrant, C.J., and Mattig, W. : 1984, in S. Keil (ed.) 'Small Scale Dynamical Processes in the Solar and Stellar Amospheres' Sacramento Peak Observatory, Sunspot, New Mexico, p. 243.

Nordlund, Å : 1982, Astron. Astrophys. 107, 1.

Nordlund, Å : 1984, in S.L. Keil (ed.), 'Small-Scale Dynamical Processes in the Solar and Stellar Atmosphere', Sacramento Peak Observatory, Sunspot.

November, L.J. and Simon, G.W. : 1988, Astrophys. J. (submitted).

Oda, N. : 1984, Solar Phys. 93, 243.

Pravdjuk, L.M. : 1982, Soln. Dann. N° 2, 103.

Rösch, J. and Hugon, M. : 1959, Comptes Rendus Acad. Sci. Paris, 249, 625

Roudier, Th. and Muller, R. : 1986, Solar Phys. 107, 11.

Schmidt, H.U., Simon, G.W., and Weiss, N.O. : 1985, Astron. Astrophys. 148, 191.

Schmidt, W., Grossmann-Doerth, U., and Schröter, E.H. : 1988, Astron. Astrophys. 197, 306.

Schröter, E.H. : 1957, Z. Astrophys. 41, 141.

Schröter, E.H. : 1962, Z. Astrophys. 56, 183.

Simon, W., Title, A.M., Topka, K.P., Tarbell, T.D., Shine, R.A., Ferguson, S.H., Zirin, H. and the SOUP Team : 1988, Astrophys. J. 327, 964.

Stenflo, J.O. : 1973, Solar Phys. 32, 41.

Title, A., Tarbell, T. and the SOUP team : 1986, in Theoretical Problems in High Resolution Solar Physics, II. Boulder, Co. p. 55

Title, A.M., Tarbell, T.O. and Topka, K.P. : 1987, Astrophys. J. 316,892.

Voight, H.H. : 1956, Z. Astrophys. 40, 157.

Wiehr, E. and Kneer, F. : 1988, Astron. Astrophys. 195, 310.

Zahn, J.P. : 1987, 5th European Meeting on Solar and Stellar Physics. Titisee (in press).

Zwaan, c. : 1985, Solar Phys. 100, 397.

Discussion

TITLE — We do not see the break at 1.37 arcsec in either the number density or the area perimeter ratio.

FOX — Given that the cyclic variation of granules was derived from the last solar maximum, which was the second largest ever, do the observers have any indication of granular changes with the current cycle as it approaches what may be the largest maximum ever?

MÜLLER — We have been observing the granulation continuously since 1978; however, the data taken after 1984 have not yet been analysed.
We are planning to observe the granulation at the Pic du Midi Observatory at least until the next maximum.

WEISS — I find it hard to accept that the granulation is the inertial range of a turbulent cascade driven by larger scale supergranulation or mesogranulation. The development of an exploding granule seems to be initially a thermal phenomenon (though the fragmentation is probably an inertial instability), and it seems much more likely that granules have a convective origin.

MÜLLER — It is troublesome that, from the power spectrum, the inertial range would start at 3 arcsec, while from the fractal analysis it would start at 1.5 arcsec.

DRAVINS — A general remark: it may be important to separate clearly what is *really observed*, and what is *merely inferred* from the data. Many granulation phenomena may be too complex for straightforward interpretation, and then the confrontation with theory should come only after detailed numerical modeling.

MÜLLER — You are right. But in order to attract the attention of theoreticians we have to be somewhat provocative. However, it is troublesome that even with a straightforward interpretation several different results favour a turbulent origin, at least of the smaller granules.

BECKERS — I would like to add one other item to your list of needs for future observations, and that is for granulation observations at the wavelength of the H⁻ opacity minimum at 1.6 μm. Although this radiation is formed only 25–50 km deeper in the atmosphere than the visible, one looks at that wavelength into the convection zone rather than in the convective overshoot region. Things may look quite different there.

LINE SHIFTS IN THE INFRARED SPECTRA OF LATE-TYPE STARS

Daniel Nadeau[1], Jean Bédard[1] and Jean-Pierre Maillard[2]

[1] *Université de Montréal, Canada*
[2] *Institut d'Astrophysique de Paris, France*

1 Introduction

The direct study of stellar convection is of considerable interest for stellar astrophysics, because it provides a diagnostic of the stellar structure and magnetic field generating mechanisms below the photosphere (Sofia and Endal 1987), and because of its role in the energy transport through the atmosphere. The study of convective motions in red giants in particular, on which little has been done, could help to understand such phenomena as the isotopic enrichment of the atmosphere by elements synthesized within the star, the excitation of chromospheric gases, the onset of strong stellar winds in the later-type stars, the photometric variability and other non-thermal phenomena. Detailed hydrodynamical computer simulations are restricted by our very limited knowledge of the parameters of convective motions in these stars until these parameters (granulation scale, horizontal temperature inhomogeneity, relative velocity and areal ratio of the upward and downward moving streams) are determined by observational studies and empirical multi-stream models. The study of this fundamental stellar activity has received impetus from the application of modern high-resolution, high-signal-to-noise spectrometers.

2 The Observability Of Stellar Convection

The evidence of convective motions in stellar atmospheres has been detected first in the Sun by high spatial resolution images of the solar disk. The accurate analysis of small asymmetries in selected metallic line profiles, induced by the solar granulation, has provided a spectroscopic criterion. The usual method involves finding the mid-points of the horizontal segments bounded by the sides of the line profile. The resulting line bisector pattern exhibits a typical C-shape which can be interpreted as the combination of a blueshifted component originating from the hot rising granules and a component shifted in the opposite direction due to the cooler falling material. The relative area of the bright granules and intergranular lanes and their temperature difference lead to a net blueshift of the lines (typically $\sim$300 m/s). A three-dimensional detailed simulation of this mechanism has been performed by Nordlund (1982, 1984), who has been able to reconstruct the observed line profiles and shifts.

R. J. Rutten and G. Severino (eds.), Solar and Stellar Granulation, 125–134.
© 1989 by Kluwer Academic Publishers.

The bisector method has been extended successfully to stars and provides an effective way of characterizing the stellar granulation on objects other than the Sun, even if the asymmetry of the profiles is less pronounced when averaged over the full disk. However, the precise determination of the line bisector requires very high resolution spectra (at least 10^5) with a signal-to-noise ratio of the order of 500. Therefore, only the brightest stars can be observed, and only the lines of isotopically pure elements that appear unblended in the spectrum should be selected. The accuracy is improved by averaging the bisectors from several lines, but since high resolution is generally incompatible with a large spectral range, the number of usable lines is small. Nevertheless, the method has been applied to a few main-sequence to supergiant stars of spectral types F5 to K2 (Dravins *et al.* 1981; Gray 1982; Gray and Toner 1986), providing unique information on the variation of stellar granulation with spectral type and luminosity class.

The increasing strength of molecular bands in spectra of later type stars prevents the use of the visible metallic lines of Fe I, Fe II, Ca I, etc, which are generally used for the study of bisectors in hotter stars. The extension of the bisector method to late K- and M-type stars can be pursued in the infrared part of the spectrum by using CO lines in the 2.2μm region. The relative blending of lines is compensated by the access to a large number of lines. However, a high resolution, a high signal-to-noise ratio, and a large spectral coverage are required. A Fourier Transform Spectrometer (FTS) is the only instrument offering all these advantages. With such an instrument, set for a resolving power of 160,000 at 4000 cm^{-1}, Ridgway and Friel (1981) have detected a redshift of the wings of CO lines with respect to their core (average 200 m/s) in the spectra of eight K- and M-type giants. They obtained a 6 km/s minimum velocity separation of the two components from an analogy with the bisector shapes that were measured in other stars and reproduced by a two-stream model. Individual stars, particularly of the M type, show large variations in their bisector characteristics, with core to wing displacements varying from 200 to 500 m/s.

In order to compare unambiguously the bisectors of different lines, it is necessary to determine their relative line shifts. The measurement of the relative shifts at line bottom can in fact be used by itself as a diagnostic of convective motions. The profile at line bottom is less sensitive to blending than in the line wings, and the spectral resolution requirements are not as stringent for the measurement of line shifts as for those of line asymmetry. Simple simulations show that the asymmetry of a line—produced by the sum of two components—is reduced much faster than its shift at line bottom when the line is convolved with an instrumental profile (a sinc function) of increasing width. For the typical width of the CO lines in M-giant stars, a resolution of the order of 75,000 at 4000 cm^{-1} (or $\simeq 4$ km/s) is adequate to keep the information on lineshift. This is half the resolution used on the line bisector determinations, allowing a gain of a factor 4 in observing time for the same signal-to-noise ratio on sources of similar brightness. This gain makes it easier to obtain data on a significant sample of stars, from which it is possible to determine the parameters of convective motions as a function of spectral type, the variations among objects of the same spectral type, and possibly the variability of these motions in individual stars.

The FTS technique is well suited for the measurements of relative line shifts. This instrument gives access to a large spectral range, covering a full atmospheric window

(4000 - 5200 cm^{-1}, 5500 - 7000 cm^{-1}). A large number of lines of several species (CO, OH, and metals), having different strengths and excitation energies, and probing various parts of the stellar atmosphere, can be measured simultaneously. The FTS is capable of a high spectral resolution, leading to a precise determination of the position of the bottom of a line. The frequency of an unblended line observed with a resolution limit of 3 km/s, and a S/N ratio of 100, can be found to an accuracy of ± 0.03 km/s. The main advantage offered by a FTS, however, is the uniformity of the frequency scale over the whole spectral range measured, enabling the measurement of very accurate relative frequencies. The determination of absolute line frequencies depends only on the calibration of the reference line used in the instrument (generally a laser line) and it is very good (10^{-6} or better), even without any special correction. The uncertainties are nonetheless much larger than those affecting relative line shifts within a single spectral range, and in this project we do not seek to obtain the absolute frequency shifts of the lines.

Velocity differences between CO lines of the first overtone band and lines of neutral metals have since long been reported in the atmosphere of a few late-type supergiants such as α Boo, α Her, α Ori, α Sco, from the early spectra obtained with a FTS (Bopp and Edmonds 1970; Maillard 1974; Brooke et al. 1974). These differences were described as the result of a differential expansion of photospheric layers related to the mass-loss in the atmosphere of these late-type stars. After analysis, this assumption appeared invalid, the resulting mass-loss rate being too high. Observational evidence of mass-loss in similar spectra came from the detection of distinct components in the lines of the fundamental band of CO (80 times stronger than its first overtone), which could be attributed to colder circumstellar CO (Bernat 1981). The weaker CO lines of the first overtone band must sample cyclic motions of the atmospheric gases, and their line shifts with respect to the metallic lines are most likely caused by convective motions.

The reliability of these early detections was beset, however, by the lack of accurate laboratory spectra that ate calibrated with respect to a common standard. An absolute calibration $\Delta\sigma/\sigma$ better than 10^{-6} must be achieved to be able to detect a typical convective blueshift of 300 m/s. Only the laboratory spectra obtained by the FTS technique have this accuracy. A number of molecular and atomic spectra of astrophysical interest, obtained by this method, are now available (v.g. Mantz and Maillard 1974; Guelachvili et al. 1983; Biémont et al. 1985; Davis 1987). No interferometric spectra are available, however, for common elements like Ca I and Si I, and their lines, although easily detectable, are almost unusable, the uncertainty on their line positions being larger than the effect to be measured. In addition to using a FTS, the source of the spectrum must be carefully designed to limit the effects of pressure shifts.

We have determined that the lines of CO and Fe I have an accuracy of 0.001 cm^{-1} with respect to a common standard (Nadeau 1988; Nadeau and Maillard 1988), and that the lines of OH, which were calibrated with respect to CO, should have a similar accuracy (Maillard et al. 1976). These species can be used to determine differential velocity shifts for lines of different strengths and excitation energies in the infrared spectra of K and M giants to an accuracy of the order of 100 m/sec.

3 Results

We have observed 11 stars in the infrared K-band (4000-5200 cm^{-1}), among which 7 were also observed in the H-band (5500-7000 cm^{-1}). All these stars are of spectral type M0-M5 III (except HR7635: K5.6) and show little variability, thus providing a priori a reasonably homogeneous sample of "normal" stars. They were also selected for their brightness in order to obtain a high signal-to-noise ratio; the S/N ratio of the spectra obtained so far varies between 40 and 70, and it is apparent that this ratio dominates the uncertainties in our data.

The method used for detecting and identifying the spectral lines is described in detail in Nadeau and Maillard (1988). The identification of stellar lines with laboratory lines is based on the agreement of their frequencies within a velocity range chosen small enough to limit the noise from spurious identifications but large enough not to bias the results. Since the amplitude of molecular lines varies smoothly with the energy levels involved in the transition, those lines identified with CO and OH transitions, of which the amplitude is inconsistent with this smooth variation, are considered contaminated and therefore eliminated.

The K-band spectra contain strong CO (and ^{13}CO) lines of the first overtone band ($\Delta v = 2$). A number of lines of Fe I, Ca I, and Si I are also detected. A median velocity is obtained for all the CO lines having an excitation energy less than 1 eV, for the CO lines with excitation energy between 1 eV and 2.5 eV, and for the Fe I lines, which have an average excitation energy of 5.5 eV. The laboratory frequencies of the Ca I and Si I lines are too uncertain to be used for the measurement of line shifts. This grouping of the lines serves the purpose of reducing the scatter due to the S/N ratio and possible line contamination. The resulting median velocities for each group of lines are shown in Figure 1, together with the differential velocities found for the Sun (at disk center) using lines of Fe I with excitation energies and line depths that correspond to the stellar data (Nadeau 1988). The frequency shifts of the low energy lines of CO in the Sun agree with the shifts of the Fe I lines of corresponding excitation energy and line depth to within 40 m/sec.

The data in Figure 1 show large variations of the differential shifts even among stars of similar spectral types. From the scatter of the observed individual line shifts we estimate the uncertainty on the value of the median velocity shifts to be ±100 m/sec. The stars HR0045 and HR3950 have been observed at two different epochs, and the median differential shifts measured in the two stars are reproducible within ±30 m/sec. The variations of the shifts among the different stars appear to be real features of their spectra. The general trends are that little shift is seen between the low and high energy lines of CO for the early M stars (there may be a relative redshift of the high energy lines in HR7635, although the uncertainties are somewhat larger for this star), while a blueshift of the high energy CO lines is seen in the spectra of stars of type M2 and later. The lines of Fe I are consistently blueshifted with respect to the lines of CO for the stars of spectral type from M0 to M3. The three stars later than M3 do show a blueshift of the Fe I lines with respect to the low energy CO lines, but little or no shift with respect to the high energy CO lines.

The stars of type M0 to M3 make up a reasonably uniform group in terms of their differential shifts. It is noticeable that for these stars the blueshift of the Fe I lines with

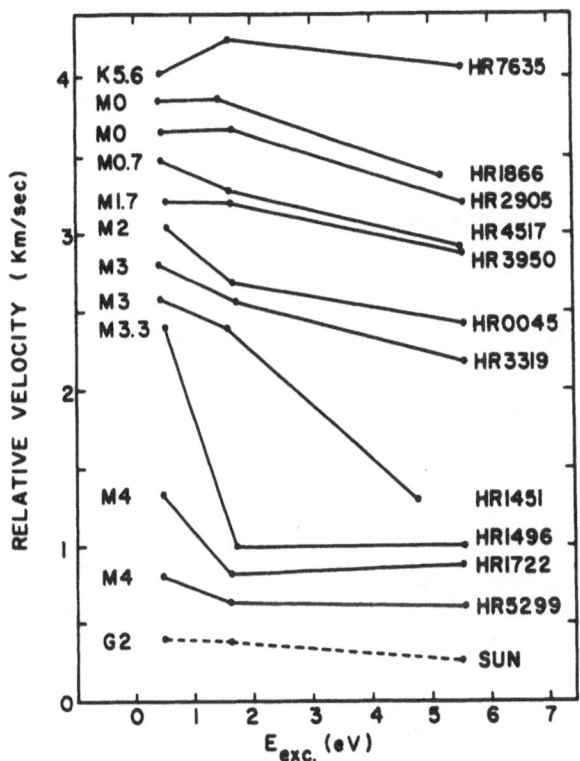

Figure 1: The differential shifts between the CO lines with energy below 1 eV, the CO lines with energy between 1 and 2.5 eV, and the lines of Fe I. The relative velocity scale shows the amplitude of the velocity shifts between the three groups of lines. The uncertainty on the plotted velocities is estimated to be ±100 m/sec. The shifts of solar lines observed at disk center are shown for comparison.

respect to the CO lines is of the order of 500 m/sec, while the corresponding blueshift in the Sun is about 150 m/sec, even though the stellar lines are integrated over the disk of the star, which has been shown, in the case of the Sun, to decrease the line shifts by a factor of 2.

There is a strong correlation between excitation energy, line depth, and species in the K-band data so that it is difficult to separate the dependence of the line shifts on each of these factors. The spectra obtained in the H-band contain numerous lines of Fe I with excitation energies between 5 and 6 eV, and covering a large range of line depths. The relative shifts of these lines are plotted in Figure 2 as a function of line depth for four of the stars observed.

The relative blueshift of the weaker lines is qualitatively similar to that found in the Sun but the velocity shift is, on average, 3 times as large in the stars as in the Sun for the same change in line depth. The median velocity and line depth of the lines of CO

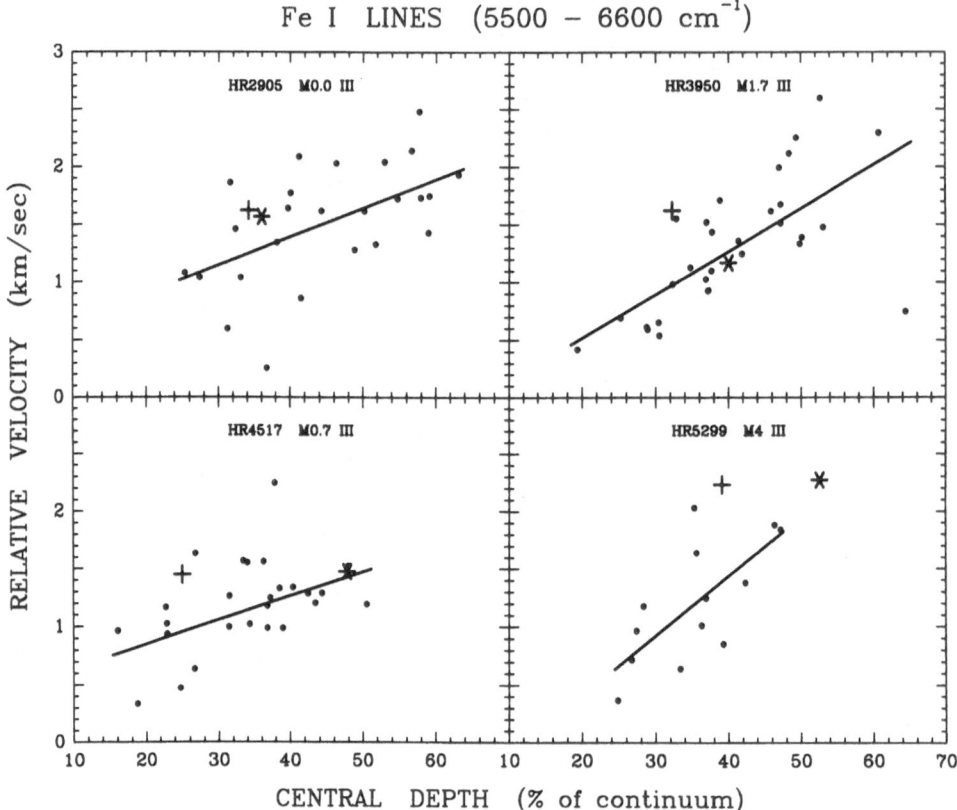

Figure 2: The relative velocity shifts of lines observed in the H-band. The dots represent the individual lines of Fe I with excitation energy between 5 and 6 eV. The cross and the star indicate the median velocity and line depth of the lines of CO and OH respectively. Straight lines are fit to the Fe I data.

($\Delta v = 3$) and OH observed in the H-band are plotted in Figure 2 as well. The CO lines have excitation energies between 0 and 1 eV and the OH lines have energies near 1 eV. In general, the OH lines are deeper than the CO lines, but they have a similar velocity, indicating a smaller dependence of the line shift on line depth for the low energy lines than for the lines of Fe I. The star HR4517 has a spectral type intermediate between those of HR2905 and HR3950 but the depth of its CO and OH lines are significantly different from those of the other two stars. This indicates that the spectral classification may not describe M stars uniquely. Uncertainties in the spectral classification may also explain some of the variations observed in the K-band among stars of similar spectral types. The CO lines are not as strongly redshifted with respect to the Fe I lines in the H-band as they are in the K-band, and the Fe I lines are stronger with respect to the CO lines in the H-band than in the K-band. This is another indication of the dependence of the line shifts on line depth.

It is interesting to compare the data for HR5299 in Figure 1 and Figure 2. In the K-band a relatively small blueshift is observed between the lines of CO and Fe I, while in the H-band HR5299 is the only star that shows all the Fe I lines to be blueshifted with respect to the CO lines, and it is the star for which the dependence of the shifts on line depth is strongest. It is also the star for which the CO lines in the H-band have the largest depth. Considering the fact that the $\Delta v = 2$ CO band has a strength that is more than an order of magnitude higher than the $\Delta v = 3$ band, the differences seen in the K-band and H-band data may be due to the sampling of different regions of the atmosphere. Tsuji (1987) has found evidence of a circumstellar shell component in the low energy K-band lines of CO for stars of type M5 and later. A similar phenomenon may be affecting the later stars in our sample.

4   Comparison of Stellar and Solar Line Shifts

As mentioned in §2, the observed differential line shifts cannot be due to mass-loss because they would imply mass-loss rates a few orders of magnitude larger than those expected for the stars in our sample. The observed velocities must be related to oscillatory or cyclic motions. Global oscillations with successive outward and inward motions are ruled out on statistical grounds by the fact that the great majority of stars show similar shifts, and because the two stars observed at different epochs show no change of the differential shifts with the epoch.

There are striking analogies between the line shifts observed in the giant stars and the shifts of the Fe I lines observed in the Sun. The lines of Fe I in the Sun show a blueshift of the weaker lines with respect to the stronger ones (Dravins *et al.* 1981; Nadeau 1988). This dependence of the blueshift on line depth becomes steeper with increasing excitation energy of the lines, and the deeper high energy lines are in fact redshifted with respect to the low energy lines. Similar trends can be observed in the stellar data. The weaker Fe I lines observed in the H-band are blueshifted with respect to the stronger ones. Essentially no dependence of the velocity on line depth is seen between the low energy CO and OH lines. The weak high energy Fe I lines are generally blueshifted with respect to the strong CO lines in the K-band, and those high energy Fe I lines which have a central depth of more than 50% are redshifted with respect to the low energy CO lines in the H-band.

The differential lineshifts of the Fe I lines observed in the Sun are well reproduced by models of the solar granulation. We conclude from the qualitative agreement between the shifts observed as a function of excitation energy and central line depth in the solar and late-type giants spectra, that convective motions are the main cause of the line shifts observed in the giant stars.

5   Future Developments

Important progress in the study of convective motions in late-type stars should be possible in the near future, in the first place by acquiring higher signal-to-noise spectra for a larger number of stars. The S/N ratio of the stellar spectra obtained so far is the dominant factor limiting the accuracy of the frequency shift determinations in the CO,

OH, and Fe I lines. Because of the large differences observed between stars of similar spectral types, it is important that a larger sample of stars be observed in order to determine trends as a function of spectral type. The methods of spectral classification of the late-type giants should also be looked at carefully, since there may be inconsistencies in the present classification, as shown by the OH/CO line ratios in our H-band spectra.

The determination of accurate reference frequencies for additional species is also crucial in order to measure the shifts of lines with different strengths and excitation energy, and hence obtain better sampling of the different parts of the atmosphere. It may be possible to circumvent the limitations of the laboratory spectra of species such as Ca I and Si I by using their solar frequencies corrected with the help of a model of the solar line shifts, but the urgent need for laboratory spectra with an absolute uncertainty $\Delta\sigma/\sigma \sim 10^{-7}$ should be stressed.

By a careful selection of uncontaminated lines, it should be possible to obtain a large amount of information from the bisectors of a few lines, since each bisector samples a whole range of depths into the atmosphere.

A necessary complement to the observational studies is the application of empirical 2-stream or multi-stream models to late-type stellar atmosphere models in order to determine the velocity and temperature contrast and the stream areal ratios, and to find the correlation between the horizontal inhomogeneities in velocity and temperature as a function of height in the atmosphere. The determination of these parameters may ultimately will ultimately enable the detailed hydrodynamical modelling of the convective layer. The most important parameter for these calculations, however, is probably the granulation scale. It may be possible to estimate this scale from the height of penetration of the convective overshoot into the atmosphere of the giant stars, since observational and theoretical studies of the Sun and solar type stars may well establish a relation between the two quantities. This would be another demonstration of the fruitful interaction between studies of the solar and stellar granulation.

6 References

Bernat, A. P. 1981, *Ap. J.*, **246**, 184.

Biémont, E., Brault, J. W., Delbouille, L., and Roland, G. 1985, *Astr. Ap. Suppl.*, **61**, 107.

Bopp, B. W., and Edmonds, F. N., Jr. 1970, *Pub. A.S.P.*, **82**, 299.

Brooke, A. L., Lambert, D. L., and Barnes, T. G., III. 1974, *Pub. A.S.P.*, **86**, 419.

Davis, S. P. 1987, *Pub. A.S.P*, **99**, 1105.

Dravins, D., Lindegren, L., and Nordlund, Å. 1981, *Astr. Ap.*, **96**, 345.

Gray, D. F. 1982, *Ap. J.*, **255**, 200.

Gray, D. F., and Toner, C. G. 1986, *Pub. A.S.P.*, **98**, 499.

Guelachvili, G., de Villeneuve, D., Farrenq, R., Urban, W., and Vergès, J. 1983, *J. Molec. Spectrosc.*, **98**, 64.

Maillard, J.-P. 1974, *Highlights Astr.*, **3**, 269.

Maillard, J.-P., Chauville, J., and Mantz, A. W. 1976, *J. Molec. Spectrosc.*, **63**, 120.

Maillard, J.-P., and Nadeau, D. 1988, in *IAU Symposium 132, The Impact of Very High S/N Spectroscopy on Stellar Physics*, ed. G. Cayrel de Strobel and M. Spite

(Dordrecht: Kluwer), p. 249.

Mantz, A. W., and Maillard, J.-P. 1974, *J. Molec. Spectrosc.*, **53**, 466.

Nadeau, D., 1988, *Ap. J.*, **325**, 480.

Nadeau, D., and Maillard, J.-P. 1988, *Ap. J.*, **327**, 321.

Nordlund, Å. 1982, *Astr. Ap.*, **107**, 1.

Nordlund, Å. 1984, in *Small-Scale Dynamical Processes in Quiet Stellar Atmospheres*, ed. S. L. Keil (Sacramento Peak, NM: National Solar Observatory), p. 181.

Ridgway, S. T., and Friel, E. D. 1981, in *IAU Colloquium 59, Effects of Mass Loss on Stellar Evolution*, ed. C. Chiosi and R. Stalio (Dordrecht: Reidel), p. 119.

Sofia, S., and Endal, A. S. 1987, *Pub. A.S.P.*, **99**, 1241.

Tsuji, T. 1987, in *IAU Symposium 120, Astrochemistry*, ed. M. S. Vardya and S. P. Tarafdar (Dordrecht: Reidel), p. 409.

Discussion

MARMOLINO — What are your reasons to exclude other dynamical phenomena (e.g. gravity waves) to explain some of the trends in the differential velocities you observe with line depth and with line excitation potential?

NADEAU — We cannot exclude the presence of gravity waves or other oscillatory phenomena in the atmospheres of red giant stars. In view of the complexity of the motions shown by the solar atmosphere the presence of similar phenomena in red giants would not be surprising, but the limited amount of data available in the stellar case precludes the testing of a complex model with numerous free parameters.

In analogy with the solar case, and knowing from theoretical models that convection plays an important role in the atmospheres of red giants, we expect that convective motions are the dominant cause of the observed line shifts. More detailed studies that we are pursuing now may show, however, that some of our data are inconsistent with pure convective motions, and that other phenomena must indeed be included.

VINCE — Did you check the influence of the line shifts caused by collisions between the absorbers and perturbers?

NADEAU — No, but I think that the pressure broadening shift is negligible in red giant atmospheres.

ASYMMETRY OF ABSORPTION LINES IN THE SOLAR AND PROCYON SPECTRA

I.N.Atroshchenko, A.S.Gadun, R.I.Kostik
The Main Astronomical Observatory of the
Academy of Sciences of the Ukr.SSR
252127 Kiev Goloseevo, USSR

ABSTRACT. The comparative analysis of solar and Procyon spectral line bisectors is carried out. The interpretation of the basic features of solar and Procyon spectral line bisectors is given on the basis of the theoretical inhomogeneous models of their atmospheres and also on the basis of stochastic approximation theory of spectral line formation. The conclusion is made that the main differences between the Sun and Procyon are caused by the differences of correlation between of the temperature and vertical velocity fluctuations in their atmospheres.

At present we have extensive data on the bisectors of absorption lines for the stars of different spectral classes. Detailed analysis of those data was carried out by Dravins(1987). The continuous sequence of bisectors for F-G supergiants was given by Gray and Torner (1986).

In our study we compare the solar and Procyon spectral line bisectors. Solar bisectors were obtained for the disc center from the Jungfraujoch atlas (Delbouille et al.,1973), for the limb (μ =0.2) from the Kitt Peak atlas (the version on the magnetic tape) and for the Sun as a star - from the R.Kurucz et al. atlas (1984). Bisectors of Procyon lines were obtained from the R.Griffin and R.Griffin atlas data (1979). Note, that earlier Procyon line asymmetries were analysed by D.Dravins (1987), who used the same atlas. However, in addition to the work by D.Dravins, besides the iron lines we used in our study the lines of six other elements. Preliminarily 145 lines of various elements were selected. But finally we left only 87 lines as the most suitable for the solution of this task. The list of lines used is given in Table 1, the bisectors of lines of iron and other elements are shown in Fig.1 and 2 (the numbers of lines is specified in brackets). These bisectors were derived by averaging of several line bisectors without accounting for absolute shifts.

As the majority of lines investigated are iron lines (59) we have divided FeI and FeII lines for detailed study into several groups, that depends on their strength. In addition, FeI lines were divided into the groups by excitation potetial. From the analysis of bisector shapes we

R. J. Rutten and G. Severino (eds.), Solar and Stellar Granulation, 135-143.
© *1989 by Kluwer Academic Publishers.*

TABLE 1

nm	el.	nm	el.	nm	el.	nm	el.
450.8286	Fe11	534.8319	Cr1	576.0829	Ni1	616.2182	Ca1
454.5960	Cr1	537.3706	Fe1	577.5083	Fe1	616.5364	Fe1
455.4042	Ba11	537.9574	Fe1	579.3918	Fe1	617.3344	Fe1
457.4220	Fe1	538.6330	Fe1	580.9216	Fe1	620.0320	Fe1
457.6338	Fe11	539.8280	Fe1	585.2219	Fe1	621.3436	Fe1
460.2006	Fe1	540.1265	Fe1	585.6090	Fe1	624.6328	Fe1
460.2948	Fe1	541.7036	Fe1	585.9590	Fe1	629.7802	Fe1
462.0516	Fe11	542.5259	Fe11	586.2364	Fe1	633.5336	Fe1
463.0128	Fe1	543.5856	Ni1	594.8544	Si1	641.1660	Fe1
471.8424	Cr1	546.1546	Fe1	598.3686	Fe1	643.0854	Fe1
478.8761	Fe1	552.6813	Sc11	598.4823	Fe1	643.2883	Fe11
495.0112	Fe1	554.3936	Fe1	598.7066	Fe1	675.0158	Fe1
495.3212	Ni1	556.0209	Fe1	600.8561	Fe1	675.2712	Fe1
506.7152	Fe1	557.8722	Ni1	602.7055	Fe1	676.7781	Ni1
509.4410	Ni1	558.8760	Ca1	605.6008	Fe1	677.2321	Ni1
510.9648	Fe1	559.3733	Ni1	606.5492	Fe1	684.3660	Fe1
514.1742	Fe1	561.8634	Fe1	608.4106	Fe11	685.8154	Fe1
515.5126	Ni1	563.5821	Fe1	609.3647	Fe1	740.5788	Si1
519.8712	Fe1	563.8266	Fe1	610.8126	Ni1	742.2985	Ni1
521.7392	Fe1	565.1469	Fe1	614.9248	Fe11	744.5742	Fe1
528.8528	Fe1	567.9032	Fe1	615.1622	Fe1		
529.5306	Fe1	573.1763	Fe1				
532.5548	Fe11	575.2038	Fe1				

may list the following basic reduliarities (most of them were also noticed previously by other authors):
1. The shape of bisectors is determined mainly by line strength.
2. The dependence of bisector shape on the excitation potential is not
   clearly visible (within the limits of errors of bisector calculation).
3. Moderate-strong and strong lines for the center of the solar disc and
   for the Sun as a star reveal a specific "C"-shape profile.
4. All lines at the solar limb indicate a negative (violet) asymmetry.
5. The asymmetry of lines in the spectrum of the Sun as a star is smaller
   than the asymmetry for the disc center.
6. Contrary to the Sun "C"-shape of bisector is absent even for very strong
   Procyon lines. This is clearly visible for the bisectors of Ba11 and
   Ca1 lines.

Let us consider the spectral lines of the solar disc center which
have characteristic "C"-shape bisector. At the certain point the bisector changes its direction, so that its upper part has the slope to the
red and its lower part has the slope to the violet. For some groups of
lines we have estimated the geometrical depth of the layer in which the
bisector changes its direction(on the basis of depression contribution
function in LTE approach). The results are listed in Table 2, where the

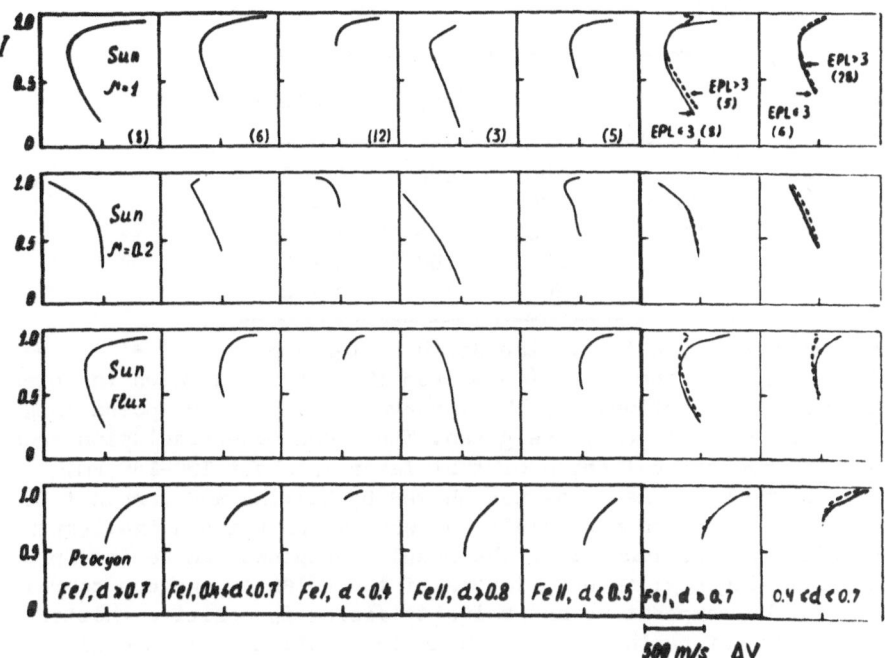

Figure 1. Bisectors of FeI and FeII lines.

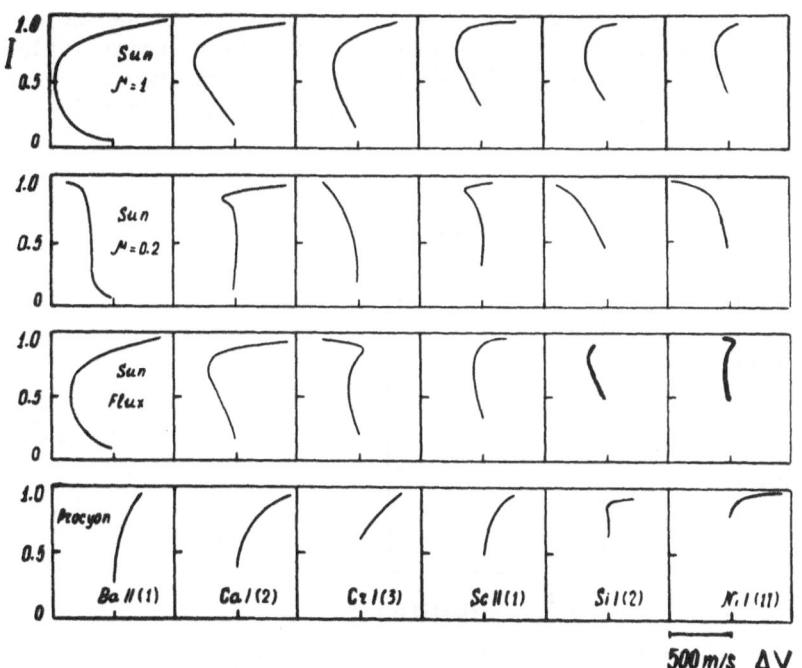

Figure 2. Bisectors of BaII, CaI, CrI, ScII, SiI, NiI lines.

TABLE 2

d	d'	h(km)	χ (eV)
0.800	0.260	130	0 - 3
0.700	0.280	95	3 - 9
0.600	0.2500	95	3 - 9
0.550	0.250	150	0 - 3
0.200	0.200	120	0 - 3

Central line depth (d), the depth of turning point (d'), the geometrical height of the layer (h - from the HOLMU) in which the bisector changes its direction, and the interval of excitation potentials for the groups of lines (χ) are given. Thus, one make conclusion from Table2 that the geometrical depth of that layer is about 100-140 km.

On the other hand, as was showed by Kaising and Durrant(1982) bisector shape is determined mainly by correlation between the temperature and velocity fluctuations. Thus the results obtained can be interpreted in the following way: in deep layers of the solar atmosphere correlation between the temperature and velocity fields is positive (hotter matter rises), therefore the bisector near the continuum has the slope to the red; starting from the certain depth the correlation between the temperature and velocity fluctuations changes its sign (cooler matter rises), which leads to the violet asymmetry. Theoretical nonhomogeneous models (Gadun,1986) show that at the height $\sim$ 50 km above the photosphere correlation between the temperature and vertical velocity fields rapidly decreases and at the height about 120-150 km changes its sign. It is clearly seen in Fig. 3 taken from the paper Gadun (1986).

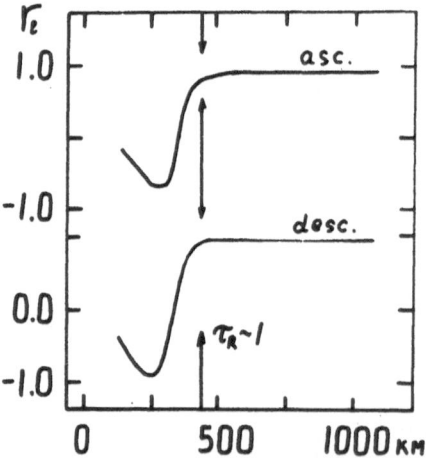

Figure 3. Correlation between the temperature and vertical velocity fluctuations in the solar photosphere (theoretical models).

The same fact is confirmed by the relation between the characteristic sizes and velocity RMS fluctuations of nonhomogeneous formations in the solar photosphere.

Applying stochastic approximation for the theory of spectral line formation and assuming that the velocities of turbulent elements have Gaussian distribution, their sizes being determined according to Poisson low (Kubo-Anderson process; Auvergne M., et al.,1973), we can find the sizes of turbulent elements (L) from the comparison of observed and calculated central line intensites, taking velocity RMS fluctuations $V_{rms} = \sqrt{(V_{mi}^2 + V_{ma}^2)/2}$. V_{mi} and V_{ma} were determed for the solar center and limb (μ =0.2) using 47 FeI lines with Oxford oscillator strengths. Final values of V_{rms} for the solar center and limb are shown in Fig.4 in comparison with the values obtained by other authors from Doppler line shifts.

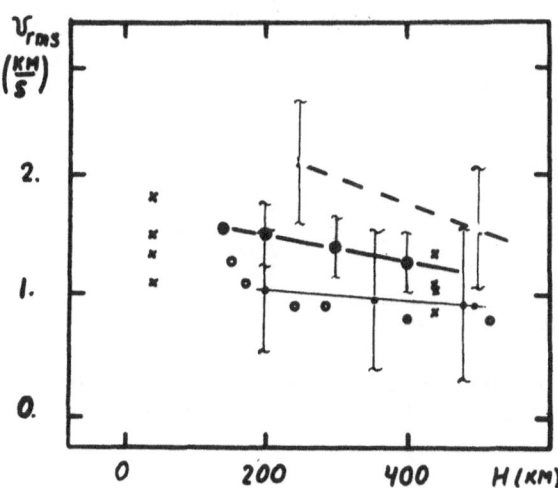

Figure 4. The results of determination V_{rms}, x –Bassgen,Deubner,1982; o – Canfield,Mehltretter,1973; ● – Durrant et al.,1979; ● – present work,--- – V_{rms}^{hor} , present work.

The results of calculation of turbulent element sizes depending on V_{rms} are presented in Fig. 5 a,b and lead to the following conclusions:
1. Turbulent elements in the photosphere are very deformation: in the lower photosphere the ratio L^{hor}/L^{ver} =1.5-3, but in the upper photo - sphere the ratio is 10 - 50 (Fig.6).
2. We can subdivide the photosphere into 3 layers according to the kind of physical processes dominated in the generation of nonhomogeneous formations:
   "A"-is the region of penetrative convection (H $\leqslant$ 180 km). Actually, as it was shown by Hollandskiy (1972), Kolmogorovspectrum of turbulence in the presence of radiation field is $\varepsilon(k) \sim k^{-7}$ for vortices which luminate in the whole volume and $\varepsilon(k) \sim k^{-3}$ for partial lumination. Then the value of V_{rms} is given by $V_{rms} \sim L^n$, n= 3 - 1.

140

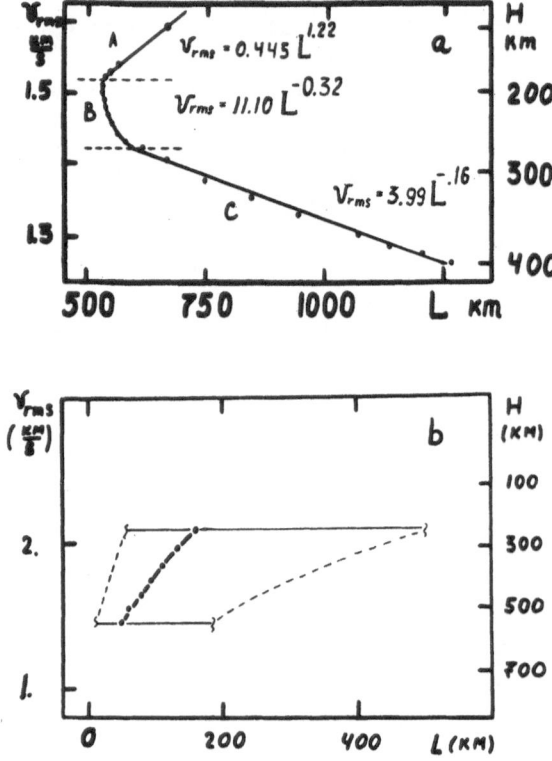

Figure 5. The dependence of the turbulent element motion velocity upon its sizes (a – the disk center, b – the limb (μ =0.2)).

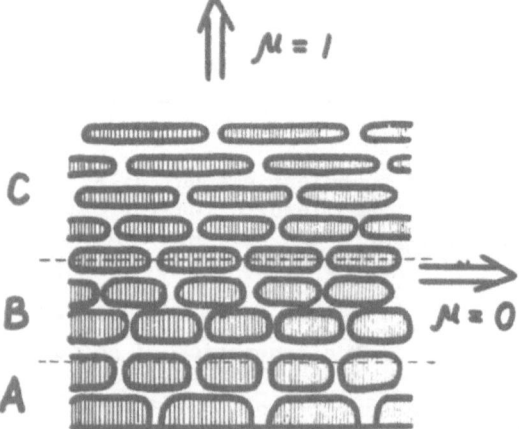

Figure 6. Schematic shape of nonhomogeneous formations in the solar photosphere.

"B"-is the transitional region (180 < H < 270 km). In this region
turbulence elements (due to the penetrative convection) dissipate.
At the same time wave motions accelarate. The influence of both
processes on line formation is approximately the same.
"C"-is the region of the upper photosphere (H ⩾ 270 km). Wave motions
play here the basic role.

Thus, moderate and weak line profiles with clear red asymmetry are
formed in the penetrative convection region with positive (or closely
to zero) coefficient of correlation between the temperature and velocity
fields.

Contrary to the Sun "C"-shape of bisectors is absent for the Procyon. Beginning from the continuum they don't change their direction
even for very strong lines. Therefore we can assume that the correlation
between the temperature and velocity fields remains positive in the
whole height interval where spectral lines were formed. Actually, the
calculated dynamical Procyon models indicate the positive correlation
between the temperature and vertical velocity fluctuations at all geometrical heights of modelled region (Fig. 7).

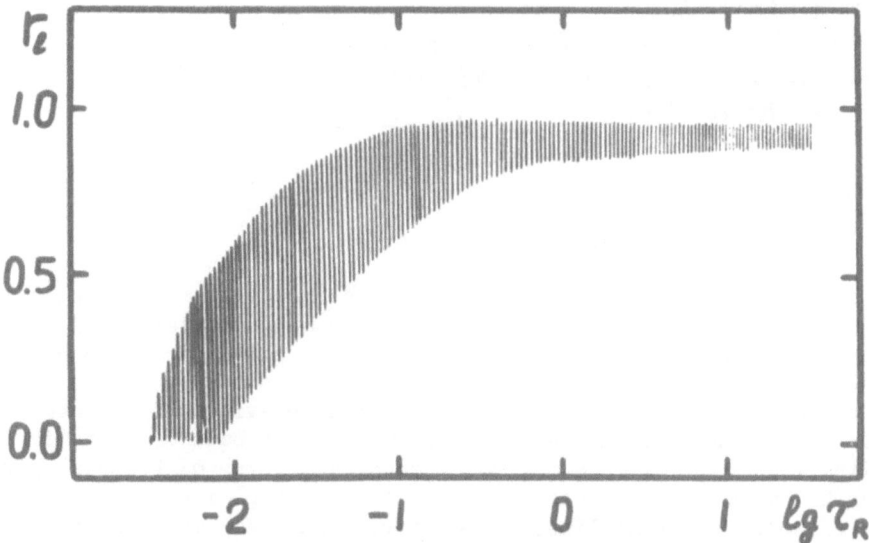

Figure 7. The correlation coefficient between the temperature and
vertical velocity fluctuations in the Procyon photosphere.

Spectral line profiles were calculated on the basis of these models
of the Procyon photosphere and also on the basis of three-dimensional
nonhomogeneous solar photosphere models (Gadun,1986). They are shown in
Fig. 8. Solar models well reproduce observed bisectors in the disc center; for the limb the fitting is worse that leads to the worse agreement
for line bisectors in the spectrum of the Sun as a star.

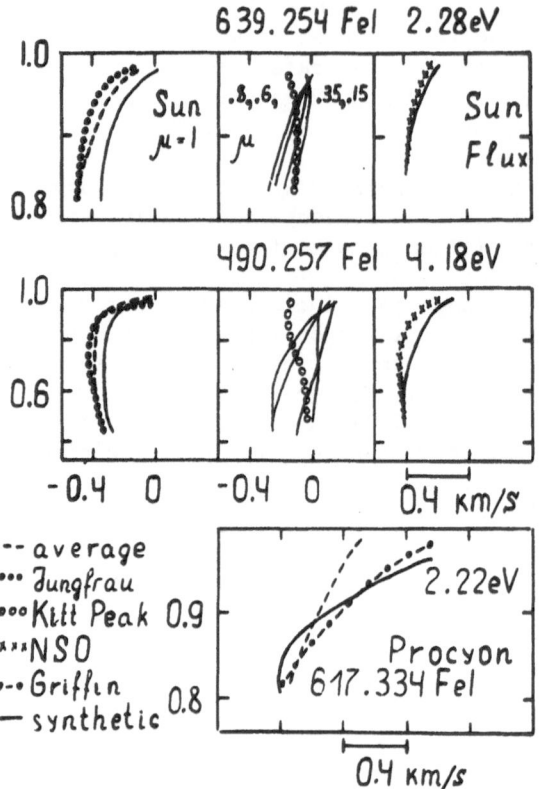

Figure 8. The solar and Procyon line bisectors.

The presence of slight violet asymmetry in the core of calculated line is the main difference between the observed and calculated FeI 617.334 nm line bisector of Procyon. However, this effect may be due to influence of upper boundary conditions.

One can make the following conclusions. The behaviour of the absorption line bisectors observed in the Procyon spectrum both quantitatively and qualitatively differs from the solar ones. Asymmetry of Procyon spectral lines has monotonous character and does not change its sign, the asymmetry of solar spectral lines is positive (red) near the continuum and changes negative (violet) in the cores of strong and moderate lines. Such behaviour of the bisectors is due mainly to the shape of correlation relations of temperature and velocity fields in the atmospheres of these stars. Rapid decrease of the correlations in the penetrative convection region, their absence in the region of mechanical motion dissipation and reversal of correlation in the upper photosphere is typical for the solar photosphere. Smooth monotonous decrease of positive correlation between the temperature and vertical velocity fields in the spectral line formation region is characteristic for the Procyon photosphere.

We stress that when considering the correlation between the temperature and velocity fields the averaging procedure was not strickly defined. We assume that for the averaging over time and area it is important that time and area intervals were not smaller than characteristic time development and scale of nonhomogeneous formations which are main contributors to the spectral line profile formation.

Disk averaged correlation between temperature and radial velocity fluctuations must be consider for Procyon. But the central parts of the disc give the màin contribution, therefore we can consider temperature and vertical velocity fields only.

We thank Dr. R.Griffin and Dr. R.Griffin for making available the version of the Procyon atlas on the magnetic tape, Dr. H.Batcher and Dr. R.Rutten for the possibility to use the version of solar limb atlas of Kitt Peak Observatory. Especialy we wish to acknowledge R.Rutten for help in the practical using all versions of atlases on the magnetic tapes.

REFERENCES

Auvergne M.,Frish N.,Froeschle C.,Pouque A. 1973, Astron.Astrophys., 29,93

Eassgen M.,Deubner F.L. 1982, Astron.Astrophys.,111,L1

Canfield R.C.,Mehltretter J.P. 1973, Solar Phys., 33,33

Delbouille L.,Neven L.,Roland C. 1973, Photometric Atlas of Solar Spectrum from 3000 to 10000 A, Liege

Dravins D. 1987, Astron. Astrophys., 172,211

Durrant C.J.,Matting W.,Nesis A.,Reiss G.,Schmidt W. 1979,Solar Phys,, 81,251

Gadun A.S. 1986, Preprint ITP of Ac.Sc.Ukr.SSR,ITP-86-106R

Gray D.F.,Torner C.G. 1986, Publ.Astron.Soc.Pacific,98,499

Griffin R.,Griffin R. 1979, A Photometric Atlas of the Spectrum of Procyon 3140 - 7470 A, Cambridge

Hollandskiy O.P. 1972, Izv.Krymsk.Astrofiz.Obs.,56,83

Kaising M.,Durrant C.J. 1982, Astron.Astrophys.,116,332

Kurucz R.L.,Furenlid J.,Erault J.,Testerman L. 1984, Solar Flux Atlas from 290 to 1300 nm, Harvard University

LINE ASYMMETRIES IN LATE-TYPE DWARF PHOTOSPHERES

David H. Bruning[1]* and Steven H. Saar[2]*

[1]*Dept. of Physics, University of Louisville, Louisville, KY 40292, USA*
[2]*Harvard-Smithsonian Center for Astrophysics, Cambridge, MA 02138, USA*

Abstract

We have begun a program to observe photospheric line asymmetries in late-type dwarfs. Analysis of spectral line profiles of 61 Cyg A suggests that the combination of line blends and the line asymmetry may give rise to false magnetic field detection. The effects of rotational velocity and excitation potential on line asymmetry are shown for several stars.

1 Introduction

The study of stellar photospheric line asymmetries was begun a short time ago by Gray (1980; 1982) and by Dravins (1987b). Their observations have largely concentrated on F, G and early K spectral types of luminosity classes I-V. Recent work has focussed on the interpretation of stellar line asymmetries as the correlation between granular velocities and intensities (Gray and Toner, 1985; Dravins, 1987c), extending the explanation of solar line asymmetries to stars (see review by Dravins, 1982).

Livingston (1982) has shown that the asymmetry of solar spectral lines is different in magnetic regions and in non-magnetic regions, and that the line asymmetry appears to vary with the solar cycle (Livingston, 1983). Bruning and LaBonte (1986) describe this cyclic variation not in terms of a change in the global characteristics of surface convection, but rather as a change in the amount of surface covered by magnetic field. Temporal variations in line asymmetries for ξ Boo A have been observed by Toner and Gray (1988) which suggests that regions of stronger and weaker granulation are being rotationally modulated. Although Toner and Gray define the modulation in terms of a new phenomenon they call a "starpatch", the variations are perhaps due to magnetic fields (Saar et al., 1987), similar to the case for the sun. Temporal variations of line asymmetries may prove to be a powerful diagnostic for the analysis of magnetic field distributions since field distributions are presently found from simultaneous measurements of Zeeman broadening and linear polarization, two difficult measurements.

---

*Visiting Astronomer, National Solar Observatory, operated by the Association of Universities for Research in Astronomy, Inc. under contract with the National Science Foundation.

R. J. Rutten and G. Severino (eds.), Solar and Stellar Granulation, 145–151.
© 1989 by Kluwer Academic Publishers.

Despite the potential of the study of stellar line asymmetries to tell us about surface convection in stars and, perhaps, about magnetic fields, few stars have actually been observed. This paper describes an on-going program to observe late-type dwarfs and the possibility of using the infrared for these observations.

2   Observations

A series of optical measurements of stellar line bisectors has begun at the McMath telescope of the National Solar Observatory. The TI4 CCD camera is used along with the echelle grating to obtain spectra near 6250 Å with a signal-to-noise ratio of several hundred and a spectral resolution of 80000. As pointed out by Dravins (1987a) and Livingston and Huang (1986), line asymmetries are degraded at spectral resolutions below 100,000. To provide higher resolution at the McMath telescope, a new image slicer is being built (completion Fall 1988) which should permit a spectral resolution of 160,000 with no loss in the signal-to-noise ratio.

Observations from stellar magnetic field studies have also been pressed into service for our study of stellar line asymmetries. Several K and M dwarfs were observed at the 2.5 m Hooker telescope at Mt. Wilson, several FTS observations of late-type stars were made at the 4 m telescope at Kitt Peak, and several late-type stars have been observed at the ESO Coudé Auxiliary Telescope. A total of 17 dwarfs with spectral types ranging from F5 to M0.5 have been observed, and 4 giants have been studied for comparison of our data with that of other authors.

3   Results

The initial observing run at the McMath was exploratory in nature to determine whether line asymmetries could be effectively observed with this telescope. The line asymmetries we observe for Arcturus and Procyon are similar to those reported by Gray (1982, 1983) and Dravins (1987b), and the variation of the line bisector slope with excitation potential (Dravins, 1982) is also seen (Figure 1).

Comparison of the line bisectors for β CVn ($v \sin i = 1.8$ km s^{-1}), β Com ($v \sin i = 4.3$ km s^{-1}) and χ^1 Ori ($v \sin i = 9.4$ km s^{-1}), three G0 V stars, reveals the so-called rotation effect whereby stellar rotation enhances the line asymmetry (Figure 2). The enhancement observed for these three stars is different from that predicted by Gray (1986) and Smith, Huang and Livingston (1987). Smith, Huang and Livingston predict a blueward asymmetry, increasing as the square of the rotational velocity. Gray predicts that rotation will produce a pronounced redward asymmetry. The observed bisectors show an enhanced redward asymmetry that increases with increasing rotational velocity, but the bisector changes less quickly than predicted by Gray or Smith, Huang and Livingston.

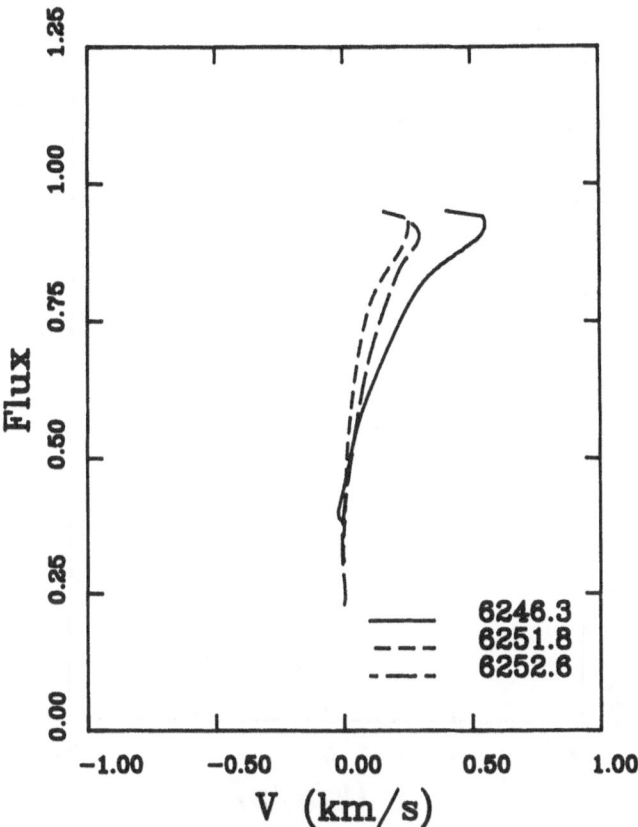

Figure 1: The variation of line asymmetry as a function of excitation potential for Arcturus. The excitation potentials are: 0.29 eV (6251.8 Å), 2.40 eV (6252.6 Å) and 3.60 eV (6246.3 Å).

The Mt. Wilson data were obtained at 6173 Å, which is not an optimum region for observing line asymmetries. However, one interesting result has come from these observations. Saar's (1987) observation of 61 Cyg A in the infrared fails to detect a magnetic field for this star, while measurements of the Fe I 6173 line show strong field strengths with 30 percent filling factors (Marcy, 1984; Bruning, Chenoweth and Marcy, 1987). If the optical measurements are corrected for the line asymmetry, the filling factor and field strength for 61 Cyg A are decreased radically. Since line blends are also a problem at λ6173 (Hartmann, 1987; Saar, 1987), the discrepancy between the optical and infrared magnetic field measurements appears to result from a combination of the line asymmetry and line blends for the optical lines. Future magnetic field measurements should, therefore, take the line asymmetry into account.

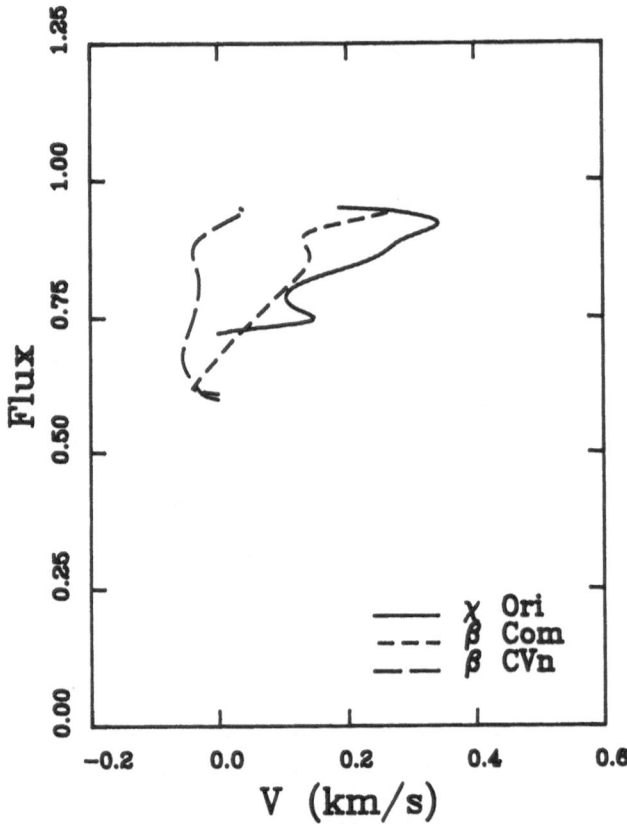

Figure 2: Increasing rotational velocity enhances the line asymmetry as shown for three G0 V stars: β CVn (1.8 km/s), β Com (4.3 km/s) and χ^1 Ori (9.4 km/s). The Fe I line at 6246.3 Åis shown for all three stars.

The infrared is intriguing for the study of line asymmetries since line blends are less severe than at $\lambda 6200$. It is difficult, however, to obtain the high spectral resolution necessary to properly observe the line shape. Our present observations obtained a spectral resolution of 45,000 at the expense of long integration times with a 4 meter telescope. Higher spectral resolution measurements of dwarf stars, at least at 2.2 microns, do not appear to be practical at this time.

4   Interpretation of Line Bisectors

Several models are being used to interpret the observed line bisectors in terms of stellar granulation. Our present model uses either three or four streams and includes properties from the two-stream model by Gray and Toner (1985) and the multi-stream model of Dravins (1987c). Three or four streams permit the use of one or two velocities for the granule, one for the neutral regions and one for the inter-granular lanes.

In addition, models of solar line asymmetry by Beckers and Nelson (1978) and observations by Balthasar (1985) suggest that horizontal motions are important, and these motions will therefore be included in our model.

As shown by Bruning (1984), convolutions are not an appropriate way to treat rotation for slowly rotating stars. Rotation is included in our model as an explicit disk integration which also permits the inclusion of limb darkening and the variation of line shape from disk-center to the limb. Input line profiles are calculated from LTE model atmospheres. Future models will include magnetic fields and line blends for direct comparison with the observations.

Acknowledgements

This work has been supported in part by an NSF EPSCoR grant and by the Harvard-Smithsonian Postdoctoral Fellowship program.

References

Balthasar, H.: 1985, *Solar Phys.* **99**, 31.

Beckers, J.M. and Nelson, G.D.: 1978, *Solar Phys.* **58**, 243.

Bruning, D.H.: 1984, *Astrophys. J.* **281**, 830.

Bruning, D.H., Chenoweth, R.E. and Marcy, G.W.: 1987, in J.L. Linsky and R.E. Stencel (eds.) *Cool Stars, Stellar Systems, and the Sun*, p. 36.

Bruning, D.H. and LaBonte, B.J.: 1985, *Solar Phys.* **97**, 1.

Dravins, D.: 1982, *Ann. Rev. Astr. Astrophys.* **20**, 61.

Dravins, D.: 1987a, *Astr. Astrophys.* **172**, 200.

Dravins, D.: 1987b, *Astr. Astrophys.* **172**, 211.

Dravins, D.: 1987c, preprint.

Gray, D.F.: 1982, *Astrophys. J.* **255**, 200.

Gray, D.F.: 1983, *Pub. Astr. Soc. Pacific* **95**, 252.

Gray, D.F.: 1986, *Pub. Astr. Soc. Pacific* **98**, 319.

Gray, D.F. and Toner, C.G.: 1985, *Pub. Astr. Soc. Pacific* **97**, 543.

Hartmann, L.: 1987, in J.L. Linsky and R.E. Stencel (eds.) *Cool Stars, Stellar Systems, and the Sun*, p. 1.

Livingston, W.C.: 1982, *Nature* **297**, 208.

Livingston, W.C.: 1983, in J.O. Stenflo (ed.) *Solar and Stellar Magnetic Fields: Origins and Coronal Effects*, p. 149.

Livingston, W.C. and Huang, Y.-R.: 1986, in M. S. Giampapa (ed.) *The SHIRSOG Workshop*, p. 1.

Marcy, G.W.: 1984, *Astrophys. J.* **277**, 640.

Saar, S.H.: 1987, in J.L. Linsky and R.E. Stencel (eds.) *Cool Saars, Stellar Systems, and the Sun*, p. 10.

Saar, S.H., Huovelin, J., Giampapa, M.S., Linsky, J.L., and Jordan, C.: 1987, presented at *Activity in Cool Star Envelopes*, Tromsø, Norway, July 1-8,1987.

Smith, M.A., Huang, Y.-R., and Livingston, W.: 1987, *Pub. Astr. Soc. Pacific* **98**, 297.

Toner, C.G. and Gray, D.F.: 1988, *Astrophys. J.*, in press.

Discussion

MAILLARD — With respect to other line bisector measurements, you used a somewhat lower resolution than what seems to be required. Can you comment on that?

BRUNING — The Mt Wilson and Kitt Peak 4m observations were taken for magnetic field analyses, where lower spectral resolution is permissible. We thought it would be interesting to use these data for line bisector studies in spite of the low resolution, since it would extend our knowledge to cooler dwarfs.

CHALLENGES AND OPPORTUNITIES IN STELLAR GRANULATION OBSERVATIONS

DAINIS DRAVINS
Lund Observatory
Box 43
S-22100 Lund
Sweden

ABSTRACT. Effects of stellar granulation are observed in photospheric line asymmetries and wavelength shifts. Challenges for the future include achieving sufficiently high spectral resolution to fully resolve stellar line asymmetries; methods to identify granulation signatures also in heavily blended spectra; measuring differential wavelength shifts between groups of different lines in the same star; obtaining very accurate laboratory wavelengths; measuring different convective lineshifts in different stars; understanding activity—cycle lineshift variations that may hinder the detection of extrasolar planets; and ultimately spectroscopic and other studies of spatially resolved granulation structure across stellar disks.

1. Observing Line Asymmetries

The faithful observation of the subtle photospheric line asymmetries caused by stellar granulation is a more demanding task than most other stellar spectroscopy applications. In particular one can point to the issues of:

1.1. FINITE SPECTRAL RESOLUTION

The asymmetries of spectral lines are often represented by their bisectors. In order to "fully" resolve the bisector shape may require some $4 - 5$ independent points on the curve. Since the relevant width for most photospheric lines in ordinary stars is on the order of $8 - 10$ km/s, and each bisector point is obtained from *two* intensity measurements in opposite flanks of the line, this demands $8 - 10$ points across the line, corresponding to a resolution of $\simeq 1$ km/s ($\Delta\lambda/\lambda \simeq 300,000$). Numerical simulations of how bisector shapes are distorted by lower resolution clearly illustrate this point (Dravins, 1987; Livingston and Huang, 1986).

Such resolutions are normally not available in stellar spectrometers, where $\Delta\lambda/\lambda \simeq 100,000$ is more common. Although such a resolution might be adequate to detect the *presence* of line asymmetries, the study of their detailed shapes or the detection of any differences between different types of lines becomes marginal.

Of particular importance for exact comparisons with theoretical data is the availability of an accurately measured instrumental profile (including its possible asymmetry), and the carefully measured spectral degradation in the instrument (including the scattered light). One way of illustrating the spectrometer

153

R. J. Rutten and G. Severino (eds.), Solar and Stellar Granulation, 153–160.
© *1989 by Kluwer Academic Publishers.*

performance is by recording spectra of the sunlit Moon, which to a good approximation equals disk–integrated sunlight, and thus can be readily compared to solar spectrum atlases. Due to incoherent scattering in the Earth's atmosphere, however, *skylight* may be too different from integrated sunlight to be reliably usable for high–resolution work (Lind and Dravins, 1980).

The fundamental problem is *not* the lack of spectroscopic instruments of sufficient performance, but the fundamentally poor light efficiency of those that are technically sufficient. For example, Fourier transform spectrometers (FTS) offer fully adequate spectral resolution, a good instrumental profile, possibilities for accurate wavelength determination, and other valuable features. The problem is that an FTS, scanning successive Fourier components of absorption–line spectra in the optical, is photon–noise limited, analogous to a wavelength–scanning single–channel spectrometer. (This is because all the light in the instrument falls onto the detector, and contributes photon noise from all wavelengths all the time, while the desired signal is only a slight modulation on top of this.) In order to use FTS's for stellar work one is forced either to accept a lower spectral resolution or to introduce rather narrow bandpass–limiting wavelength filters. Even so, the observation of only a few spectral lines in very bright stars with a large telescope may require several hours (e.g. Wayte and Ring, 1977; Nadeau and Maillard, 1988). Double–pass scanning spectrometers have very much the same limitations (Dravins, 1987). Their wavelength scanning interval may be easier to define, but it is awkward to increase the spectral resolution towards 10^6, say.

Fortunately, there is reason for optimism for the future due to the current wave of construction of very large telescopes in different parts of the world. Their greatly increased light collecting power will overcome the photon limits in today's "large" telescopes, and make Fourier transform spectrometers generally available for stellar spectroscopy. An estimate of the likely performance of an FTS on an $8 - 10$ meter class telescope shows that, at spectral resolving power $\simeq 10^6$, a signal–to–noise ratio of 100 should be reached for an $m_v = 3$ star in one night's integration over a bandpass of 10 nm (100 Å). Although some astronomers might still think of $m_v = 3$ as representing rather bright objects, the number of stars whose spectra could be studied in great detail will now be counted in the hundreds rather than the handful of the very brightest ones that at present are barely accessible to such studies.

1.2. ASTROPHYSICAL PROBLEMS IN STELLAR SPECTRA

Although spectral lines to be studied obviously are selected to be as undisturbed as possible, they are almost never completely unblended. This makes it awkward to draw conclusions on line asymmetries from measurements of very few lines only: how is one to know if an asymmetry is intrinsic to the stellar photosphere or due to a blend? These problems become especially pronounced for measurements close to the continuum: in weak lines or in the wings of stronger ones. Since one can expect blending lines to be randomly positioned in the wings of different primary lines, one solution is to average the asymmetries of several similar lines.

If very high spectral resolutions are required, this is not a trivial requirement, since each line to be observed might require hours of observing time. In cooler and/or more rapidly rotating stars it may in practice be impossible to find any significant number of sufficiently undisturbed lines. Solutions could possibly include going to the infrared (where line densities are lower), or face the problem of observing (and modeling!) not individual spectral lines, but rather *line complexes*. For very cool stars this could even be the only alternative. Such complexes of blended lines will carry much the same information as individual lines, but the observational possibilities of averaging bisectors of many similar lines will disappear. Likewise, accurate laboratory wavelengths will be needed for all components making up the blend since average values over similar lines will not suffice.

2. Determining Wavelength Shifts

Since stellar spectral lines are intrinsically asymmetric, their "wavelengths" can not simply be defined as single quantities. Ideally, line profiles and their bisectors should be measured on an *absolute* wavelength scale that allows the determination of *convective lineshifts* in different parts of the line, relative to the line's laboratory wavelength (corrected for the stellar radial velocity and its gravitational redshift). Such accurate wavelengths in spectral lines form important constraints on different theoretical models, which sometimes may even predict the same line asymmetry, but different wavelength shifts. This observational task presents several challenges:

2.1. WAVELENGTH DEPENDENCE ON SPECTRAL RESOLUTION

The finite spectral resolution in stellar spectrometers causes wavelength shifts of stellar lines. The instrumental convolution of an asymmetric stellar line profile with even a perfectly symmetric instrumental one results in a line profile of different asymmetry, where different parts of the bisector now are at different wavelengths. Since e.g. differently strong lines have different intrinsic asymmetries, they will be differently affected by the same spectral resolution, possibly mimicking the astrophysically expected behavior between different groups of lines. For solar–type asymmetries, such instrumentally induced shifts may reach 100 m/s for resolutions $\Delta\lambda/\lambda \simeq 100,000$ (Bray and Loughhead, 1978; Dravins, 1987).

The best solution would again be to achieve very high spectral resolution. Since this could be difficult in practice, the next best thing might be to accept the available resolution, but assure that the instrumental profile and the wavelength scale of the spectrometer is well determined from laboratory sources. For broader spectral intervals, the best wavelength calibrations are probably available in Fourier transform spectrometers. However, a word of caution may be warranted concerning the practice to use the commonly available He–Ne laser for wavelength calibration also for wavelengths far away from its 632.8 nm line. Such practice involves the implicit assumption that the interferometer cavity length and the (microscopic) depth of penetration of the light into the metal overcoatings of the optical elements is equal for different wavelengths. This assumption may not be valid if the highest accuracies are required. In any case, it is advisable to check the wavelength scales against *different* laboratory sources.

2.2. WAVELENGTH SHIFTS BETWEEN DIFFERENT LINES IN THE SAME STAR

Convective wavelength shifts are expected to be similar for groups of lines with common properties. Since only *statistical data* for groups of lines are required, it is *not* really necessary to record the full stellar spectrum. Rather, it could suffice with data from radial velocity measuring machines, utilizing the cross–correlation of the stellar spectrum with some spectrum template. The well–known instruments of the Griffin– and CORAVEL–type employ hardware masks and reach precisions on the order of 100 m/s, adequate to begin searches for signatures from stellar granulation.

With different masks, e.g. such preferentially selecting high— or low—excitation lines, one could detect *differential* radial velocities between groups of different lines in the same star. Next—generation radial velocity instruments are likely to avoid mechanical masks in order to improve light efficiency and allow the integration of the full spectrum all the time. Their spectrum templates will then be defined in software by selecting features in the spectrum for cross—correlation. Such designs appear very suitable for our purposes. Although there are likely to be many practical hurdles, it should in principle be straightforward to define different templates by selecting interesting groups of lines (from either observed or synthetic; stellar or laboratory spectra) to search for signatures from stellar granulation.

2.3. THE NEED FOR VERY ACCURATE LABORATORY WAVELENGTHS

The most useful atomic species for convective lineshift studies appears to be iron. It has high atomic mass (minimizing the thermal broadening of the stellar lines), its hyperfine and isotope splitting has few complications from atomic and isotope structure, and it has a rich and well—studied spectrum. In the Sun and solar—type stars, there are approximately 500 "unblended" Fe I and 50 Fe II lines in the visual spectrum, with another 100 or so lines in the infrared. Reasonable laboratory wavelengths exist, and convective lineshifts have been studied in the visual solar spectrum for Fe I and Fe II (Dravins et al., 1981; 1986) and in the infrared for Fe I (Nadeau, 1988).

Although significant effects are visible for line—group averages, the laboratory wavelength accuracies are *not adequate* to test different granulation models using measurements of *individual* lines only. The noise may be $\simeq 100$ m/s for stronger Fe I lines, but rapidly gets worse for weaker lines, for high—excitation ones, for Fe II, and for lines in the infrared. Indeed, the lack of sufficiently accurate laboratory data is now the main limiting factor in these studies. Since astronomers do not ordinarily provide such laboratory data, we have to direct our requests to atomic physics groups.

2.4. THE ROLE OF LABORATORY ASTROPHYSICS

Several institutes in the world work with problems in atomic spectroscopy of immediate astrophysical relevance. From time to time, their projects are redefined in response to new scientific problems. Over the last several years, much of the effort in several atomic physics groups has been the identification of lines in the ultraviolet and/or from different ionized species. This work has been rewarding thanks to the availability of ultraviolet spectrometers, observing high—temperature astronomical objects. By contrast, there has been less motivation to pursue detailed determinations of the exact energy levels in ordinary atoms.

The determination of very accurate laboratory wavelengths is a challenging task and such projects are unlikely to be pursued unless there is felt to be an important application of such data. Further, since the data needed are for *ordinary* atomic species such as neutral iron, the intrinsic atomic physics value of such studies might be limited (in contrast to, perhaps, the study of the complicated energy levels in multiply ionized rare—earth elements). In order to motivate atomic physicists to take up such work (and convince their funding agencies of adequate support), it is necessary to increase the awareness of the need for very accurate wavelengths. As a step in this direction, a resolution could be taken by the participants of this meeting and forwarded to IAU Commission 14 on *Atomic and Molecular Data.*

2.5. DIFFERENT WAVELENGTH SHIFTS IN DIFFERENT STARS

Stars of different temperature and luminosity are expected to have different amounts of convective lineshift. Significant new information could be obtained if such shifts were measured in different stars. The vigorous granulation in a $T_{eff} = 6600$ K model of the F5 star Procyon causes convective blueshifts in ordinary Fe I lines of $\simeq 1000$ m/s, while the shifts in a cooler $T_{eff} = 5200$ K model of a K1 dwarf amount to only $\simeq 200$ m/s (Dravins and Nordlund, 1989). Observed solar values fall in between, typically around 400 m/s. Understanding such effects is important not only for the study of stellar photospheres *per se*, but is also required for the accurate determination of stellar radial velocities, in particular in systems with small internal velocity dispersions, e.g. open galactic clusters.

To separate convective shifts from shifts due to stellar motion, requires one to somehow determine the absolute or relative stellar radial velocity without using the spectral lines. (One also needs to correct for different gravitational redshifts in different stars, but stellar models allow this to be done with good accuracy.)

Differential convective lineshifts should be measurable between stars that share the same space velocity (even if the exact amount of this velocity is not known). The components of *binary stars* must share the same system velocity when averaged over their orbits. Thus, visual binaries with not too long periods would be suitable objects to search for spectral–type dependent differences in orbit–averaged apparent radial velocities.

From galactic dynamics arguments, the velocity dispersion of stars in young galactic clusters is expected to be only a fraction of one km/s: less than expected differences in convective lineshift between different spectral types. Thus, if one could identify systematic differences in apparent velocities between different classes of cluster members, this would be evidence for different convective lineshifts in different stars.

To determine *absolute lineshifts* is more challenging since it requires the accurate determination of stellar radial velocities without using any spectral lines nor invoking the Doppler principle. This is possible for the Sun, where the solar motion is well determined from planetary system dynamics (rather than from the apparent Doppler shifts of spectral lines), and consequently solar wavelengths can be corrected for the Sun–Earth motion and compared to laboratory values.

In principle, *astrometric* measurements could do the same for stars: if one could accurately determine a star's three–dimensional position in space at different times, the difference in position would yield its space velocity. Unfortunately, current astrometric accuracies are insufficient for such direct measurements of individual stars. However, the advent of space astrometry promises to make at least some classes of related measurements possible. One of these concerns young galactic clusters (such as the Hyades), whose member stars share a common space velocity. A classical method of determining the distance to such moving clusters involves measuring the radial velocities across the cluster, and combining this with the measured proper motions towards the cluster apex. Since the cluster subtends a certain angle in the sky, the velocity component in the radial direction changes across the cluster, and the geometry of the situation then permits a determination of the cluster distance. The milliarcsecond accuracies expected from space astrometry will make it possible to directly measure the distance from trigonometric parallaxes. It then becomes possible to solve the *inverse* problem, i.e. determine the heliocentric radial velocities from astrometric data only. Numerical simulations of likely measuring errors, combined with plausible values for the cluster dynamics, suggest that accuracies of such *astrometric radial velocities* could reach $\simeq 400$ m/s for stars in the Hyades, with a potential for higher accuracy in the more nearby Ursa Major cluster. Such an observing program has been approved for the ESA astrometry satellite HIPPARCOS (Dravins and Lindegren, 1982).

3. Cyclic Changes of Line Asymmetries and Wavelength Shifts

A temporal variation in the granulation pattern may change the convective wavelength shifts, and thus mimic a varying stellar radial velocity. To understand such effects could be very important for one very challenging astronomical problem: the search for possible extrasolar planets.

3.1. THE DETECTION OF PLANETS AROUND OTHER STARS

Among plausible means of detecting such planets, one of the most promising seems to be the long–term monitoring of the radial velocity of the parent star. If the star is moving in conjunction with an unseen planet, one expects a cyclic change in its velocity. For the Sun–Jupiter system, the amplitude of the solar velocity due to its motion around their common center of gravity is 13 m/s, with a period of 12 years. There now exist stellar radial velocity instruments with measuring precisions that are sufficient for this task, but there remains the problem of separating cyclic changes due to stellar motion from those due to changes in convective lineshifts.

Solar granulation structure, bisector curvature and amount of lineshift are observed to vary between solar active regions and quiet ones. Consequently, the different area coverage of active regions in different phases of the solar 11–year cycle must lead to changes in the bisector curvature and wavelength shift also in integrated sunlight. Such changes have been studied by Livingston (1983), Deming et al. (1987), Jiménez et al. (1988) and Wallace et al. (1988).

The exact effect on the apparent radial velocity depends upon precisely how the line is measured, but may correspond to an amplitude of $\simeq 30$ m/s over the solar cycle. Such a magnitude is consistent with observed differences between active and quiet regions, and their different area coverages in different years of the solar cycle. Since many stars possess activity cycles, one should expect qualitatively similar effects in other stars. Clearly, if the Sun, seen from afar, displays an apparent velocity variation of perhaps 30 m/s with a period of 11 years, that will not be simple to disentangle from the 13 m/s amplitude over 12 years, induced by Jupiter.

To overcome the problem requires a better understanding of stellar granulation properties. Of great importance would be a systematic long–term monitoring of stellar line asymmetries and accurately measured wavelength shifts in different stars. Granulation changes correlate with active region coverage, and a measure of that is given by the Ca II K chromospheric emission intensity. This should also be monitored in order to identify the phase and period of possible stellar activity cycles. True velocity changes must affect all spectral lines, while granulation changes should affect different lines differently. Variations on the shorter timescales of stellar rotation might identify active region patches of significantly modified granulation (Toner and Gray, 1988), while variations over several years might identify the changing area coverages of such features.

4. Observing Structures On Stellar Disks

Much of the progress in astronomy is based upon improved spatial resolution. Our understanding of planets, nebulae or galaxies would have been very limited, if we only had observed them as point sources. One of the major aims of stellar granulation studies must be to ultimately observe the fine structure of stellar surfaces and to enable the study also of stars as extended objects. Along with this go related aims, such as understanding the physics of stellar line formation in different atmospheric inhomogeneities. The observational problems are of course caused by the small angular extent of stellar disks – no more than a few tens of milliarcseconds even for the largest stars. Nevertheless, with proper techniques, also the fine structure on stellar disks should be accessible for observation.

4.1. INDIRECT METHODS: DEDUCING CENTER–TO–LIMB LINE PROFILE CHANGES

Line profiles in disk–integrated starlight are built up by contributions from different disk positions. Since both line asymmetry, convective lineshift and continuum brightness depend on the disk position, the integrated line profile incorporates these effects in a complex manner. What is needed is a tool to disentangle the different quantities. Such a tool could be available in *stellar rotation*.

The line profile contributions from different center–to–limb positions, $\cos \theta = \mu$, are *not* equal for stars of different rotational velocities $V \sin i$. Increased rotation changes the Doppler shift at each μ due to the increased projected velocity, and since the contributions from each μ have a different line asymmetry, these will be differently distorted by different rotational broadening (Gray and Toner, 1985; Smith et al., 1987).

Numerical simulations of granulation now permit the computation of synthetic line profiles for different center–to–limb positions in different stars (Dravins and Nordlund, 1989). From such data, synthetic full–disk line profiles are obtained for different $V \sin i$. What is still lacking is the observational counterpart: a sequence of line profile, line asymmetry and wavelength shift measurements for groups of stars of the same spectral type, only differing by successively more rapid rotation. Such data could be an effective constraint on stellar granulation models and permit the analysis of line profile changes across stellar surfaces.

4.2. DIRECT METHODS: INTERFEROMETRIC IMAGING OF STELLAR SURFACES

Although indirect methods can be quite powerful, they do not replace the ultimate need for direct methods to obtain *images* and *spatially resolved spectra of stellar surfaces*. The angular (diffraction–limited) resolution of present telescopes in the 4–6 meter class is around 20 milliarcseconds in the visual. Using e.g. speckle interferometry, this allows one to resolve perhaps 10 surface elements on the largest red giants. The forthcoming very large telescopes in the 8–10 m class will significantly improve the situation. Of particular promise is the potential of active optics, which could allow the stable imaging of a stellar disk onto a Fourier transform spectrometer for two–dimensional high–resolution spectral observations with good wavelength calibration. Such data of the center–to–limb changes of stellar line profiles and wavelength shifts are precisely what is required to unravel the physics of stellar line formation.

Optical interferometers with baselines on the order of 100 m are now becoming operational, and will offer resolutions around 1 milliarcsecond. That is sufficient for thousands of resolution elements on the disks of red giants, and for resolving nearby main–sequence stars. (The baseline requirements for resolving different classes of stars are discussed by e.g. Dupree et al., 1984.) Since granules on some giant stars might subtend a significant fraction of a stellar diameter, their appearance and spectral features might soon become detectable through e.g. speckle spectroscopy. Solar–type granules, however, have sizes only about one thousandth of the stellar diameter, and their imaging requires baselines a thousand times longer than those required to resolve the stellar disk. To achieve this may require kilometric arrays of space–based optical phase interferometers, and their feasibility is now under active study by different groups in the world. These frontiers in stellar granulation research indeed offer interesting challenges for the future.

ACKNOWLEDGEMENT

This work is supported by the Swedish Natural Science Research Council

References

Bray, R.J., Loughhead, R.E.: 1978, *Astrophys.J.* **224**, 276
Deming, D., Espenak, F., Jennings, D.E., Brault, J.W., Wagner, J.: 1987, *Astrophys.J.* **316**, 771
Dravins, D.: 1987, *Astron.Astrophys.* **172**, 200
Dravins, D., Larsson, B., Nordlund, Å.: 1986, *Astron.Astrophys.* **158**, 83
Dravins, D., Lindegren, L.: 1982, *Convection in Stellar Atmospheres*, observing program for the ESA astrometry satellite HIPPARCOS
Dravins, D., Lindegren, L., Nordlund, Å.: 1981, *Astron.Astrophys.* **96**, 345
Dravins, D., Nordlund, Å.: 1989, *Astron.Astrophys.*, to be submitted
Dupree, A.K., Baliunas, S.L., Guinan, E.F.: 1984, *Bull.AAS* **16**, 797
Gray, D.F., Toner, C.G.: 1985, *Publ.Astron.Soc.Pacific* **97**, 543
Jiménez, A., Pallé, P.L., Régulo, C., Roca Cortés, T., Elsworth, Y.P., Isaak, G.R., Jefferies, S.M., McLeod, C.P., New, R., van der Raay, H.B.: 1988, in J.Christensen–Dalsgaard, S. Frandsen, eds. *Advances in Helio– and Asteroseismology*, IAU symp **123**, p. 215 (Reidel, Dordrecht)
Lind, J., Dravins,D.: 1980, *Astron.Astrophys.* **90**, 151
Livingston, W.C.: 1983, in J.O.Stenflo, ed. *Solar and Stellar Magnetic Fields: Origins and Coronal Effects*, IAU symp. **102**, p.149 (Reidel, Dordrecht)
Livingston, W., Huang, Y.R.: 1986, in M.S.Giampapa, ed. *The SHIRSHOG Workshop*, National Solar Observatory, Tucson, p.1
Nadeau, D.: 1988, *Astrophys.J.* **325**, 480
Nadeau, D., Maillard, J.P.: 1988, *Astrophys.J.* **327**, 321
Smith, M.A., Huang, Y.R., Livingston, W.: 1987, *Publ.Astron.Soc.Pacific* **99**, 297
Toner, C.G., Gray, D.F.: 1988, *Astrophys.J.*, submitted
Wallace, L., Huang, Y.R., Livingston, W.: 1988, *Astrophys.J.* **327**, 399

HIGH RESOLUTION GRANULATION OBSERVATIONS FROM LA PALMA: TECHNIQUES AND FIRST RESULTS

G.B. SCHARMER
Royal Swedish Academy of Sciences
Stockholm Observatory
S-133 00 Saltsjöbaden
Sweden

ABSTRACT. The paper reviews the design of the Swedish vacuum solar telescope, emphasizing the importance of its aerodynamic design and high optical quality, and the new techniques used for real-time frame selection and data reduction.

Recent results from the reduction of a 79 minute sequence of quiet sun granulation data involve the discovery of a stable vortex structure, which dominates the development of nearby granules and the measurement of instantaneous and average flow fields related to granule dynamics and mesogranulation.

1. Introduction

The Swedish Vacuum Solar Telescope is located at the island of La Palma in the Canary Islands. The succes of this telescope is due to the excellent seeing, the very high quality of the telescope optics, a powerful image acquisition system for making CCD observations and the reduction software, originally developed by Lockheed for the SOUP experiment on Spacelab. The telescope and its equipment is dedicated primarily for time studies of fine-structures in the Solar Atmosphere.

This paper describes a partly new approach to high-resolution ground-based solar observations and the first results of such studies of granulation dynamics.

2. Solar Telescope

The telescope is described in detail in a paper by Scharmer et. al. (1985). Here I shall mention only the most important properties of the telescope.

The telescope uses a turret design, similar to that of the Sacramento Peak Vacuum Telescope (Dunn 1969). The outstanding advantage of this design is that dome seeing is completely eliminated and that due to the telescopes aerodynamic shape, air can smoothly flow over the top of the tower and around the telescope, causing a constant removal of any heated air created in the vicinity of the telescope. Furthermore, the mirrors are located inside vacuum thus eliminating convection which would otherwise occur near the heated mirror surfaces. In spite of the advantages of this design we have seen occasional signs of local heating, by making knife edge tests on the limb of the sun, which emphasizes the importance of the problem of local heating.

R. J. Rutten and G. Severino (eds.), Solar and Stellar Granulation, 161–171.

The telescope optics consists of a classicaly designed doublet objective which forms the image 22.35 meters behind the lens, of three flat folding mirrors and an exit vacuum window. We do not think that the excellent image quality is related to the fact that we use a lens to form the image rather than a mirror, except that possibly irregularities in a transmitting surface scatter much less light than those of a reflecting surface. Much more significant is that the optical quality of all components is very high. Furthermore, there is no central obscuration in the beam, which would cause additional scatter and reduced contrast. The annular diaphragma in front of the objective effectively prevents the heating of the lens cell which would lead to radial temperature gradients in the objective and spherical aberration (Mehltretter 1979). Finally, the mirror cells are excellent and give no measurable astigmatism. Interferometric tests on stars indicate that the peak-to-peak wave-front error of the entire optical system is approximately 1/15 wave, corresponding to 1/40 wave rms. In spite of the modest apperture, which is only 48 cm, this telescope gives excellent images of solar fine structures. Figure 1 shows an example of a granulation picture obtained on March 18, 1986 through a filter centered on 468.6 nm.

The drawback of the optical design is that because of the colour curve of the objective, spectra at two or more wavelengths can generally not be obtained strictly simultaneously.

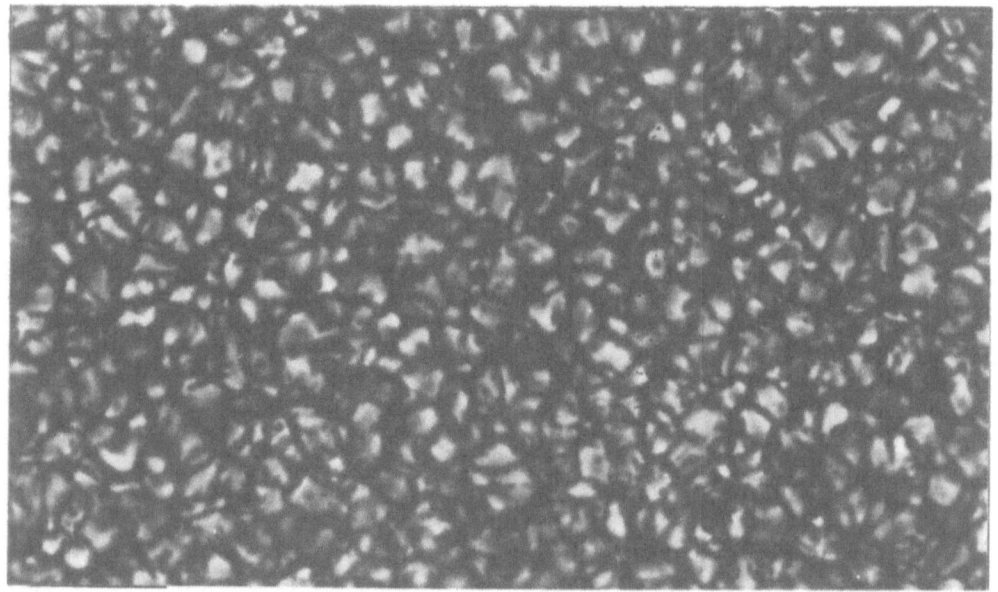

Figure 1. Granulation picture obtained at 468.6 nm on March 18, 1986

2.1 AUXILIARY INSTRUMENTATION

Most results reported in this paper have been obtained by direct imaging using a simple CCD-camera. We have also made observations with a short Littrow spectrograph (see Lites, Scharmer and Nordlund, these proceedings). The spectrograph is equipped with a holographic grating (2400 lines/mm) which gives a very low level of scattered light. The length of the grating is 22 cm and the resolution of the grating is 480,000. The spectrograph optics consists of a cemented aspherical doublet, which is corrected for spherical aberration and coma. The optical quality of the lens is approximately 1/20 wave ptp when used in single transmission. The lens is anti-reflection coated to minimize scatter from the surfaces.

We use a slit width of 25 μm, corresponding to 0.22 arc seconds. With this slit width, the practical resolution of the spectrograph is approximately 240,000 at 590 nm and 170,000 at 480 nm. The efficiency of the grating is 50%.

The advantage of the spectrograph is that because of the high optical quality and because there is only one reflecting surface there will be almost no degradation in the image quality. Spectra with 1/4 arc second resolution are therefore possible. The relatively low dispersion gives high intensity and enable exposure times of 1/25 second with acceptable S/N using a good CCD. The spectrograph is intended for programmes requiring very high spatial resolution but is not optimal whenever very high spectral resolution is required. Observations are only possible at one wavelength at the time.

3. Image Acquisition Techniques

For ground-based studies of solar fine structures at the resolution limit of the telescope the seeing will distort the image and cause a significant - and often unknown - reduction of the contrast. In view of this it does not seem meaningful to strive for extremely high accuracy in measurements of intensities, but rather to minimize the effects of seeing. To do this we have constructed a seeing monitor that measures the image sharpness in real time and sends commands to an image acquisition system that grabs the sharpest image during a specified time interval. Second, a technique for measuring wave-front errors in front of the objective is currently beeing developed by Högbom (these proceedings). This technique may allow the full compensation of the degradation caused by seeing, even for large-amplitude phase fluctuations.

3.1 IMAGE ACQUISITION SYSTEM

For real-time image selection it is essential to read out images at 25 - 60 Hz or faster. This was regarded to be of much higher importance than obtaining images with very high photometric accuracy since seeing causes a severe limitation in the measurement of accurate contrasts of small-scale solar structures. Because of this we have installed a fast 8 bit image acquisition system (DEC/IPS) from Kontron in Munich rather than a conventional slow12 bit CCD system. The system digitizes the video signal from PAL or NTSC CCD-cameras.

A very important property of the DEC/IPS is the possibility of simultaneous acquisition from two synchronized CCD cameras. There are several applications for which this capability is essential. Direct imaging at two wavelengths provide useful information on colour temperature or atmospheric dynamics at two different layers in the solar atmosphere. By taking images at the two different wavelengths strictly simultaneously, the differential effects of seeing are minimized. The same applies to the measurement of magnetic fields which require images taken at left and right circular polarization and subtracting the two images from each other. Since the polarization signal is usually rather weak it is important that the seeing does not distort one image relative to the other. Secondly, interpretations of spectra are greatly simplified if a slit-jaw image is taken strictly simultaneously so that the exact position of the slit as well as the quality of the seeing can be determined. Finally, the technique of Högbom (these proceedings) for measurement of and compensation for seeing requires a pair of images taken simultaneously at focus and out of focus. The capability of obtaining two simultaneous CCD images was therefore regarded as essential and ruled out the use of conventional CCD systems.

The DEC/IPS allow up to 25 (PAL) or 30 (NTSC) images per second to be stored in the image memory, which is presently 16 Mbytes large. We often use the system in non-interlaced mode which allows digitization at 50 or 60 Hz. The maximum size of the images are 768x512 pixels and any size smaller than that can be selected. The number of images that can be taken in a burst depends on the image size and is e.g equal to 64 for 512x512 images. The DEC/IPS is connected to a Micro VAX II via a DRV11-W interface.

3.2 REAL-TIME IMAGE SELECTION

The seeing monitor plays an essential role in the process of image acquisition. Even on days with excellent seeing the image quality can fluctuate from good to bad in a fraction of a second. Since for most applications it is sufficient to record an image for every 10 second time interval it follows that more than 99.5% of the images read out (at 25 or 30 Hz) are redundant. In order to improve the quality of the observations it is therefore sensible to select, in real time, the best image during every 10 second interval and to store only the best image on disk. For this purpose we have constructed a seeing monitor that measures the quality of the seeing in real time, compares with previous images and controls the DEC/IPS in the process of image acquisition.

The sharpness of the image quality is measured by high pass filtering of the video signal and integrating the contrast of the filtered signal within a window, the location and size of which can be set by potentiometers. The so-obtained sharpness value is digitized and evaluated by a Z-80 micro processor which compares the current sharpness value with the previously best value obtained within a given time interval. If the current value exceeds the previously best value the Z-80 sends a "grab" command to the DEC/IPS via the Micro VAX .

In order to eliminate time delay between the sharpness evaluation (which is done at the end of each image) and the grabbing of images the DEC/IPS continuously stores all images at a specific location in the image memory. When it receives a grab command it jumps, or toggles, from that image location to another location in the image memory, continuing to store images there. When receiving the next grab command it toggles back to the original position. The grab command therefore prevents the (currently best) image from beeing overwritten by the next image instead of grabbing the next image.

At the end of each time interval (usually 10 seconds) the seeing monitor sends a "store" command to the Micro VAX and the image is stored automatically on disk. After that the image selection process starts all over again.

We have used this system to record excellent time sequences of granulation and to produce "movies". The first results of these studies are discussed in these proceedings (see also Brandt et. al 1988) and are summarized in this paper.

Our current system suffers from a time delay, because of the slow response time of the Micro VAX. To eliminate that time delay Kontron will rebuild our system to allow the seeing monitor to talk directly to the DEC/IPS. This will significantly decrease the number of bad images stored.

4. Reduction Techniques

The first movies of granulation were made in june 1987 by the author using reduction software written for the Micro VAX and the DEC/IPS. The reductions proceed in steps: First the images are corrected for dark current and gain. Then the effects of bad tracking and seeing are removed by a cross correlation program (originally written by Andreasen) which calculates the displacements of all images relative to a reference image. Then the images are displaced according to the results of the cross correlation calculations and a sequence of images (typically 100-200) are read into the image memory of the DEC/IPS which replays the images at a certain frame rate while a U-matic video tape recorder records the images. The tape recorder is then halted by the Micro VAX and a new sequence of images are read into the image memory. In this way a movie can be made from many short segments.

During the last year a collaboration has been started with the Lockheed group (Title, Shine et. al.) who used basically similar techniques for reducing SOUP data. The reduction language developed for these and similar applications, which is called ANA, is much more extensive and flexible than the software developed for the DEC/IPS.

A major problem of making good movies based on ground-based data is that differential image motion across the field of view, caused by seeing in the upper layers of the earth's atmosphere, is very disturbing when the evolution of small-scale structures is studied even if the image motion amounts to less than 0.1" rms. To compensate for this, the Lockheed group remove differential image motion by a local cross correlation technique. A recent improvement in the image quality has also been achieved by removing frame-to-frame image quality fluctuations using FFT techniques. These consist in taking the full 3D Fourier transform of the entire movie sequence and applying a filter that removes Fourier components with super-sonic phase velocities. The local cross correlation program is also used to measure systematic flows by averaging out the seeing effects over the entire duration of the time sequence. The results of these reductions have yielded exciting new results on photospheric dynamics, as discussed below

The greater potential of the ANA software has caused us to abandon the original software developed for the DEC/IPS and to implement ANA instead. For future reductions we will use ANA installed in Stockholm and Lund on a VAX Station 3200 and a VAX Station 2000.

5. Results

The best sequence analyzed so far was obtained on June 16, 1987 and was reduced in collaboration with Brandt et. al. (1988). The sequence covers 79 minutes of time and 18 x 18 arc seconds of the quiet sun, but was reduced to 14.2 x 12.2 arc seconds to avoid problems due to telescope guiding errors. The observations were made through a 25 Å filter centered at 4686 Å using an EEV CCD, the DEC/IPS and the seeing monitor described in the previous sections. The reductions were made at Lockheed using the ANA reduction language

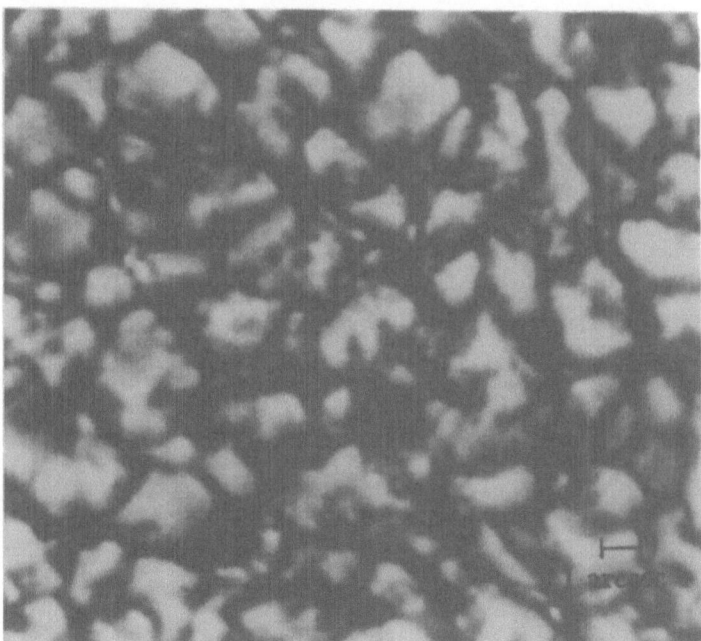

Figure 2. Snapshot from the movie sequence, showing many structures of sub-granular size, some of which may be filigree.The image has been slightly enhanced by inverse Fourier filtering to emphasize the small-scale structures.

It is likely that the area observed covers the boundary of a super granulation cell. The spatial resolution of many of the images approaches 0.25 arc seconds and several stable small-scale structures, probably filigree, can be identified near regions of converging flows (see Fig 2). The rms contrast of the individual images ranges from 8.5 to 10.6 %.

Local velocities were determined by measuring displacements on 0.8 and 0.4 arc second grids. Displacement velocities were obtained from images separated by approximately 60 seconds in time. Movies of the displacement field overlayed on the destretched images showed that the inferred velocities were qualitatively correct. The rms flow speeds, corrected for noise, in the individual flow maps was 1.6 km/s. These speeds observed in the granulation are much larger than those reported previously (Title et. al. 1987) due to the higher resolution and smaller field of view in the present data as compared to the data from SOUP. When SOUP data are reanalyzed with smaller field of view, the speed increase to values consistent with our results. By averaging the inferred displacement velocities over the entire duration of the sequence the average flow field was obtained. The speeds of 20 minute average maps are approximately 0.8 km/s.

The most spectacular discovery is that of a vortex which had a diameter of approximately 5000 km and persisted for the 1.5 hour duration of the sequence. The vortex can easily be seen in the movie and in the average flow maps. Granules are seen to spiral around and disappear in the vortex center. Some of them elongate in the streamline direction and are compressed perpendicular to the streamlines, thus resembling a solar "maelstrom". It is clear that all the granules in a radius less than 3000 km from the center of the vortex have their life histories dominated by the vortex flow. The average circulation was 4000 km^2/s, which is an order of magnitude stronger than vorticity structures observed with SOUP. Figure 2 shows the flow field obtained by averaging the displacement maps. Note the vortex in the upper-left part of the figure.

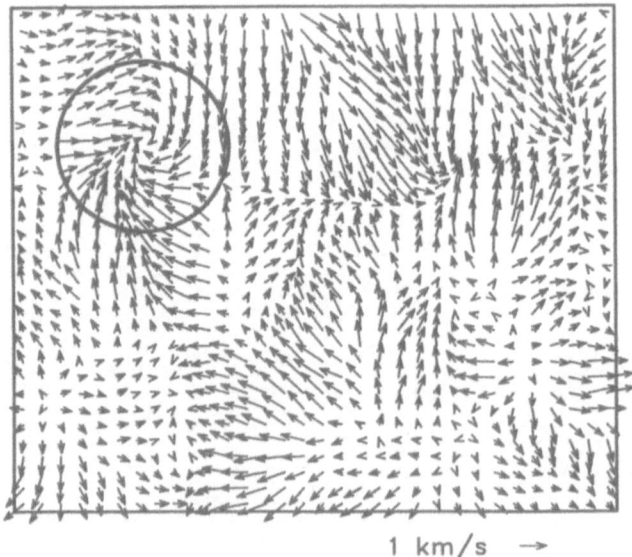

1 km/s →

Figure 3. Flow-field calculated by averaging displacement vectors over the entire 80 mins sequence. Note the vortex in the upper-left part of the figure.

There are also several strong linear downflow regions and sink regions into which granules flow but do not spiral around. Granules born near such sinks are strongly influenced by the presence of the sink. In addition there is an overall steady flow pattern with a size of 5-7 arc seconds, which is the scale of mesogranulation (Title et. al. 1986, November et. al. 1981, Koutchmy and Lebecq 1986, Oda 1984). For further details, see Brandt et. al. (1988).

Other sequences, including two sets of granulation data in active regions (obtained on August 1,1987) are presently beeing analyzed. These data will provide information on granulation dynamics and mesogranular flows in magnetic regions, their relations to filigree and information on flows near pores and sunspots. We are currently trying to obtain time sequences on sunspots.

6. Conclusions

The results from the Swedish solar telescope demonstrate what can be obtained even with a small (48 cm), but excellent, telescope on a good site. The techniques used for image acquisition, including real-time frame selection, and the software developed by the Lockheed group for reduction of data improves the quality of the data to a point where it can be superseeded only by a large (>50 cm) telescope even in space.

The technique of Högbom for measuring and correcting for seeing effects as well as the implementation of adaptive optical systems on major solar telescopes points to the possibility of obtaining at least 0.1 arc second resolution from the ground, which with the additional development of sophisticated data handling and image reduction techniques should mark the beginning of a new era in observational solar physics.

Acknowledgements. This research was supported by grants from the Crafoord foundation, the Wallenberg foundation and the Swedish Natural Science Research Council

References

Brandt, P.N., Scharmer, G.B., Ferguson, S., Shine, R.A., Tarbell, T.D., Title, A.M., Submitted to *Nature*.
Dunn, R.B., 1969, *Sky Telesc.* **38**, 368
Koutchmy, S., Lebecq, C., 1984, *Solar Phys.* **93**, 243.
Mehltretter, J.P.,1979, *J. Opt.* **10**, 93.
November, L.J., Toomre, J., Gebbie, K.B., Simon, G.W., 1981, *Astrophys. J.* **245**, L123.
Oda,N., 1984, *Solar Phys.* **93**, 243.
Title, A., Tarbell, T., Simon, G. and the SOUP Team, 1986, *Adv. Space Research* **6**, 253.
Title, A., Tarbell, T.D and the SOUP Team, 1987, in *"High Resolution Solar Physics II"* , ed G. Athay & D. Spicer, NASA Conference Publication **2483**, 55.

Discussion

(Editorial note: G. Scharmer gave two presentations, on the Swedish Solar Telescope on La Palma and on observations of the solar granulation obtained with this telescope respectively. Both topics are covered in the preceding paper. The first part of the discussion below took place after the talk about the telescope.)

MÜLLER — What is the quality of the site, and what is the stability of the quality?

SCHARMER — We have not directly measured the quality of the site as yet, but it seems excellent, with very high stability when the wind is low.

TITLE — Let me comment on Richard's question. We have measured the size of r_0 using time sequences of images from La Palma and comparison images from SOUP. From these studies we believe that r_0 is 80 cm in about 60% of the selected images taken over 82 minutes.

SCHARMER — Alan, I am not sure I believe in your results!
1). How did you determine r_0?
2). How can you determine r_0 from *selected* images?

TITLE — Images that are of high quality, as evident from their spatial power spectra, and that were taken close together in time are compared. In particular, we compare the phase shifts between the high-frequency spatial components. We assume that these phase shifts would be the same if the effects of the atmosphere were absent, so that the amplitude differences of these phase shifts are due to the atmosphere. Comparison with models that have been degraded assuming a Kolmogorov spectrum then enables us to estimate the value of r_0

KOUTCHMY — 1). What are the results of the measurements of the MTF of your telescope?
2). Did you start to measure the polarization properties of your telescope?
3). In case you have image motion produced by the instrument (vibration, etc.), did you use an active mirror? Are you planning to use the Lockheed active mirror and perhaps also their correlation tracker?

SCHARMER — Ad 1): the MTF of the telescope and the MTF of the atmosphere have been measured by David Brown *separately*, using a shearing interferometer and long exposures of stars. The telescope MTF was determined by measuring the shape of the fringes (deviations from straight lines). The atmospheric MTF was determined by measuring the contrast for several shear values. The result was that the optical quality of the telescope is approximately $\lambda/15$ peak to peak, corresponding to a nearly diffraction-limited MTF.
Ad 2): no.
Ad 3): vibrations are noticeable if the wind speed is higher than 10-15 m/s, but then the seeing is usually not very good. We will use the Lockheed active mirror with the correlation tracker in September this year, and we hope to install a similar active mirror next year.

BECKERS — It would be interesting to measure the optical quality on stars in day-time, by rapidly moving over from solar observations to observing a bright star. I would trust such results more than the results from the nighttime observations.

SCHARMER — I agree that this should be done.

(Editorial note: the discussion which took place after G. Scharmer's presentation of granulation observations now follows .)

KOUTCHMY — This is a rather naive question, addressed not only to you but also to Alan Title and George Simon. It concerns the interpretation of the pattern of horizontal 'velocities' deduced from your cork analysis, in which you find 'sinks' which are presumably locations of concentrated magnetic field, and which imply large downflows. If this is true, why do spectra taken above these regions (line gaps) not show significant red shifts? (Both high spatial resolution spectra, after removing the 5-minute oscillations, and very high spectral resolution FTS data now agree on this point!) And why do you claim high downward motions there, and not upward ones?

NORDLUND — The size of the blueshift is only about 400 m/s. You cannot see this vertical flow easily because of the noise due to the 5-minute oscillation.

TITLE — What we see is a flow pattern in the continuum. We do not know where the magnetic fields are. Continuity suggests that the matter goes down. But indeed, you can only see the pattern after the 5-minute oscillation has been removed first.

RUTTEN — From the movie itself, you cannot state whether the cork gathering locations are sinks or updrafts?

TITLE — From the movie itself, no. But if you consider the continuity equation and the fact that local correlation follows the flow, you must conclude that the corks collect at downflows.

BECKERS — There is no reason for the downflows to be confined to the 'singularity' of the final cork location. The downflow can cover a much larger or smaller area, covering as much as many granules, or as little as the intergranular lane only.

SIMON — The local correlation tracking shows several places where corks are concentrated by the mesogranular flow pattern. In the vortex center, and at the pore boundaries, you say that granules are 'sucked' into the downdrafts, while at another cork concentration you say that granules are 'squeezed out of existence'. I think it very likely that these two phenomena actually represent a single process, namely, the disappearance of matter at loci of downdrafts (sinks).

TITLE — I agree.

DRAVINS — The spatial resolution in the best images is probably close to the diffraction limit of your 50 cm telescope. In these images, one is (barely) beginning to glimpse small features, which are not visible at lower resolution. Since these are clearly not resolved, their full resolution will require telescopes several times larger. The important conclusion for the future is that a diffraction-limited telescope in the 1-meter class is *not* likely to be sufficient: rather, telescopes in (at least) the 2-3 meter class may be required.

TITLE — The photon mean free path is about 50 km, so in an optically thick medium you cannot resolve details smaller than this size.

DRAVINS — But what if there are magnetic fields present in the medium?

TITLE — The image is still set by the photon mean free path.

KNEER — One should be cautious in attributing to the photon mean free path a definite length of say 50 km. The photon mean free path changes, depending on the height in the atmosphere. It can be much smaller than 50 km in subphotospheric layers.

TITLE — Yes, I was referring to scale lengths of $\tau = 1$ in the continuum.

BECKERS — I agree with Dravins, especially when I consider observations at the H^- opacity minimum at 1.6 μm, where 0.5 arcsec seeing in the visible corresponds to $r_0 = 80$ cm. At that wavelength, a 2.4 meter telescope will often give diffraction-limited images (perhaps 10–50% of the time). A 2.4 meter aperture LEST has therefore only a modest aperture.

RUTTEN — Two comments and one question.
1). Regarding the smallest features and the photon mean free path: one should also describe the solar atmosphere with a modulation transfer function. Magnetic features with structures smaller than the photon mean free path may be visible at a few percent modulation yet.
2). The corks float at the optical depth $\tau = 1$ surface, not at the same geometrical depth. Would it be of interest to use modelling to get corks of the latter type too?
3). Is it right to say that you cannot identify the mesogranulation from the morphology of the movie alone?

TITLE — Ad 2): the height variations are small compared to the flow grid. But it may help.
Ad 3): One can see a clear non-uniformity in the evolution of granulation. You can see it too, but you need to be trained. Larry November saw it already in his PhD thesis. The 'corks', however, outline this flow in an objective way.

SIMON — You can see the mesogranulation if you spend many hours watching the movie, but it will give you a headache.

ULMSCHNEIDER — Do you have Ca K-line pictures from La Palma or from SOUP which correlate with these granulation pictures, to see what goes on at greater heights?

SCHARMER — No, but this is planned to be done soon with the Lockheed Ca K filter.

WEISS — Can one say that small granules are *never* squeezed out of existence at the centres of mesogranules, or that exploding granules *never* occur at mesogranular boundaries?

TITLE — No. *Never* is too fancy a word for such a complex flow. It is rare that exploders occur in boundaries. When several exploders occur simultaneously in a region of positive divergence, internal granules will be 'squeezed'.

SIMON — To me, the most interesting part of the movie is the long lifetime of a few tiny bright features (filigree? faculae?) that are located in intergranular lanes at a cork concentration (sink point). If they were normal (typical) granules, I would expect them to disappear quickly into the downdraft. Therefore, I suspect that these features correspond to loci of vertical magnetic flux tubes which are anchored in the boundary of a supergranule.

TITLE — That may well be. We plan to observe with our narrow-band filter and the Scharmer video system this fall. Perhaps we can then find out where the field is with respect to the small granules.

SPATIALLY RESOLVED SPECTRA OF SOLAR GRANULES

Hartmut Holweger
Institut für Theoretische Physik und Sternwarte
Universität Kiel
D-2300 Kiel, F.R.G.

Franz Kneer
Universitäts-Sternwarte
D-3400 Göttingen, F.R.G

ABSTRACT.

An observing campaign has been carried out with the Göttingen
Gregory-Coudé-Telescope at the Observatorio del Teide, Canary
Islands. Efforts have been made to record line profiles of bright
and dark parts of individual granules at high spectral resolution.
We report first results and compare them with numerical simulations.

1. INTRODUCTION

The combination of high-dispersion spectrometry with the highest
attainable spatial resolution is a challenge for observational solar
physics. It offers the most direct empirical access to basic
properties of solar convection cells, providing a crucial test for
numerical simulations of the complex interplay of hydrodynamics,
radiative transfer, and thermodynamics which characterizes the solar
granulation.

The new solar facilities on the Canary Islands are well-known for
their excellent seeing conditions. A high-resolution spectrograph is
available at the Göttingen Gregory-Coudé-Telescope of the
Observatorio del Teide, Tenerife. In the following we report some
results based on an observational campaign in August 1987, shortly
after the telescope went into regular operation. The intention was
to assess the spectroscopic and imaging capabilities of this
instrumentation. First, encouraging results were reported by Wiehr
and Kneer (1988); they are based on the statistical analysis of one
single spectrogram of excellent definition recorded at an earlier
date. In the present complementary study we focus our attention on
the line spectrum of individual structures - granules or network

173

R. J. Rutten and G. Severino (eds.), Solar and Stellar Granulation, 173–186.
© 1989 by Kluwer Academic Publishers.

bright points - that can be seen on high-quality spectrograms
selected from a fairly large number of frames obtained during
periods of very good seeing.

2. OBSERVATIONS

The 45 cm-Gregory-Coudé-Telescope (GCT) used for the present work is
located at Izaña, Tenerife, at an altitude of 2400 m. The optical
system and basic features of this instrument are outlined elsewhere
(Schmidt and Soltau, 1986; Kneer et al., 1987). The entire optical
path between the entrance aperture of the telescope and the exit
window in front of the spectrograph slit is being kept under vacuum.

The effective focal length of 25 m combined with the 10 m-Czerny-
Turner-spectrograph with a 12.8 x 25.6 cm Echelle grating leads to
an image scale of 8.2 arc sec/mm in the focal plane of the spectrograph, the diffraction-limited spatial resolution at 6150 Å being
0.34 arc sec (250 km on the Sun). The dispersion is 0.1 - 0.2 Å/mm
in the visible. This, together with the entrance slit width of 60
microns used here, gives a spectral resolution element of about
10 mÅ at the wavelengths considered. This high resolving power
ensures that solar line profiles are negligibly degraded. In the
spatial domain, 60 microns correspond to 0.5 arc sec (360 km on the
Sun). Therefore, under good seeing conditions, the spatial
resolutions in the direction of dispersion and perpendicular to it
are comparable.

The spectrograms studied here were obtained during a 12-day
observing campaign (August 21 - September 1, 1987). At the beginning
the transparency of the sky was somewhat affected by Sahara dust.
Apart from this, the sky above Izaña remained cloudless at all
times. In contrast, the sites at La Palma were frequently covered by
a cloud layer whose top reached the 2400 m level there (but not on
Tenerife).

Spectra were recorded on high-resolution Kodak SO 115 film. The
format of the individual frames, 24 x 88 mm, corresponds to a
wavelength coverage of about 15 Å and a projected slit length of
198 arc sec (144000 km at disc center). Two wavelength bands have
been recorded alternatively, centered at 6125 and 6173 Å,
respectively. Selection criteria were occurence of suitable, well-
isolated lines, and freedom of telluric contributions. Exposure
times were typically 4 sec. For calibration and test purposes we
recorded also spectra at low spatial resolution by defocussing the
telescope until the granulation became smeared out. With a grey
wedge put in front of the spectrograph slit this permitted
subsequent photometric calibration. Spectra taken without the grey
wedge served to compare our line profiles with those found in
published spectrum atlases. The areas on the Sun studied were
regions at the center of the disc well removed from sunspots. The

disc-center area projected on the polished slit jaws was recorded
with a TV camera and the enlarged image monitored visually during
periods of good seeing. Such periods, usually lasting for 1 - 2
hours in the morning, occurred on 5 of the 12 days of observation.
Each time when the observer felt that the image of the granulation
was particularly stable and well-defined, a spectrum was exposed.
Frequently individual granules could be traced for several minutes
and their evolution followed on the TV.

3. PHOTOMETRY

A total of 370 high-resolution frames have been recorded in the
manner described above. By careful visual inspection, 130 of them
were found adequate for closer scrutiny, the goal being to select a
small number of the very best spectra for more detailed analysis.

The next step in this selection procedure was to find a more
objective measure for the spatial resolution achieved in each of
these 130 spectra. As a criterion we used the granular contrast at a
fixed wavelength in the continuum. Each frame was scanned
photometrically perpendicular to the direction of dispersion, using
the PDS microdensitometer at Kiel. Inspection and intercomparison of
these 130 tracings permitted to identify about 10 frames of superior
quality, with a well-defined granulation pattern extending along
most of the slit. Further analysis was restricted to this sample.

As mentioned above, the intention of this study was to assess the
capability of the GCT for spectroscopy at high spatial resolution.
Therefore we did not attempt to make a full statistical analysis of
the frames, but focussed our attention on the spatial variation of
the profiles of a few selected spectrum lines. Initial tests
involved various lines with equivalent widths in the range
$W = 10 - 230$ mÅ. It turned out that the spectroscopic signature of
granular inhomogeneities - variations of line intensity and
wavelength - was most conspicuous for lines with equivalent widths
between about 20 and 60 mÅ. The lines studied in detail are quoted
below, together with their excitation potentials and equivalent
widths (data from Moore et al., 1966):

Ti I	6126.224 Å	1.07 eV	20 mÅ
Fe I	6127.912 Å	4.14 eV	48 mÅ
Fe I	6173.341 Å	2.22 eV	50 mÅ

In further analysis priority was given to the two stronger lines.

Under excellent seeing conditions the projection of the spectrograph
entrance slit on the center of the solar disc outlines a (nearly)
rectangular area 0.5 arc sec wide and 198 arc sec long, embracing
some small granules as a whole, and intersecting bright and dark
parts of many larger ones. Our goal was to study line profiles

emerging from individual surface elements. For a given frame we made
a sequence of microdensitometer scans across the line under
consideration, consecutive scans being displaced with respect to
each other in the slit direction by 50 microns (0.41 arc sec; 300 km
on the Sun). A 50 x 50 micron square scanning aperture was used. In
this way 6 of the very best spectra have been evaluated, taking care
to restrict each spatial sequence to those part of the frame where
the instantaneous seeing appeared to be of uniform high quality. A
total of 1081 line profiles were recorded for the three lines
mentioned above.

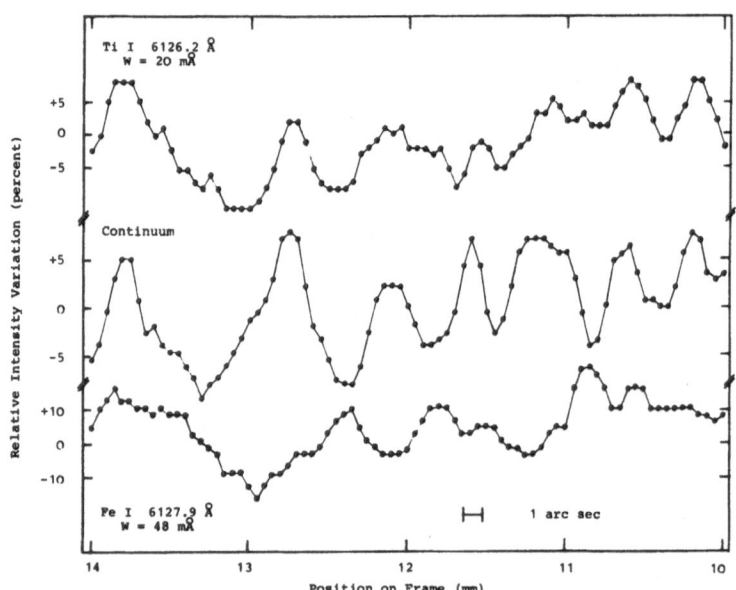

Figure 1: Example illustrating the spatial variation of intensity in
the continuum (middle curve) and in the core of two lines of
different strength (plotted above and below the continuum data to
facilitate comparison). The abscissa is the coordinate on the film
in the slit direction. The data points (connected by straight lines)
correspond to individual spectrum scans; the spacing between
consecutive scans is 50 microns (0.41 arc sec; 300 km on the Sun).
The entire spatial sequence covered 15 mm in this case.

4. RESULTS AND DISCUSSION

4.1 Continuum

Fig. 1 (center curve) illustrates the spatial resolution obtained in the high-dispersion spectrograms. In the section shown, several well defined granules show up. Obviously a sequence of granules of comparable size happened to be lined up along the slit. There are more nice sequences of this kind in our sample, but usually the spatial variation is not as regular as in Fig. 1. This is not surprising since the field-of-view defined by the slit intersects an assembly of larger and smaller granules at random positions. Nevertheless, well-defined granules are rather common.

4.2 Weak Line

The Ti I 6126.2 Å line has a mean equivalent width of 20 mÅ and a mean residual intensity of 78 percent of the continuum, according to published photometric atlases of the disc center recorded at low spatial resolution. As can be seen in Fig. 1, the core of this weak line moves up and down essentially parallel with the continuum. This pronounced correlation has been found in all spatial sequences studied. Close comparison of the line profiles shows that, in addition, lines emerging from the bright region of a granule are always blue-shifted with respect to those of adjacent dark regions, by typically 0.8 km/s. Furthermore, the intergranular line profiles are always broader. The stronger lines discussed below behave in a similar way. For a discussion we refer to section 4.4.

4.3 Stronger Lines

The two stronger lines have mean residual intensities with respect to the continuum of 53 percent (Fe I 6127.9 Å) and 38 percent (Fe I 6173.3 Å), respectively. Their spatial variation is strikingly different from that of the weak line: the correlation between residual intensity and continuum found for the latter normally turns into an anticorrelation in the case of the stronger lines. Examples can be found in Figure 1.

However, the actual situation is more complicated. Even the limited sample of structures shown in the figure contains one 'anomalous' element which stands out bright both in the continuum and in the line core; it is located at position 13.8 mm in Figure 1. Of the total of 74 well-defined bright continuum structures studied, 16 percent turned out to be anomalous in this sense, whereas 60 percent behaved 'normal' (anticorrelation between continuum and residual intensity). The remaining 24 % exhibited little variation in the line core. The anomalous structures and those showing little varia- tion in the core have one distinct property in common: a large fraction of them is exceptionally bright in the continuum. A straightforward explanation is that their vertical temperature profile differs from that of most other granules. We shall come back to this point in section 4.5, focussing our attention for the moment on normal granules like those which are the dominating features in Figure 1. Typical line profiles of adjacent bright and dark conti- nuum structures are shown in Figures 2 - 5. These examples, like all

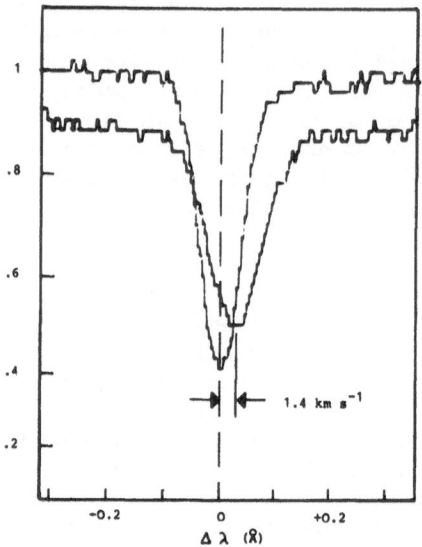

Figure 2: Profile of Fe I 6127.9 Å in the bright and dark part of a
granule.

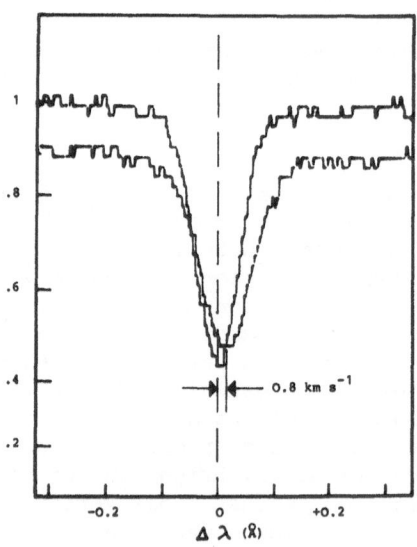

Figure 3: Same as Figure 2; another granule with smaller Doppler
shift

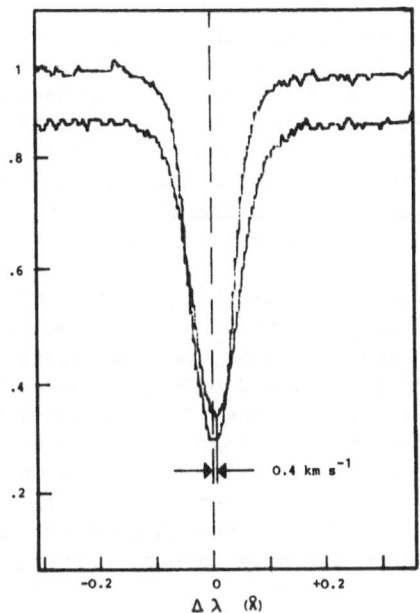

Figure 4: Profile of Fe I 6173.3 Å in the bright and dark part of a
granule. The two tracings can best be 'disentangled' by
turning the figure by 90 degrees.

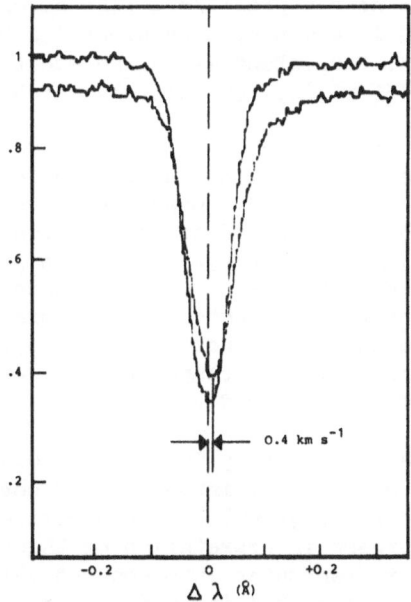

Figure 5: Same as Figure 4; another granule with smaller contiuum
contrast.

granules studied, share two properties with the weak line discussed
in section 4.2. In dark structures the line cores are always red-
shifted with respect to adjacent bright areas, and the profiles are
wider and - in some cases - perceptibly asymmetric, the red wing
being depressed. Red-shift and granular contrast are generally
larger in the 6127 Å frames as compared to 6173 Å. Since periods of
excellent seeing seldom exceeded a few seconds, this difference may
be due to the fact the the exposure time for 6127 Å (4 sec) was 20
percent smaller than for 6173 Å.

Some comments to Figures 2 - 5: these four examples are arranged in
order of decreasing Doppler shift and continuum contrast, and
represent the full range of these quantities seen in the spectra.
The quantity plotted is actually photographic density, which was
found to be an almost linear function of intensity. The intensity
scale is given in the figures. Doppler shifts were determined from
the lower part of the line profile. The distance between bright and
dark areas was typically 2.5 arc sec (1800 km), i.e. the granules
were rather large. It must be stressed that our procedure to select
well-defined structures strongly favours large granules.

4.4 Comparison with Theoretical Predictions

Synthetic line profiles derived from numerical simulations of solar
convection have been published by Dravins et al. (1981), Steffen and
Gigas (1985), and Steffen (1987, 1988). In the paper by Steffen and
Gigas, the variation of a typical Fe I line across a synthetic
granule is displayed in a manner convenient for comparison with the
present observations. Equivalent width and excitation potential of
the synthetic line (56 mÅ, 3.7 eV) closely match our two stronger
lines.

The similarity between observation and theoretical prediction is
striking. Indeed the line core is predicted to be brighter in
intergranular regions as compared to areas which are bright in the
continuum. In addition, the intergranular line profiles are red-
shifted and tend to be asymmetric, in accordance with our results.
The line shift, of course, reflects the intergranular downdraft, and
the asymmetry is caused by the increase with depth of its velocity
rather than by a correlation of temperature fluctuation and
velocity, as is commonly believed.

As mentioned, the calculations by Steffen and Gigas (1985) predict
the anticorrelation of residual intensity and continuum which is so
conspicuous in most of our granules. The numerical simulations
provide a straightforward interpretation of this phenomenon. When
hot, ascending matter penetrates into convectively stable layers, it
necessarily becomes cooler than its surroundings. Hotter than
average in deep layers means bright in the continuum: the 'granule'.
On the other hand, the residual intensity of a saturated line formed
not far from LTE is essentially given by the Planck function at line

optical depth unity; i.e. in the upper photosphere. Therefore, if the granule is observed at the wavelength of the line center, it will look darker than the intergranular region. Thus, the anticorrelation of line core intensity and continuum is a direct manifestation of convective overshooting.

Since the lateral temperature differences in a granule vary with height both with respect to magnitude and sign, the spectroscopic signature of convective overshooting will depend on line strength. Steffen (1989) gives a nice illustration of what is to be expected if one looks at one of his synthetic granules at the center wavelength of lines of different strength. In stronger lines, as we have discussed, the ascending part of the granule looks darker than the descending part. With decreasing strength the line core is formed in successively deeper layers. The granular contrast in the line core decreases until it almost disappears, then it changes sign and gradually approaches the spatial intensity profile of the continuum. Indeed, our weak line Ti I 6126.2 Å (section 4.2) confirms Steffen's prediction.

4.5 Bright Points

As reported in section 4.3 a smaller but significant fraction of the bright elements identified in the frames does not show the anticorrelation of residual intensity and continuum. Many of these structures are bright also in the line core. Two typical examples are shown in Figures 6 and 7.

As already mentioned, they appear to constitute a distinct class of features also because they are often unusually bright, as can be seen in the two examples. About 80 percent of all exceptionally bright features in the frames studied exhibit this 'anomalous' behaviour of the line core. In the light of the foregoing discussion (section 4.4) it is tempting to speculate that their vertical thermal structure differs from that of normal granules in the sense that they possess a temperature excess both in the lower and in the upper photosphere.

Most probably this class of bright features is identical with 'network bright points' (NBP), also called facular points. According to Muller and Roudier (1984), NBP can be found at all latitudes on the Sun, and their number per unit area is at maximum during sunspot minimum. NBP are known to be associated with 'line gaps', which may be nothing else than the enhancement of residual intensity seen in Figures 6 and 7. NBP tend to form aggregates ('filigree'). Indeed, in our spectra nearly 90 percent of bright points with anomalous line cores are concentrated in one third of the frames studied.

NBP are generally considered to be associated with magnetic flux tubes. No anomalous line broadening is discernible in Figures 6 and 7. This does not exclude the presence of magnetic fields because

182

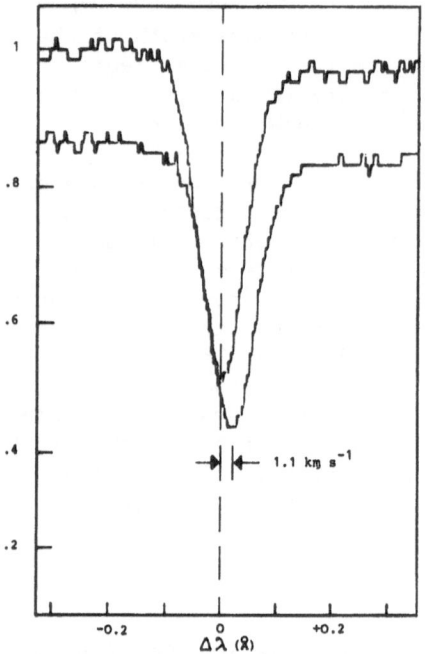

Figure 6: Profile of Fe I 6127.9 Å in a 'bright point' and in the adjacent dark area.

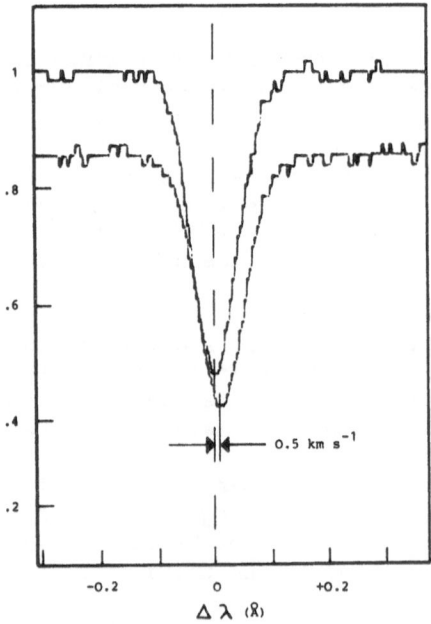

Figure 7: Same as Figure 6; another typical 'bright point'.

the Fe I 6127.9 Å line is magnetically insensitive (g_{eff} = 0.375).

5. CONCLUSIONS

The intention of this study was to test the qualification of the site at Izaña, and of the Gregory-Coudé-Telescope, for solar spectroscopy at high spatial and spectral resolution. We feel that this new facility is well-suited for future detailed investigations of small-scale features.

A few points should be emphasized:

1) High spatial and spectral resolution are crucial for the study of the thermal and kinematical structure of solar convections cells. The spatial resolution should be improved further.

2) Numerical simulations are able to reproduce essential properties of granules. Moreover, they are valuable as a theoretical guide for the physical interpretation of observations. At the outset of observations the simulations can help to select spectrum lines whose properties (equivalent width, excitation potential) qualify them as sensitive probes for the physical parameters to be investigated.

3) On the spatial scale of the granulation, bright elements can be found and studied at high resolution, whose vertical structure differs from that of normal granules. More systematic observations, employing lines of different Zeeman sensitivity, will help to decide whether these bright points represent magnetic flux tubes.

In the last two points, we caution against undifferentiated statistical analysis of spectrograms:

4) The study of individual structures, if their number is large enough, can provide more information than a 'full' statistical analysis of spectrograms. In the latter, different structures - like normal granules and bright points - will be mixed together, causing a loss of physical information.

5) The spectroscopic signature of small-scale photospheric structures strongly depends on line strength. For example, the core of a line of intermediate strength is expected to show little spatial variation across a granule. Statistical analysis of such lines will yield little or no correlation with the continuum. This must not be interpreted as a lack of physical correlation between the lower and upper photosphere. In fact, continuum and line center originate in a coherent vertical structure - the granule.

184

The spatial resolution attainable at the GCT can still be improved. For our photographic spectra we were forced to employ exposure times of 4 sec or more. Visual inspection of the spectrum in real time, by looking at the focal plane of the spectrograph with a magnifying lens, revealed moments when considerably finer structures became visible. During hours of good seeing these moments were not rare, but their duration was much less than 4 sec. For future work, use of a high-resolution CCD detector with an adequate image processing system is under consideration.

REFERENCES

Dravins, D., Lindegren, L., Nordlund, Å., 1981, Astron. Astrophys. **96**, 345.

Kneer, F., Schmidt, W., Wiehr, E., Wittmann, A.D., 1987, Mitt. Astron. Gesellsch. **68**, 181.

Moore, C.E., Minnaert, M.G.J., Houtgast, J., 1966: Nat. Bur. Stand. (US) Monograph **61**.

Muller, R., Roudier, T., 1984, Solar Phys. **94**, 33.

Schmidt, W., Soltau, D. 1986, Geowissensch. in uns. Zeit **4**, 87.

Steffen, M., Gigas, D., 1985, in: Theoretical problems in high-resolution solar physics, H.U. Schmidt (ed.), Max-Planck-Institut für Physik und Astrophysik, München, p. 95.

Steffen, M., 1987, in: The role of fine-scale magnetic fields on the structure of the solar atmosphere, E.H. Schröter, M. Vázquez, A.A. Wyller (eds.), Cambridge University Press, p. 47.

Steffen, M., 1989, this conference.

Wiehr, E., Kneer, F., 1988, Astron. Astrophys. **195**, 310

Discussion

SCHÜSSLER — Are downflows near bright points larger than in other places?

HOLWEGER — We did not look at this yet.

KOUTCHMY — 1). Concerning the 'exceptionally bright features' you analyzed, can you say what the proportion of these features is compared to the overall number of granules?
2). A question concerning possible errors in your analysis: since both the line-core brightness and the continuum intensity are affected by oscillation phenomena (5-minute etc.), how can these affect your conclusions?

HOLWEGER — Ad 1): about 16% of the well-defined continuum structures in our sample were found to be exceptionally bright.
Ad 2): the pairs of line profiles studied arise from adjacent bright and dark areas. The effect of the 5-minute oscillation has not been removed. We assume that it is similar for both areas.

DEUBNER — I had the impression that most of the redshifted profiles you have shown were distinctly wider than the other ones; would you care to comment on that?

HOLWEGER — Indeed, the profiles of both the weak and the medium strong lines are wider in the intergranular regions (where they are redshifted). Sometimes they are also slightly asymmetric, the red wing being depressed. This can be understood on the basis of the recent simulations by Gigas and Steffen: in the dark regions there is a downdraft of which the velocity increases steeply with depth. This velocity dispersion along the line of sight is causes both the asymmetry and the increase of the line width.

NORDLUND — In one case, where you find an unusual *positive* correlation between the continuum and the line core brightness, the maximum line core brightness actually occurs at the point of maximum *change* of the continuum intensity (the edge of the granule). Do you see this often?

HOLWEGER — It occurs sometimes. In 10-20% of the cases studied, the positions of the maxima are shifted slightly with respect to each other. This can easily be understood if one assumes that these structures are not perfectly vertical, but that they are slightly inclined.

STEFFEN — Another reason why you cannot expect a perfect (anti-)correlation between the continuum and line core intensities is that granulation actually is a time-dependent phenomenon. What happens in the continuum forming layers happens at a later time, further up in the atmosphere. A correlation is to be expected chiefly for long-lived, well-established granules.

EDITH MÜLLER — Do I understand correctly that you define the center of the spectral line as the wavelength where the line has its deepest point? In the core of the line the lowest part may be shifted in one direction as compared to the upper part of the line core.

HOLWEGER — The center wavelength was determined from the lower third of each line profile. We did not try to determine bisectors because of the limited accuracy one can achieve with photographic spectra.

RUTTEN — I have three points.

1). Is the normal-granule contrast reversal you see between line core and continuum the same as the old Evans-Catalano reversal of the streaks seen in the wings of Ca II H and K? If so, a comparison may be worthwhile.

2). A comment regarding NLTE effects in your lines. Assuming LTE for these Fe I lines is probably correct for their source functions, but wrong for their opacities. If you convert intensity from $\tau = 1$ into temperature at a given geometrical height, NLTE overionization must be taken into account. A comparison with the Ca II H and K line streaks can be a diagnostic to the size of this correction.

3). I gather that your exposure times represent a limiting factor. Would you be willing to sacrifice spectral resolution to reduce the exposure duration? If so, by a factor of 2? A factor of 10?

HOLWEGER — Ad 1): the contrast reversal is indeed in qualitative agreement with the statistical studies by Evans and Catalano[1] and by Canfield and Mehltretter[2]. Ad 3): the spectral resolution that we used was about 10 mÅ. This could be degraded to 20 mÅ without significant loss of spectral information. However, the spatial resolution in the direction of dispersion—as determined by the width of the spectrograph entrance slit—would then change from 0.5 arcsec to 1 arcsec, implying loss of spatial information.

RIGHINI — May I suggest to add together separately the line profiles of the intergranular, granular and bright (magnetic?) regions, in order to obtain their typical bisectors? This would be of great help in interpreting the line bisector behaviour for active regions.

KALKOFEN — For the numerical simulations by Steffen, you have shown the isotherms plotted against geometrical depth. It would be interesting to look at them plotted instead against optical depth (or depths), in order to display what one would observe.

HOLWEGER — Since the H^- opacity increases with temperature, and the temperature in the upper photosphere is higher above the intergranular region than above the granule because of overshooting, one would expect that above the intergranular region the location of optical depth unity in the line core occurs higher up in the atmosphere than above the granule. This will enhance the granular contrast with respect to the case of equal geometrical heights.

---

[1] Solar Physics **27**, 299, 1972.

[2] Solar Physics **33**, 33, 1973.

GRANULATION LINE ASYMMETRIES

W. Mattig[1], A.Hanslmeier[2], A. Nesis[1]
[1] Kiepenheuer-Institut für Sonnenphysik, Freiburg, FRG
[2] Institut für Astronomie, Univ. Graz, Austria

ABSTRACT. We give some preliminary results of spectroscopic solar granulation observations done with the Gregory-Coudé-Telescope at Izaña (Tenerife). It is clearly seen that the granular-intergranular regions are well resolved. Bisectors of some iron lines are remarkable stronger asymmetric in the spatially resolved spectra than in the unresolved spectra.

1. INTRODUCTION

More than 30 years ago Voigt (1956) and Schröter (1957) proposed inhomogeneous solar models in order to explain the center to limb variation of the asymmetries in the infrared Oxygen triple resp. the limb-effect of middle strong Fraunhoferlines.

In these models the rising and sinking elements have a strong velocity gradient leading to intense line-asymmetries. The line-asymmetry of the individual components can amount to 20 mÅ as well as blue and red shifted. Asymmetries of this magnitude have not yet been observed though Kavetsky and O'Mara (1984) find stronger asymmetries in the individual components as had been known from spatially unresolved spectra. The line asymmetries of spatially unresolved spectra (C-shape) are about 5 mÅ as for example Brandt, Schröter (1982) or Cavallini et al. (1982) showed.

Voigt (1956) and Schröter (1957) compose the measured line-asymmetries of components with a higher line-asymmetry. Dravins (1975) explains the observed asymmetry as a superposition of two symmetric line profiles thus the individual components having no velocity gradient.

Recently performed granulation model calculations by Steffen (1987) show that for spectra with high spatial resolution one would expect line-asymmetries up to 0.7 km s^{-1} (13 mÅ) blue-asymmetric and up to 2 km s^{-1} (36 mÅ) red-asymmetric. The asymmetry in the spatial unresolved spectrum is 0.2 km s^{-1} (3.6 mÅ); this was calculated for the Ti II-line (λ 5336.7 Å).

R. J. Rutten and G. Severino (eds.), Solar and Stellar Granulation, 187–193.
© 1989 by Kluwer Academic Publishers.

New observations with enhanced spatial resolution should verify if the line-asymmetries increase (Steffen) or decrease (Dravins).

2. OBSERVATIONS AND REDUCTIONS

The observations were done on July 20, 1987, with the 45 cm Gregory-Coudé-Telescope at Izaña (Tenerife). We observed four different Fe-I-lines near the disc center. There was no CaII-activity seen on the slit jaws. The wavelengths of the lines are given in Table 1 with their corresponding equivalent widths.

Table 1: Fe-I-lines used for observation

Line Nr.	Wavelength	Equivalent width
I	6494.499 Å	34 mÅ
II	6494.994	165
III	6495.740	42
IV	6496.472	69

The dispersion of the used spectrograph was 5.1 mm/Å. The spectra were photographically recorded, a typical exposure time was 3s. The frames were photoelectrically digitized with the Optronics Microdensitometer and transformed into intensities.

This paper presents preliminary results of the reduction of one spectrum in order to see, whether a distinction between granular and intergranular line asymmetries is possible. The spectrum covers a spatial region of about 140 arcsec, the 906 scans have a spatial distance of 0.16 arcsec. The bisector curves were calculated for all four lines but in the blue wing of line I and in the red wing of line III also a terrestrial line was present. In order to avoid influences on the bisectors, we could only go to a lower intensity level than for lines II and IV.

3. RESULTS

The 906 scans in the region of 140 arcsec lead to the following preliminary results:
i. We could clearly measure the variation of the residual itensities in the bright and dark regions. This was not investigated further.
ii. There is a good correlation between the relative wavelength shift and the intensity. Bright structures (granules) show a characteristic blue shift with strong asymmetries.
iii. The bisectors of the spatially resolved spectra show no C-shape as has been known from spectra of the spatially unresolved Sun.
iv. The bisectors have both a red and a blue-asymmetry (Fig. 1,2,3). Fig. 1 shows 30 spatially neighboring bisectors covering a region of 4.8 arcsec. The peak to peak velocity varies from the line core from 10 mÅ

(0.45 km s^{-1}) to 1 km s^{-1} at the line wings which confirms once again the decrease of the rms velocity fluctuation with height in the atmosphere.

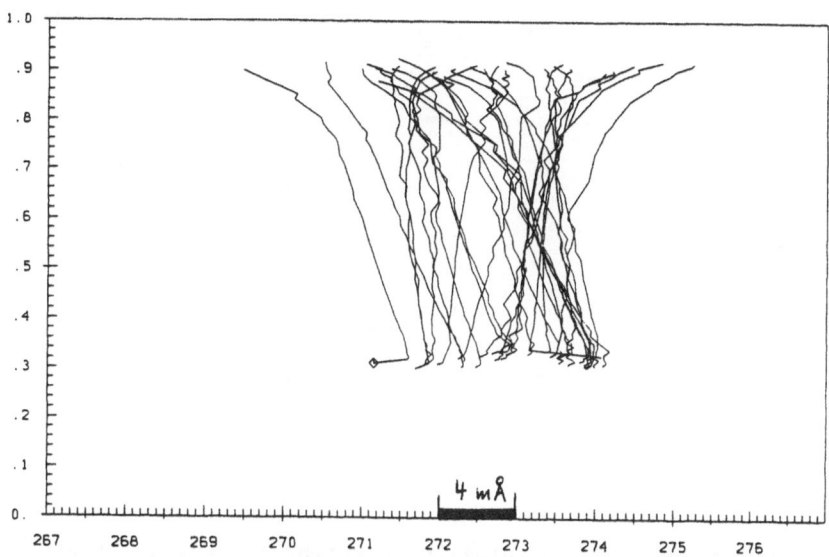

Fig. 1: 30 spatially neighboring bisectors of the line 6494.944 Å covering a region of 4.8 arcsec.

v. For considering only the asymmetries we shifted in Fig. 2 and in Fig. 3 the bisectors of 100 line profiles covering a range of 16 arcsec to the same line core position. For the strongest line investigated a blue asymmetry of max 24 mÅ and a red asymmetry of 10 mÅ were obtained. For the weakest line the maximum blue asymmetry was 16 mÅ and the red asymmetry 6 mÅ.

vi. As is evident from the figures the blue asymmetries are remarkable larger than the red asymmetries. The blue asymmetries are of the same order found by Voigt, Schröter and Steffen. The red asymmetries are considerably smaller as one would have expected. This can be explained that the integranular space is not sufficiently resolved. Wiehr (1987) claimed that the resolving power of the used spectrograph is 0.5 arcsec. For unambiguous measurements of line profiles in the intergranular space instruments with higher resolving power i.e. larger aperture will be necesarry

4. CONCLUSIONS

The presented preliminary observational results clearly demonstrate that with sufficient resolution there is a blue and a red asymmetry in the lines. These are considerably larger than for spatially unresolved spectra. This leads to the conclusion that the observed line asymmetry of

190

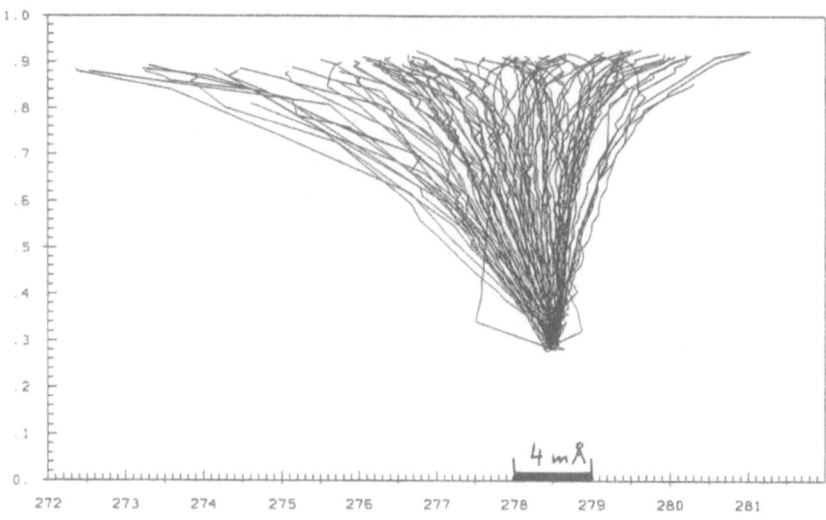

Fig. 2: bisectors of 100 line profiles covering a range of 16 arcsec.
Strong line 6494.944

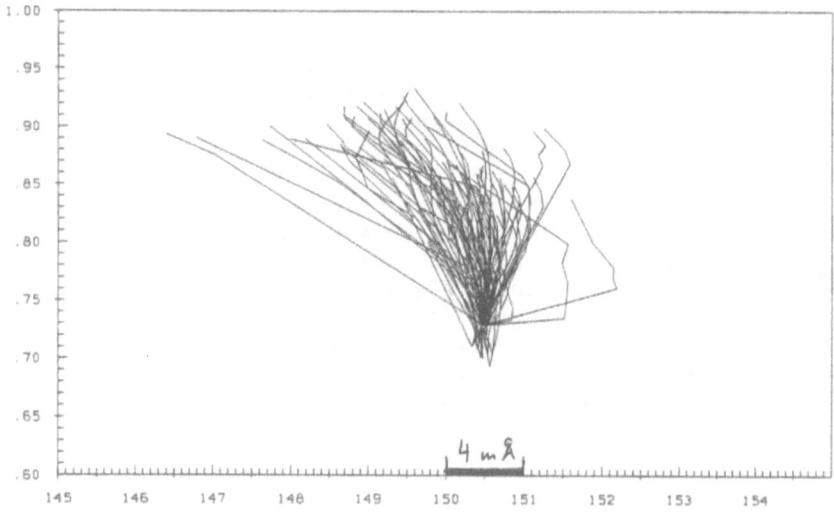

Fig. 3: bisectors of 100 line profiles covering a range of 16 arcsec.
Weak line 6494.499 Å

spatially unresolved spectra (C-shape) consists of more asymmetric profiles and not of symmetric profiles.

Since there are now observations of sufficient spatial resolution a
direct comparison between observations and model calculations should be
possible.

5. REFERENCES

Brandt, P.N., Schröter, E.H.: 1982, Solar Phys. **79**, 3.
Cavallini, F., Ceppatellli, G., Righini, A.: 1982, Astron. Astrophys.
    109, 233.
Dravins, D.: 1982, Ann. Rev. Astron. Astrophys. **20**, 61.
Kavetsky, A., O Mara, B.J.: 1984, Solar Phys., **92**, 47.
Schröter, E.H.: 1957, Z. Astrophys. **41**, 141.
Steffen, M.: 1987, in the Role of Fine-Scale Magnetic Fields on the
    Structure of the Solar Atmosphere, ed. E.H. Schröter, M. Vazquez,
    A.A. Wyller, page 47.
Wiehr, E.: 1987, in the Role of Fine-Scale Magnetic Fields on the
    Structure of the Solar Atmosphere, ed. E.H. Schröter, M. Vazquez,
    A.A. Wyller, page 93.
Voigt, H.H.: 1956, Z. Astrophys. **40**, 157.

Discussion

RIGHINI — I suggest to superimpose line bisectors by taking as zero point in the wavelength scale that point which corresponds (via the response function) to the level where the correlation between the temperature fluctuation and the velocity fluctation changes sign.

MATTIG — In this presentation we demonstrate only that the effects on the bisectors are remarkably more pronounced in the resolved spectra than in the unresolved spectra.

DRAVINS — What is your spatial resolution?

MATTIG — About 0.5 arcsec.

DRAVINS — Even if the spatial resolution in the spectra (in the half-width sense) is perhaps 0.5 arcsec, then the width at 10% or 20% must be at least about 1 arcsec. Both high-resolution images and detailed numerical simulations indicate the presence of significant structure down to a level of 0.1 arcsec. Since the observed line profiles are then spatial averages over some 100 different spatial elements, the observed line asymmetries may be largely a function of the spatial resolution, and perhaps not so much of e.g. the depth dependence of the velocity fields.

MATTIG — May be. I hope that we will have spectra with better resolution from the VTT in the near future. On the other hand, from measurements of the velocity fluctuations in different lines, we know that there is a strong depth dependence of the velocity field.

EDITH MÜLLER — None of the many bisectors you have shown, and have measured up to 10% of the continuum intensity, have a C-shape as found previously by various others. So you define the asymmetry given by a bisector as its largest extension, and then find values of up to 20 mÅ for the asymmetry. Is that right?

MATTIG — Yes, this is correct. That is the main difference between spatially resolved bisectors and the typical C-shape from unresolved line profiles.

EDITH MÜLLER — Let me comment: I have a PhD student at the University of Athens, Mrs. Mariella Stathopoulou, who has been studying the bisectors of a large number of iron lines determined on Kitt Peak FTS spectrograms observed at disk center, near the limb and in integrated solar light. At disk center, many lines have C-shaped bisectors and others do not, whereas at the limb only a few lines have C-shaped bisectors while most of the lines do not. Furthermore, the asymmetries show a clear dependence on the excitation potential of the lines. The bisectors of the lines on the integrated solar light spectrograms behave somewhat intermediate between the bisectors at the center and near the limb of the solar disk. So, according to your spatially resolved observations, our C-shaped bisectors would be a smearing over a number of differently extended and directed line bisectors.

DEUBNER — You have shown an impressive variety of line bisector shapes and a large enough scatter of their asymmetries to worry about. However, one should not forget that 10 mÅ corresponds to velocities not uncommon for p-modes. In order to separate the effects of p-mode oscillations and of granular dynamics, one would certainly need time series of profiles at different positions with respect to the outlines of granules.

MATTIG — I agree completely with you and we will observe this as soon as possible. Here, I have concentrated on the line asymmetries and not on the line shifts. We assume that p-modes more or less produce shifts only, because of the small height gradient derived from the variation between different lines. Is 10 mÅ asymmetry really a representative value? It is not observed in spatially resolved spectra.

KNEER — I think that the line asymmetries shown here are mainly of granular origin and not due to oscillations, because the oscillatory velocity increases with height and thus with line depth (grossly). But here the extensions are largest in the wings of the spectral lines.

RIGHINI — Let me comment on that. We must keep in mind that due to phase effects, we have a drastic change of bisector shape during the oscillation.

STEFFEN — I think it is not quite fair to confront the results of my numerical simulations with the impression you get from the very simple, schematic illustration given by Dravins et al.[1]. I wonder whether there is still a contradiction if you use the more detailed results from the granulation models of Nordlund et al.?

MATTIG — I have not compared our data with the models of Nordlund.

BECKERS — Steffen's models also predict that regions on the sun with large red line asymmetry have also shallower lines. I do not see this in your measurements. Comment!

MATTIG — That is correct, but in this presentation we have only discussed the line asymmetries and not the line profiles in detail.

[1]D. Dravins, L. Lindegren and Å. Nordlund, 1981, *Astron. Astrophys.* **96**, 345.

GRANULATION AND WAVES?

Franz − L. Deubner

Institut für Astronomie und Astrophysik
der Universität Würzburg
Am Hubland, D−8700 Würzburg, Germany F.R.

Abstract

At the photospheric level convective motions generated in the interior of the sun interact with the visible layers of the atmosphere. The physics of this interaction and the observed photospheric dynamical phenomena are not yet fully understood.

This article briefly reviews recent observational work in this field, mostly spectroscopic, as well as some pertinent older results. The emphasis is on the dynamical character of small scale granulation, and on mesogranulation as a genuine photospheric convective regime.

Introduction

S i m u l a t i o n of convective motions in stellar atmospheres has come to a state of perfection (see Stein, Nordlund and Kuhn, this Conference), where the results are expected to reproduce in great detail the fine scale phenomena observed in the solar photosphere. In particular we have here in mind the occurrence of waves generated by the interaction of convective motions with the ambient medium in the convection zone itself, or within the stably stratified atmosphere on top of it. A typical example of this kind of interaction, involving a rather extended range of heights of the atmosphere, may be found in Deubner and Laufer (1983), who describe the excitation of short period acoustic pulses by concomitant motions of convection, and of photospheric as well as chromospheric p—modes. Previous anelastic approximation models were fundamentally inapt to describe this type of dynamical coupling which may be of importance for the chromospheric heating problem.

A special aspect of (one—dimensional) simulation calculations in our context is singled out by the work on spectral line bisector diagnostics, and the effects of thermal relaxation, in the presence of oscillations, on the intensity fluctuations measured in the two wings of a Fraunhofer line. The Firenze and the Napoli group have been particularly active in this field (see e.g. Alamanni et al., 1989; Cavallini

R. J. Rutten and G. Severino (eds.), Solar and Stellar Granulation, 195–205.
© 1989 by Kluwer Academic Publishers.

et al., 1986; Marmolino et al., 1988; Severino et al., 1986). In several papers, e.g. by Marmolino et al., 1987, and by Gomez et al., 1987, the model dependence of simulated line shifts has been investigated. The results prove that different parameters taken for "granular" or "intergranular" model atmospheres have a much smaller effect on the fluctuations in the wings of a spectral line than the well known effects of thermal relaxation (Schmieder, 1976).

In the following, we shall leave simulation studies aside, and concentrate on o b s e - r v a t i o n a l topics.

In Part I we shall consider granulation, the types of waves it might generate in the stably stratified layers on top, and the effects these waves have on the interpretation of spatial coherence spectra used to study the physical properites of granular flow in the photosphere.

In Part II we shift to larger scales, and highlight some properties of the mesoscale flow as observed in the photosphere. We shall pay particular attention to temporal phase relations among the velocity and brightness fluctuations in different spectral lines, in order to improve our understanding of the dynamics of the flow.

Before taking up our first subject, we might step back for a minute, and look at this report from a different angle.

It is relatively easy for an observer to explain to his colleagues the tools he is using to reduce his data, so that they become interpretable. In the solar case we usually treat the observed motions and brightness fluctuations as linear modes; we compute spatial and temporal Fourier transforms to deduce power spectra, to carry out coherence and phase analyses, to obtain $k-\omega$ diagrams for comparison with theoretical dispersion relations; by theory we are guided to distinguish among stable and unstable layers of the atmosphere; and with all these nice tools at hand one would expect that it is not all that difficult to sort things out, and to arrive at an acceptable i n t e r p r e t a t i o n. Unfortuantely, the contrary is true:

Our confusion about the important physical processes involved in a dynamical picture of the solar atmosphere seems to be still increasing in spite of a steadily increasing data base. Due to considerable overlap of frequency ranges and spatial scales of convective modes, and waves of different kind, or even turbulent motions, quick conclusions based on only one particular type of data set can be misleading. Both, granulation and mesogranulation are typical examples which deserve further attention.

Part I

The spatial coherence of granular velocity and brightnes fluctuations.

In a recent paper, Roudier and Muller (1987) investigated the horizontal structure of granulation in white light. By displaying the area of granules as a function of the perimeter they found a typical size, of 1."37, of the granule diameter, where the complexity of the shape of the granules changes in the sense that larger granules typically have a more complex horizontal structure. The authors find this particular size of ~1."3 confirmed by a histogram of granule areas, where the rate of decrease of numbers with increasing area changes abruptly in a log–log presentation at a diameter of 1."31. The authors conclude, that smaller size elements probably have a different origin, and based on the gradient of the slope they have determined in the granular spatial power spectrum Roudier and Muller suggest that turbulent decay of the larger elements into smaller ones may be the relevant process.

Figure 1, which is taken from Deubner (1988), where the Figure is explained in more detail, aimes at an independent search for such a characteristic size in spectral brightness (I) as well as in the velocity (V) distribution near the center of the solar disc. S p a t i a l coherence and phase are shown as function of the spatial wavenumber k_x. Before computing the spatial Fourier transforms, the velocity and brightness data obtained from a 32 min series of spectra of superior quality were conditioned in two different ways. (a) Only the best 10% of the original frames selected for high r.m.s. velocity contrast in the range from 1" to 3" were retained for the spatial coherence analysis (square dots in Fig. 1); (b) The raw data of the complete set of 320 frames were subjected to a subsonic filter to reject oscillations and high frequency noise (open dots).

This study which concentrates on the lower photosphere (K for continuum, C for C I 5380, F for Fe I 5383) arrives at the following conclusions:
1)   Subsonic filtering is useful, at all levels, to increase spatial coherence in the critical wavenumber range.
2)   In the higher line (Fe I 5383) the coherence between continuum brightness and intensity in the core of the line is low (see also Nesis et al., 1988), except for the main spatial scale of granulation (2 to 4 Mm^{-1}) where anticorrelation is observed ($\phi = 180°$) — a consequence of overshoot in the higher, stable layers.
3)   Good correlation ($\phi = 0°$) is observed for all the other combinations, i.e. up to 11 Mm^{-1} (~ 0."8) for continuum intensity and C I 5380 line velocity, and up to 9 Mm^{-1} (~1."0) for continuum intensity and Fe I 5383 line velocity, as well as for the C and Fe line velocities.
4)   No change of the coherence is indicated at 6 Mm^{-1}, corresponding to 1."3.

Even if we can not find in these data specific support for a characteristic scale as claimed by Roudier and Muller (1987), we certainly should like to know whether the breakdown of coherence at 9(11) Mm^{-1} has to be attributed to solar "turbulence", seeing or other effects.

At this point the possibility that gravity waves interfere with convective motions and destroy the phase coherence at levels close to $\tau_{5000} = 1$ needs to be looked into.

Gravity waves exist without any doubt in the stable layers of the atmosphere. In addition to the inverted brightnes contrast of granules in the upper layers of the photosphere, phase investigations of small scale photospheric brightness and velocity fluctuations have shown ever more convincingly (Schmieder, 1976; Altrock et al., 1984; Staiger et al., 1984), that the dynamic behaviour of the photosphere at frequencies below 2.5 mHz follows closely the predicitions of the Souffrin – Schmieder theory of hydrodynamic waves including thermal relaxation effects. A detailed comparison (Deubner and Fleck, 1988) of the wavenumber and frequency dependence of the observed phase differences with theory leaves no more doubt, that gravity waves do exist at least down to the level of Fe I 5929, and probably even as low as C I 5380 (see also discussion of Fig. 2).

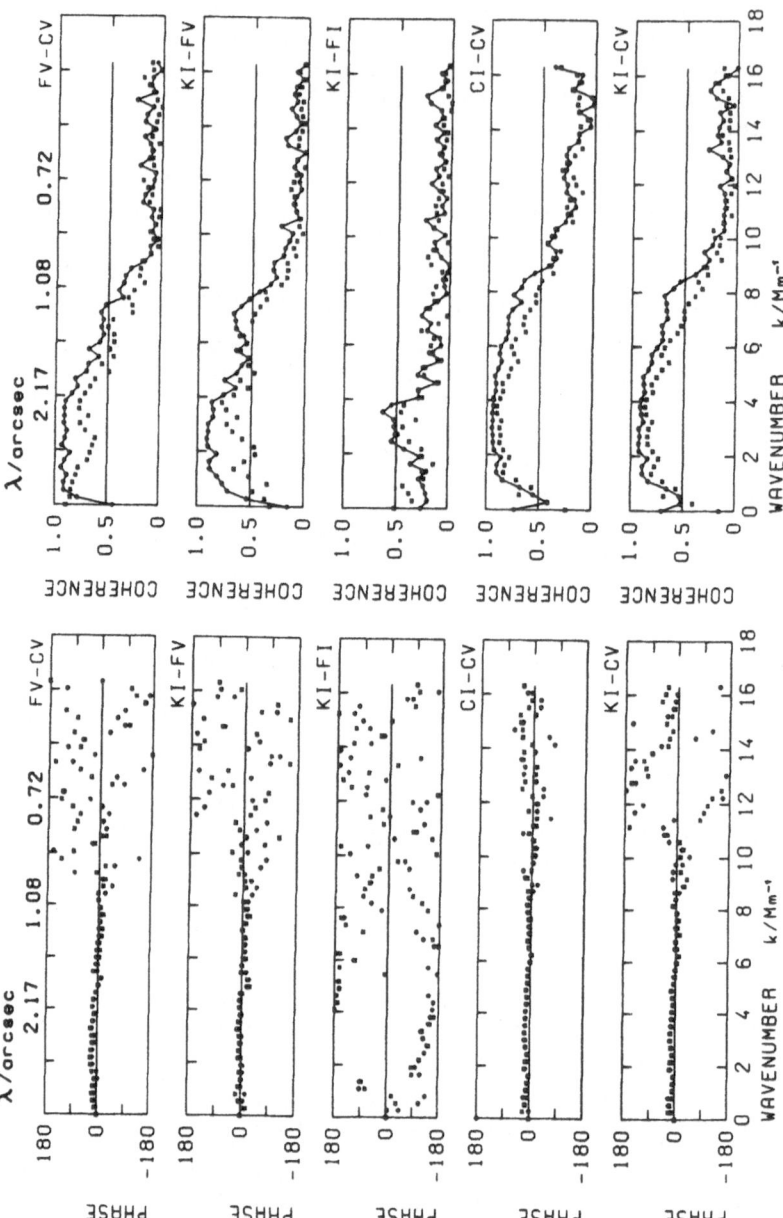

Fig. 1. Spatial coherence spectra of photospheric velocity (V) and brightness (I)
fluctuations in the continuum and in the C I 5380 and Fe I 5383 lines. Spatial
phase and coherence are plotted in the left rsp. right column. The spectra are
derived using a time series of 320 exposures after suppressing non-convective
motions by filtering in k, ω space (circles), and a subset of 32 frames selected
for best quality (squares).

Since at the high wavenumbers, and at frequencies comparable to granular life times brightness and velocity get very much out of phase ($\gtrsim \pi/2$) it seems quite possible

that even those spatial phase spectra which have been subjected to subsonic filtering become contaminated by the effects of gravity waves at horizontal wavenumbers higher than 6 Mm^{-1}. It appears that the phenomenological description of the vertical coherence of small scale vertical and horizontal motions given by Nesis et al. (1988) fits quite well with our picture, suggesting that convective energy is quickly dispersed with height by gravity waves in the stably stratified layer ("loss of memory"); in particular, the rapid loss of coherence of the horizontal motions can easily be accounted for by the interference of wave trains from adjacent sources.

We have given reasons, why spatial coherence spectra of small scale photospheric brightness and velocity fluctuations cannot be used prima vista in support of suggestions of a turbulent decay of the granular flow. But, of course, we are still left with the problem to decide which process (seeing, turbulence, waves) determines ultimately the observed coherence spectra. Better seeing than the one prevailing during the observations leading to Figure 1 is hardly available for extended periods of time from the ground. Therefore, spectral observations from space will eventually be necessary to arrive at a clear answer. Note also, that only observations of rather weak, high excitation lines like those of C I are comparatively little affected by gravity waves, and some further improvement should still be gained from spectra taken at $\lambda \sim 1.6\mu$, formed somewhat deeper than the C I lines in the visible.

Part II

Mesogranulation as a convective phenomenon.

The physics of mesoscale flow patterns appears to be even more confusing than that of the smaller scale motions we have just discussed.

On the one hand, there are the observations of November and coworkers (1981) who detected in 1 hour averages of Mg I Doppler heliograms (i.e. near the temperature minimum) a quasistationary velocity distribution with a typical horizontal scale of 5" to 10" and vertical velocities of ~60 ms^{-1}.

These observations resemble in a strange way the patterns shown by Damé and Martic (Damé, 1985), who apply various filters in the frequency domain of p–mode oscillations to time series of chromospheric spectroheliograms. Damé and Martic (see also this Conference) deduce more or less the same spatial pattern from their own observations (slightly dependent on the frequency filter); their interpretation, however, rests on the instantaneous horizontal phase distribution of a (standing?) oscillatory field rather than on a stationary velocity component.

On the other hand there is a class of white light or low photospheric spectral observations (Deubner, 1974, 1977; Oda, 1984; Koutchmy and Lebecq, 1986) which clearly point to the existence of a stationary distribution, in the 5" to 10" range, of brightness, as well as vertical and horizontal velocities near the $\tau = 1$ level. Recent observations of the divergence of granular proper motions (November and Simon, 1988; Brandt et al., 1988) have added a rather convincing piece of independent information supporting these earlier photospheric results.

If "mesogranulation" really merits its name, i.e. if it is a convective phenomenon linked by its spatial scale to the He I ionization zone (Simon and Leighton, 1964; November et al., 1981), then the dynamical properties of the flow should testify to its convective origin as well as the beautiful horizontal flow patterns revealed by the recent high resolution white light observations.

Figure 2 exhibits results from the same data set as the one used to prepare Figure 1; only, the resolution in horizontal wavenumber k_x is increased by a factor of 6, and the phase difference displayed is temporal rather than spatial. The data are filtered in ω to reject acoustic and evanescent oscillations. The cut–off frequency is 1/8 min for all values of k_x. In addition to the coherence and phase spectra of the low frequency component of C I and Fe I intensity and velocity fluctuations, the spatial power spectra of the filtered data are displayed for comparison. The diagrams in the left column refer to the center of the sun, the right column to cos $\Theta = 0.8$, where already 60% of the horizontal velocity component parallel to the radius vector on the disc contributes to the measured velocity power.

While interpreting the power, coherence and phase spectra one must remember, that they are one–dimensional in space and have not been corrected for spatial isotropy of the wavefront vectors, in order to retain the phases. Therefore, the peaks of the power and coherence distributions tend to be smeared out in the direction of smaller wave numbers, and the valleys in between to be filled in. As a result the contrast of every spectral feature is diminished considerably, and the phases may be biased by higher spatial frequency components.

A detailed description of the following discussion is in preparation. Let us summarize the salient points:

1)   From the diagram in Figure 2 there emerges a regime with enhanced power and temporal coherence in the range of ~1 to 2 Mm^{-1} (4."5 to 9").
2)   The enhanced power is much more pronounced at $\mu = 0.8$ than in the disc center, and is best visible in CV, CI, FV; at these different levels of the atmosphere the spatial structure appears rather similar in each data set (arrows).
3)   Near the maxima of the spectral distributions the power is by a factor of ~ 2 enhanced over the adjacent minima; the maxima of the velocity power spectra at $\mu = 0.8$ are higher than those at disc center by a factor of 3 (dashed lines).

These three points confirm the existence of a distinct flow pattern with a typical scale in between granulation and supergranulation, which is seated in the photosphere and transports thermal energy. The maximum horizontal component of the flow exceeds the vertical by at least a factor of 2.6.

4)   In the disc center the V–V phase at the mesoscale scatters around zero (dropping to about −10° at granular scales) whereas the V–I phase at the lower level is close to zero at the granular scale, and increases slightly but steadily to 10° or 15° at the mesoscale; at the upper level it falls from values around − 140° at higher wavenumbers to about −100° at the mesoscale.

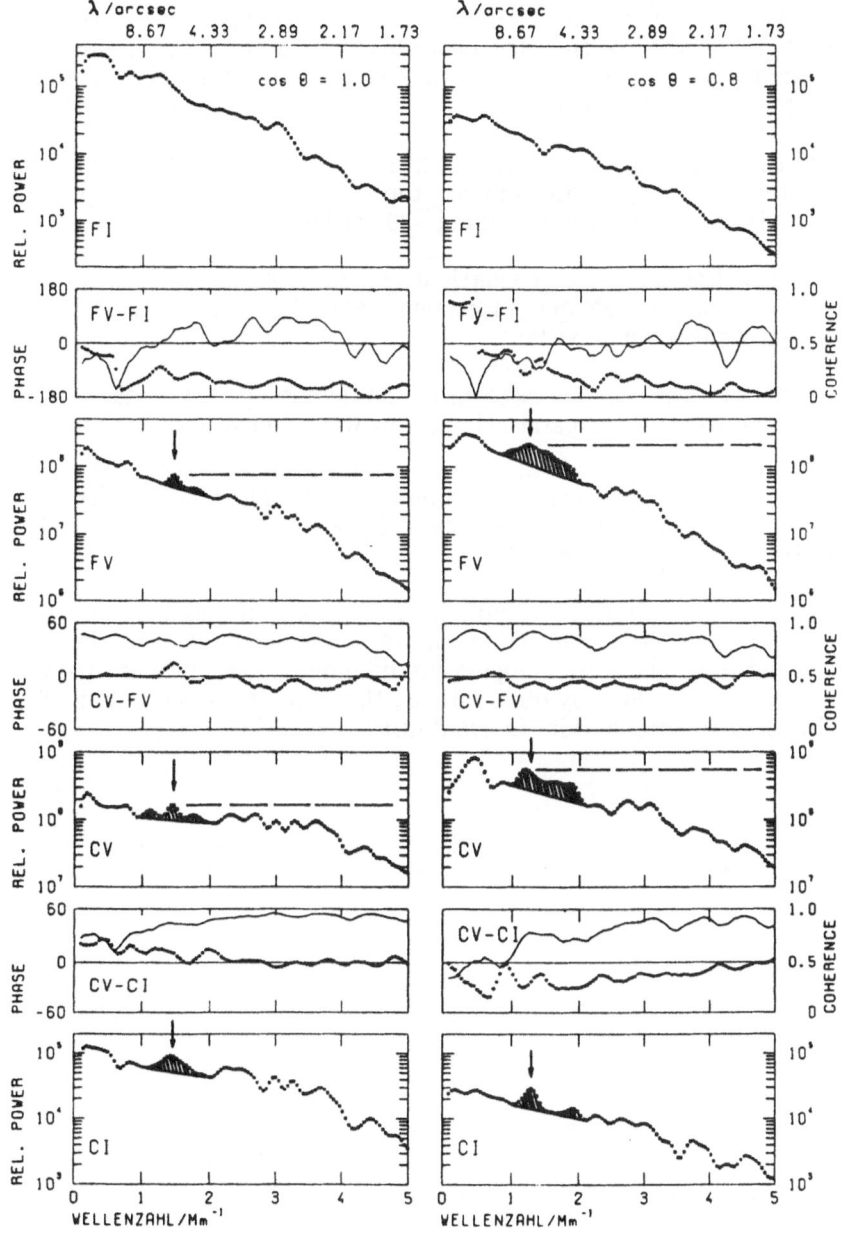

Fig. 2. Spatial spectra of power (squares), temporal coherence (solid line) and phase (diamonds) of photospheric velocity and brightness fluctuations as described in the text. Left column: center of the disc, right column: $\cos \Theta = 0.8$. The spectra are derived using a time series of 320 exposures after suppressing non-convective motions by filtering in ω space.

A simple model explaining the increase of V–I phases of convective motions with increasing scale, as observed at the C I level, has been suggested by Deubner and Fleck (1988). Else, the wavenumber dependence of the phase diagrams is well in line with common ideas about convective overshoot at the upper level, and the onset of gravity waves in interjacent medium.

5)   Near the solar limb, the phase spectra look very much the same, with one exception: the V–I spectrum for the C I line. There the phase drops distinctly to negative values around −30° at low wavenumbers.

Here, the effects of an increased geometric height ($\mu = 0.8$) favouring gravity waves over convection, and of oblique projection enhancing the contribution of the horizontal motions become important.

In conclusion, we seem to be discovering an almost perfect sosie of ordinary granulation, centered at a comparatively narrow wavenumber range around $k_x \simeq 1.5$ Mm⁻¹. We observe motions convective in character near the bottom of the photosphere; overshoot and gravity waves on top of it. The phases differ only little between the two scales, and there is a smooth monotonic transition of all phases across the borderline close to 2 Mm⁻¹.

Measurements to confirm these results with spatially two–dimensional data are underway. The physical connection between the mesoscale motions in the photosphere and those above the temperature minimum is yet unclear. Until such a relation has been established by observations, the term "meso–granulation" should be applied with great care in order to avoid further confusion.

References

Alamanni, N., Bertello, L., Cavallini, F., Ceppatelli, G., Righini, A.: 1989, Astron. Astrophys., in print.

Altrock, R., Musman, S., Cook, M.C.: 1984, in Small Scale Dynamical Processes, S.L. Keil ed., Sunspot, NM 88349, p. 130.

Brandt, P.N., Scharmer, G.B., Ferguson, S., Shine, R.A., Tarbell, T.D., Title, A.M.: 1988, in print.

Cavallini, F., Ceppatelli, G., Righini, A., Alamanni, N.: 1986, Astron. Astrophys. 173, 161.

Damé, L.: 1985, in Theoretical Problems in High Resolution Solar Physics, H.U. Schmidt ed., Max Planck Institut für Astrophysik, München. p. 244.

Damé, L. Martic, M.: 1989, this Conference.

Deubner, F.–L.: 1974, Solar Phys. 36, 299.

————: 1977, Mem. Soc. Astr. Italiana 48, 499.

————: 1988, Astron. Astrophys., in print.

Deubner, F.–L., Fleck, B.: Astron. Astrophys., in print.

Deubner, F.–L., Laufer, J.: 1983, Solar Phys. 82, 151.

Gomez, M.T., Marmolino, C., Roberti, G., Severino, G.: 1987, Astron. Astrophys. 188, 169.

Koutchmy, S., Lebecq, D.: 1986, Astron. Astrophys. 169, 323.

Marmolino, C., Roberti, G., Severino, G.: 1987, Solar Phys. 108, 21.

————: 1988, in Physics of Formation of Fe II Lines Outside LTE, R. Viotti et al. eds., D. Reidel Publ. Comp. p. 217.

Nesis, A., Durrant, C.J., Mattig, W.: 1988, Astron. Astrophys., in print.

November, L.J., Simon, G.W.: 1988, Astrophys. J. 333, in print.

November, L.J., Toomre, J. Gebbie, K.B., Simon, G.W.: 1981, Astrophys. J. 245, L 123.

Oda, N.: 1984, Solar Phys. 93, 243.

Roudier, Th., Muller, R.: 1987, Solar Phys. 107, 11.

Schmieder, B.: 1976, Solar Phys. 47, 435.

204

Severino, G., Roberti, G., Marmolino, C., Gomez, M.T.: 1986, Solar Phys. 104, 259.

Simon, G.W., Leighton, R.B.: 1964, Astrophys. J. 140, 1120.

Staiger, J., Schmieder, B., Deubner, F.–L., Mattig. W.: 1984, Mem. Soc. Astr. Italiana 55, 147.

Stein, R., Nordlund, A., Kuhn, J.: 1989, this Conference.

Discussion

MÜLLER — You did not tell us where the data were obtained which you compared with our analysis of Pic du Midi white-light photographs.

DEUBNER — These were photographic spectra taken in September 1971, with the Vacuum Tower Telescope of the Sacramento Peak Observatory under excellent seeing conditions.

HIGH RESOLUTION DIAGNOSTIC OF THE MESOCELLS IN THE SOLAR TEMPERATURE MINIMUM REGION

M. MARTIC
Laboratoire d'Astrophysique Théorique du Collège de France
Institut d'Astrophysique de Paris
98 bis Bd Arago, Paris 75014, FRANCE

and

L. DAMÉ
Laboratoire de Physique Stellaire et Planétaire
Institut d'Astrophysique Spatiale d'Orsay
BP n° 10, Verrières-le-Buisson 91371 Cedex, FRANCE

ABSTRACT. The Transition Region Camera during its fourth rocket flight provided us with a set of UV filtergrams at a spatial resolution sufficient to resolve fine structures in the Tmin region. Seeing free and undistorted, these high resolution images of the solar surface provide a means for studing the weak mesogranular flow. Spatio-temporal analysis allowed to precisely determine the geometrical characteristics and evolution of the mesocells. The problem of the origin of the mesogranulation phenomenon is rised.

1. Introduction

The mesogranulation structure has been recently discovered and got rising interest in the studies of the solar atmospheric structures. Following the first observations of this secondary scale of the atmospheric organization by November et al. (1981), several related observations were carried using diferent techniques either through intensity measurements, Doppler shifts or by measurements of granular flows and, in all these studies the mesogranulation was evidenced as a cellular flow of characteristic 10-12 arcsec spatial extent. However, the evidence of the meso-scale structure in the photosphere and chromosphere (maybe even in the Transition zone at the CIV level, Dere et al., 1986) without subsequent scaling with

R. J. Rutten and G. Severino (eds.), Solar and Stellar Granulation, 207–215.
© *1989 by Kluwer Academic Publishers.*

height, and a possible influence of the magnetic field (Simon et al., 1988), rise the question of what determines the dominant horizontal scale of the pattern. Moreover the mesoscale flow is much weaker than the granular convective flow (thus explaining its later discovery and the necessity for high resolution, seeing free observations, e.g. Spaselab2/SOUP, Title et al., 1986) and the origin of the mesogranulation could not be just associated to another scale in the cascade of solar turbulent convection. In this paper we are further studing this pecular phenomena and its morphological and dynamical properties.

2. Observations

A sequence of solar UV images with an effective resolution of 0.8 arcsec was obtained during the fourth rocket flight of the Transition Region Camera (Damé et al., 1986). Detailed description of the instrument was given in Foing and Bonnet (1984) and in references therein. For this study we have used the filtergram sequence formed around the Temperature minimum region through the 156 and 169 nm interferential filters. However the limited time of the flight sequence (4.5 min), the calibration procedure (characteristic curve determined by two-filter method) permitted to use the whole sequence of 31 filtergrams (spaced in time by a maximum of 10 sec). The high dynamic of the unblurred images and the temperature sensitivity of the brightness in the far ultraviolet allowed to discerne even the faintest structures of the solar field (Fig.1).

3. Meso-scale Evidence in the Tmin Region

Visual inspection of the images reveals the weak structures in the interior of the supergranular cells. A spatial frequency analysis indeed confirmed the existence of the secondary (meso-scale) structure : power spectra show, beside the characteristic dimension of the supergranular cell, a clearly resolved (FWHM 4 Mm) peak at 8 Mm. This agrees well with previous analysis of the mesogranulation spatial extent obtained by a one dimension auto-correlation technique (Foing and Bonnet, 1984) or by direct displacement measurements of the granules (Title et al., 1986). The resulting spatial power spectra of the individual image and time sequence

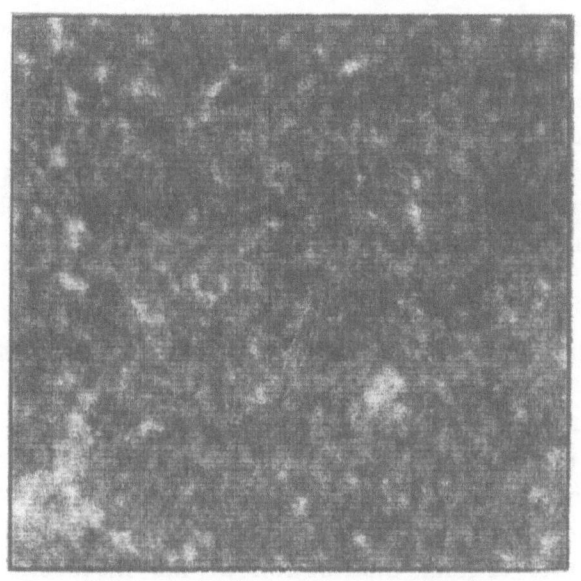

Fig. 1 - Quiet sun area of 105 x 105 arcsec^2 used for the present analysis (filtergram obtained through the filter centered at 156 nm).

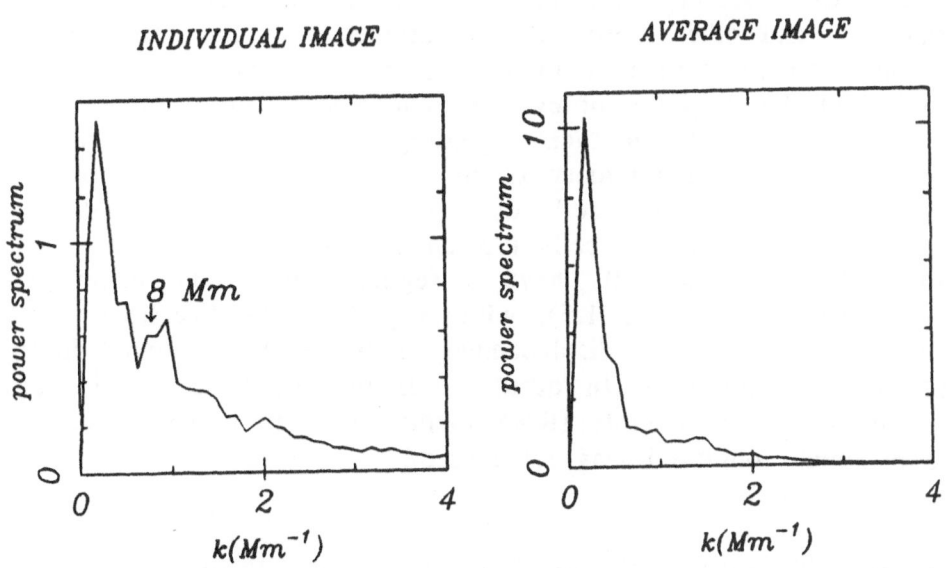

Fig. 2 - Spatial power spectra of the solar field; (a) for an individual filtergram; (b) for the time sequence average

average are shown on Fig. 2a and 2b. One can notice the disappearing of the mesostructure spatial signature on the average image. This time sequence (~ 5 min) is too short to indicate the meaning of absence of the 8 Mm peak but previous studies done on a 50 min time sequence of Ca II K filtergrams showed the same phenomenon (Damé and Martic, 1988). In other words the meso-structures present a time evolution that prevent their observation on a long time average.

4. Geometrical Properties and Distribution of the Mesocells

The topology of the mesocells is evidenced through a selective spatial filtering of the images. The convolution with bandpass spatial filter has been done in Fourier domain due to the defined bandwidth from the spatial frequencies spectrum. The filtered image shown in Fig. 3 illustrates rather regular distribution of the mesocells. In order to determine the geometrical properties and to quantify the boundaries of the mesocells and consequently to follow the time evolution of each cell the pattern recognition algorithm was applied on every filtergram of the sequence. The method of "Moments of Inertia" was used to calculate the excentricity, center of gravity, orientation and other morphological parameters of the detected cells. We have to mention that in our analysis we faced the standard problem of the detection of evolving phenomena (e.g identification of the granules known for their repeating fragmentation or grouping) and that using just intensity tresholding in the algorithm prevents the identification of the cells due to their subsequent disappearing. The analysis of detected cells (75 on the 105 x 105 arcsec^2 field) shows that the mesocells have a regular shape (the excentricity mostly between 0.5 and 1.5), without preferential directions of the axis and no significant displacement of the gravity centers through out the time sequence. In other words the brightening and fading cells do not move lateraly (RMS displacement is ± 0.36 arcsec and the maximum observed was 1.1 arcsec).

5. Dynamic Nature of the Brightness Fluctuation

The previous analysis showed that the mesocells have a regular shape, narrow distribution of the sizes and stable spatial location but

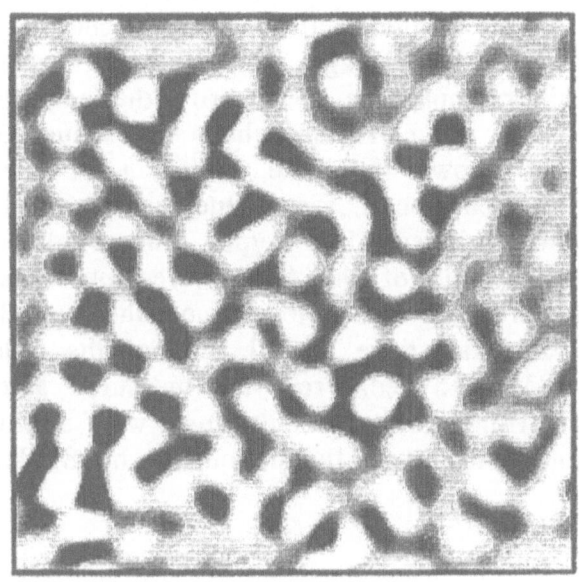

Fig. 3 - Illustration of the mesocells distribution on the filtered image (FWHM of the filter : 6 - 10 Mm)

MESOCELL TIME EVOLUTION

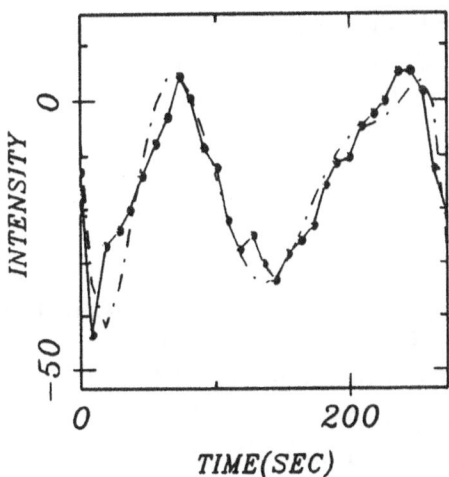

Fig. 4 - Time evolution of the intensity of a standard mesocell, integrated over a 5.5 arcsec diameter area (half wavelength of the observed phenomenon). The regularity of the oscillation is common to most identified cells.

also important variations of their intensity in time. Our sequence is short but consists of 31 calibrated filtergrams spaced in time by a maximum of 10 sec and the images are coaligned together better than 0.2 arcsec. This prompted us to quantify the amplitude of the brightness fluctuation of the cells throughout the sequence. Selected curve in the Fig. 4 shows the time variation of the intensity integrated over a reference cell area. One can notice periodical brightening of almost sinusoid-like type. The evolution curves established for all the cells of the field were approximated by a polynome of fourth order in order to deduce the period of these regular oscillations. Statistics done for all cells of the field gave an average period of 183 ± 15 sec. From the measured amplitudes of the intensity variations on the curves we evaluated the corresponding average temperature fluctuations to be 81 K.

6. Conclusion

The results of a morphological study on the evolution of mesocells presented above confirm the conclusions of the previous analysis in Damé and Martic (1987), based mainly on the evidence of the mesostructure by wave analysis at the chromospheric level. Our new results provide more precise informations (e.g stability of the center of the gravity within $\sigma = \pm 0.36$ arcsec and well defined 180 sec period of the oscillation of the detected cells) for theoretical interpretation of the meso-scale phenomenon (Martic and Damé, 1988)

7. References

Damé, L., Foing, B., Martic, M., Bruner, M.E., Brown, W.A., Decaudin, M. and Bonnet, R.M. : 1986, *Adv. Space Res.* 6(8), 272
Damé, L. and Martic, M. : 1987, *Astrophys. J.* **314**, L 15
Damé, L. and Martic, M. : 1988, in *Advances in Helio- and Asteroseismology*, eds. J. Christensen-Dalsgaard and S. Frandsen, IAU *Symp.* **123**, 433
Dere, K.P., Bartoe, J.-D. F. and Brueckner, G. E. : 1986, *Astrophys. J.* **305**, 947
Foing, B.H. and Bonnet, R.M. : 1984, *Astrophys. J.* **279**, 848

Martic, M. and Damé, L. : 1988, *Astron. Astrophys.*, submitted
November, L.J., Toomre, J., Gebbie, K.B. and Simon, G.W. : 1981, *Astrophys. J.* **245**, L23
Simon, G.W., Title, A.M., Topka, K.P., Tarbell, T.D., Shine, R.A., Ferguson, S.H., Zirin, H. and the SOUP team : 1988, *Astrophys. J.*, accepted for 15 April
Title, A.M., Tarbell, T.D., Simon, G.W., and the SOUP Team : 1986, *Adv. Space Research* **6**(8), 253

Discussion

KALKOFEN — What is the spatial scale of the 3-minute variations at the temperature minumum?

MARTIČ — We have not measured it directly at the temperature minimum level since the time sequence was too short for complete wave analysis, but we have studied the spatial phase coherence at the chromospheric level (Ca II K) using a longer filtergram sequence, and found a coherency extent of the phase cells of 8 Mm associated with the 3-minute oscillations[1].

SIMON — You have described intensity patterns (network) in the chromosphere, and you call them mesocells or mesogranules probably because the sizes of these structures are the same as the mesogranular velocity structures seen lower in the atmosphere. But these structures do not represent the same phenomenon. The structures that you see are undoubtedly co-spatial with magnetic structures which have been pushed to the boundaries of mesogranules and supergranules by the convective flows lower in the atmosphere. These magnetic features are generally vertical structures, and they extend upwards into the chromosphere where you detect them as loci of local heating and hence enhanced brightness.
I also find it rather surprising that the predominant scale size which you find is mesogranular (5–10 arcsec) rather than supergranular (25–35 arcsec). I wonder whether you have simultaneous Ca II K or H_α data, and whether they show the same spatial and temporal power spectra?

MARTIČ — The power spectra showed that the dominant scale is indeed the supergranular one. The meso-scale peak is a secondary one, a smaller but clearly resolved peak centered at 8 Mm (FWHM 6–10 Mm). The meso-pattern that I described has nothing to do with the chromospheric network. It is a smooth pattern isotropically distributed over the solar surface.
In fact, we have simultaneous Ca II K observations taken at Sac Peak during a rocket flight of the TRC, and they show the same spatial spectra. Moreover, we have studied an active solar region (plage) and got the same uniform distribution of the meso-cells (indicating that perhaps the magnetic field at this particular size is co-spatial with the meso-patterns) with enhanced brightness, but still showing regular, periodic variations of the intensity.

DEUBNER — The similarity of the spatial scale of mesogranulation in the photosphere and of the standing wave pattern above the temperature minimum may be purely accidental; but probably the upper layers do not see the granulation and the 5 arcsec size is just the smallest scale it senses from down below. As a resonant cavity it can only react with its own eigenfrequency to whatever happens underneath.

MARTIČ — It is probably true that the oscillations we observe on mesogranular scales are the response of the atmosphere to excitation down below in the photosphere, and that the 3-minute oscillations are characteristic of a chromospheric resonant

---

[1]L. Damé and M. Martič, 1987, Ap. J. 314, L15.

cavity coupled with a photosperic one.

However, (active) emerging granules at the center of mesocells (5 minute lifetime) are clearly linked with the 5-minute photospheric oscillation, of which the coherent extent may well be 8 Mm.

NORDLUND — There is a very simple explanation for these observations. The p-modes at $\nu \simeq 5$ MHz propagate just above the acoustic cut-off frequency. The propagation is then very sensitive to small changes in the properties of the surface layers. These properties are modulated by meso-scale convection, which consequently leaves an 'imprint' on the amplitude measured in the temperature minimum and chromospheric regions.

KNEER — The observations presented here are really admirable for their high spatial resolution and their excellent definition. However, before going into too much speculation I would suggest to perform a longer time sequence observation of 1 to 2 hours (the instrument must then of course fly on a satellite). I say so because our Ca II K filtergram observations (and similar ones) did show at first sight the chromospheric 3-minute oscillation, whereas the analysis of longer time sequences showed that the 3-minute oscillation 'dissolved' into a superposition of the 5-minute modes and acoustic waves detectable in our data at periods as short as 60 seconds.

DAMÉ — Let me comment: we have also used longer time sequences higher in the chromosphere of Ca II K (20 minutes and 50 minutes). They lead to the same kind of results.

The advantage of the TRC rocket flight data is their cleanliness, wich produces clear evidence of this prominent phenomenon of a 3-minute oscillation at a mesostructure of 8 Mm, and allows us to quantify the mesocell displacements without any ambiguity, and the amplitude of the intensity fluctuations as well. This does not require a longer time sequence (although it would be interesting), since the p-mode spectrum is not what we are looking for. The aim of using the TRC data was to quantify properly the dynamic properties of the mesocells, which clearly indicate their strong coupling with the oscillation phenomenon.

Pole-Equator-Difference of the Size of the Chromospheric Ca II - K - Network in Quiet and Active Solar Regions

H. Münzer[1],   A. Hanslmeier[2],   E.H. Schröter[1]   and   H. Wöhl[1]

[1] Kiepenheuer-Institut für Sonnenphysik, Schöneckstr.6,
    D-7800 Freiburg, Federal Republic of Germany
[2] Institut für Astronomie, Universität Graz, Universitätsplatz 5,
    A-8010 Graz, Austria

ABSTRACT.   The dependence of the size of chromospheric network cells on latitude was investigated for quiet and active solar regions. Calibrated photographic Ca II K - filtergrams were taken at the Schauinsland observatory of the Kiepenheuer-Institut during the period 1982 until 1984. Six overlapping images on 24 mm * 36 mm film were taken from the solar equatorial and meridional belts, respectively. A total of 53 series could be reduced.
The images were digitized with a stepwidth of about 1500 km in the solar center and 2400 km near to the limb. After transforming the densities into intensities a smoothing with a 4th order polynomial was performed. The deconvolution of the geometric shortening was done using a cylindric projection, which preserves the area.
Two-dimensional Fourier transformations were performed on subimages of 20 deg * 20 deg heliographical sizes to obtain the network cell sizes. The power spectra were corrected by a filtering for influences of seeing effects.
The two-dimensional power spectra were integrated over the azimuthal angles to obtain mean cell sizes. The mean at the equator being 28 700 (± 200) km from the east to west scans. Above 60 deg in latitude the cell sizes were found to be about 10 % smaller than at the equator. From the integration parallel and normal to the solar rotation it was found that the cell sizes are about 5 % larger perpendicular to the solar rotation than parallel to it.

Using the image intensity in the Ca II K line as a measure for the local solar activity the subimages were divided into subgroups with non active, low active and strong active regions. An increase of the cell sizes with locally increasing Ca II K intensity was found. The differences of cell sizes between non active and strong active regions amount to about 10 %.

The paper abstracted here is submitted to ASTRONOMY AND ASTROPHYSICS.

R. J. Rutten and G. Severino (eds.), Solar and Stellar Granulation, 217–218.

Discussion

FOX — Given that these observations are at a time of decreasing activity, have you attempted to obtain and compare the torsional oscillation data (shear pattern) as a function of time, to see whether any correlation exists between the supergranular scale flows and the true indicators of solar activity?

WÖHL — No, we have not done that.

SIMON — Your observation that the size of the network structure is larger in active regions seems to contradict the extensive earlier work of Zwaan[1], and also our small set of observations from Spacelab 2, which suggest that supergranules are smaller in active regions.

A reason for this may be that if the activity is strong enough, the convection cell can be completely suppressed. So you may be including regions in your measurements where there are 'missing' cells, and therefore your cells appear larger because they contain areas *without* a cell.

WÖHL — Perhaps we miss some cell borders which might happen in regions of low activity, e.g. in polar regions. Within active regions, however, the cell borders are pronounced, and the Ca II K network structure is more clearly detectable.

SCHRÖTER — I would like to add a comment.

The mean cell size was defined as the center of gravity of the power spectrum. Being aware that changes of the tail of the power spectrum at high k-numbers due to seeing etc. may shift the center of gravity, we used only those spectra for further reduction (filtering etc.) wich had similar high-frequency tails.

TITLE — How can you reconcile your result with that of Zwaan (*ibid.*), who finds diameters of $12–15.10^3$ km for network cells in plages?

WÖHL — There is a continuous increase of the Ca II K network cell sizes with increasing local activity. The strongest activity nevertheless shows only an increase by 10% to 15%, and not a variation by a factor two or so.

---

[1]C. Zwaan, 1978, Solar Phys. **60**, 213.

TEMPERATURE WAVES AND SOLAR GRANULATION

Y. D. Zhugzhda
IZMIRAN, USSR Academy of Sciences
Troitsk, Moscow Region
142092 U.S.S.R.

ABSTRACT. The problem of propagation and linear interaction of atmospheric and temperature waves in an atmosphere with stratified heat exchange has been considered. It is noted that atmospheric waves, for example five-minute oscillations, in an stratified atmosphere must generate temperature waves. It is suggested that temperature waves have been observed by Hill et al. Granulation must also generate temperature waves.

An adiabatic approximation, as a rule, is used by the theory of hydrodynamic waves in photosphere. Periods of photospheric waves are about or less than 5 minutes. Temperature disturbances relaxation time varies within the range of tens to hundreds seconds in photospheric layers. Consequently, an adiabatic approximation can't be applied to the theory of photospheric waves.

An isothermal approximation presuppose an instant relaxation of temperature disturbances ($\delta T = 0$). In this case gravity waves are impossible. For acoustic waves the specific-heat ratio equals 1 in this approximation. Hence using the both approximations one can't take into account two main properties of nonadiabatic waves in the solar photosphere.

The First property. The equation of nonadiabatic waves has second solution except that the heat exchange follows the Newton law. This solution discribes propagation of temperature waves (Landay, Lifshic, Hydrodynamics).

The Second property. The heat exchange in photosphere and convection zone varies considerably with depth. In the case of stratified heat exchange the linear interaction of hydrodynamic and temperature waves appears across the path propagation through atmosphere. Consequently, hydrodynamic waves in solar photosphere must generate temperature waves.

R. J. Rutten and G. Severino (eds.), Solar and Stellar Granulation, 219–223.
© 1989 by Kluwer Academic Publishers.

1. THEORY OF NONADIABATIC OSCILLATIONS

The study of wave propagation in an isothermal atmosphere plays a Key role in the theory of atmospheric waves. It is the consideration of an isothermal atmosphere that the classification and theory of adiabatic atmospheric waves are based upon. Souffrin (1966) considered the case of height-independent relaxation time of temperature disturbance in an optically thin atmosphere. In this case the coefficients in the wave equation are also constant and the problem is reduced to an analysis of a dispersion equation. Souffrin can't investigate the temperature waves and waves's interaction.

The case of depth-dependent relaxation time is considered by Zhugzhda (1983). The equation of non-adiabatic oscillations in an isothermal atmosphere is

$$\frac{d^2}{dz^2}\left(\frac{\delta Q}{\rho_o}\right) + \frac{1}{H}\frac{d}{dz}\left(\frac{\delta Q}{\rho_o}\right) + \left(\frac{\omega^2}{c_*^2} - k_\perp^2\right)\frac{\delta Q}{\rho_o} -$$

$$-\frac{i\gamma T}{(\gamma-1)T}\left[\frac{d^2\delta T}{dz^2} + \frac{1}{H}\frac{d\delta T}{dz} + \left(\frac{\omega^2}{c_o^2} + \frac{(\gamma-1)g^2 k_\perp^2}{\omega^2 c_o^2}\right)\delta T\right] = 0, \tag{1}$$

where c_o and c_* are the adiabatic and isothermal sound velocity respectively.

In the case of optically thin temperature disturbances

$$\delta Q = -\frac{\rho_o c_v}{t_R}\delta T , \tag{2}$$

where t_R - relaxation time.

In the case of optically thick disturbance

$$\delta Q = \lambda\left(\frac{d^2\delta T}{dz^2} - k_\perp^2 \delta T\right) + \frac{d\lambda}{dz}\frac{d\delta T}{dz} = div\,\lambda\nabla\delta T, \tag{3}$$

where λ is radiative conductivity.

Equation (1) has been solved for exponential dependence of relaxation time on depth. In this case oscillations are quasidiabatic in deep layers of the atmosphere and quasiisothermal in high layers. Temperature waves are generated in the both layers. The dispersion equation of the temperature waves in a quasiadiabatic area is

$$(k_z H_R)^2 = -i\gamma\omega t_R , \tag{4}$$

where H_R is scale of relaxation time depth-variation, t_R is relaxation time of temperature disturbance measures H, H is the scale of pressure. These waves are damped on the scale of a few wavelength.

The dispersion equation of the temperature waves in quasiisothermal area is

$$k_z^2 - k_z \left(1 - \frac{H_R}{H}\right)\frac{1}{H} - \left(k_\perp \frac{H_R}{H}\right)^2 = 0. \qquad (5)$$

These waves are surface temperature waves, which propagate horizontally and vary exponentially with height. For $k_\perp H \ll 1$ and $H_R \ll H$, $k_z \cong 1/H$.

Linear interaction of hydrodynamic and temperature waves is responsible for the generation of temperature waves by hydrodynamic waves as they pass through non-adiabatic area ($\gamma \omega t_R \sim 1$). Temperature waves penetrate into quasiadiabatic ($\gamma \omega t_R \gg 1$) and quasiisothermal areas ($\gamma \omega t_R \ll 1$).

Hydrodynamic and temperature waves in non-adiabatic layers ($\gamma \omega t_R \sim 1$) constitute an indivisible mixture. Neither hydrodynamic nor temperature waves can exist without each other. This property is shown in Figure.

Actually the boundaries of interaction layer are not sharp as shown in Figure. Interaction of waves decreases with the distance from layer of $\gamma \omega t_R = 1$.

Figure. Generation of evanscent (et) and damping (dt) temperature waves by acoustic waves (a) in the atmosphere with stratified heat exchange.

2. FIVE-MINUTE TEMPERATURE WAVES IN PHOTOSPHERE

The theory of five-minute oscillations has not so far taken into account interaction with temperature waves. The minimum relaxation time in photosphere is about tens seconds. Consequently, five-minute oscillations must generate temperature waves in photosphere. The distinguishing feature of the temperature waves as opposed to the five--minute oscillations in photosphere is travelling waves. It is likely that temperature waves in photosphere have been observed by Hill, Goode and Stebbins (1982). They observed travelling five-minute waves in photosphere. Hill et al interpreted this wave as gravity mode. It is difficult to agree to this interpretation because the gravity waves of such periods are unlikely to exist in photosphere due to fast heat exchange.

3. GRANULATION AND TEMPERATURE WAVES

If granulation is not a stationary convection, a generation of temperature waves must occur. Moreover, in this case granulation is an indivisible mixture of hydrodynamic and temperature disturbances. Consequently, correlation between granular velocities on different layers and intensity may be different from unity.

In any cases the temperature waves make a contribution to the velocity and intencity pattern of photosphere.

REFERENCES

Hill H.A., Goode P.R. and Stebbins R.T., 1982, Ap. J., 256, L17.
Souffrin P., 1966, Ann. Astrophys., 28, 463.
Zhugzhda Y.D., 1983, Astrophysics and Space Sci., 95, 255.

Discussion

WEISS — Have you considered the effect of perturbing a thermally stratified layer? In particular, will acoustic waves and temperature waves be destabilized by sub-adiabatic as well as super-adiabatic stratifications?

ZHUGZHDA — I consider only an isothermal atmosphere. In this case there are no instabilities. Instabilities must occur, however, in the case of a non-isothermal atmosphere.

DEUBNER — Can you provide order of magnitude estimates of the effects you are describing, in relation to those which follow from simpler, say adiabatic, calculations?

ZHUGZHDA — This is a general theory. An analysis of the observational effects that are connected with temperature waves is indeed required. I plan to do this investigation.

ULMSCHNEIDER — Mathematically, this is a nice derivation which is valid for small-amplitude waves. However, there are now nonlinear time-dependent large-amplitude acoustic wave calculations available, which are not restricted by Newton's law of cooling, which is valid only in the optically thin limit, but that employ the diffusion approximation which is valid just at the low heights where the radiative relaxation time t_{rad} is smallest. To treat cases where $t_{rad} << P$ (with P the wave period) with Newton's law of cooling is very dangerous, whereas at these low heights the diffusion approximation is valid. Moreover, I am not so happy with calling the case $t_{rad} << P$ a temperature wave. In calculations of acoustic waves in O-type stars, where the radiation damping is extremely large ($t_{rad} << P$), my student Wolf has shown that the wave has no temperature oscillation at all, but only pressure and velocity oscillations. Only when the wave passes the radiation damping zone does the temperature oscillation grow and does the wave develop into a strong shock, with a 3.10^6 K temperature jump.

ZHUGZHDA — I agree that the considered model is oversimplified. We cannot include radiation in its full form, and we cannot include the nonlinear effects. However, such general considerations help us to study *possible* effects of temperature waves in the non-adiabatic layer. When using non-analytical methods to solve the problem (i.e. by computation), it is possible to miss some effects, and it is also possible that such effects are not strong, and therefore not easily noted. Further computations are needed.

VAN BALLEGOOIJEN — What about gravity waves?

ZHUGZHDA — We only studied acoustic waves.

FLOWS, RANDOM MOTIONS AND OSCILLATIONS IN SOLAR GRANULATION DERIVED FROM THE SOUP INSTRUMENT ON SPACELAB 2

A.M. Title, T.D. Tarbell, K.P. Topka, S.H. Ferguson, R.A. Shine and the SOUP Team[1]

Lockheed Palo Alto Research Laboratory

ABSTRACT

The Solar Optical Universal Polarimeter (SOUP) on Spacelab 2 collected movies of solar granulation that are completely free from the distortion and blurring introduced by the Earth's atmosphere. Individual images in the movies are diffraction-limited (30 cm aperture) and are not degraded by pointing jitter (the pointing stability was 0.003" root mean square). The movies illustrate that the solar five minute oscillation has a major role in the appearance of solar granulation and that exploding granules are a common feature of the granule evolution. Using 3-D Fourier filtering techniques, we have been able to remove the oscillations and demonstrate that they dominate the temporal autocorrelation functions (ACF) of the granulation pattern. When the oscillations are removed the autocorrelation lifetime of granulation is a factor of two greater in magnetic field regions than in field-free quiet sun. Using a technique called local correlation tracking we have been able to measure horizontal velocities and observe flow patterns on the scale of meso- and supergranulation. In quiet regions the mean flow velocity is 370 m/s while in magnetic regions it is about 125 m/s. We have also found that the root mean square (rms) fluctuating horizontal velocity field in quiet regions increases from 0.45 to 1.4 km/s and in strong magnetic field regions it increases from 0.3 to 0.75 km/s as the measuring aperture decreases from 4 to 1 arc seconds. Combining the results from temporal and spatial ACF's with the velocity measurements, we conclude that the decay of the temporal ACF is due as much to motion and distortion of granules as to the lifetimes of elements in the pattern. By superimposing the location of exploding granules on the average flow maps we find that they appear almost exclusively in the center of mesogranulation size flow cells. The density of exploding granules is sufficient for their expansion fronts to cover a

---

[1](L. Acton, D. Duncan, M. Finch, Z. Frank, G. Kelly, R. Lindgren, M. Morrill, N. Ogle (deceased), T. Pope, R. Reeves, R. Rehse, R. Wallace, Lockheed Palo Alto Research Laboratory; J. Harvey, J. Leibacher, W. Livingston, L. November, H. Ramsey, National Solar Observatory; G. Simon, Air Force Geophysics Laboratory)

R. J. Rutten and G. Severino (eds.), Solar and Stellar Granulation, 225–252.

typical mesogranule in 900 seconds. Because of the non-uniformity of the distribution of exploding granules, the evolution of the granulation pattern in mesogranule cell centers and boundaries differs fundamentally. It is clear from this study that there is neither a typical granule nor a typical granule evolution. Even after the solar oscillations have been removed, a granule's evolution is dependent on the local magnetic flux density, on its position with respect to the active region plage, on its position in the mesogranulation pattern, and on the evolution of granules in its immediate neighborhood.

1   Introduction

In recent years there have been excellent reviews of granulation by Wittmann (1979), Beckers (1981) and Bray, Loughhead, and Durrant (1984). However, a number of what are thought to be significant properties of the granulation, for example, lifetime, size, and evolution (Kitai and Kawaguchi 1979, Oda 1984, Kawaguchi 1980, Roudier and Muller 1987) are still under active discussion. Although the majority of astronomers believe that granules are thermal plumes overshooting from the convection zone into the photosphere (Bray *et al.* 1984, Nordlund 1985), there is some evidence that they are turbulent eddies (Zahn 1987). The relative contribution of wave and convective motion to the velocity field of the photosphere is still controversial (Beckers 1981, Nordlund 1985, Cox 1987). At present it is not possible to answer the question: "what is a typical granule?".

The lack of good movies has aggravated the objective definition of what constitutes a granule and granule evolution. The traditional statistical tools for characterizing intensity fluctuations in the photosphere side-step this problem. Unfortunately, they also include other physical processes and instrumental effects unrelated to granules. Spatial autocorrelation functions (ACF's), their Fourier transforms and intensity power spectra are all compromised by lack of knowledge of the combined modulation transfer function of the atmosphere, telescope, and detector system (Bahng and Schwarzschild 1962, Schmidt *et al.* 1981, Nordlund 1984). The SOUP data are only partially free from these problems. Temporal ACF's do not separate decorrelations due to size and shape changes from proper motions, nor do they distinguish changes due to the convective flows from other patterns, such as oscillations and waves.

A great deal of insight into defining granulation has been gained from the SOUP movies which are free from the blurring and distortions introduced by the Earth's atmospheres. They are qualitatively different from any previous groundbased movies. From the first viewings, it became clear that oscillations played a major role in the appearance of the photospheric surface, and that exploding granules were much more common than had been suggested previously (Title *et al.* 1987), and also that there are regions of ordered flow (November *et al.* 1987) and that magnetic fields affected the time history of granulation (Simon *et al.* 1988).

There is no evidence from these movies that a single "typical" granulation evolution pattern exists. Rather, the surface of the Sun has brightness fluctuations which are caused by a combination of convection, turbulence, waves, and magnetic fields. It is impossible to separate these effects in a single image, and the first steps in effecting the separation in movie data are reported here.

We have been able (section 2) to remove a great deal of the solar oscillations by 3-D Fourier filtering. When these oscillations are removed, quiet and magnetic regions differ significantly in temporal history. The differences between quiet and magnetic regions are also visible in their average intensities (section 3), in their root mean square (rms) intensity fluctuations, and in their quasi-steady horizontal velocities (section 4). The accurate estimate of the autocorrelation lifetime obtained from Fourier filtered images coupled with horizontal velocities obtained from local correlation tracking show that the temporal change in the temporal ACF is dominated by the displacements rather than by the lifetimes of the individual granules.

Once the oscillations are removed the role of exploding granules becomes much more evident. In the SOUP data at least half of the solar surface is affected by the circle of expansion of an exploding granule (section 5). Often, a granule is terminated (e.g. squashed) by a neighboring exploder. Further, these exploders are concentrated in meso- and supergranule interiors, so that their effects are not uniform. Due to this fundamental non-uniformity of distribution, lifetimes of individual granules are often determined by events in their local neighborhood, which in turn depends on their position in the mesogranulation pattern.

2 Temporal Autocorrelation Measurements

We define the correlation lifetime of granulation by the $1/e$ width of temporal autocorrelation function (ACF), where the temporal ACF of the intensity is defined as

$$ACF(\tau) = \frac{\langle \delta I(x,y,t_0)\, \delta I(x,y,t_0+\tau) \rangle}{\langle \delta I^2(x,y,t_0) \rangle}$$

where x and y are the spatial coordinates, t_0 is the time, and τ the time separation. The brackets indicate an average over space and time, and

$$\delta I(x,y,t) = I(x,y,t) - \langle I(x,y,t) \rangle.$$

The ACF has been calculated for granulation in quiet sun regions that are essentially free from strong magnetic fields, and for granulation in the magnetic field regions in the vicinity of the active region. The locations of quiet and magnetic areas were determined with the BBSO magnetograms by carefully registering features visible on both the SOUP and BBSO images. Small image areas (24×24 pixels, called boxes) were then chosen to be either outside (quiet sun) or inside (magnetic areas) of the magnetic contours. The sunspots and pores were avoided. Results for larger areas were obtained by averaging ACF's from several boxes. Figure 1 shows (a), a quiet sun image, and (b), a pore region image. The sample boxes used to calculate the ACF and the BBSO magnetogram contours are indicated as well.

The ACF's from two quiet sun sequences are shown in figure 2. The quiet sun correlation lifetime is about 5 minutes. Measurements from a number of other studies (Mehltretter 1978) are also included on figure 2. As can be seen from the figure, both the shape and the lifetime of the SOUP data are in agreement with the previous values.

The ACF's from the individual boxes indicated in figure 1, which cover only 36 square arc seconds, are qualitatively different from their sum. Figure 3 (solid) shows the ACF's generated from four of these boxes in regions of quiet sun. These ACF's do not drop monotonically, but they rather exhibit oscillations with periods of 3 to 6 minutes (depending on the sample). This suggests that solar oscillations are affecting the correlation lifetime, which is not too surprising as they are strongly visible in the granulation movies.

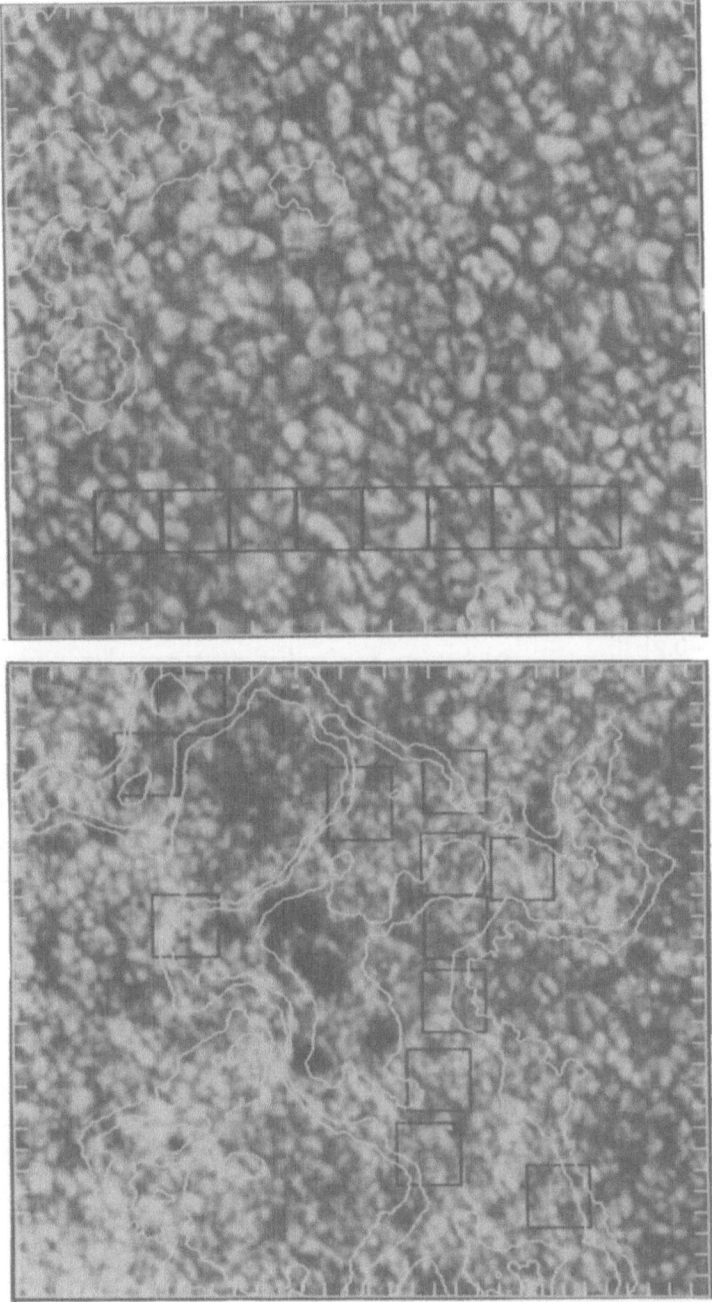

Figure 1: SOUP images of (a) a quiet sun region and (b) a region surrounding pores. The boxes define the areas for which the temporal autocorrelation function was calculated. Iso-gauss contours from a registered Big Bear Solar Observatory magnetogram (contours approximately ±50 G, ±100 G, and ±200 G)are also shown. The tick marks are 2″ apart.

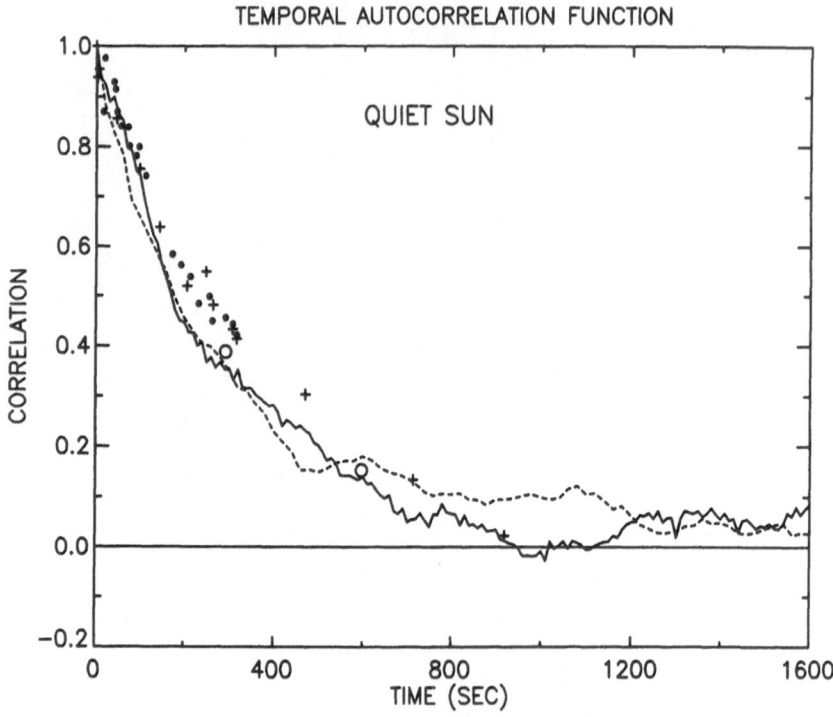

Figure 2: Autocorrelation functions from 2 quiet sun regions, original data. Solid line: 0.161″/pixel image sequence, dashed line 0.134″/pixel image sequence. Superimposed are autocorrelation measurements from previously published results for comparison (dots, crosses, and open circles).

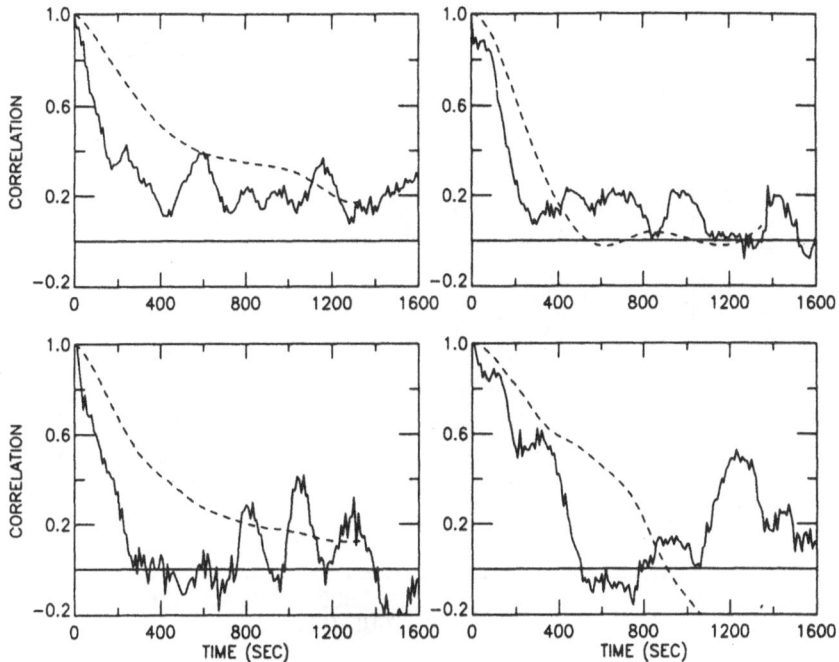

Figure 3: Autocorrelation functions in four $6'' \times 6''$ boxes of quiet sun from an original image sequence (solid), showing the effects of intensity oscillations. The dashed curves are the results for the areas outlined by the boxes in fig.1, after 3-D Fourier filtering of the image sequence which removes the effects of oscillations.

To remove the effect of oscillations, we have applied what we call a subsonic Fourier filter. A raw image sequence, which can be considered to be a single 3-D function of intensity versus x, y, and t, is Fourier transformed into a function of k_x, k_y, and ω. The subsonic filter is defined by a cone

$$v_p = \frac{\omega}{k}$$

in $\vec{k}-\omega$ space, where $\vec{k}$ and ω are spatial and temporal frequencies and v_p is the maximum phase velocity. All Fourier components inside the cone (i.e., with phase velocities less than v_p) are retained, while all those outside are set to zero. Then a new sequence of images is calculated by an inverse Fourier transform. For the subsonic filters, values of v_p of 3, 4, and 5 km/s, well below the 7 km/s sound speed were used. For these values of v_p, the velocity cone is totally inside the region occupied by the solar f- and p-modes.

A subsonic filtered movie exhibits very little solar oscillation, and the same is true for the ACF's formed from subsonic filtered data. The dashed curves in figure 3 are for the same areas as the solid curves, but they have been derived from subsonically filtered ($v_p = 3$ km/s) data rather than the original data.

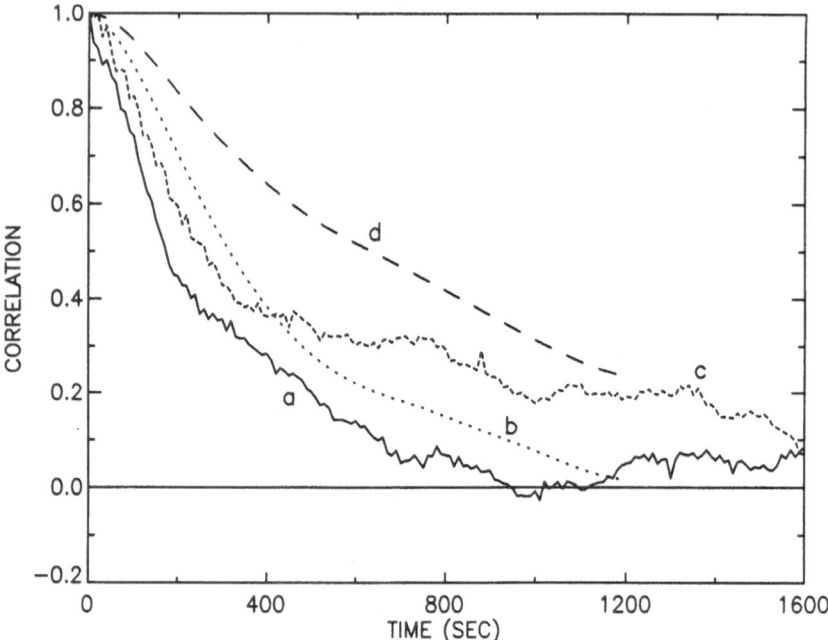

Figure 4: Autocorrelation functions for (a) quiet sun original data and for (b) the same data Fourier filtered to remove the effects of solar oscillations. Figure 1 (c) is the autocorrelation function for magnetic region original and (d) shows the same data as (c) after Fourier filtering.

Figure 4 shows a comparison between the ACF from 300 square arc seconds of quiet sun for (a), the original, and (b), the subsonic filtered image sequences. The quiet sun ACF lifetime of the subsonic data is one and a half times greater than the original—a lifetime of 410 versus 270 seconds.

We have investigated the effects of our choice of subsonic filter, and whether or not an abrupt cutoff has any important consequences. We also repeated the subsonic filter and ACF calculations for values of v_p of 4 km/s and 5 km/s. Regardless of the filter selection, the ACF lifetime that results is within 10% of the original 3 km/s filter value.

Figure 5 shows how the ACF of the original data changes with spatial resolution. The solid curve (a) is the original data ACF, and the dashed curve (c) the ACF calculated from data degraded in resolution to $1''$ by smearing the original frames. Smearing was performed by first calculating the Fourier transform of each image in a sequence, followed by filtering in the spatial frequency domain with the appropriate circular aperture diffraction pattern, and then performing the inverse Fourier transform. The insignificant difference caused by loss in resolution probably explains the consistancy of the ACF results obtained by different observers at many observing sites (see figure 2).

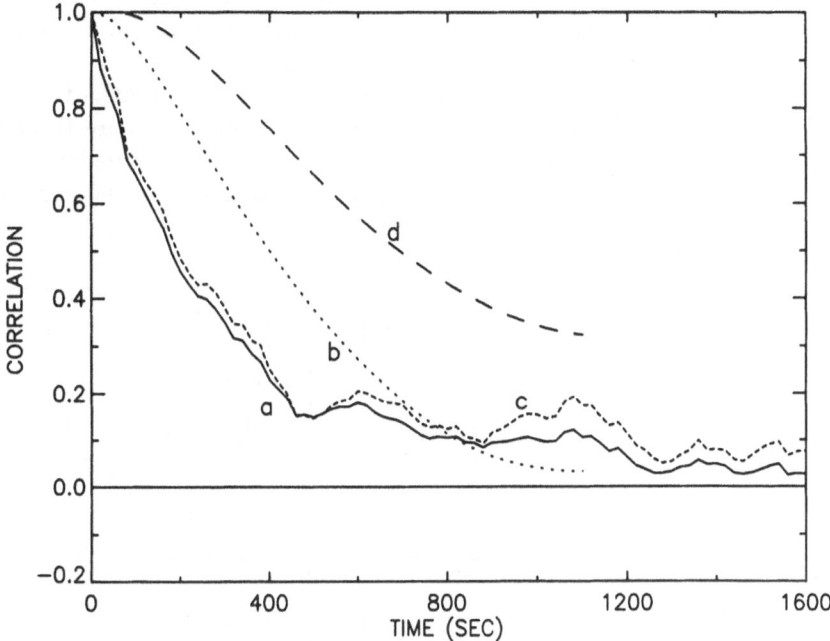

Figure 5: Autocorrelation functions from a $o.143''$/pixel quiet sun sequence for (a) original data, (b) the same data Fourier filtered to remove the effects of solar oscillations, (c) the same data as (a) but smeared to $1''$ resolution, and (d) the same data as in (b) but smeared to $1''$ resolution.

Because neither the p- or the f-modes have significant power on the scale of an arc second, blurring the data should have little effect if the ACF's are dominated by solar oscillations.

On the other hand, granulation does have significant power at the scale of an arc second, so that blurring subsonic data should have a significant effect. Figure 5 (d) also shows ACF's that are calculated from blurred subsonic filtered data. Here, the lifetime of the degraded data (dashed) (d) is doubled in comparison with the unblurred subsonic data (dotted) (b). Taken together, figures 1 through 5 demonstrate that the solar oscillations add a dominant contribution to the temporal ACF's and that the characteristics of the granules only become apparent when the oscillations are removed. It also suggests that the differences between quiet and magnetic sun may be masked by the oscillations present in unfiltered data.

For a long time, there have been suggestions that granulation in magnetic regions is different. Indications of this difference that have been put forward are a reduction in contrast in magnetic regions—"abnormal granulation" (Dunn and Zirker 1973), an increase of the granule number density in magnetic regions (Macris 1951), and differences in the shape of the line bisector in photospheric lines in magnetic regions as compared to quiet regions (Bonet et al. 1984). However, this evidence has not been convincing because of the variations of granule properties reported by various observers.

There is also evidence that the mean bisector shape varies with solar cycle (Livingston 1984), and that perhaps the granule density changes with solar cycle (Macris and Rosch 1983).

To observe the effect of magnetic fields on the granulation pattern ACF, twelve boxes were selected within the 70 gauss contour of the magnetograms (see figure 1b), but well outside of pores. The magnetic flux contained within these 12 regions varied from 3×10^{19} Mx to 8×10^{19} Mx. The ACF's formed from original (solid) and subsonic filtered (dashed) data for four of these boxes are shown in figure 6 .

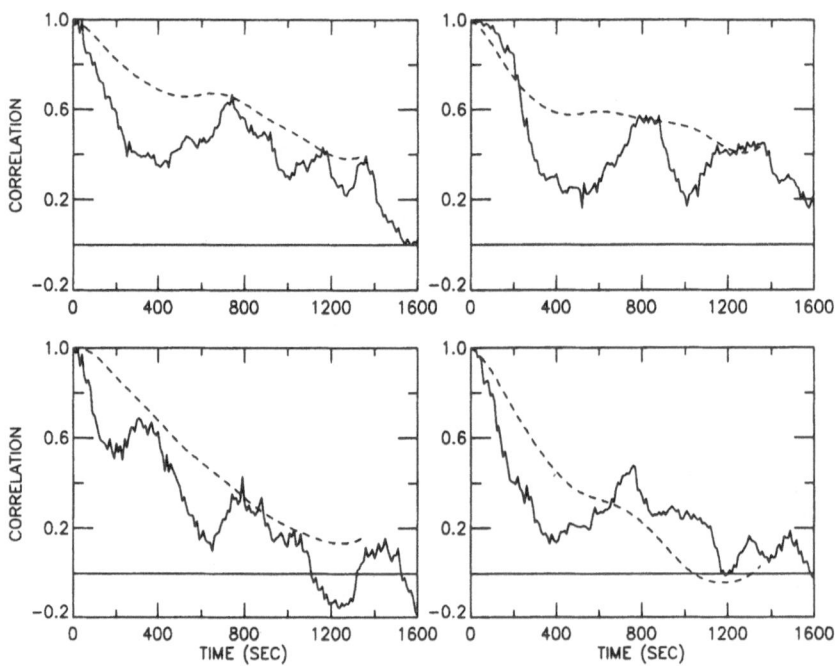

Figure 6: Autocorrelation functions in four 6×6 arc second regions of magnetic regions from original (solid) and Fourier filtered (dashed) to remove the effects of solar oscillations.

Figure 4 shows a comparison of the summed (all 12 regions) ACF's from original (c) and subsonic filtered (d) data in the magnetic areas. The lifetime in a magnetic region is more than a factor of two greater in subsonic than original data—890 versus 420 seconds. Bahng and Schwarzschild (1962) also compared the ACF's in quiet and magnetic areas, but saw no difference.

Figure 4 summarizes the effects of oscillations and magnetic fields on the temporal ACF. The original data (a) in quiet sun and (c) in magnetic regions differ somewhat, while the subsonic filtered data show substantial differences between (b) the quiet sun and (d) magnetic areas. The results are independent of the details of the Fourier filter used.

3 Intensity Fluctuations

The mean intensity of granules in the SOUP passband increases in magnetic field areas, while both oscillations and granule scale intensity fluctuations are reduced in time. The mean intensity and the rms intensity fluctuation over the 28 minute observing interval have been computed in the pore and quiet sun sequences, for both original and subsonic filtered data. The mean intensity image for the pore region clearly shows areas where the granules are brighter, and comparison with a registered BBSO magnetogram taken simultaneously reveals that in these areas magnetic fields were present.

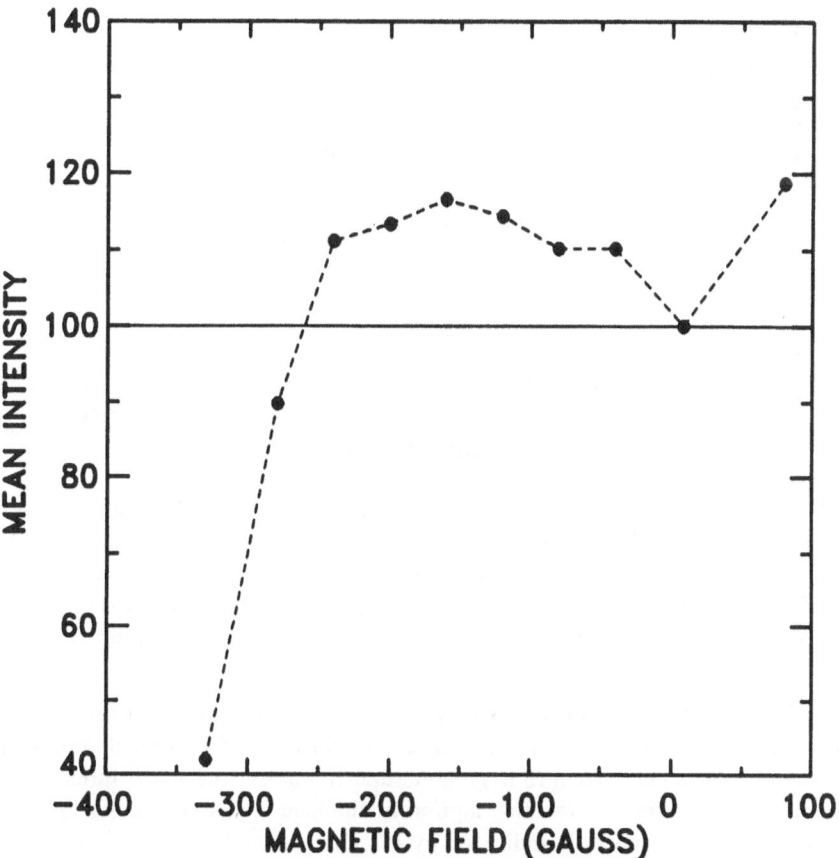

Figure 7: Plot of the mean value of intensity in the photosphere as a function of magnetic field strength. The intensities are from an average over 83 images of the pore region spanning 28 minutes. The magnetic field strength is from a registered BBSO magnetogram.

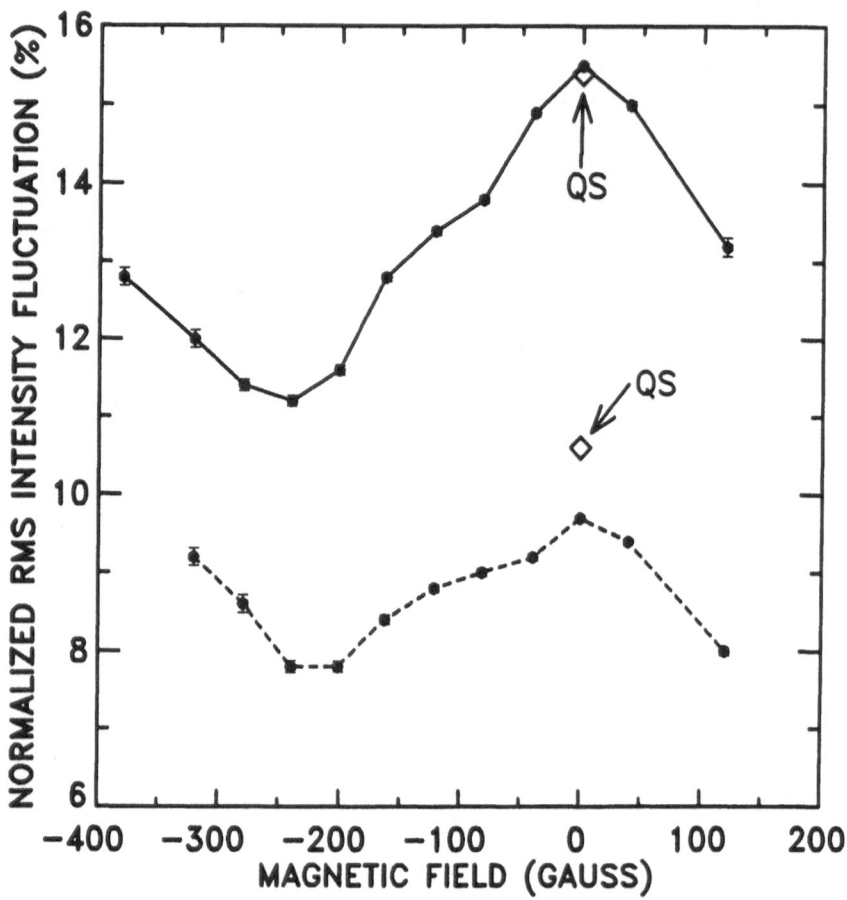

Figure 8: Plot of the mean value of the rms intensity fluctuation in the photosphere as a function of magnetic field strength, for original data (solid), and for Fourier filtered data (dashed). As in figure 7, the rms intensity fluctuations were calculated from a 28-minute long SOUP image sequence centered on the pores. The two diamonds labelled 'QS' indicate the mean rms fluctuation at zero magnetic field for a SOUP quiet sun region.

Figure 7 quantifies this result by showing the mean intensity versus the magnetic field strength. The mean intensity at zero field has been normalized to 100. Significantly, as soon as the magnetic field strength is greater than the noise, the mean intensity of the granules clearly increases and remains high until the field is strong enough for pores to form. The formation of pores is indicated by the sharp drop in mean intensity at -275 G.

Comparison of the rms intensity fluctuations from the original and subsonic filtered image sequences shows that intensity fluctuations are considerably reduced in the subsonic filtered images. Figure 8 shows the mean normalized rms intensity fluctuation versus magnetic field strength for original data (solid) and subsonic filtered data (dashed). The values of the rms fluctuations in both sets of quiet sun data are also shown (diamonds). The rms intensity fluctuation is always largest at zero magnetic field, and then it declines with increasing field strength until pores form. Note that the point at which figure 8 shows an upturn at the magnetic field value where figure 7 shows a downturn. Pores show significant intensity fluctuations because any change in a pore's size or shape will cause an intensity fluctuation due to the contrast between the pore itself and the surrounding bright granules. These variations within the pore region are consistent with our subjective impression of the "vigor" of the granulation in the movies.

The BBSO magnetogram used in the construction of figures 7 and 8 has a resolution somewhat worse than 2″. This limits both the accuracy of its registration to the SOUP images and our knowledge of the spatial distribution and strength of the magnetic flux. The existence of the relationships shown in figures 7 and 8 therefore suggests that the presence of a magnetic flux tube affects an area of granulation that is considerably larger than the diameter of the flux tube.

The diamonds in figure 8 indicate the values of the rms intensity fluctuations for quiet sun. The average fluctuation for quiet sun (subsonic filtered) is about 10% higher than for the corresponding value from the zero field areas in the plage regions surrounding.the pores. Therefore, even the granulation in zero field regions in plage differs from field free quiet sun. Magnetic effects are not limited to filligree in intergranular lanes.

4 Horizontal Flows via Local Correlation Tracking

It is apparent from the movies that granules both move randomly and are advected by meso- and supergranulation scale flows. We have measured the horizontal flow velocities by a technique we call local correlation tracking (November et al. 1987). Briefly, a coarse rectangular grid of points is chosen which covers the image field-of-view. A spatial cross-correlation between two images (typically 60 seconds apart in time) is computed in the neighborhood of each grid point. The position of maximum cross-correlation yields a local displacement vector for each grid point. We interpret these displacements as horizontal velocities. The neighborhood of each grid point used for the comparison is defined by a gaussian apodizing function, and the full width at half maximum (FWHM) of this gaussian represents the spatial resolution of the velocity measurement.

We have produced movies of the flow field in this manner by applying local correlation tracking to a time series of image pairs. Composite movies have been created by overlaying the flow field on the original and subsonic filtered data. These movies clearly show that granules are advected by the flow field, that magnetic regions have smaller steady flows than quiet areas, and that meso- and supergranular scales are visible in the displacement field of the granules (Simon et al., 1988). These larger scales became clear by averaging the velocity over the duration of the movie, since it suppresses motions of individual granules and effects of five-minute oscillations. Figures 9 and 10 show the average velocity field overlayed on (a) an image and (b) the divergence of the average velocity field for the quiet sun and the pore region respectively. The two-dimensional divergence of the horizontal flow is roughly proportional to the vertical flow (November et al. 1987).

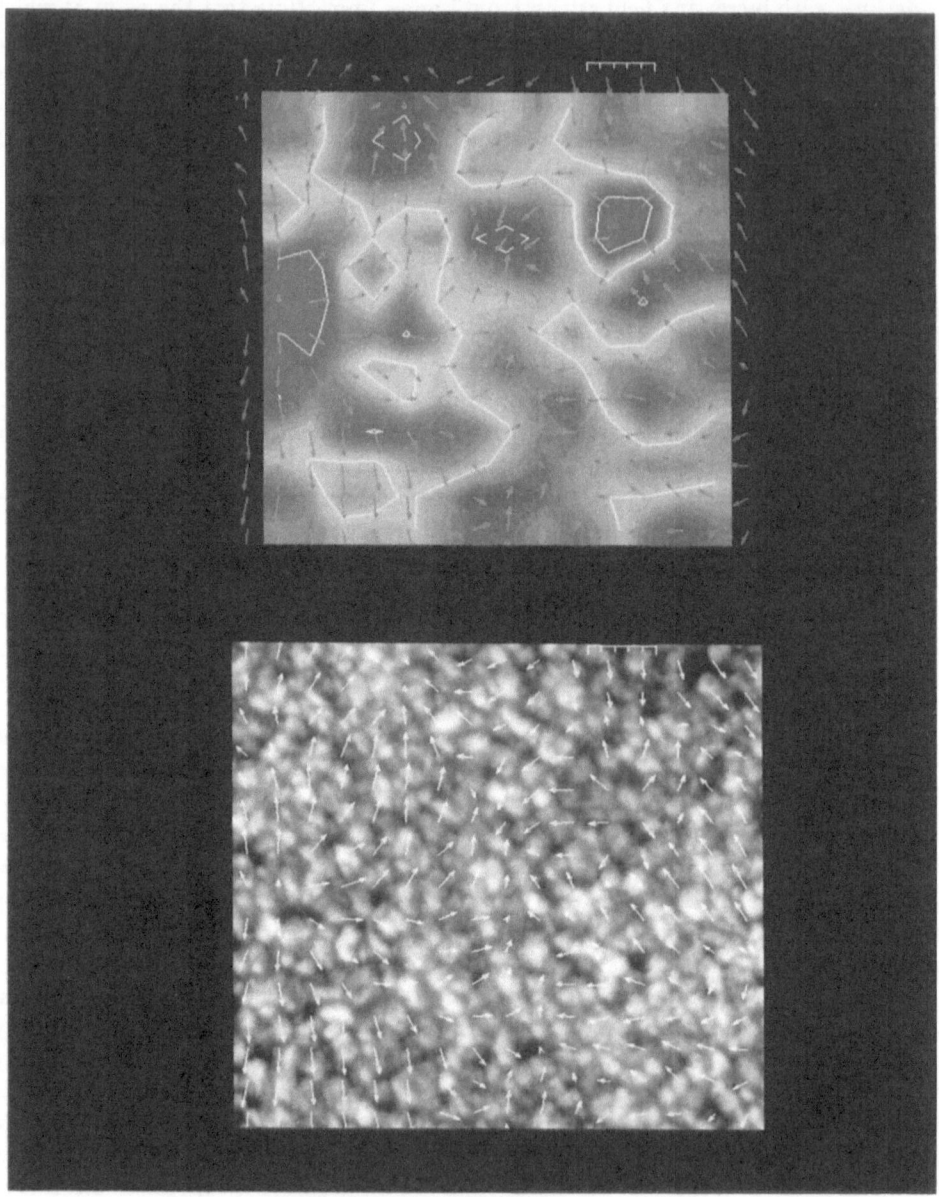

Figure 9: Image from quiet sun sequence (0.161″/pixel) with superimposed arrows (a) indicating direction and magnitude of surface flows measured from local correlation tracking (see text). The divergence of the flow field is shown in (b). Note that the color convention used (red for outflowing cells, blue for inflowing regions) is opposite to what is traditionally used for Doppler flows.

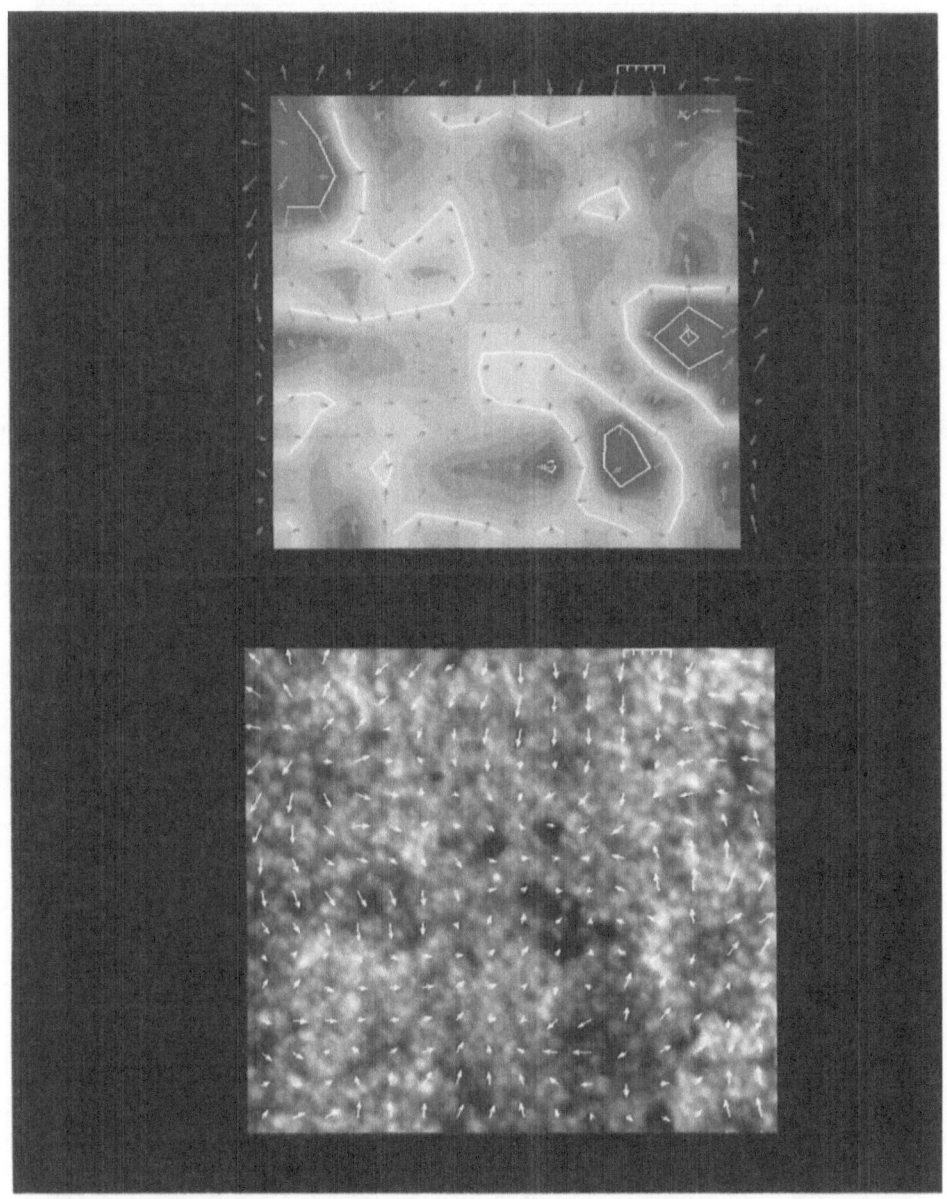

Figure 10: Pore region image (a-0.234″/pixel), with average horizontal flow field super-imposed, and (b) the divergence of this flow field with the average horizontal flow field superimposed. The flow speeds are slower in magnetic areas than in quiet sun.

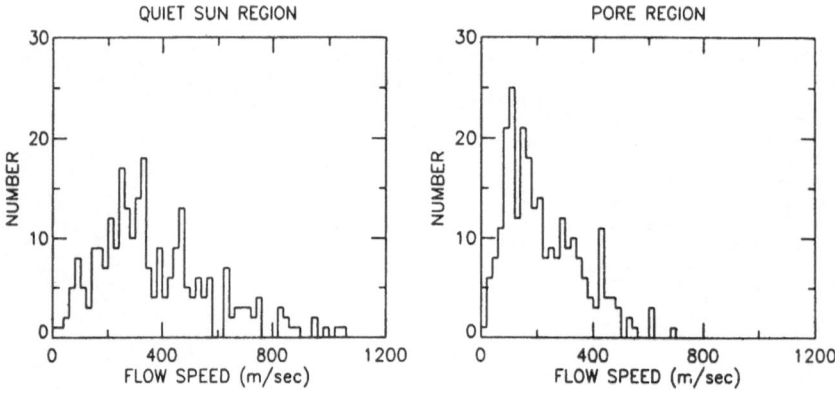

Figure 11: **Histograms** showing the distribution of measured flow speeds in (a) the quiet sun region and (b) the pore region.

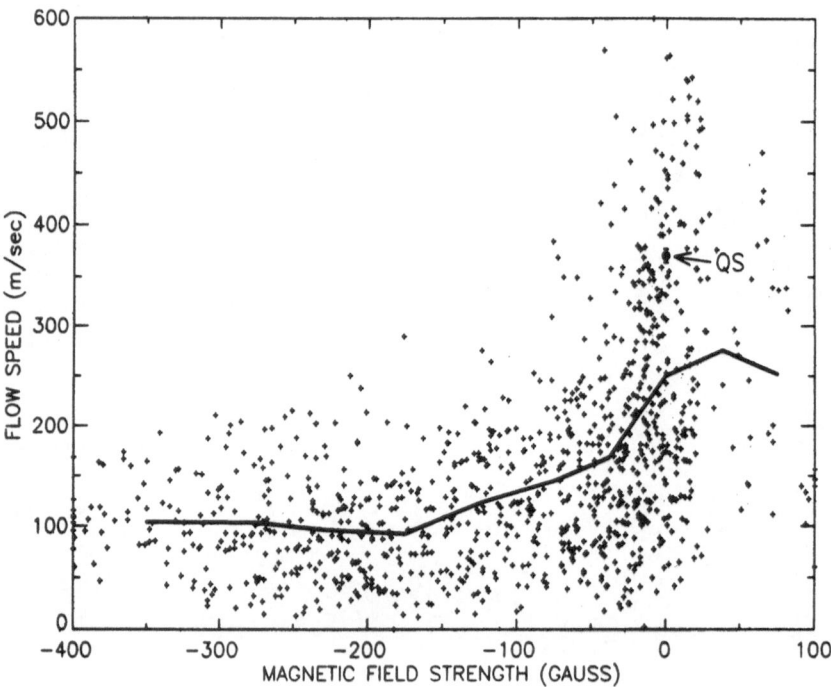

Figure 12: Scatter plot showing the flow speed versus local magnetic field strength in the pore region (0.134″/pixel sequence). Also indicated are the mean flow speed (solid line) as a function of field strength and the mean value from the quiet sun region (dot labelled 'QS' at 0 gauss).

The mean flow speed (magnitude of the velocity vector) is 370 and 163 m/s in the quiet sun and pore region sequences, respectively. Figure 11 shows the histogram of the speeds in (a) the quiet sun and in (b) the pore region. Figure 12 shows a scatter diagram of the flow speed versus magnetic field strength. The velocities for figures 9 through 12 were calculated using a gaussian with FWHM of 4″ and a grid spacing of 2.5″. This mask size was chosen to suppress the local expansion and collapse of individual granules, but still show meso- and supergranule scale flows. The magnetic field strength is the average from an 8 × 8 pixel area centered at the location where each horizontal velocity vector was calculated. Also marked in figure 12 is the mean flow speed (solid) as a function of magnetic field and the average speed in quiet sun (dot). The mean speed is strongly dependent upon magnetic field; it varies from 275 km/s at zero field to 100 m/s for field strengths greater than 175 G. Significantly, the quiet sun average flow speed is 35% greater than the flow speed in zero field areas near the pores. This implies that the magnetic field can affect surface flows for a considerable distance outside the flux tubes themselves. Figures 11 and 12 show that there is a difference between quasi-steady flows in quiet and magnetic regions, and that, like the intensity fluctuations, these flows are also different in granulation in zero or low field areas in plage compared with field-free quiet sun.

Local correlation tracking cannot only be used to measure quasi-steady flow fields, but also to estimate averages of the instantaneous velocities of the granules. The horizontal velocity vector at a single grid point shows fluctuations with time due to granulation, oscillations, waves, and random noise. To separate the noise from the solar signals, we have calculated the Fourier power spectrum of the x and y components of the velocity in all of the points in the image. From the average power spectrum, the white noise level is easily identified; the rms fluctuating solar velocity and the rms noise can then be calculated. We have measured the velocities using a set of gaussian masks of FWHM of 4″, 2″, 1.5″ and 1″. The smaller masks show more of the random motions and of the rates of expansion and the collapse of individual granules than the larger masks. Unfortunately, the rms noise increases as the width of the mask decreases, so that it is not possible to measure rms velocities at the resolution limit of the images. Plots of rms speed as a function of gaussian FWHM are shown in figure 13 for quiet region, non-magnetic areas of the pore region (magnetogram < 40 Gauss), and medium and strong magnetic areas of the pore region (100 to 400 Gauss and > 400 Gauss respectively).

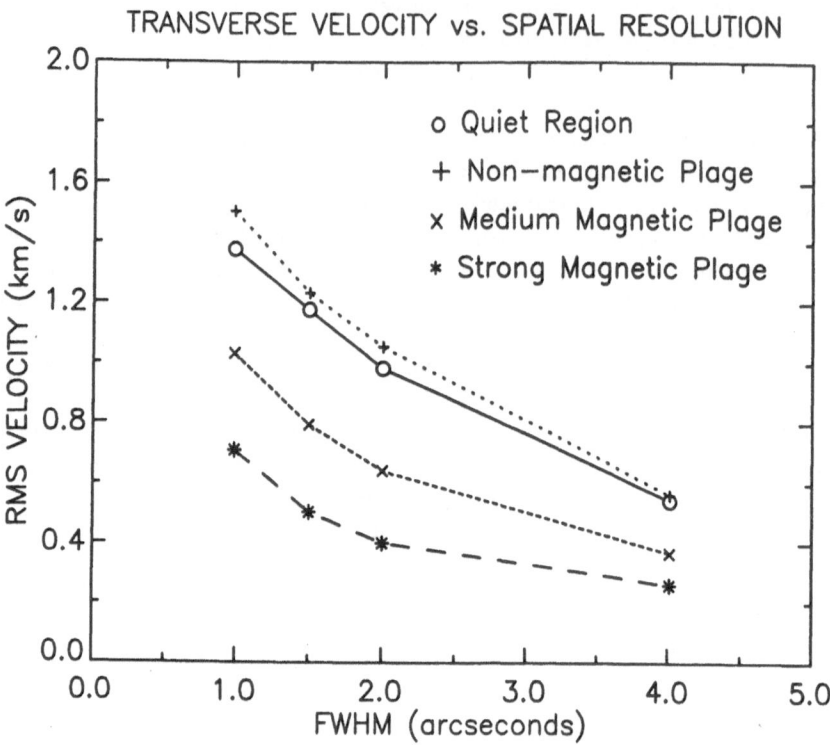

Figure 13: Plot showing RMS velocity versus the FWHM of the gaussian distribution used in the local correlation technique. See text. Data for different regions are designated by open circles (quiet sun), plusses (plage), crosses (medium magnetic plage), and asterisks (high strength magnetic plage).

The rms velocities plotted in figure 13 represent the fluctuating flow at a single point, as measured in different regions and with different spatial resolutions. The sharp increases at finer resolution must be due to granulation, since most power in the five minute oscillations occurs at larger spatial scales. This is borne out by movies of the flow vectors overlaid on the images, in which exploding granules are especially obvious. In strongly magnetic areas, the velocities are systematically 45% less than quiet sun values. Thus, magnetic fields strongly suppress the vigor of the granulation on 1″ scales. Non-magnetic locations within the plage are essentially the same as quiet sun far from strong fields, which implies that the magnetic fields do not effect the vigor of granular flows more than a few seconds beyond the boundaries of the flux tubes themselves.

From figure 13 the rms speed in quiet sun is at least 1.4 km/s. This is quite high compared to previous measurements of the vertical component of photospheric velocities (Beckers 1981).

It is possible that the local correlation tracking technique is measuring phase velocities as well as mass motion. However, the convective component of the flow is driven by
thermal buoyancy, and radiative losses should cause the brightness front to lag behind
the gas motion. Numerical simulations by Nordlund (private communication) verify that
brightness fronts lag behind material flows. Thus, our velocities may even be underestimated. Details of the relations between brightness and mass flows will be discussed in a
separate paper. Nonetheless, the speeds measured here furnish a description of how the
temporal ACF's evolve.

The temporal ACF's compare sets of spatially aligned data as a function of time and
yield information related to feature lifetime. Spatial ACF's compare the same image
with different spatial offsets and yield information about the spatial scale of the image.
The spatial ACF is defined as

$$ACF_s(\Delta) = \frac{\langle \delta I(x,y,t)\, \delta I(x+\Delta,y,t) \rangle}{\langle \delta I^2(x,y,t) \rangle}$$

where Δ is the spatial offset. The spatial ACF can be used to estimate the effect of flows
on the autocorrelation lifetimes because the displacements can be considered to be the
results of a flow. A velocity, v, which lasts of a time, t, will cause a displacement, $\Delta = v \times t$. We define a displacement lifetime, T_d, as the time required for the rms flow v_{rms}
to cause the spatial ACF to drop to $1/e$:

$$ACF_s(v_{rms}/T_d) = 1/e.$$

Figure 14 shows (a), the spatial ACF for subsonic filtered quiet sun, and (b), the
spatial ACF for magnetic regions. From the figure the $1/e$ width is 450 km and 500 km for
quiet and magnetic regions respectively, which is consistent with previous measurements.
On average, structures in the quiet and magnetic regions are about the same size in the
SOUP images. While we have shown that granules in the magnetic region are different
from quiet granules, the spatial ACF is not a good discriminator of that difference.
Nevertheless, the eye can see a subtly different texture in the magnetic parts of the pore
region.

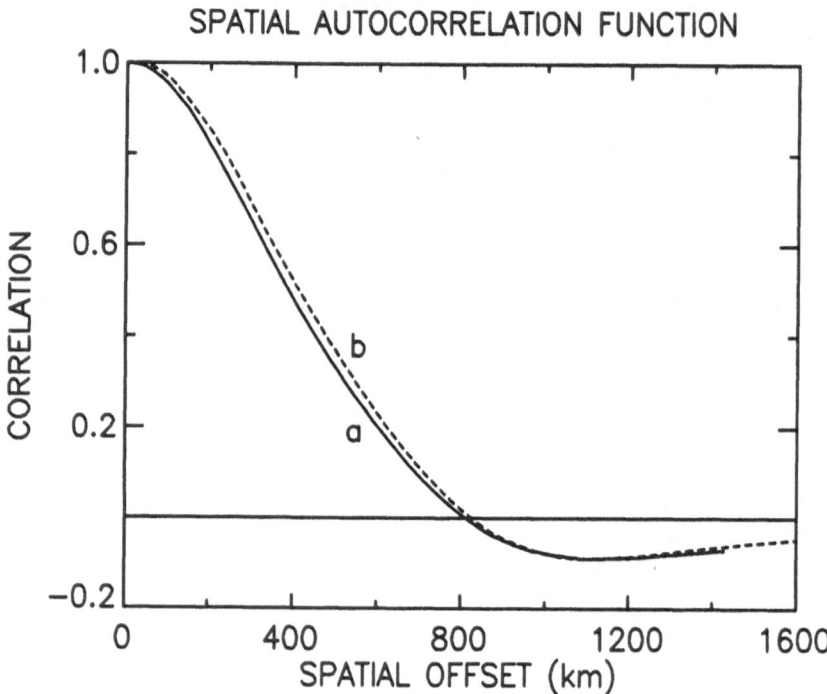

Figure 14: Spatial autocorrelation measurements for (a) subsonic filtered quiet sun and (b) for subsonic filtered magnetic region.

Either random or locally ordered motions will cause the lifetime measured from the temporal ACF to underestimate the lifetime of the individual features in the granulation field. Using a velocity of 1.4 km/sec for quiet sun, the displacement lifetime T_d is 320 seconds, as compared to an autocorrelation lifetime of 270 seconds. In the medium magnetic plage (see figure 13), we use 1.0 km/sec which yields a T_d of 500 seconds, compared with 420 seconds from the ACF. Therefore, the flows are sufficient in both cases to cause much of the decay in the temporal ACF. Note that this flow includes not only the displacements of entire granules but also some of the distortions and explosions which are seen in the local correlation tracking with 1″ FWHM gaussian. The temporal ACF does not simply reflect the lifetime of isolated intensity features in granulation: motion and distortion of the features have a large effect.

5   Exploding Granules

Granules that develop a dark central region and expand radially—exploding granules— were discovered by Rösch and his co-workers (Carlier *et al.* 1968). A prototypical example is shown in figure 15.

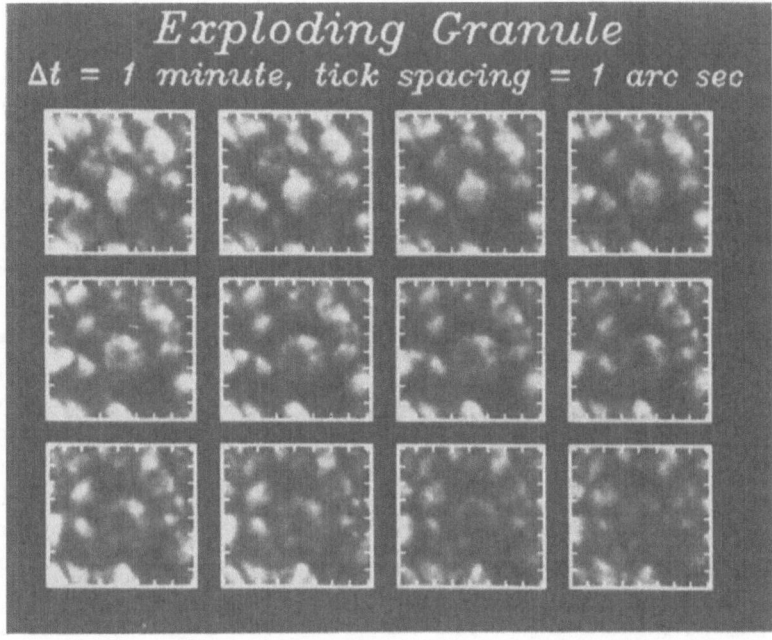

Figure 15: Prototypical exploding granule time history. Tick marks are at 1″ intervals and the time between the frames is one minute.

Usually, the expanding ring is not a complete symmetrical structure seen in the figure, but rather an ensemble of local brightenings which are all traveling outward from a center. They are somewhat rare in earlier granulation data (Mehltretter 1978, Namba and van Rijsbergen 1977), but in the SOUP data they are common. Recently, Steffen (1988) has calculated that the "exploding" configuration is the steady state form of granulation.

The radial expansion velocities of the SOUP events are in the range 1 to 2 km/s, which is consistent with values measured by Musman (1972) and Namba (1986). More interesting is the number and distribution of the exploders. We have found 41 clear examples in the central 32″ × 32″ of the 40″ × 40″ 1360 second quiet sun movie. The central region was used to avoid mis-identifying exploding granules occurring at the edges of the region. The average diameter of the cells at maximum expansion is about 4.2″. Because it is hard to recognize an event unless the initial center or the final ring can be seen, about 360 seconds are lost for detection at the beginning and end of the movie. This suggests that at least 4.0 exploding granules are born in a 10″ × 10″ area during the movie, if they were distributed uniformly. This is a lower limit since, due to the relatively high density of exploders, some events will not be recognized because of overlap and interference.

Allen and Musman (1973) and Oda (1984) have shown that exploding granules are not statistically associated with the supergranulation network. Oda (1984) found some evidence that growing granules may be associated with mesogranulation. Figure 16 shows the map of the divergence of the horizontal flow pattern overlayed by the location of exploding granules. Only one exploding granule is well inside a region of negative divergence. The strong tendency of exploders to occur in mesogranules at a high rate is probably the reason why other observers have noted that exploding granules repeat and spawn new exploders (Carlier et al. 1968, Kawaguchi 1980).

Since regions of positive and negative divergence have nearly equal areas, about 8 exploding granules per $10'' \times 10''$ area per 1000 sec are born in cell-center regions during the duration of the movie. The mean area of maximum expansion of an exploding granule is about 14 square arc seconds, hence 150 square arc seconds of the average $10'' \times 10''$ area are impacted by exploders during the 1360 seconds of the movie. Therefore, every 900 seconds or less such regions are swept by the expansion fronts of exploding granules. It is hard to avoid the conclusion that virtually all the granules that appear in the centers of mesogranulation cells either explode and are pushed aside, or are terminated by an exploding granule in their vicinity. In short, explosions are a major factor in determining the lifetime of granules in quiet sun. We can also conclude that the granules born in the mesogranulation lanes evolve without exploding. Both these conclusion are verified by observations of the SOUP and La Palma movies.

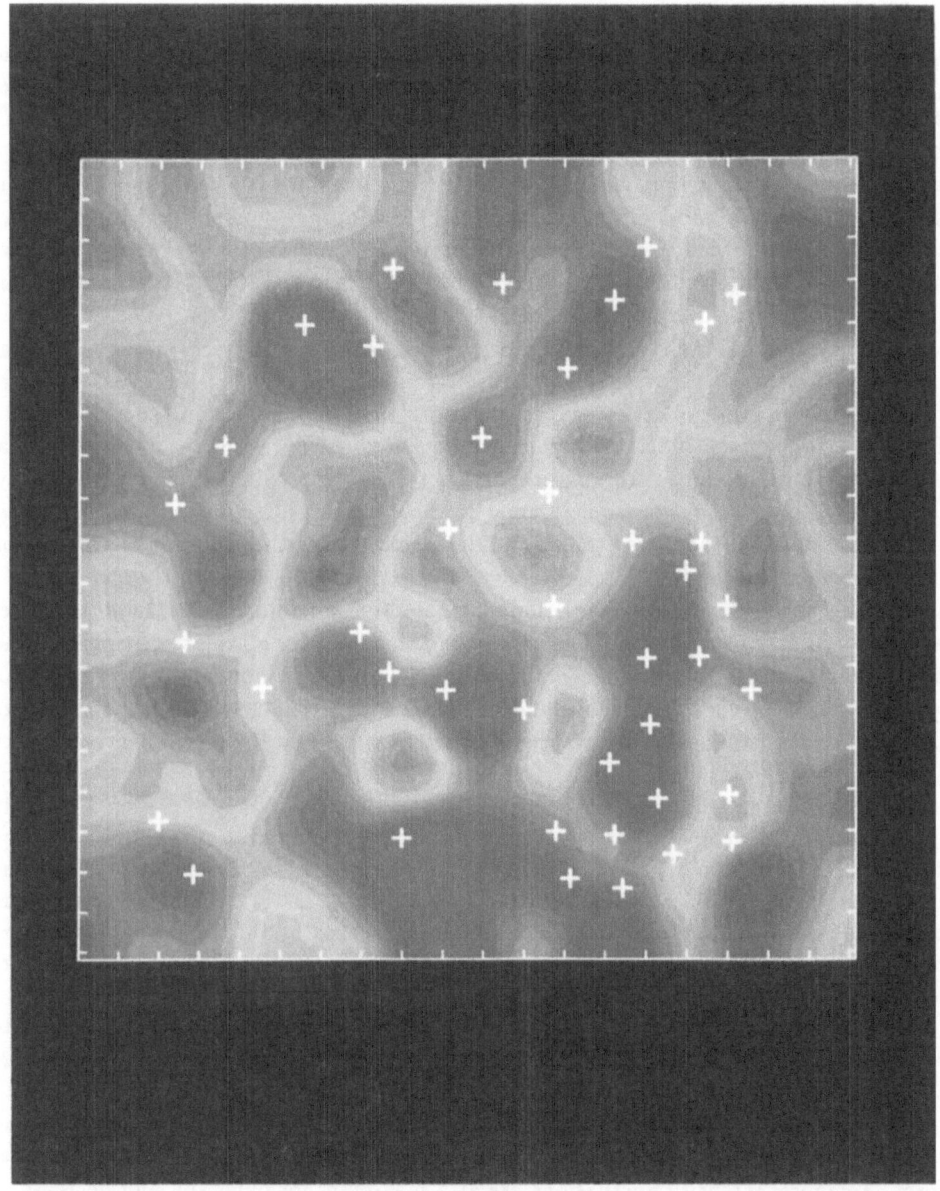

Figure 16: Divergence of the flow field in quiet sun with the location of 41 exploding granules overlaid. Red areas are positive and blue areas are negative divergence (not the same convention as Doppler shifts). White plusses mark the locations of the center of every exploding granule found during the 28 minute observing period.

6 Discussion

We have attempted in this paper to use 3-D Fourier filtering and statistical techniques to identify, measure, and quantify properties of solar granulation. Such a plan is complicated by the richness of the phenomena in the solar surface. The intensity pattern of the solar photosphere is due to granulation (convective overshoot and turbulent flows), oscillations (p- and f-modes), local internal gravity waves, and magnetic fields. From a single photograph it is impossible to separate these phenomena. With the SOUP time sequences of images, most of the effects of global oscillations and some of the effects of waves can be separated from the intensity pattern. Simultaneous magnetograms have allowed us to isolate some of the effects of magnetic fields as well. In spite of various difficulties, the initial plan has been successful and we have found that:

1) The temporal ACF of the photosphere is dominated by the presence of solar oscillations, rather than by the evolution of granulation. If sufficiently small areas of the photosphere are studied, the autocorrelation function oscillates with periods of 3 to 6 minutes. When the solar oscillations are removed by Fourier filtering, the temporal autocorrelation function of the photosphere shows an approximate exponential decay. The decay lifetime in magnetic field areas is about twice as long as in quiet sun areas, which indicates that there are significant differences between the evolution of granulation in quiet regions on the sun.

2) The quasi steady (average) component of the horizontal velocities of granules averages 370 m/s in quiet sun. The patterns in this flow are on the scale of meso- and supergranules. Magnetic fields strongly inhibit the average horizontal flow speeds of granules to 275 m/s for weak fields, and to 100 m/s for strong fields. In plage regions of nearly zero magnetic field, the quasi steady flows are down by 10 percent compared with the field-free quiet sun.

3) The rms (fluctuating) component of the horizontal velocities are also strongly dependent on the type of solar region and the size of the aperture used for correlation tracking. In quiet sun these velocities go from about 0.45 to 1.4 km/s and in strong magnetic regions the increase is from 0.3 to 0.75 km/s as the measurement aperture decreases from 4 to 1 arc seconds. The velocities measured by the smallest apertures are on the order of velocities required for models of solar line widths. The rms velocities are lower in magnetic field areas by about 45%

4) The expansion, collapse, and horizontal motions of granules alone are sufficient to explain much of the decay of the temporal ACF. Thus, temporal ACF's do not measure the lifetime of granules, but rather evolution of the patterns in granulation.

5) On average, granules are brighter in magnetic field areas than in quiet sun and the magnitude of the intensity increase is nearly independent of field strength until pores form. Temporal intensity fluctuations are at a maximum for quiet sun and decrease as magnetic flux in the local neighborhood increases. The fluctuations decrease monotonically with increasing field strength until pores form. The effects of magnetic fields on granulation extend well beyond (perhaps $2'' - 4''$ or more) the boundaries of the individual magnetic flux tubes.

6) Exploding granules tend to occur inside of mesogranules, i.e. in regions of positive divergence of the horizontal flow. Every 900 seconds, all of the surface within these cells has been affected at least once by the expansion fronts of an exploding granule. This

also means that granules that are located within the network between mesogranules differ much in their evolution from those within the cells.

The mesoscale advective flow(November *et al.* 1981) may be sufficient to account for the exploders, or it may be that a nonuniform energy input associated with the meso-granules causes the thermal plumes in the centers of these cells to be more vigorous. Recent hydrodynamic calculations by Nordlund and Stein (1988) and Merryfield and Toomre (1988) show that there are indeed very strong stable downdrafts, with a sepa-ration typical of mesogranulation. These calculations also show that the upward flow is expanding rapidly as it rises. It may well be that the penetrating plumes dominate the form and the evolution of granulation and so create the mesoscale.

In 1962 Leighton noted that there is an asymmetry between the granules and inter-granular lanes visible on the best quality photographs. He then argued that granulation consists of convective cells rather than turbulent flows. In 1987 Roudier and Müller de-termined that the number of granules increases monotonically with decreasing size, and they argue that this provides evidence for a turbulent origin for smaller granules. Our observational results agree substantially with those of these authors.

It is easy to imagine that granulation results from random vertical flows of thermal plumes that penetrate upwards into the photosphere, and that when they reach the convectively stable layer, they begin to expand horizontally, then become unstable, and eventually break up into a turbulent cascade with a Kolmogorov spectrum. However, we abstain here from drawing conclusions. Rather, we will await further reports on the details of the behavior of granules in regions of convergence and divergence, based on groundbased observations at the new Swedish Observatory on La Palma.

When this work started, we hoped to write the definitive life history of a typical granule. We now know that this is not possible. The evolution of a granule is strongly influenced by its environment, namely magnetic fields in its vicinity and its location with respect to meso- and supergranular flows. Many, if not most, granules do not evolve to completion, but rather are ended, probably randomly, by the radial expansion of another granule in their vicinity.

Small granules or fragments are another problem. Is there an objective way to handle them? Is there a size spectrum, perhaps turbulent, of small granules? Are the small granules the result of turbulent breakup of larger ones? Clearly, some smaller fragments are generated during the final stages of the exploding granule process. Are other small granules the signature of a turbulent convective layer? Are some just small convective cells? Here again we are tempted to make suggestions, but will await the results of the new studies.

Finally, we re-emphasize the importance of the digital movie techniques for studying and guiding data processing of images of granulation or any complex evolving system. Raw movies are so chaotic at high resolution as to be nearly incomprehensible. The 3-D Fourier filtering allows us to concentrate on any phenomenon of interest which can be isolated in the $k - \omega$ plane. The all-pervading nature of exploding granules only becomes evident when the oscillations are removed, and the differences between quiet and magnetic granulation also become clearer. The ability to overlay the flow arrows and divergence maps on movie sequences was crucial to understanding the role of these flows. The eye-brain system is also very sensitive to time evolution and motion when it

250

is the relatively narrow range the system expects, so that it is critical to view movies at the proper "brain speed", both in forward and reverse order. Hopefully, the next few years will see the development of image processing procedures that may measure more accurately what we can see. Until that time, we will have to attend workshops and meetings to see some of the remarkable movies generated from observations and simulations.

Special thanks are extended to the crew of Spacelab 2 and the controllers and planners on the ground who worked so hard to get these observations. Data reduction and analysis at Lockheed was assisted by Zoe Frank, Mike Levay, Michael Morrill, and Karen Chen. We would also like to acknowledge the groundbased observers around the world who supported both the Naval Research Laboratory High Resolution Telescope Spectrograph and the SOUP experiment during the flight of Spacelab 2. We benefited from useful discussions with Juri Toomre, Göran Scharmer, Åke Nordlund, Aad van Ballegooijen, and Peter Brandt. And special thanks go to Hal Zirin of Big Bear Solar Observatory and Jack Harvey at the National Solar Observatory for providing magnetograms. The SOUP instrument development and data analysis were supported by NASA under contract NAS8-32805. The CCD camera used to digitize all SOUP data was developed under contract NAS5-26813 for the Coordinated Instrument Package of the High Resolution Solar Observatory. The image processing systems for movie data analysis using laser optical disks have been supported by Lockheed Independent Research funds.

REFERENCES

Allen, M. S. and Musman, S., 1973, *Solar Phys.*, **32**, 311.

Bahng, J., and Schwarzschild, M., 1962 *Ap. J.*, **134**, 312.

Beckers, J.M., 1981, in *The Sun as a Star*, ed. Stuart Jordan (NASA SP-450), 11.

Bonet, J. A., Marquez,T., Roca-Cortez, T., Vazquez, M., Wohl, H., Wittmann, A., 1984, in *Small Scale Dynamical Processes in Quiet Stellar Atmospheres*, ed. S. Keil, NSO conf., 323.

Bray, R.J., Loughhead, R.E., and Durrant, C.J, 1984: *"The Solar Granulation,"* Cambridge University Press.

Carlier, A. Chauveau, R., Hugon, M., and Rosch, J., 1968, *C. R. Acad. Sc. Paris*, **226**, 199.

Cox, A., 1987, *Ap. J.*, submitted.

Dunn, R. B. and Zirker, J. B., 1973, *Solar Phys.*, **33**, 281.

Karpinsky, N., 1980, *Soln. Dannye*, **7**, 94.

Kawaguchi, I., 1980, *Solar Phys.*, **65**,207.

Kitai, R., Kawaguchi, I., 1979, *Solar Phys.*, **64**, 3.

Leighton, R.B., 1962, *Ann. Rev. Astron. Astrophys.*, **1**, 19.

Livingston, W., 1984, in *Small Scale Dynamical Processes in Quiet Stellar Atmospheres*, ed. S. Keil, NSO conf., 330.

Macris, C.J., 1951, *The Observatory*, **75**, 122.

Macris, C.J. and Rosch, J., 1983, *C. R. Acad. Sc. Paris*, **296**, 268.

Mehltretter, J.P., 1978, *Astr. Ap.*, **62**, 311.

Merryfield, W.J.and Toorme, J., 1988, *B.A.A.S*, **20**, 702.

Musman, S., 1972, *Solar Phys.*, **26**, 290.

Namba, O., Diemel, W. E., 1969, *Solar Phys.*, **26**, 290.

Namba, O., van Rijsbergen, R., 1977, in *Problems of Stellar Convection, IAU Coll.38*, ed. E.A. Spiegel and J.P. Zahn, 119.

Namba, O., 1986, *Astr. Ap.*, **161**, 31.

Nordlund, A., 1984, in *Small Scale Dynamical Processes in Quiet Stellar Atmospheres*, ed. S. Keil, NSO conf., 174.

Nordlund, Å., 1985, *Solar Phys.*, **100**, 209.

Nordlund, Å. and Stein, R.F., 1988, *B.A.A.S*, **20**, 702.

November, L. J., Toomre, J., Gebbie, K. B., and Simon, G. W., 1981, *Ap. J.* **245**, L123.

November, L., Simon, G., Tarbell, T., Title, A., and Ferguson, S., 1987, in *High Resolution Solar Physics II*, ed. G. Athay and D. Spicer, NASA Conference Publication 2483, 121.

Oda, A., 1984, *Solar Phys.*, **93**, 243.

Roudier, Th. and Muller, R., 1987, *Solar Phys.*, **107**, 11.

Simon, G. W., Title, A. M., Tarbell, T. D., Topka, K. P., and the SOUP Team, 1988, *Ap. J.*, in press.

Schmidt, W., Knolker, M., and Schroter, E.H., 1981, *Solar Phys.*, **73**, 217.

Steffen, M., Ludwig, H.G., and Kruss, A., 1988, *Astron. Astrophys.*, submitted.

Title, A., Tarbell, T., Simon, G., and the SOUP Team, 1986, *Adv. Space Research*, **6**, 253.

Discussion

VAN BALLEGOOIJEN — How does your lane-finding method work?

TITLE — It examines *all* points in the image to find valleys, which are local minima in any direction. This forms an incomplete net. Then, net cells are completed, starting by filling in those nets, which were most nearly complete in the previous pass of the process.

MÜLLER — The difference between the size histogram you have shown and the one we published[1] can be due to the fact that we presented a histogram of 'granule' sizes while you present a histogram of cell sizes (granule + intergranule). In fact, we have shown in our paper that histograms of granule sizes and of intergranular distances (which are related to the 'cells') have very different shapes.

DAMÉ — A comment: your results show some aspects of the strong coupling between granulation and oscillation. Let me point out:
– individual granules last 5 minutes and no more;
– their initial collapse is very fast (i.e. non-linear in oscillation terms);
– new (or exploding) granules appear in the center of mesoscale patterns (positive divergence regions); they correspond to the active granule (maximum oscillation?) motion;
– fragments (or flows) are slower than the 5-minute lifetime; perhaps they are pushed away by the oscillation in the center of the mesocell (they move even slower if magnetic field is present).

TITLE — What you say is correct, but we need to establish the connection between the observed flow field in the continuum and the oscillation in the temperature minimum.

SCHRÖTER — A remark: Dr. A. Title's statement that the measured surface density of granules varies from author to author by almost a factor two is not correct. Bray, Loughhead and Durrant[2] have summarized in their monograph such investigations. Their table 2.4 shows that the lowest number is 29.5 (per 10 arcsec x 10 arcsec) and the highest number is 33.5. The only exception is the result of Namba and Diemel[3] with 51.3–58.0 based on the Stratoscope experiment. Schröter used the same material and arrived at 31.9. For this discrepancy see Namba and Diemel (*ibid.*, p. 168); it is just a matter of definition.

VAN BALLEGOOIJEN — For the purpose of detecting solar-cycle variations of granulation, it would be useful to use well-defined properties such as temporal or spatial correlation functions, rather than algorithm-dependent quantities such as the number of granules per unit area.

TITLE — Yes. One of the major problems is to define good statistical properties. That is, properties that are not dependent on the 'recognition' algorithm.

---

[1]Th. Roudier and R. Müller, 1986, Solar Phys. **107**, 11.

[2]R.J. Bray, R.E. Loughhead and C.J. Durrant, 1984, *The Solar Granulation*, Cambridge University Press, 2nd edition.

[3]O. Namba and W.E. Diemel, 1969, Solar Phys. **7**, 167.

GRANULATION IN AND OUT OF MAGNETIC REGION

Serge Koutchmy
AFGL/NSO–Sac Peak Observatory
Sunspot, NM 88349 (USA) and
Paris Institut d'Astrophysique, CNRS (France)

ABSTRACT: We performed an analysis of several granulation pictures obtained in both a quiet region and a plage or magnetic region. Filtergrams of ultra–narrow passband and free of line–blocking were obtained with the Sac Peak Vacuum Telescope with an average 0.5 arcsec spatial resolution. Results of a statistical comparative study are presented: amplitude spectra, RMS and histograms of intensity fluctuations. The precise location of a large sample of magnetic field structures identified with MgI b_1 + 0.4A filtergrams was analyzed; 20.4% of these structures overlap or are inserted in granules, which gives a natural explanation of the high value of the RMS in the magnetic region. We used selected very high spatial resolution broad–band short exposure time pictures to show crinkles–type structures at the center of granules that we identify with the abnormal granulation and "magnetic" granules.

I.   INTRODUCTION

Since the discovery by Dunn and Zirker, 1973, of solar filigrees, many attempts have been made to explain the high contrast of facular and network structures observed in the continuum by Mehltretter, 1974, Koutchmy, 1977 and Muller and Keil, 1983, at the center of the Solar disc or at optical depth near unity. Moreover a well–established correlation between these structures and the concentration of magnetic flux as detected in photospheric lines (10^{-1} to 10^{-3} optical depths) exists, see, e.g. Dunn et al. 1974. When the granulation in magnetic regions is considered, a large confusion seems to emerge from an inspection of the literature. Although the term "abnormal granulation" has been introduced, beginning with the first paper of Dunn and Zirker, 1973 few authors indeed have analyzed the behaviour of granulation in a magnetic region, preferring to "forget" that filigrees in form of "crinkles" interact with granules, as clearly stated, e.g., by Mehltretter, 1974. This last author did not however identify the phenomena with the facular granules analyzed later on by Muller, 1977. Further, Muller, 1983 claimed that facular structures (presumably filigree structures) at the disk center are made of points and, additionally, that these "facular points" occur <u>only</u> between granules. High resolution pictures from Sac Peak Observatory show clearly filigree structures not only inserted between granules, but also structures well connected with granules ("overlapping" them), see figure 1 and even, more rarely, at the center of granules! Obviously, it is far more easy to see a very small bright "crinkle" over the dark intergranular lanes background than over a bright granules, so the analysis of these structures is greatly biased by the inability of the best instruments, (not to mention the seeing effects) to correctly

R. J. Rutten and G. Severino (eds.), Solar and Stellar Granulation, 253–271.

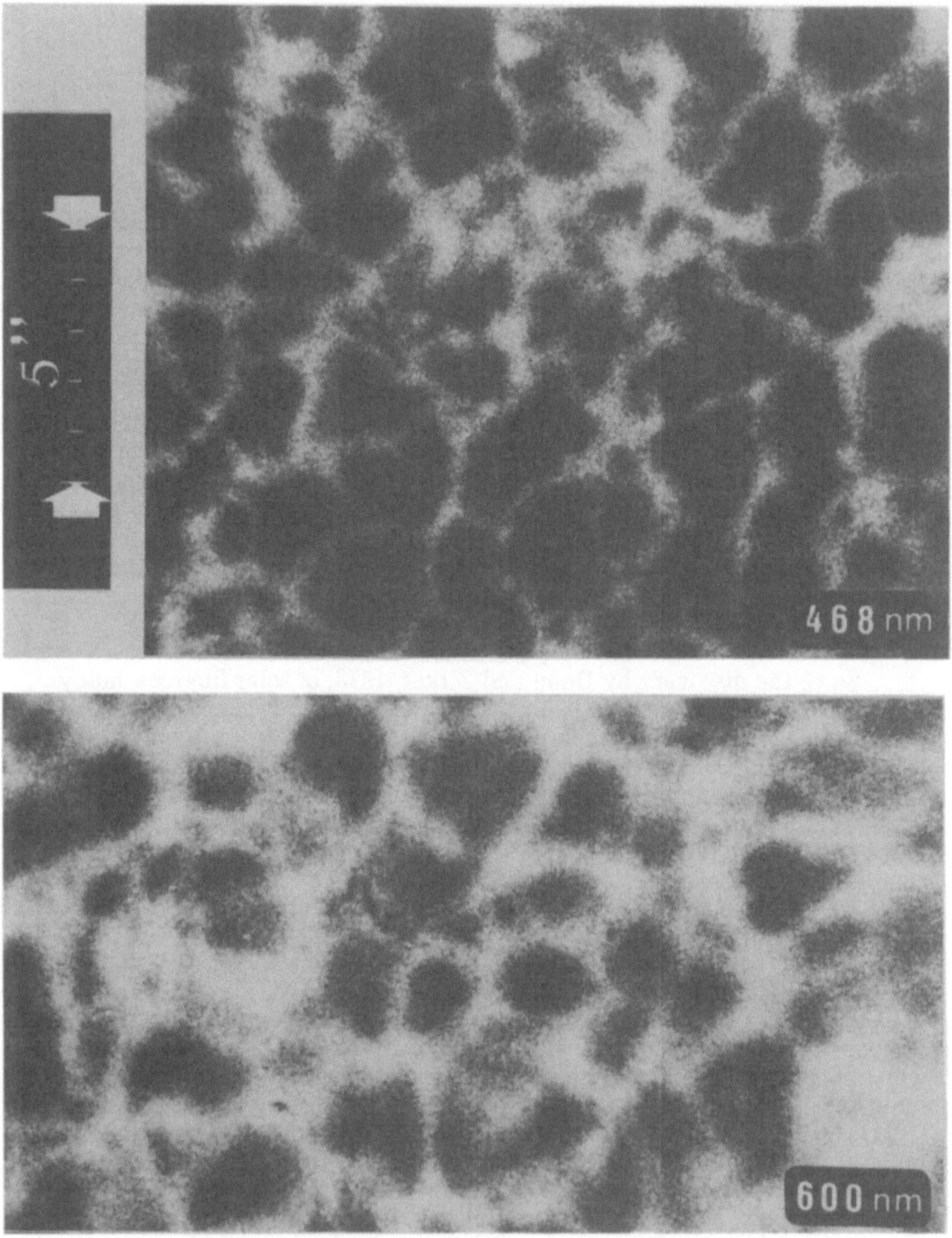

254

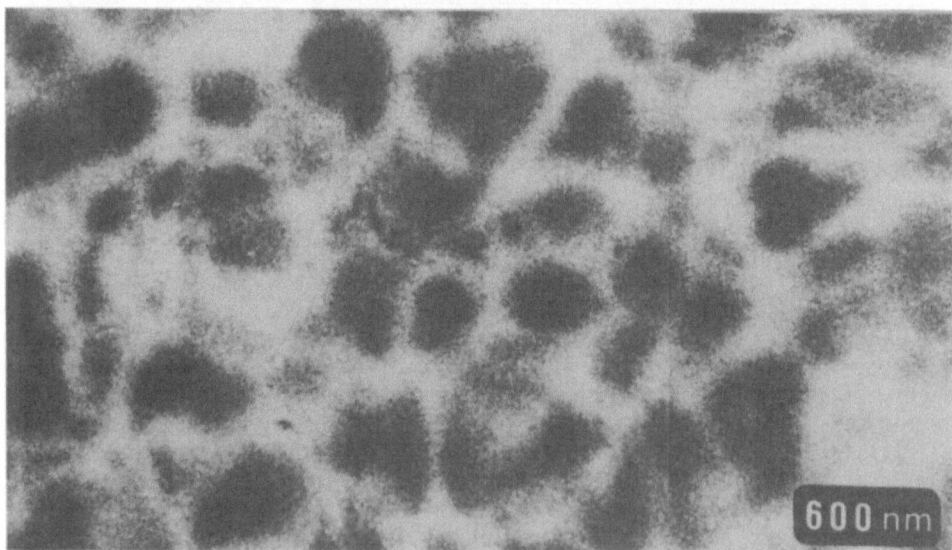

Figure 1. Typical filigree structure observed at 468nm with a passband of 3.5nm (line–blocking effect 11%) and at 600nm with 6nm passband (ℓ–blocking 4%), and a 5msec exposure time, at the prime focus of the SP/VTT (original scale 3.75 arcsec/mm on Kodak microfilm 5460). Note the morphology of the "crinkle" structures suggesting that at least a part of them is overlapping the edge of an "abnormal" granule.

transfer the intensity modulations corresponding to these very small features. Fortunately, magnetic structures can be easily identified in photospheric lines or wings of strong lines, so filtergrams can be used. The granulation is observed at the best in the continuum, avoiding line blocking effects; then very narrow band filtergrams are required. From the theoretical point of view, models of flux tubes have been promoted to explain observations. There again a confusion exists, as different authors use different, sometimes incompatible, approaches. The popular "hot—walls" effect, Spruit 1976, is still competing with the "hot—cloud" model, Rogerson, 1961 and Chapman, 1970, to explain line profiles and center—limb contrasts. However, apparently no arguments have been brought in favor or against the interaction and even the co—existence of a magnetic element and of a granule...The Nordlund (1983) numerical simulations suggest a relation between the concentrationof magnetic field and the downdraft velocities (which are observed typically in intergranular lanes). However, both high spatial resolution spectroscopic observations (Koutchmy and Stellmacher, 1978) and V—Stokes precise profile measurements (Stenflo et al. 1987) agree in showing that no large downdrafts are present in flux tube regions; moreover, the 5—min. oscillations are present, see Dara et al. 1987, making the interpretation of instantaneous observations of velocities rather delicate (saying nothing, again, of problems produced by seeing effects). Finally, let us notice that since downdrafts being systematically observed in intergranular lanes and filigrees observed only in few special locations, at least one additional condition for the "production" of flux tubes should exist, even if the connection downdraft/magnetic structure exists. Here, we will leave this question open and will rather concentrate on the question of "co—spatiallity" of the magnetic field region and the granulation field, as evidenced by the comparison of different filtergrams. This question was also addressed in the recent paper of Title et al. 1987, also performed using the Sac Peak Vacuum Tower Telescope (SP—VTT) with a slightly different technique. Additionally, this latest work gives a comprehensive presentation of the "abnormal granulation" phenomenon; the main conclusion of these authors is that the magnetic field structures map a larger area than just the filigree and especially the "bright—point" locations, a conclusion in agreement with the early finding of Simon and Zirker, 1974, in a plage region and of Koutchmy and Stellmacher, 1978, in a quiet network region.

II. OVERVIEW OF THE METHODS USED AND APPLICATIONS

It is well known that magnetic structures are difficult to map at very high spatial resolution because the signal to noise ratio is rather low and integration is required; then, the spatial resolution is lost due to the variation of image quality, especially image motions and distortion. On the other hand, frame selection of short exposure time pictures is a very effective method, as it has been demonstrated in the Rosch's and Muller's analysis of the Pic du Midi observations. At the SP—VTT, using frame selection, it is also possible to obtain diffraction limited pictures, provided a sufficiently short exposure time is used (recent measurements performed by P. Brandt at the SP—VTT confirmed that moments of seeing quality with r_0 at least equal to the radius of the entrance effective aperture of 76 cm are not infrequent). So, we decided to use this method applied to sequences of filtergrams made with the UBF of the SP—VTT, see e.g. Beckers, 1976, in order to preserve the best possible spatial resolution. However, to map the magnetic structures, we used their thermodynamic properties, namely the greatly enhanced temperature in these regions, at heights of photospheric line formation. Further, instead of making the difficult choice of the best suited line, we decided to use the line which is best suited for UBF observations, with its limited spectral range and its limited spectral resolution. Figure 2 shows the computed profile of the b_2–line of MgI (the b_1 line of MgI shows almost exactly the same behaviour) obtained by G. Stellmacher from the model of Koutchmy and Stellmacher, 1978 shown on Figure 3. This last figure compares different models of magnetic field structures. The temperature effect is always present at the heights of formation of the wing of the b_1 line of MgI and we found, following Beckers 1976, that the positions at plus or minus 0.4A from the line center give a good contrast for magnetic structures, as it is clear also from the inspection of Figure 2 (see also Dara and Koutchmy, 1983). Let us notice, however, that although a general agreement exists concerning the temperature effects at these heights (roughly $\tau_5 = 10^{-3}$), it is not yet clear if the magnetic field is concentrated in the tiny crinkles only or not. Figure 4 shows the distribution, at high spatial resolution, of the I and the V–Stokes parameter over a magnetic structure. It demonstrates again that at the continuum level, no peculiar brightening is observed, suggesting that at this level the magnetic structure is more extended than just the tiny crinkles (not resolved, evidently, on the spectrum). This deduction does not take into account the possible effect of a large constriction of the magnetic field in the low photosphere.

III   OBSERVATIONS

All reported observations were made with the SP–VTT during moments of superb seeing. Our spectroscopic observations were already reported in several papers, see e.g. Koutchmy and Stellmacher, 1978 and Dara et al. 1987. Figure 4 illustrates what can be extracted from these sequences. The results motivated the new observations performed with the UBF in order to cover simultaneously a large magnetic region. Figure 5 shows the whole region observed, around a small sunspot, as well as the two selected regions where a photometric analysis was performed with the fast microphotometer of SPO. The sequence selected is made of four pictures taken in the true continuum of the blue, green and red parts of the solar spectrum, see Table I. These selected pictures are separated by a time interval not exceeding 5 sec of time. The exposure time was chosen to give an optimum photographic density (nearly 2) for the photosphere; however, umbral dots are well exposed in the core of the sunspot on the continuum pictures, confirming their high quality. The wavelength in each spectral region has been carefully chosen, after the inspection of several solar spectrum atlases and, also, the J. Harvey sunspot spectrum atlas. The calibration has been made with an out of focus Sun and a specially designed space–qualified sensitometer with 36 neutral steps. Flat fields were also photographed to take into account the noise of the very fine grain and the vignetting of the optical system. Matrices of 256 x 256 points were recorded with a sample interval equivalent to .128 arcsec with a .35 arcsec FWHM gaussian pin–hole. Figure 6 shows the reproduced from the video display matrices of both the quiet photosphere and the plage region in true continuum emission of different spectral regions.

Finally, observations were also made with broad bands, see Table 1, in order to have very short exposure times and a great number of pictures with a good scale for

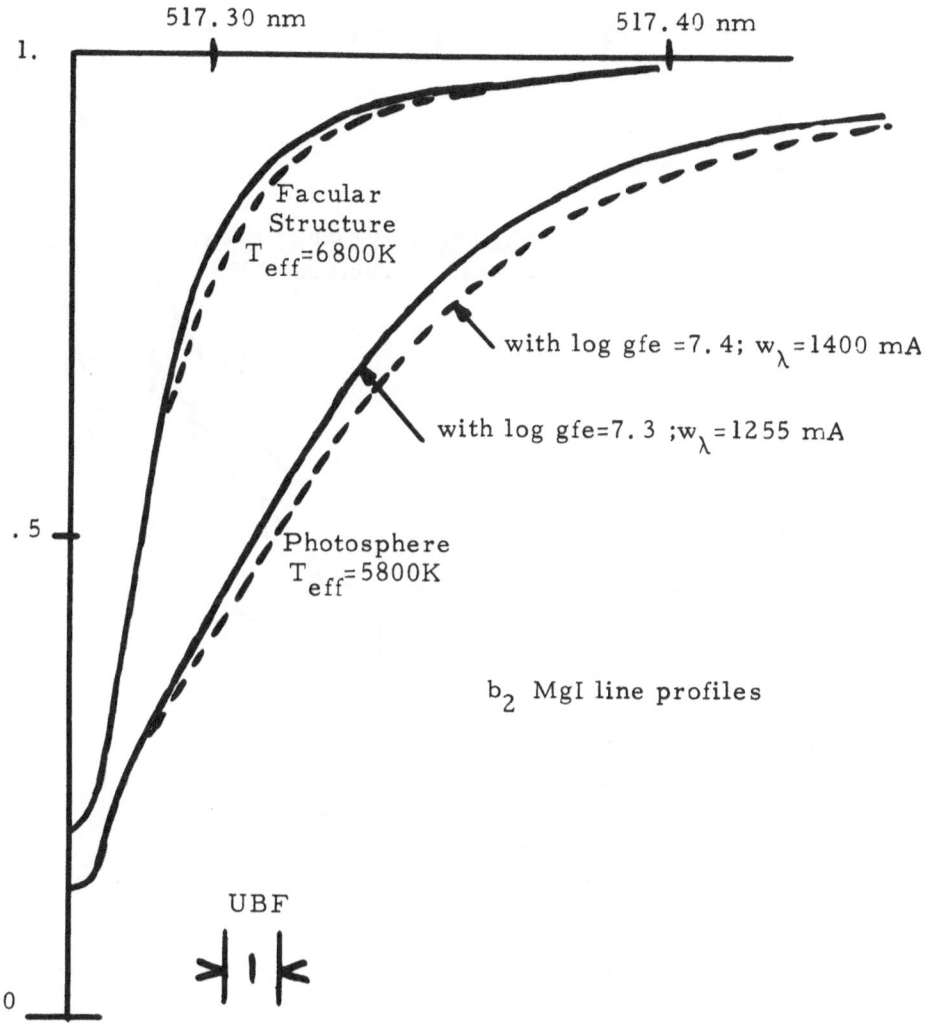

517.30 nm   517.40 nm

Facular
Structure
T_{eff}=6800K

with log gfe =7.4; w_λ =1400 mA

with log gfe=7.3 ;w_λ =1255 mA

Photosphere
T_{eff}=5800K

b_2 MgI line profiles

UBF

Figure 2. Profile of the Mg b₂ line, calculated for the homogeneous photosphere and for a bright element of a magnetic field structure with the model of Koutchmy and Stellmacher, 1978. The position of the UBF passband used to make filtergrams showing the magnetic structure is indicated.

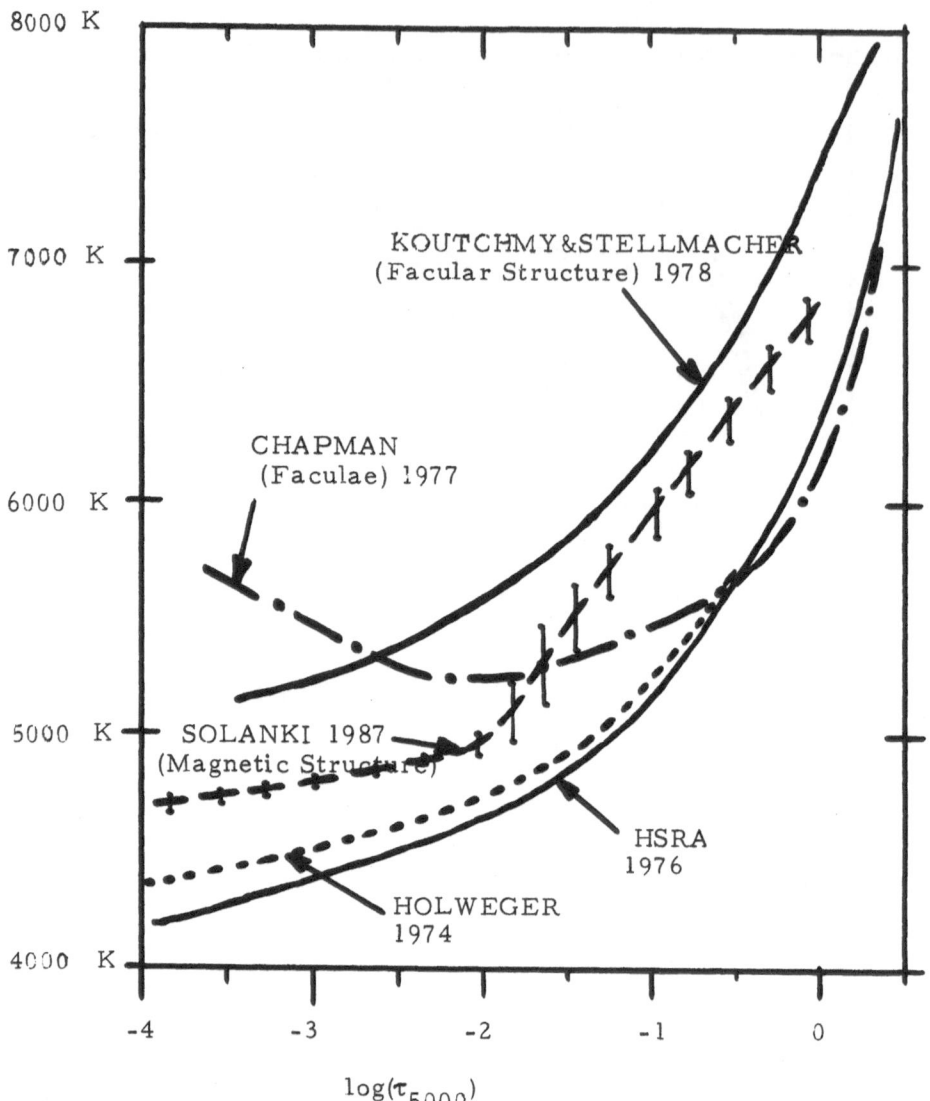

Figure 3. Different temperature models of photospheric atmospheres. Note that the model of Koutchmy and Stellmacher, 1978, corresponds to the hot component alone of the magnetic structure which is more extended than just the bright structure. The Chapman's model does not reproduce observations of a bright structure at the low photospheric level; the Solanki's model is based on the interpretation of the V–Stokes whole profile of several photospheric magnetic lines in magnetic regions; it incorporates a partially physical model of flux tubes.

259

Table 1 Parameters of the Analyzed Observations (VTT/SPO)

Spectral Region	Wavelength λo (nm)	FWHM(nm)or dispersion	τ (Sec)	Film Type (Kodak)	Contrast	Resolution	Scale (arcsec/mm)
B	445.1240	0.009	0.5	2415	$\gamma = 3.6$	0".6	10
V	525.635	0.013	0.1	2415	$\gamma = 3.8$	0".5	10
R	606.960	0.018	0.1	2415	$\gamma = 3.8$	0".4	10
Mgb$_1$ +0.4	518.405	0·012	0.3	2415	$\gamma = 3.8$	0".5	10
V.L.	4680	3.5	0.004	5460	$\gamma = 3.5$	0"17	3.75
V.L.	6000	6	0.004	5460	$\gamma = 3.6$	0"21	3.75
Spectra	6302.5	8mm/0.1nm	1.5sec	50392	$\gamma = 3.6$	0"7	3.75

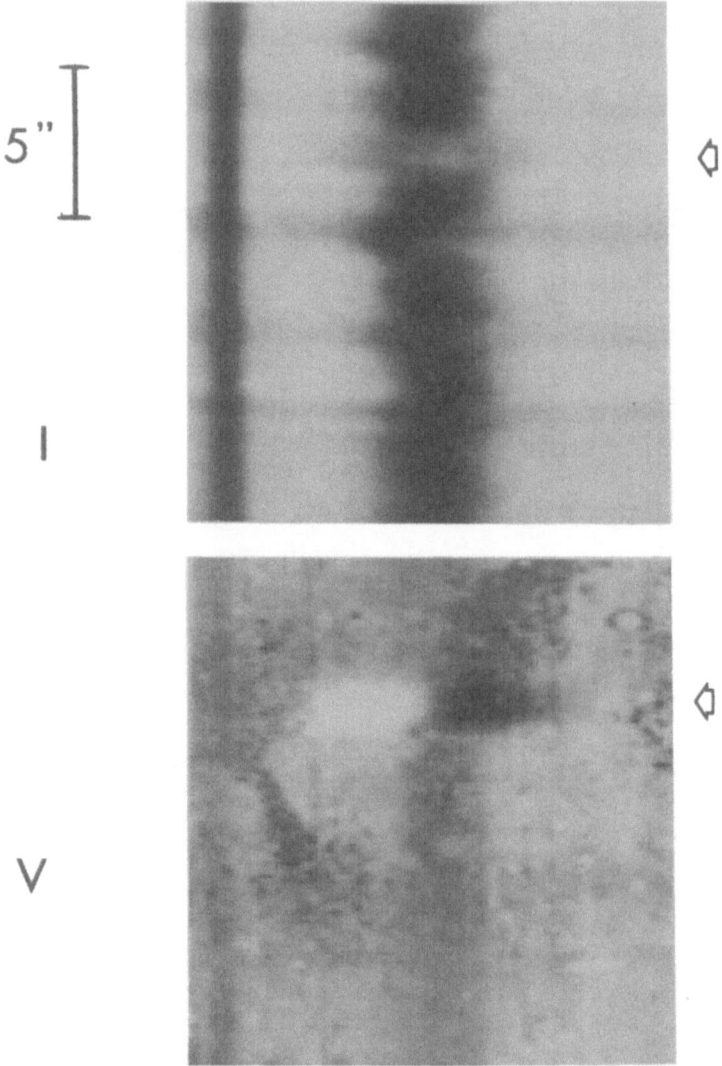

Figure 4. One–dimensional distribution of the I and of the V–Stokes parameters observed around a magnetic element of the chromospheric network in the 630.25nm line of FeI (Landé factor: 2.5). A typical 1 KG magnetic field is observed directly (splitting) over an area of 1.5 to 2 arcsec, showing no prominent brightening (nor darkening) on the continuum.

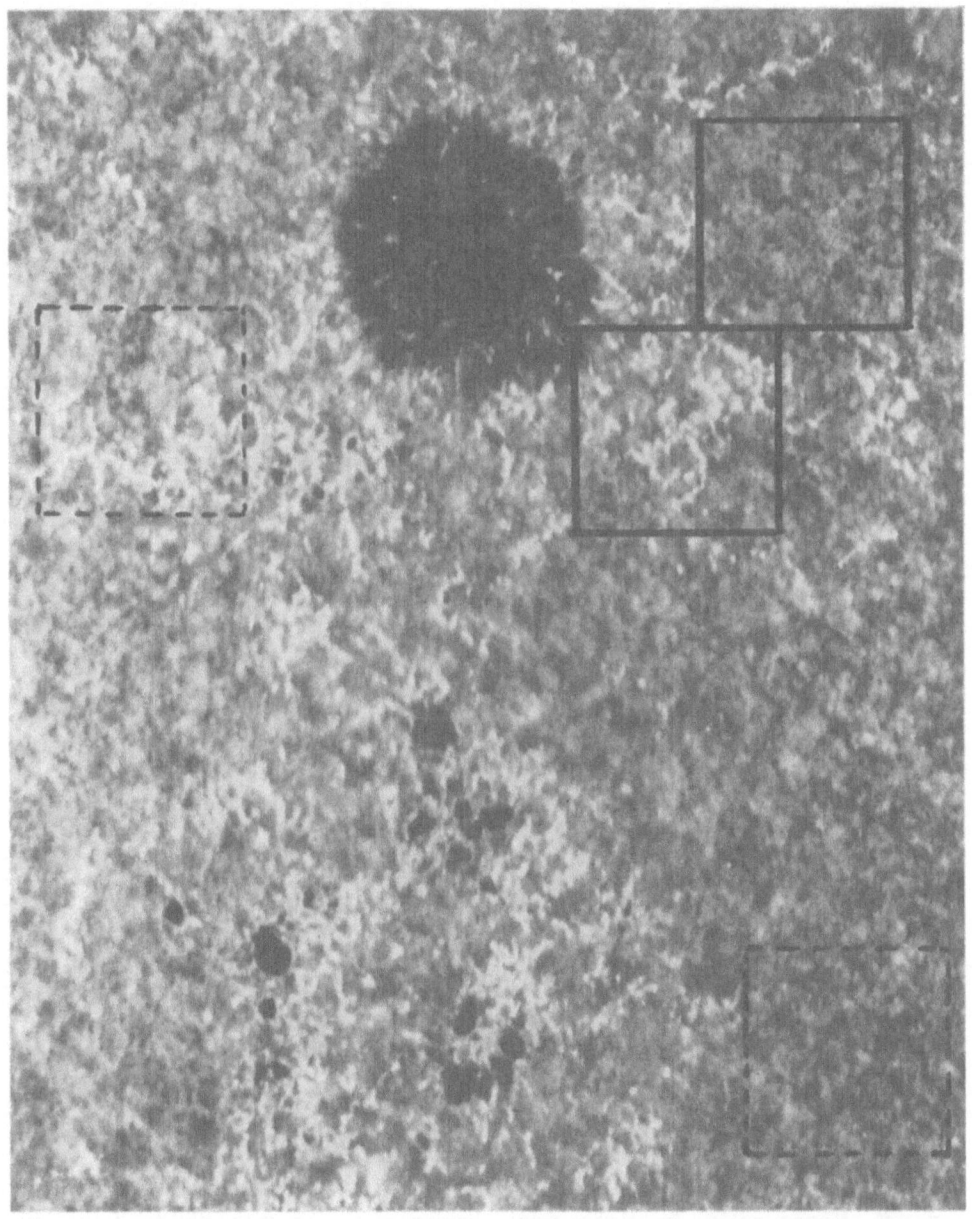

Figure 5. Print of a large portion of the original field of view selected for the study of the granulation in and out of the magnetic regions. The drawn squares represent the region selected for the microphotometric analysis. Their size is 32.6 x 32.6 arcsec^2.

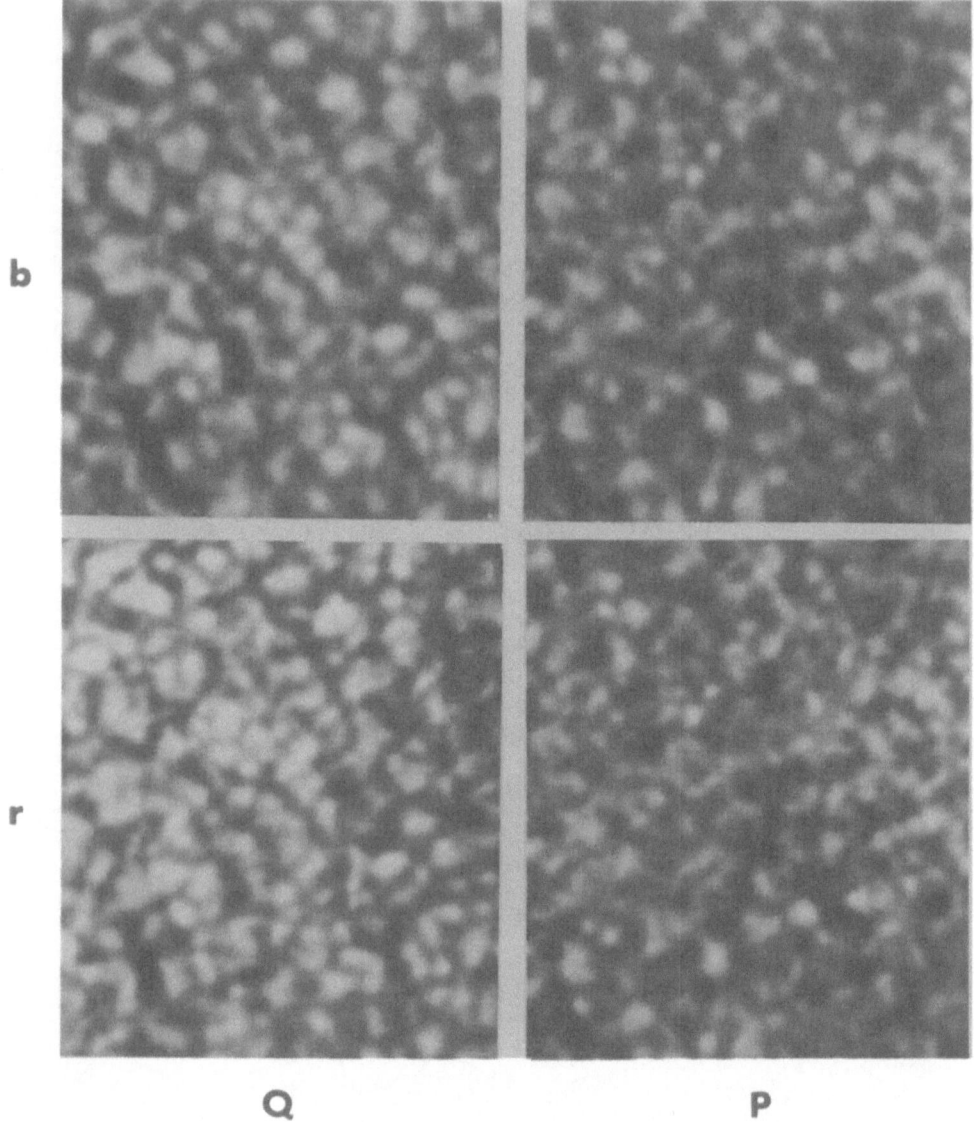

b

r

Q   P

Figure 6. Reproduction from the video–display of the recorded fields of the quiet region (Q) and of the plage region (P), after calibration and removal of the vignetting effect. Note the slight difference in image quality from the blue (b) to the red (r) and the appearance of the "abnormal" granulation.

reducing the noise due to the photographic grain. Figure 1 give a sample of these pictures, showing a very small part of the photographed field of view. Thousands of photographic pictures are available for the analysis; some of them were selected to make a movie. We also tried, without success, to use different video CCD cameras. Results on very fine grain film are still the best.

IV.   ANALYSIS, RESULTS AND DISCUSSIONS

a.   The results of the statistical analysis of the intensity modulations measured on the filtergrams taken in the true continuum are shown in Table 2 and Figure 7, which represent only a sample of the results. Surprisingly, the RMS of intensity fluctuation of the plage region is found higher than the RMS of the quiet photosphere. This result could be especially convincing if we take into account the fact that the "blue" pictures show the best contrast, although the "red" pictures show the best resolution as expected from the consideration of exposure time and seeing effects. It needs confirmation.
Figure 7 compares the radial amplitude spectra (square root of the 2 dimensional power spectra per interval of frequency) observed on the "red" pictures (after removing the effect of the noise); no attempt is made to remove the instrumental attenuation of the spectral distribution, see Koutchmy 1977. However, it seems that more spectral power is present in the high frequency part of the spectrum, a result which is well understood in term of small scale fluctuations produce by the influence of filigrees.

Table II. Values of the RMS of intensity variations observed over the recorded pictures in different windows of the true continuum of the solar spectrum (after removal of the DC component row per row, in 2 directions).

Spectral Regions (nm)	RMS (±%) Quiet Region	RMS (±%) Plage Region
445.1240	7.02	7.48
525.635	6.73	6.74
606.960	5.37	5.86

This result is partially substantiated by the analysis of histograms of intensity fluctuations, showing a definite difference, when the two regions are compared. On the plage (magnetic) region, histograms are more extended; addionally more "bright" pixels are observed and less "dark" ones, for intensities around the average.
Further,using the properly calibrated intensity matrices in different colors, a color index mapping was attempted. Up to now, mixed success was obtained because the differential distortion effect on pictures produces a rather important noise. Only a part of the fields shown on figure 5 and 6 can be properly cross—correlated. A destretching program could be tried also. Preliminary results indicate few new features seems to appear: the map of color

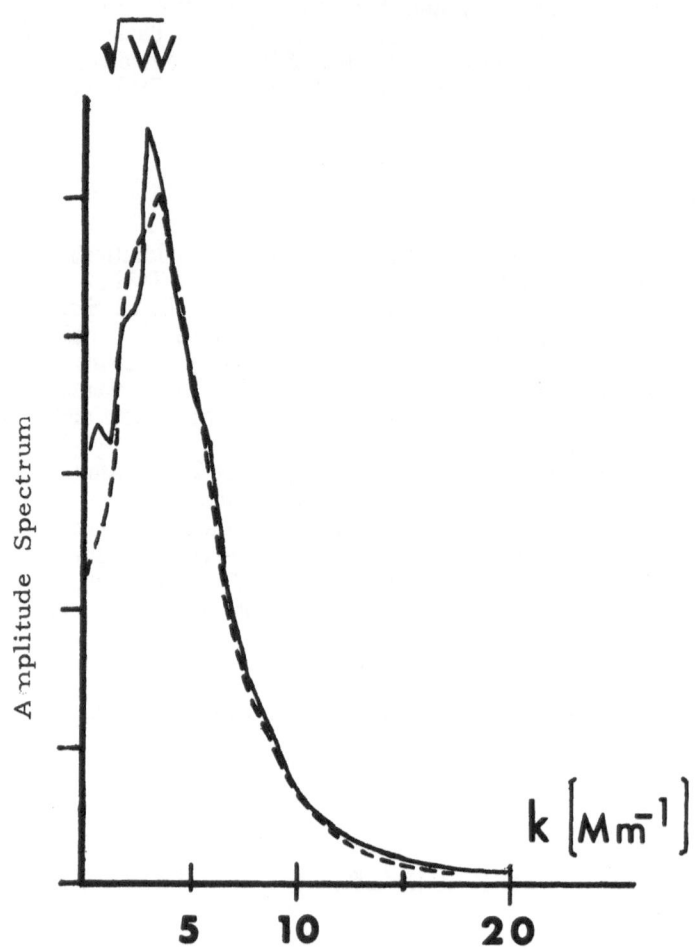

Figure 7. Amplitude radial spectra obtained from the study of the intensity fluctuations (at 606.95nm over the quiet photosphere (dashed line) and the plage region (full line). Note that the plage (or magnetic) region gives slightly higher amplitudes in the high frequency part of the spectrum.

index distribution (ratio of intensities of different color) is showing structures definitely different from just the 2D–intensity distribution over the field.

b. Further, we looked at the locations of magnetic field structures, as deduced from our best MgI b_1 + 0.4A filtergram, with respect to the location of granules and intergranular lanes. Figure 8 shows the distribution of magnetic field structures outside the sunspot and outside the pores (which are well seen on the continuum pictures. We chose to sample the bright elements at 0"77 arcsec intervals and 1244 elements were identified over the whole field. Furthermore, a count was made of magnetic elements per unit area. We took as a unit area 60 arcsec^2, so 207 areas were analyzed (48% of the overall surface of the field). The distribution function of the number of magnetic elements per unit area is shown on Figure 9. Finally, a careful identification of magnetic elements location was made by superposing the map shown on Figure 8 on the continuum pictures. In order to reduce the noise in making the identification, we considered only cases when the magnetic element is located "well" inside a bright granule (we did not count cases when the magnetic element is near the edge of the granule or evidently, when it is in darkintergranular lane). The result of this analysis is shown on figure 10; it shows maximum well displaced with respect to the position of the maximum on figure 9. We first notice that the total number of magnetic elements inside a granule (254) represents 20.4 % of the total number of identified magnetic elements over the picture. The density distribution of the magnetic elements (Figure 9) seems to suggest that an "optimum" density exists;more importantly, when only magnetic elements located in a granule are considered, (Figure 10) an "optimum" density of magnetic elements exists which corresponds to maximum occurence of magnetic elements located in a granule.

V.   CONCLUSIONS

Taking into account the results presented in part 4, we conclude:

a. abnormal granulation (granulation in plage region) does not show an RMS of intensity fluctuations smaller than in the normal granulation, as it was expected if bright filigree features were only "filling" the intergranular space. Evidence was shown for the reverse.

b.   Amplitude spectrum, color index and histogram analysis made in different spectral regions of the true continuum suggest complex relations which need further investigations.

c.   A visual correlation analysis of the location of magnetic elements, performed over a statistically significant number of elements, show convincingly that 20.4% of magnetic elements are located in a granule and, additionally, these "magnetic granules" are more numerous for a high enough number of magnetic elements per unit area, they should be also brighter.

d.   We did not try in this study to deduce any value for the strength of the magnetic field structures. We, however, noticed a slight tendancy for magnetic elements located in a granule to be less bright in MgIb_1 + 0.4, which could mean a smaller amplitude of B in the high photosphere. The existence of a moderate strength magnetic field component in magnetic structures is still an open question, see Semel, 1986, so no

Figure 8. Location of magnetic field structures obtained from the visual inspection of the best MgI b_1 + 0.4 filtergram. Locations of the sunspot and of the pores are also indicated. Magnetic elements were sampled by hand over a greatly magnified print, using a threshold of local brightness.

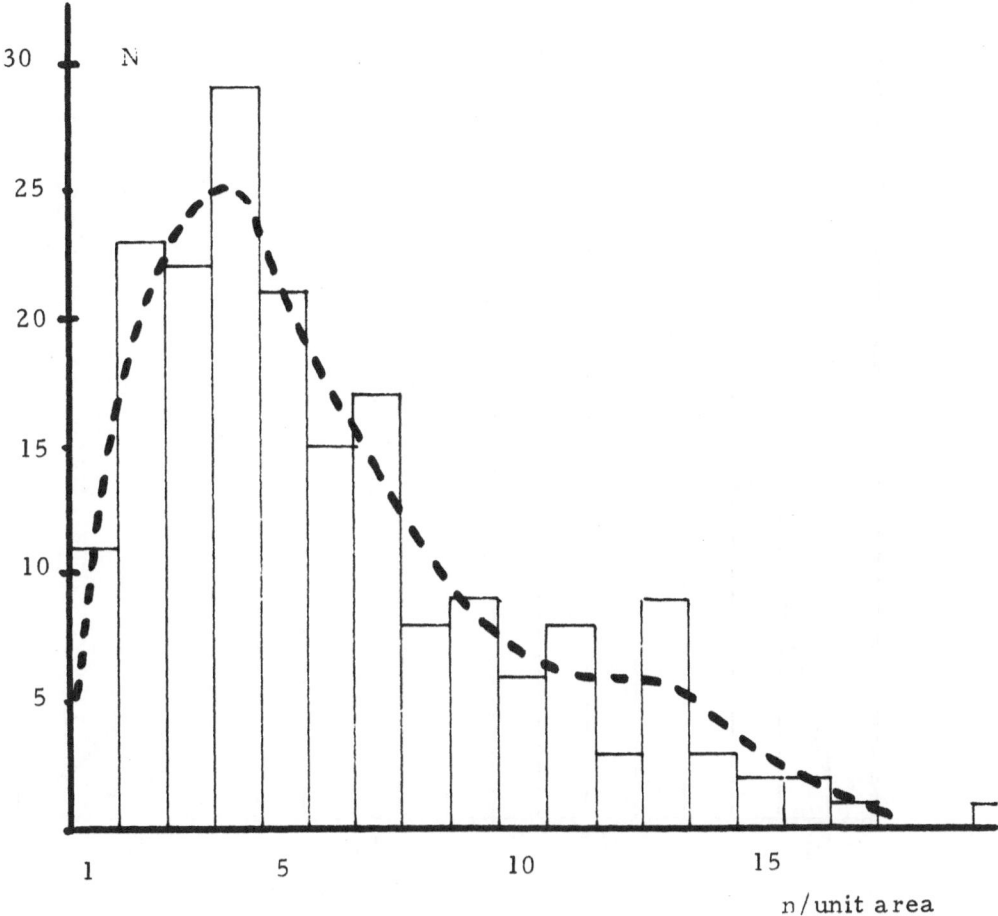

Figure 9. Distribution function (deduced from the histogram) of the number of magnetic elements per unit area. 1244 elements were taken into account; the unit area is 60 arcsec² and the sample "length" of each magnetic elements is 0.77 arcsec.

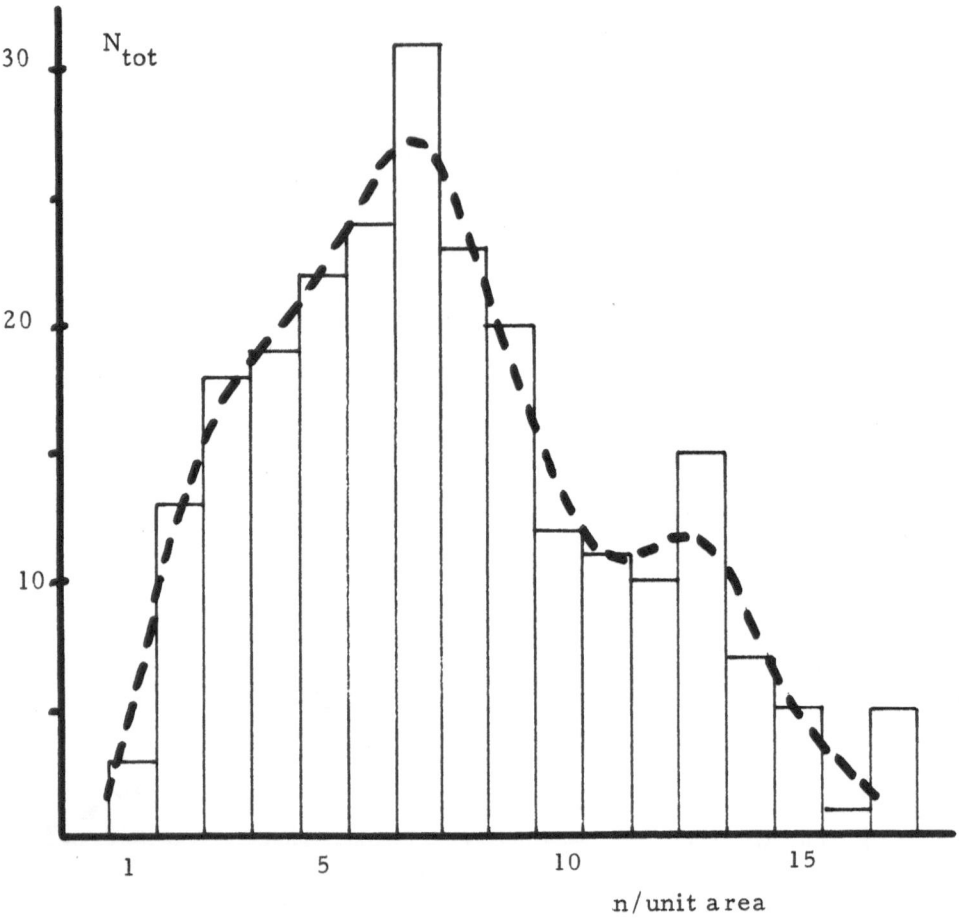

Figure 10. Distribution function of the total number of magnetic elements located in granules as a function of the number of magnetic elements per unit area.

definite conclusions can be made.

      e.   The picture of abnormal granulation coming from this analysis should be incorporated in a dynamical context. Taking a typical life–time of a granule 7 m in, our statistical analysis implied that magnetic elements live at least 7 min x $0.204^{-1} \simeq 34.3$ min. These results compare rather favorably with the theoretical predictions, see e.g. Schmidt et al. 1985.

References

Beckers, J.M. 1976, 'Magnetic Fields in the Solar Atmosphere,' AFGL–TR–76–0131, Environment Research Papers No. 568, Air Force Systems Command, USAF.

Chapman, G.A.: 1970, *Solar Phys.* **14**, 315.

Dara, H. and Koutchmy, S.: 1983, *Astron. Astrophys.* **125**, 280.

Dara, H.C. Alissandrakis, C.E. and Koutchmy, S.: 1987, *Solar Phys.* **109**, 19.

Dunn, R.B. and Zirker;, J.B.: 1973, *Solar Phys.* **33**, 281.

Dunn, R.B., Zirker, J.B. and Becker, J.M.: 1974, in 'Chromospheric Fine Structure,' (R.G. Athay), Ed. Reidel Pub. Company, IAU, p. 45.

Harvey, J.W.: 'A Photographic Spectral Atlas of a Sunspot Between 3813 and 918 Angstroms', Kitt Peak Obs.

Koutchmy, S.: 1977, *Astron. Astrophys.* **61**, 397.

Koutchmy, S. and Stellmacher, G.: 1978, **67**, 93.

Mehltretter, J.P.: 1974 *Solar Phys.* **38**, 43.

Muller, R.: 1977, *Solar Physics*, **52**, 249.

Muller, R.: 1983, *Solar Physics*, **85**, 113.

Muller, R. and Keil, S.L.: 1983, *Solar Phys.* **87**, 243.

Nordlund, A.: 1983, in 'Solar and Stellar Magnetic Fields,' IAU Symp. **102**, (J.O. Stenflo, Ed.) 79.

Rogerson, J.B.: 1961, *Astrophys. J.* **134**, 331.

Schmidt, H.U., Simon, G.W. and Weiss, N.O.: 1985, *Astron. Astrophys.* **148**, 191.

Semel, M.: 1986; in 'Small Scale Magnetic Flux Concentrations in the Solar Photosphere,' Proceedings ed. by Deinzer, Knölker and Voight, Göttingen, V & R Ed., p. 39.

Simon, G.W. and Zirker, J.B.: 1974, *Solar Phys.* **35**, 331.

Spruit, H.C.: 1976, *Solar Phys.* **50**, 269.

Stenflo, J.O., Solanki, S.K. and Harvey, J.W.: 1987, **Astron. Astrophys. 173**, 167.

Title, A.M., Tarbell, T.D. and Topka, K.P.: 1987, *Astrophys. J.* **317**, 892.

Discussion

BECKERS — The Mg I observations which identify the magnetic elements are formed higher in the solar atmosphere than the continuum observations. If magnetic elements are tilted, then this height difference might affect your statistics on the number of magnetic elements inside granules. True?

KOUTCHMY — That is correct. However, the contribution function corresponding to intensities recorded with the UBF at the wavelength we used ($\triangle\lambda = 0.4$ Å from the center of the Mg I b_1 line), has its maximum at a rather small height of 200–300 km, which is smaller than the size of the magnetic elements which we picked up (0.77 arcsec). Additionally, I should mention that these filtergrams were taken practically at the center of the disk.

NORDLUND — Your results are instrument dependent, so you should be careful in interpreting them!

KOUTCHMY — We try to avoid instrumental effects. For example, instead of using white light observations that are polluted by line blocking, we select a segment of true continuum by using a very narrow passband. Also, the results of statistical analyses with large samples are less sensitive to instrumental effects. However, let me point out that the results presented here concern plages; for the quiet Sun the conclusions may not be valid.

TITLE — I am not sure that the line blocking effect is important in our SOUP observations, and whether this can explain the discrepancy in the rms results.

KOUTCHMY — This should be investigated. Concerning the rms results, I should point out that these are clearly affected by errors of statistical origin. The behaviour of the power spectra seems to be more significant: plages give more power at high frequencies and less power at low frequencies than quiet regions. Subtracting the effects of the 5–minute oscillations won't change this conclusion, since these effects are concentrated in the low-frequency range. Finally, our results concern instantaneous rms values computed over the field of view, not over a time sequence (I thank T. Tarbell for pointing this out).

OBSERVATION AND INTERPRETATION OF PHOTOSPHERIC LINE ASYMMETRY CHANGES NEAR ACTIVE REGIONS

Stephen L. Keil[1]
Thiery Roudier[1]
Estelle Cambell[1],
Bon Chul Koo[1]
Air Force Geophysics Laboratory
National Solar Observatory[2], Sacramento Peak
Sunspot, New Mexico, 88349

Ciro Marmolino[3]
Dipartimento di Fisica, Universita di Napoli
Mostra d'Oltremare, Pad. 19-20 - I 80125 Napoli

ABSTRACT: Changes in spectral line asymmetries between regions of quiet sun and regions containing varying amounts of magnetic flux were observed. The observed asymmetry changes are used to deduce parameters that describe the local state of convective overshoot. From these parameters we deduce how the local convective energy flux depends on magnetic fields within the observed regions. Observations were obtained during five separate observing runs between July 1983 and February 1987. We observe FeI 5434 and FeI 6302. Fe 5434 is not split in magnetic regions while 6302 is used to map out the longitudinal component of the field. Additionally we obtain a K-line slit jaw image to locate areas of plage. The observations indicate that magnetic fields decrease the flux carried by convective elements deep in the photosphere, but permit the elements to overshoot higher into the photosphere and temperature minimum regions. The quantitative effect on the convective transport depends strongly on the strength of the magnetic field.

[1] Work performed while holding a National Solar Observatory, Summer Research Associateship

[2] Operated by the Association of Universities for Research in Astronomy, Inc. under contract AST 78-17292 with the National Science Foundation.

[3] Work Completed while an NAS/NRC Research Fellow at Sacramento Peak.

R. J. Rutten and G. Severino (eds.), Solar and Stellar Granulation, 273–281.
© 1989 by Kluwer Academic Publishers.

1. INTRODUCTION

The correlation between velocity and temperature fluctuations associated with fine-scale solar granulation produces the well-known convective blue-shift of photospheric lines, when the lines are observed with insufficient spatial resolution to resolve the granulation (Beckers and Nelson, 1978; Dravins et. al., 1981). It is also well established that the characteristic C-shape of bisectors of spatially unresolved photospheric lines results from this granular velocity-temperature correlation (Voigt, 1956; Schroter, 1957; Nordlund, 1979). Reviews of observations and models are available in Magnan and Pecker (1974) for earlier work and Dravins (1982) for more recent contributions.

Changes in the structure of the solar granulation caused by magnetic fields produce wavelength shifts of photospheric lines and changes in their shapes (Livingston 1982, 1984; Kaisig and Schroter, 1983; Miller, Foukal and Keil, 1984; Righini et. al., 1984; Cavallini et. al., 1985, 1987). Because of these shifts and shape changes induced by changes in the granulation, measurements of flows associated with magnetic regions need very careful interpretation.

Livingston (1982) measured the profile of Fe 5250 (Landre g = 1.5) from areas of the sun having high and low magnetic flux and found the bisector to be red-shifted in the magnetic region with-respect-to the non-magnetic region. The position of the line core changed very little between the two regions. Livingston found similar results for Fe I 5576 (g = 0), indicating that the Landre g-factor does not influence the measurement. Miller, Foukal and Keil (1984) measured three other FeI lines, 5434, 4065, and 5233 and obtained very similar results between network and cell profiles, where network profiles represent averages over regions exhibiting high Ca II K-line emission and cell profiles are averages over regions showing low Ca II K-line emission. They found the network profiles were red-shifted with-respect-to the cell bisectors. Fe 5434 and 4065 showed small core blue-shifts in the network, but the shifts (50km/sec) were not much larger than the uncertainty in the measurements. Their observations were made in regions of the quiet Calcium network (away from sources of activity).

Kaisig and Schroter (1983) measured profiles for six FeI lines formed in both plage regions near sunspots and quiet regions of the disk. They find strong blue-shifts in the cores of Fe 5434, 5635, 5395 and 5576 (> 200 m/sec) and somewhat weaker blue-shifts in Fe 5855 and 5560. Their results for 5434 can be compared directly with those of Miller, Foukal and Keil. The observed changes between plage and non-plage regions are remarkably different in the two papers. The difference in the results could be interpreted as differences in magnetic strength near sunspots and in the quiet network. Cavallini et. al. observed Fe I 6301.5, 6302.5 and 6297.8 in three different active regions of varying magnetic strength. They find that, near disk center, the active region bisector is always red-shifted with-respect-to the quiet sun profile and that the amount of red-shift is directly proportional to the field strength. At about 0.56 solar radii from disk center they observe a slight blue shift of the magnetic profile. They comment that larger shifts correspond to stronger fields and that there is a correlation between field strength and bisector shape.

Both the Cavallini *et. al.* measurements and the Kaisig and Schroter measurements depend on temporal averaging of the data to cancel the effects of the five-minute oscillations. Both papers use ten manifestations of the profile to compute a mean and thereby claim this cancels the shifts due to oscillations. However, the amplitudes of the velocity associated with each cycle of the oscillation is not constant. We could expect the residual error in the mean defined over ten cycles to be about 30%. More recent work by Immerschitt and Schroter (1986) indicates that the earlier work of Kaisig and Schroter contains errors, and that the large observed blue-shift is probably not real.

Because of the discrepancies in the observations mentioned above and the lack of statistical significance, we felt in worthwhile to collect a larger set of data and to investigate differences between various types of plage regions, those near sunspots, near isolated filaments, and in quiet regions. This talk is a preliminary report on some of this data.

2. OBSERVATIONS AND REDUCTION

We obtained observations during five different time periods. These were 2-4 July 1983, 9-14 November 1984, 3-8 November 1985, 2 Dec 1986 and 13-15 February 1987. The first two sets of observations were made using film as the detector and the latter observations use CCD arrays. Here we will only report on the CCD data. A complete analysis of all of the data is in preparation.

We use two CCD arrays; FeI 5434 spectrograms are recorded on one array. The other array is split into three spectral regions, one containing FeI 5576, the other two image the right and left circularly polarized components of FeI 6302 as well as a neighboring terrestrial line that is used as a velocity reference. The spectrograph slit was scanned across the solar disk in 200, 0.5" steps. The spatial resolution along the slit is 0.33"/pixel and the slit width was 0.5". Spectral resolution is 9 mA/pixel in 5534 and 6.5 mA/pixel in 6302.

Before beginning a scan we make a set of dark current and gain tables. Several tests were conducted to verify the linearity of the gain corrections. Exposures of a flat field were made using a series of exposure times, differing slit widths, and neutral density filters. We found small deviations from linearity (2-3%) for light levels near the center of the CCD's range. At low light levels the deviations could be large (80-100%) for some of the pixels. Linear correction factors for the gain tables were derived from the exposures with variable slit width.

The FeI 6302 images are used to make a map the line-of-sight component of the magnetic field. The magnetic field was measured using both the line core position and the center-of-mass of the left and right circularly polarized components. For each region scanned we then have $B_z(x,y)$, where z is the direction along the line-of-sight, x the direction along the slit, and y the direction in which the slit is stepped. Similarly, for each region scanned, the FeI 5434 spectrograms are used to generate a velocity map (line core position), $V_z(x,y)$, and intensity maps in both the line core and continuum, $I_0(x,y)$ and $I_c(x,y)$.

The line profile at each x,y point is averaged into a bin that depends on the strength of the magnetic field at that point. The individual profiles are referenced to the average core position over the scan:

$$V_{ref} = <V_z(x,y)>_{x,y}$$

and divided by a reference continuum intensity obtain by fitting a two-dimensional, second order polynomial to the continuum intensity map before adding them into a bin. Bins are created for both positive and negative polarities. The line center and continuum intensity maps are used to exclude profiles formed in sunspots or dark pores from the bin averages, since these profiles are much weaker and usually shifted with-respect to those formed in the surrounding gas.

3. OBSERVATIONAL RESULTS

The observed profile bisectors vary considerably from region to region, but show several consistent properties as a function of magnetic activity.

On November 6, 1985 we scanned USAF\NOAA region 4700. This region emerged on the 5th and continued to grow. Cosine of the heliocentric position angles was $\mu = 0.83$. The region was scanned several times with the scans spaced at 12.5 minute intervals to aid in removing the effects of the five minute oscillations. A small active region at $\mu = 0.82$ was scanned on December 2, 1986. On February 13, 1987 we observed a small plage area at $\mu = 1.0$.

Using the magnetic maps for the regions we have averaged the profiles into bins that depend on the field strength of the emitting region. In Figure 1 bisectors obtained by combining the data from active region 4700 on November 6, 1985, the plage region from December 2, 1986, and the plage region from February 13, 1987 are plotted. The profiles have been averaged into 200G bins. Profiles show a clear separation between negative and positive polarities, with the positive polarities primarily red-shifted and the negative polarities blue-shifted.

In Figure 2 we replot the data, but we do not distinguish between positive and negative polarity. This give a figure that can be easily

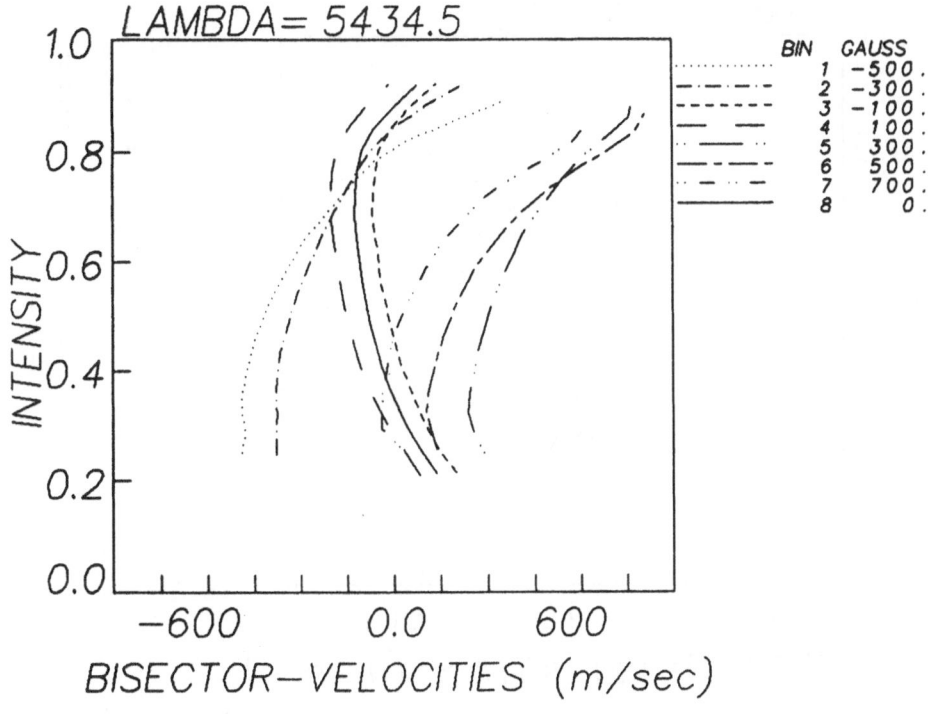

Figure 1. Bisectors of FeI 5434, generated by averaging all of the data from the scans made on November 6, 1985, December 2, 1986 and February 13, 1987, are shown as a function of magnetic field strength. Profiles were averaged into bins that depended on the strength of the local magnetic field. The bins each covered a range of 200 Gauss centered at -500, -300, -100, 100, 300, 500, and 700 Gauss respectively.

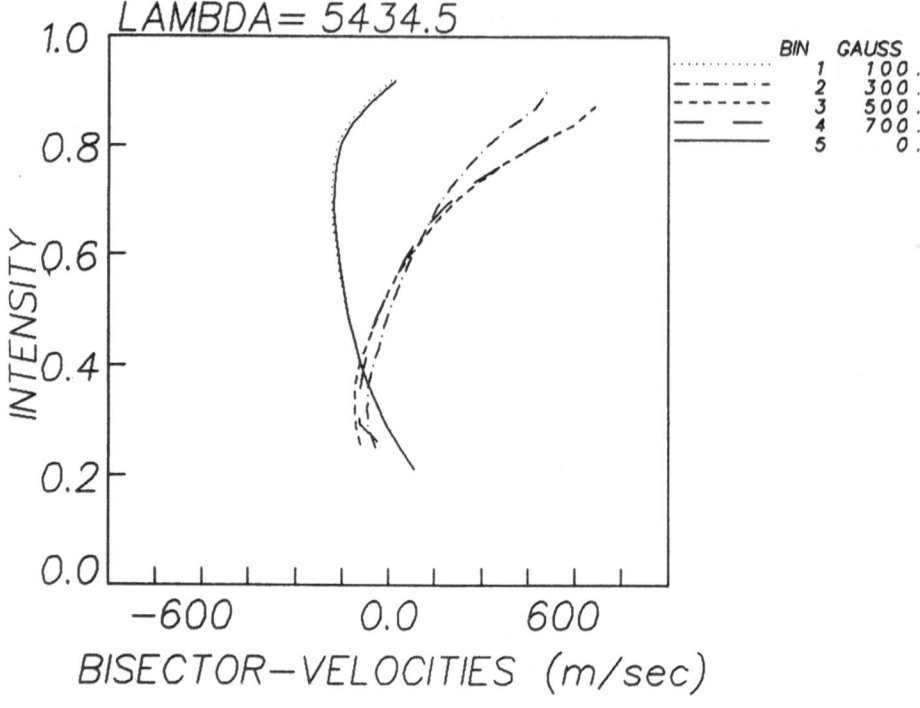

Figure 2.   Bisectors as in Figure 1, however, the sign of the local field was ignored in generating the average for each bin.

compared with earlier work that only separated profiles into plage and non-plage categories. The profiles show a slight core blue-shift and supression of the wings. The bisector of lines from regions exhibiting stronger fields show a greater supression of the red wing.

4. DISCUSSION

Our results can be compared with those of Cavallini *et.al.* (1985). They observe three FeI lines, all of which have Landre g-factors greater than 1, while we observe FeI 5434 with a g-factor of zero. Their results for active regions near disk center (their Figure 1) can be compared to our Figure 2. Both sets of observations show a strong red-shift in the wings of the line. However, we do not observe the consistent red-shift of the core that they report. Our observations show a slight blue-shift of the core. All of our data was obtained either very near disk center or slightly towards the west limb of the sun.

The data presented here indicate that the line asymmetries vary as a function of field strength and perhaps even polarity of the field. Since the bulk of the data in Figure 1 comes from positions on the disk westward of the central meridian, the separation between polarities could be caused by flows along the fields that project differently for each polarity. For example, if one polarity is consistenly closer to the limb and a downflow along the field line exists, it would have a component towards us for one polarity and away from us for the other.

It is clear from the observations that further work is needed to systematically determine the effects of magnetic fields on convection. The observations are consistent with the picture that magnetic fields suppress the amplitude of convective velocities and permit convective elements to penetrate higher into the atmosphere. Currently, we are developing dynamical models to explain the individual bisector shapes and will present these results along with a more systematic analysis of the bisectors from a wide variety of regions, both east and west of the central meridian.

5. REFERENCES

Beckers, J. and Nelson, G.: 1978, *Solar Phys.* **58**, 243.

Cavallini, F., Ceppatelli, G., Righini, A.: 1985, *Astron. Astrophys.* **143**, 116

_____: 1987, *Astron. Astrophys.*, (in press)

Debouille, L., Roland, G., Neven, L.: 1973, *Photometric Atlas of the Solar Spectrum,* Inst. d'Astrophysique, Univ. Liege

Dravins, D., Lindegren, L., Nordlund, A.: 1981, *Astron. Astrophys.* 96, 395.

Dravins, D.: 1982, *Ann.Rev.Aston.Astrophys.* 20, 61.

Immerschitt, S., Schroter, E.: 1986, Proceedings of the conference on *The Role of Fine-Scale Magnetic Fields on the Structure of the Solar Atmosphere*, Tenerife, Canary Islands, (in press)

Kaisig, M., Schroter, E.: 1983, *Astron.Astrophys.* 117, 305.

Righini, A., Gavallini, F., Geppatelli, G., : 1984, in *Small-Scale Dynamical Processes in Quiet Stellar Atmospheres*, (ed. S.L.Keil), National Solar Observatory, Sunspot, NM, page 300

Livingston, W.C.: 1982, Nature, 297, 208.

Magnan, C., Pecker, J.C.: 1974, *Highlights of Astron.* 3, 171.

Miller, P., Foukal, P., Keil, S.: 1984, *Solar Phys.* 92, 33.

Schroter, E.: 1957, *Z.Astrophys.* 41, 141.

Voigt, H.: 1956, Z.Astrophys. 40, 157.

Discussion

SCHRÖTER — Let me add a remark: Presently,Immerschitt and myself have a paper in press with exactly the same type of observations and almost the same results. The only difference is that we used the brightness of Ca^+ mottles as indicator of the magnetic flux. However, our model calculations show that in order to fit the observed changes of the line profile and the C-shape with increasing magnetic activity, one has to diminish the number of ascending granules with constant intergranular area and fill the lack by "static" bright elements ("filigree"?). The comparison with your model calculations shows that there is some ambiguity in such model calculations.

KEIL — The number of free parameters in empirical models of the granulation permits us to match any line bisector shape. These computations can only serve as a guide to the range of variations within the convective overshoot region.

MATTIG — What is the spatial resolution of your data? You also find real C-shapes, but in my data presented this morning, I never found any C-shape but only straight lines.

KEIL — All of the spectra which I showed represent averages over large spatial areas and long temporal periods. All the profiles coming from areas within a set of magnetic contours were averaged, except for those showing strong darkening in the continuum (spots and pores) and also for those showing strong brightness enhancements in the core (line gaps due to facular points). Thus, the data have no spatial or temporal resolution at all. However, the "resolved" granular profiles which I have published before (with Frank Yackovitch[1] and in Keil 1984[2]) also show "straight" bisectors.

MÜLLER — Using line asymmetry observations performed at the Hida Observatory in Japan and high resolution photographs obtained from Pic du Midi near the same sunspot, we have found[3] that both the line asymmetry and the size of granules decrease when approaching the sunspot. Our observations also show that very few facular elements are present, the filling factor being less than 1%; they cannot contribute much to the observed line asymmetry. Thus, the decrease of the line asymmetry is correlated with the granule size decrease, similarly to the joint variation of line asymmetry and granule size over the solar cycle.

[1]S.L. Keil and F.H. Yackovitch, 1981, Solar Phys. **69**, 213

[2]S.L. Keil, 1984, in S.L. Keil (Ed.),*Small-Scale Dynamical Processes in Quiet Stellar Atmospheres*, Proceedings NSO Conference, Sacramento Peak Observatory, Sunspot, p. 148

[3]K. Ichimoto, R. Kitai, R. Müller and Th. Roudier, to be submitted to Solar Physics.

LINE BISECTORS IN AND OUT MAGNETIC REGIONS

F. Cavallini, G. Ceppatelli
Arcetri Astrophysical Observatory
Largo E. Fermi 5 - I-50125 Firenze, Italy
A. Righini
Institute of Astronomy, University of Florence
Largo E. Fermi 5 - I-50125 Firenze, Italy

ABSTRACT. Some photospheric lines have been recently observed in active regions at different distances from the solar disk center and the asymmetry and shift dependence on the magnetic field has been evaluated. In particular, near the disk center, we find that the red-shift of the line bisectors increases with the field and decreases towards the core, while, at the bottom of a strong line, a small blue-shift is observed. Measurements performed between 0.5 and 0.8 solar radii from the disk center show that the line bisectors are red-shifted near the continuum and blue-shifted at the bottom. These results are qualitatively interpreted as due to the magnetic inhibition of the vertical and horizontal component of the convective motions.

1.   INTRODUCTION

Previous observations by Cavallini et al. (1985a) have shown that in solar active regions, up to about 0.6 solar radii from the disk center, the line bisectors are red-shifted, while, at greater distances, their lower part is blue-shifted. However, these preliminary measurements only refer to three Fe I lines and to eight active regions, and they lack any information about the error in assigning the bisector wavelength position.

To confirm the observed phenomenon and to gain a better insight into the interaction between turbulence, magnetic field and line formation processes, new observations in solar active regions, at different distances from the disk center, have been performed with the spectro-interferometer installed at the Arcetri Solar Tower (Cavallini et al., 1988a).

In this paper the results of these measurements will be discussed.

2.   OBSERVATIONS AND DATA REDUCTION

In order to cover a wide range of formation heights and to investigate a possibile dependence of the observed phenomenon on the Landé g factor,

R. J. Rutten and G. Severino (eds.), Solar and Stellar Granulation, 283–287.
© *1989 by Kluwer Academic Publishers.*

the following lines have been choosen: 5569.6 Å FeI (g=0.75), 5576.1 Å
FeI (g=0), 6149.2 Å FeII (g=1.33), 6162.2 Å CaI (g=1.25). The effects
of the magnetic field on the line bisectors have been evaluated by alternately observing the active region, selected by visual inspection
through a Zeiss H-alpha filter, and the corresponding quiet area at the
same heliographic longitude but at opposite latitude. The pointing was
mantained, during all the observing time, by automatically compensating
the solar rotation. In the active and quiet reference regions the Sun-
observer component of the differential rotation is in principle the same
and will be eliminated in evaluating the relative wavelength position
of the observed lines. However, since the two considered regions are
not generally at the same distance from the disk center, the line wavelengths and central intensities have been corrected for the center-to-
limb effect, by using independent observations carried out across the
solar disk (Cavallini et al., 1988b).

To disentangle the effects produced on line asymmetry by the magnetic field from those due to geometrical effects, we have selected two
areas on the disk. The first one ranges from the disk center up to
about 0.5 solar radii, the second one ranges from 0.5 to about 0.8 solar
radii. These two regions have been choosen to minimize the bisector
shape variation due to the geometry.

About the magnetic field, we have no direct information of its intensity. however, as shown by Brandt and Solanki (1987) and confirmed
by Livingston (1987), the line weakening may be assumed as an indirect
measurement of the field. Therefore, the line bisectors obtained in
active regions, in each of the two selected solar areas, have been grouped and averaged, by putting together bisectors with similar weakening.

To directly evaluate experimental errors, the line profiles have
been observed in couples of quiet regions, following the same procedures
adopted for the observations in active areas. We find that the central
intensity dispersion is about 0.3% of the continuum, independent of the
position, while the standard deviation of the shift is about 1 mÅ up to
0.4 solar radii, increasing towards the limb. However, in some cases,
shifts reaching 5 mÅ have been observed, most likely due to supergranulation.

3. DISCUSSION AND CONCLUSIONS

If we first consider the bisectors observed in active regions near the
disk center, we find that the red-shift increases with the field and
decreases at lower relative intensities. These experimental results
may be qualitatively interpreted by assuming that: i) the correlation
between brightness and velocity fluctuations in the solar photosphere
produces the well known blue-shift of the Fraunhofer lines relative to
laboratory sources (Beckers and Nelson, 1978). This correlation decreases with increasing photospheric height and vanishes at about 300 km
(Canfield and Meheltretter, 1973), ii) the magnetic field partially
inhibits the convective motions (e.g. Livingston, 1982; Miller et al.,
1984; Cavallini et al., 1985b). On these assumptions, an increasing
magnetic field will produce an increasing red-shift of the solar lines,

as confirmed by the observations. Moreover, the red-shift will be smaller when observing higher in the photosphere and null at about 300 km. In fact, assuming as formation layer, the layer where the velocity Response Function (Beckers and Milkey, 1975; Caccin et al., 1977) is maximum, we find that the Fe II line core (formed at about 100 km), shows a red-shift. This red-shift however vanishes at the bottom of the two Fe I lines and at 30% of the Ca I line (formed at about 300 km).

At residual intensities lower than 30% in the Ca I line the red-shift changes in a blue-shift amounting to about 1 mÅ at the line core. This effect, comparable to the relative positioning error of the mean bisectors (about 0.5 mÅ), might be differently explained. The core of the Ca I line is formed in the high photosphere (about 500 km), where some observations (Canfield and Meheltretter, 1973) and some numerical simulations (Nordlund, 1984) suggest that the granulation pattern is inverted, while the velocity pattern is well correlated with that of the deep photosphere. Therefore, the lower part of a strong line should be convectively red-shifted and then show a blue-shift in presence of a magnetic field. If this hypothesis is correct, we expect to find strong solar lines red-shifted with respect to laboratory ones. Such a test leads however to contradictory results (Balthasar, 1985; Cavallini et al., 1986), showing that the wavelength data are not yet fully reliable for this purpose. However, the observed effect could be alternatively interpreted as due to the horizontal component of the convective motions, producing a mean red-shift of about 200 m/s at 500 km (Balthasar, 1985). If we assume that the considered bisectors have been observed, on average, at about 0.25 solar radii from the disk center, the magnetic inhibition of the horizontal motions should produce on the Ca I line core a blue-shift of just 1 mÅ. To test which one of these hypothesis is correct, new measurements of strong lines are needed in active regions very near the disk center. Finally, the very similar behaviour of the two Fe I lines seems to confirm that the Landé g factor does not appreciably affect the observed phenomenon (e.g. Livingston, 1982).

If we consider now the results obtained in the external region of the solar disk (0.5-0.8 solar radii from the center), we find that, for strong magnetic fields, the bisectors are red-shifted near the continuum and blue-shifted on the lines bottom. This behaviour produces a crossing of the "quiet" and "active" bisectors, confirming the findings by Cavallini et al. (1985a). In particular, the "crossing point" in the Ca I line rises from 30% to about 50%, as expected for a depth dependent phenomenon. The magnetic inhibition of the horizontal convective motions, showing Doppler shifts towards the observer in deep photospheric layers and away from the observer in high layers (Balthasar, 1985), is most likely relevant in producing this result. However, the behaviour of the two Fe I lines, showing a red-shift for medium-strong magnetic fields, is rather puzzling. More observations are needed, considering that, at these distances from the disk center, the standard deviation of the wavelength shift amounts to about 2.5 mÅ.

REFERENCES

Balthasar, H.:   1985, Solar Phys. 99, 31.
Beckers, J.M., Milkey, R.W.:   1975, Solar Phys. 43, 289.
Beckers, J.M., Nelson, G.D.:   1978, Solar Phys. 58, 243.
Brandt, P.N., Solanki, S.K.:   1987, in "The role of fine scale magnetic
    fields on the structure of the solar atmosphere", E.H. Schroter
    and M. Vazquez eds., Cambridge University Press (in press).
Caccin, B., Gomez, M.T., Marmolino, C., Severino, G.:   1977, Astron.
    Astrophys. 54, 227.
Canfield, R., Meheltretter, J.P.:   1973, Solar Phys. 33, 33.
Cavallini, F., Ceppatelli, G., Righini, A.:   1985a, Astron. Astrophys.
    143, 116.
Cavallini, F., Ceppatelli, G., Righini, A.:   1985b, Astron. Astrophys.
    150, 256.
Cavallini, F., Ceppatelli, G., Righini, A.:   1986, Astron. Astrophys.
    163, 219.
Cavallini, F., Ceppatelli, G., Righini, A.:   1988a, Astron. Astrophys.
    (in press).
Cavallini, F., Ceppatelli, G., Righini, A.:   1988b, (in preparation).
Livingston, W.C.:   1982, Nature 297, 208.
Livingston, W.C.:   1987, (private communication).
Miller, P., Foukal, P., Keil, S.:   1984, Solar Phys. 92, 33.
Nordlund, A.:   1984, in "Small-Scale Dynamical Processes in Quiet Stel-
    lar Atmospheres", S. Keil ed., Sacramento Peak National Observa-
    tory, p. 181.

Discussion

VAN BALLEGOOIJEN — Observations of spectral profiles in circular polarization have shown the existence of V-profile asymmetries, which are interpreted as correlations of magnetic and velocity fields along the line of sight. Is there any connection between these V-profile asymmetries and the line bisectors in magnetic regions that you report on?

CAVALLINI — Our instrument does not provide any information on the circular polarization of the observed spectral profiles. However, we cannot exclude that this effect could contribute to the behaviour of the bisector shape and shift that we observe in active regions.

STEFFEN — You conclude that in the presence of inverted temperature fluctuations in the overshooting regions (as found in the numerical simulations) the cores of strong lines are redshifted (in the quiet sun). It seems to me that this conclusion is incorrect. In fact, we find that the opposite is the case: the temperature fluctuations imply an additional blueshift. So probably you can abandon the first of the two alternative explanations that you mentioned.

CAVALLINI — May be. I think, however, that more reliable laboratory wavelengths are needed to ascertain the convective redshift sometimes observed in strong lines.

RMS VELOCITIES IN SOLAR ACTIVE REGIONS [1]

A. Nesis, K.-H. Fleig, and W. Mattig

Kiepenheuer-Institut für Sonnenphysik, Freiburg, FRG

ABSTRACT. Spectrograms taken at Sacramento Peak were analyzed by means of power analysis. The vertical and non-vertical rms granular velocity field in active Solar regions shows a significant reduction compared to that of the non-active regions. We explain this reduction in terms of enhanced magnetic fields in active regions and the resulting inhibition of small-scale velocity fields.

The interaction of the small-scale velocity fluctuations with the magnetic field in the overshoot layers prove to be of central significance for the physics of the solar atmosphere: the heating of the chromosphere and the bright points (filigree) in the photosphere are two different phenomena which can be considered as manifestations of this interaction.

To determine the influence of the magnetic field on the velocity fluctuations in the photosphere, we compare the power spectra of the velocity fields of active with those of non-active regions. Here, the non-active regions are regarded as being free from magnetic flux and, therefore, can be used as reference for this comparison.

Mattig and Nesis (1974) showed that there is a difference between the power spectra of active and non-active regions; they furthermore investigated the height dependence of the granular rms velocity of active and non-active regions by means of the line core and line wings of the absorption line. Mattig and Nesis (1976) repeated this analysis with new material but the results seemed to be in conflict to those of the previous study. Now, about ten years later, we analysed new spectrograms obtained at Sac Peak observatory at two different positions on the disc of the Sun, at $\cos\theta = 1.0$ and $\cos\theta = 0.8$. Thus, we can compare active and non-active regions and their variation with height in the photosphere for the vertical ($\cos\theta = 1.0$) as well as for the non-vertical ($\cos\theta = 0.8$) velocity fields. In this investigation the variation of the velocity fields with height is calculated by means of the absorption lines 6496 Å and 5576 Å. Using the intensity contribution functions of these lines we found the line formation to be at a height of 175 km and 325 km above $\tau_{5000} = 1.0$, respectively. The layers at about 325 km level are in the following denoted as the higher layers, those at about 125 km as the deeper photospheric layers.

The calculated power spectra of the active and non-active regions are shown in Figs. 1 and 2 for $\cos\theta = 1.0$ and $\cos\theta = 0.8$, respectively. Here, we realize that the power spectra of the active regions (A.R) show smaller values than the power spectra of the non-active regions (N.A.R), regardless of the heights in the photosphere and the position on the

---

[1] Mitteilungen aus dem Kiepenheuer-Institut Nr. 294

R. J. Rutten and G. Severino (eds.), Solar and Stellar Granulation, 289–294.

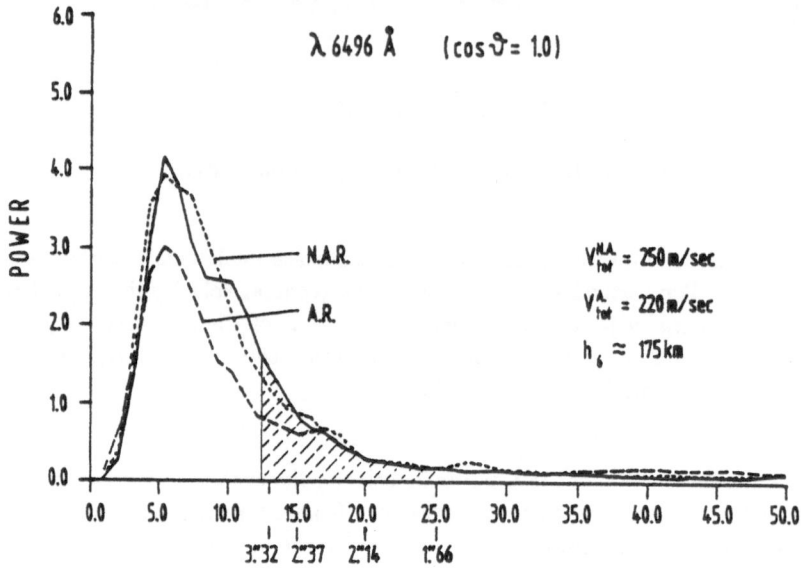

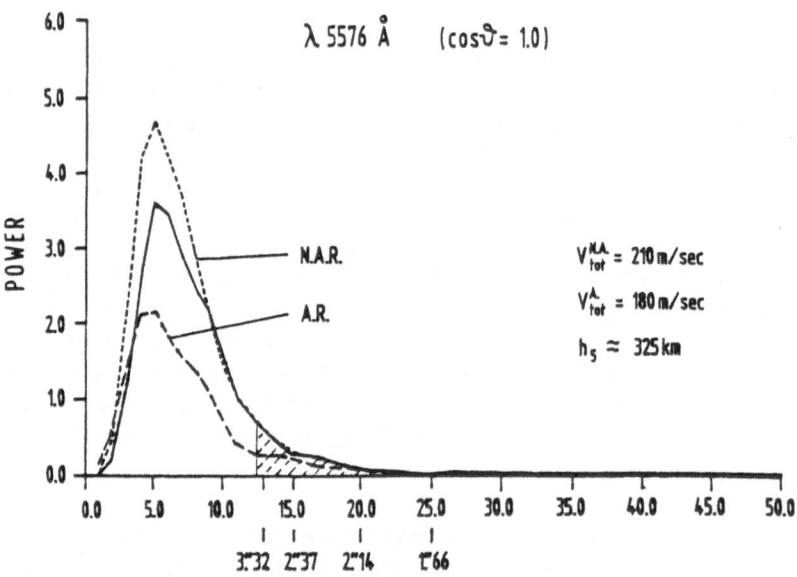

Figure 1: Velocity power spectrum

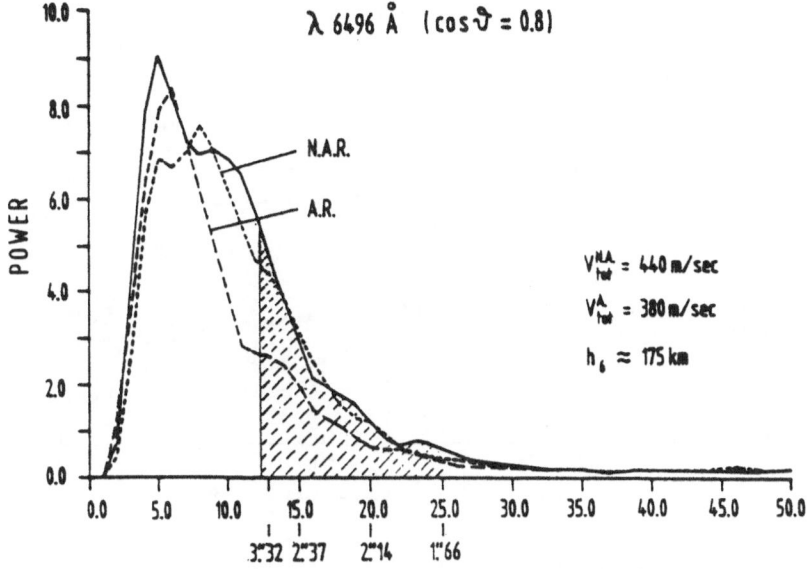

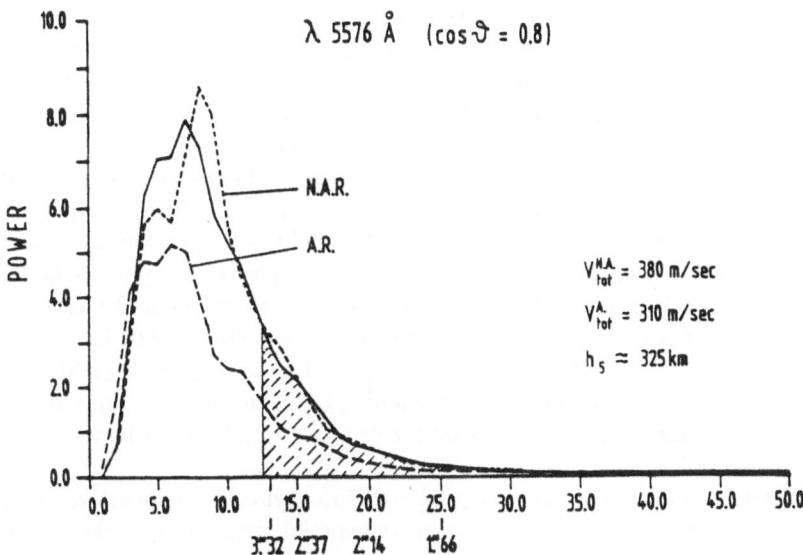

Figure 2: Velocity power spectrum

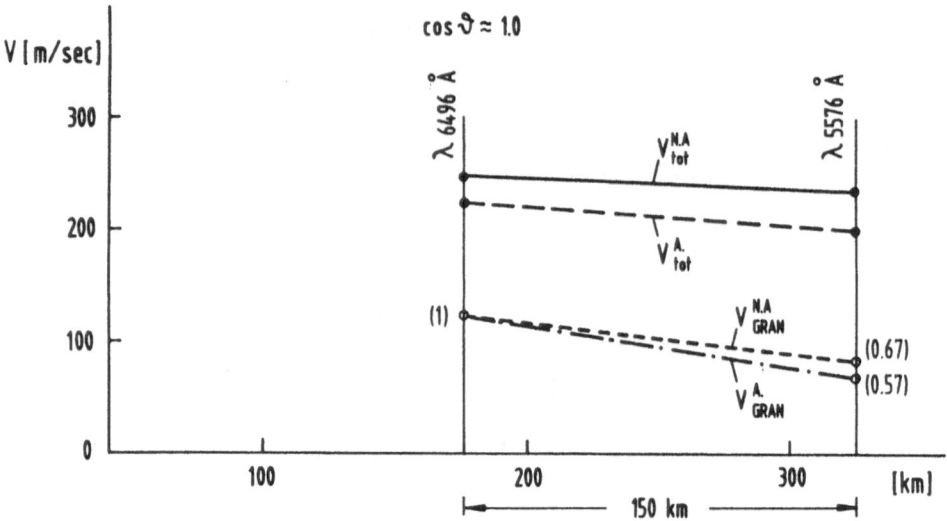

Figure 3: Velocity variation with height $\cos\theta = 1.0$

solar disc. Using these power spectra we calculated the V_{tot}, the total rms velocity. The values are given in Figs. 1 and 2.

To determine if there is a difference between the velocity varitions with height in the active and the non-active regions we calculated $V_{tot}^{A}/V_{tot}^{N.A}$ in the deeper and higher layers. We find that the value of $V_{tot}^{A}/V_{tot}^{N.A}$ does not change with height, which demonstrates the similarity of the variation of V_{tot} in the active and non-active regions. This does not depend on the position on the solar disc. In Figs. 3 and 4 the height dependence of the velocity V_{tot} is shown for the active and non-active regions.

The range of small-scale rms velocity V_{gran} in the total velocity V_{tot} is marked by the shaded areas in the power spectra shown in the Figs. 1 and 2. V_{gran} was not calculated directly. To estimate the variation of V_{gran} with height, we compared the shaded areas in the deeper layers (for 6496 Å, Figs. 1 and 2, above) with those of the higher layers (for 5576 Å, Figs. 1 and 2, below). Normalizing the values of the deeper layers to 1, we found a decrease of V_{gran} at the higher layers by a factor of 0.67 in the non-active regions, and by 0.57 in the active regions for $\cos\theta = 1.0$, and by factors of 0.84 and 0.69, respectively, for $\cos\theta = 0.8$. The comparison of V_{gran} in the active regions at $\cos\theta = 1.0$ and $\cos\theta = 0.8$ (Figs. 3 and 4) shows that the decrease with height is larger at $\cos\theta = 0.8$ than at $\cos\theta = 1.0$.

Summarized, we can say that the gradient of the velocity in the active and those in the non-active regions depends sensibly on the spatial structuring of the velocity field. In the case in which the velocities correspond to a velocity field which is represented by all spatial wave numbers there is no difference in the gradients between active and

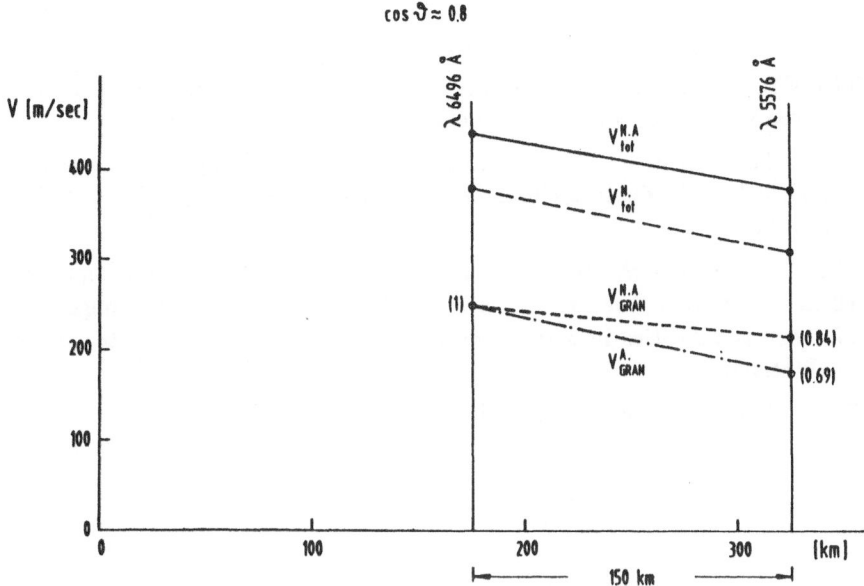

Figure 4: Velocity variation with height $\cos\theta = 0.8$

non-active regions neither for the vertical nor for the non-vertical velocity fields (Figs. 3 and 4). However, this is not the case for the small-scale (3.″5–1.″6) velocity field. Here, we observed a difference between the velocity gradients in the active and the non-active regions (0.57 and 0.67, respectively). This difference is enhanced if we consider the velocity gradients calculated for the case of $\cos\theta = 0.8$ (0.85 and 0.69, respectively).

Now, the faster decay of the small-scale velocity in the active regions can be interpreted as an emerged energy deficit in the higher layers, which seems to be larger at $\cos\theta = 0.8$ than at $\cos\theta = 1.0$. We interpret this finding as an enhanced loss of kinetic energy with respect to the non-active region. A first estimation of the variation of the kinetic energy (ρv^2) with height in the active regions reveals a value that is 30% smaller than the corresponding value of the non-active region. This energy deficit could be due to the influence of the horizontal velocity. This has to be shown by further measurements towards to the limb of the solar disc.

References

Mattig, W., Nesis, A.: 1974, *Solar Phys.* **38**, 337
Mattig, W., Nesis, A.: 1976, *Solar Phys.* **50**, 255

Discussion

KOUTCHMY — Could you specify the method used to measure velocities, and also how you remove the 5-minute oscillation?

NESIS — The velocities correspond to the measured rms Dopplershift in the core of the line. The 5-minute oscillations were removed by filtering the spatial power spectrum.

DEUBNER — A remark: since motions across magnetic field lines are impeded more efficiently at lower β-values, i.e. at greater heights, the relatively stronger gradient of the rms velocities in active regions at $\mu = 0.8$ appears easy to understand qualitatively.

NESIS — This is a possible explanation.

REPORT ON THE SESSION: POSITION AND DISCUSSION: SOLAR DATA

A. Righini[1], E.H. Schröter[2]
[1]*Osservatorio Astrofisico di Arcetri, Firence, Italy*
[2]*Kiepenheuer-Institut, Freiburg, FRG*

1   Introduction

The authors have been asked by the scientific organizing committee to initiate and to guide a round table discussion on solar granulation data presented during the workshop. They understood their task to provoke discussions by listing a number of questions which arose from the preceding oral presentations and from posters. An extended and lively discussion followed after the presentation of such questions, which lasted almost twice as long as foreseen in the program.

2   Topics of discussion

Among others the following topics were discussed to a higher extent:

a) Aspects arising from observational and data-reduction procedures.

What is meant when the achieved spatial resolution is described by, say: X arcsec? Is this description still adequate in the new era of sub-arcsec solar research from space and ground? Can observations of solar granulation at $\lambda = 1600$ nm give us a new and better insight into the processes in deeper photospheric layers, as proposed e.g. by Beckers and Koutchmy (with lower granular contrast and worse MTF as compared to the visible range)? Is instrumental and atmospheric scattered light really a crucial point in context with observations of granulation? In the framework of somewhat controversial results presented by the Pic du Midi team and the Spacelab 2 group regarding the granular cell-size the question came-up: do modern image processing methods really make the reduction of granulation data more objective? If this is not completely the case, should we aim for a unification of procedures and definitions?

b) Benefit from data of moderate spatial resolution.

Can granulation data with moderate spatial resolution ($\Delta x \geq 1$ arcsec) still add to our knowledge of the physics of granulation or not? Do C-shape and line shift investigations help us to reconstruct the temperature and velocity structure in the overshoot region? The behaviour of the C-shape and line shift from spectra of rather high spatial resolution, recently obtained on Tenerife (see contribution by Mattig et al. in this volume),

R. J. Rutten and G. Severino (eds.), Solar and Stellar Granulation, 295–297.
© *1989 by Kluwer Academic Publishers.*

demonstrate what complex superposition of very local variations makes up the C-shape and line shift as observed with low spatial resolution. The steep ΔT-gradient seems to be established by several investigations and from theoretical models; the same applies to the gradient of the rms-value of the vertical granular velocities. But, there is still a controversy between observations and model-calculations regarding the gradient of the horizontal velocities.

c) The morphology of granulation; mesogranulation.

In context with the discrepancy between the observed granular contrast ($\simeq 14\%$ at $\lambda = 500$ nm) and model-calculations leading to 24% (e.g. Nordlund) a discussion arose how much tiny unresolved granules and intergranular lanes may affect the measured contrast? What can be the reason that Pic du Midi observers find a distribution of granules as function of their area, which yields no "characteristic" scale, and a change of the slope in the power-spectrum, which is in contradiction to the results from the Spacelab 2 experiment? Here, once again, the problem of data handling and of definitions of parameters came up. Connected with this problem area the discussion passed over to the cell-size of granules in the vicinity of sunspots (or more general in active regions). Several observers found the mean cell-size of granulation surrounding a sunspot to be smaller by $\sim 20\%$ whereas others could not confirm it. Subject to definition or depending on evolutionary stage of the sunspot?

Evoked by A. Title's presentation it was speculated whether stream vortices as observed by the SOUP-group and recently also on La Palma by Brandt et al. are a common or rare phenomenon. Evidently very high spatial resolution for hours and sophisticated reduction methods are needed to detect such phenomena. It was considered very unfortunate, that in both cases simultaneous Ca^+-, $H\alpha$-pictures and magnetograms were not available for comparison.

Finally the authors raised the question: what is common to the following observations:

i) Kawaguchi's "families of large granules" (members having a diameter of 3–5 arcsec, the lifetime of the family being ~ 40 minutes)

ii) long lived cells of large granules, as observed by Dieletis et al. with a lifetime of more than 20 minutes (see also the poster at this workshop) and

iii) the mesogranulation detected by November et al. in 1981 as a cellular structure in the upper photosphere on a scale of 5–10 Mm and a lifetime of ~ 2 hours.

Although there was no consent whether all 3 observations describe but the same phenomenon this ambiguity was used to extensively rediscuss the reality of mesogranulation, its observed properties and its possible origin. As to the first point, it was mentioned that already thirty years ago, observers have found "conglomerates of granules" on a scale of 5–10 arcsec (see the first edition of Bray's and Loughhead's "Solar granulation"). Deubner claimed to have found more than ten years ago in his powerspectra peaks on the scale of mesogranulation but did not give recognition to them. Some speakers doubted that was observed by Damé and Martić with a space-born camera at $\lambda=190$ nm as intensity fluctuations on a scale of 8 Mm is to be identified as mesogranulation above temperature minimum. As to the origin two possible processes have been considered: convective origin due to ionization of He^+ (about 7 Mm deep in the convection zone) and active granule grouping maintained by the structure of the magnetic fields. For the phenomenon described by Damé and Martić it has been argued that it may represent either the response of the atmosphere to 5 min-oscillations or the result of coupling of chromospheric spicules to their photospheric foot points.

THE LIMB EFFECT OF THE KI 769.9 NM LINE IN QUIET REGIONS

J A Bonet, I Marquez, M Vazquez
Instituto de Astrofisica de Canarias,
38200 - La Laguna, Tenerife, Spain.
H Wöhl
Kiepenheuer Institut für Sonnenphysik,
7800 Freiburg, FRG.

1. INTRODUCTION

The limb effect of photospheric spectral lines has been used rather frequently in the last few years as a tool to learn about the parameters characterizing solar granulation. Dravins et al. (1981), Balthasar (1984) and Nadeau (1988) presented observations for a large sample of lines.

In fact the KI 769.9 nm line has been used as a representative of the upper photospheric layers, where the influence of the convective motions is reduced. Gasanalizade (1979) and Roca Cortés et al. (1983) did not find any significant center-to-limb variation of the line's minimum position. On the contrary, Andersen et al. (1985), Lo Presto and Pierce (1985) and Anguera et al. (1987) reported values around 100 m/sec.

In this work we present a preliminary study of the KI limb effect at different residual intensities expressed in an absolute wavelength scale based on the nearby terrestrial lines. A numerical simulation tries to fit the observations using a standard two-component model.

2. OBSERVATIONS AND DATA REDUCTION

The KI profiles were obtained during several observing runs along the years 1978, 1980, 1981 with the Gregory Telescope of the Universitäts-Sternwarte Göttingen placed at Locarno (see Roca Cortés et al., 1983) and in 1983 using the MacMath Telescope at Kitt Peak (Bonet et al., 1988).

The data reduction procedure is described in Bonet et al. (1988). The Locarno data were reanalyzed with the same computer code.

R. J. Rutten and G. Severino (eds.), Solar and Stellar Granulation, 299–303.
© *1989 by Kluwer Academic Publishers.*

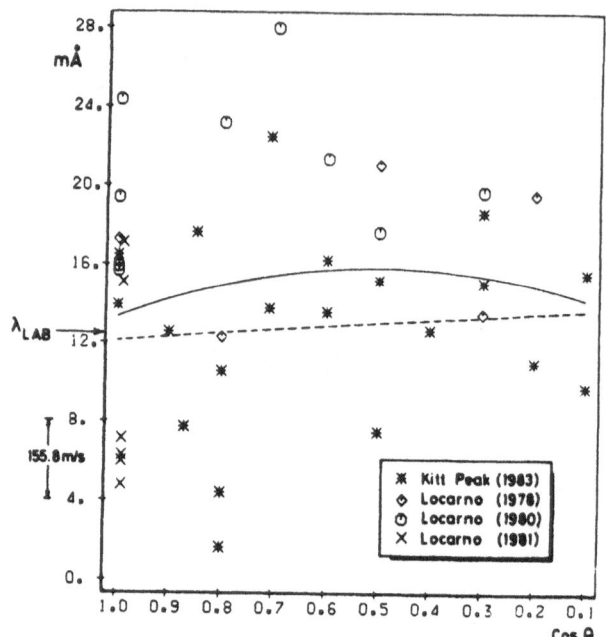

Figure 1. Limb effect of the KI 769.9 nm line and the parabolae fitted to all data (full line) and to Kitt Peak data only (dashed line).

3. RESULTS

Figure 1 shows the measured positions, for different cosθ values, of the line minimum in an absolute wavelength scale. A second order polinomial $(\Delta\lambda(\theta)=a+b(1-\cos\theta) + c(1-\cos\theta)^2$ is fitted to all data (full line) and to Kitt Peak data only (dashed line). Several points are worth outlining:

(a) As far as we are aware, the laboratory wavelength of the KI line is known with rather large uncertainty (769.8959 ± 0.0006 nm; Risberg, 1956) since the potassium is not isotopically pure.

(b) The scatter in the data points is produced mainly by supergranulation velocity fields. In most of the cases the measurements are averages of periods longer than 45 min, cancelling most of the influence of solar oscillations.

(c) There are indications of a temporal variation of the limb effect, which changes from one year to another; this variation can be partly attributed to the use of different instrumentation and different oxygen lines to determine the dispersion of the spectrograph.

(d) The Kitt Peak data give a value of the limb effect similar, but somewhat smaller (70 m/sec), than those published by Andersen et al. (1985), Lo Presto and Pierce (1985) and

301

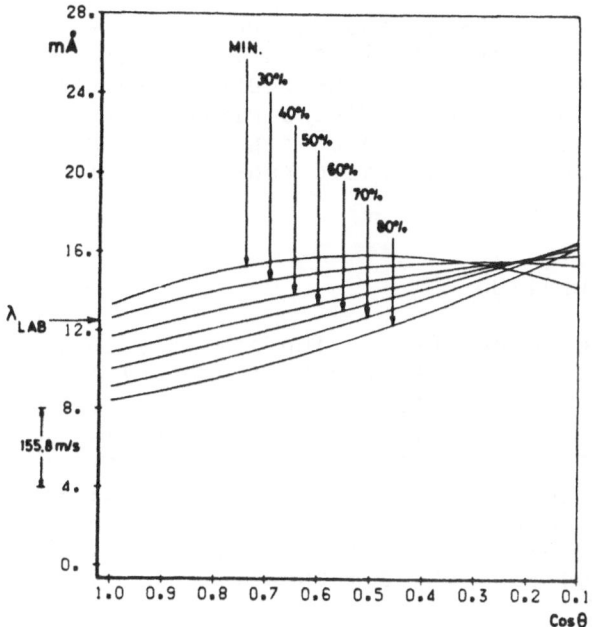

Figure 2.   Center-to-limb variation of the KI 769.9 nm line
bisector at several levels; parabolae fitted to both the
Locarno and Kitt Peak data.

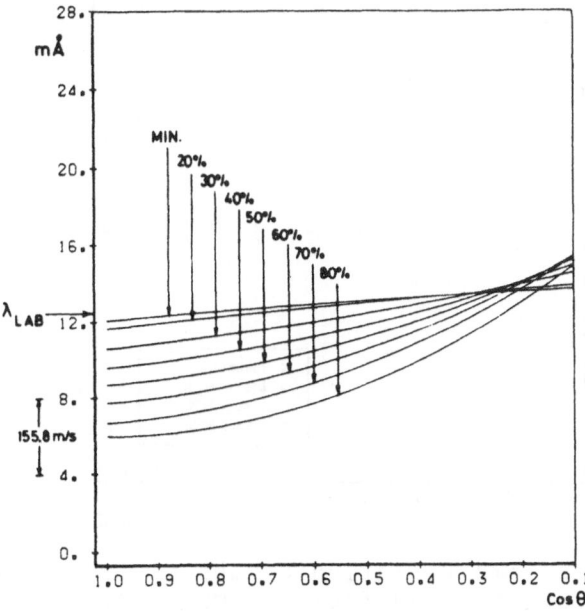

Figure 3.   The same as in Figure 2 but for Kitt Peak data
only.

302

Anguera et al. (1987).
    Figure 2 (for all data) and Figure 3 (for Kitt Peak
data only) show the center to limb variation of the
bisector at different levels. The represented curves are
parabolae fitted to the data. The amplitude of this
variation increases towards the continuum as expected from
a convective origin. An intrinsic redshift is found near
the limb.

4. NUMERICAL SIMULATION

We assume that the solar granulation consists in two average
component atmospheres represented by the model of Nelson
(1978). We consider the same area for the hot and cool
components. Vertical velocities are positive in the hot
(rising) component and negative in the cool (descending)
component. Although we know that the horizontal velocities
play an important role in the limb effect (Balthasar, 1985)
in our preliminary simplified simulation these velocities
are zero on average for each component.

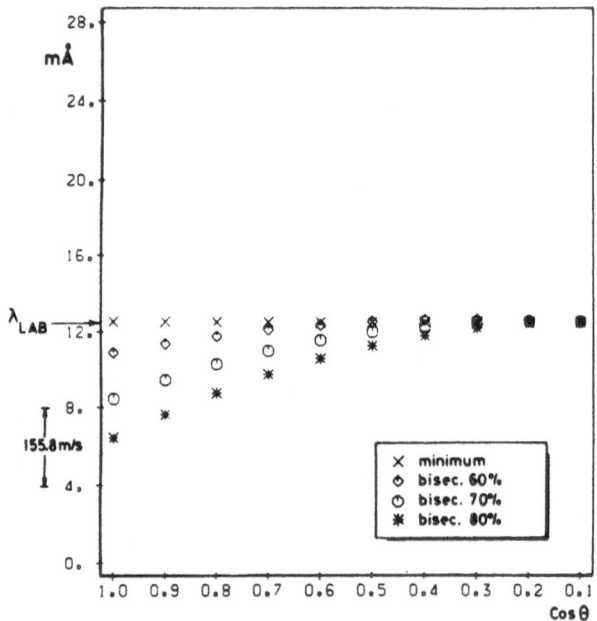

Figure 4. Center-to-limb variation of the KI line bisector
at different levels, obtained from a simplified numerical
model.

The theoretical profiles of the KI lines in both components are computed solving the transfer equation via the Feautrier method (Mihalas, 1978). We assume LTE, but to improve the line depth we have corrected the line source function using a ratio given by Marmolino et al. (1984).

Figure 4 shows the obtained center to limb variation of the bisector at different levels. We have multiplied by a factor 1.5 the velocities in order to reproduce the bisector at the 80% level (the more blue shifted) in the solar center. A microturbulence value of 1.2 km/sec is used to reproduce the equivalent width and the FWHM at the center of the disk.

REFERENCES

Anguera M., Palle P.L., Regulo C., Roca Cortés T., Isaak G.R., McLeod C.P., Van Der Raay H.B.: 1987, in 'The Role of Fine-Scale Magnetic Fields on the Structure of the Solar Atmosphere'. Cambridge University Press.
Andersen B.N., Barth S., Hansteen V., Leifsen T., Lilje P.B., Vikanes F.: 1985, Solar Phys. **99**, 17.
Balthasar H.: 1984, Solar Phys. **93**, 219.
Balthasar H.: 1985, Solar Phys. **99**, 31.
Bonet J.A., Marquez I., Vazquez M., Wöhl H.: 1988, Astron. Astrophys. **198**, 322.
Dravins D., Lindegren L., Nordlund A.: 1981, Astron. Astrophys. **96**, 345.
Gasanalizade A.G.: 1979, Soln. Dannye No. 7. pg. 85.
Lo Presto J.C., Pierce A.K.: 1985, Solar Phys. **102**, 21.
Marmolino C., Roberti G., Severino G., Vazquez M., Wöhl H.: 1984, "The Hydromagnetics of the Sun", ESA SP-220 pg. 191.
Mihalas D.: 1978, Stellar Atmospheres, W.H. Freeman.
Nadeau D.: 1988, Astrophys. J. **325**, 480.
Nelson G.D.: 1978, Solar Phys. **60**, 5.
Risberg P.: 1956, Arkiv. Phys. **10**, 583.
Roca Cortés T., Vazquez M., Wöhl H.: 1983, Solar Phys. **88**, 1.

Vortex Motion of the Solar Granulation

P.N. Brandt[1], G.B. Scharmer[2], S.H. Ferguson[3], R.A. Shine[3], T.D. Tarbell[3], and A.M. Title[3]

[1]*Kiepenheuer-Institut, Freiburg*
[2]*Stockholms Observatorium, Saltsjöbaden*
[3]*Lockheed Palo Alto Research Laboratory, Palo Alto*

Summary

A-79 minute series of 388 CCD granulation frames of sub-arcsec resolution was observed with the Swedish 50 cm vacuum telescope at La Palma in June 1987. The TV system was equipped with analogue real-time image selection circuitry. Horizontal flow fields of the granulation were determined with a local correlation technique. Their vectorial time average shows a pattern of converging and diverging flows of average speed $v = 0.7$ km/s, corresponding to the mesogranulation. Moreover, a clear example of vortex motion was detected, which visibly dominates the motion of the granules in its neighbourhood. The vortex had a diameter of ≈ 5000km and a vorticity at the center of 1.4×10^{-3} s^{-1}.

1 Observations

Object:	quiet sun granulation near disc center at $\lambda = 4696 \pm 27$ Å
Telescope:	Swedish solar vacuum telescope, La Palma (Canary Islands)
Time:	June 16, 1987, 12:41 - 14:00 h
Technique:	commercial CCD TV camera with real time image selection
Field of view:	14.2 by 12.2 arcsec^2
Pixel size:	0.07 arcsec
Series:	388 frames at approx. 12.2 sec time intervals, total duration 79 min
Intensity contrast:	8.5 to 10.6 % rms, uncorrected (example Fig. 1)

305

R. J. Rutten and G. Severino (eds.), Solar and Stellar Granulation, 305–310.
© 1989 by Kluwer Academic Publishers.

306

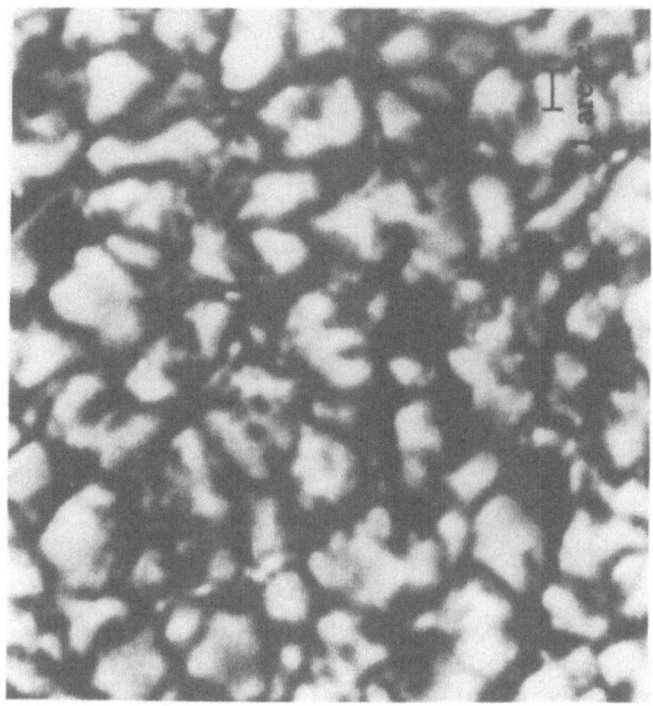

Figure 1: Granulation image taken with CCD video camera and "image grabbing" system at the 50 cm Swedish solar vacuum telescope at La Palma (Canary Islands). Field of view 14.2 by 12.2 arcsec; $\lambda = 4696 \pm 27$ Å.

2 Data analysis

After background subtraction and flat-fielding of the single frames, horizontal motions of the granulation pattern were determined using a "local correlation technique" (November *et al.*, 1987). This was done by dividing each frame into sub-images of approximately 0.45 by 0.45 arcsec^2 width and searching for the best cross-correlation of the intensity patterns in pairs of frames taken at intervals of 61 seconds. Velocity maps were generated from the shifts determined from these cross-correlations and the known time difference. Vectorial averaging over these velocity maps yielded average flow maps of the granulation pattern.

Video movies of granular evolution and motion were produced. The effects of seeing were reduced by "de-stretching" the images frame by frame (Topka *et al.*, 1986). This de-stretching essentially removes differential image distortion by cross correlating 2.4 x 2.4 arcsec^2 sub-images in subsequent frames and reshifting them by an interpolation scheme.

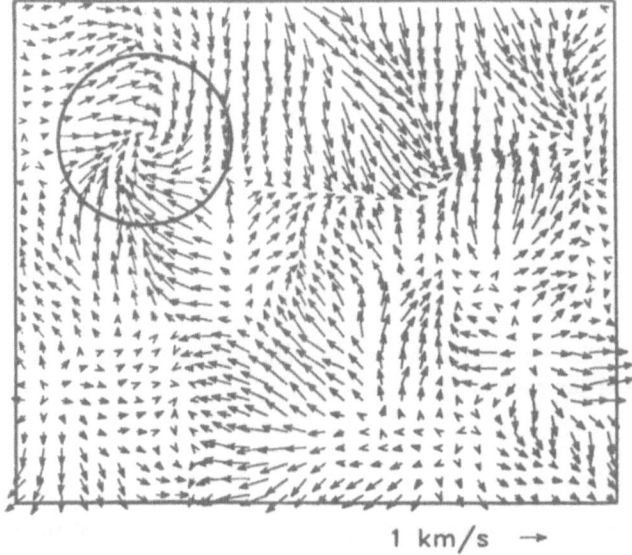

1 km/s →

Figure 2: Horizontal velocity field (14.2 by 12.2 arcsec) derived from granule motions averaged over 79 minutes. The vortex structure is circled. The lengths of the arrows are proportional to the velocity. A reference arrow indicating 1 km/s is shown below the box.

3 Results

3.1 Flow fields

Vectorial averaging of the flow fields over the 79 minute observing period shows that the granules are not moving at random, but that they follow a consistent large scale pattern of motion (Fig. 2) which can be identified as "mesogranulation". Its pattern size is 5 to 7 arcsec.

Flow speeds in the 79 minute average map range from a few hundred m/s to 1.2 km/s with an average of 0.67 km/s.

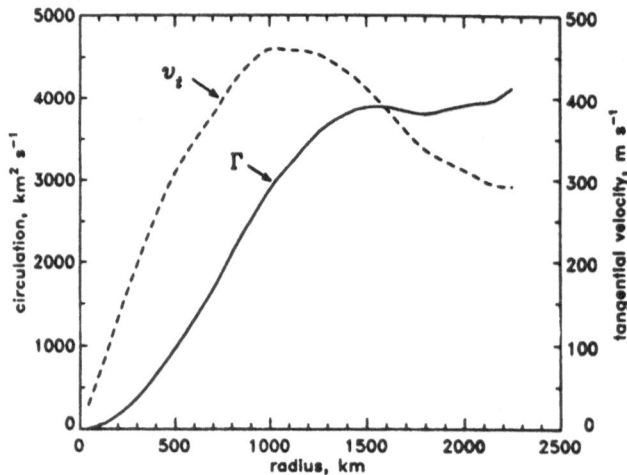

Figure 3: The circulation (solid line) and the tangential velocity v_t (dashed line) as functions of the radius.

3.2 Vortex flow

A conspicuous feature was seen in the upper left corner of the frames: granules spiral around, and disappear at, a certain point, thus forming a vortex. The vortex had a diameter of > 5000 km and the vorticity ($curl_z \, v_{hor}$) at the center was $1.4 \times 10^{-3} \, s^{-1}$. Its tangential velocity and circulation

$$\Gamma = \int_a curl_z(\vec{v})da = \int_c \vec{v}d\vec{l}$$

as function of the radius are shown in Fig. 3. The paths of test particles drifting in the average flow field are shown in Fig. 4. The vortex motion persisted over the entire 79 minute observing period.

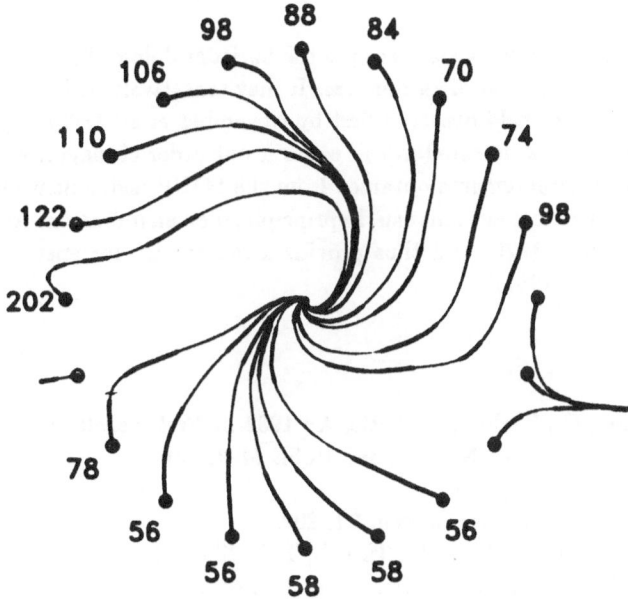

Figure 4: The paths of test particles from the circumference of a 2000 km radius circle centered on the vortex. The starting positions are labeled with the travel times to the center in minutes. The path lines change between thick and thin every 15 minutes.

4 Conclusion

A flow pattern of mesogranular scale is seen in the horizontal flow of granules, persisting over the entire observing period of 79 minutes. It may tentatively be identified with the mesoscale vertical velocity field observed first by November *et al.* (1981).

The vorticity of the flow detected in this series is one order of magnitude larger than seen in previous flow measurements obtained from the SOUP instrument on Spacelab 2 (Simon *et al.*, 1988). Vortical motions can, in principle, play an important role in twisting magnetic fields (Sakurai, 1979) and thus provide a mechanism for energy storage and coronal heating (Parker, 1987).

References

November, L., Simon, G., Tarbell, T., Title, A.: 1987 in *High Resolution Solar Physics*, ed. G. Athay & D. Spicer, NASA Conf. Publ. 2483, 121

Parker, E.: 1987, *Ap. J.*, **318**, 876

Sakurai, T.: 1979, *Publ. Astr. Soc. Japan*, **31**, 209

Simon, G.W., Title, A.M., Topka, K.P., Tarbell, T.D., Shine, R.A, Ferguson, S.H., Zirin, H., and the SOUP Team: 1988, *Astrophys. J.*, **327**, 964

Topka, K. P., Tarbell, T. D., Title, A. M.: 1986, *Astrophys. J.*, **306**, 304

THE GRANULATION SENSITIVITY OF NEUTRAL METAL LINES

J.H.M.J. Bruls, H. Uitenbroek, R.J. Rutten

Sterrekundig Instituut Utrecht, The Netherlands

Abstract. We discuss the sensitivity of neutral metal lines to the temperature variations imposed by the granulation on the solar atmosphere. We concentrate on the Ni I 676.78 nm line, which the Global Oscillation Network Group (GONG) has chosen as Doppler sensor for helioseismological study of the sun. We have studied the NLTE formation of this line, in particular its sensitivity to granular temperature fluctuations, because these may represent a major noise source in the oscillation signal. A more detailed account will be published elsewhere.

The GONG line turns out to be a typical metal line from a minority ionization stage. As in the case of Fe I (Lites 1972, see Rutten 1988), the most important NLTE aspect of its formation is its sensitivity to the hot ultraviolet radiation field. This causes two effects which affect the observed line core.

The first effect is the large overionization due to the ultraviolet imbalance $J_\nu > B_\nu$. This overionization empties all Ni I levels in the upper photosphere considerably. It results in an appreciable rescaling of the optical depth scale for all Ni I lines, giving a lower height of formation (for a given transition probability) than LTE modeling of the ionization equilibrium would indicate. This ultraviolet overionization effect is not compensated by overrecombination to high levels in the red (where $J_\nu < B_\nu$) as it is in the case of alkalids as K I (Gomez *et al.*, these proceedings), because the levels that are high enough in the Ni I term diagram to feel the red imbalance are too high to affect the overall populations.

The second effect is overexcitation in the upper photosphere, again due to the ultraviolet imbalance $J_\nu > B_\nu$. This effect has been studied for Fe II (Cram *et al.* 1980; Watanabe and Steenbock 1986), but it appears in Ni I and probably Fe I as well: the numerous ultraviolet lines overexcite their upper levels, and these overpopulations produce enhanced source functions in the upper photosphere for subordinate lines at longer wavelengths such as the GONG line (see Rutten 1988).

How do these NLTE formation effects interact with the granular temperature modulation for the GONG line? We have studied this in detail, using the Carlsson radiative transfer code (Carlsson 1986, Scharmer and Carlsson 1985) for granulation models that were kindly supplied by M. Steffen from his simulation (Steffen and Muchmore 1988).

The computations show that although Steffen's granules have only a very small (and reversed) temperature contrast at the height where the core of the GONG line is formed, it is nevertheless sensitive to what happens in the granulation. This is so because the ultraviolet radiation carries the large temperature contrasts present in the granulation

R. J. Rutten and G. Severino (eds.), Solar and Stellar Granulation, 311–312.

in deep layers all the way up to the height of formation of the line core, and there affects both the line opacity and the line source function through the two NLTE mechanisms described above. As a result, the line core brightens appreciably above a hot granule, although in Steffen's results the temperature at that height is actually slightly lower than above an intergranular lane.

In conclusion: the core of the GONG line shows granular intensity variations due to NLTE coupling with the ultraviolet radiation from deeper layers, even if the granulation temperature contrast has vanished at its formation height.

This behaviour is not typical for this specific Ni I line, and we may expect that other comparable minority-species metal lines such as Fe I lines behave similarly.

References

Carlsson, M., 1986, Uppsala Astronomical Observatory Report **33**

Cram, L.E, Rutten, R.J. and Lites, B.W., 1980, Astrophys. J. **241**, 374

Lites, B.W., 1972, *NCAR Cooperative Thesis* **28**, University of Colorado, Boulder

Rutten, R.J., 1988, in: R. Viotti, A. Vittone, M. Friedjung (Eds.), *Physics of Formation of FeII lines Outside LTE*, Proceedings IAU Colloquium 94, Reidel, Dordrecht, p. 185

Scharmer, G.B. and Carlsson, M., 1985, J. Comput. Phys. **59**, 56

Steffen, M, and Muchmore, D., 1988, Astron. Astrophys. **193**, 281

Watanabe, T. and Steenbock, W., 1986, Astron. Astrophys. **165**, 163

GRANULE LIFETIMES AND DIMENSIONS AS FUNCTION OF DISTANCE FROM A SOLAR PORE REGION

Tron Andre Darvann[1] and Ulf Kusoffsky[2]

[1]Institute of Theoretical Astrophysics
University of Oslo, P.O. Box 1029, Blindern
N-0315 OSLO 3, Norway

[2]Grupo Sueco, Observatorio del Roque de los Muchachos
Apartado 66, Santa Cruz de La Palma
Islas Canarias, Spain

ABSTRACT. The lifetimes of photospheric granules have been determined from cross-correlation of granulation images (von der Lühe 1983, Andreassen et al 1984, Darvann 1986). The measurements were carried out on a 19 minute long time series of solar granulation obtained with the Swedish 50 cm Vacuum Tower Telescope at La Palma (Wyller and Scharmer 1985). A rapid decrease of the granule lifetimes, from about 12 minutes close to a pore region to 5 minutes at a distance of about 10 arcseconds from the pores, is found. A morphological study of the granulation (i.e. measurement of the average granule size, number density and fractal dimension) in the same pore region has been carried out, and the results are discussed.

1   Introduction

The lifetime and dimension of granules of the solar photosphere are important parameters for the understanding of solar convection. Most measurements have been carried out on normal granulation in quiet areas of the photosphere. Alissandrakis et al (1986) and Mehltretter (1978) describe various methods for measurement of lifetimes that have been used. Results from earlier studies of granule lifetimes are summarized by Wittmann (1979), and by Bray et al (1984). The values range from 3 to 16 minutes.

The lifetime of granules in strong magnetic regions may be different from the lifetime in other parts of the photosphere (Bray et al 1984). This was observed by Schröter (1962) to be the case for granules close to a large sunspot, where he found a typiical lifetime of 12 minutes.

Several studies have been carried out to measure average size and number density of granules close to and away from sunspots and active regions. The results are conflicting, as pointed out by Collados et al (1987). They find no dependence of the granule size on their distance from sunspots, while Schmidt et al (1988), and Macris et al (1989)

313

R. J. Rutten and G. Severino (eds.), Solar and Stellar Granulation, 313–323.

find on the average smaller granules close to sunspots. Schmidt et al (1988) derive a smaller RMS intensity contrast close to sunspots, which they attribute to smearing of small granules by seeing. On the other hand, Koutchmy (1989) does not find a smaller contrast in magnetic regions (plage regions).

Title et al (1987) have studied the correlation between magnetic fields and Doppler motion in the photosphere, and found a smaller downflow in the intergranular lanes where the magnetic fields are strong. Title et al (1986) used "seeing-free" images recorded with SOUP and found the average granule lifetime in magnetic regions to be 16 minutes, and 8 minutes in non-magnetic regions. They also found that granules are smaller in magnetic regions than in non-magnetic regions. Dialetis et al (1989a,b) showed that long-lived granules, both in active and non-active regions, form a network comparable to the mesogranulation. Müller et al (1989) found a network on a scale somewhat smaller than the mesogranulation (6 - 7 arcsec) consisting of granules with size smaller than average.

We have applied the method of cross correlation of granulation images to measure average granule lifetimes. In a pore region we found that the lifetime close to the pores was more than twice (12 min) the lifetime further away (5 min).

2   Observations and data processing techniques

The observations used in the present study were obtained on July 4, 1986, with the 50 cm f/45 Swedish Vacuum Tower Telescope at La Palma (Wyller and Scharmer 1985). We used secondary optics to form a 42 cm diameter solar image. The exposure time was 1/1000 s on Kodak Technical Pan, centered at λ 4680 Å and with bandwidth 50 Å. The young pores of the sunspot group 4734, which was in the process of forming (Solar Geophysical Data), were observed. The resolution was better than 0.4 arcseconds in the 17 negatives selected for the lifetime study. The images were digitized using the scanning microphotometer of Institute of Theoretical Astrophysics, University of Oslo. The resulting image scale is 0.11 arcsec/pixel. The time series of granulation images is 19 minutes long, and the time between each image is approximately 70 s. A rotation algorithm using cubic spline interpolation has been applied to each image in order to compensate for the rotation of the field caused by the heliostat of the telescope. Median filtering was applied to the images to remove noise and dust spikes. The images were transformed to a linear intensity scale by standard calibration of the film and determination of the characteristic curve.

Granule sizes were computed using an algorithm which will be explained below. Other algorithms have been applied by Title et al (1987), Title et al (1986), Roudier and Müller (1986) and Collados et al (1986). The method presented here is based on the following definition: a point in the image is part of a granule if the intensity is higher than the intensity along a curve through the inflection points in the image. This definition is objective, independent of large-scale intensity gradients, and corresponds well to the human visual impression of sizes and forms in the granulation pattern (Figure 1a). Figure 1b shows the gradient image corresponding to the granulation pattern of Figure 1a. Contour lines of the granules correspond to the locations of maximum intensity gradients.

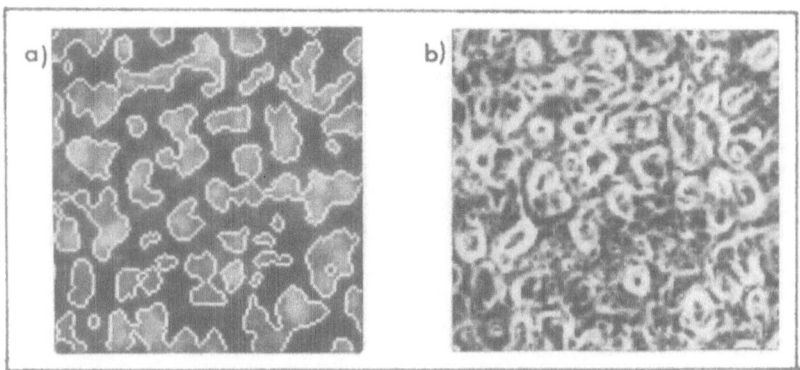

Figure 1: a) Contours around granules drawn by use of the granule finding algorithm. b) Gradient image corresponding to the granulation in Figure a).

3   Measurement of granule lifetimes as function of position

The method of cross correlation (CC) (see von der Lühe 1983, Andreassen et al 1984, Darvann 1986) was applied to measure the lifetime of granules in 9 selected subframe areas in the 17 images of the 19 minute long timeseries. Figure 2 shows the position of the subframes (each being 7.0 x 7.0 arcsec^2) in the full reference (time=0) image. The subframes are labeled with letters from A to I with increasing distance from the pore region, the largest distance being approximately 20 arcseconds. All subframes are carefully positioned in order to avoid pores in the field of view.

Each subframe in the timeseries was compared to the corresponding reference (time=0) subframe by computing the CC coefficient. The decrease of the CC with increasing time separation between the frames is the effect of the granule decay. The time for the CC to fall to 1/e is taken as the average granule lifetime. A linear least square fit of these points for CC values greater than 0.4 was obtained for 5 decay curves for the same subframe position, using different reference images. The average values of the 5 cases were taken as the average granule lifetime for the subframes of a given position in the field. Figures 3a and 3b show examples of decay curves corresponding to lifetimes of 7 ± 1 minutes and 15 ± 3 minutes, respectively. The decay curves for all of the 9 subframe areas yield lifetimes as a function of distance from the pore region (Figure 4). The squares in the figure are values obtained by using a 7x7 arcsec^2 subframe size, and the crosses give the results from 9x9 arcsec^2 subframes centered at the same position. (Subframe A had to be displaced 2 arcseconds to avoid one of the pores.) The results obtained from the two different frame sizes are not significantly different. The general trend of the plot is quite clear. The areas A to D, which are close to the pore region (or perhaps rather inside the pore region, see Figure 2), have a significantly higher lifetime (about 12 minutes) as compared to the areas E to I, which are further away from the pore. In the areas E to I the average lifetime is 5.3 minutes, which is the same as the lifetime of granules in quiet regions of the photosphere measured by Mehltretter (1978).

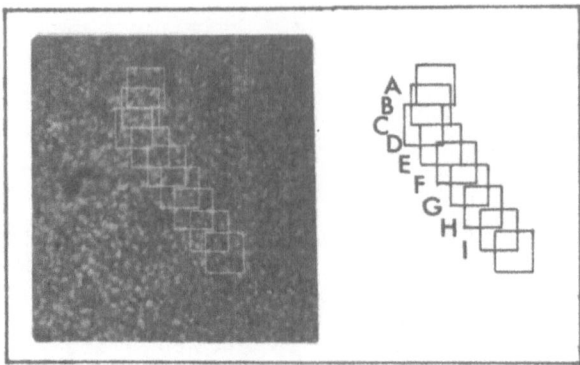

Figure 2: The pore region used for the study in this article. The boxes (7x7 arcsec^2) show the subframes used for lifetime measurements.

Our result confirms earlier measurements showing that granule lifetimes in active regions are much longer than in non-active regions.

4 Granule size as function of distance from pore region

Having found a longer granule lifetime close to the pore region, we started to look for possible variations in the average granule size and number density and the fractal dimension of the granulation also. The granule finding algorithm was applied to the same pore region as the one used for lifetime studies. The positions of the subframes in the images were the same as in Figure 2, but the areas were enlarged to cover 11x11 arcsec^2 which resulted in about 30 granules per subframe. By using all 17 images of the time series the total number of granules used for all nine subframe positions (A to I in Figure 2) were between 500 and 600.

The average granule size, and number of granules per 100 arcsec^2, were derived, and the results plotted as a function of distance from the pores (Figures 5a and 5b). The plotted values represent the mean values of 17 different times, ± standard deviation. Seeing fluctuations are probably responsible for most of the variation in the data at each of the nine positions. Figure 5a indicates that the average number of granules per 100 arcsec^2 is somewhat less inside the pore region (areas A - C) than outside (areas F - I). The average granule size (Figure 5b) shows no variation with distance from the pores except for a slight increase at the position of subframes D and E. This result could imply that the intensity profile of the granules varies systematically with distance from the pores. Close to the pores a larger fraction of the total area consists of intergranular lanes. Note that the changes in the average size and number density occur in the regions C - E (between 5 and 10 arcsec from the pores), and that this corresponds to the position of the abrupt change in lifetime (see Figure 5). At this apparent "boundary" of the pore region, the granules are larger and have a lower number density.

The fractal dimension D of granulation was derived as a function of position and time. This parameter is obtained from the relation $P \sim A^{D/2}$, where P is the perimeter

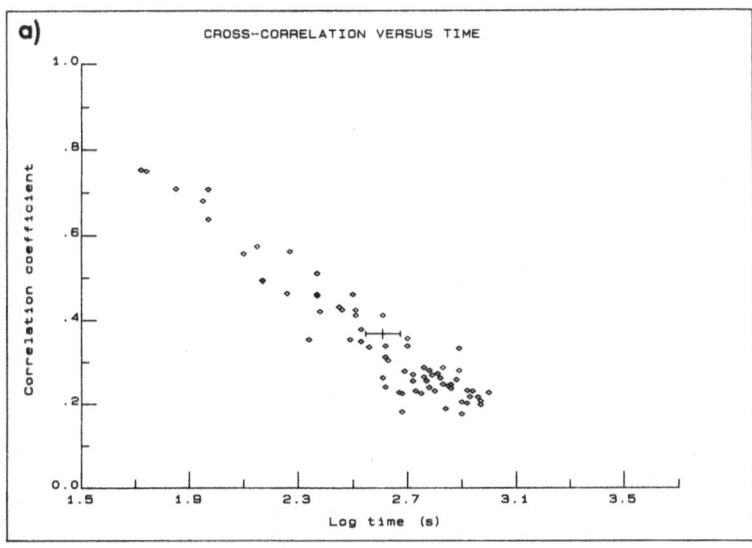

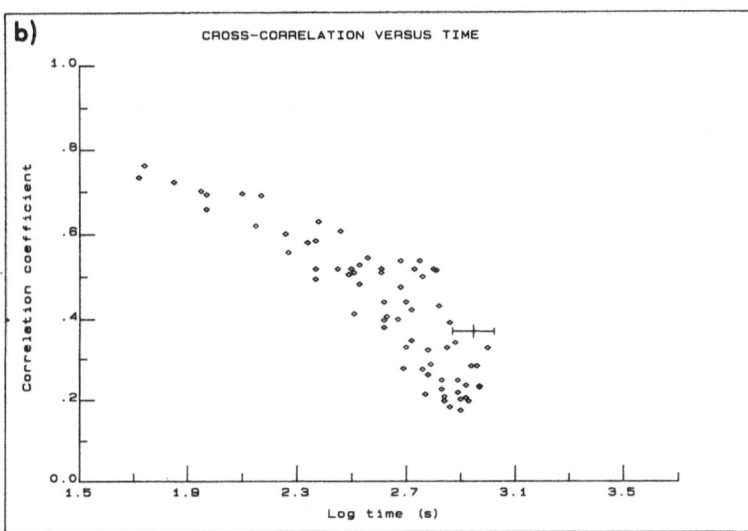

Figure 3: Cross correlation coefficient (CC) as a function of time for two different sub-frame positions in the granulation time series. Lifetimes, i.e. the time it takes for the CC to decrease to 1/e : a) 7±1 minutes, b) 15±3 minutes.

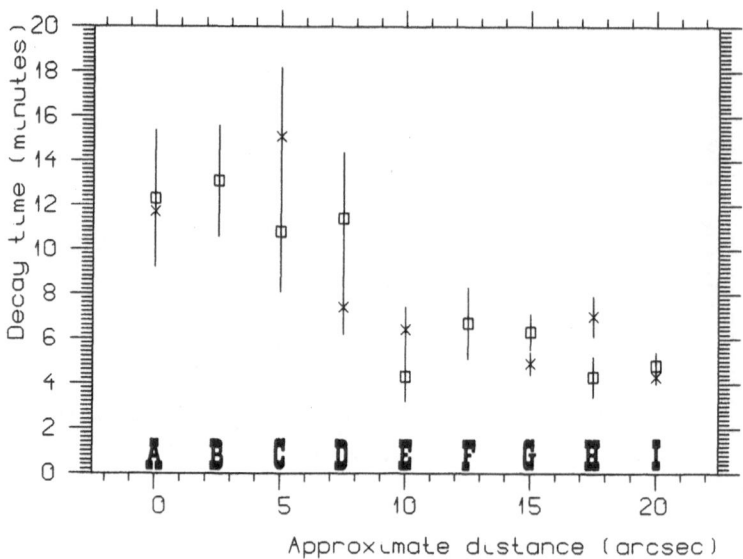

Figure 4: Granule lifetime as function of distance from pores. Labels A to I refer to the areas in Figure 2. Squares and crosses are the results for subframe areas 7x7 arcsec^2 and 9x9 arcsec^2 respectively.

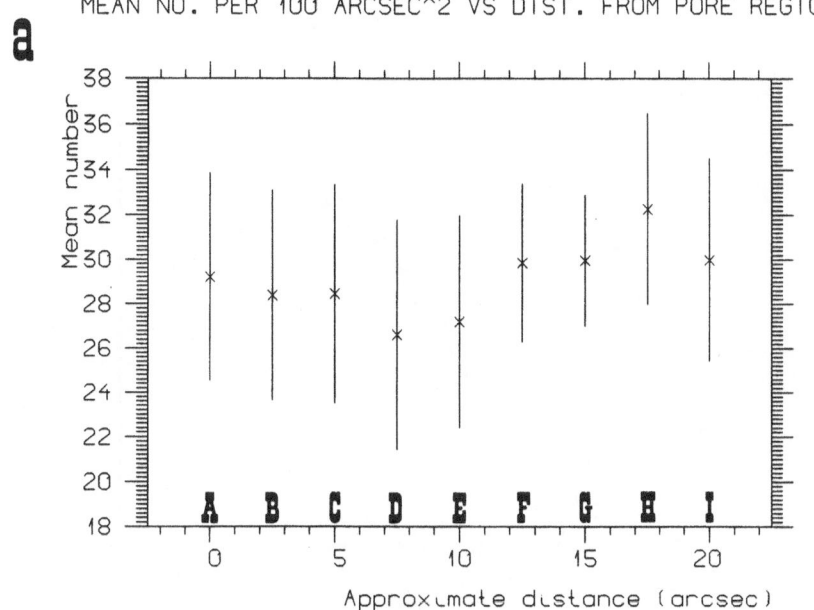

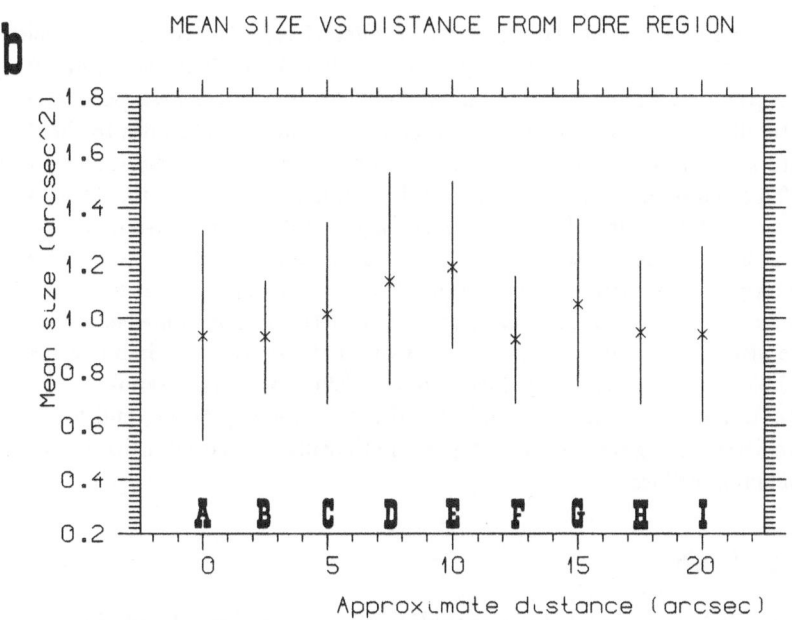

Figure 5: Granule parameters as a function of distance from the pore region: a) the average number density, b) the average size of granules. Labels A to I refer to the areas in Figure 2. Error bars show rms deviation.

320

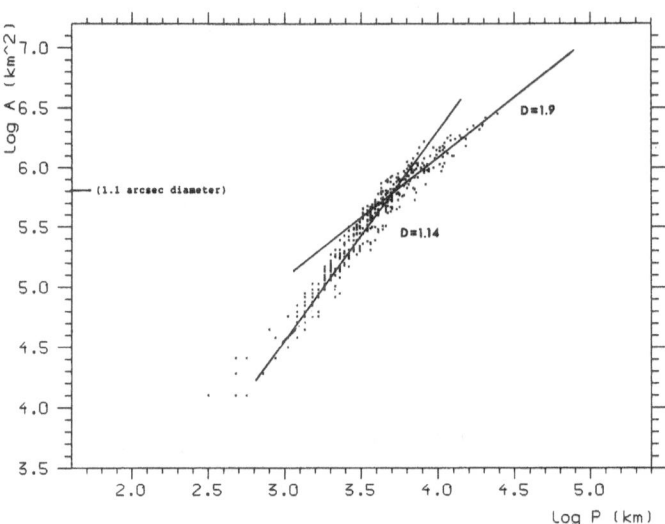

Figure 6: Area-perimeter relation. D is the derived fractal dimension of the granulation.

and A the area of a granule (Roudier and Müller 1986). The fractal dimension in our data was found to be 1.14 ± 0.05 for granules smaller than 1.2 arcsec^2, and 1.9 ± 0.1 for the larger granules. Figure 6 shows an example of a plot of the area-perimeter relation. The fractal dimension with deviation is found by fitting straight lines to the data.

Figure 7 shows the fractal dimension plotted as a function of distance from the pore region. This parameter does not vary with distance from the pore. More than 5000 measurements of granules of different sizes, ages, and distances from the pore region, have been used in the analysis. The fractal dimension found by Roudier and Müller (2.15 for large, 1.25 for small granules) (see Figure 11 in their paper) is somewhat larger than our values. The granule diameter at which the fractal dimension changes (1.37 arcsec) is also different from ours (1.1 arcsec). This discrepancy is not a result of the specific choice of the intensity level used for the definition of the granules. We computed the fractal dimension using both higher and lower intensity levels, and found that the results depicted in Figure 7 do not depend noticeably on the intensity level chosen for the granule borderlines.

5 Conclusion

We have studied solar photospheric granulation in the neighbourhood of one pore region and found larger granule lifetimes closer to the pores. The lifetime changes suddenly

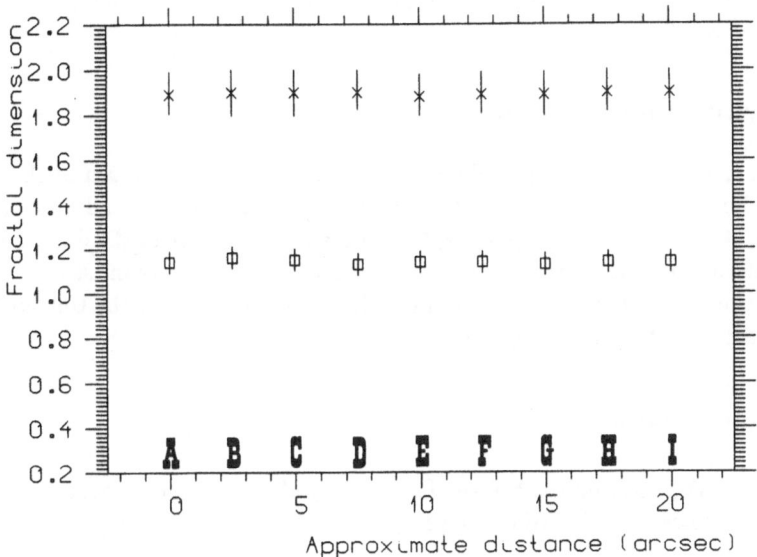

Figure 7: Fractal dimension as a function of distance from the pore region. The crosses represent granules with diameter smaller than 1.1 arcsec, squares are from measurements on granules larger than 1.1 arcsec. Error bars show rms deviation.

322

at the "border" of the pore region. The average size and the number density of the granules also change slightly at the "border". The intergranular area is larger inside the pore region than outside. The fractal dimension of granulation is found to be independent of the distance from a pore.

6 Acknowledgements

We wish to acknowledge helpful discussions with Øyvind Andreassen of the Norwegian Defence Research Establishment, and with Bjørn Håkon Granslo and Jan Karud of the Institute of Theoretical Astrophysics, University of Oslo. Part of the data reduction has been carried out at the Image Processing Laboratory, University of Oslo.

The present contribution is part of the thesis work of TAD at the University of Oslo, (supervisor Dr. Oddbjørn Engvold).

7 References

Andreassen, Ø., Engvold O., Müller, R.: 1984, *High Resolution in Solar Physics*, Proc., Toulouse, France, Ed.: R. Müller, p.91.

Alissandrakis, C.E., Dialetis, D., Tsiropoula, G.: 1987, *Astron. Astrophys.* **174**, 275.

Bray, R.J., Loughhead, R.E., Durrant, C.J.: 1984, *The solar granulation.* Cambridge University Press, 2nd. edition, p.49, p.86.

Darvann, T.A.: 1986, *Proc. of the Inaugural Workshop at Tenerife,* Oct. 1986.

Dialetis, D., Macris, C., Müller, R., Prokakis, Th.: 1989a, *(this volume)*.

Dialetis, D., Macris, C., Müller, R., Prokakis, Th.: 1989b, *(this volume),* also submitted to *Astron. Astrophys.*

Collados, M., Marco, J., Del Toro, J.C., Vazquez, M.: 1986, *Solar. Phys.* **105**, 17.

Koutchmy, S.: 1989, *(this volume)*.

Macris, C., Prokakis, Th., Dialetis, D.: 1989, *(this volume)*.

Mehltretter, J.P.: 1978, *Astron. Astrophys.* **62**, 311.

Müller, R., Roudier, Th., Vigneau: 1989, *(this volume)*.

Roudier, Th., Müller, R.: 1986, *Solar Phys.* **107**, 11.

Schröter, E.H.: 1962, *Z. Astrophys.* **56**, 183.

Schmidt, W., Grossmann-Doerth, U., Schröter, E.H.: 1988, *Astron. Astrophys.* **197**, 306.

Solar Geophysical Data prompt reports, September 1986.

Title, A.M., Tarbell, T., and the SOUP team.: 1986, *Proceedings, Boulder, Colorado 1986.*

Title, A.M., Tarbell, T.D., Topka, K.P.: 1987, *Ap.J.* **317**, 892.

von der Lühe, O.: 1983, *Astron. Astrophys.* **119**, 85.

Wittmann, A.: 1978, *Small scale motions on the Sun, Mitteilungen aus den Kiepenheuer-Institut* Nr. **179**, 1979.

Wyller, A.A., Scharmer, G.B.: 1985, *Vistas in Astronomy* **28**, 467.

SPATIO-TEMPORAL ANALYSIS OF WAVES IN THE SOLAR ATMOSPHERE

F.-L. Deubner and B. Fleck
Institut für Astronomie und Astrophysik der Universität
Am Hubland
D - 8700 Würzburg, F.R.Germany

ABSTRACT. A time series of simultaneously measured velocity (V) and intensity (I) fluctuations of different spectral lines formed in the photosphere and low chromosphere is analysed. In order to separate the different patterns of motions (granulation, mesogranulation, gravity waves, acoustic waves) V-V and V-I phase difference spectra have been computed not only with respect to frequency ν but also to the horizontal wavenumber k_x. The V-V phase difference spectra of the Na D_1 line and two photospheric Fe I lines (5929.7Å, 5930.2Å) clearly show the existence of gravity waves in the solar atmosphere. The V-I phase spectra also exhibit the behaviour of gravity waves at high wavenumbers ($k_x \gtrsim 1.7$ Mm^{-1}). At low wavenumbers one finds positive V-I phase differences. This we take as an indication for convective motions with a typical spatial scale of more than 5" (mesogranulation).

The paper will be published in Astronomy and Astrophysics.

R. J. Rutten and G. Severino (eds.), Solar and Stellar Granulation, 325.
© 1989 by Kluwer Academic Publishers.

ON THE GRANULE LIFETIME NEAR AND FAR AWAY FROM SUNSPOTS

D. Dialetis*, C. Macris*, R. Muller**, Th. Prokakis*.

  * National Observatory of Athens, Athens 11810,
    P. O. Box 20048, Greece.
 ** Observatoires du Pic-du-Midi et de Toulouse,
    65200, Bagneres-de-Bigorre, France.

ABSTRACT. We have studied the spatial distribution of all the granules (516) located in a photospheric region 45"x34", according their life-times. The location of this region, near a large and a small sunspot, gave us the possibility to examine possible differences in the granule lifetime in magnetic and non-magnetic regions. We conclude that the presence of sunspot do not disturb the network structure defined by the long-lived granules (Dialetis et al., 1988). Our study is based on an excellent sequence of photos taken at the Pic-du-Midi Observatory on May 11, 1979.

1. INTRODUCTION

   In a previous paper (Dialetis et al. 1988, hereafter referred to as Paper I) we have shown that the long-lived photospheric granules are not randomly distributed, but they are located in cellular regions that cov-ered ≈35% of the whole area of the region under study (37"x37"). The size and the morphology of these structures are similar to the   spatial scale of mesogranulation (November et al, 1981), the spatial scale of structures formed by the brightness fluctuations (Oda, 1984), as well as to the cells of quasi-stationary intensity modulations at the mesogranu-lation scale (Koutchmy and Lebecq, 1986).
   The greater regions were oblong with length greater than 15" and a mean width ≈ 5". The mean distance between them was ≈ 5". The smaller regions seem to be the remnants of greater ones which fade and vanish. But this was an impression caused by the morphology of these regions.
   The region that we have studied was located in a quiet area of the center of the solar disk. Our study was based on an exceptional sequence of photos taken at the Pic-du-Midi Observatory on May 16, 1979 (solar maximum).
   We have pointed out that this result must be confirmed with more observational data for different periods referred to the solar cycle and for different positions on the solar disk.
   In the present work we examine the variation of the lifetime of solar granules as a function of their location by studying a greater solar region (45"x34") located near a large and a small sunspot. Our

327

R. J. Rutten and G. Severino (eds.), Solar and Stellar Granulation, 327–331.
© *1989 by Kluwer Academic Publishers.*

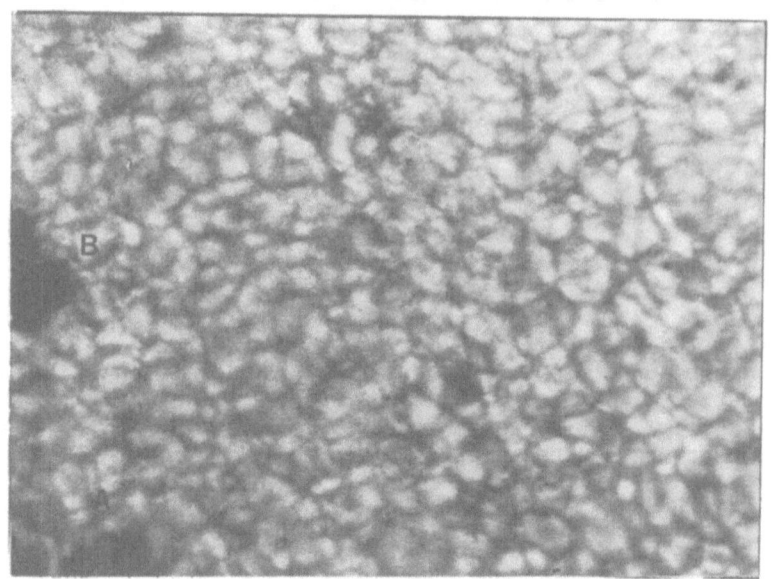

10"

Figure 1. The photospheric region under study. (A) and (B) are parts of the penumbrae of the two spots.

purpose was to confirm the existence of regions of long-lived granules and to study the possible influence of sunspots in the morphology of these regions. Our study is based on an excellent sequence of photos taken by one of us (R.M.) with the 50 cm refractor at the Pic-du-Midi Observatory on May 11, 1979. The duration of the observational sequence was 73 min with a time difference of ≈1 min between successive pictures.

2. REGIONS OF LONG-LIVED GRANULES

In order to examine the relation between lifetime and location of solar granules, we have determined the lifetime of all granules (n=516) located in an area 45" x 34" (Figure 1). The area was selected in a "master" frame at the middle of the sequence, and every granule with diameter >0.4" was traced back to formation and forward to disappearance. See paper I for more details about this size-criterion, as well as about the adopted criteria for the determination of the "formation" and "disappearance" of a granule. The mean lifetime of the population of 516 granules is 14 min (with standard deviation ≈ 6 min). Exactly same value was found in our first study when we have used for the determination of the mean lifetime the same technique (Technique A, Paper I). A determination of the mean lifetime by using the "survival" function is possible to yield a value about 15 min (Alissandrakis et al., 1987). It is interesting to note that for an active region the mean value of the granule lifetime is the same with the mean value find for a quiet region (Paper I) some days later .

In order to construct the map of the region with contours of equal lifetime, for every granule we have determined the lifetime and the coordinates of its center. These irregularly spaced data were placed in

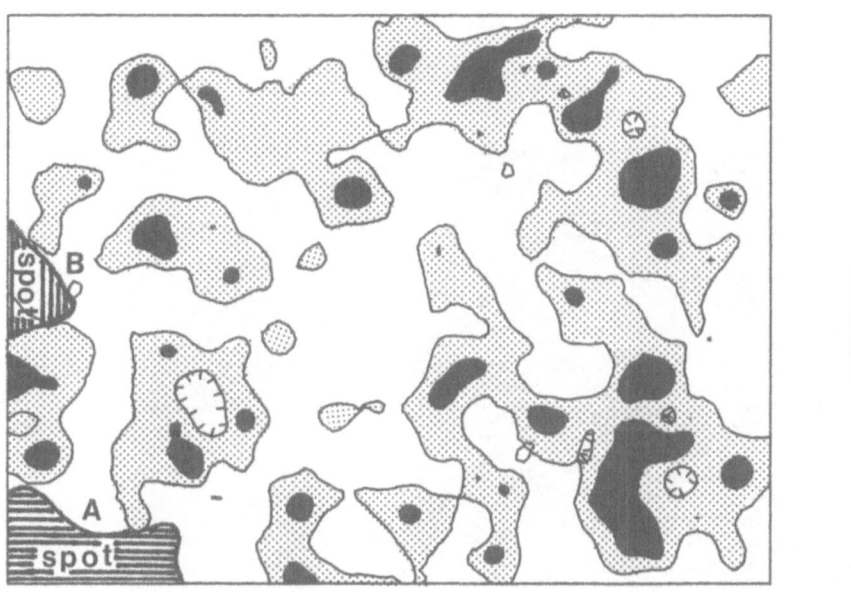

Figure 2. Regions with long-lived granules. Dotted regions correspond to lifetimes >15 min and black regions to lifetimes >21 min.

a regularly spaced rectangular grid; each lifetime value was given a weight proportional to the inverse of the square of the distance of the granule from the grid point.

As in Paper I we considered as long-lived granules the granules with lifetimes greater than 15 min. These granules are grouped in well defined regions (Figure 1) that cover ≈ 40 % of the whole area. These regions are similar to the regions presented in paper I. The mean length is 15" and the mean width 6". The location of these structures create the impression that these regions are distributed near the outer edge of a cell with dimensions ≈30" x 35". The location of the very long-lived granules (lifetime >21 min) give also this impression (Figure 1). The central part of this cell is characterized by granules with lifetimes < 15 min. In our previous paper we have not notice the existence of such a structure composed by the regions of the long-lived granules. But a second glance at the Figure 1 of Paper I shows that in the upper part of the figure we can distinguish the possible existence of a similar structure with comparable dimensions.

Finally we tried to identify this greater cell in observations taken simultaneously at the Ca$^+$K line, where the bright points of the photospheric network are very prominent. There is a relatively good corre-

spondence between the bright areas in the central part of the region and
the structures of the short-lived granules.

As it concerns the existence of regions composed by granules with
longer lifetimes near the spots, we have not found essential differences. The regions near the spots (A) and (B) are occupied by structures
of long-lived granules and structures with short-lived granules located
radially around them.

10"

Figure 3. A picture of the region at the Ca^+K line. There is a
relatively good correspondence between the bright areas in the central
part of the region and the structures of the short-lived granules.

3. RESULTS AND DISCUSSION

The results of this study confirm without any doubt the existence of
cellular regions composed by the long-lived granules. The size and the
morphology of these regions are similar to the mesogranulation cellular
patterns discovered by November et al (1981). It is interesting to study
in the future if the pattern composed by the long-lived granules is a
fairly stationary pattern like the mesogranulation (lifetime $\approx$ 2 hr.).
It is also very interesting to study if the position of these regions
correspond to the cells of horizontal flow mentioned by Title et al
(1986).

The morphological characteristics of these regions as well as the
mean granule lifetime seem to be the same in quiet and active regions.
We have not found any evidence that near the spots and the pores the

mean lifetime of granules changes dramatically.

We believe that we dispose a first observational indication that these structure are not randomly distributed. They compose a larger cell with dimension ≈ 30". The central part of this cell is composed by short-lived granules. The relatively restricted dimensions of the region under study does not give the possibility to confirm if this greater cell is a pattern repeated systematically.

Acknowledgements.

We wish to thank the post graduate student E. Bratsolis for his kind help in the determination of the lifetimes.

References.

Alissandrakis C. E., Dialetis D., Tsiropoula G.: 1987, Astron. Astrophys., **174**, 275.
Dialetis D., Macris C., Muller R., Prokakis Th.: 1988, Astron. Astrophys. in press.
Koutchmy S., Lebecq C.: 1986, Astron. Astrophys., **169**, 323.
November L. J., Toomre J., Gebbie K. B., Simon G. W.: 1981, Astrophys. J., **245**, L123.
Oda N.: Solar Phys., **93**, 243.
Title A. M., Tarbell T. D., Simon G. W. and the SOUP Team : 1986, Adv. Space Res., vol. **6**, No. 8, 253.

OBSERVATIONS OF HIGH FREQUENCY WAVES IN THE SOLAR ATMOSPHERE

Stephen L. Keil
Air Force Geophysics Laboratory
Sunspot, NM 88349

Amy Mossman
Wellesley, College
Wellesley, Massachusetts

ABSTRACT. An agile (tiltable) mirror was used in conjunction with the echelle spectrograph of the Vacuum Tower Telescope at Sacramento Peak to measure propagating acoustic waves in the solar atmosphere. The agile mirror permits the removal of image motion caused by the earth's atmosphere before feeding the image into the spectrograph. A CCD array was used to record 132 spectral points across the photospheric line Fe I 5434 every 1.5 seconds at 100 spatial points along the spectrograph slit. Spatial resolution along the slit was 0.33" and the spectral resolution is 9.2 mA. The images were stabilized using the signal from a quad-cell locked onto small sunspots or pores. Several, approximately one-hour-long observing runs were made with the slit either crossing the spot or displaced several arc-seconds from the spot. Observations were made at disk center and at a heliocentric angle of 37^{0} ($\cos(\theta)= 0.8$). We have computed phase differences between velocities in the line core and at various bisector levels in the line as a function of frequency. The image stabilization permits us to see wave propagation out to about 24-28 mHz without any further correction for seeing. After correction for radiative transfer effects, we find a flux of energy in the propagating waves that is sufficient to heat the chromosphere.

1. INTRODUCTION.

A quantitative determination of the flux of propagating acoustic waves in the solar atmosphere is of interest for several reasons. The problem of heating of the outer layers of the atmosphere has not yet been solved. Propagating waves at frequencies above the acoustic cutoff might be a possible source of this heating, at least in the upper photosphere and lower chromosphere. These waves might provide a physical explanation for

R. J. Rutten and G. Severino (eds.), Solar and Stellar Granulation, 333–345.

the excess broadening of spectral lines formed in the solar atmosphere
without having to invoke arbitrary velocity fields such as micro- or macroturbulence. The propagating waves may have effects on spectral lines that
can cause errors in abundance determinations (Marmolino, 1987). The waves
can introduce line asymmetries that must be accounted for to correctly
interpret the shapes of photospheric lines (Cram *et. al.*, 1979, Keil and
Marmolino, 1987). Finally, the interaction between waves and magnetic
flux tubes could generate MHD or magneto-acoustic waves that can contribute
to the heating in the corona.

Earlier attempts to measure the acoustic flux have generally used
spectral time sequences and relied on post-facto techniques to remove the
effects of image motion introduced by turbulence in the earth's atmosphere.
When the solar image moves the region on the sun being observed by the
spectrograph is shifted, this can in-turn result in an arbitrary change in
the observed velocity.   These random changes in the velocity make it
difficult to measure the power and phases of the velocity fluctuations
needed to determine the flux and propagation characteristics of the wave.
Techniques to correct for velocity fluctuations introduced by seeing have
been developed by Deubner and his colleagues (Endler and Deubner, 1983,
Deubner, Endler, and Staiger, 1984, and Deubner, Reichling and Langhanki,
1987). The techniques are based on assumptions about the properties of the
earth's atmosphere and are fairly complex to apply. By stabilizing the
image before sending it to the spectrograph slit, we can substantially
reduce the magnitude of these corrections. It should be noted however,
that although the agile mirror removes gross image motion within a certain
radius from the object being used for stabilization, it does not correct
for differential image motions. To minimize the effects of differential
motions we observed only when seeing conditions were very good.
Furthermore, as the distance away from the stabilized feature is increased
beyond a certain radius the agile mirror can actually increase the
differential image motion.

2. OBSERVATIONS WITH THE AGILE MIRROR

The observations were obtained with the echelle spectrograph of the Vacuum
Tower Telescope at Sacramento Peak Observatory. Before light was passed
into the spectrograph slit, it was sent through a box containing an agile
mirror and a quad-cell detector. When a solar feature is imaged on the
quad-cell, the output from the cell can be used to steer a tiltable or
agile mirror to keep the feature locked in a fixed position. A complete
description of the stabilization system is given in von der Luhe (1988,
available at this meeting). The stabilized image is then passed into the
spectrograph. Figure 1 shows the error signals generated by the quad-cell
that are used to drive the agile mirror as well as their power spectra.
Typical driver signals gave rms image motions of about 0.3"-0.5".

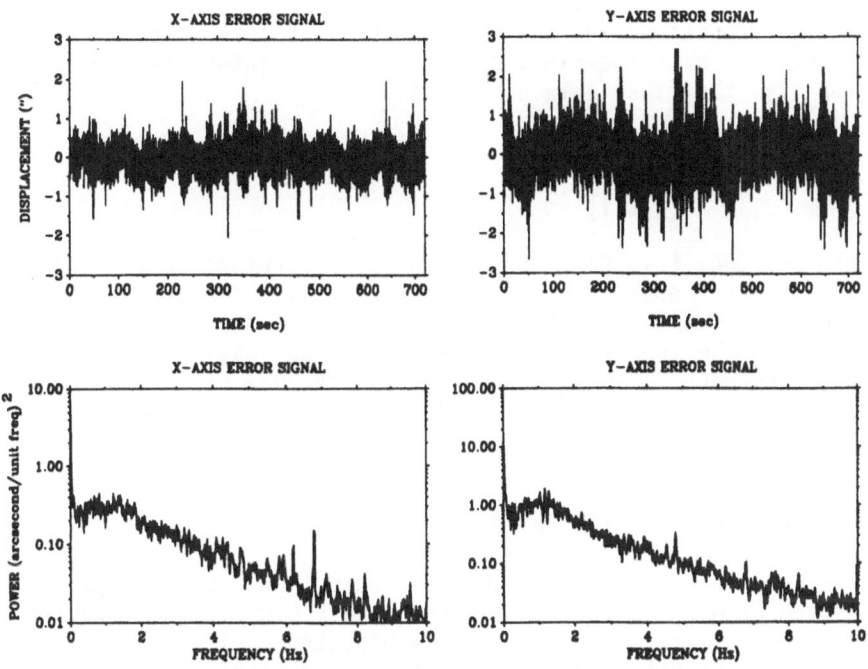

Figure 1. Mirror correction signals. The upper curves show the mirror displacement signals converted to arc seconds for the x and y offsets. The lower curves show the power spectra of the fluctuations. The mirror was locked onto a small sunspot while the error signals were recorded.

We observed FeI 5434, which is a non-magnetically sensitive line (g=0) formed in the upper photosphere. The line was recorded on a CCD with 132 pixels in the spectral direction and 100 pixels in the spatial direction. Spectral resolution was 9.15 mA/pixel and the spatial resolution was 0.33"/pixel. With the slit locked onto a small sunspot or pore, we recorded a spectrogram every 1.5 seconds for periods of about one hour. The width of the slit corresponded to 0.50" on the sun.

At each point along the slit, we extract the FeI 5434 profile, compute its fourier transform, apply an optimum filter to the transform (Brault and White, 1972), extend the transform with zero's and finally retransform to obtain an interpolated, smoothed line profile. Sixteen points are interpolated for every CCD pixel. The final dispersion is 0.57 mA per interpolated pixel. The interpolated, smoothed profiles are used to determine the line center position and the position of bisectors at 9 points equally spaced in intensity between the line core and the continuum. We also record the intensity in the line core and the continuum. This process is repeated for each point along the slit and for each time step. Letting V represent the position of the line core or bisectors after calibration for dispersion, I_c the continuum intensity and I_0 the line core intensity, the final data output is:

$$I_c(x,t)$$
$$I_0(x,t)$$
$$V(x,t,i)$$

where x=1,100; t=1,2048, and i=1,10 where 1 refers to the line core and 2,10 are the 9 equally spaced bisectors. All velocities are referenced to the mean velocity along the slit. From these data, we compute power spectra of the temporal fluctuations at each spatial point, and phase spectra between various pairs of data. Because of the presence of the sunspot or pore in the data, we do not attempt to compute the spatial transform of the data since the spot would dominate the k_x spectra.

3. VELOCITY FLUCTUATION POWER SPECTRA

Power spectra of temporal velocity fluctuations are plotted in Figures 2 (measured at disk center) and Figure 3 ($\mu = 0.8$) at four levels in the line, the core (V_0), and at the 20% bisector (V_2), the 50% bisector (V_5) and the 80% bisector (V_8) levels. The dotted line shows the noise level determined by averaging the high frequency tail of the spectra (from 200-333 mHz). Each spectrum was generated by averaging the individual spectra at each of the 100 spatial points along slit. Power above the assumed white noise level is present out to frequencies above 100 mHz. Leakage of power from the 5 minute band can probably only account for about 20% of this high frequency power (see Deubner, 1976 and the argument presented in section 5). The shift to increased convective (low frequency) power is obvious in the spectra from the bisector levels near the line wings. These spectra should be compared with those of Melroy and Keil (1984) and those of Deubner (1976). The periodic maxima and minima reported by Deubner are visible in the spectra of the bisector fluctuations. Our maxima at 22 mHz and 31 mHz are in fair agreement with those observed by Deubner in FeI 5383. The presence of these cascades in the power spectra has been interpret by Deubner as due to interference in the line formation region (Deubner, 1976) and by Deubner, Durrant and Kaltenbacher (1982) as due to non-linear process in the transfer of the Doppler signal.

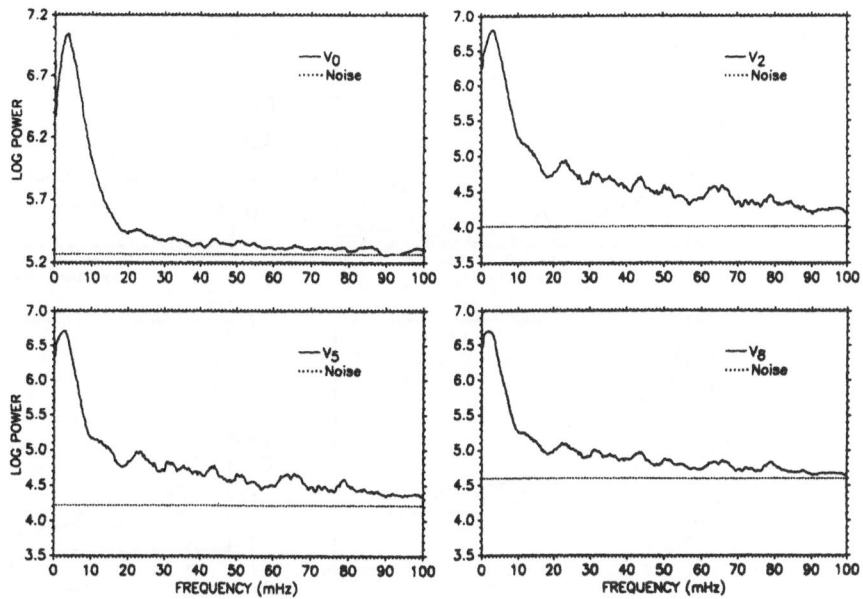

Figure 2.   Power Spectra at Disk center. Power spectra of velocity fluctuations measured at disk center (V_0) and at the 20% (V_2), 50% (V_5) and 80% (V_8) bisector levels are shown.

338

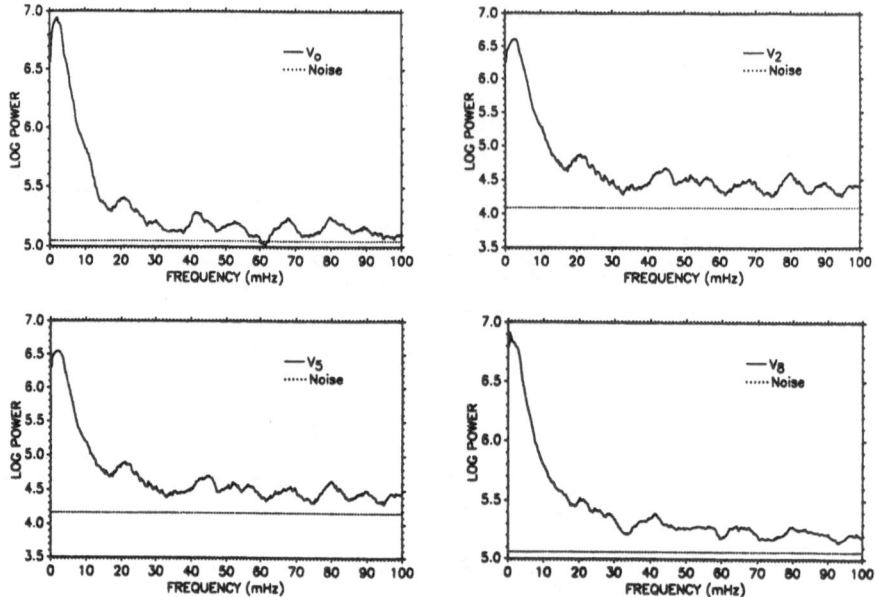

Figure 3.   Power Spectra a μ = 0.8.   Power spectra at several bisector levels as in Figure 2.

4. PHASE SPECTRA

Phase difference spectra between fluctuations observed in the line core and fluctuations in the 50% bisector level are plotted in Figure 4 (disk center) and Figure 5 (μ = 0.8) for various sets of spatial points. All 100 points along the slit have been plotted in the upper left diagram of each figure while various subsets of the data representing varying distances from the sunspot or pore are plotted in the other diagrams. The scatter in phase results from real differences on the sun, instrumental and atmospheric noise, and the fact that regions near the spot are corrected better than regions further from the spot.

At disk center, Figure 4, spatial points 1 through 20 are plotted separately in the upper right diagram. These points are all over 20" from the sunspot. In the lower-right only spatial points 40 through 60 are plotted, consisting of points between 7" and 13" from the sunspot. In the lower left spatial points 70 through 90 are plotted. This region contains the sunspot on which we are stabilizing. Points 40-60 clearly show wave propagation at frequencies above 6 mHz. At frequencies above 28 mHz the diagrams become more confusing and the propagation characteristic is lost. For points 1-20, which are further from the sunspot on which the image is being stabilized, the phase spectra are much noisier and propagation of the waves is not clearly seen.

At cos(θ) = 0.8, Figure 5, points 50-70 contain the small sunspot used to stabilize the image. Although the propagation characteristics of the spectra are not as strong as those at disk center, they are still recognizable, especially for those points near the stabilized spot (31-45). At a heliocentric angle of 38° the line of sight shifts horizontally about 100 km between the height of formation of the line core and the 50% bisector level. Thus, the horizontal wavelength plays a role. In addition, horizontal velocity fluctuations contribute to the measured signal.

Mean phase differences between line center velocity fluctuations and fluctuations at bisector levels 10%-60% above the line core are plotted in Figure 6.. The phases found at spatial points 40-60 were averaged to obtain the means at disk center (upper diagram) and those at spatial points 31-45 were averaged at μ = 0.8. These represent points near enough to the spot to be well stabilized, but not containing the spot. The increased slope of the differences with increased frequency agrees well with the interpretation of an upward propagating wave.

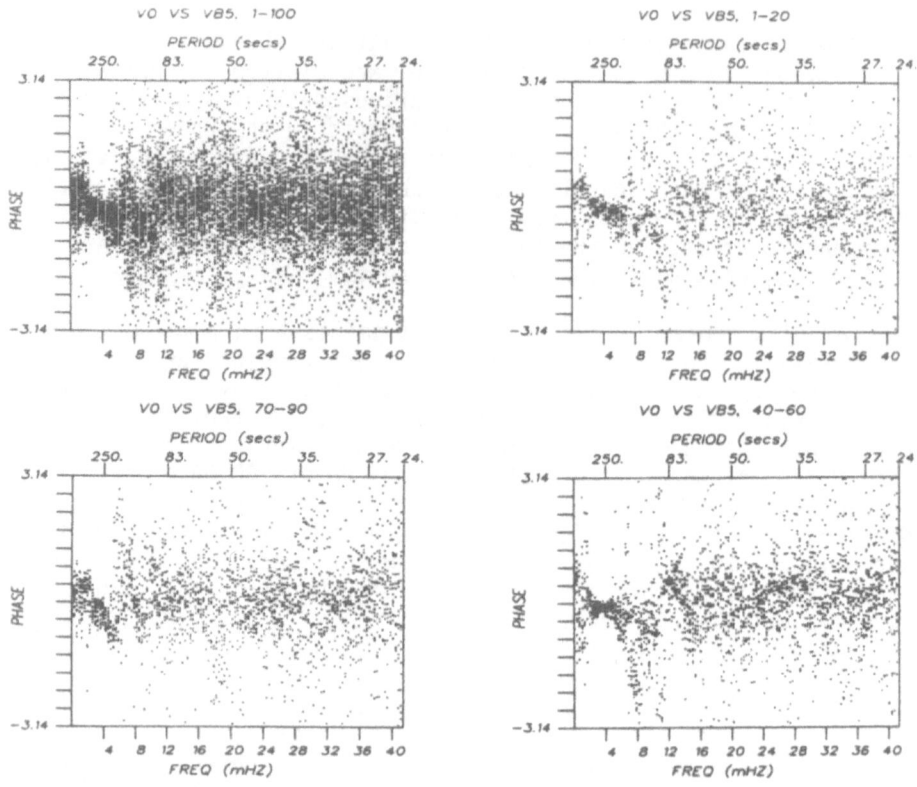

Figure 4. Phase spectra at disk center.   Phase difference spectra between fluctuations observed at disk center in the line core and fluctuations in the 50% bisector level for all 100 spatial points are plotted in the upper left. At any given frequency, there are 100 dots, representing the phase difference at each of the 100 points along the slit. In the upper right, spatial points 1 through 20 are plotted separately. These points are all over 20" from the sunspot. In the lower right only spatial points 40 through 60 are plotted, consisting of points between 7" and 13" from the sunspot. In the lower left spatial points 70 through 90 are plotted. This region contains the sunspot on which we are stabilizing.

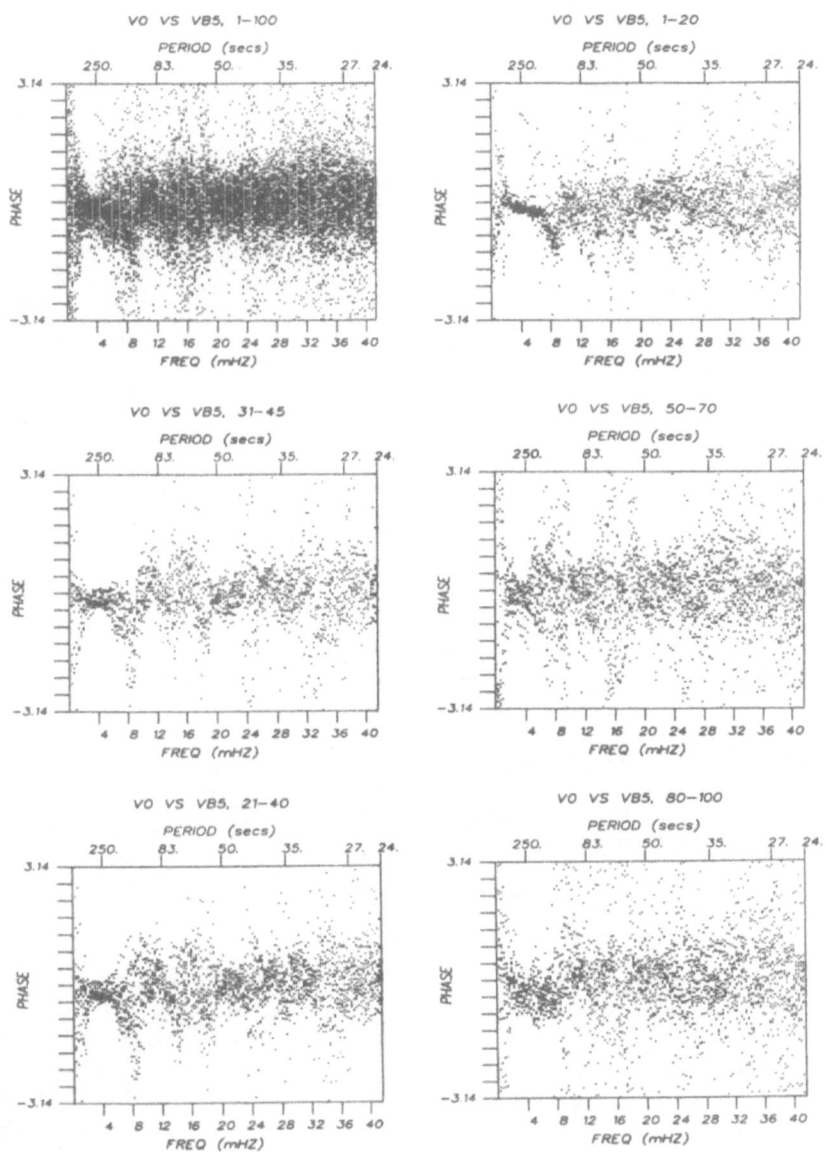

Figure 5. Phase spectra at μ = 0.8.   As in Figure 4, phase difference spectra between fluctuations observed at $\cos(\theta)$ = 0.8 in the line core and fluctuations in the 50% bisector level for all 100 spatial points are plotted in the upper left. The differences observed in selected spatial bands are plotted in the other figures. Points 50-70 contain the small sunspot used to stabilize the image.

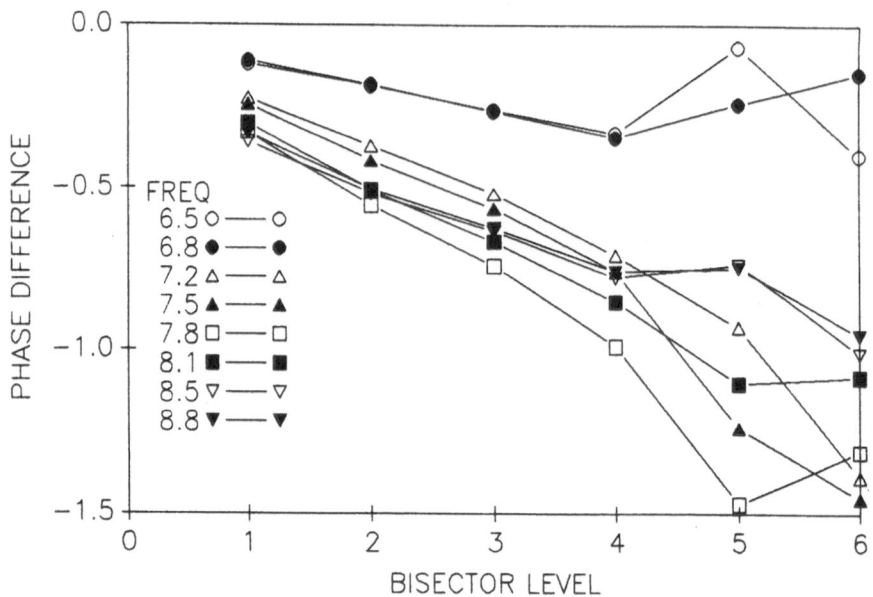

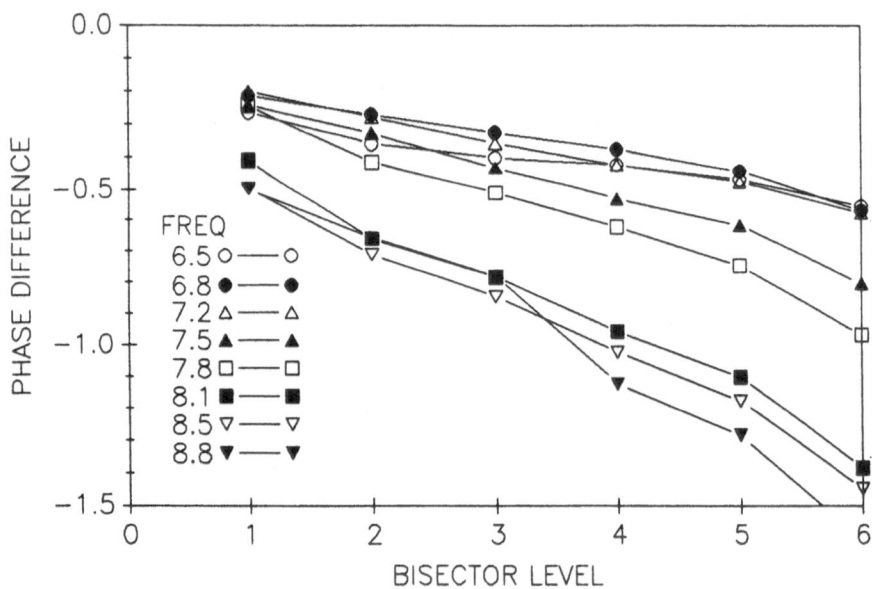

Figure 6. Mean phase differences. Mean phase differences between line center velocity fluctuations and fluctuations at bisector levels 10%-60% above the line core are plotted. The upper plot is for disk center and the lower plot for $\mu = 0.8$. The phases found at spatial points 40-60 were averaged to obtain the means at disk center and at points 31-45 at $\mu = 0.8$.

5. MECHANICAL FLUX AND ATMOSPHERIC HEATING

Figure 7 shows the flux of mechanical energy propagating through the atmosphere as well as the ratio of the flux at $\mu = 0.8$ to the flux at disk center. The flux of propagating energy is estimated from the power spectra of each bisector level and is shown as a function of bisector level (height in the solar atmosphere), at disk center (top) and at $\cos(\theta) = 0.8$ (middle). Heights for the formation of each bisector level were estimated from the phase spectra and from line synthesis calculations of Fe I 5434. Power spectra were averaged over those spatial points that showed strong propagation signals in the phase difference spectra. The mechanical flux is computed from:

$$F_{mech} = \rho c_s <v>^2$$

where

$$<v>^2 = \int_{\omega 1}^{\omega 2} (P(\omega) - P_{noise} - P_{osc}(\omega))\, d\omega$$

is the rms velocity, c_s is the local sound speed, ρ is the local density, and $\omega 1$ and $\omega 2$ are determined from the phase spectra. $P_{osc}(\omega)$ is the power in the 5 minute band and we assumed it is given by:

$$P_{osc}(\omega) = P_0 e^{-(\omega-\omega_0)^2/\Delta\omega^2}$$

where P_0 is determined from the data, $\omega_0 = 3.3$ mHz and $\Delta\omega = 12$ mHz. This probably is a slight overestimation of the power that leaks into the high frequency bands. The solid circles represent the observed values of F_{mech}. The open triangles are "corrected values", where the magnitude of the correction was obtained from the calculation of Keil and Marmolino (1987), who compute line shift fluctuations produced by propagating acoustic waves. The filled square is the value reported by Deubner (1976) after estimating the correction for radiative transfer effects. The unfilled square is the value he measured. The dotted line is the level of flux that would be required to heat the chromosphere. The bottom plot show the ratio of the fluxes observed at $\cos(\theta) = 0.8$ and disk center. This curve indicates that the waves have a strong horizontal component deep in the atmosphere which appears to weaken with height.

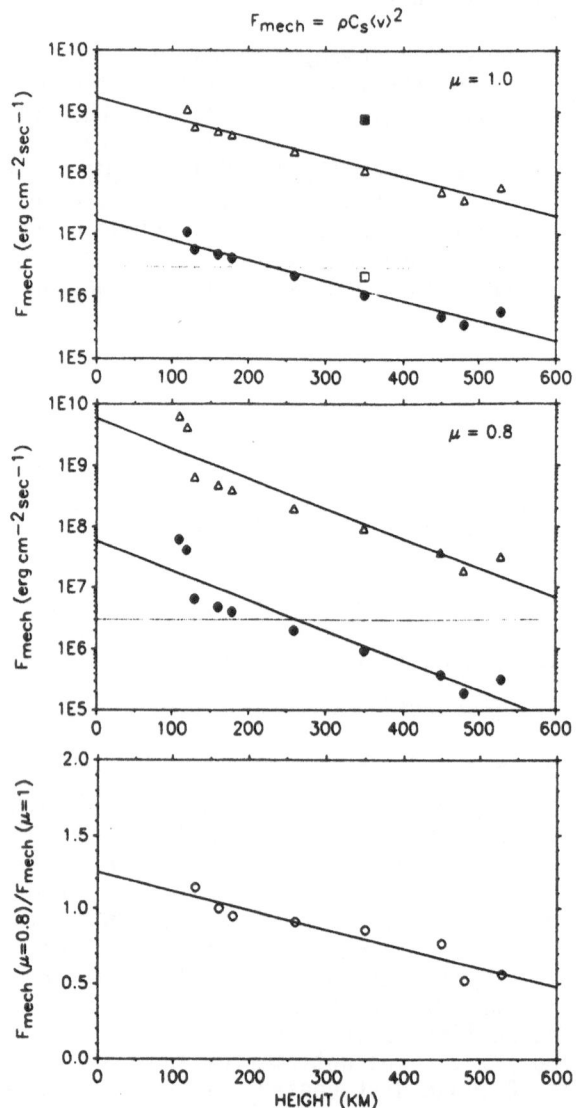

Figure 7. Mechanical flux vs. height in the atmosphere. The upper plot
shows the flux at disk center estimated from the high frequency part of the
power spectra at each bisector level. Filled circles are the observed
values. Open triangles represent an estimate of the correction required
because of radiative transfer effects. The squares represent observed
(open) and corrected (filled) values measured by Deubner (1976). The
middle plot is the same as the upper except it is computed for the data at
$\mu = 0.8$. The bottom plot is the ratio of the flux at $\mu = 0.8$ to the disk
center flux.

6. CONCLUSIONS

The agile mirror has permitted us to observe wave propagation out to 24-28 mHz, without making corrections for seeing. Previous attempts were limited to frequencies below about 12 mHz unless substantial seeing corrections were made. These corrections lead to large uncertainties in the observed flux. With further refinements, it should be possible to measure the acoustic energy flux out to about 30 mHz using an agile or active mirror in combination with a spectrograph and CCD detectors, and to place an accurate lower limit on the total flux of acoustic energy in the photosphere and temperature minimum regions.

7. REFERENCES

Brault, J. W. , White, O. R.: 1971, *Astron. Astrophys.*, **13**, 169.

Cram, L. E., Keil, S. L., Uhlmschneider, P.: 1979, *Astrophys. J.*, **234**, 768.

Deubner, F.-L.: 1976, *Astron. & Astrophys.*, **51**, 189.

Deubner, F.-L., Durrant, C., Kaltenbacher, J.: 1982, *Astron. & Astrophys.*, **114**, 85.

Deubner, F.-L., Endler, F., Staiger, J.: 1984, *Mem. S.A.It.*, **55**, 135.

Deubner, F.-L., Reichling, M., Langhanki, R.: 1987, paper presented at Aarhus meeting.

Endler, F., Deubner, F.-L.: 1983, *Astron. & Astrophys.*, **121**, 291.

Keil, S. L., Marmolino, C.: 1987, *Astrophys. J.*, **310**, 912.

Marmolino, C.,: 1988, *Solar Phys.* **112**, 211.

Melroy, P., Keil, S. L.: 1984, in *Proc. Workshop on Small-Scale Dynamical Processes in Quiet Stellar Atmospheres*, ed. S. L. Keil (Sunspot, NM: Natioanl Solar Observatory), p. 19.

von der Luhe, O.: 1988, submitted to *Astron. $ Astrophys.*

PHOTOELECTRIC ANALYSIS OF THE SOLAR GRANULATION IN IR

Serge Koutchmy

AFGL/NSO/SP, Sunspot, New Mexico 88349 (USA)
and
Paris Institut d'Astrophysique, CNRS. France

ABSTRACT. Results of both image analysis and statistical analysis of 1D scans of the IR granulation observed in the opacity minimum region are briefly presented. The same technique was applied at 600 nm and results partially presented by Koutchmy and Lebecq, 1986, *Astron. Astrophys.* **169**, 323. All observations were collected at the prime focus of the Sac Peak VTT using high signal–to–noise ratio measurements with a specially–designed pinhole PbS spectro–photometer; imaging has been made using both 2D scanning of the telescope and processed video–scans of an IR–vidicon. Images of the IR granulation and of sunspot umbral dots and penumbral filaments are presented for the first time. The main results are:

A. 1. Uncorrected for the smearing RMS of intensity fluctuation at 1.75 μm are typically ± 2.1%; scans performed at $\cos \Theta = 1$, .8, and .7, corrected for the foreshortening, show a slight decrease to the limb.

   2. Power spectrum analysis of the IR scans shows evidence for more power at higher frequencies, and, conversely, less power in the low frequency tail.

   3. Histograms of intensity fluctuations show a more pronounced than in the optical region "bi–distribution" of intensity variations corresponding to the dark intergranular lanes (with a larger number) and to the bright granules.

   4. Cross–correlation analysis of the scans shows:

      a. A typical lifetime of granules of 3.25 min instead of 6 min (decrement value) found at 600 nm with the same method.

      b. Displacements of granules in rough agreement with the solar rotation.

   5. Image reconstruction of a region observed around a sunspot at $\cos \Theta = 0.71$ shows a definite positive contrast of facular features at low spatial resolution.

B. 1. Time sequences of IR diffraction–limited granulation images were obtained

347

R. J. Rutten and G. Severino (eds.), Solar and Stellar Granulation, 347–348.
© *1989 by Kluwer Academic Publishers.*

at a slow rate of 2 images of 512x512 per sec (limited by the speed of the computer processing). Umbral dots were observed for the first time in the opacity minimum region with improved quality as far as image blurring and image motion are concerned when comparison is made with the visible image obtained simultaneously at 800 nm.

2. Although the ultimate spatial resolution achieved is far from what can be currently obtained with high speed photography in the optical region, the great advantages offered by working in IR (especially the gain offered in Zeeman sensitivity) make this technology very promising for sunspot and magnetic elements studies, including out–of–disc analysis.

CONSTRAINTS IMPOSED BY VERY HIGH RESOLUTION SPECTRA AND IMAGES ON THEORETICAL SIMULATIONS OF GRANULAR CONVECTION

B. W. Lites[1], Å. Nordlund[2], and G. B. Scharmer[3]

[1]*HAO/NCAR*[1]
[2]*JILA, University of Colorado*
[3]*Royal Swedish Academy of Sciences*

Abstract

We have obtained a sequence of simultaneous granulation spectra of the line Fe I 6302.5 Å and narrow band slit-jaw images using the Swedish Vacuum Solar Telescope (SVST) on La Palma, Canary Islands, Spain. Simultaneous 0.02s CCD exposures of both the spectrum and the slit-jaw image effectively 'freeze' the atmospheric seeing motions and permit unambiguous identification of the spectral features with the corresponding structures of the granular convection.

The unique combination of simultaneous images and high resolution spectra places significant constraints on theoretical simulations of compressible granular convection. These observational constraints have led to the following comparisons: 1) The simulations predict spatially averaged line widths very close to that observed, thereby validating the mean vertical velocity amplitudes of the numerical simulations. 2) The ratio $\sigma_{Irms}/\sigma_{Vrms}$ in the simulated granulation is in agreement with that inferred from the observations. However, the observed intensity and velocity fluctuations are systematically less than in the simulations, suggesting a substantial level of "scattering" from both the seeing and the telescope. We are able to give quantitative estimates of the effects of seeing by using the fact that the total velocity amplitude is *known* (from the broadening of spatially unresolved spectral lines). 3) Comparison of the observed and simulated granular images reveal the size of convective elements to be systematically larger in the simulations. Reduction of the size of these simulated convective elements was accomplished through a decrease in the adopted value of numerical viscosity. However, even with this reduced viscosity, small scale structures are more prevalent in the observations.

1 Data Acquisition and Analysis

Shown in Figure 1 is a spectrum of Fe I 6302.5 Å obtained at about local noon on 1987

---

[1]The National Center for Atmospheric Research is sponsored by the National Science Foundation.

R. J. Rutten and G. Severino (eds.), Solar and Stellar Granulation, 349–357.

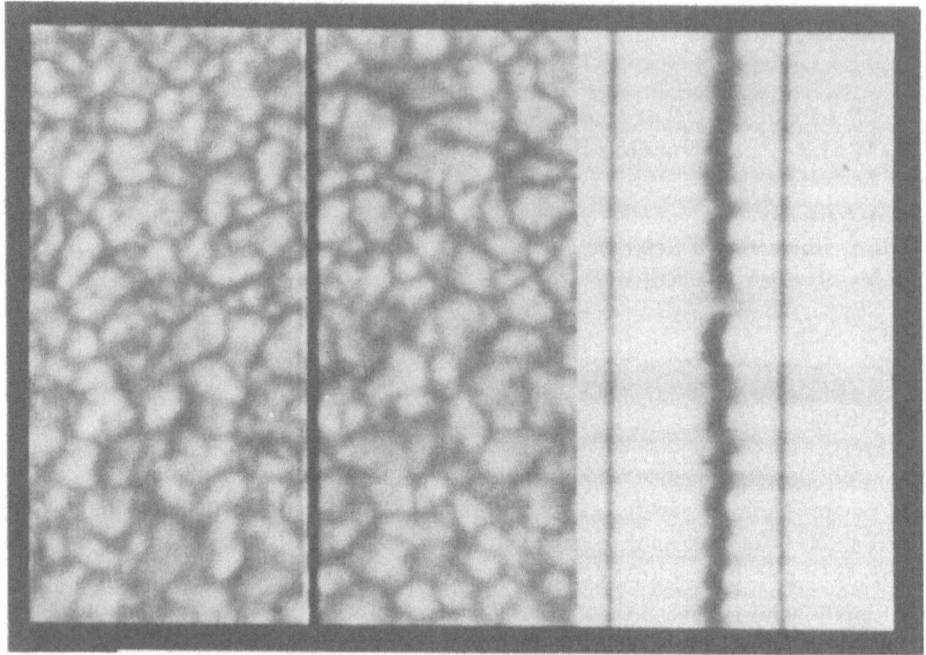

Figure 1: A slit-jaw image (left), the corresponding spectrum (right) of Fe I 6302.5 Åare shown for measurements of a quiet region near disk center at $11h51m17s$ UT on 1987 October 29. The apparent slit width is greater than its actual width. The length along the slit corresponds to 19.3 Mm at the surface of the Sun. This image had the highest rms velocity fluctuation (0.68 km/sec) of any of the spectra analyzed. However, the rms intensity contrast (0.056) for the slit jaw image is about average.

October 29 in a quiet region near disk center. The slit-jaw image on the left is recorded simultaneously with the accompanying spectrum. The very short exposure time (0.02 sec) of this digital image, the excellent optical quality of the telescope, the high efficiency of the spectrograph, and the excellent seeing conditions of the La Palma site permit the acquisition of highly dispersed spectra with angular resolution approaching that of the telescope itself (0.3"). The spectrograph has a measured FWHM of 0.025 Å at this wavelength. The observed range of intensity fluctuations ($0.066 \leq \sigma_{Irms} \leq 0.091$) and vertical Dopper velocities ($0.42 \leq \sigma_{Vrms} \leq 0.68$ km/sec after removal of most of the p-mode oscillatory signal) attest to the exceptional quality of these observations.

Video synchronization of two CCD cameras assured simultaneity of both the spectrum and slit-jaw images. The video signals from each of the cameras were combined with a video signal mixer, and the resultant video signal was 'grabbed' and processed by the image processing system (Scharmer 1987). In the present analysis we consider 11 individual frames (of high rms intensity contrast in the slit-jaw images) from 8 bursts of four images each. In these bursts, the time interval between the starts of the video scans of the successive images is 0.04 sec.

The digital images were corrected for dark-level offset. Observations of a defocused quiet Sun provided the basis for flat-field corrections (c.f. Lites and Scharmer 1988). These corrections greatly reduce the wavelength-dependent fringing in the CCD device used to record the spectrum. After Thomas et al. (1987), we adopt the centroid method of determining Doppler shifts for this line.

2 Fully Compressible Simulations of Granular Convection

The method of numerical simulation of the granular convection (as well as p-mode resonant oscillation of the subsurface cavity), is being described by Stein et al. (1989) and Nordlund & Stein (1989) in papers presented at this symposium. These simulations allow a time history of monochromatic images (including, for example, selected wavelengths within an absorption line) to be generated. The convection is computed for a 6×6 Mm area of the surface, with periodic boundary conditions. Figure 2 presents three such images. The simulated continuum images at 6300 Å are shown beside the simulated spectrum of Fe I 6302.5 Å. These images denote: (left) the unsmoothed image with the resolution of the numerical model (0.097 Mm/pixel), (middle) the image smoothed to mimic the diffraction-limited resolution of the SVST, and (right) the image smoothed with both the telescope and a seeing modulation transfer function (MTF) of exponential form (e.g. Nordlund 1984). It should be noted that this exponential form is not unlike that adopted by Schmidt, Knölker, and Schröter (1981) for small wavenumber k; but, unlike their theoretical telescope MTF, it does not vanish at a finite k. The Doppler shifts inferred from the unsmoothed and smoothed spectra are shown in Figure 3, along with Doppler shifts from the observed spectrum of Figure 1. An identical analysis procedure was used to extract Doppler shifts from the simulated and observed spectra. The oscillatory component of the Doppler shift has been removed from the curves. Note the discontinuities in simulated Doppler shift due to the image boundaries at 6 and 12 Mm. For the lower curve, Doppler shifts are presented relative to average shift with redshift positive, the two lowest spatial frequencies representing p-modes with wavelengths of $\approx$ 20 and 10 Mm have been filtered, high frequency noise has also been removed, and 1.5 Mm on each end of this curve has been apodized. One feature in the spectrum (labeled "a") shows a particularly large shear in the vertical velocity. It is associated with the border of the large granule nearly bisected by the slit near the center of the image in Figure 1.

The combined telescope and atmospheric MTFs adopted for the image on the right of Figure 2 yield rms velocities (after removing the average Doppler shifts due to p-mode oscillations) approximately equal to the average for the observed spectra. The derived exponent (see caption, Figure 2) increases the half-width of the telescope point-spread function (PSF) by about 15%, although it contributes a significant "scattered light" component to the PSF. The effect of this smoothing is to decrease the contrast of the features, yet allow visibility of small features down to the diffraction limit (as also seen in the observed image of Figure 1). The contrast and physical dimensions are identical in all images of Figures 1, 2, and 5.

The main uncertainty in the simulations is probably the one due to the approximate

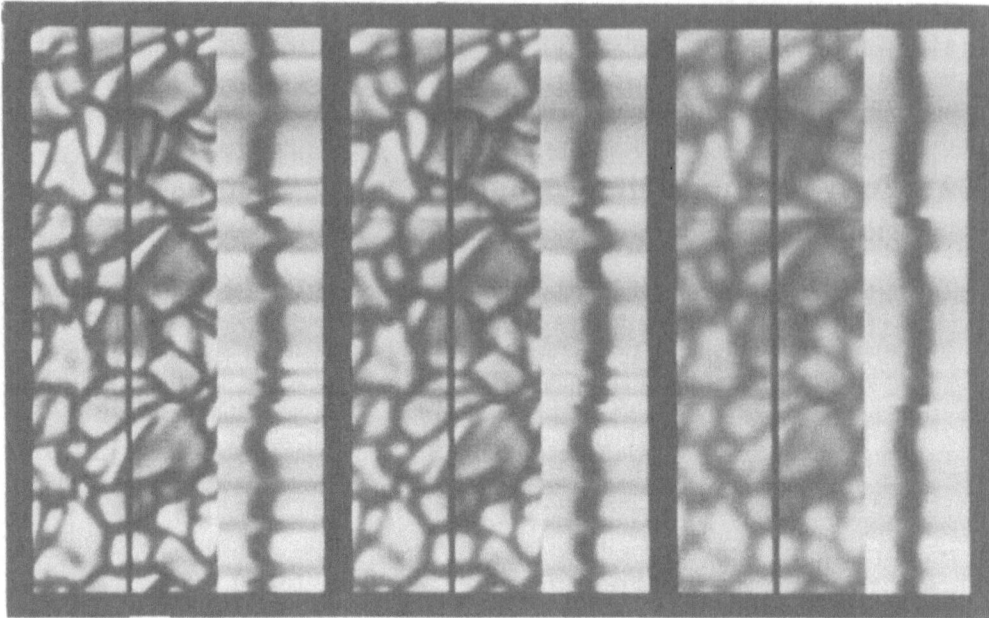

Figure 2: Simulated continuum images at 6300 Å and adjoining spectra of Fe I 6302.5 Å are shown. Each vertical granulation image is a mosaic of three time steps of a convection sequence, separated from each other by 4 minutes. The sequential images are stacked vertically (with the first image at the top) so as to simulate the longer slit sample of the observed spectra. A vertical line designating the artificial "slit" of width 0.19 Mm (approximately equal to the actual slit width of the observations) shows the locations from which the simulated spectra were derived. The image to the left is unsmoothed, the middle image has been smoothed with a diffraction pattern corresponding to the SVST, and the image to the right has been convolved with both the telescope diffraction MTF and an exponential function of the form $exp(-0.13k/k_0)$, where $k_0 = 1.05$ $Mm-1$.

treatment of the non-gray radiative transfer, with the influence of millions of spectral lines approximated by averaging the source function into just four bins (representing the continuum, weak lines, intermediate lines, and strong lines). Care has been taken to use a good approximation to the true Rosseland mean in the continuum forming layers, while using an intensity weighted mean in the optically thin layers. However, the position of the "surface" (defined as the location of maximum radiative cooling) depends sensitively on the value of the continuum opacity, and therefore uncertainties in this opacity *relative to the monochromatic opacity used when calculating the synthetic spectra* do influence to what extent the temperature fluctuations are visible; i.e., the σ_{Irms}. We have estimated the effect of a fairly drastic change of the continuum bin opacity (a factor of 100.2, probably an upper limit to the uncertainty), and we find that the ratio of velocity to intensity rms changes by about 10% (cf. Table below).

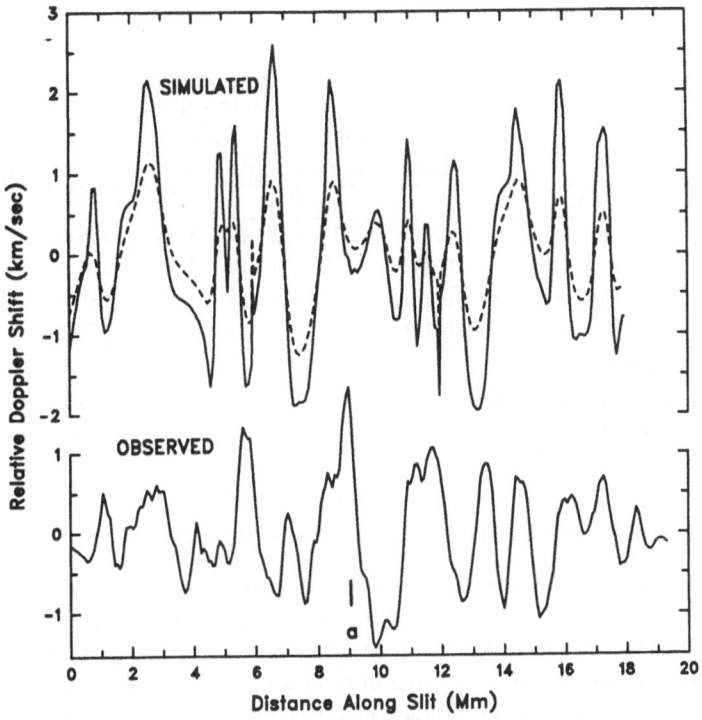

Figure 3: The top curves represent the Doppler shifts from the unsmoothed simulations (solid – left image in Figure 2), and fully smoothed (dashed – right image in Figure 2). The bottom curve represents Doppler shifts inferred from the observations shown in Figure 1. The scale in Mm along the slit is shown, with left and right corresponding to top and bottom, respectively, in Figures 1 and 2.

3 Results

3.1 Statistical Properties of Observed and Simulated Granular Convection

Table 1 summarizes the statistical properties of the granulation from both the observations and the simulations. For the observations we consider one frame from each of the 8 separate bursts. The rms intensity fluctuation σ_{Irms} of the observations is given for both the slit-jaw images and the spectra. The observed vertical velocity fluctuations σ_{Vrms} are limited in sample to the area admitted by the slit, so that the variance of these measurements from frame to frame is much larger relative to that of the intensity fluctuation. In the simulations, the numerical diffusion was changed in such a way as to keep the numerical energy diffusion approximately constant, while decreasing the ratio of momentum to energy diffusion (the Prandtl number P). The simulated σ_{rms} are derived from 10 solar minute averages over 16 different slit positions, except for the $P=1$ case, which is an average over 50 solar minutes. The simulations show a variance similar to that of the observations for comparable "slit" widths.

Table 1

Case	No Smearing		Telescope Only		Telescope and Seeing	
	Irms	Vrms	Irms	Vrms	Irms	Vrms
Simulations						
P=5	0.175	1.13	0.145	0.95	0.094	0.58
P=2	0.175	1.13	0.142	0.94	0.089	0.57
P=2 Modified Opac.	0.144	0.97	0.113	0.80	0.068	0.49
P=1	0.166	1.08	0.133	0.88	0.081	0.52
Observations						
Spectral Continuum	-	-	-	-	0.080	0.52
Slit Jaw Image	-	-	-	-	0.056	-

3.2 Line Width Constraint on the rms Velocity of Granular Convection: Calibration of the Smoothing due to Seeing

The simultaneous observations of velocity and intensity fluctuations presented here place important constraints on the numerical models of solar granulation. Whereas interpretations of observed intensity fluctuations alone are always subject to uncertainties as well as to the amount of light scattered beyond a few arcseconds (Nordlund 1984), such ambiguities do not arise in the present analysis since, unlike the intensity fluctuation, the rms velocity fluctuation arising from granular convection is reasonably well-known from line widths of spatially averaged line profiles. Thus we have another independent parameter constraining the permissible amount of smoothing due to seeing.

Since the convection model yields approximately the observed average line width (FWHM = 0.118Å simulated, 0.114Å observed), we believe the velocity distribution at the levels of formation of the Fe I line in the simulations to be roughly correct. Figure 4 compares the FTS profile (Livingston 1987) with the average of the simulated profiles. However, since the width of this medium strong line depends sensitively on the (uncertain) Van der Waals damping, weaker lines should be used to put more stringent constraints on the velocity field amplitude.

3.3 Compatibility of the rms Intensity and velocity Fluctuations

The difference between measured σ_{Irms} of the slit-jaw image and that of the spectral continuum is of some concern. The slit-jaw reflection introduces some scattering, and the optics used in these measurements to relay the slit-jaw image to the camera are also known to introduce some scattering and degradation of the image. The rms granular contrast measured (at another occasion) with this telescope at 4700 Å without these relay optics show $\sigma_{Irms} \simeq 9 - 10\%$, which corresponds to an rms contrast of $\approx 7\%$ at 6300 Å. Thus, even though there remains some uncertainty in the dark level of the spectrum observations (Lites and Scharmer 1988), the σ_{Irms} from the spectral continuum probably represents a more accurate mean of the true intensity fluctuations.

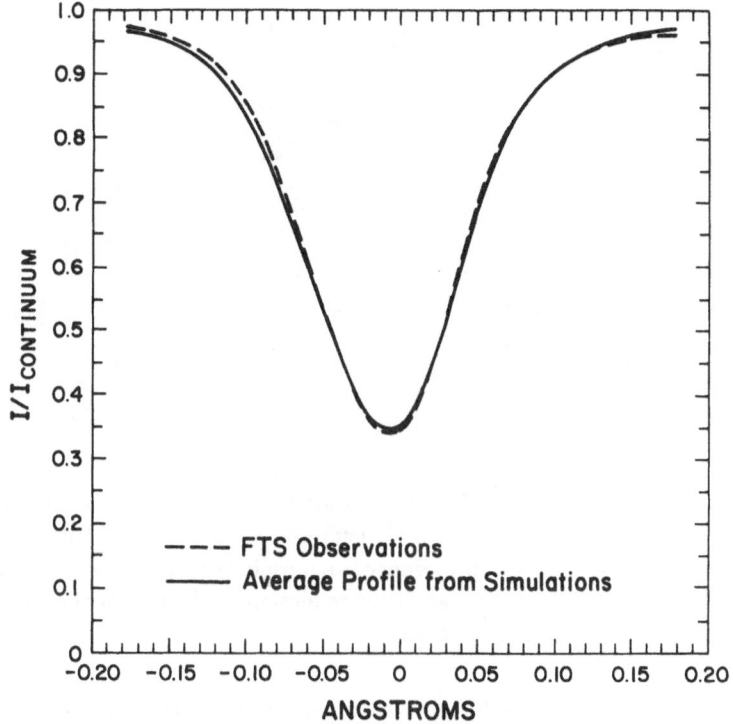

Figure 4: The spatial and temporal average of the Fe I 6302.5 Å line from the simulations (solid curve) is compared with disk-center FTS observations (dashed curve).

Using Wien's law, the simulated $\sigma_{I rms}$ of 16.6% in Table 1 corresponds to $\approx$ 20% at 5200 Å, which is in disagreement with the 13% contrast found by von der Lühe and Dunn (1987). If it can be shown that the atmosphere and telescope are not responsible for all of the "scattering" needed to bring our simulations and observations into agreement, a very special error would have to exist in our numerical model. This error would have to be such that its removal would preserve both the total velocity amplitude and the ratio of measured velocity and intensity amplitudes, while at the same time decreasing the measured intensity rms by a sizeable factor. Even if we disregard the difficulty of finding any possible physical or numerical cause for such an error, it would seem to be hard to adjust two quantities (velocity and intensity fluctuation) in such a way as to preserve the amplitude of one quantity, the observed ratio between them, while still reducing the other by a large factor.

3.4 Observed Size of Granules: Constraints on Numerical Viscosity

The improvement of the correspondence between the size of the largest granules in the observations and simulations upon reducing the numerical viscosity indicates that a numerical viscosity of some 3×10^{12} cm²s⁻¹ is marginally small enough to represent

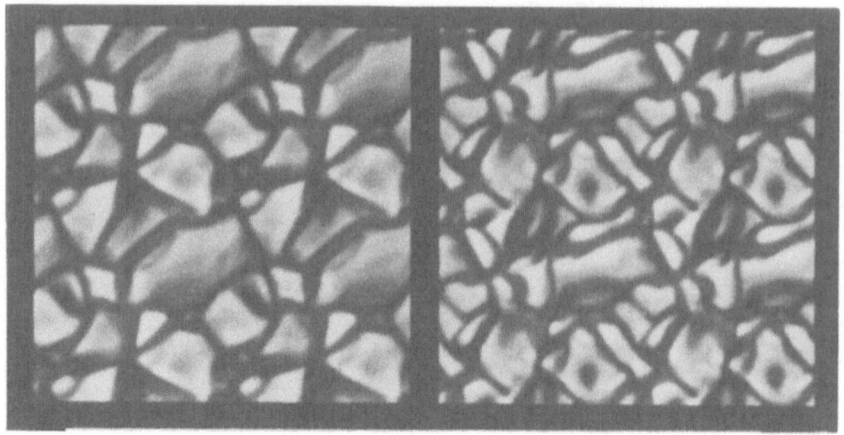

Figure 5: Granulation images are compared after 20 minutes of simulated solar time for the cases of large ($\approx 9 \times 10^{12}$ cm^{2}s^{-1}, left) and small ($\approx 3.5 \times 10^{12}$ cm^{2}s^{-1}, right) numerical viscosity. Each 12×12 Mm image is a mosaic of four identical 6×6 Mm images with periodic structure. Note the significant reduction of the size of the convective cells upon reduction of the numerical viscosity.

large granules well. The actual diffusivities in our numerical simulations are functions of the local velocity and velocity differences, so we cannot provide more accurate limits. In order to improve the similarity between observations and numerical models on smaller scales, the effective numerical viscosity needs to be further reduced. This might perhaps be accomplished by using a pseudo-spectral representation (as in Nordlund 1982), instead of finite differences.

4   Conclusions

Comparison of Observations and Simulations:

1. The average width and central depth of the simulated Fe I 6302.5 Å line agree remarkably well with the FTS observations. This is evidence that the average vertical velocity in the simulations is nearly correct.

2. The ratio of the intensity to the velocity fluctuations agrees in the simulations and observations. Both the intensity and velocity fluctuations of the simulated granulation are larger than observed. By smearing the simulated images with the theoretical telescope modulation transfer function, and an exponential MTF similar in shape to empirically determined seeing MTFs (Nordlund 1984), the simulated fluctuations can be reduced so as to agree with the observed ones. The atmospheric smearing is dominant. It allows one to see features near the diffraction limit of the telescope, but it greatly reduces the contrast of those features. The simulated contrast of 17% disagrees with the 13% contrast found by von der Lühe and Dunn (1987).

3. There is apparently more power in the real granulation at small size scales than in the simulations, even with the numerical viscosity reduced to the lowest value permitted.

Future Work:

1. Spectral observations covering a larger area of the solar surface, and comprising weaker spectral lines (less affected by damping) from both neutral and ionized species, will help to constrain further the thermodynamics and vertical motions of solar granulation.

2. The question of the amount of "scattering" caused by atmospheric seeing will be further investigated. In particular, the central intensity observed in small sunspots and pores might be used to check the level of "scattering" present in the observed σ_{Irms}.

3. Numerical experiments with the viscosity will continue in order to improve the resolution of small scale features of the granulation.

Acknowledgments

We thank U. Kusoffsky and R. Kever for assistance with the observations and processing of the data at the Swedish Vacuum Solar Telescope. We also thank W. Livingston of the National Solar Observatory for kindly providing us with the FTS measurements shown in Figure 4; and M. Knölker for reading the manuscript under very severe time constraint. Å. N. gratefully acknowledges the hospitality of the faculty and staff at JILA during a one year Visiting Fellowship period. He also acknowledges support from the Danish Natural Science Research Council and the Carlsberg Foundation.

References

Lites, B. W., and Scharmer, G. B. 1988 "High Resolution Spectra of Solar Magnetic Features. I. Analysis of Penumbral Fine Structure," (in preparation).
Livingston, W. C. 1987 (private communication).
von der Lühe, O., and Dunn, R. B. 1987, *Astron. Astrophys.*, **177**, 265.
Nordlund, Å. 1982, *Astron. Astrophys.*, **107**, 1.
Nordlund, Å. 1984, in *Small-Scale Dynamical Processes in Quiet Stellar Atmospheres*, ed. S. L. Keil (National Solar Observatory, Sunspot, N. M.), p. 174.
Nordlund, Å., and Stein, R. F. 1989 (in these proceedings).
Scharmer, G. B. 1987, in *The Role of Fine-Scale Magnetic Fields on the Structure of the Solar Atmosphere*, ed E. Schröter, M. Vázquez, and A. Wyller (Cambridge: Cambridge Univ. Press), p. 349.
Schmidt, W., Knölker, M., and Schröter, E. H. 1981, *Solar Phys.*, **73**, 217.
Stein, R. F., Kuhn, J.R., and Nordlund, A. 1989 (in these proceedings).
Thomas, J. H., Lites, B. W., Gurman, J. B., and Ladd, E. F. 1987, *Ap. J.*, **312**, 457.

THE VARIATION OF THE MEAN SIZE OF THE PHOTOSPHERIC GRANULES NEAR AND FAR FROM A SUNSPOT

C. Macris*, Th. Prokakis*, D. Dialetis*, R. Muller**.

* Astronomical Institute, National Observatory of Athens, Athens 11810, P.O. Box 20048, Greece.
** Observatoire du Pic du Midi, F-65200 Bagneres de Bigorre, France.

ABSTRACT. We study the mean size of granules as a function of the distance from the boundaries of the sunspot penumbra. We use for the determination of the mean size two different methods, a visual and a photometric one. In all cases the mean diameter of the granules remote from the spot was greater than the mean diameter of the granules in the neighbourhood of the penumbra. Our study is based on an excellent sequence of photos, taken by one of us (R.M.) at the Pic du Midi Observatory in May 11, 1979.

1. INTRODUCTION

A very interesting and controversial question concerning the solar granulation is the possible variation of mean diameters of the photospheric granules near the sunspots. According to Macris (1949, 1953), Macris and Prokakis (1962) and Schroter (1962) the mean diameter of the photospheric granules near the spot's penumbra are smaller than the ones of the granules found far from the spots. Bray and Loughhead (1964) found that only in very few cases this difference has been noticed in the appearance of the granulation around a spot. Finally in a recent paper Collados et al (1986) expressed the view that the difference between the mean diameters of the granules far from and near the spot regions is not significant.

The aim of the present work is to find if, with the use of better observational data, there exists indeed such a difference in the granular size near and away of sunspots penumbra.

2. METHODS OF MEASUREMENTS AND RESULTS

The observational material used in this work was taken at the Pic du Midi observatory by R. Muller, in May 11, 1979 during a day of excellent seeing conditions (Figure 1). In the sunspot group (Mc Math 15990) existed some small spots and a bigger one.

R. J. Rutten and G. Severino (eds.), Solar and Stellar Granulation, 359–363.
© *1989 by Kluwer Academic Publishers.*

For our measurements we have considered as granules only the larger structures (D>0.3") and we have ignored all the traces of structure and all the little fragments within the granules and the intergranular lanes, because it is not evident that all these small structures that we can identify as individual features at a given moment are individual granules.

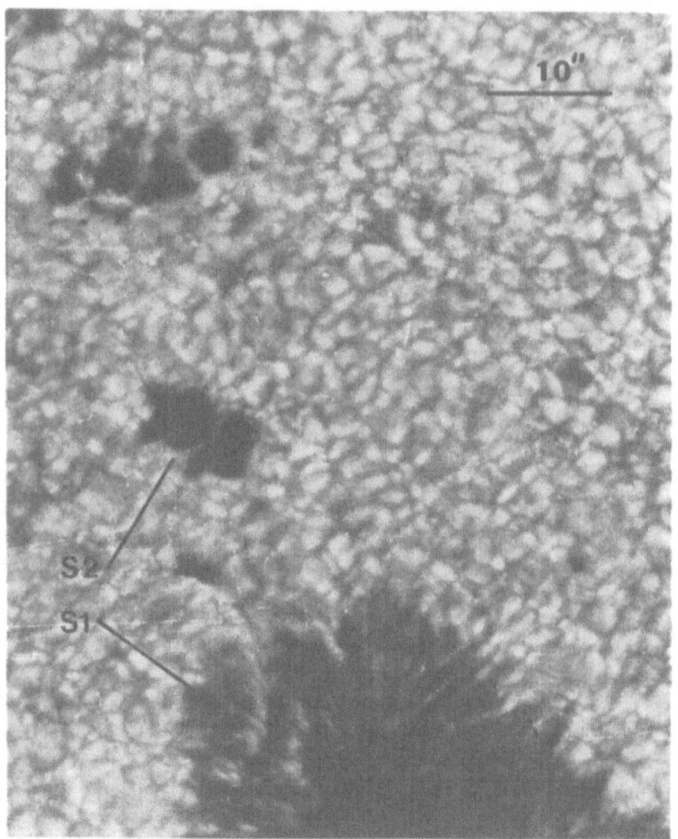

Figure 1. General view of the studied photospheric region.

For the determination of the mean diameter we have used two different methods.

2.1. The visual method.

We have selected two different groups of granules. The first group

was composed of 200 granules located near the outer edge of the penumbra
(maximum distance < 4"). The second group was also composed of 200 granules with a mean distance of 39" from the outer edge of the penumbra.
The diameter of granules was determined by using an ocular micrometer.
Each granule was measured twice in two perpendicular directions. We consider as granule diameter the mean value of these two diameters (Macris
1979). The results are the following:

Mean diameter of the granules located near   1.12" ± 0.33"
the spot (n= 200)

Mean diameter of the granules located far   1.54" ± 0.44"
from the spot (n=200)

Under the assumption of a Gaussian distribution of the diameters we can
be 99% confident that the two means (near and far from the penumbra) are
different.

2.2. The photometric method.

The determination of the granule diameter in a visual way is
affected by uncertainties such as apparent contrast of the picture in
the photographic paper, the subjective criterion defining the granule-
intergranule boundary etc. The only way to minimize these uncertainties
is to use photometric methods for the determination of the granule
diameter. For this reason a photometric determination of the granule
size was used.

The pictures were digitized using the Joyce-Loebl microdensitometer
of the National Observatory of Athens (Figure 2). A photometric criterion gave the possibility to define objectively the granular limits. By
using a random process we have selected 443 granules located in different distances from the main spot of the group. For every granule we have
measured the area and the distances from the two spots (S1 and S2). As
diameter of the granule we have considered the diameter of a circle with
equal area (Karpinsky, 1980). We have considered as granules located
near the sunspot penumbra all granules with distance from the outer edge
of the penumbra <6".

The results obtained by the second method are the following:

Mean diameter of granules located near
the penumbra of sunspot S1 (n=85)   1.30" ± 0.32"

Mean diameter of granules located far
from the penumbra of sunspot S1 (n=358)   1.53" ± 0.44"

Mean diameter of granules located near
the penumbra of sunspot S2 (n=50)   1.35" ± 0.40"

Mean diameter of granules located far
from the penumbra of sunspot S2 (n=393)   1.51" ± 0.43"

Mean diameter of all the measured granules (n=443)	1.49" ± 0.43"

Under the assumption of a Gaussian distribution of the diameters we can be 99% confident that the two means (near and far from the penumbra) are different for both sunspots.

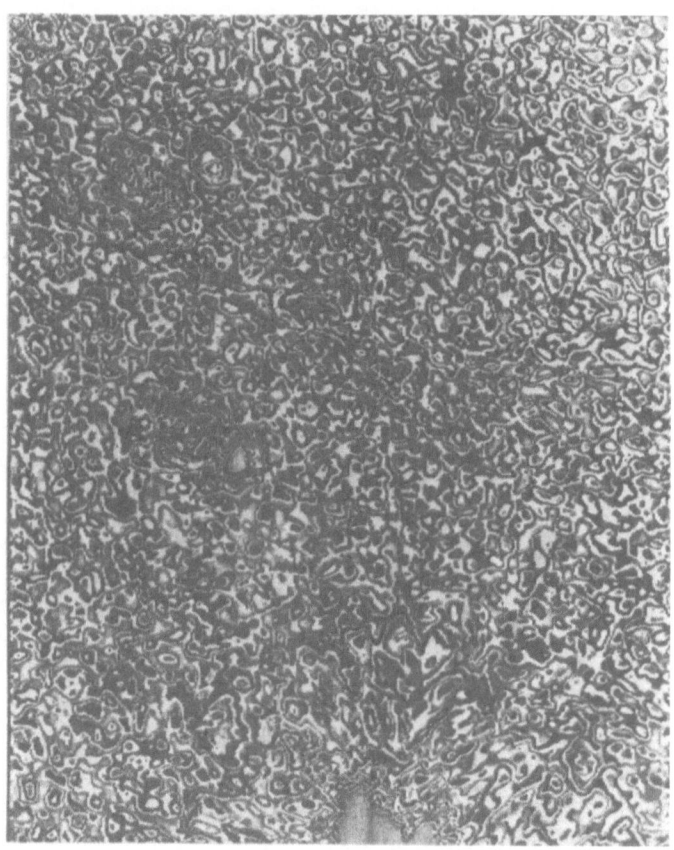

Figure 2. Isophotometric map of the studied photospheric region.

3. DISCUSSION

The results coming from the two methods confirm without any doubt the existing size difference for the granules located near and far from the sunspot penumbra. Comparing these results we found a large size difference by using the first method. We cannot conclude whether that is due to a possible systematic bias introduced by the visual method, or to the fact that we have measured all the granules located very close (d<4") to the outer edge of the penumbra.

Using the results of the photometric method we had also the possibility to study the granule diameter as a function of the distance of a granule from the penumbra. We give in Figure 3 the results for the sunspot S1. It is interesting to note that there is a linear increasing of the mean diameter from the outer edge of the penumbra to a distance of ≃ 16". In greater distances it is clear that the mean diameter vary with the location, but in any case is greater than the mean diameter of the granules located near the penumbra. This phenomenon is very prominent

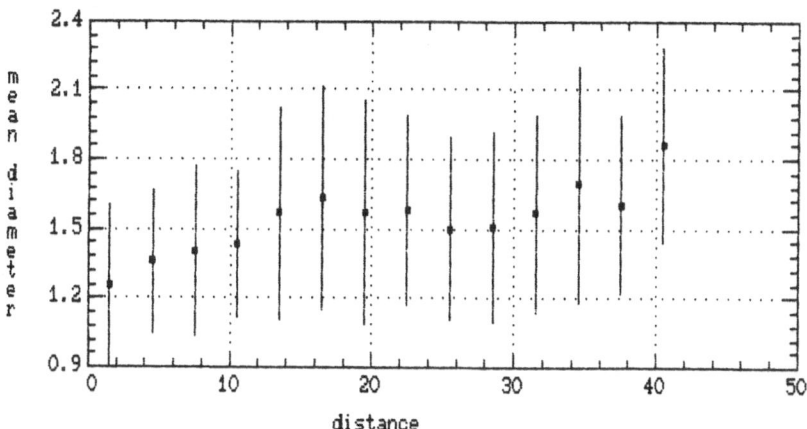

Figure 3. Variation of the mean granule diameter as a function of the distance from the outer edge of the penumbra.

for the sunspot S1 but not for the much smaller spot S2.

This result must be confirmed with more observational data for different spots. If it is true we must accept that the deformations of granules near the penumbra boundary do not depend directly to the strength of the magnetic field at sunspot boundary, because it dropped off to very small values at very short distance from the penumbra (Beckers and Schroter, 1969; Wiehr et al., 1986).

References

Beckers J.M., Schroter E.H. : 1969, Solar Phys., 10, 384.
Bray R.J., Loughhead R.E. : 1964, 'Sunspots', Chapman and Hall, London.
Collados M., Marco E., Toro del C.J., Vazquez M. : 1986, Solar Phys., 105, 17.
Karpinsky V.N. : 1980, Solnechnye Dannye Biulleten, 7, 94.
Macris C. : 1949, Compt. Rend. Acad. Sci. Paris, 228, 2010.
Macris C. : 1953, Ann. Astrophys., 16, 19.
Macris C. : 1979, Astron. Astrophys., 78, 186.
Macris C., Prokakis Th. : 1962, Compt. Rend. Acad. Sci. Paris, 255,1862.
Schroter E.H. : 1962, Zeit. fur Astrophysik, 56, 183.
Wiehr E., Stellmacher G., Knolker M., Grosser H: 1986, Astron. Astrophys., 155, 402.

ABNORMAL GRANULATION

R. MULLER
Pic du Midi Observatory
65200 Bagnères-de-Bigorre
France

ABSTRACT

When observed under very high resolution (0".25) the abnormal granulation appears to be formed with facular points filling intergranular
lanes. A slight seeing degradation is sufficient to smear the area.

In some places the granulation appears smeared and of low contrast
and called abnormal granulation (Dunn and Zirker, 1973). These places
coïncide with facular areas or with remnants of active regions having
a "filigree" appearance when observed in the wings of Hα. The question
is to know whether abnormal granulation is a kind of inhibited granulation or a granular pattern in which intergranular lanes are filled
with facular points, smeared by seeing. In order to solve this problem
we performed high resolution observations of a remnant of an old active
region near the disc center with the 50 cm refractor of the Pic du
Midi Observatory. The observations consist of pictures taken at
λ 4308 Å, which show facular points brighter than the surrounding
photosphere, and of quasi-simultaneous white-light pictures of the same
regions taken under different seeing conditions (Figure 11, in the
review paper 'Solar Granulation : Overview', by Muller, in this volume).
In the white-light picture taken under moderate seeing conditions
(Figure 11 b), the granulation appears smeared in the regions corresponding to facular areas. However in Figure 11 c, taken under superb
seeing conditions, close to the diffraction limit of the telescope
(0".25), facular points can be distinguished in the intergranular
lanes between granules. This observations strongly suggests that the
abnormal granulation is a region where many facular points fill the
intergranular lanes. A slight seeing degradation is sufficient to give
a smeared appearance to this region. However the granules seem to be
smaller than normal.

REFERENCE

Dunn, R.B. and Zirker, J.B. : 1973, Solar Phys., 33, 281.

365

SUPERGRANULAR PATTERN FORMED BY THE LARGE GRANULES

R. Muller[1], Rh. Roudier[2], J. Vigneau[2].

1 . Pic du Midi Observatory, 65200 Bagnères-de-Bigorre, France
2 . Pic du Midi Observatory, 31500 Toulouse, France.

ABSTRACT

The granules larger (or smaller) than the critical size (1".4) form a
cellular pattern of mean size 6"-7". This pattern could be related to
the mesogranulation.

1. INTRODUCTION

When trying to measure the scale of the granulation with uni-dimensio-
nal autocorrelation function, we were surprised to find that in some
areas, the autocorrelation exhibits a well defined secondary maximum,
whilst in other areas it does not. It was then apparent that in the
first kind of areas, large granules are dominant and that small granules
are dominant in the second kind of areas, where the autocorrelation
function does not exhibit a secondary maximum. We are thus incited to
look more carefully at the spatial distribution of large and small
granules.

2. SPATIAL DISTRIBUTION OF LARGE AND SMALL GRANULES.

For this purpose we used the processed granulation image shown in
Roudier and Muller (1986, Figure 4). The image processing provides the
position, the shape and the area of every granule. In this paper we
found that the granules have a critical size of 1".37. We then plotted
maps containing only the granules of area A larger than this critical
size ($A = \pi d^2/4$), Figure 1a, and maps containing only the granules
smaller than the critical size, Figure 1b. It is clear from those figures that either the large or the small granules form a cellular pattern.
From an autocorrelation analysis, the characteristic size of these cells
is found to be 6"-7". In order to check whether this granule cellular
pattern could result from a 'chance grouping' of granules distributed
at random, we performed a pattern of synthetic granules. For this
purpose we used the granule size distribution and the number density
published by Roudier and Muller (1986); the granules were then placed

367

R. J. Rutten and G. Severino (eds.), Solar and Stellar Granulation, 367–369.
© 1989 by Kluwer Academic Publishers.

368

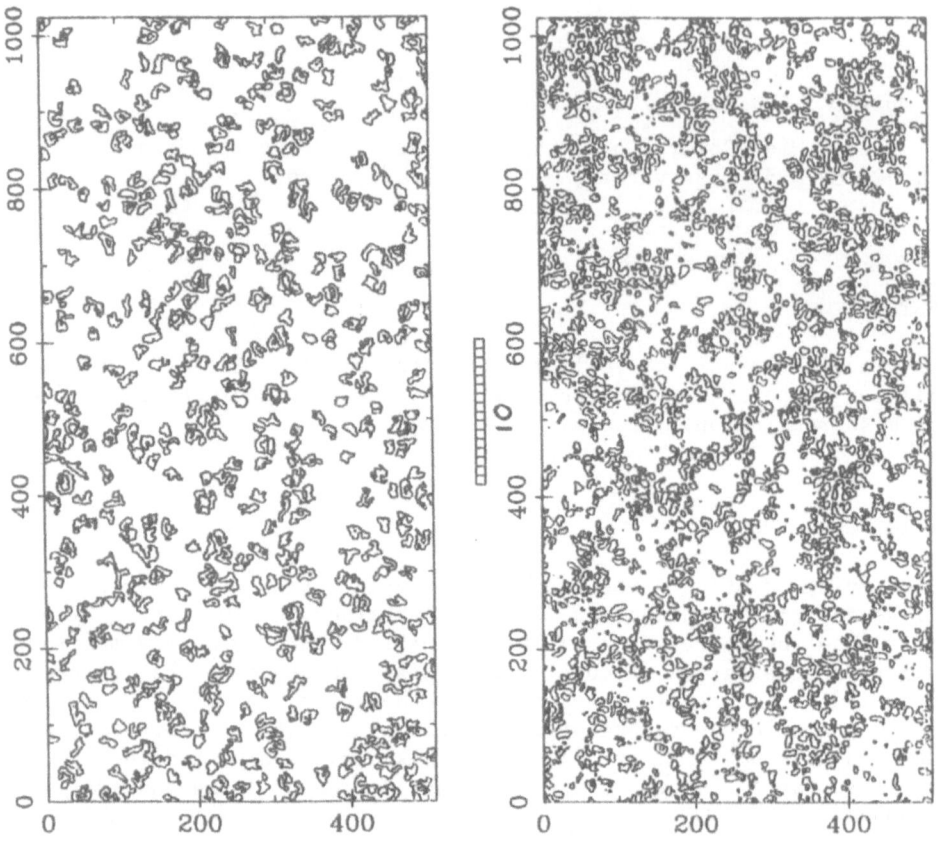

Fig. 1. a (left side): pattern of the granules larger than 1".37;
b (right side) : pattern of the granules smaller than 1".37.

at random. The autocorrelation of such a field of synthetic granules spatially distributed at random does not exhibit any secondary maximum. Consequently real granules are not distributed at random on the surface of the sun; granules larger than the critical size (1".37) as well as smaller granules appear to be grouped, forming a cellular pattern of average size 6"-7".

3. DISCUSSION

A cellular pattern was previously revealed by the active granules (those which fragment or merge, Oda, 1984) and by the position of the long-lived granules (Dialetis et al., 1988). The mean cell size of this pattern (11"-12") strongly suggests that it is associated with the mesogranulation discovered by November et al. (1981) on dopplergrams. Large scale brightness fluctuations were also found at the mesogranule scale (Oda, 1984; Koutchmy and Lebecq, 1986).This is not very suprising because the granules which fragment are the brightest and have the longest lifetimes; they are also the largest, thus suggesting that the pattern of large (or small) granules is the same as those quoted above; however its size is found to be significantly smaller, for a yet unknown reason.

REFERENCES

Dialetis, D., Macris, C., Muller? R., and Prokakis, Th. : 1988, Astron. Astrophys.(in press).
Koutchmy, S. and Lebecq, Ch. : 1986, Astron. Astrophys. 169, 323.
November, L.J., Toomre, J., Gebbie, K., and Simon, G.: 1981, Ap.J. (Letters), 245, L123.
Oda, N. : 1984, Solar Phys. 93, 243.
Roudier, Th. and Muller, R. : 1986, Solar Phys. 107, 11.

DETAILS OF LARGE SCALE SOLAR MOTIONS REVEALED BY GRANULATION TEST PARTICLES

G.W. Simon[1], L.J. November[2], S. Ferguson[3], R. Shine[3], T. Tarbell[3],
A. Title[3], T. Topka[3], H.Zirin[4]

[1] Air Force Geophysics Laboratory National Solar Observatory/Sacramento
Peak*, Sunspot, NM, USA
[2] National Solar Observatory/Sacramento Peak*, Sunspot, NM, USA
[3] Lockheed Palo Alto Research Laboratory, Palo Alto, CA, USA
[4] California Institute of Technology, Pasadena, CA, USA

Abstract. We present details of large scale horizontal (mesogranular, supergranular) flows which have been discovered by measuring the motions of granules with local correlation techniques. Not only do the flows expel magnetic field from the centers of source regions to their boundaries, thus forming the well-known supergranular network, but they also exhibit convergence into sinks, and significant vorticity. The displacement, distortion, and twisting of magnetic field by such motions may lead to a better understanding of solar flare buildup processes, and an increased ability to predict the occurrence of flares.

Keywords. Granulation, mesogranulation, supergranulation, atmospheric motions, large scale flows, divergence, vorticity, magnetic fields.

1 Introduction

During the Spacelab 2 (STS-19) mission of the Space Shuttle, the SOUP (Solar Optical Universal Polarimeter) experiment built by the Lockheed Palo Alto Research Laboratory (Title *et al.* 1986) obtained a 27.5 min movie of solar granulation during orbit 110 (19.10–19.38 UT 5 August 1985). Because the images are free of distortion caused by the earth's atmosphere, it is possible to detect very small motions (less than 5 km) of solar intensity features, using the method of local correlation tracking (November 1986, November and Simon 1988). This analysis has shown that solar granules move like test particles ('corks') on top of larger scale flow patterns in the solar atmosphere. These patterns have mainly mesogranular (6″–10″) and supergranular (20″–40″) scales. By correlating these flows with simultaneous observations of the solar magnetic field it was found that magnetic flux elements move in patterns which are in almost perfect agreement with those of the granular corks (Simon et al 1988): granules and magnetic features are expelled from the centers of mesogranules, moving first to the mesogranular boundaries and ending up at the boundaries (network) of the larger supergranules. The test particles of the motion then slowly concentrate into small downdrafts (sinks) over a period of one to two days,

*Operated by the Association of Universities for Research in Astronomy, Inc., under contract with the National Science Foundation; partial support for the National Solar Observatory is provided by the USAF under a Memorandum of Understanding with the NSF.

R. J. Rutten and G. Severino (eds.), Solar and Stellar Granulation, 371–377.
© 1989 by Kluwer Academic Publishers.

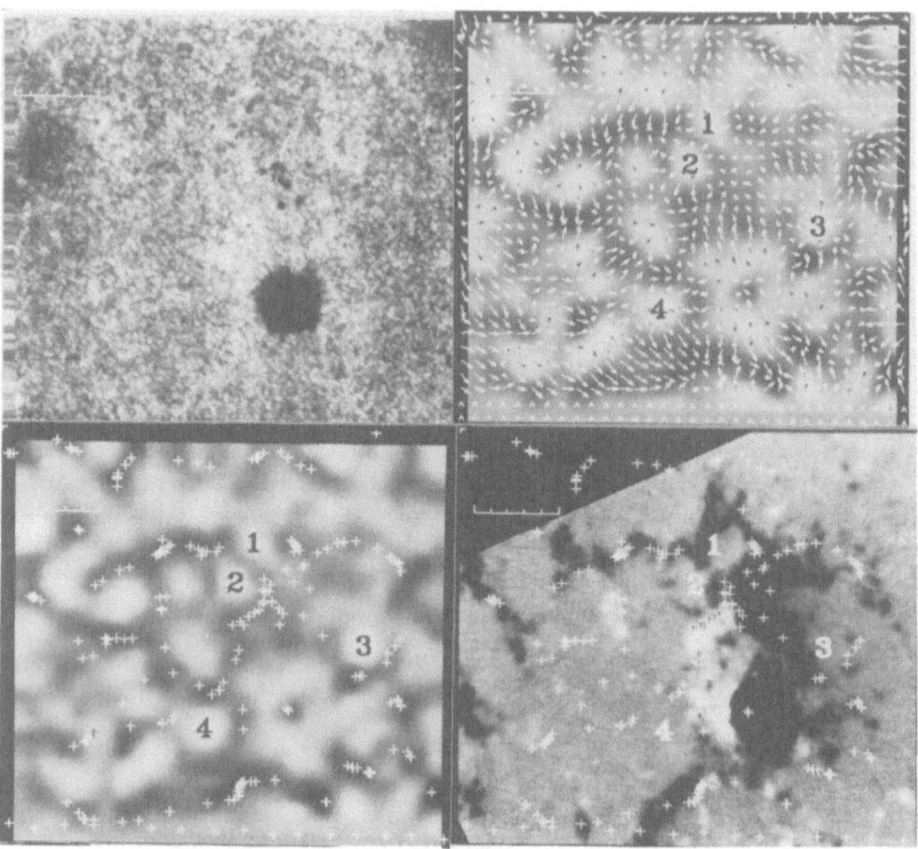

Figure 1: (a) (upper left) A white light photograph of the 131″ by 119″ area of the SOUP frame from orbit 110 on 5 August 1985 which was analysed in this work. A scale with 5″ tik marks is shown. (b) (upper right) SOUP flow vectors superposed on a gray scale image of the flow divergence (with bright = sources and dark = sinks). Four prominent flow cells are identified in the figure. (c) (lower left) Positions of corks 12 hr after starting from a uniform distribution, superposed on the divergence figure, and (d) (lower right) superposition of the 12 hr corks and the magnetic field structure as observed simultaneously at Big Bear Solar Observatory.

although the magnetic elements apparently prefer to remain part of a network pattern. These results are shown most effectively by high speed (300–1000 times acceleration) cinematography, but they are summarized in figure 1. In this paper we present some details of the observed horizontal flow patterns.

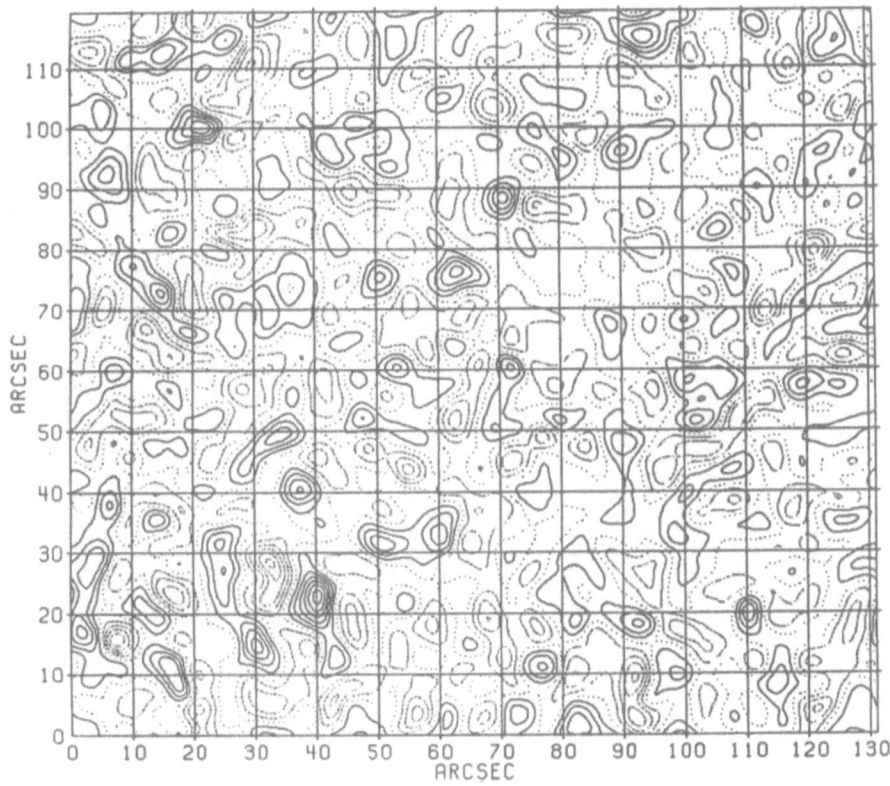

Figure 2: Contour plot of the divergence field for the same area shown in figures 1b and 1c. Solid contours (sources) correspond to the bright regions in figures 1b and 1c.

2 Results

From the components v_x and v_y of the horizontal velocity vector v it is possible to calculate the divergence D and vertical component E of the vorticity at each point in the field of view:

$$D = \frac{\delta v_x}{\delta x} + \frac{\delta v_y}{\delta y} \qquad E = \frac{\delta v_y}{\delta y} - \frac{\delta v_x}{\delta x} \tag{1}$$

In figures 2 and 3 we show contour maps of the quantitites D and E respectively, derived from the vectors of figure 1(b), in a 119″ by 131″ area of the SOUP image. There appear to be about 40 local maxima (sources) and a roughly equal number of minima (sinks) of the divergence in an area of 10000 square arcsec. The vorticity map has a very similar appearance, and contains about the same number of clockwise and counterclockwise vortices. Typical values for the local maxima are $10^{-4}s^{-1}$.

374

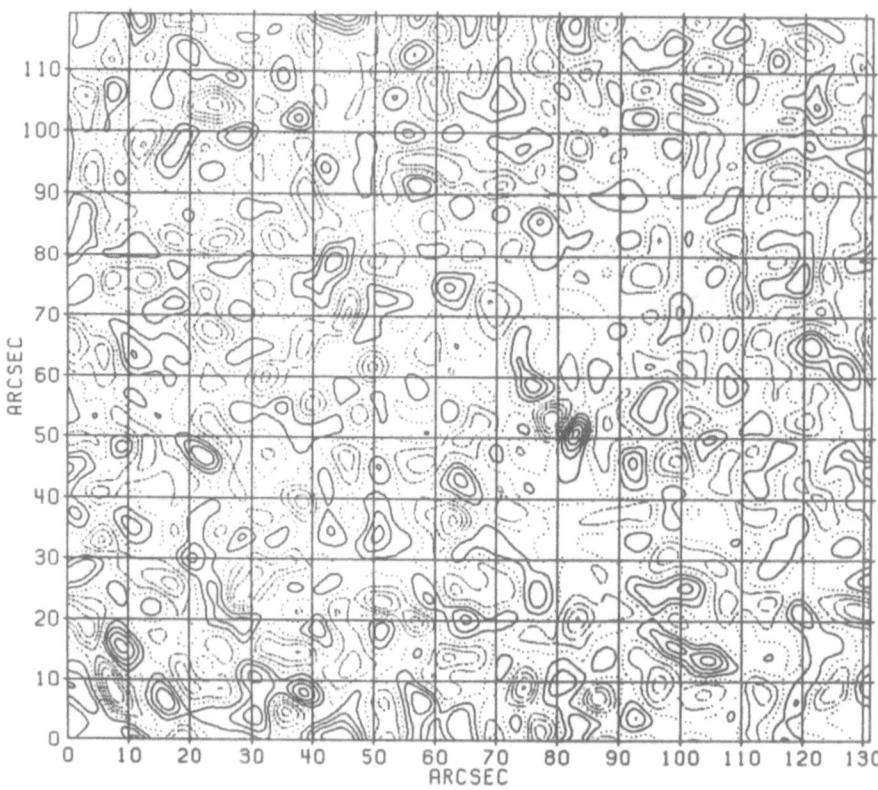

Figure 3: Contour plot of the vertical component of vorticity for the same area shown in figure 2.

An interesting observation is that the maxima and minima of vorticity occur only rarely at the same loci as the maxima or minima of divergence. This disagrees with one's intuitive expectation that the flow would probably follow a vortical path as it is sucked into a sink. However, observation of a particularly large vortex by Brandt et al. (1988) does, indeed, show the expected behavior. Since the method of local correlation tracking has been applied to only a few observations of divergence and vorticity in the solar atmosphere, further data are needed to clarify this situation.

In figure 4 we show twelve particularly good examples of positive and negative divergence extracted from figure 2. The left-hand six examples are of positive divergence (sources), while the right-hand six show convergent flows at sinks. In the same manner we show in figure 5 six cases of clockwise vorticity on the left, and six counterclockwise flows on the right. These examples can all be found in the large vorticity map in figure 3. Two interesting regions of multiple and opposite vorticity are shown in figure 6: a triple vortex on the left, and a double vortex on the right.

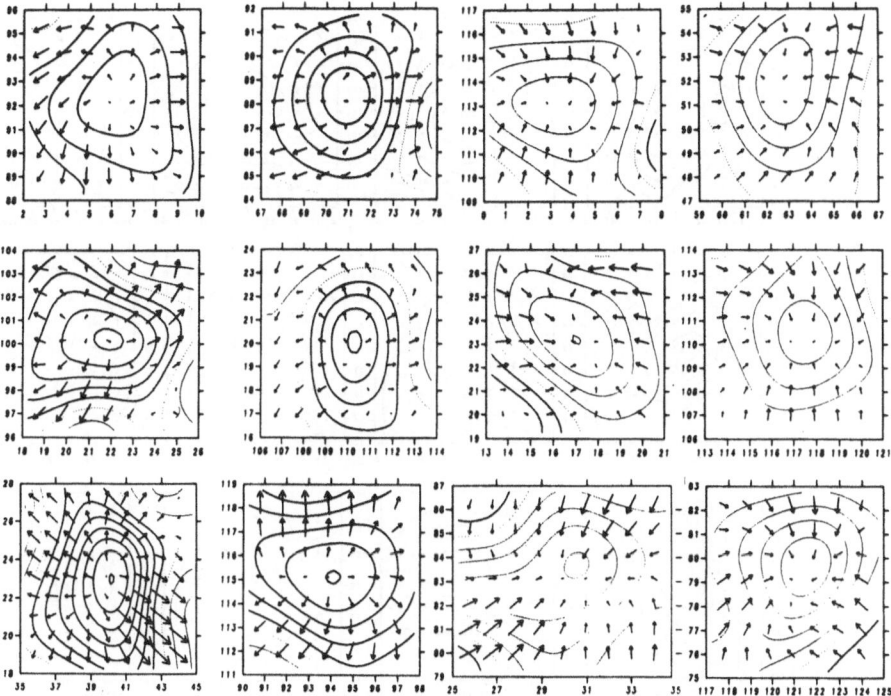

Figure 4: Twelve examples of diverging and converging flows. The six left-hand images are sources; the six right-hand ones are sinks. Numbers along the abscissa and ordinate of each small figure correspond to the same numbers in figure 2, and can thus be used to locate each source (and sink) in the large image.

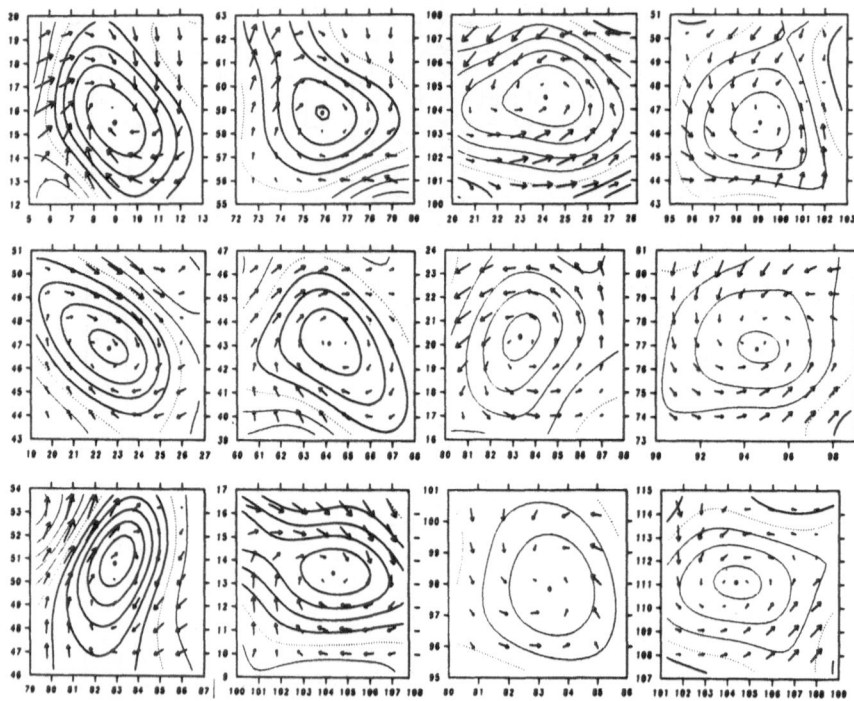

Figure 5: Twelve examples of vertical vorticity. Six clockwise flows are shown on the left, and six counterclockwise on the right. These examples can all be located in figure 3.

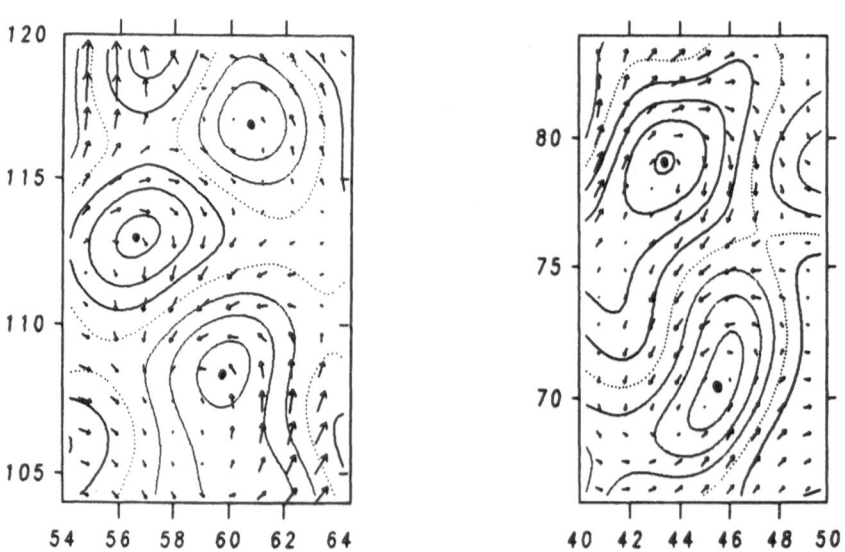

Figure 6: Illustrations of more complex vortex patterns. On the left a triple vortex, on the right a double vortex. Adjacent vorticies flow in opposite directions. These vorticies are taken from figure 3.

3 Conclusions

These observations have important implications for the buildup of magnetic activity on the sun. We have pointed out that the structure of the sun's surface magnetic field closely mimics that of test particles which are driven by mesogranular and supergranular flow patterns. Thus the behavior of such corks should help one to predict rather accurately where magnetic fields will pile up and undergo stress. If one adds to that the observed vorticities which are large enough to twist a magnetic flux tube by a full 360 degrees in less than three hours, it should become possible to pinpoint loci of magnetic mixing and twisting, and thus to increase one's ability to predict the occurrence of solar flares, coronal heating, and mass ejections.

Both at the Sacramento Peak Observatory and the Swedish Solar Observatory, techniques have been developed and observations are in progress to measure these large scale flows on a regular basis over large areas of the solar disk. (Note that such proper motion measurements are usable from disk center out to a radius vector of about 0.9, while traditional Doppler techniques work only in the range of about 0.5 to 0.8 and are more difficult to interpret.)

One attempt has already been made and is also being presented at this conference (Simon and Weiss 1989, **1988**) to model the large scale outflows by axisymmetric potential functions which preserve mass continuity. The remarkable success of this simple kinematical model in simulating the actual observed behavior of the magnetic field and the associated cork patterns of the SOUP data, gives real promise that models will eventually predict the effects of surface flows on subsequent magnetic phenomena.

References

Brandt, P., Scharmer, G., Ferguson, S., Shine, R., Tarbell, T., and Title, A. 1988, Nature, in press.

November, L. 1986, Appl. Optics **25**, 392.

November, L., Simon, G., Tarbell, T., Title, A., and Ferguson, S. 1987, in Proc. 2nd Workshop on Theoretical Problems in High Resolution Solar Physics, NASA Conf. Pub. 2483 (Washington, DC: NASA), ed. G. Athay, p.121.

November, L. and Simon, G. 1988, Ap. J. 333, in press (1 October).

Simon, G., Title, A., Topka, K., Tarbell, T., Shine, R., Ferguson, S., Zirin, H. and the SOUP Team 1988, Ap. J. **327**, 964.

Simon, G. and Weiss, N. 1989, in Proc. Workshop on Solar and Stellar Granulation, (Dordrecht, Holland: Kluwer), ed. R. Rutten and G. Severino, in press.

Simon, G. and Weiss, N. 1988, Ap. J., submitted.

Title, A., Tarbell, T., Simon, G. and the SOUP Team 1986, Adv. Space Res. **6**, 253.

Part 3
Modelling

CONVECTION AND WAVES

R.F. Stein[1], Å. Nordlund[2] and J.R. Kuhn[1]

[1] *Dept. of Physics and Astronomy, Michigan State University, U.S.A.*
[2] *JILA, University of Colorado, U.S.A.*

1   Introduction

We believe that computer simulations are a powerful tool for developing a qualitative understanding of solar physics. To this end, we have written a three-dimensional, compressible, magneto-hydrodynamic code, with a sophisticated equation of state and including radiative energy transport, to produce such simulations. In this talk we discuss how we use simulations of solar convection to study the interaction between the 'five minute' global solar oscillations and convection. In a second talk (Nordlund and Stein, 1989) we discuss using computer simulations of magneto-convection to investigate how pores and sunspots develop. Of course, with our simulations we can do much more. We have already started looking at what controls the scale of structures in the solar convection zone. For the future, we plan to investigate the generation and propagation of waves, and the role of convection and magnetic fields in heating the solar chromosphere and corona.

Global solar oscillations are resonant acoustic modes of a cavity in the solar interior (Ulrich, 1970; Leibacher and Stein, 1971). The upper boundary of the cavity is produced by the steep density gradient at the cool solar surface, and the lower boundary is produced by refraction due to increasing sound speed with depth. Since modes of differing frequency and horizontal wavelength penetrate to different depths in the sun, comparison of observed mode frequencies with those calculated from a solar model enables us to probe the internal structure of the sun. The oscillations are an invaluable tool for unraveling the nature of the solar interior. Using them is practical because of the unusual precision (by astrophysical standards) with which they are observed. To employ the oscillations for probing the solar interior also requires comparable precision in the calculated eigenmodes.

We were motivated to investigate the interaction between the oscillations and convection by the discrepancy that exists between the observed and the best theoretical eigenfrequencies. Christensen-Dalsgaard (1984, see also Christensen-Dalsgaard *et al* 1985) has pointed out that the general trend of the difference $\Delta\nu = \nu_{observation} - \nu_{theory}$ is a function of frequency but independent of horizontal wavelength (for $k_H < 0.3 Mm^{-1}$), which indicates that the major source of error is near the solar surface. Another way of revealing the problem was discovered by Duvall (1982). He found that all the modes map

381

R. J. Rutten and G. Severino (eds.), Solar and Stellar Granulation, 381–399.

onto a single line when plotted as $(n + \epsilon)\pi/\omega$ vs. ω/k_H, which indicates that the eigen-frequencies are a function primarily of the run of sound speed with depth. The phase ϵ is due to the penetration of the eigenfunctions beyond the boundary of the cavity. The observed phase is 1.5, the theoretical phase is 1.3. Again, an effect of the boundary layer is indicated. We anticipate that the outer boundary layer of the sun will have a sig-nificant effect on the eigenfrequencies of the global solar oscillations, because the waves spend a significant fraction of their period in the low sound speed, cool, outer regions of their cavity. We expect that the inhomogeneous temperature, density, velocity and magnetic field in the superadiabatic upper region of the convection zone will produce a mean structure different from that calculated by mixing length theory, as well as directly modifying the eigenfrequencies.

Can global solar oscillations be studied by computer simulations, particularly ones covering only 1% of the solar convection zone? Typical lifetimes of the solar modes are hundreds of hours, which would require weeks of super-computer time. How could a mode maintain its coherence over such long times, given the inevitable numerical noise in the simulations? Also, typical mode velocity amplitudes are a fraction of a m/s. How could they be seen above the much larger chaotic convective motions? And, how can one investigate frequency discrepancies of a few μHz between the observed frequencies and those calculated from one-dimensional spherically symmetric models, which would again require simulating days of solar time to obtain such resolution? This freqency difference is about 0.1% of the frequency itself. Since multi-dimensional numerical hydrodynamic calculations often have discretization errors of a few percent, how could one expect to say anything about such small discrepancies. Finally, even if we should see some oscillations in our simulations, will they have any relevance to the real sun? We shall see.

First, in Section 2, we introduce you to the equations and the code we use to solve them. Section 3 presents some preliminary results which show that oscillations do exist in our simulation and that they have relevance for the real sun. In Section 4 we briefly discuss the theory of the interaction of waves and p-modes with convection. And, we recapitulate and look toward the future in Section 5.

2 Code

To solve a fluid dynamics problem, we must solve the fluid dynamic equations – the equations for the conservation of mass, momentum and energy. There is no one best way to solve these equations; the nature of the problems we wish to tackle controls the method to be used. Convection is essentially a three-dimensional phenomenon, so we must solve the equations in three dimensions. The solar atmosphere is stratified, and we wish to include the region from the chromosphere to a depth in the convection zone comparable with the horizontal meso-scale. This covers a range of 15 pressure scale heights, so there are large changes in the pressure and density. We therefore pre-condition the equations we solve in order to reduce the magnitude of the derivatives of these variables. The fluid equations are pre-conditioned for a stratified atmosphere by using the logarithm of the density, $\ln \rho$, the velocity, u, and the internal energy per unit mass, e, rather than the per unit volume quantities ρ, ρu, and ρe. The equations we solve are: the equation of mass conservation

$$\frac{\partial \ln \rho}{\partial t} = -\mathbf{u} \cdot \nabla \ln \rho - \nabla \cdot \mathbf{u} \ , \tag{1}$$

the equation of momentum conservation

$$\frac{\partial \mathbf{u}}{\partial t} = -\mathbf{u} \cdot \nabla \mathbf{u} + \mathbf{g} - \frac{P}{\rho} \nabla \ln P \ , \tag{2}$$

and, the equation of energy conservation

$$\frac{\partial e}{\partial t} = -\mathbf{u} \cdot \nabla e - \frac{P}{\rho} \nabla \cdot \mathbf{u} + Q_{rad} + Q_{visc} \ . \tag{3}$$

The pressure is found from a tabular equation of state

$$P = P(\rho, e) \ . \tag{4}$$

The viscous dissipation is

$$Q_{visc} = \nu \sum_{i,j} (\frac{\partial u_i}{\partial x_j})^2 \ , \tag{5}$$

and the radiative energy exchange rate is

$$Q_{rad} = \int_\lambda \int_\Omega \kappa_\lambda (I_{\lambda,\Omega} - S_\lambda) d\Omega d\lambda \ , \tag{6}$$

where the specific radiation intensity $I_{\lambda,\Omega}$ (at wavelength λ, in direction Ω) is related to the source function S_λ by

$$\frac{dI_{\lambda,\Omega}}{d\tau_\lambda} = I_{\lambda,\Omega} - S_\lambda \ . \tag{7}$$

$d\tau_\lambda$ is the optical depth increment, defined by

$$d\tau_\lambda = \rho \kappa_\lambda ds \ ,$$

where ds is the path length along the straight ray and κ_λ is the absorption coefficient per unit mass.

An essential aspect of a successful complex code is that it be very modular. We have done this by isolating different operations in different subroutines. We solve for the time derivatives of our fundamental variables from the equations (1)-(3) in vector form, as shown above, a separate routine for each equation. These routines contain only vector operations, and are independent of the geometry of the problem. Routines are provided to perform each vector operation, e.g. grad, div and scalar product. The vector operation routines are common to all the equations, and contain all the geometry. They depend, in turn, on lower level routines that calculate the derivatives in each coordinate direction using spline fits to the function. Other terms are also evaluated in separate subroutines, for instance, the radiative heating (6) (which calls other routines to solve the radiative transfer equations (7)) and the viscous dissipation (5). Such modularity greatly improves the readability and reliability of the code. Once a routine is tested and found to work, it can be employed elsewhere with confidence.

By preconditioning these equations for a stratified atmosphere, we have greatly increased the smoothness of the derivatives. Replacing the gradients of density and pressure by the gradients of their logarithms, means that instead of the derivative of a variable that changes by five orders of magnitude, we have the derivative of a variable that changes by only one order of magnitude. Further, the coefficient of the pressure gradient in the momentum equation is now $P/\rho \propto T$, which varies by only one order of magnitude, instead of five for ρ^{-1}.

One must also be concerned with the stability of a numerical code. We advance the variables in time by the second-order Adams-Bashforth algorithm, with unequal time steps, which is mildly unstable. The evaluation of the spatial derivatives, by centered differences, introduces errors, in the form of one-zone spatial oscillations. Since this is a compressible flow, shocks can occur. These instabilities are removed and the code is stabilized by applying artificial diffusion to all the fluid equations. The diffusion term is

$$\nu\nabla^2 f + (\nabla\nu + \nu\nabla\ln\rho) \cdot \nabla f \tag{8}$$

for each variable f. The diffusion coefficient, ν, has three types of contributions: (1) a term proportional to the fluid velocity, to prevent ringing at sharp changes in advected quantities, (2) a term proportional to the sound speed, to smooth centered differences and to stabilize weak waves and the Adams-Bashforth time stepping, and (3) a term proportional to the finite difference velocity convergence (where positive) to stabilize shock fronts.

It is not possible (because of limitations in computer processing power, memory and storage) to model the entire convection zone surrounded by stable layers and still resolve the small scale granulation structure that is an essential part of the convective phenomena. Hence, we simulate only a thin layer of the upper convection zone and some of the atmosphere above it. The region we simulate is coupled to an external medium about which we have no information, and is, in reality, influenced by what happens in those external regions. Boundary conditions on the computational box then become a thorny issue. In keeping with our modular style, we isolate them so that we can easily experiment with them. Since we have no information on what is occurring outside our computational domain, we must therefore impose boundary conditions on our computational box that can not possibly be correct. The best that can be achieved are boundary conditions that are stable and do as little violence to the real situation as possbile. In practice, this means that all information interior to the computational domain that is being carried out across the boundaries should be permitted to do so with minimal reflections, while everywhere information is being carried into the computational domain must be kept stable. In the horizontal directions the problem is easy – we assume periodic boundary conditions – what goes out at one side comes in at the other. At the top, what we have found works reasonably well is to take an extra large boundary zone ($\geq$ scale height). In this zone we impose the conditions that the amplitude of the velocity and the density fluctuations remain constant, while the energy density at the boundary is fixed at its initial average value. At the bottom: we impose constant pressure by adjusting the density; we impose $\partial\mathbf{u}/\partial z = 0$; and, we specify the internal energy density of inflowing material, while allowing outgoing material to carry out whatever energy density it has. This keeps the mean convective flux constant, so the mean state of the model does not drift. Since convection is driven from the superadiabatic region at the top of the convection zone, we expect that the dominant causal influence is from inside our computational box to the outside. That is, we expect what is going on outside our computational box to really have only a small effect on what happens inside. Therefore, these boundary conditions should be physically realistic.

Another important part of our simulation is that we try to make the physics as realistic as we can. This means that we use a realistic equation of state, and include radiative energy transport. The equation of state includes ionization and excitation of hydrogen and other abundant atoms, and formation of H_2 molecules. Radiative energy exchange plays a crucial role in determining the structure of the upper convection zone. Indeed, the observed granulation pattern is due to the radiation of heat carried to the surface by ascending hot gas in the convective flow. The net outward flux at any wavelength originates at the layer where the monochromatic optical depth at that wavelength is of order unity. According to the Eddington-Barbier relations,

$$F_\lambda \simeq B_\lambda(\tau_\lambda = 2/3) \ .$$

Hence cooling is strong where the continuum optical depth is of order unity. Since the opacity is a rapidly increasing function of temperature, this cooling layer is very thin. The transfer equation is solved along straight rays, after averaging the Planck function into four bins by wavelength sorted according to opacity (cf. Nordlund, 1982). The reference opacity and averaged Planck functions are pre-calculated and stored in a table as a function of energy and $\ln \rho$. The Uppsala package is used to calculate these values (Gustafsson, 1973).

3 Simulation Results

We simulate a layer of the Sun extending from the temperature minimum $(-0.5Mm)$ to a depth of $2.5Mm$, with horizontal extent $6 \times 6Mm$. To date we have simulated 3 solar hours, which has required approximately 600 cpu hours on an Alliant FX/8 computer.

What did our simulations reveal? We see convection, of course. A granulation pattern is visible at the surface, with granule sizes and velocities in agreement with observations. This pattern is produced by the sudden release of radiation at optical depth unity by ascending hot gas in disconnected cells. The hot gas ascends slowly over a broad region and is nearly structureless below the surface. The upward flow diverges and turns over into cool descending gas in connected intergranule lanes. This descending flow converges into thin filaments below the surface, which then converge into larger plumes that plunge downward (Nordlund and Stein, 1988). Above the visible surface horizontally propagating shocks occasionally develop, particularly around bright granules that are being squeezed out of existence by their neighbors.

We also find that oscillations are excited in our simulation. Figure 1 shows the mass flux, normalized by $< \rho >^{1/2}$, averaged over a horizontal plane, as a function of time, at the surface and a depth of 1 Mm. Do these oscillations have any relation to the global solar oscillations? We shall see.

Notice how the oscillation amplitude grows linearly with time, and saturates after only 130 minutes or 17 periods. How is it we can see this growth and saturation after only

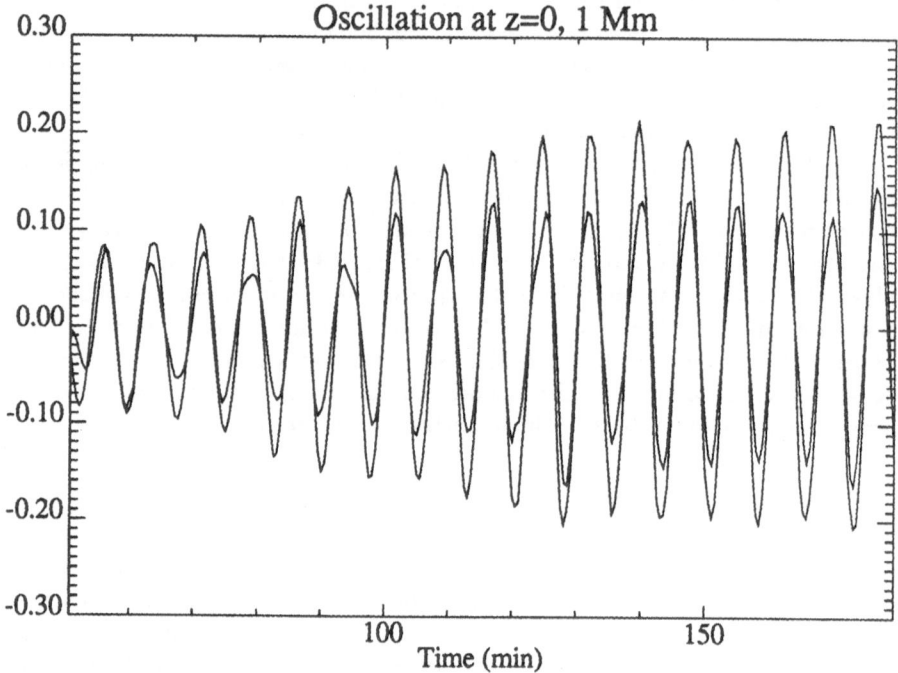

Figure 1: The horizontally averaged mass flux $< \rho u >$ (divided by $< \rho >^{1/2}$), as a function of time, at $z=0$ and 1 Mm depth. The larger amplitude trace is for $z = 1Mm$, near where the first harmonic has a node.

17 periods? What happened to the mode lifetime of days? This boon is a consequence of our shallow computational box. In the Sun, each global mode has an enormous inertia. In our simulation, the modes occupy only the outer 1% of the convection zone. Since they don't extend deep into the Sun, they have much less inertia than the solar oscillation modes, and therefore have short lifetimes and large growth rates. This is one advantage to having a shallow box, and there are others, as discussed below. The fact that the velocity seems to grow linearly, not exponentially, with time suggests that the modes are excited by stochastic interaction with convection, not by an overstability.

The modes are also clearly revealed as peaks in velocity power spectra. Spectra for modes with $k_H = 0$ (horizontally uniform) and $1\ Mm^{-1}$ (one horizontal cycle) are shown in figure 2. Is there any other evidence that convective motions drive the oscillations? The velocity amplitudes also point in that direction. They are enormous, of order 100 m/s per mode, not 0.1 m/s per mode as observed. We can understand the mode amplitudes in our box from a simple heuristic model. If the driving is due to some sort of random kick by the convection, then the effect of many kicks is a random walk in amplitude, with the expectation value of the velocity, V, growing as

$$V \simeq N^{1/2} \Delta V ,$$

where N is the number of kicks of magnitude ΔV. The final amplitude depends on how

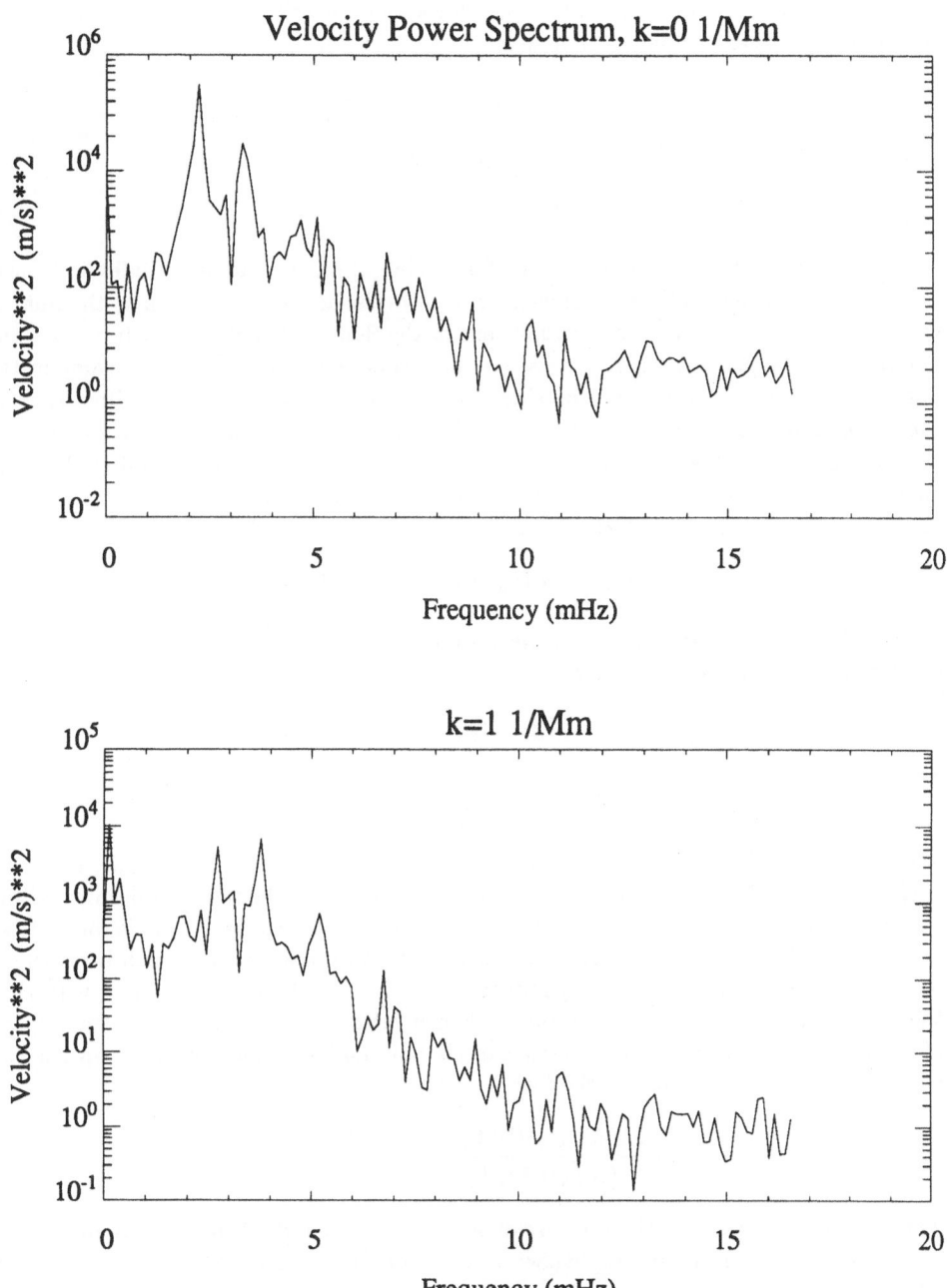

Figure 2: Velocity power spectra for modes with horizontal wavenumber 0 and 1 Mm^{-1}, at 0.3 Mm above the surface.

rapidly the modes are damped. They grow for approximately one damping time, τ, and the number of kicks per damping time is

$$N \simeq \dot{N}\tau .$$

Thus, the saturated amplitude will be approximately

$$V \simeq (\dot{N}\tau)^{1/2}\Delta V .$$

To compare our simulation box and the Sun, think of our box with periodic horizontal boundary conditions replicated sufficiently to have the same surface area as the Sun. If convective kicks drive the global oscillations in the Sun as they do in our box, then the number of kicks per unit surface area, per unit time is approximately the same in the Sun and in our computational box, since the kicks probably come predominantly from the superadiabatic layer where the convective velocity is largest, and our simulations include that layer. However, in the Sun the kicks are randomly distributed in time and space, while in our replicated boxes, with their periodic horizontal boundary conditions, many come in unison. If M kicks occur in unison, the expected amplitude is

$$V \simeq (\dot{N}\tau/M)^{1/2}M\Delta V = (M\dot{N}\tau)^{1/2}\Delta V ,$$

where M (the replication factor) is the ratio of the surface area of the Sun relative to one box of horizontal dimension L:

$$M = 4\pi R_\odot^2/L^2 .$$

The damping time is

$$\tau \simeq E/\dot{E} ,$$

where E is the mode energy. If the damping occurs by interaction with convection and radiative losses, it occurs predominantly near the surface, in the region contained within our simulation box. Then we expect the damping times to scale as the mode mass. If we assume that the energy input per kick is independent of mode mass (which is necessary for the mode amplitude to be independent of ℓ for a given frequency as observed), then the velocity amplitude imparted to the mode is inversely proportional to the square root of mode mass. The mode mass then cancels out, and the ratio of mode amplitudes expected for our box to that in the Sun is

$$\frac{V_{box}}{V_\odot} \simeq \frac{(M\dot{N}\tau_{box})^{1/2}\Delta V_{box}}{(\dot{N}\tau_\odot)^{1/2}\Delta V_\odot} = \frac{(4\pi)^{1/2}R_\odot}{L} \simeq 400 ,$$

where L (=6 Mm) is the horizontal size of our box. This is indeed the order of magnitude of the ratio between mode amplitudes in our box and in the Sun ($\approx$ 100 m/s vs. 0.2 m/s), and provides more indirect evidence for stochastic convective driving of the oscillations.

We can find the modes in our simulation from the peaks in the power spectrum for each horizontal wave number. Each mode can be checked by comparing its eigenfunction with that calculated for the mean atmosphere. As in the observational case, we have plotted the $k-\omega$ diagram for our modes in figure 3. They are shown together with two sets of modes calculated for the 1-D ensemble averaged mean atmosphere of our simulation (averaged over horizontal planes and time). The mean atmosphere eigenmodes were computed using two sets of boundary conditions. Both have zero displacement at the upper boundary; one also has zero displacement at the lower boundary, while the other has zero derivative of the displacement there. The mode structure has a similar shape to the Sun's, but is much sparser. This is another consequence of the shallow depth of our computational box. As in the Sun, the modes only have significant amplitude up to the acoustic cutoff frequency of the temperature minimum ($\nu_{acoustic} = 5.3 mHz$ in our model), and have a significantly smaller amplitude as the cutoff is approached. The minimum frequency of the modes in this simulation is determined by the depth of our box.

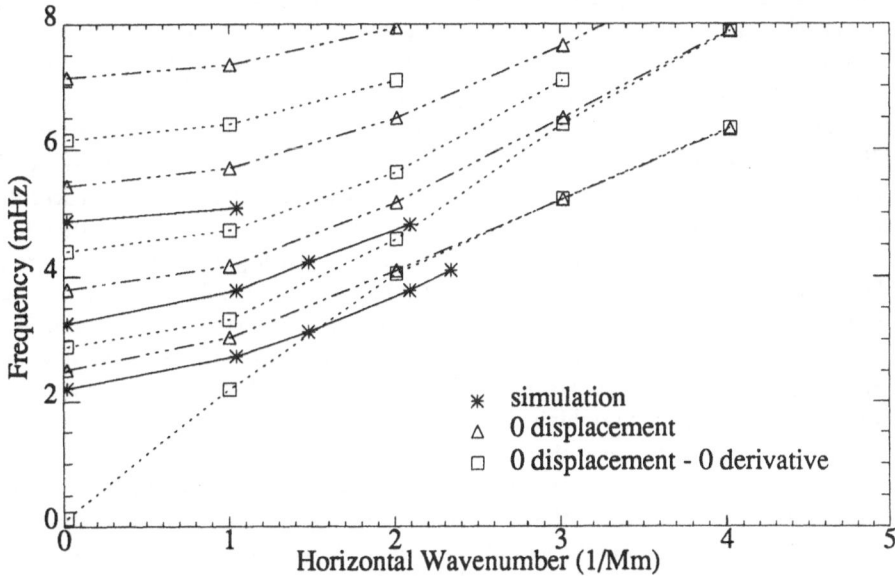

Figure 3: $k - \omega$ diagram. The simulation modes are connected by solid lines. The mean atmosphere modes, for different lower boundary conditions, are connected by dotted and dot-dashed lines.

The dependence of the mode frequencies on boundary conditions is clearly revealed by comparing the eigenfrequencies of the mean atmosphere modes calculated with different lower boundary conditions. Boundary effects are severe for small horizontal wave number and/or high frequency. Towards larger horizontal wave number and lower frequency, the dependence on the lower boundary condition disappears. The reason for this is that the total internal reflection point for the modes of small horizontal wave number and/or high frequency occurs at a higher sound speed than occurs in our box. Hence, these waves are propagating at the lower boundary. How many wavelengths can fit into the box then depends on the boundary condition and determines the mode frequency. Modes with large horizontal wave number and low frequency have a turning point within our computational box, so they are evanescent at the lower boundary, and hence are insensitive to the lower boundary condition. At the upper boundary all the modes with frequency less than the acoustic cutoff are evanescent, so the effect of the boundary condition is small. We want to investigate the differences in mode frequencies between 1-D mean atmosphere and fully convective atmospheres. To achieve this, it is necessary to remove the dependence on the boundary conditions. Luckily, it is easy to extend the computational box to greater depth. Because of the high temperature in the interior the scale height is large, so a 400% increase in depth requires only a 30% increase in computational levels. Such an extension, which we plan to implement, will put the turning points of all but the $k_H = 0$ modes within the box. The price we will pay, of course, is greater mode inertia and slower growth rates.

We have used comparisons of the eigenfunctions for the simulation and mean atmosphere modes to help us determine the eigenfrequencies for modes corresponding to small peaks in the power spectrum and to further clarify the dependence of the mode frequencies on the boundary conditions (figures 4 and 5). From the curvature of the eigenfunctions we can see that the waves are evanescent near the upper boundary, and presumably insensitive to the boundary condition used there. The similarity of simulation and mean atmosphere eigenfunctions near the upper boundary shows that the zero displacement boundary condition closely matches the simulation results there. We see that the character of the effective lower boundary condition changes from zero displacement like for n=1 and small horizontal wave number, to zero derivative like for $n > 1$ and large horizontal wave number. We intend to investigate whether a more general lower boundary condition $a\xi + bd\xi/dz = 0$, where ξ is the displacement, can be adjusted to better match the eigenfunctions and frequencies of the simulation modes. By better matching the effective lower boundary condition our objective is to be able to isolate and study the effects of convection. Modes with $k_H = 1 Mm^{-1}$ correspond to a modes with $\ell \sim 700$, which is very high. The $k_H = 0$ modes we expect are representative of all of the low ℓ modes because they all have about the same structure in this upper part of the convection zone. We should therefore be able to provide information even on the low ℓ modes from our simulations, if we can treat the bottom boundary condition properly, since those low ℓ modes will always penetrate through the bottom boundary. By extending the depth of our box, we will include several radial nodes of the low ℓ modes within the computational domain, which should reduce the dependence on the bottom boundary condition.

As is clear from the above discussion, we can not directly compare our simulation modes with those of the Sun. What we can do, is compare our simulation modes to those of the 1-D mean atmosphere and look for differential effects. We can compare these differential effects with those between the solar observations and the 1-D spherically symmetric calculations. Notice that the differences in the frequencies between are simulation modes and the mean atmosphere modes are substantial, even where the boundary condition effects have become small. They are not the order of μHz, but rather of $10^{-1} mHz$. Hence we don't have to run our simulation for days of solar time to resolve the discrepancies. Again, this is a consequence of the shallowness of our computational domain. The modes are much sparser than in the Sun. We have also tried to make some of the same comparisons that have been made in the Sun. We have calculated the frequency difference $\Delta\nu = \nu_{simulation} - \nu_{mean}$ vs. ν for the zero-displacement boundary

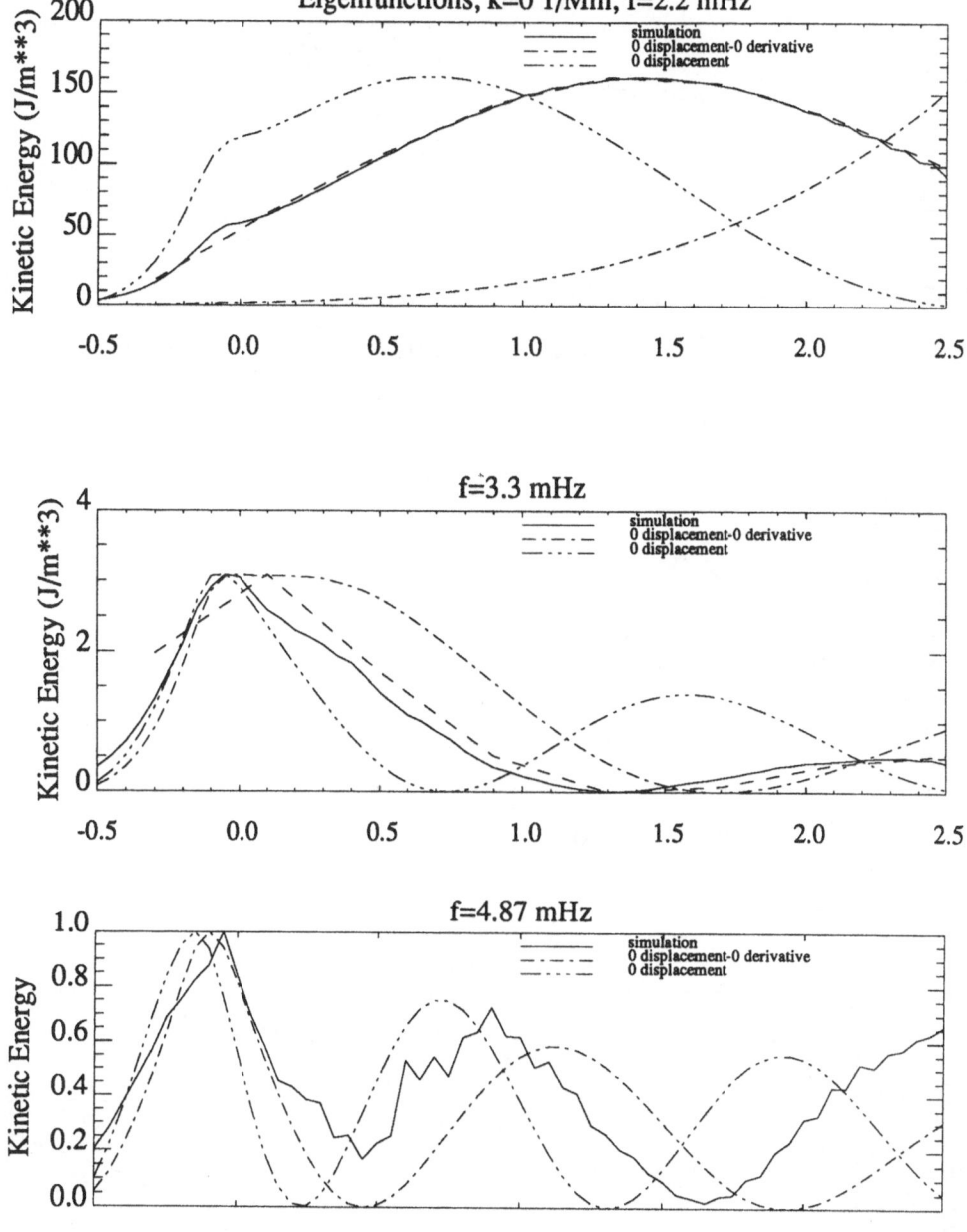

Figure 4: Eigenfunctions for $k_H = 0$. Solid and dashed lines both represent the simulation eigenfunctions, calculated differently. The dashed curve has coarse resolution, (only 8 levels).

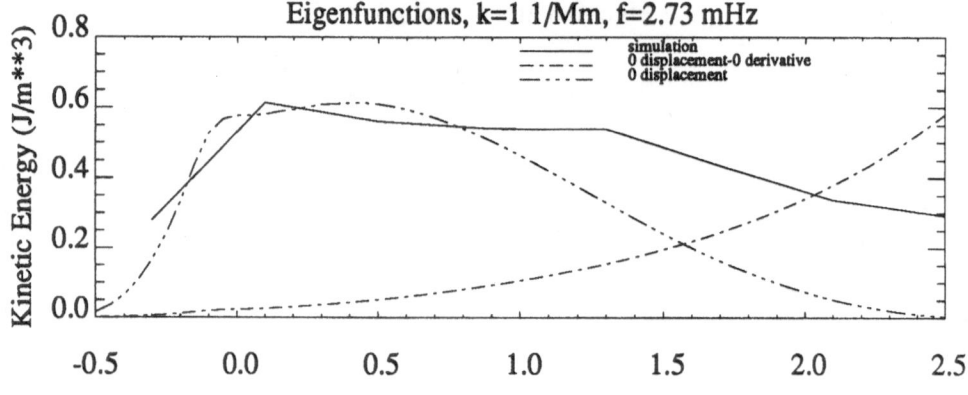

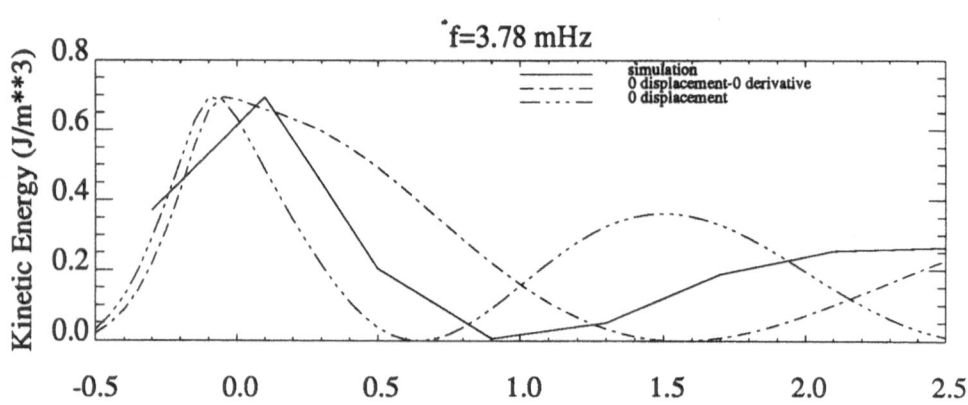

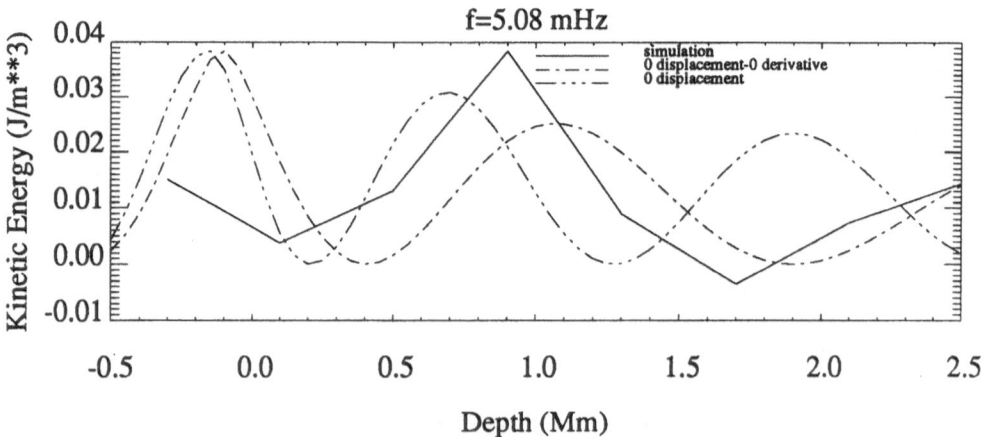

Figure 5: Eigenfunctions for $k_H = 1 Mm^{-1}$.

condition mean atmosphere modes. The general trend is $\Delta\nu$ decreasing with increasing ν, similar to what Christensen-Dalsgaard (1984) found for the Sun, but with all $\Delta\nu < 0$, rather than $\Delta\nu > 0$ at low frequency and $\Delta\nu < 0$ at high frequency, as occurs in the Sun. We have also made a Duvall (1982) diagram of $(n+\epsilon)/2\nu$ vs. ν/k_H. The modes do collapse to a single line, but the results for the phase ϵ were inconclusive. A more careful analysis, using the results from the deeper box with its greater number of modes, will be needed to determine the difference in phase ϵ between the simulation and mean model modes.

4 Interaction of Convection and Waves

Our goals are to figure out how our 3-D simulation behaves differently from a 1-D mean atmosphere, and to understand the processes that drive and damp the oscillations. There are three interaction process between waves and convection: emission, absorption, and scattering and refraction.

4.1 Emission

The sources of sound are (Lighthill, 1952; Howe, 1975):
vorticity
$$div(\mathbf{u} \times curl\mathbf{u}) ,$$
entropy fluctuations
$$div(TgradS) ,$$
and fluctuating buoyancy force
$$\dot{\rho}g .$$

The primary location where all three of these source terms are significant is in the downward plunging plumes. An alternative scenario is that when the dense head of a plume passes through the lower boundary, the weight of the overlying material is reduced, and the pressure, which is kept constant at the boundary, produces an upward pulse. Goldreich and Kumar (1988) have emphasized the fluctuating buoyancy force as a source, because it has a dipole, rather than a quadrupole, nature, and therefore will be larger at low Mach numbers (a phenomenon first pointed out be Unno (1964) and Stein (1967)). The difference is that Goldreich and Kumar assume $\delta\rho \sim u$, while the earlier work assumed $\delta\rho \sim u^2$. The horizontal vorticity is concentrated in the intergranule lanes at the surface and about the downward plunging plumes below the surface. There is a large horizontal vorticity at the edges of the plumes, because the velocity rapidly changes from a small upward flow to a large downward flow there. There is also vertical vorticity because the downward plunging plumes are rotating, but that is a smaller effect. The entropy in the plumes is less and the density greater than the surroundings, because of their low temperature, but as they plunge downward they entrain higher entropy, lower density fluid from their surroundings, so the contrast gradually decreases with depth. These sound sources are largest at the heads of the downward plunging plumes, and that is where all three of the sound sources listed above will be greatest.

The fact that the amplitude grows linearly with time (figure 1) means that the driving is not stochastic (which would produce a linear growth in the energy, not the velocity), but requires phase coherence between the driving mechanism and the oscillation mode. That is, the oscillations must be modifying the driving. One possibility is that there is an asymmetry in the driving by the downward plunging plumes, depending on whether they plunge down during a compression or rarefaction phase of the oscillation. For instance, in the scenario where the downward plunging plumes are thought of as removing mass from the atmosphere, they remove more mass during the higher density compressive phase than the low density rarified phase of an oscillation.

The observed mode amplitudes are independent of k_H and only a function of frequency for small k_H (Libbrecht, et. al. 1986). This implies that excitation and damping occur in the same region for all the modes, which must therefore be a thin layer near the top of the convection zone. In our model, where k_H is large, the amplitude is a function of k_H. Our driving region is comparable with the depth of our box and different modes sample it differently.

4.2 Absorption

Bogdan (1988) has pointed out the possibility of a resonant interaction between waves and vortices at the surfaces where the wave angular frequency matches a harmonic of the vortex angular velocity, $\omega_{wave} = m\Omega_{vortex}$. At such a surface the group velocity becomes parallel to the vortex tube, so wave energy is deflected to propagate along the vortex tube. Energy propagates into the resonant layer and piles up there. As the energy grows there, it eventually either couples to some other, shorter wavelength, modes, or it is damped by viscosity.

Goldreich and Kumar (1988) consider three possible damping processes: (i) eddies scattering off each other in the presence of waves. The doppler shift produced by the wave changes the eddy energy, which gives an absorption coefficient

$$\alpha(\omega) \sim \frac{M^2}{\tau_H} .$$

(ii) eddy-wave interaction in the presence of a fluctuating external force, which gives similar absorption, and (iii) three-mode coupling which can transfer energy to higher frequency modes that can escape up into the atmosphere. The mode amplitude is determined by the balance between emission and absorption:

$$\epsilon(\omega) = \alpha(\omega)E(\omega)\frac{dn}{d\omega} ,$$

from which Goldreich and Kumar (1988) find that

$$E(\omega) = \rho_o H^3 c^2 / (\omega\tau_H)^{13/2} , 1 \le \omega\tau_H \le M^{-2} .$$

Many people have focused on their result that for $\omega\tau_H \le 1$ the denominator is absent, and all oscillation modes have energies comparable to the thermal energy of the dominant eddies. However, on the Sun, the oscillation periods are smaller than the granulation- and certainly the meso-granulation- time scales, so this regime does not occur. In the observed oscillation frequency regime, they predict an inverse frequency dependence of the mode energy.

The important question we face is how to get information from our simulation to choose among these various excitation and damping scenarios. How do we take this data set, where we have a lot of information, and analyze it or do experiments on it in order to find the dominant driving and damping processes.

4.3 Scattering and Refraction

The oscillation waves are propagating through an inhomogeneous medium, where there are temperature fluctuations which produce fluctuations in the sound speed. Therefore, there will be refraction and scattering. In addition, on the resonant surfaces of the vortices, there will also be very large scattering, with a cross section (Bogdan, 1988)

$$\sigma \sim \pi a^2 (ka)^4 \ ,$$

which will be a function of frequency for small k_H and both frequency and k_H for large k_H. Refraction and scattering alter the wave travel time and so change the mode frequency, which broadens the mode.

5 Concluding Remarks

We want to utilize our simulations to increase our qualitative understanding of what modifies the mode frequencies and what drives and damps the oscillations. Our simulations are not a model of the Sun. We have a computational box with properties representing the outer layers of the solar convection zone and the photosphere. Hence, we must make differential comparisons between our dynamic simulation and other ways of looking at the properties of our box, e.g. mean and static representations of our box. Then we must generalize these differential comparisons to the Sun.

At present, most of the oscillation modes in our simulation depend sensitively on the lower boundary condition, because the modes are still propagating at a depth of $2.5 Mm$. In the future, we intend to extend our box to greater depths (and heights) to make all the $k_H \neq 0$ modes become evanescent inside our computational box, and reduce the dependence on the boundary conditions. This will also increase the number of modes excited in the $2 - 5$mHz range, which will enable us to investigate the mode amplitude behavior better. Unfortunately, it will also increase the mode inertia and so will reduce the growth rates, necessitating longer simulation runs.

We want to isolate the dominant effects on the mode frequencies. To this end, we have already compared modes of our convecting box with modes of the ensemble average mean atmosphere. The convective fluctuations alter the mean atmospheric structure. We intend to investigate the sensitivity of the eigenfrequencies to changes in the mean atmosphere by calculating the variation in the mode frequencies for a sequence of instantaneous mean atmospheres, and also for a mixing length atmosphere, constructed using the same opacity table as the full simulation. Such tests should illuminate the extent to which departure of the mean solar atmosphere from the model calculated with mixing length theory accounts for the discrepancy between observed and theoretical mode frequencies. We would also like to make comparisons with an instantaneously frozen snapshot of the atmosphere, in order to isolate the effects of inhomogeneities from the effects of the altered mean atmosphere and from the effects of the dynamics. To do this by brute force is a daunting prospect. We are seeking simpler, approximate methods that will allow us to study the modes of interest.

In order to elucidate the driving and damping processes, we intend to evaluate the terms in the momentum and energy equations and the work terms. The individual mode frequencies will be isolated, and we will look for the dominant contributions and their location, during both the mode growth and saturation eras.

The overall moral is: Computer simulations have a tremendous potential for increasing our qualitative understanding of complex processes, such as discussed here. But, we can not just look at the results of such simulations. Rather, we must use the results of the simulation as a data set to experiment upon. Simulation results can supply a much more complete set of information than is available from astronomical observations. Guided by the observations, we can test the feasibility of many different scenarios.

Acknowledgments

This work was supported in part by grants from the National Science Foundation to R.F.S (AST 83-16231) and J.R.K (ATM 86-04400). They wish to express their appreciation to N.S.F. Å. N. gratefully acknowledges the hospitality of the faculty and staff at JILA during a one year Visiting Fellowship period. He also acknowledges support from the Danish Natural Science Research Council and the Carlsberg Foundation. The authors would like to thank Nigel Weiss for stimulating discussions.

References

Bogdan, T. J. (1988), *Astrophys. J.* (in press).
Christensen-Dalsgaard, J. (1984), *The Hydromagnetics of the Sun*, esa **sp-220**, 3.
Christensen-Dalsgaard, J., Gough, D.O., Toomre, J. (1985), *Science*, **229**, 923.
Duvall, T. L. Jr. (1982), *Nature*, **300**, 242.
Goldreich, P. and Kumar, P. (1988), *Astrophys. J.* **326**, 462.
Gustafsson, B. (1973), *Uppsala Astron. Obs.* Ann 5, No 6.
Howe, M. S. (1975), *J. Fluid Mech.* **71**, 625.
Leibacher, J. W. and Stein, R. F . (1971), *Astrophys. Lett.* **7**, 191.
Lighthill, M. J. (1952), *Proc. Roy. Soc.* **A 211**, 564.
Libbrecht, K. G., Popp, B. D., Kaufman, J. M. and Penn, M. J. (1986) *Nature*, **323**, 235.
Nordlund, Å (1982), *Astron. and Astrophys.* **107**, 1.
Nordlund, Å, and Stein, R.F. (1988), *Bull. Amer. Astron. Soc.* **20**, 702.
—— (1989) (this volume).
Stein, R. F. (1967), *Solar Phys.* **2**, 385.
Ulrich, R. K. (1970), *Astrophys. J.* **162**, 993.
Unno W. (1964), *Trans. I.A.U.* XIIB, Academic Press, NY, 555.

Discussion

KOUTHCHMY — Looking at the results of your simulations of the temperature maps (contrasts in intensities?) at different levels, I am surprised to find a rather mixed agreement with observations of the granulation, namely:

1). I miss the well-established reversing of contrasts at the 200 km level in most of the cases at the scale of granules.

2). When one observes in the deepest observable layers observed at the 1.6 μm opacity minimum, which have a narrow contribution function, the granulation looks more regular than it looks a little bit higher. In your simulations, however, it seems that the inhomogeneities become concentrated in a few points only. Is this correct?

STEIN — We do in fact see a clear reversal of contrast a few hunderd kilometers above the visible surface. The granules become dark and the intergranular lanes become bright. This is due to the adiabatic cooling of the upward moving plasma above the granules and the adiabatic heating of the downflowing plasma above the intergranular lanes.

NORDLUND — The concentration of the downflow into a few narrow plumes occurs at a greater depth than can be seen at any wavelength, and thus the plumes are not directly observable. They become quite noticeable at 500 km below the visible surface. However, their effects are clearly visible in Title's movie of Scharmer's observations.

FOX — I would like to add support to these general results by mentioning that my forthcoming 2-D and 3-D calculations with Chan and Sofia exhibit the same velocity oscillations, also depending on the domain size. In addition, when a polytropic model is used (ideal gas), then these oscillations are still apparent, giving weight to the argument that they are a consequence of the statistical flow and not a thermal effect (beneath $\tau_{5000} = 1$).

Have you estimated the influence of numerical truncation on the modes of oscillation that you find?

Have you removed the granular (rms) velocity component from the oscillatory plots, i.e. is the high velocity solely due to domain size?

STEIN — We have looked at all of the modes for the mean atmosphere. The modes we see in the simulation correspond to the lowest modes in the box. All the other modes have frequencies above the acoustic cut-off frequency at the temperature minimum, which for our atmosphere is 5.3 mHz. We expect not to see them because our upper boundary condition is, we hope, transmitting, and the modes are damped by the waves running out at the top.

We do subtract the background rms velocity. It only has an effect for the smaller peaks in the power spectrum. The larger peaks are an order of magnitude above the background.

HOLWEGER — It is nice to see that your 3-D calculations lead to the same kind of oscillations as Steffen got with his 2-D code some time ago. In the proceedings of the Åarhus meeting, there is a figure in Steffen's contribution which shows these oscillations beautifully, much like the figure you have shown.

Do your oscillations show an asymptotic behaviour, i.e. does their amplitude become constant after a long enough time?

STEIN — Over the period over which we have run our simulation (approximately 4 solar hours) the higher modes seem to have saturated, while the lowest mode is growing.

DEUBNER — Talking about testing the simulation scenarios, and the generation of acoustic oscillations in particular, I should like to mention that J. Laufer has pointed out[1] that bursts of high-frequency acoustic oscillations are observed preferentially at locations of granular downdrafts, and during those time intervals when chromospheric 3-minute modes and photospheric 5-minute modes coincide to produce a coherent downward motion. This finding appears to tie up nicely with the ideas you have developed.

WEISS — These are impressive simulations. If I understand them correctly, the scale of motion near the base of the box is determined by the dimensions of the region studied. Near the top, the scale is reduced owing to the effects of an unstable thermal boundary layer, producing the familiar granulation. How far does this process depend on the artificial diffusion introduced into your equations? Is it a natural feature of high Reynolds number convection, or merely a consequence of the particular closure introduced by your procedure?

STEIN — Our results are sensitive to the magnitude of the artificial diffusion we use. We reduced the diffusion coefficient at one point, and found that the granulation pattern became noticeably smaller. We are going to have to increase the resolution at some time to check whether our results are still sensitive to the diffusion, or whether we have reached the point where some other, physical, process has taken over controlling the large scales, and all the diffusion is doing is just wiping out the small scales. We cannot answer this question yet. However, after the diffusion was reduced, the granulation scales matched the size observed in the best observations.

KALKOFEN — You have listed three ways of generating waves (four ways if you separate $\sim \sigma$, $\sim \sigma^2$). Can you comment on the relative importance of the driving terms?

STEIN — Not yet.

ZHUGZHDA — I agree that the 5-minute oscillations are generated by turbulence. Let us consider the excitation of the fundamental mode in the nonadiabatic case. In the general case, without any restrictions on the atmosphere and on the law of heat exchange, instabilities of the fundamental mode are absent. Now, a ridge is present for the fundamental mode on observational $k - \omega$ diagrams of the 5-minute oscillation. Hence, overstability can supply some energy to p-modes because, as can be seen, the fundamental mode is weaker than the p-mode ridges (reference: Astronomical Journal USSR Letters, 8, p. 562–565).

VAN BALLEGOOIJEN — What about gravity waves and short-period acoustic waves ($P < 100$ sec)?

---

[1] F.-L. Deubner and J. Laufer, 1983 *Solar Phys.* **82**, 151.

STEIN — We cannot study acoustic waves with periods less than 100 sec, because we do not have the spatial resolution. We do see some events that we believe may be gravity waves, and we will check their properties to verify their nature.

SCHÜSSLER — Is the last word really spoken concerning the driving of the oscillations that you find? From the spatial structure of the oscillating flow in Matthias Steffen's calculation and also from the observations that Franz Deubner mentioned, it appears as if the oscillations have their origin in the downdrafts, exactly at those places where you have the maximum horizontal radiative exchange which is the driver for overstable thermal oscillations. If you consider a distribution of such structures which start oscillating with random phases, don't you think it would be possible to have a linear growth of amplitude of the *global* oscillations?

STEIN — The downdrafts are also the locations of maximum divergence of vorticity and of maximum density and entropy fluctuations. Hence, all possible source terms are at a maximum there. The linear growth of amplitude indicates that the driving must be locked in phase with the amplitude and does not occur with random phases, because then the amplitude would grow as $t^{1/2}$.

HOW MUCH CAN THEORETICAL MODELS OF COMPRESSIBLE CONVECTION TELL US ABOUT SOLAR GRANULATION?

Peter A. Fox

Center for Solar and Space Research, Yale University

Abstract. Many hydrodynamic models of solar convection have been developed to study and discuss granulation and the interaction of convection and magnetic fields. They have not, however, been extensively applied to mesogranulation, supergranulation or oscillations in sunspots. In this paper both the modal expansion technique and finite difference models are examined in the application of the fully compressible equations of hydrodynamics to solar granulation. The linear theory of convective motions only gives limited information and in particular does not give the preferred spatial scales of motion that are observed. The non-linear integrations are able to reproduce the dynamic characteristics of granulation such as vertical and horizontal velocities, intensity fluctuations, energy transport and the lifetime to a good degree. This agreement is over a large range of cell sizes. Particular attention will be paid to how well each method treats each of the observed features.

1 Physical effects and Motivation

The primary goal of modelling convection in the Sun is to reproduce surface layer observations (those below the temperature minimum) on varying spatial scales. Features such as granules, fibrils, pores, oscillations, mesogranules, plages, supergranules and sunspots demand an explanation in terms of existing convection theory. Some of the characteristics to be studied are sizes, lifetimes, RMS velocities, intensity fluctuations, fluxes, periods, magnetic field geometries and strengths, and more.

To do stellar convection however, it is necessary to constrain the theory much more closely with less dependence on parameters. In the stellar case there are limited detailed observations to calibrate parameters unlike the amount that exist for the Sun.

It is also useful to further develop theories of convective and turbulent energy transport for stellar structure calculations as a replacement for the current methods (e.g. mixing length theory and its non-local extenstions) and develop a better understanding of convective instability especially in the interface regions where convective and radiative energy transport are comparable.

Then, using the accumulated findings of models of granulation, establish the relationship of other phenomena to granular convection. This includes the interaction with radiative layers (overshooting and penetration), coupling with acoustic oscillations in the 3 to 5 minute range, their influence on large scale flows, contribution (if any) to torsional

401

R. J. Rutten and G. Severino (eds.), Solar and Stellar Granulation, 401–413.

oscillations, meso/super granular interactions, the solar cycle, interaction with magnetic fields (global and fibril) and magnetic (intra-, inter- and chromospheric) networks.

2 Theoretical aspects

2.1 Hydrodynamics

In ascertaining the amount of information convection models may provide about granulation, it is necessary to briefly examine the underlying physics of the outer layers of the Sun. The medium is highly compressible and the region of interest contains many pressure scale heights which vary rapidly. It also seems reasonable that convection interacts in some way with the observed non-radial pulsations. The medium is turbulent, characterized by short correlation lengths and high RMS velocities even though the turbulent pressure is not very large relative to the total pressure. The dynamic viscosity is supplemented by turbulent effects and is usually represented by an effective viscosity, μ, given by

$$\mu = \mu_s + \mu_t + \mu_m$$

where μ_s is a modelling of the sub-grid scale action of the turbulence, μ_t is some 'bulk' viscosity of the mixing length type and μ_m is the molecular viscosity. The understanding of the first two terms relies in the interpretations of incompressible turbulence, atmospheric (terrestial) calibration and geometric arguments.

The main progress made in the treatment of hydrodynamics has been to make use of the fully compressible equations and in the testing and calibration of the expressions for μ_s and μ_t. Little in the basic interpretation of turbulent viscosity has changed in the last ten years and that situation must be due to change soon. As far as granulation models are concerned, sensitivity tests relating to the turbulent transport coefficients are urgently required.

2.2 Thermodynamics

The equation of state of the (real) gas is that of a partially ionized plasma of which the opacity changes rapidly (reflecting the steep gradient in temperature) near unit optical depth. Around this level the outward transport of energy changes from convective to radiative. This means that radiative transfer must be modelled, at least, in the diffusion approximation and preferably in the Eddington approximation to treat the "overshoot" region and the upper boundary correctly, let alone make useful comparisons with atmospheric observations. In this area, much more progress has been achieved especially in the inclusion of radiative transfer not using the diffusion approximation (Nordlund 1982, Stefanik et. al. 1984). In addition, there are groups working actively on stellar opacities which should lead to even more reliable treatments.

The comparison of theoretical spectral line asymmetries and wavelength shifts to observation (see Dravins et. al. 1981) of spatially unresolved granulation is a very important test for models. No further discussion of these observations will be made, since neither of the techniques to be discussed in this paper makes the required detailed treatment of radiative transfer.

The usual thermal conductivity, K, is also enhanced by turbulent and radiative effects in a similar way to the viscosity and may be written

$$K = K_r + K_s + K_t + K_c$$

$$K = \frac{4acT^3}{3k\rho} + \frac{\mu_s c_p}{\sigma} + K_t + K_c$$

where K_r is the radiative conductivity in the diffusion approximation, K_s is the counterpart of the sub-grid scale viscosity (σ is the Prandtl number), K_t is the bulk conductivity, used to transport the heat flux using a mixing length scheme and K_c is the molecular conductivity. Usually one of the terms dominates, but near optical depth unity K_r and K_s or K_t are comparable.

On the stellar structure side it appears that continued diagnostics such as those of Chan and Sofia (1989) should lead to improved treatments of convection zones.

2.3 Magnetodynamics

The influence of solar magnetic fields on granular convection is quite variable depending on the strength of the local field. The main difficulty is that the interaction between convection and magnetic fields often transforms weak, uniformly distributed, fields into regions of intense fields whose resulting contribution to the pressure significantly affect the surrounding medium. Fortunately magnetic fields in the Sun may be treated in the MHD limit.

Whilst the coefficients of viscosity and conductivity are *somewhat* understood in a turbulent medium, the magnetic (resistivity) diffusivity is not so well known. In particular there are reasons for considering it to be anisotropic, depending on the magnetic field strength and the current density as well as the local velocity (shear) field. The effective magnetic diffusivity, η, may then be written,

$$\eta = \eta_t + \eta_m$$

where η_t is the turbulent contribution and η_m is the molecular contribution. In most current models η_t is simply a mixing length expression that is ad-hoc at best, including no variation with position. A slightly better expression would involve its relation to the viscosity using the *magnetic* Prandtl number r, with $\eta_t = r\mu_s/\rho\sigma$ although this introduces another scaling parameter too. Clearly, very little progress has been made in this area.

2.4 Boundary conditions

Since observations of granular convection are made in a region where considerable uncertainties in the treatment of turbulence and radiation exist, the treatment of surface boundary conditions becomes of utmost importance. The interface to the radiative regions may be treated in several ways. This region is difficult to include in the integration, since the pressure scale height falls off rapidly in this region and much higher resolution is required. It should be noted that stellar structure calculations use a black-body radiation law boundary condition with effective temperature ≈ 5770K at optical depth 2/3.

Boundaries where the convection must interface to a convective region below the domain of integration can cause serious problems. In particular, stresses imposed by a rigid boundary can result in quite different flows to those with other boundary conditions. It is of prime importance for convection models of the upper layers to determine the influence of this lower boundary specification, *especially* if convective cells transverse the entire integration depth. A large advantage would be gained if the 'boundary scale height' (the distance over which the boundary is felt) is small compared to the layer depth. This is mainly due to the restriction current convection models face with respect to the number of pressure scale heights they can include in their domain. Since there is no chance of integrating over the entire convection zone in the very near future, an enormous benefit would be gained if there were a natural order of vertical scales into which the domain could be split.

The lateral boundaries of a domain, however, are also important in determining the influence on interior and surface flows. Since the horizontal thermodynamic quantities do not vary dramatically at a given level (as shown by the intensity pattern of granulation—except perhaps in regions of intense magnetic fields), one solution is to make the domain wider than the region of interest. This easy solution is possible because the resolution required is smaller than for the vertical variations (but it may also make the interpretation of the flows harder—see later).

The treatment of magnetic fields at the boundaries of a domain can be somewhat difficult depending on the particular numerical method and formulation of the magnetic field components. Usually it is necessary to ensure continuity of the field and its derivatives (the current) over the boundaries. Another interesting possibility is to use strong magnetic fields at or near the lateral boundaries of a large domain. This would simulate a barrier (perhaps impenetrable) to convective flow, much like the observed magnetic networks at supergranule boundaries, but still ensure a self consistent solution of the dynamic equations. In this case, the velocity flow field may not be an artifact of poorly determined boundary conditions.

The adequate treatment of boundary conditions (perhaps insensitivity as well) is vital to a reliable description of any surface feature but especially so for granulation.

2.5   Initial conditions

Since the task of calculating compressible convection is difficult enough without having to determine the static thermodynamic structure, it is important to have a reasonable representation to start with. In particular, the existence of a steep gradient in temperature near unit optical depth and correct stratifications in density, pressure etc. should be included in the initial model. This gives added incentive to the improvement of prescriptions of convective transport for stellar structure calculations as highlighted by Sofia and Chan (1984).

It is now well established that numerical models of convection for the Sun must be integrated for a thermal relaxation time in order to remove all traces on the initial (hydrodynamic) conditions, this means that perturbations of the non-thermodynamic quantities are not important. The statistically stationary state that ensues is the one from which meaningful results can be derived (taking into account all the limitations of the modelling). The problem with this restriction is that the thermal relaxation time depends on the number of pressure scale heights and the mass contained within the layer; hence integrating to large depths implies not only a larger number of grid points but also a longer integration time. The important question in this case is whether the statistical equilibrium can be attained in another manner? This requires a much deeper understanding of solar turbulence and a workable prescription for it.

The initial magnetic field is a little more complex since it is really an external influence that can fulfil different roles, depending on its strength and configuration, in the statistical equilibrium that exists in the Sun. This is an area that is still not well understood.

3 Methodology

3.1 Introduction

Modal expansion and finite difference techniques are two methods that are used in modelling compressible convection. In this section a brief description will be given on each before their ability to represent granular convection and their relative usefulness is examined.

3.2 Modal representation

The modal expansion (ME) technique (Unno 1969; Van der Borght 1971) is motivated by the cellular structure of granulation and other convective features on the solar surface. Their purpose is to treat horizontal surfaces in some finite statistical sense (like a horizontal template) while retaining the physical dimension in the vertical direction, where most changes in the structure occur. The planform function which describes the horizontal dependence satisfies a two–dimensional Helmholtz equation,

$$\left(\nabla^2 + a_i^2\right) f_i = 0$$

and for many cases hexagonal shaped cells are used (see Bisshopp 1960). The usual expressions for velocity, density, temperature etc. are expanded as a mean vertical component that varies in time and a fluctuating part which contains the horizontal dependence. For example, the temperature in an N-mode expansion is expressed as

$$T(x, y, z, t) = T_0(z, t) + \sum_{i=1}^{N} F_i(z, t) f_i(x, y).$$

Most studies restrict their attention to the most severe truncation, that of one mode, however, some remarkably good agreement of some observed features in still obtained.

This formulation makes it easy to include within it the effects of rotation, interaction with pulsations, magnetic fields and ionization but not necessarily treat them completely accurately (or realistically). The main advantage is that it is possible to model different spatial scales of motion with a small number of 'modes', but still retain coupling between the scales.

To study overshooting and penetrative motions the restriction of the horizontal dependence must be remembered, especially since motions extending into the stable radiative outer layers of the Sun are unlikely to be long lived or cellular in appearance.

3.3 Finite difference

The use of finite difference methods in both two and three dimensions (e.g. Graham 1975, and the ADISM - Alternating Direction Implicit on a Staggered Mesh method of Chan and Wolff 1982) has enabled the analysis on convective flows without the concern of mode truncation. The emphasis has shifted to one of numerical accuracy with fewer parameters to be determined. The flows are certainly more difficult to interpret but have set the stage for tackling complex physical systems such as the solar convection zone. The last few years have been a learning period in which the validity of certain techniques and detailed analysis of the flows has emerged.

The ADISM technique which incorporates a staggered mesh for thermodynamic and hydrodynamic quantities, implicit time stepping and a sub grid scale (SGS) viscosity will be referenced in this paper as representative of what most finite difference (and similar) techniques can provide in the treatment of solar granulation.

4 What information to extract

4.1 How to extract information

As previously mentioned, results from any model must be obtained after all traces of initial conditions have been removed. At this stage the time accuracy of a scheme becomes important, as it is essential that many cell turnover times are used in order to get the full character of the solutions. This is true not only for doing averages and statistics but for flow studies as well, in order to handle any range in timescales that may be present.

When treating large scale and combined scale flows, this is a little harder because statistics and even the eye will miss many salient features of the flows. Some methods handle these situations more easily than others. The modal technique, for example, allows simple recognition of separate spatial and time scale flows.

4.2 Comparisons

The exact information to be extracted from a model depends on which feature is being examined. At the same time, though, it is also necessary to validate any assumptions that have been made and try to extract other diagnostic information that may aid the investigation of related features.

4.3 Diagnostics

The validity of the anelastic approximation for granular motions has long been doubted. With the advent of more powerful computers and better numerical techniques it seems that this restriction has been removed from most models. For an examination of the anelastic approximation, see Van der Borght and Fox (1983) and Fox (1985).

Another consideration is the influence of the flow on the hydrostatic structure. Although models of the solar convection zone are quite reliable (due to better calibration), the detailed structure of stellar convection zones is still quite uncertain.

A restriction of the modal formulation which has significant consequences when relating the model to observations is due to the imposition of the horizontal functional form. This implies, for the one-mode case, that the entire surface on the Sun appears to behave in the same manner. This template effect means that all cells know the exact behaviour of their neighbours, a situation which is far from true on the solar surface. The incorporation of more modes in the expansion can allow for more inhomogeniety but can have other damaging effects on the properties of the mean flow (Marcus 1980).

4.4 Statistics and mean quantities

In analyzing the models of compressible convection and the enormous amount of data that is generated it is essential that a conservative view of the results be taken. In all techniques, the calculation of statistical quantities provides a good measure of both the characteristics of granular convection and the general behaviour of highly compressible turbulent flow. Of course observers have known these techniques for many years, and it is only fitting that theoreticians use the same or similar techniques.

Some of the averages that are useful include the combined time (over many turnover times) and horizontal average ($\theta_1 \leq \theta \leq \theta_2, \phi_1 \leq \phi \leq \phi_2$) of a quantity q at a given mean pressure $\bar{p}$ is

$$\bar{q}\,(\bar{p}) = \frac{1}{t_2 - t_1} \int_{t_1}^{t_2} dt \left[\frac{1}{(\cos\theta_1 - \cos\theta_2)\,(\phi_2 - \phi_1)} \int q \sin\theta d\theta d\phi \right].$$

The Root–Mean–Square (RMS) of a quantity is then

$$\mathrm{rms}\,[q]\,(\bar{p}) \equiv \left(\overline{q^2}\right)^{\frac{1}{2}}.$$

The correlation of two quantities q_1 and q_2 at two levels $\bar{p}_1$ and $\bar{p}_2$ is

$$C\,[q_1, q_2]\,(\bar{p}_1, \bar{p}_2) = \frac{\overline{q_1 q_2}}{\mathrm{rms}\,[q_1]\,\mathrm{rms}\,[q_2]}.$$

The autocorrelation of a quantity q, at two levels is

$$A\,(\bar{p}_1, \bar{p}_2) = C\,[q, q]\,(\bar{p}_1, \bar{p}_2),$$

an example of the quantity q is the radial velocity v_r and the autocorrelation over a range of depths gives an estimate of the "mixing length" (the distance over which one element *feels* its neighbours) of the motion (Chan and Sofia 1987).

One correlation that is particularly significant for granulation is $C\,[v_r, T']$, the RMS vertical velocity – temperature fluctuation relation. There is almost a 100% correlation between the direction of upward moving flows and temperature excesses (Bray et. al. 1976). The actual correlation calculations from observation must be carefully extracted since the high correlation is actually between small scale (granular - < 5000 km) velocity fields and the temperature. Larger scale effects significantly lower (smear) the correlation—hence the technique of *moving averages* by Stuart and Rush (1954), which selects certain small scale regions to do the correlations, and which can change a low value of -0.30 (Richardson and Schwarzschild 1950) into a high value of -0.68 on the same set of data. Very similar effects are true for theoretical models as well. The correlations demonstrated by Chan and Sofia (1986) for comparable width to depth domain ratios reflect a high correlation for this quantity. However, when the domain is extended in the horizontal direction (Fox, Chan and Sofia 1989), the relationship is decreased dramatically unless small sub-regions are examined. The modal technique actually handles this aspect very well since the scales of interest are usually represented by an individual mode. This provides the ability to distinguish between granular and larger (mesogranular and supergranular) scales easily.

Other useful calculations include the power spectrum (both one and two-dimensional), which describes the contribution of each spatial scale to the total fluctuation of the brightness, and the quadratic coherence (Jenkins and Watts 1969) which determines a linear correlation coefficient filtering out structures both larger and smaller than some band centered on a sample wavenumber k. It is related to the two-dimensional power spectrum.

4.5 Cell sizes

Granulation seems to have an upper limit on its horizontal extent of about 4000 km between centers of cells. The modal expansion technique requires information about the size (both depth and width) of a cell. Hence, examination of the variation in the size distribution requires many calculations and the comparison of the granular characteristics (RMS velocities etc.) to observation. The finite difference technique simply requires a large horizontal domain size, and, provided that the upper boundary is treated adequately, the range in scales will be reproduced by analysis of the flows remembering the possible restrictions (regarding large domains) mentioned earlier.

The depth penetration of granular cells is not relevant for observational comparison provided that the granular characteristics are reproduced but it does provide a diagnostic of our understanding of turbulent convection and possible clues to the interaction with other scales of motion.

In the case of modal convection placement or determination of the layer, depth is *very* important in reproducing granular motions. Examination of the marginal stability using the linearized hydrodynamic equations often provides information on the preferred scales of motion based on the principle of maximum instability or growth rate of convective motion. In the solar case, however, extensive calculations using the linearized modal equations (Fox 1985), have not indicated any cut-off in horizontal scales or preferred depths for granular convection. Indeed, they suggest that for a large range in cell widths a great range in cell depths is allowed (equally preferred). This result is not surprising in view of the statistical nature of solar granulation which is unlikely to be described by conditions of marginal stability.

The non-linear calculations of modal convection (Fox 1985,1989) do place quite good limits on the width and depth scales of granular convection by noting which granular characteristics (velocities, intensities, granular lifetimes etc.) start to depart markedly from observational limits. The suggested scales are less than $\approx$ 4000 km and indicate that many of the variations in surface features are explained by depth penetration and lifetime variations.

Two-dimensional finite difference calculations for granulation (Fox, Chan and Sofia 1989) show a range of horizontal cell sizes from 500 km up to about 3500 km over a total domain size of 20000 km and several vertical scales. The restriction to two dimensions is enforced by the large region of integration and thus should only be considered a guide to what three dimensional calculations would show.

4.6   Velocities

Both modal and finite difference models have no trouble in reproducing RMS velocities (both horizontal and vertical) in good agreement with the solar values. Since the comparisons are made at or near the upper boundary, again care should taken. In particular, most models indicate that the vertical velocity vanishes at the upper boundary and hence the magnitude is usually compared just 'below' the surface since the opacity is increasing rapidly. Models which have peaks in the depth distribution of the RMS vertical velocity close to the surface seem to be favoured, otherwise the predicted velocity would be quite small. Both modal and finite difference models exhibit this behaviour of being strongly peaked near the upper boundary. Once again the subsurface velocities are only good for diagnostic purposes (analyzing convection etc.) and determining the influence of boundary conditions.

The existence of motions that "overshoot" into the stable radiative layer above the convection zone also places constraints on granular models. Both the modal and finite difference techniques can place their upper boundaries within this overshoot region providing they take radiation into account. The modal models predict an overshoot distance of about 50 km since at that level the granular characteristics begin to depart from observations (Fox 1985). The restriction of the horizontal dependence makes this some average value. At this time the ADISM models have not been applied to study overshooting at the solar surface.

Other velocity features that must be reproduced are the fast intergranular downdrafts of a few km sec^{-1}. Modal models, again with the horizontal restriction, actually give good estimates of this feature. The ADISM results clearly show the downdrafts in two and three-dimensions, but care must be taken since these flows are often induced by the lateral boundaries of a domain. Further studies of vertical and horizontal flow velocities are required especially for larger scale features such as mesogranulation and supergranulation.

4.7   Intensity fluctuations

The existence of a contrast between centers and edges of granules is perhaps their most notable feature. Modal models are ideally suited to estimate the difference in brightness as the horizontal dependence from center to edge is known (not the magntitude), even on varying spatial scales. They give values well within the 7–19% range that is observed, and in fact are used as an estimate of the lifetime of granules (see next section). The fluctuations predicted are *very* sensitive to the temperature boundary treatment (a black body law gives the best agreement), and they also imply an assembly of cells all the same.

When examining the finite difference models, individual cells must be treated and an estimate of their 'center' and 'edge' (or average thereof) must be defined, and certainly not be averaged over many turnover times. Except when extracting the continuum brightness, this is one case where horizontal averaging must be used carefully, as it is very difficult to analyze individual intensity fluctuations. There is much room for improvement in this area.

The contributing RMS fluxes (radiative, convective etc.) are important in the very upper layers in determining the transport processes and their effect on the flows, while The depth distribution of fluxes is important for the study of turbulent convection, in particular when searching for a representation of the flux in terms of local quantities (advantageous in stellar structure calculations - see Chan and Sofia 1989). It is also useful when looking at the reductions in intensity of pores, sunspots etc.

4.8 Lifetime

The estimate of granulation lifetimes (coherent surface flows) is really one of pattern recognition, and certainly difficult. Clearly, the statistical nature of the flow does not indicate a single lifetime, but more likely a range depending on many aspects of the flow.

The modal models (one-mode) cannot reproduce the lifetime distribution of granulation in the absence of external influences (rotation and magnetic fields) since they are only limited by non-linear (inertial) interactions, and every neighbour is identical to them. This represents a problem which can only be resolved by estimating the lifetime by some other observational constraint. The technique adopted by Fox (1985) is to *define* the lifetime of a granule as the time from which a cell takes to become visible, observationally, (this is usually when its intensity fluctuation exceeds 2% of the background value) until it reaches an intensity fluctuation of an average granule (usually between 7 and 19%), say 15%. At this point the other granular characteristics are compared to observations. Remarkably enough, this scheme provides good general agreement for granular motions over a wide range of scales, lifetimes, and velocities.

Finite difference models again suffer from the need to follow individual cells, and perhaps a technique can be developed to determine the timescale for coherent flows much in the way that extra contributions (e.g. oscillatory) to granule velocities are extracted in observations. The actual rich array of time dependence, exploding, merging etc. is something that shows up very well in these calculations. Some estimates of lifetime can easily be made by studying the flow fields (see next section).

4.9 Flow fields

The time evolution of the flows (both dynamic and thermal) is by far one of the most interesting and informative diagnostics for convection models . The time evolution of RMS averaged vertical fields can also provide very important information on the energy carrying scales. In three-dimensional models, horizontal plane views, especially those at the surface, are very important in determining the lifetimes and intensity fluctuations. Since the modal models have a known horizontal dependence, only multi-mode formulations may need to reconstruct the total horizontal field. Analysis of the horizontal field at various depths is also very informative, especially in determining the changing nature of spatial (horizontal) scales for deeper layers (downdrafts, swirls etc.). Vertical cross sections can also be useful, but they usually have to be chosen very carefully when interpreting any flows.

4.10 Other scales

The interaction of granules and other horizontal scales (mesogranules, supergranules etc.) has received far less attention than granulation on its own. The evidence for interactions or dependencies is quite strong (see Zahn 1987), even though the characteristics displayed are of a different nature (e.g. granules are an intensity pattern and supergranules are a velocity pattern).

The modal technique can perform these multi-scale calculations relatively easily (so called multimode), and preliminary work (Van der Borght and Fox 1984; Fox 1985) has indicated that coupling between granular and supergranular scales occurs. The ADISM technique (as mentioned previously) also shows signs of giving mesogranular and granular type flows a single domain, but a significant amount of work (analysis) remains to be done.

5 Summary

5.1 Conclusion and prospects

The representation of granulation by models of compressible convection has advanced rapidly in the last few years, which is mainly due to the emergence of numerical techniques and fast computers that can perform the detailed calculations. In this paper the emphasis has been placed on extracting certain types of information from theoretical models taking into account limitations of the technique, implementation of the boundary conditions, and the validity of approximations relating to the equations.

The necessary comparison with observed features is an important tool in tuning convection models. The analysis of other, non-observable, quantities using statistical techniques is providing a much better understanding of highly compressible, turbulent convection. Thus, the limitations of detailed granular models that are applied to, or related to, other phenomena, may be understood.

The modal expansion technique (one-mode) has proved a flexible and reliable way of examining granulation. Several inherent limitations make its further applicability to granulation unattractive. The use of multiple modes, however, still can be a useful method in anaylzing multiple scale flows.

The finite difference technique will continue to provide insight into many aspects of convection as long as its computational requirements are satisfied – such is the price of a more general approach.

5.2 Future

In the not too distant future (and even at present), models of compressible convection will be applied to small and large scale magnetic field interactions, multi-scale surface flows and interactions with non-radial pulsations. The existence of stellar granulation is also an important area in which current convection models can be truly tested. There is still a great deal of analysis of convective flow to be done, in particular those associated with magnetic fields. Finally, the treatment of truly fundamental problems in solar physics, such as the global solar dynamo and solar cycle, may be one of the end products of a systematic treatment of solar convection.

This work is supported under a grant from NASA (NAGW-778). The author would like to acknowledge collaboration with R. Van der Borght, S. Sofia and K. Chan.

6   References

Bisshopp, F. E. 1960, *J. Math. Anal. and Applic.*, **1**, 373.

Bray, R. J., Loughhead, R. E. and Tappere, E. J. 1976, *Solar Phys.*, **49**, 3.

Chan, K. L. and Sofia, S. 1986, *Ap. J.*, **307**, 222.

Chan, K. L. and Sofia, S. 1987, *Science*, **235**, 465.

Chan, K. L. and Sofia, S. 1989, *Ap. J.* in press.

Chan, K. L. and Wolff, C. L. 1982, *J. Comp. Phys.*, **47**, 109.

Dravins, D., Lindegren, L. and Nordlund, Å. 1981, *Astron. Astrophys.*, **96**, 345.

Fox, P. A. 1985, *Ph. D. thesis*, Monash University, Australia.

Fox, P. A. 1989, to be submitted to *Ap. J.*

Fox, P. A., Chan, K. L. and Sofia, S. 1989, in preparation.

Graham, E. 1975, *J. Fluid Mech.*, **70**, 689.

Jenkins, G. M. and Watts, D. G. 1969, Spectral Analysis and its Applications. San Francisco, Holden-Day.

Marcus, P. S. 1980, *J. Fluid Mech.*, **103**, 241.

Nordlund, Å. 1982, *Astron. Astrophys.*, **107**, 1.

Richardson, R. S. and Schwarzschild, M. 1950, *Ap. J.*, **111**, 351.

Sofia, S. and Chan, K. L. 1984, *Ap. J.*, **282**, 550.

Stefanik, R. P., Ulmschneider, P., Hammer, R. and Durrant, C. J. 1984, *Astron. Astrophys.*, **134**, 77.

Stuart, F. E. and Rush,J. H. 1954, *Ap. J.*, **120**, 245.

Unno, W. 1969, *Publ. Astron. Soc. Japan*, **21**, 240.

Van der Borght, R. F. E. 1971, *Publ. Astron. Soc. Japan*, **23**, 539.

Van der Borght, R. F. E. and Fox, P. A. 1983, *Proc. Astron. Soc. Austr.*, **5**, 170.

Van der Borght, R. F. E. and Fox, P. A. 1984, in *The Big Bang and Georges Lemaître*, Ed. A. Berger, Holland, Reidel, p269.

Zahn, J.-P. 1987, in *Solar and Stellar Physics*, Eds. E. H. Schröter and M. Schüssler, Lecture Notes in Physics, Springer.

Discussion

TITLE — Lifetimes and other statistical parameters must be defined carefully because granules interact. Pattern recognition techniques are required.

FOX — I entirely agree, and hope that I did not give the impression of using statistical techniques for all comparisons.

STEIN — How many modes do you use in your modal calculations? One mode has only one length scale and is quantitatively different from convection. Two modes can only have two scales, one being twice as large as the other. It is only from three scales onwards that you can get chaotic behaviour. Even then, the behaviour is very different from real turbulence, as Phil Marcus has shown. Modal calculations are best suited for studying wave propagation. They are inappropriate for convection.

FOX — In many of the calculations I have done, only one mode was used, which actually describes granulation very well. In particular, single-mode calculations conserve the mean and rms properties of the flow fields very well. I agree that the use of multiple modes may change the conservation properties of the flows, but multiple modes are necessary to obtain the stochastic behaviour we see in granulation, which is not apparent in the single-mode calculations. The multi-mode calculations are very useful in modelling different spatial and temporal scales of motion (e.g. granules and supergranules). I agree that modal techniques do have limitations for the highly time-dependent granulation flow, but I disagree that they are inappropriate for convection.

TWO AND THREE-DIMENSIONAL SIMULATIONS OF COMPRESSIBLE CONVECTION

Fausto Cattaneo, Neal E. Hurlburt and Juri Toomre

Joint Institute for Laboratory Astrophysics, University of Colorado

ABSTRACT. We present some results of numerical studies of compressible, nonlinear convection in highly stratified fluids. Both two and three-dimensional flows are considered. In three dimensions the flow forms an irregular cellular pattern near the surface with downflows at the cellular boundaries and upflows at the cellular centers. At greater depth the downflows collapse into isolated columns surrounded by gentler upflows. The two dimensional simulations show that at high Rayleigh numbers convection becomes supersonic near the upper boundary. Non-stationary shock systems appear which interact with the thermal boundary layer giving rise to a vigorous time dependence.

1   Introduction

The theoretical study of turbulent convective motions, such as we find in the solar convection zone, is a task of formidable complexity. The difficulties in obtaining an analytical description of the flows in the highly nonlinear regime have forced the study of convection to rely heavily on numerical simulations (Graham 1975, 1977; Gilman and Glatzmaier 1981; Glatzmaier and Gilman 1981; Chan *et al.* 1982; Nordlund 1982, 1983, 1984, 1985; Hurlburt *et al.* 1984, 1986; Sofia and Chan 1984; Yamaguchi 1984; Chan and Sofia 1986, 1987). With few exceptions numerical studies of convection have dealt with idealised model problems. Although such models cannot be related to stellar convection directly, they provide valuable insight into the physics of convective flows. It is this insight that is then used to further our understanding of convection in realistic situations. In this spirit we present here the results of two families of numerical experiments describing convective flows in two and three dimensions.

Because of the differences in geometry and resolution the two studies address different aspects of the problem of convection. In three dimensions we consider flows with moderate Rayleigh numbers where we seek to understand the general structure and topology of the flows, the role of stratification in producing up-down asymmetries and the relation between the generation of vertical vorticity and the time dependence of the horizontal planforms. Although the observed motions are vigorously time dependent, the convection in these three-dimensional simulations is far from turbulent. In an attempt to approach a description of turbulent convection we also consider flows at large Rayleigh numbers. In order to describe accurately the small scale motions that develop at such regimes we limit these studies to two dimensional flows where we can achieve a much

R. J. Rutten and G. Severino (eds.), Solar and Stellar Granulation, 415–423.

higher numerical resolution. The emphasis of these calculations is on the generation of small-scale structures by secondary flow instabilities. In these simulations we observe transsonic flows in the upper thermal boundary layer which lead to the formation of non-stationary shocks. A striking property of these shock systems is their variability on very short time scales which might bear on the problem of non-thermal line broadening.

We provide a brief description of the mathematical model and the numerical methods in the next section. In sections three and four we present the results of the simulations. Section five is devoted to a discussion of the results, and section four contains suggestions for further investigations.

2 Model and method

We study an idealised model of thermally driven, compressible convection. We consider a layer of perfect gas confined between two horizontal, stress-free boundaries. A constant heat flux is prescribed on the lower boundary while the upper boundary is kept at a fixed temperature. The problem is simplified considerably by assuming that the shear viscosity and the thermal conductivity of the fluid are constant. At present we make no provision to describe subgrid effects such as transport by unresolved turbulent scales. The field variables are assumed to be periodic in the horizontal directions.

The evolution equations, continuity, momentum and energy are solved numerically by a hybrid, semi-implicit scheme. We found it advantageous to both computational speed and accuracy to adopt a pseudo-spectral representation in the horizontal coordinates while treating vertical variations by finite differences. Further details on the numerical methods can be found in Cattaneo and Hughes (1988).

The calculations were started from an initially static polytropic state with a polytropic index of unity and a density contrast of 11. All the Rayleigh numbers quoted in this paper are measured relative to this initial state and are evaluated at midlayer depth. In order to minimize the effects of the initial conditions on the final states, particularly in three-dimensional simulations where the solution can "lock" into one of many stationary patterns, we triggered the motions by a small, white-noise perturbation of the temperature field consistent with the boundary conditions. The properties of the flows discussed in the rest of this paper describe solutions in a statistically steady state.

3 Three-dimensional flows

With the present resolution of 64^3 collocation points, we were able to extend our calculations to Rayleigh numbers of approximately 6×10^4 in a domain with an aspect ratio of $1 : 8 : 8$. We now summarize some important aspects of a representative solution.

3.1 Flow pattern and variability

The motions develop into a coherent circulation after a few sound crossing times while the mean stratification relaxes to a statistically steady state on a much longer timescale comparable to the thermal relaxation time. The typical appearance of the flow after the initial phases have occurred can be seen in figure 1 which shows a perspective view of the vertical velocity planforms at two depths. The convection forms an irregular pattern of cells with a peak Mach number of 0.5 and an aspect ratio between 3 and 4. Near the upper boundary the cellular lanes, defined by the downflow regions, curve and slide along each other on timescales of the order of the turnover time. This continuous readjustment of the flow pattern is associated with the generation of vertical vorticity by the curvature of the intercellular lanes. As the adjustment progresses cells are occasionally destroyed or created. These events occur through the coalescence of two distinct vertical flows into a single one or through the splitting of a single vertical flow region into two. Typically flow merging or division is initiated at some depth and then proceeds vertically until it is completed (like closing or opening a zipper). The surface manifestation of these events is of a cell being squeezed out of existence by its expanding neighbours or of a downflow lane splitting and forming a new cell.

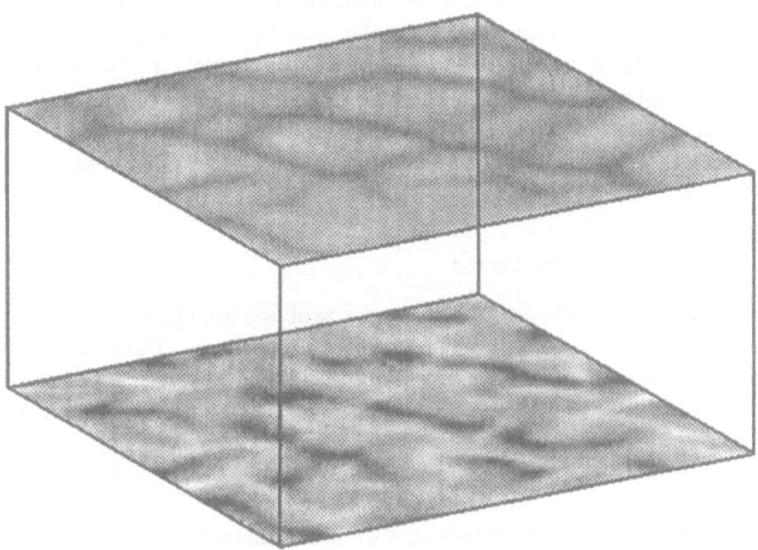

Figure 1: The vertical velocity at two depths $z = 0.18$ and $z = 0.87$ reveal that three-dimensional compressible convection forms a network of descending fluid (dark) near the top of the layer which breaks into narrow plumes near the base. The ascending fluid forms a similar network at the base of the layer but forms broad regions of gentler flow near the top. Hence the flow possesses topological symmetry in the vertical but lacks geometrical symmetry.

3.2   Absence of topological asymmetries

Topological asymmetries arise in convection where the regions of upflow and downflow have different connectedness properties. A common example of a circulating flow with this type of asymmetries in convection with hexagonal planforms. In this case the downflow regions, say, coincide with the cellular boundaries and are therefore connected, while the upflows are sited at the cellular centers and are consequently isolated. It is well known that flows with this property give rise to a mean advection of passive vector fields (topological pumping) along the direction of the connected flows (Drobyshevski and Yuferev, 1974; Arter, 1983; Moffatt, 1983; Cattaneo et al., 1988). The existence of connected flow regions also provides the basis for our intuitive notion of a convective cell.

In our solutions the downflows form a connected network near the surface, with cold, dense fluid sinking at the cellular boundaries and hot, lighter fluid rising in the centers. At greater depths, however, the downflows break up into sheets and at greater depths still, into isolated columns. The collapse of sheets into columns is likely to be due to the contraction experienced by cold, dense fluid elements as they descend through an almost adiabatically stratified medium. Near the lower boundary the descending columns coincide with the cellular centers of a different network which is now connected along the upflow regions (see figure 1). Thus the idea of a flow network is seen to be a shallow phenomenon confined to the vicinity of the horizontal boundaries. Throughout the bulk of the convection the upflows and downflows do not exhibit any particular connectedness property.

3.3   Geometrical asymmetries

Geometrical asymmetries between upflows and downflows are caused by buoyancy braking and buoyancy enhancement in some regions of the flow. These asymmetries are peculiar to convection in highly stratified fluids. When the horizontal pressure and temperature fluctuations are correlated the density deficit of a rising fluid element is decreased, thus reducing the net effect of buoyancy (buoyancy braking). A negative correlation between pressure and temperature fluctuations has the opposite effect (buoyancy enhancement).

In our simulations the dynamical effect of pressure fluctuations is to organise the flow into concentrated downward directed plumes surrounded by broader regions of gentler upflow. An indicator of the existence of geometrical asymmetries in our solutions is the horizontal average of the kinetic energy flux which can become a significant fraction of the total energy flux. Throughout most of the convection we observe the kinetic energy flux to be directed downwards in agreement with previous two-dimensional studies (Hurlburt et al., 1984). The existence of asymmetric flows with downward directed plumes has also been reported in other three-dimensional studies of convection by Nordlund and Chan and Sofia (private communication).

4   Two-dimensional flows

Due to the lighter computational effort required by two-dimensional simulations we were able to study convective flows with Rayleigh numbers exceeding 10^7 in domains with aspect ratio of 1 : 4. In order to resolve the flows a grid with 512×128 collocation points was used.

For Rayleigh numbers between 10^5 and 10^6 the convection consists of a single cell filling the entire domain with a strong concentrated downflow and broader upflows. The motions are almost steady except for a slow drift of the downflow site. For Rayleigh numbers above 10^6 the character of the motion changes dramatically. The horizontal flows near the upper boundary become supersonic and shock systems appear; the convection cell breaks into two new cells with an aspect ratio of 2 and the flow becomes vigorously time dependent. All three changes are related to the transition to supersonic flow.

After the transition to a smaller aspect ratio has occurred the following sequence of events is observed. Shocks form near the upper boundary on either side of the downflows and propagate upstream, leaving behind a layer of dense fluid moving subsonically. The shocks weaken as they propagate outward towards regions moving at lower Mach number until they eventually disappear. The region of subsonic horizontal flow downstream of the shocks becomes unstable to Rayleigh-Taylor instabilities, and a heavy blob of cold, dense fluid falls thus generating a strong horizontal pressure gradient. The horizontal flows become once again supersonic, and two new shocks are formed. This entire cycle repeats over and over with a period comparable to the sound crossing time.

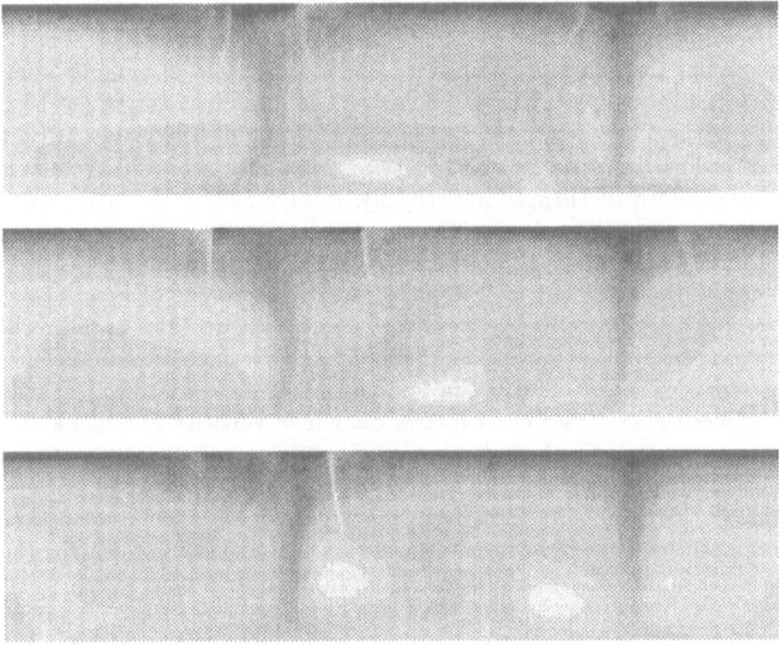

Figure 2: Shock waves form near the regions of downflow in two dimensional simulations once the convection become sufficiently nonlinear. Here the Rayleigh number is 10^7 or about a factor of 10^4 larger than its critical value. The shocks form as the supersonic flows are turned downward. They then propagate away from the downward plumes until they reach subsonic regions.

Figure 2 shows the entropy distribution throughout the layer at three different times. Dark regions, corresponding to low entropy, coincide with the upper thermal boundary layer and with the downflows. The shocks appear as sharp, bright, almost vertical features near the upper boundary. The sequence of events just described can, to some extent, be followed in figure 2. Going from top to bottom, a pair of shock forms near the downflow on the left, the shocks propagate outwards and are eventually replaced by a new pair. An interesting feature of this "pulsating" flow is that the shocks systems often do not propagate symmetrically with respect to the downflows. When one shock, for example the right one, propagates faster than the left one, the thermal boundary layer thickens asymmetrically, causing the downflow site to move rapidly (in this particular case to the right).

In the simulation shown in figure 2 the shocks have a peak Mach number around 2. The density and temperature increase dramatically across the shock and the entropy reaches a local maximum consistent with the theory of viscous, thermally conducting shocks (see for instance Mihalas and Mihalas, 1984). The occurrence of a transition to supersonic flow in two dimensional simulations of convection at high Rayleigh numbers has also been observed by Porter and Woodward (private communication).

5 Conclusion

The purpose of numerical studies is to provide physical insight. The results of our simulations prompt the following comments. Our intuitive idea of granular convection, suggested by observations, as a connected network of downflows might not be entirely representative of the motions throughout the bulk of the fluid. The picture we have from our numerical studies seems to be that of a series of strong, isolated downflow columns whose location and identity changes quite rapidly. It is not clear at the moment over what depth these columns remain coherent. In our simulations the vertical motions extend over the entire depth of the layer. However, we anticipate that for flows with a higher Reynolds number the coherence will be disrupted by the onset of shear instabilities. There are some serious difficulties involved with a direct study of this problem since it is known that shear instabilities develop much more readily in three-dimensional flows than in two-dimensional ones. Furthermore, in two dimensions the conservation of energy and of entropy implies the existence of an inverse cascade of vorticity to larger scales which seems to have no analogue in three dimensions. In other words, the study of convection with shear instabilities is intrinsically a three-dimensional problem. We estimate that the resolution required to study these flows adequately is 128×256^2 collocation points which, at present, is outside our computational reach. Simulations of this size, however, might be possible on massively parallel machines in the near future.

The appearance of supersonic flows with shocks also prompts some interesting speculations. In three dimensions the shocks would form predominantly in the vicinity of cellular corners where the horizontal flows are strongest. Shocks would appear horizontally as segments of arc propagating away from the downflows. The lifetime of a shock system in the granular flow should be of the order of one minute with the shocks propagating over a distance of a few hundred kilometers. The rapid fluttering of the downflow sites induced by the shocks would give rise to some interesting hydrodynamical phenomena on timescales of only a fraction of the turnover time. A worthwhile extension of the present work is to consider supersonic convection with penetrative geometry, where the upper boundary is replaced by a stable fluid layer. The dynamics of shocks in this layer is likely to induce turbulent motions on spatial scales of only tens of kilometers and on temporal scales of less than a minute. The local hydrodynamics of convectively driven shocks is probably relevant to the problem of non-thermal line broadening, however, such study would involve a more realistic treatment of radiative transfer effects. The coupling of shocks to the acoustic field is also a subject that deserves further investigation.

This research was supported by the National Aeronautics and Space Administration under grant NAGW-91. The three-dimensional simulations were carried out at the John von Neumann National Supercomputer Center funded by the National Science Foundation.

REFERENCES.
Arter, W. 1983, *J. Fluid Mech.*, **132**, 25.
Cattaneo, F. and Hughes, D. W. 1988 *J. Fluid Mech.*, in press.
Cattaneo, F., Hughes, D. W. and Proctor, M. R. E. 1988, *Geophys. Astrophys. Fluid Dyn.* **41**, 335.
Chan, K. L. and Sofia, S. 1986, *Ap. J.*, **307**, 222.
Chan, K. L. and Sofia, S. 1987, *Science*, **235**, 465.
Chan, K. L., Sofia, S. and Wolff, C. L. 1982, *Ap. J.*, **263**, 935.
Drobyshevski, E. M. and Yuferev, V. S. 1974, *J. Fluid Mech.*, **65**, 33.
Gilman, P. A. and Glatzmaier, G. A. 1981, *Ap. J. Suppl.*, **45**, 351.
Glatzmaier, G. A. and Gilman, P. A. 1981, *Ap. J. Suppl.*, **45**, 335.
Graham, E. 1975, *J. Fluid Mech.*, **70**, 689.
Graham, E. 1977, in *IAU Colloquium 38, Problem of Stellar Convection*, ed. E.A. Spiegel and J.-P. Zahn (Berlin: Springer-Verlag), p. 151.
Moffatt, H. K. 1983, *Rep. Prog. Phys.*, **46**, 621.
Hurlburt, N. E., Toomre, J. and Massaguer, J. M. 1984, *Ap. J.*, **282**, 557.
Hurlburt, N. E., Toomre, J. and Massaguer, J. M. 1986, *Ap. J.*, **311** , 563.
Mihalas, D. and Mihalas, W. B. 1984, *Foundations of Radiation Hydrodynamics*, (New York: Oxford university press).
Nordlund, Å. 1982, *Astr. Ap.*, **107**, 1.
Nordlund, Å. 1983, in *IAU Symposium 102, Solar and Stellar Magnetic Fields: Origins and Coronal Effects*, ed. J. O. Stenflo, (Dordrecht: Reidel), p. 79.
Nordlund, Å. 1984, in it The Hydromagnetics of the Sun, ed. T. D. Guyenne (ESA Sp-220; Noordwijkerhout: ESA), p. 37.
Nordlund, Å. 1985, in *Theoretical Problems in High Resolution Solar Physics*, ed. H. U. Schmitt (Munich: Max-Plank-Institute fur Astrophysik), p. 1.
Sofia, S. and Chan, K. L. 1984, *Ap. J.*, **282**, 550.
Yamaguchi, S. 1984, *Publ. Astron. Soc Japan*, **36**, 613.

Discussion

DAMÉ — In your movie, in Mach number and in vorticity, we observe an oscillation of the granular/downflow lanes, going back and forth horizontally driving shocks. What is the period of this phenomenon? Is there an observational counterpart?

CATTANEO — The shocks propagate roughly sonically, therefore we expect the 'oscillation' to have a period of about one minute for a flow of granular size . The surface manifestation of the boundary layer instability is to shift the position of downflow sites laterally by about 100 km on a time scale of the order of one minute.

KOUTCHMY — One of the interesting and new results of these wonderful numerical simulations is the discovery of shocks in the granulation layers, propagating in horizontal directions. Before trying to compare this prediction with solar observations in order to find what could be the signature of the phenomenon, let me ask you respectfully, is it possible that such shocks are an artifact of the method you used? Will they be present in the case of a 3-D simulation, and will they be present if you take into account features like photospheric magnetic fields and NLTE radiation effects?

CATTANEO — It is always possible that any numerical result is bogus[1]. In defence of the credibility of our results, I should point out that similar behaviour, that is, a transition to supersonic flow with shocks, has been observed in independent 2-D simulations by Woodward et al[2] and by Porter and Woodward[3]. R. Stein has also shown the existence of shock systems in 3-D simulations with penetrative layers and radiative effectsin his talk. I think that if the dissipation is low enough, shocks will be present quite generally.

STEIN — Shocks are not merely a 2-D artifact. They also occur in 3-D flows with overshoot and radiation, as I have shown in my talk.

NORDLUND — A comment: there is actually (indirect) observational confirmation of supersonic flows in stellar atmospheres. In anelastic numerical simulations, where the flow is subsonic by definition, we obtain synthetic spectral line profiles which are too narrow, especially in the line wings, for stars with higher T_{eff} and lower $\log g$ (e.g. Procyon). Thus, there must be (quite common) regions of supersonic flows to explain the observed line profiles, since the simulated flows are already close to sonic.

HOLWEGER — You recommend to revise the idea of cellular flow. But if I recall the beautiful movies by Title, it seems to me that these 'exploding' granules still show a lot of cellular topology.

---

[1] bogus = bad output generated using a supercomputer (Ed.).

[2] P.R. Woodward, D.H. Porter, M. Ondrechen, J. Pedelty, K.-H. Winkler, J.W. Chalmers, W. Hodson and N.J. Zabusky, 1987, Science and Engineering on Cray Supercomputers, Cray Research, Minneapolis.

[3] D.H. Porter and P.R. Woodward, 1986, in Proceedings of the Conference on Scientific Applications and Algorithm Design for High Speed Computing, National Center for Supercomputer Applications, Champaign, Illinois.

CATTANEO — Observations show us the surface appearance of convective flows. Simulations, in principle, extend our knowledge of the flow structure to regions well within the bulk of the motion, which cannot be reached directly by observations. The evidence from the simulations of Stein and Nordlund and from our simulations is that the connectivity of downflows, i.e. the existence of a *cellular network* is only a shallow phenomenon. Throughout most of the convection the upflows and downflows are disconnected. The appearance of the flow below the observable surface is that of a series of isolated downwards-directed columns, and broader regions of gentler upflow. The idea of cellular flow, which was developed mainly to describe connected flows, is not so appropriate to describe what we see.

DRAVINS — Different 3-D simulations have now confirmed the structural change from the apparently cellular granulation patterns on the surface to a pattern of narrow channels with sinking matter beneath the surface. An early suggestion of this nature of convection was made by Fred Hoyle *et al.*[4] already some 25 years ago. The presence of isolated downflows was predicted from the high opacity of the cool material, which isolates it from the surrounding hot material.

CATTANEO — Thanks to Dr. Dravins for pointing out this early work.

VAN BALLEGOOIJEN — You point out that the downflows are concentrated into disconnected columns. Some observational indication that this is indeed the case comes from the 'cork movies' derived from the recent granulation data.

FOX — I would like to add that compressible calculations of modal convection also indicate a depth cut-off for the depth penetration of cellular granular convection, even though the horizontal structure that is imposed would prefer cellular motions to large depths.

NESIS — The investigation of Spectrostratoscope data, especially for the horizontal velocities $(\cos\theta = 0.2)$ at small scales in the photosphere and their variation with height supports the results of your 3-D simulations.

---

[4]J. Faulkner, K. Griffiths and F. Hoyle, 1965, Mon. Not. Roy. Astr. Soc. **129**, 363.

Spectroscopic Properties of Solar Granulation obtained from 2-D Numerical Simulations

Matthias Steffen[1,2]

[1]*Institut für Theoretische Physik u. Sternwarte,
Universität Kiel, Federal Republic of Germany*

[2]*National Solar Observatory, Tucson, USA*

1 Introduction

The theoretical understanding of the phenomenon of solar granulation is one of the topics of present-day solar physics. Despite extensive observational and theoretical efforts our knowledge about the dynamical and thermal effects of granulation on the solar photosphere is still incomplete. Commonly, granulation is interpreted as a pattern of surface convection cells at the boundary between the solar convection zone and the photosphere, but from time to time even the convective character of granulation is questioned, as in a recent study by Roudier and Muller (1986), who suggest that granules smaller than ≈ 1000 km might actually be turbulent eddies.

It has become clear over the years that it is impossible to obtain a consistent, quantitative picture of the conditions prevailing in the solar granulation layers from a purely empirical analysis of existing observational material (e.g. Bray et al., 1984). To make progress, theoretical granulation models having the potential to provide data that can be compared directly to solar observations are indispensable.

Notoriously, the calculation of convection in stellar atmospheres is a complex problem, requiring the application of time-dependent, non-linear hydrodynamics to a highly turbulent flow that, to complicate the situation, strongly interacts with the radiation field. Although in principle the problem is well defined by a few differential equations, an analytical solution is impossible in practice. Relevant results can only be obtained from numerical simulations on powerful computers, allowing realistic background physics to be included.

The first model calculations of this kind were successfully performed by Nordlund (1982, 1984a). His 3-dimensional, time-dependent models are capable of reproducing essential dynamical and spectroscopic features of the solar granulation (Wöhl and Nordlund, 1985; Dravins et al, 1981). Recently, similar calculations were reported by Kostik and Gadun (preprint; this conference). While their results seem to be similar to Nordlund's at least in some respects, a more detailed comparison is needed to assess the overall consistency of the two sets of models.

The purpose of this contribution is to demonstrate what kind of results can be obtained from 2-dimensional models of granular convection cells. In contrast to the full

R. J. Rutten and G. Severino (eds.), Solar and Stellar Granulation, 425–439.

3-dimensional case, our models are comparatively simple, especially if only steady state solutions are considered. Under these circumstances it is more readily possible to study in some detail the physical mechanisms governing granular convection.

By varying the model diameter we are able to simulate convection cells of different horizontal dimension. We find that the properties of granulation depend systematically on its horizontal scale. Spectroscopic properties derived from our models are generally in good agreement with solar observations. Remaining discrepancies may partially be explained by the unrealistic temperature structure of the upper photosphere resulting from the hydrodynamical calculations as a consequence of treating radiative transfer in the grey approximation.

Finally, we briefly discuss existing inconsistencies between our models and those of Nordlund (and Kostik and Gadun). In a joint effort it should be possible to locate the basic reason for the significant disagreement by performing appropriate test calculations. Ultimately, spectroscopic observations with very high spatial resolution ($\approx 0.1\,''$), which should become possible with future instrumentation, must provide a crucial check of the basic validity of the theoretical models.

2 Calculations

The numerical simulations referred to in this work are restricted to 2 dimensions in cylindrical symmetry. They have a higher spatial resolution than Nordlund's models, and avoid the anelastic approximation to retain all effects of compressibility. Thermodynamics include ionization of hydrogen. The radiation field is calculated by solving the equation of radiative transfer along a large number of rays crossing the model at various inclinations and directions. In this way non-local radiative exchange is taken into account both vertically and horizontally. Using a realistic Rosseland mean opacity, the grey approximation in LTE has been adopted so far. Magnetic fields are not included. The resulting set of differential equations is solved numerically using a Characteristics method. The final results are self-consistent, physical models of granular convection cells.

Details about the physical assumptions, numerical scheme, model parameters and boundary conditions of the hydrodynamical calculations have been given elsewhere (Steffen, 1987; Steffen and Muchmore, 1988; Steffen et al., 1988; Stefanik et al., 1984).

3 General Results

In this section we briefly summarize the basic characteristics of the solutions obtained from the numerical simulations.

We find that models with a diameter exceeding a critical upper limit of approximately 2000 km exhibit a truly non-stationary behaviour, never reaching a steady state. The flow changes significantly on time scales of the order of 10 minutes, comparable to typical granular life times. In this paper, however, we primarily analyze the spectroscopic properties of the smaller, steady state convection cells. Typical features of these solutions are narrow downdrafts with high velocity and broader ascending regions with lower flow velocities. The thermal structure is characterized by large horizontal temperature differences, and by steep temperature gradients in the hot, rising parts of the granules. Convective motions extend a few hundred km into the stable layers above large granules, causing an inversion of the temperature fluctuations in the upper photosphere. Overshooting becomes progressively weaker with decreasing cell size. At the same time the continuum intensity contrast drops sharply. For a detailed study of the calculated velocity- and temperature fields as a function of horizontal scale see Steffen et al. (1988). Finally, it is worth mentioning that oscillations, superimposed on the convective flow, seem to be ubiquitous. Typically, their periods range near 250 sec. (Steffen, 1988a)

4 Spectroscopic properties of granular convection derived from the numerical simulations

4.1 Continuum

Although in the numerical simulations described above radiative transfer is treated in the grey approximation, it is possible (somewhat inconsistently) to use the resulting 2-dimensional hydrodynamical model atmospheres for detailed line formation calculations to obtain basic spectroscopic properties from the models. For this purpose modified versions of the computer codes ATMOS and LINFOR, originally developed by the Kiel group for the fine analysis of stellar spectra, were employed, assuming LTE and pure absorption. The results reported in the following refer to solar disc-center.

For the largest steady state models with diameters between 1000 and 2000 km, the rms contrast of the 2-dimensional intensity pattern ranges between 11 and 13% in the continuum at λ 5380 Å. From the models with even larger horizontal dimension, which are strongly non-stationary, we obtain a time average of the continuum intensity contrast that is not significantly different, maximum values at individual instants in time still lying well below 20%. Towards smaller granular scales the amplitude of the horizontal intensity fluctuations declines strongly. As discussed in detail by Steffen et al. (1988), the main reason for this behavior is that horizontal radiative exchange becomes increasingly more efficient with decreasing cell size, reducing horizontal temperature fluctuations particularly in the continuum forming layers ($\tau=1$). In reasonable agreement with observations, the continuum intensity contrast derived from the numerical simulations depends on wavelength roughly as λ^{-1}, essentially reflecting the wavelength dependence of the Planck Function (cf. Wittmann, 1979).

4.2 Line Spectrum

The line spectrum can be calculated with a spatial resolution that corresponds to the horizontal grid distance of the hydrodynamical calculations (typically 10.....40 km). Although currently no spectroscopic observation is capable of such an extremely high spatial resolution, it is nevertheless instructive to look at the predicted individual line profiles. The general picture emerging from the numerical simulations may be summarized as follows:

The cores of absorption lines originating from the bright granular regions are blue-shifted (relative to the laboratory wavelength), and the blue wing of the corresponding line profile is depressed relative to the red wing, resulting in a considerable line asymmetry. Lines formed in the dark intergranular lanes exhibit an even stronger asymmetry,

428

but in the opposite sense, their cores being red-shifted. The horizontally averaged line profile, obtained as a superposition of the spatially resolved profiles, turns out to be much less asymmetrical than most of the line profiles seen at high spatial resolution.

The corresponding line bisectors are inclined to the blue in the bright parts of the granulation (the top portion near the continuum being blue-shifted relative to the line core). In the dark intergranular regions the line bisectors are inclined even stronger, but to the red. The different slopes of the bisectors of the spatially resolved line profiles reflect the different depth-dependence of temperature and convective velocity at the various horizontal positions within a granular convection cell. Notably, the bisector of the horizontally averaged line profile exhibits the typical C-shape in close agreement with observation. Opposite asymmetries of the spatially resolved profiles cancel to a large degree when the spectrum is averaged over the granulation pattern. For illustrations and a more detailed discussion of these results see Steffen and Gigas (1985), and Steffen (1987). It is encouraging to see that recent spectroscopic observations with high spatial resolution carried out at the Observatorio del Teide, Tenerife (Wiehr and Kneer (1988); Mattig, this conference) indeed seem to confirm the general spectroscopic characteristics predicted by our hydrodynamical models.

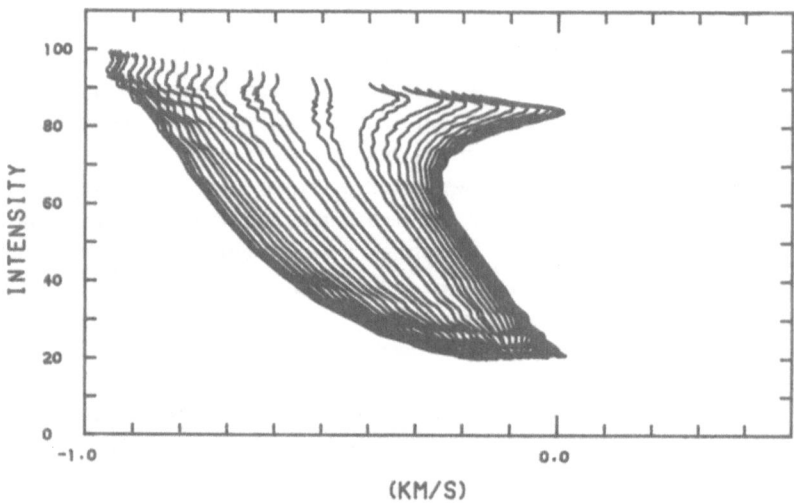

Fig. 1a: Series of computed bisectors of a strong Fe II line (λ 5197.6 Å, $\chi = 3.23$ eV) showing the cumulative effect of decreasing spatial resolution. From left to right, the surface area over which the spectrum is integrated is expanded in such a way as to progressively include contributions of darker and darker granulation elements.

The calculated effect of spatial resolution on the resulting bisectors of a strong Fe II line is shown in Fig. 1a. The leftmost bisector displayed corresponds to an imaginary spectroscopic observation with extremely high spatial resolution centered on a bright granule. Advancing to the right, we see the cumulative effect of progressively allowing darker and darker parts of the granulation to enter the aperture of the spectrograph, simulating a gradual decrease of spatial resolution. The rightmost bisector in Fig. 1a, therefore, corresponds to that of the horizontally averaged line profile observable at low spatial resolution at solar disc-center. This sequence demonstrates that it is the presence of the dark intergranular downflows that makes the bisector turn to the red as it approaches the continuum. The results of a similar experiment, but now beginning with the bisector found in the very darkest parts of the granulation pattern, are displayed in Fig. 1b.

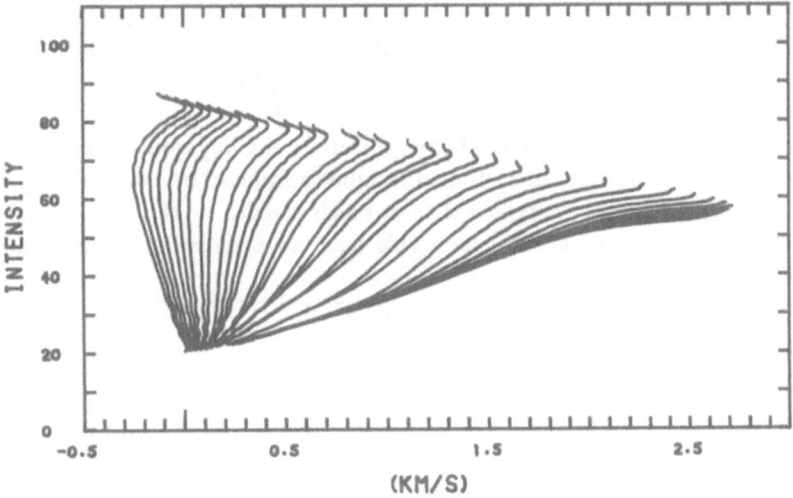

Fig. 1b: Similar to Fig. 1a, but starting with the bisector found in the very darkest parts of the granulation pattern (seen at extreme right). Advancing to the left, the area of horizontal averaging is expanded to include increasingly brighter surface elements. By definition, the rightmost bisector in Fig. 1a is identical to the leftmost bisector in Fig. 1b. The velocity scale used in Figs. 1a, 1b gives the line shift relative to the laboratory position (corrected for gravitational redshift, orbital motion, etc.).

Since the temperature fluctuations in the upper photospheric layers of our models become negligible (Steffen and Muchmore, 1988; Steffen, 1988b), the residual intensity in the core of very strong spectral lines is found to be essentially independent of the horizontal position relative to the granulation pattern. For lines of moderate strength the central intensity tends to be even lower in the ascending granular flows where the continuum intensity is high than it is in the cores of lines originating from the the dark intergranular regions. This is due to the fact that temperature fluctuations change sign in the overshooting layers as convective motions penetrate into a stably stratified atmosphere from below. Indeed, observational evidence for this behavior has been found from spectrograms of high spatial resolution (cf. Holweger, this conference).

Although under these circumstances it is true that the intensity in the cores of the blue-shifted lines is less than in the red-shifted line cores, this does not imply that horizontal averaging produces a net red shift of the cores of strong to medium strong spectral lines, as has sometimes been incorrectly concluded. Actually, the resulting net blue shift is even enhanced in the presence of a temperature field in the upper photosphere that is inverted relative to that of the continuum forming layers, because its effect is to strengthen the blue-shifted lines while weakening the red-shifted contribution. As may be seen from Fig. 1, the horizontally averaged profiles of strong lines are essentially unshifted. Weak lines, however, are blue-shifted up to ≈ 400 m/s.

Intuitively it is usually expected that the way the strength of a particular spectral line changes over the granulation pattern is primarily determined by the temperature dependence of the population density of the respective atomic species. This would be true, for example, if hot and cool parts of the granulation could be described as an ensemble of independent standard (1-dimensional) atmospheres of different effective temperature arranged side by side. However, this is not the situation found in our 2-dimensional models, where conditions are homogeneous in the uppermost photospheric layers. The thermal structure obtained from the models implies an average temperature gradient in the line formation regions that is much steeper in the hot, granular regions than in the cool, intergranular lanes. This leads to a somewhat surprising result: referred

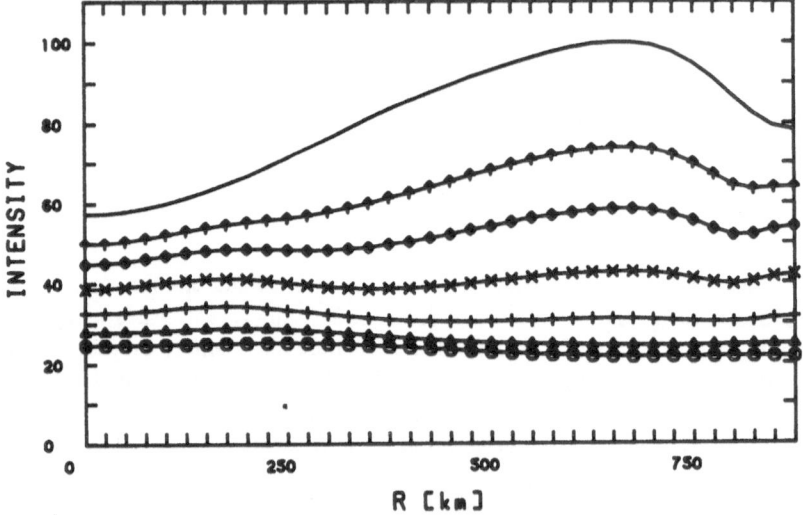

Fig. 2a: Continuum intensity (solid line) and residual intensity in the cores of several fictitious Fe I lines ($\chi = 0.0$ eV) of different strength (symbols) in the visible spectral region (λ 5380 Å) calculated for various horizontal positions across a typical model granule as seen at disc-center.

to the local continuum, all lines are found to be stronger in the bright granular surface elements as compared to the dark lanes, irrespective of the particular temperature sensitivity of the spectral line considered. It is the dominating effect of the temperature gradient that is responsible for this behavior.

Fig. 2a shows an example of how the continuum intensity and the intensity in the cores of several fictitious Fe I lines (λ 5380 Å, $\chi = 0.0$ eV) of different strengths vary across a typical steady state granular convection cell obtained from the numerical simulations. The corresponding results for Fe II (λ 5380 Å, $\chi = 2.8$ eV) are displayed in Fig. 2b. Note that the behavior of the temperature sensitive Fe I lines is qualitatively very similar to that of the much more temperature insensitive Fe II lines. Fig. 3 shows the dependence of the equivalent width (referred to the local continuum) of various different spectral lines on the horizontal position across the same model granule as used in Fig. 2. Again we see no qualitative difference between lines of neutral and ionized elements, all lines attain their maximum strength in the brightest granular structure. Strong Fe II lines show the smallest spatial variation, while C I (λ 5380 Å, $\chi = 7.68$ eV) is predicted to weaken considerably in the dark regions.

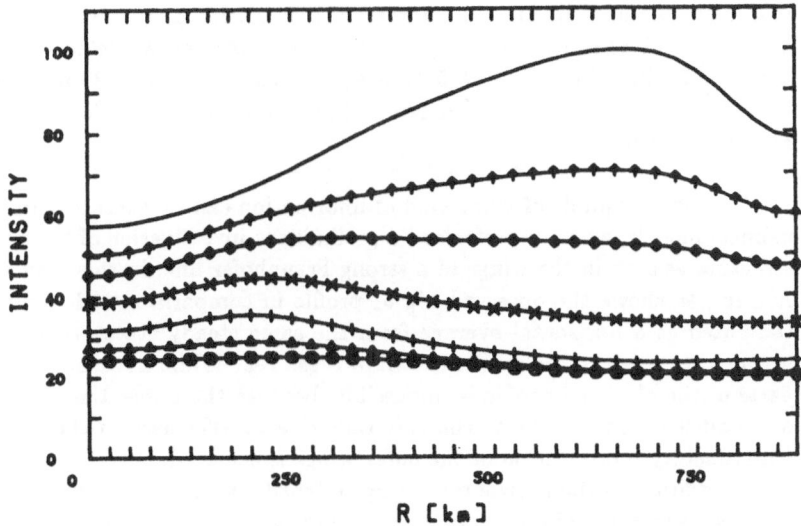

Fig. 2b: As Fig. 2a, but for a series of fictitious Fe II lines ($\chi = 2.8$ eV). Note that the behavior of the temperature sensitive Fe I lines is qualitatively very similar to that of the much less temperature sensitive Fe II lines.

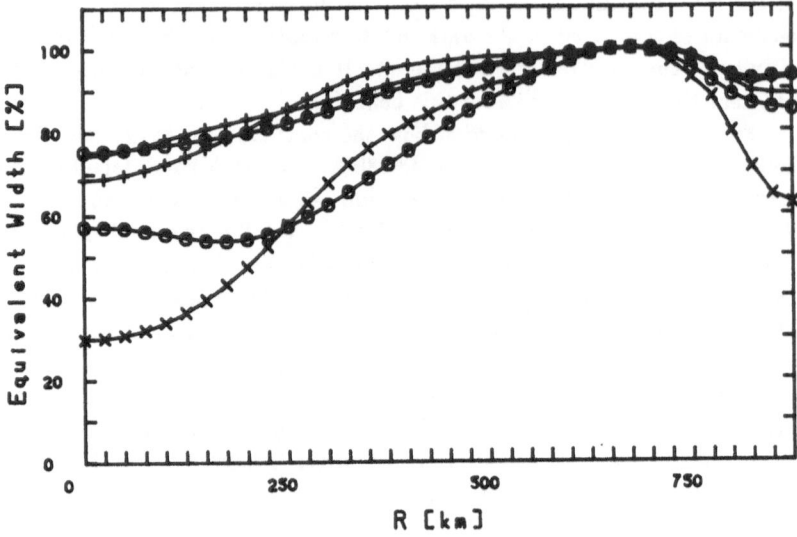

Fig. 3: Relative change in equivalent width of various spectral lines over the same model granule used in Fig. 2. Crosses: C I (λ 5380.3 Å, $\chi = 7.68$ eV, $W_\lambda = 22$ mÅ), circles: Fe I (λ 5127.7 Å, $\chi = 0.05$ eV, $W_\lambda = 20$ mÅ, and λ 5242.5 Å, $\chi = 3.63$ eV, $W_\lambda = 99$ mÅ), plus signs: Fe II (λ 5256.9 Å, $\chi = 2.89$ eV, $W_\lambda = 20$ mÅ, and λ 5197.6 Å, $\chi = 3.23$ eV, $W_\lambda = 87$ mÅ). For Fe I and Fe II, the weak lines show the larger variation across the model.

As a final example of what kind of information can be extracted from the hydrodynamical models, we have performed a preliminary investigation of the granular intensity structure as seen in the wings of a strong Fraunhofer line, in this case Mg b_2 at λ 5173 Å. Fig. 4a shows the observed Mg b_2 profile in comparison with the synthetic profile computed as a horizontal average from the same steady state granulation model that was used to produce the results shown in Figs. 1–3. While a reasonable fit of the central parts of the observed profile is impossible, because the upper temperature structure of our models is too hot due to the grey radiative transfer used in the model calculations, a satisfactory reproduction of the outer wings is achieved.

In addition to the horizontal average intensity, we have calculated the rms intensity contrast (a measure of the *relative* power of the granular intensity fluctuations) as a function of wavelength across the synthetic Mg b_2 line profile. Similarly, we have obtained the linear correlation coefficient measuring the coherence between the intensity fluctuations in the continuum and those in the line at the respective wavelength position along the profile (Fig. 4b). We see that the intensity contrast is found to decrease sharply as we proceed from the outer wings towards the line center, reaching a deep minimum at about 0.4 Å from the line core. At the same position the correlation between continuum and line intensity essentially disappears. Remarkably, the amplitude of the intensity fluctuations strongly increases again towards the line center, reflecting a corresponding

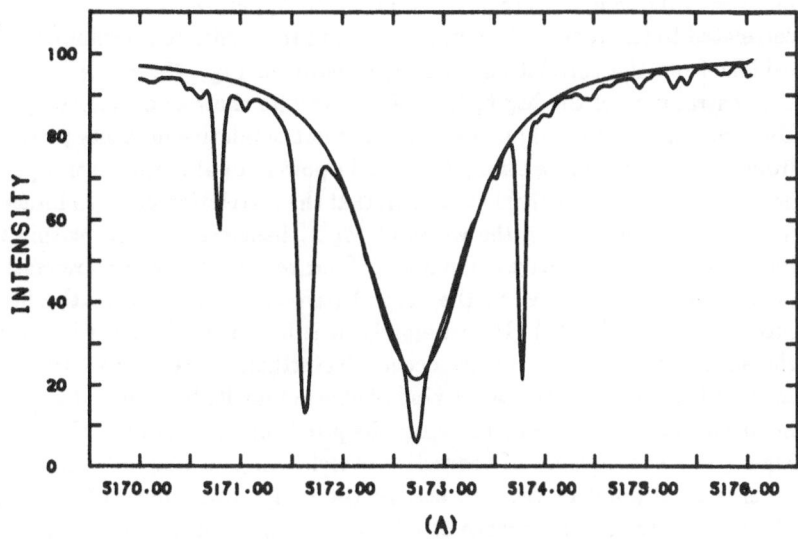

Fig. 4a: Mg b_2 λ 5173 Å, as observed at solar disc-center, compared with a synthetic profile calculated as a horizontal average over a typical model granule.

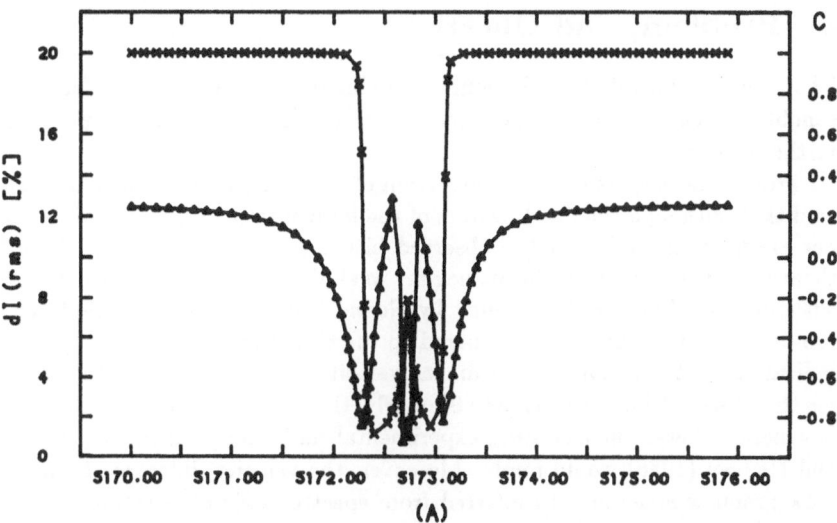

Fig. 4b: Relative rms intensity contrast (triangles, left scale) and linear correlation coefficient, between intensity fluctuations in the line and in the continuum (crosses, right scale), as a function of wavelength along the Mg b_2 profile obtained from the same hydrodynamical granulation model as used in Fig. 4a.

increase of the temperature fluctuations in the upper overshooting layers, which are anti-correlated to the temperature fluctuations in the continuum forming layers. This is also evident from the correlation coefficient shown in Fig. 4b, which is clearly negative in the central part of the Mg b_2 line. These results show some similarity to observations. Durrant and Nesis (1981) find a minimum in the relative intensity power at about 0.5 Å from the line core and a strong increase in power towards the central part of the profile. Evans and Catalano (1972) have shown that the correlation between intensity fluctuations in the continuum and in the wings of Mg b_1 becomes negative within 0.3 Å from the line center. A quantitative comparison of our results with solar observations is currently of limited value, at least in the central part of the line where the fit of the average profile is poor (Fig. 4a). Unfortunately, it is here that the most interesting features in the spatially resolved spectrum occur. Nevertheless, we suggest that our calculations show at least qualitatively what kind of effects may be be expected in the analysis of the intensity fluctuations observed across the profile of a strong solar absorption line, such as Mg b_2, as a consequence of convective overshooting. It may be tempting, from a purely observational point of view, to ascribe the temperature fluctuations seen in the upper photosphere to the appearance of a 'second component' of different physical origin (e.g. gravity waves) unrelated to granulation (Kneer et al., 1981, Durrant and Nesis, 1981). However, our results indicate that convective overshooting alone produces a temperature field with complex properties whose interpretation is by no means straightforward.

5 Problems and Questions

The results obtained from the numerical simulations are generally found to be in reasonable agreement with observations. However, there are certainly points that deserve further scrutiny.

From a more systematical comparison of observation and theory it has been discovered that, although the line bisectors of the horizontally averaged profiles resulting from our granulation models fit the observed bisectors very well, the fit of the line profiles themselves is less perfect. In particular, weak lines of Fe I as well as of Fe II, computed without any additional broadening by classical micro- or macroturbulence, tend to be too deep and too narrow as compared to the observed profiles (disc-center). This is an indication that the computed granular velocity field is somewhat too weak to account for the observed line broadening. Similarly, the resulting net blueshift in the line cores is generally lower than existing experimental data compiled by Dravins et al. (1981) and Nadeau (1988) would imply. Moreover, the velocity difference between bright and dark granular structures as inferred from spectroscopic observations with high spatial resolution (Wiehr and Kneer, 1988; Holweger, this conference), is more extreme than predicted by the numerical simulations.

Another problem that may be related to these inconsistencies refers to the comparison with model calculations by different authors. The most prominent difference between our results and those of Nordlund (1982, 1984a, 1984b) concerns the continuum intensity contrast. While our models yield a rms intensity contrast of typically 13 % at λ 5000 Å, the corresponding value resulting from Nordlund's simulations is of the order of 25 - 30 % in the same spectral region. Similar results have recently been reported by

Kostik and Gadun (preprint; this conference). As discussed in more detail by Steffen (1988b), the reason for this disagreement is not fully understood at the moment. It is clear, however, that the temperature fluctuations in the continuum forming layers must be significantly different in the two kinds of models. Since we employ the grey approximation, whereas Nordlund uses a more complex, frequency-dependent radiative transfer, one may speculate that this might be a basic reason for the existing inconsistencies. However, this seems not very likely in view of the high intensity contrast obtained by Kostik and Gadun despite treating radiative transfer in the grey Eddington approximation. Likewise, it seems unlikely that the anelastic approximation used in Nordlund's calculations produces a spuriously high intensity contrast, because the compressible simulations by Kostik and Gadun give similar results. Of course, these conclusions must remain very preliminary until a more extensive documentation of the calculations of the latter authors becomes available.

As mentioned in Section 4.1, the emergent spectrum resulting from our horizontally more extended, strongly non-stationary models is not significantly different, averaged in time, from the results derived from the steady state solutions. We conclude that the time-dependent character of the flow is not responsible for an increased intensity contrast. Thus the question arises whether the disagreement between the different sets of models is an indication that (granular) convection behaves basically different in 3 dimensions than in 2 dimensions.

Finally it should be mentioned that the physical mechanism responsible for the oscillations found in our numerical simulations is not yet understood. In particular, is is not clear what determines the period of ≈ 250 sec., and to what extent the results are relevant to the understanding of the solar oscillations observed in the 5 minute band. Interestingly, similar oscillations have recently been found in the fully compressible 3-D simulations of solar granulation reported by Nordlund and Stein (cf. Stein, this conference).

6 Conclusions and Suggestions

The results presented in this contribution suggest that basic properties of granular convection can be derived from 2-dimensional steady state models, requiring considerably less computational effort than the calculation of fully time-dependent, 3-dimensional models. There are, however, significant differences between our models and comparable models of different authors that need to be explained. The fundamental reason for this disagreement is not known at the present. Suitable test calculations are indicated to make progress. For example, it would be of great interest to see how Nordlund's results would change if he used grey radiative transfer and/or would restrict his simulations to 2 dimensions. In a joint effort it should be possible to settle this issue before the theoretical models will be confronted with unprecedented observations of high spatial resolution expected from future generation instruments.

Acknowledgements. I would like to thank H. Holweger, D. Gigas and W. Livingston for valuable discussions. Part of the research presented here has been carried out at the National Solar Observatory (NSO) in Tucson, USA, providing computational and typesettig facilities. The participation in this workshop was made possible in part by a travel grant from NSO. This work was supported by grant Ste/427-1 from the Deutsche Forschungsgemeinschaft.

7 References

Bray, R.J., Loughhead, R.E., Durrant, C.J.: 1984,
   The Solar Granulation, Cambridge University press (2nd ed., p. 189)
Dravins, D., Lindegren, L., Nordlund, Å.: 1981, Astron. Astrophys. **96**, 345
Durrant, C.J., Nesis, A.: 1981, Astron. Astrophys. **95**, 221
Evans, J.W., Catalano, C.P.: 1972, Solar Phys. **27**, 299
Kneer, F., Mattig, W., Nesis, A., Werner, W.: 1981, Solar Phys. **68**, 31
Nadeau, F.: 1988, Astrophys. J. **325**, 480
Nordlund, Å., 1982, Astron. Astrophys. **107**, 1
Nordlund, Å.: 1984a, in *Small Scale Dynamical Processes in Quiet Stellar Atmospheres*, ed. S.L. Keil, (Sunspot, N.M. 88349), p. 181
Nordlund, Å.: 1984b, in *Small Scale Dynamical Processes in Quiet Stellar Atmospheres*, ed. S.L. Keil, (Sunspot, N.M. 88349), p. 174
Roudier, Th., Muller, R.: 1986, Solar Phys. **107**, 11
Stefanik, R.P., Ulmschneider, P., Hammer, R., Durrant, C.J.: 1984, Astron. Astrophys. **134**, 77
Steffen, M., Gigas, D.: 1985, in Proc. *Workshop on Theoretical Problems in High resolution Solar Physics*, ed. H.U. Schmidt, MPA **212**, p. 95
Steffen, M.: 1987, in *The Role of Fine-Scale Magnetic Fields on the Structure of the Solar Atmosphere*, eds. E.-H. Schröter, M. Vazquez, A.A. Wyller, Cambridge University Press, p. 47
Steffen, M., Muchmore, D.: 1988, Astron. Astrophys. **193**, 281
Steffen, M.: 1988a, in *Advances in Helio- and Asteroseismology*, eds. J.C. Dalsgaard, S. Frandsen, IAU Symposium No. 123, p. 379
Steffen, M.: 1988b, in *JOSO Report* (in press)
Steffen, M., Ludwig, H.-G., Krüß, A.: 1988, Astron. and Astrophys. (submitted)
Wiehr, E., Kneer, F.: 1988, Astron. Astrophys. **195**, 310
Wittman, A.: 1979, in *Mitteilungen aus dem Kiepenheuer-Institut* **179**, p. 29
Wöhl, H., Nordlund, Å., 1985, Solar Phys. **97**, 213

Discussion

KOUTCHMY — When you predict the value of rms intensity fluctuations, do you take into account the effect of the integration along the line of sight due to the full extension of the contribution function?

STEFFEN — Yes, we do the full radiative transfer, although we do assume LTE.

DAMÉ — You presented vertical oscillations in the downflows. Are they fixed at a given location (downflows can move laterally back and forth, cf. Cattaneo's talk)? And what amplitude do they have?

STEFFEN — What I showed was the vertical velocity of a fixed position within the rising flow. The amplitude depends on the particular model, and ranges between $\approx$ 50–500 m/s.

WEISS — Two questions, to you and to Åke Nordlund respectively.
1). How does the downflow at the axis of your cells vary when the diameter is increased? Can it be related to the development of an exploding granule?
2). The same effect was present in Nordlund's anelastic simulations; do the new results of Nordlund and Stein add any further information?

STEFFEN — Ad 1): the downflow becomes longer and faster as the model diameter is increased. There is a critical upper limit, beyond which the flow reorganizes into smaller convection cells. This behaviour may well be related to the expansion and fragmentation of an exploding granule.

MÜLLER — Your simulation mainly concerns large cells; you compare the rms intensity contrasts and line profiles with observations which include all granular scales. Consequently, the comparison between your simulation and the observations is not straightforward. The presence of smaller granules, even if their contribution is quite low, may reduce the contrast because they radiate in the intergranular lanes.

MAILLARD — In the results of your simulation, you mention a difference of the calculated global line shifts with respect to the observed ones. How important is the discrepancy? Did you test that on several lines? Do you have some explanation?

STEFFEN — We find from the models that strong lines are essentially unshifted, while weak lines are blueshifted up to $\approx$ 500 m/s. On the average, the blueshift we obtain is too small by $\approx$ 100–200 m/s. This is an indication that the simulated velocity field is somewhat too weak, probably because we did not take into account the time dependence of the flow.

MATTIG — Are you able to estimate the center-to-limb rms intensity contrast from your calculation?
Let me add a remark: during the last partial eclipse at the Sacramento Peak Observatory, we have observed the full center-to-limb variation of the rms intensity contrast. Taking into account all the uncertainties of the atmospheric MTF, we never found an rms contrast larger than 18%.

STEFFEN — In principle we do have the information to calculate the center-to-limb variation of the rms intensity contrast. Work is in progress to obtain the center-to-limb variation of the emergent spectrum from the hydrodynamical models.

The rms contrast value you report seems in reasonable agreement with the model results.

VAN BALLEGOOIJEN — I would like to draw a comparison between your 2-D simulations and the 3-D simulations by Nordlund and Stein. One interpretation of the occurrence of downflow on the axis of your model is that it corresponds to an exploding granule. But another interpretation is that it corresponds to the concentrated plumes of downflow found in the 3-D simulations.

STEFFEN — It seems to me that both interpretations may be true at the same time. At the center of an exploding granule, a concentrated downflow could develop after some time.

TITLE — Scharmer's data shows bright small granules. SOUP data have shown that not all granules develop dark centers, and that granules that explode are unstable.

STEFFEN — From the models, I would expect the intensity contrast to become significantly lower for granules smaller than 0.5 arcsec which is at the limit of Scharmer's spatial resolution.

The development of a dark center is probably found mainly for granules that have time to grow large enough under sufficiently symmetrical conditions. Our steady-state models can only represent a certain phase in the time evolution of an exploding granule. Using appropriate initial conditions, we could try to simulate the time evolution (expansion) of an exploding granule in future calculations. The simulation's cylindrical symmetry, however, will prevent the breaking up of the ring structure in the way it is observed.

VAN BALLEGOOIJEN — What is the reason for the small temperature fluctuations higher up in the photosphere, compared with Nordlund's simulations?

STEFFEN — Part of the differences may be explained by the fact that Nordlund's simulations are fully time-dependent, while the results that I have shown refer to a steady-state solution which represents an average state of the atmosphere.

However, there may be a more fundamental reason for the difference between these two kinds of model, however, that is at present not understood.

RUTTEN — Can you show the transparency of this difference which you have shown at the JOSO Workshop in Freiburg last fall? (This is the figure reproduced at the end of this discussion—Ed.).

The intensity contrast difference originates in the deeper layers at left, but the disagreement is even larger at right where the lines are formed. I think this difference is due to the differences in the radiative transfer approximations used in the two codes (Rutten 1988)[1]. It would be worthwhile to do experiments on the effects of the ultraviolet line haze, not only on the effects due to scattering in the line source functions as Nordlund has described in the München proceedings[2], but also on the effects of the ultraviolet line opacities, for example by decreasing their extinction coefficients by a factor of ten in the higher photosphere, as can be the case for Fe I due to overionization.

---

[1] R.J. Rutten, 1988, in R. Viotti, A. Vittone, M. Friedjung (Eds.), *Physics of Formation of Fe II Lines outside LTE*, IAU Coll. 94, p. 203ff.

[2] Å. Nordlund, 1985, in H.U. Schmidt, *Theoretical Problems in High Resolution Solar Physics*, München

STEFFEN — It is certainly worthwhile to investigate whether this speculation (that the differences in radiative transfer are responsible for the difference in the size of the temperature fluctuations in the upper photosphere) is correct. A simple way to do this would be to adopt grey radiative transfer in Nordlund's simulations.

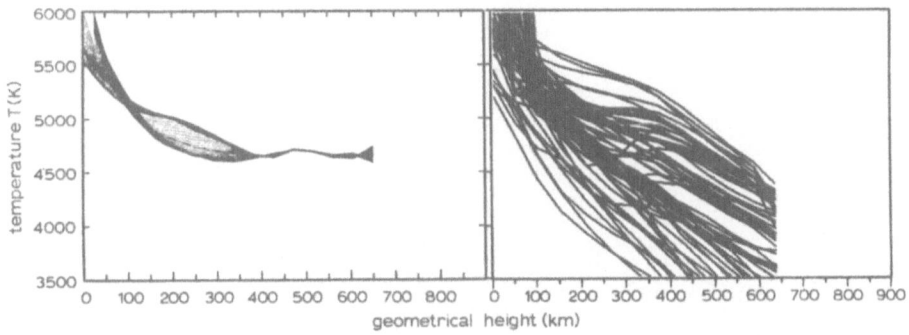

Simulation contrast of granulation contrast.
The lefthand panel shows the temperature contrast as a function of height obtained from Steffen's simulation, while the righthand panel shows, at the same scales, the contrast obtained from Nordlund's simulation as published in H.U. Schmidt (Ed.), *Theoretical Problems in High Resolution Solar Physics*, München 1985, p. 19. Figure taken from M. Steffen, 1988, JOSO Annual Report 1987, Kiepenheuer Institut, Freiburg, in press.

ON THE INFLUENCE OF OPACITY FLUCTUATIONS
ON THE ENERGY TRANSFER BY RADIATION

J. Trujillo-Bueno   and   F. Kneer
Universitäts Sternwarte
Geismarlandstrasse 11
3400 Göttingen
F. R. G.

ABSTRACT. This paper aims at gaining knowledge especially on the multi-dimensional radiative transfer effects of opacity fluctuations in stellar atmospheres. First, by means of analytic calculations, some properties of the non-local transfer effects of such fluctuations are pointed out. The potential importance of opacity fluctuations for driving radiative instabilities is demonstrated through the establishment of the possible types of instabilities in grey radiative equilibrium atmospheres. Numerical linear calculations for grey continuum radiation are then presented. The influence of the transfer effects of opacity fluctuations on the radiative relaxation time is discussed as a function of the characteristic length of the atmospheric structures.

1. INTRODUCTION

Observations of the atmospheres of the Sun and of other stars lead to the discovery of time-dependent spatial structures. The second law of thermodynamics implies that the pattern which the astrophysicist is able to scrutinize on the solar surface can only be formed and maintained through the effect of exchange of energy and matter in situations far from equilibrium. From this point of view the solar granulation constitutes a fascinating field of research. In investigating such a complicated non-linear problem, the astrophysicist is led inevitably to perform extended numerical simulations. However, one should be completely aware of the impossibility of reaching satisfactory physical understanding by means of simply increasing the physical ingredients of the calculations. Even the most skillful numerical experiment would only lead to a "reproduction" of nature, with more or less similarity. For this reason great importance is given to basic investigations where a particular physical mechanism is isolated in order to analyze in depth its physical consequences.

441

R. J. Rutten and G. Severino (eds.), Solar and Stellar Granulation, 441–452.
© *1989 by Kluwer Academic Publishers.*

The present contribution belongs to this last type of investigations. The specific physical mechanism to be isolated here is radiative
transport of energy. The aim is to obtain knowledge on the radiative relaxation and amplification processes. The strategy chosen for reaching
this objective consists of investigating the multi-dimensional radiative
transfer effects of Planck-function (B) and opacity (χ) fluctuations.
For this purpose small horizontal harmonic temperature fluctuations of
various structural lengths will be applied to a schematic grey, radiative equilibrium (RE) solar model atmosphere in LTE. (The assumptions of
greyness and of LTE are generally adequate in the solar photosphere
where the H^- ion makes the primary contribution to the opacity, and
where the use of mean opacities has proven to be fruitful).

The time behaviour of the imposed temperature fluctuations will be
assumed to be governed by the following radiative heat equation(cf. Unno
and Spiegel, 1966):

$$\rho c_p \frac{\partial T}{\partial t} = 4\pi\chi(J-B) \quad , \tag{1}$$

where J is the frequency-integrated mean intensity of the radiation
field, $B=(\sigma/\pi)T^4$ is the frequency-integrated Planck function, with σ
the Stefan-Boltzmann constant, ρ the density, and c_p the specific heat
at constant pressure for which the value 1.6×10^8 erg $K^{-1}g^{-1}$ is adopted throughout the atmosphere.

The question of whether or not such a perturbed system always re-
turns to its initial RE configuration has already led us to investigate
the dependence of the radiative relaxation time on the structural
lengths of the perturbations (cf. Trujillo-Bueno and Kneer, 1987; Kneer
and Trujillo-Bueno, 1987). The possibility of also having a negative
answer to the above question will lead us in the following sections to
establish the possible types of radiative instabilities and the conditions under which these can actually occur.

2.   FLUCTUATING RADIATION FIELD

Once temperature fluctuations are applied to a stellar atmosphere model
in RE both the opacity χ and the Planck function B deviate from their
RE values, and therefore the radiative energy balance is altered. There
exists, however, a fundamental difference between the radiative transfer
effects of B and χ fluctuations: whereas B fluctuations only give rise
to an alteration of the emissivity of the stellar gas, χ fluctuations
lead both to a change of emissivity and to a change of extinction
throughout the atmosphere. It is thus intuitively obvious that the radiative transfer effects of B fluctuations should generally play a stabilizing role, while possible destabilizing effects of χ fluctuations
depend on an intricate competition between the non-local transfer
effects of the altered emissivities and those concerning the change
in extinction which an opacity perturbation produces.

One way of calculating the response of the specific intensity I of the radiation field to small B and χ fluctuations is to deal with the following equation which results from the linearization of the radiative transfer equation:

$$\frac{d}{ds} \delta I = \bar{\chi} (\delta S^{eff} - \delta I) \; , \tag{2}$$

with

$$\delta S^{eff} = \delta B + \frac{\delta \chi}{\bar{\chi}}(\bar{B} - \bar{I}) . \tag{3}$$

Equation (2) governs the variation of the fluctuating specific intensity δI along an arbitrary direction in a three-dimensional medium. Equation (3) gives the expression for the "effective source function" which determines the solution of Eqn (2). The notational convention adopted for an arbitrary function "f" is that "$\bar{f}$" represents the unperturbed quantity, while "δf" represents its first-order fluctuation. For harmonic perturbations the symbol "Δf" will be used below to indicate the amplitude of the fluctuating quantity "δf".

As seen in Eqns (2) and (3), for a given average opacity $\bar{\chi}$, the effects of B perturbations are completely determined by the spatial variation of the perturbation itself, whilst for χ fluctuations the angle-dependent and non-local unperturbed function $(\bar{B}-\bar{I})$ comes also into play. The very existence of this property suggests the following two broad categories of physical systems:

a) Systems in Thermodynamic Equilibrium: χ fluctuations have no effect

Under thermodynamic equilibrium conditions $\bar{I}=\bar{B}$. Thus, when both B and χ perturbations are applied to these physical systems only B fluctuations may have an effect. For χ perturbations the change in extinction is just cancelled by the altered re-emission, a result first found by Spiegel (1957) for an infinite homogeneous medium in LTE and with grey absorption.

b) Systems not in Thermodynamic Equilibrium: χ fluctuations have an effect

In non-thermodynamic equilibrium situations the specific intensity is not equal to the Planck function. Therefore, according to Eqn (3) opacity fluctuations can have an effect.

For the time being consider fluctuations in the diffusion region of stellar atmospheres. In such a region the specific intensity in a direction defined by the unit vector $\vec{\Omega}$ is given by (cf. Unno and Spiegel, 1966) $\bar{I}(\vec{r},\vec{\Omega})= \bar{B}(\vec{r})+(3/4\pi)\vec{\Omega}.\vec{F}$, with $\vec{F}$ the radiation flux vector of the unperturbed medium. Consequently, the effective source function (3) reads

$$\delta S^{eff} = \delta B -\frac{3}{4\pi} \frac{\delta \chi}{\bar{\chi}} \vec{\Omega} . \vec{F} \tag{4}$$

This last equation implies that a necessary condition for opacity fluctuations to have an effect is the existence of an average radiative flux.

One may now use expression (4) and the formal solution of the transfer equation (2) with the purpose of analytically calculating the response of the radiation field to harmonic B and X fluctuations. The effort would be worthwhile, because one would then end up finding interesting properties concerning the response of the radiation field to the above-mentioned fluctuations (see Rybicki, 1965). One of these interesting properties is that the response of the mean intensity to radial opacity fluctuations in the diffusion region of stellar atmospheres is $\pi/2$ out of phase. Taking into account this peculiarity when solving the radiative heat equation (1), one can then show that a possible consequence of the transfer effects of radial opacity fluctuations is to give rise to a periodic interchange of energy between the radiation field and the stellar gas with a characteristic period inversely proportional to $|\bar{F}|\partial\ln\bar{\chi}/\partial\ln\bar{T}$. The details of this analytical study can be seen in Trujillo-Bueno (1988). We now turn our attention to investigating the radiative transfer effects of horizontal fluctuations in a semi-infinite medium.

3. HORIZONTAL FLUCTUATIONS IN A GREY RE ATMOSPHERE

Consider the unperturbed system to be a grey RE atmosphere in LTE, which implies $\bar{J}=\bar{B}$. It is a well known result of radiative transfer theory that the thermal-radiative state of this system can be conveniently described by invoking the plane-parallel Eddington approximation, which is equivalent to just taking $\mu =\pm 1/\sqrt{3}$ (μ is the cosine of the angle between the direction of the beam and the normal to the surface) when solving the one-dimensional radiative transfer equation for the unperturbed radiation field. This translates into simple analytic expressions for $\bar{B}(\bar{\tau})$ and $\bar{I}(\bar{\tau},\mu)$ (see e.g. Chandrasekhar,1960; Chapter 11), thus leading to the following simple expression for the effective source function (3):

$$\delta S^{eff}= \delta B - \frac{3}{4\pi} \frac{\delta\chi}{\bar{\chi}} \bar{F} \mu \ , \tag{5}$$

where $\mu =1/\sqrt{3}$ has to be used for calculating outgoing specific intensities, and $\mu =-1/\sqrt{3}$ for the incoming.

With this approximate effective source function we will first obtain information on the response of the radiation field to opacity changes. This information will then lead us straightforwardly to establish the possible types of radiative instabilities.

3.1. Response of the Radiation Field to Opacity Changes

It is interesting to note that, considering only opacity fluctuations (i.e. $\Delta B=0$) with $\alpha =\Delta\chi/\bar{\chi} >0$, Eqn (5) predicts negative outgoing, but positive incoming fluctuating specific intensities. Accordingly, the sign of the amplitude ΔJ of the fluctuating mean intensity would be negative if the outgoing intensity outweighs the incoming intensity, but positive in the opposite case.

For simplicity assume a one-dimensional opacity perturbation with a height-independent and positive α (i.e. $\delta\chi/\bar{\chi} = \Delta\chi/\bar{\chi} = \alpha$). Then, the effective source function (5) is also independent of the atmospheric height, being equal to $-\alpha\sqrt{3}\bar{F}/4\pi$ for outgoing rays, but equal to $\alpha\sqrt{3}\bar{F}/4\pi$ for incoming rays. The corresponding radiative transfer problem is thus very similar to that of a plane-parallel isothermal atmosphere; the only novelty being that one has to worry about the two possible signs of the source function. With these considerations in mind it is then easy to show that the amplitude of the fluctuating mean intensity can be expressed in terms of the second exponential integral as follows:

$$\Delta J(\bar{\tau}) = -\frac{\sqrt{3}}{8\pi} \alpha \bar{F} E_2(\bar{\tau}) \quad . \tag{6}$$

From the asymptotic properties of $E_2(\bar{\tau})$ (see e.g. Abramowitz and Stegun, 1965), one arrives at the conclusion that, for a positive height-independent α, ΔJ remains everywhere negative. ΔJ is equal to $-\sqrt{3}\alpha\bar{F}/8\pi$ at the surface, and approaches zero as $\exp(-\bar{\tau})/\bar{\tau}$ for $\bar{\tau} \gg 1$. Indeed, this is as it should be, since as long as one stays deeply within the atmosphere the positive incoming intensity becomes isotropic and perfectly balances the negative outgoing intensities. It becomes then evident that if $\alpha(z)$ decreases outwards ΔJ is also negative at all atmospheric heights. However, if $\alpha(z)$ increases outwards ΔJ is positive below a certain height in the atmosphere lying close to $\bar{\tau}=1$; the gradient of α determining how large the positive values of ΔJ can be.

In order to demonstrate the validity of this latter conclusion for arbitrary wavelengths Λ of horizontal cosinusoidal fluctuations, consider a grey solar RE model ($T_{eff}=5800$ K) with gravitational stratification prescribed by the following expressions for the opacity χ and the density ρ :

$$\bar{\chi} = \bar{\chi}_0 \exp(-z/H) \quad ,$$

$$\bar{\rho} = \bar{\rho}_0 \exp(-z/2H) \quad .$$

Choosing $\bar{\chi}_0$ such that the optical depth $\bar{\tau}$ at $z=0$ is unity, and measuring geometrical distances in units of the opacity scale height $H(\approx 60$ Km for H^- absorption in the solar photosphere), one has $\bar{\tau}=\exp(-z)$. Assume that opacity fluctuations of the form $\chi = \bar{\chi}(1 + \alpha \cos kx)$, with $k=2\pi/\Lambda$ and $\alpha = \Delta\chi/\bar{\chi}$, are applied to this RE solar model atmosphere. If these opacity fluctuations are the result of temperature fluctuations alone, one has

$$\alpha = \frac{\partial \ln\bar{\chi}}{\partial \ln\bar{T}} \frac{\Delta T}{\bar{T}(\bar{\tau})} = \frac{\partial \ln\bar{\chi}}{\partial \ln\bar{T}} \frac{\Delta B}{4\,\bar{B}(\bar{\tau})} \quad , \tag{7}$$

where $\Delta B = 4(\sigma/\pi)\bar{T}^3 \Delta T$ is the amplitude of the ensuing Planck-function fluctuations. Choosing $\partial\ln \bar{\chi}/\partial\ln\bar{T}=1$ and $\Delta B=1$, Eqn (7) defines an outwardly increasing positive α . These B and χ fluctuations correspond to the numerical calculations below. The calculations have been performed applying the numerical method of Kneer and Heasley (1979).

Figure 1. shows the response of ΔJ to the above-mentioned opacity variations in units of ΔB for various values of the wavenumber k. As expected, $\Delta J>0$ at depths in the atmosphere greater than $\bar{\tau} \approx 1$. Due to horizontal radiative transfer the maxima there decrease in value and are displaced towards greater depths as k increases.

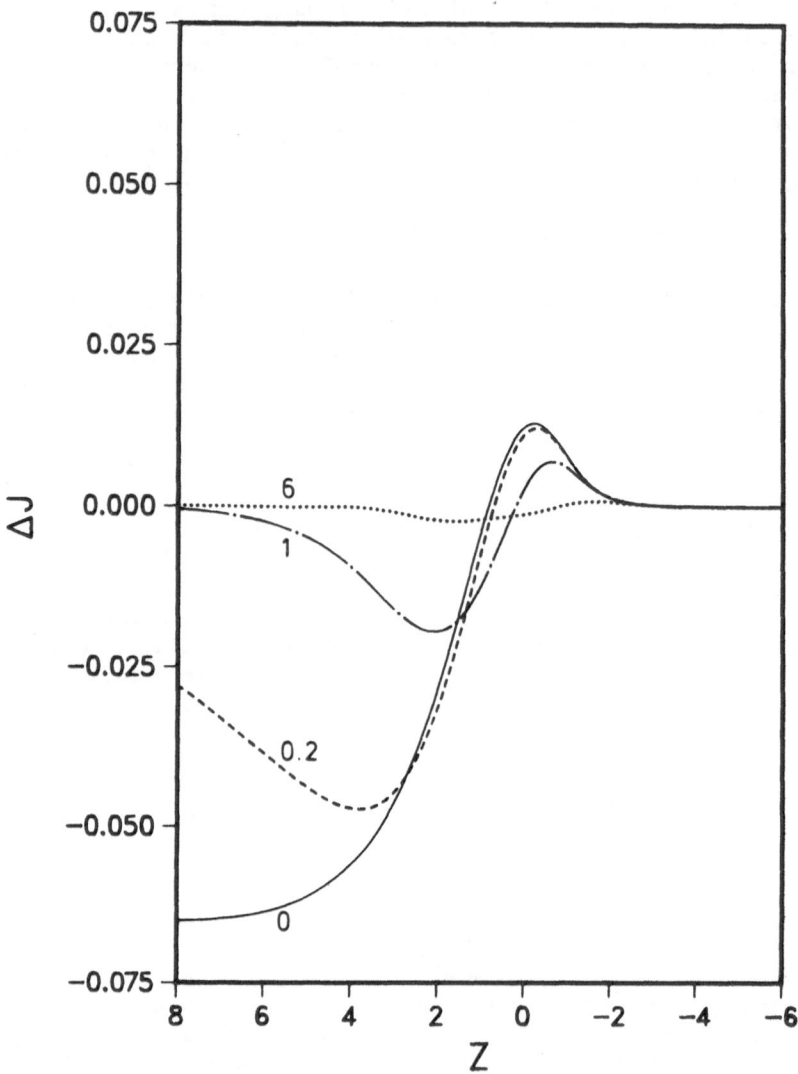

Figure 1. ΔJ in units of ΔB for horizontal opacity
fluctuations with α as given by Eqn (7). Parameter is the wavenumber k. Geometrical distances are measured in units of the opacity scale
heigth. $\bar{\tau}$ = exp(-z). The signs would have to
be reversed if $\partial \ln \bar{\chi} / \partial \ln \bar{T}$ were -1 instead of 1.

3.2 Types of Radiative Instabilities

Linearizing the radiative heat equation (1) one is left with the following equation which governs the time evolution of ΔT:

$$\frac{\partial}{\partial t}\ \Delta T =- (\eta_{\Delta B}(z,k)+ \eta_{\Delta\chi}(z,k))\Delta T \ . \tag{8}$$

The local growth rates $\eta_{\Delta B}$ and $\eta_{\Delta\chi}$ are due respectively to the transfer
effects of B and χ fluctuations. They are given by

$$\eta_{\Delta B} = \frac{16\sigma\bar{T}^3}{c_p}\ \frac{\bar\chi}{\bar\rho}\ (1-\Delta J/\Delta B) \quad , \tag{9}$$

$$\eta_{\Delta\chi} = \frac{4\pi}{c_p}\ \frac{\bar\chi}{\bar\rho}\ \frac{\bar{F}}{\bar{T}}\ \frac{\partial\ell n\bar\chi}{\partial\ell n\bar{T}}\ (- \frac{\Delta J}{\alpha\bar{F}}) \tag{10}$$

Although it is true that steep gradients of the Planck function
perturbation may give rise to values of ΔJ greater than ΔB, for reasonable spatial variations of ΔB $\eta_{\Delta B}$ >0. Thus, one is left with the transfer
effects of opacity fluctuations as the principal mechanism responsible
for the appearence of local radiative instabilities. As these instabili-
ties will occur if $\eta = \eta_{\Delta B}+\eta_{\Delta\chi}$ <0, the establishment of the possible
types of radiative instabilities follows after studying the possible
ways of obtaining negative values of $\eta_{\Delta\chi}$. Taking into account the previous results concerning the response of the radiation field to opacity
changes, the following conclusions can be reached straightforwardly.

(i) If $\partial|\alpha|/\partial z \leq 0$ the amplitude ΔJ of the mean intensity fluctuation has
the opposite sign to the change in opacity. Under these circumstan-
ces the only possibility which will give a negative growth rate $\eta_{\Delta\chi}$
is via negative values of $\partial\ell n\bar\chi/\partial\ell n\bar{T}$ (e.g. molecules; see Kneer,1983)
Whether or not radiative instabilities are possible depends on how
large the effective temperature and the thermal sensitivity of the
opacity are (see Eqn (10)).

(ii) On the other hand if $\partial|\alpha|/\partial z >0$, ΔJ changes sign at a height in the
atmosphere close to $\bar\tau =1$. If $\partial\ell n\bar\chi/\partial\ell n\bar{T}>0$ the atmosphere can in principle be unstable below this critical height, i.e. for $\bar\tau \geq 1$. However,
if $\partial\ell n\bar\chi/\partial\ell n\bar{T}<0$ radiative instabilities would be possible above such
a height, i.e. for $\bar\tau \leq 1$. As above the crucial parameters determining
whether or not such radiative instabilities actually occur are the
effective temperature and the absolute values of $\partial\ell n\bar\chi/\partial\ell n\bar{T}$.

3.3 Results from Numerical Calculations

In which extent can the radiative transfer effects of χ fluctuations influence the radiative relaxation of temperature fluctuations? Are local
radiative instabilities to be expected in the solar photosphere? Within
the framework of our assumptions and modelling we may provide answers
to these questions by comparing the growth rates $\eta_{\Delta B}$ and $\eta_{\Delta\chi}$ for the

particular case of temperature fluctuations with ΔB height-independent.

In our investigation on radiative relaxation in small-scale structures (cf. Trujillo-Bueno and Kneer, 1987), we showed a plot of $\eta_{\Delta B}(z)$ for various wavenumbers k of the thermal perturbations. Figure 1. of this latter work shows that the radiative relaxation time $t_r = 1/\eta_{\Delta B}$ is smallest near $\bar{\tau} = 1$. The growth rate $\eta_{\Delta B}$ was shown to increase as k varies from zero (plane-parallel geometry) to infinite (optically thin limit); the corresponding radiative relaxation times near $\bar{\tau} = 1$ diminishing from approximately 15 s to 1 s between these two limits.

The growth rate $\eta_{\Delta \chi}$ is shown in Fig. 2. It has been obtained from Eqn (10) by using $\partial \ell n \bar{\chi} / \partial \ell n \bar{T} = 1$ and the mean intensity amplitudes already shown in Figure 1. The solid lines in Fig. 2. represent positive values, whilst dotted lines indicate negative values. (Note that we are dealing here with the type (ii) of radiative instabilities, i.e. $\alpha(z)$ positive and increasing outwards). The critical height in the atmosphere where $\eta_{\Delta \chi}$ changes sign diminishes as the wavenumber k increases, and varies between z=1 and z=-1 for the values of k plotted. This shows that if instabilities were to occur, they would start at the height in the atmosphere where the perturbation changes from behaving optically thickly to optically thinly.

Generally $\eta_{\Delta \chi}$ decreases with increasing k due to the corresponding decrease in ΔJ. A comparison between $\eta_{\Delta \chi}$ and $\eta_{\Delta B}$ leads to the conclusion that (even with values of $\partial \ell n \bar{\chi} / \partial \ell n \bar{T}$ as large as 10) the radiative transfer effects of B fluctuations would dominate the radiative relaxation of atmospheric structures with structural lengths smaller than or equal to about 6 opacity scale heights, which corresponds approximately to 360 Km on the Sun. For increasingly larger structures one may expect χ fluctuations to become increasingly important for determining the radiative relaxation time $t_r = 1/\eta$ provided $\partial \ell n \bar{\chi} / \partial \ell n \bar{T}$ is large enough (i.e.~10). Radiative instabilities could be possible near z=-1 (i.e. $\bar{\tau} \approx 2$). However, in these atmospheric layers other modes of energy transport, i.e. convection, become relevant. Time-dependent non-linear dynamic calculations would be required to ascertain whether radiative transfer plays a significant role for establishing the dynamic behaviour (see e.g. Stein, Steffen; these proceedings).

4. DISCUSSION

What are the consequences of the previous results? To provide a theoretical explanation of the complex dynamic behaviour of stellar atmospheres via the identification of instability mechanisms requires deciding what will be the initial unperturbed configuration of the plasma which becomes the observed solar (stellar) atmosphere through the non-linear amplification of the inevitable fluctuations to which it is subject. At first sight, the fact that the previous types of radiative instabilities have been established for grey atmospheres in RE may lead to considering them only as a mere illustration of the potential destabilizing effects of opacity fluctuations. However, there exist some reasons for considering the instability mechanism related with the case where $\partial |\alpha| / \partial z > 0$ as a possible means by which stellar modes of oscillation may extract ener-

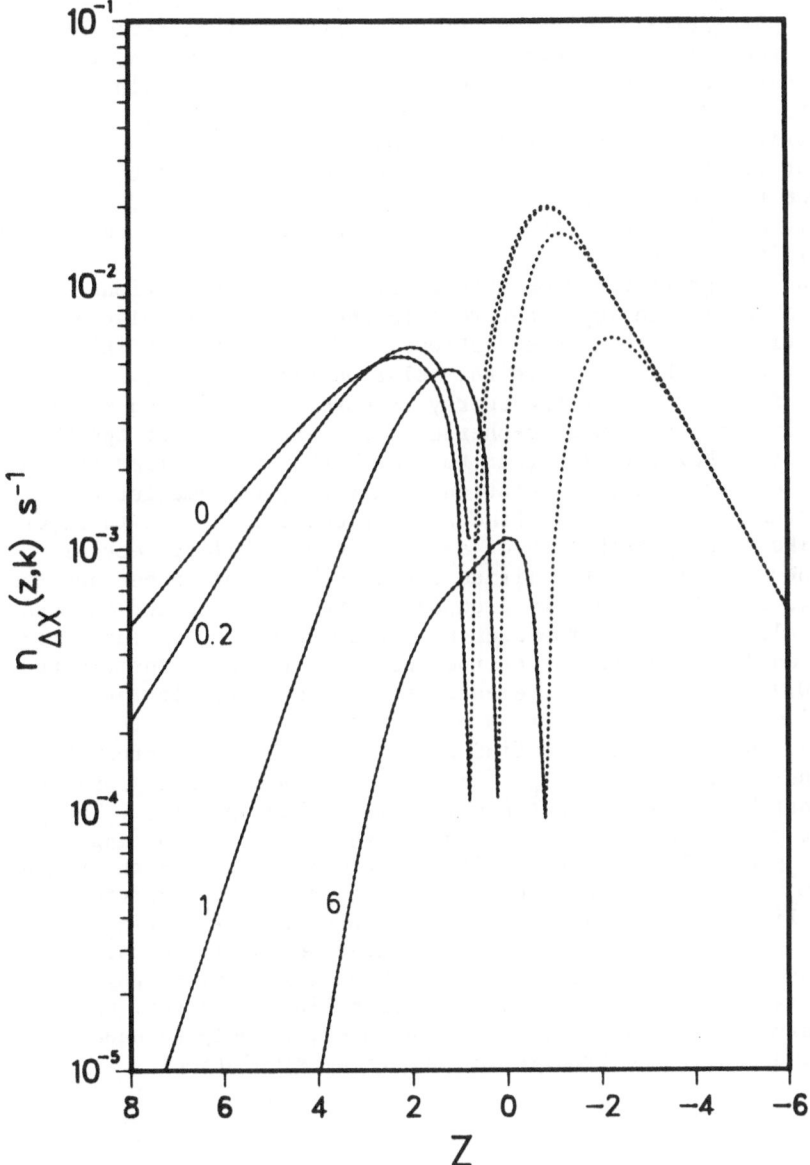

Figure 2. Growth rates as a function of heigth z for
horizontal opacity fluctuations with α as given
by Eqn (7). Parameter is the wavenumber k.
Solid lines: positive values. Dotted lines:
negative values. The signs would have to be
reversed if $\partial \ln \bar{\chi} / \partial \ln \bar{T}$ were −1 instead of 1.

gy from the radiation field. For example, it is known that the actual value of $\partial\ln\chi/\partial\ln T$ just below optical depth unity in semi-empirical models of the solar atmosphere strongly increases outwards, while in the photosphere it decreases with height starting from a value slightly larger than 10 (see e.g. Fig. 2. of Ando and Osaki, 1975). From Eqn (7) one then arrives at the conclusion that $\alpha(z)$ should actually increase outwards at least below $\tau \approx 1$ in such solar models. It is thus tempting to conclude that the κ-like overstability found by Ando and Osaki (1975) for some of the acoustic solar modes owes its existence to this type of instability mechanism.

The radiative transfer effects of opacity fluctuations are also expected to play an important role in the context of solar and stellar granulation. With the above-mentioned values of the opacity's thermal sensitivity of the H^- ion in the solar photosphere, the influence of opacity fluctuations on the energy transfer by radiation should be particularly important in atmospheric layers situated near optical depth unity. It is however important to remark that the smaller the characteristic lengths of the granulation structures, the smaller the importance of the radiative transfer effects of χ fluctuations, and therefore the less likely the possibility of a κ-like overstability mechanism.

The radiative instabilities considered in this paper owe their existence to transfer of energy processes between the matter and the radiation field, the actual appearence of them requiring of large radiative fluxes and/or large values of the opacity's thermal sensitivity (cf. Eqn (10)). Between the extremes of hot and cool stellar atmospheres with radiative forces or convection practically determining the atmospheric motions, one may expect to find a region in the Hertzsprung-Russel diagram where the values of T_{eff} and $\partial\ln\chi/\partial\ln T$ make the appearence of radiative instabilities of the types discussed above possible. The possibility that our nearest star is in such a region is fascinating. In particular the great thermal sensitivity of CO molecules near the temperature minimum region makes them an excellent candidate for driving radiative instabilities (Kneer, 1983). The purpose of this investigation is not to give concrete answers to some of these particular questions. Its real value does not, however, simply lie in having provided physical insight on the radiative relaxation and amplification processes. On the contrary its usefulness would be demonstrated should it help to understand physically the results od "realistic" numerical simulations.

ACKNOWLEDGMENTS

J. T. B. would like to thank Drs. E. Landi Degl'Innocenti and A. Righini for having invited him to visit the Osservatorio Astrofisico di Arcetri prior to this workshop. He is particularly grateful to Dr. E. Landi Degl'Innocenti for valuable discussions on the field of Radiative Transfer and for his kind hospitality. This work has been supported by the Deutsche Forschungsgemeinschaft through grant Kn 152/3.

REFERENCES

Abramowitz, M., Stegun, I.: 1965, Handbook of Mathematical Functions,
    Dover, New York.
Ando, H., Osaki, Y.: 1975, Publ. Astron. Soc. Japan 27, 95.
Chandrasekhar, S.: 1960, Radiative Transfer, Dover, New York.
Kneer, F.: 1983, Astron. Astrophys. 128, 311.
Kneer, F., Heasley, J.N.: 1979, Astron. Astrophys. 79, 14.
Kneer, F., Trujillo-Bueno, J.: 1987, Astron. Astrophys. 183, 91.
Rybicki, G. B.: 1965, Ph. D. Thesis, Harvard University.
Spiegel, E. A.: 1957, Astrophys. J., 126, 202.
Trujillo-Bueno, J.: 1988, Ph. D. Thesis, University of Göttingen.
Trujillo-Bueno, J., Kneer, F.: 1987, Astron. Astrophys., 174, 183.
Trujillo-Bueno, J., Kneer, F.: 1987, in The Role of Fine-Scale Magnetic
    Fields on the Structure of the Solar Atmosphere, E.H. Schröter,
    M. Vazquez, A.A. Wyller eds., Cambridge University Press, p. 281.
Unno, W., Spiegel, E. A.: 1966, Publ. Astron. Soc. Japan 18, 85.

Discussion

RUTTEN — I would like to comment that the situation for lines is yet more complex, because the interlocking effects *across* the spectrum can be quite large. For example, the Fe I opacities are all set in the ultraviolet due to overionization. And also the source functions of many Fe II lines are set in the ultraviolet due to interlocking effects. So, in both the Fe I opacities and the Fe II source functions, you can have ultraviolet Planck function sensitivity to temperature variations across the whole spectrum[1].

TRUJILLO BUENO — We agree completely with you that nature is much more complex than what our simple line-transfer calculations imply. Our aim has been twofold. Firstly, to point out that the radiative transfer effects of line opacity fluctuations may be highly destabilizing for some relevant spectral lines (e.g. the infrared bands of CO; see Kneer 1983[2]).

Secondly, to show that (already at structural lengths of only four scale heights) the contribution of a strong spectral line to the local energy balance may deviate from the plane-parallel case by an order of magnitude, and that the height in the atmosphere where the line's energetic role changes sign may be shifted several scale heights. Therefore, the consideration of multi-dimensional radiative transfer is of crucial importance when aiming at a reliable investigation of small-scale atmospheric processes. Details concerning our line-transfer calculations will be published elsewhere.

FOX — Can this method be used to examine the fluctuations in intensity over granules, which have an order of magnitude variation in opacity?

TRUJILLO BUENO — In this work we have been interested in the *linear* response of the radiation field to opacity fluctuations. For this reason we have solved the linearized radiative transfer equation (cf. Eqn. 2 above). This has allowed us to perform a local stability analysis.

The response of the radiation field to typical solar temperature inhomogeneities is non-linear. Some illustrative calculations showing this, taking radiative transfer effects fully into account, can be found in Trujillo Bueno (1988)[3]. With the same numerical method used in such non-linear calculations, one can also examine the fluctuations in intensity over granules.

---

[1]R.J. Rutten 1988, in: R. Viotti, A. Vittone and M. Friedjung (Ed.), *Physics of Formation of Fe II Lines Outside LTE*, Proceedings IAU Colloquium 94, Reidel, Dordrecht, p. 185.

[2]F. Kneer, 1983, *Astron. Astrophys* **128**, 311.

[3]J. Trujillo Bueno, 1988, *Multi- Dimensional Energy Transfer by Radiation in the Solar Atmosphere*, PhD thesis, Göttingen.

SIMULATING MAGNETOCONVECTION

Å. Nordlund[1] and R.F. Stein[2]

[1] *JILA, University of Colorado*
[2] *Dept. of Physics and Astronomy, Michigan State University*

1 Introduction

In this talk we discuss how to use computer simulations of magnetoconvection to test scenarios for sunspots and small scale magnetic field concentrations in the solar photosphere. As we see it, the main purpose of running solar mhd simulations is to obtain a qualitative understanding of solar mhd-processes; i.e., to support or refute pet ideas about how the Sun works, much as observations can and should be used for the same purpose.

Personally, we are biased towards a belief that the three-dimensional aspects of solar phenomena are absolutely essential: with three dimensions to play with, the Sun is capable of putting on an endless show of intricate dynamics – non-equilibrium, non-stationary, chaotic, even explosive. Anything beyond the most elementary explanations similarly requires three-dimensional scenarios. This was part of the rationale in constructing a computer code for three-dimensional compressible magneto-hydrodynamics, to use as a "test-bed" for conceptual scenarios. In this paper, we present the mhd aspects of this code (cf. also Stein et al., this volume). We present some rudimentary scenarios, discuss to what extent these may be tested with numerical simulations, and provide some examples of results from preliminary experiments we have performed with the code.

This is not a review paper, and we have not attempted to include extensive references to previous work on the subject. Rather, we offer here a mixture of ideas and scenarios, old and new, our own and others, with little or no attempt to assign credits or claim originality. These are presented as a basis for discussion at this meeting, as well as for testing against reality (be it observed or numerically simulated).

In Section 2 we introduce the equations in the form they are treated in the code, in Section 3 we discuss aspects of the code particular to magnetohydrodynamics, and in Section 4 we discuss some scenarios which we expect to be testable with this code now or in the forseeable future. In Section 5 we present some preliminary results.

2 Equations

As discussed elsewhere in this volume (Stein et al.), we write the equations of magneto-hydrodynamics (with radiation) in a form which is particularly well suited for a strongly

R. J. Rutten and G. Severino (eds.), Solar and Stellar Granulation, 453–470.

stratified medium, where the pressure and density may vary over many orders of magnitude. In this form, the equations describe the time rate of change of the logarithm of the density $\ln\rho$, the velocity $\mathbf{u}$, the internal energy per unit mass e, and the magnetic flux density $\mathbf{B}$. We write the continuity equation as

$$\frac{\partial \ln\rho}{\partial t} = -\mathbf{u}\cdot\nabla\ln\rho - \nabla\cdot\mathbf{u} \;, \tag{1}$$

the equation of motion as

$$\frac{\partial\mathbf{u}}{\partial t} = -\mathbf{u}\cdot\nabla\mathbf{u} + \mathbf{g} - \frac{P}{\rho}\nabla\ln P + \frac{1}{\rho}\mathbf{J}\times\mathbf{B} \;, \tag{2}$$

the energy equation as

$$\frac{\partial e}{\partial t} = -\mathbf{u}\cdot\nabla e - \frac{P}{\rho}\nabla\cdot\mathbf{u} + Q_{rad} + Q_{Joule} + Q_{visc} \;, \tag{3}$$

and the induction equation as

$$\frac{\partial\mathbf{B}}{\partial t} = -\nabla\times\mathbf{E} \;. \tag{4}$$

The electric current is

$$\mathbf{J} = \frac{1}{\mu_0}\nabla\times\mathbf{B} \;, \tag{5}$$

the electric field is

$$\mathbf{E} = -\mathbf{u}\times\mathbf{B} + \eta\mathbf{J} \;, \tag{6}$$

and the Joule dissipation is

$$Q_{Joule} = \eta J^2 \;. \tag{7}$$

The equations describing the radiative energy transport are given in the accompanying paper in this volume.

3   Code

As is the case with the pure hydrodynamics problem, one of the main difficulties with applying these equations to a specific simulation problem is the treatment of the "virtual boundaries"; i.e., boundaries of the computational domain which do not correspond to real boundaries, but just delimit the volume we chose (or can afford) to simulate. We treat the density, velocity, and energy at the lower boundary in the same way as in the pure hydrodynamic case, except now we keep the *total* pressure (magnetic pressure plus gas pressure) constant at the lower boundary. We adopt $\partial\mathbf{B}/\partial z = 0$ as the lower boundary condition for the magnetic flux density B.

At the upper boundary, we again adopt the same boundary conditions as in the pure hydrodynamic case for the density, velocity and internal energy. For the magnetic flux density, we assume that the field is potential at the upper boundary. Actually, we sometimes allow the region where the field is potential to extend over several depth points near the boundary. In this potential region, we force the velocity to be exactly parallel to the magnetic field. This approach is useful when the model develops a region near the upper boundary where the density is very low, and hence the Alfvén speed is very high, and the true field is very nearly potential. In such cases, the requirement that information must not be allowed to propagate over more than a fraction of a grid cell per timestep (the CFL condition) would force a very small time step, although the phenomena of interest (in deeper layers of the model) actually happen on a much longer time scale. Adopting a potential field in such cases effectively excludes Alfven waves from the potential region, and thus allow time steps to be tuned to the (slower) subsurface phenomena.

Special consideration has to be given to the condition

$$\nabla \cdot \mathbf{B} = 0 \tag{8}$$

which is in principle automatically ensured by the induction equation, since $div(curl()) \equiv 0$. In practice, however, the numerical operators do not necessarily commute (this is the property that ensures that $div(curl()) \equiv 0$). To be specific, let us look briefly at the induction equation. If we could evaluate the induction equation as it stands, we would indeed find that the numerical div and $curl$ operators could be made to commute exactly (special consideration has to be given to the boundaries). However, the central differences used for the div and $curl$ operators allow a non-propagating (non-physical) sawtooth mode to exist, and in order to damp this mode, we must evaluate the induction equation with an *explicit* resistive term:

$$\frac{\partial \mathbf{B}}{\partial t} = -\nabla \times (-\mathbf{u} \times \mathbf{B}) + \frac{\eta}{\mu_0} \nabla^2 \mathbf{B} - \nabla \eta \times \mathbf{J}. \tag{9}$$

where the ∇^2 term is evaluated as a numerical second derivative directly from the spline representation. This term is then effective in damping a sawtooth type wave. Mathematically, the divergence of the sum of this term and the last term is zero (since they are $div(curl(\eta \mathbf{J}))$, but numerically it is not. In order to keep $div(\mathbf{B}) = 0$, we must therefore explicitly enforce the divergence of the sum of the two last terms in the above induction equation to be zero. We do this by invoking a numerical procedure which modifies slightly the vertical component of a given vector with approximately zero divergence, in such a way as to make the *numerical* divergence of the vector identically equal to zero. The procedure evaluates the contribution to the divergence from the horizontal components of the vector, and then uses the inverse of the numerical vertical derivative operator to modify the vertical component of the vector.

The resistive terms in the induction equation are of a diffusive type, and the time step has to be checked against the appropriate magnetic diffusion limit, as well as against the viscous and radiative diffusion limits, and the advective time step limit. We do this by evaluating each of these four time step limits in the routines that calculate the respective diffusion terms, and then check against all of them immediately *before* the time stepping routine. The time stepping is presently done with the second-order Adams-Bashforth algorithm, allowing for varying time steps.

4 Scenarios

In this section we discuss scenarios for the various aspects of magnetic fields as they appear at the surface of the Sun: the magnetic network, plages, sunspot umbrae and sunspot penumbrae. We stress again that we do not expect all of these scenarios necessarily to be correct, but present them as a basis for discussion at this meeting, and as a provocation for numerical and observational work. Most of these scenarios are testable by direct numerical simulation with the code we presented in the previous section. In the next section we discuss in more detail how this may be done, and also present some preliminary results of such simulations. A firm understanding of the properties of magnetic fields at the surface of the Sun is also likely to inspire new thoughts about the global properties of magnetic fields on the Sun: the activity cycle and the dynamo problem.

4.1 Network and Plages

Plages are observed to have rather well defined boundaries (of the order of $50 - 100$ G, cf. Schrijver 1987). In a recent paper, Schrijver (1988) has suggested a simple mechanism, based on the topology of the paths available for horizontal flux transport, which explains how a rather sharp boundary between a plage and the surrounding network may be maintained. The mechanism depends on the much smaller scale of the magnetic network inside plages; i.e., on the absence of large supergranular openings in the main body of a plage. This allows flux elements traveling along network boundaries more paths of access to the plage area than paths of exits, which produces an effect similar to that of a semi-permeable membrane.

This explanation shifts the attention to the question why the magnetic network size is smaller in plage areas. There are at least two possible explanations: Either the large-scale velocity field is suppressed by the presence of a sufficiently large average magnetic field, or the large-scale velocity field is still there, but the motion of the magnetic flux elements is suppressed when they become sufficiently packed, as in a plage. Also, the presence of a larger number of "trace particles" may make a smaller scale structure in the velocity field more visible (cf. Simon et al. 1988).

The supergranular velocities are similar in the quiet sun and in the enhanced network (Wang 1988). The observations of Simon et al. (1988) show that the motion of magnetic flux elements agrees well with the horizontal motion of granulation cells, as determined by local autocorrelation tracking. These two observations indicate that, for the average flux densities of the normal and enhanced network, the magnetic flux elements (at least in the cell interiors) are free to move with the fluid, and hence do not exert any significant drag on the large-scale flow.

At a certain average flux density (similar to that in network boundaries ?), the magnetic field of adjacent flux elements starts to merge above the surface. This stops further concentration of the flux elements, but presumably does not significantly influence the large-scale flow at and below the surface, since the topology of the field at and below the surface is still disconnected, which makes it possible for the flow to slip easily past the vertical "trunks" of magnetic field.

However, at a sufficiently high average flux density (more than that of the normal and enhanced network), the topology of the magnetic field changes *qualitatively*: For average field strengths of 100 G or more, the magnetic field fills essentially all available intergranular lanes at the surface and hence becomes topologically connected in the horizontal plane, although at some depth below the surface the field is still concentrated into "clumps" and topologically disconnected (cf. Nordlund 1986, Figs. 8-9). For such flux densities, the large scale surface flows must be efficiently suppressed, and indeed, a strong suppression of the horizontal flows (as measured by local autocorrelation tracking of granules) has been observed by Title *et al.* (1988) for average flux densities above some 50 G.

4.2 Pores

Pores form when enough magnetic flux has accumulated in a limited area. Presumably, the accumulation of flux is controlled by large-scale flows, though here we are basically concerned with what happens to the magnetic field in an area with a given magnetic flux density; how the presence of a fairly large flux density influences the convection, and how the topology of the magnetic field changes with time.

As discussed in the previous subsection, the merging magnetic field lines above the surface provide a repelling force which tries to distribute the magnetic flux evenly. The pressure confinement at the surface level, made possible by the strong surface cooling combined with the suppression of the horizontal advection of heat, constrains the location of the flux at the surface level to the boundaries between convection "cells".

It is important to remember that the "cells" are not closed convection cells. Rather they consist of strongly divergent ascending flows, surrounded by descending flows which form topologically connected "lanes" at the surface, but which turn into concentrated, filamentary downdrafts below the surface (Nordlund & Stein 1988a, 1988b). Thus, only at the surface there is a topologically connected "network" of favorable sites for the magnetic flux; below the surface the larger scale flows tend to push the magnetic field towards the sites of the filamentary downdrafts. Thus we might expect a slow evolution of the topology of the magnetic field, on the timescale of meso-granulation (tens of minutes and up), where the magnetic field is systematically concentrated at preferred locations which, if we could look below the surface, coincide with convergence points for the larger-scale subsurface flows.

We may expect that the larger the flux concentration, the more energetically isolated the interior will be; radiative heat exchange with the field free surroundings is reduced, and so is the importance of diffusive processes at the interface between the flux concentration and its surroundings. Thus the (Wilson) depression of the surface of optical depth unity, which is caused by the temperature deficiency in the interior of the flux concentration, should increase with increasing size of the flux concentration. Likewise, the temperature at the surface of optical depth unity – and hence the surface brightness – should decrease with increasing size of the flux concentration.

In summary, if the average magnetic flux density in an area is large enough (a few hundred Gauss), the magnetic field may concentrate in locations which are singled out by the topology of the large (meso-) scale convective flows, and there form flux concentrations large enough to appear as dark patches in surface brightness: pores.

4.3 Umbra

An umbra scenario should address at least two questions: What are the major energy transport mechanisms that supply energy to the umbral surface, and what is the topology of the magnetic field below the visible surface. Obviously, these two questions are interrelated in that the topology of the magnetic field is probably both a consequence of and a major factor influencing the subsurface energy transport.

Clearly the normal mode of energy supply to the solar surface, overturning convection, is strongly suppressed in the surface layer of a sunspot, but there is still a significant energy loss at the surface of optical depth unity (of the order of 20% of the normal solar flux in large spots, probably more in smaller spots and pores).

The generic umbra scenario involves some kind of modified convective energy transport below the surface, and radiative energy transport in a thin surface region (cf. Parker 1979ab). The main observational constraints are the small convective blue-shifts and vertical velocities in umbrae (Beckers 1977, 1980; Zwaan 1981), which implies that convective energy transport does not quite reach the umbral surface.

The instabilities of a stratified medium with a uniform vertical magnetic field have recently been investigated by Moreno-Insertis & Spruit (1988), who solve for adiabatic linear eigenmode structure and growth rates for given stratifications. As pointed out by them, the subsurface stratification, although in principle unknown, in practice is significantly constrained by the requirement that the entropy and pressure approach with depth the surrounding convection zone values. They find linearly unstable convective modes using stratifications which should be qualitatively realistic for sunspot umbrae. Little can be said about the structure of fully developed nonlinear convection from these results, but at least they establish that convection is likely to be present below the surface of umbrae.

Moreno-Insertis & Spruit also point out that a shortcoming of the generic umbra scenario is the fine tuning apparently necessary to make the convection end just below the visible surface. Recent results from simulations of stellar granulation (Nordlund & Dravins 1988, cf. also Dravins, this volume) may shed some light on this: For stars of varying effective temperature, the "granulation" (i.e. strong cellular temperature fluctuations associated with the convection) occurs at *nearly the same temperatures* in different stars. The reason is the strong dependency of opacity and internal energy on temperature in the temperature range where hydrogen is ionizing (actually where hydrogen is the main electron contributor). As a result, for stars with lower effective temperature than the Sun the granulation is essentially "hidden" below the surface; i.e., only rather small temperature and velocity fluctuations are visible at the surface, with resulting small blue-shifts of spectral lines (Dravins & Nordlund 1988).

A similar situation exists in umbrae; apart from the presence of a magnetic field, the umbral atmosphere is essentially the atmosphere of a star with appreciably lower effective temperature. Thus we would actually *expect* the large amplitude fluctuations to end just below the surface, with radiative energy transfer becoming sufficiently efficient to take over the energy transport. Some preliminary results confirming this picture are presented below (section 5.3).

4.4 Penumbra

A penumbra scenario should explain (1) why the brightness of the penumbra is larger than that of the umbra, but still smaller than that of the surrounding photosphere, (2) why the penumbra is separated from the umbra by a sharp boundary, (3) why the penumbra is separated from the photosphere by a (fairly) sharp boundary, (4) why the penumbra consists of bright and dark filaments, (5) why there is a predominantly inward motion of bright penumbral features, and (6) why there is an outward directed Evershed flow, while (7) the motions observed in $H\alpha$ are again predominantly inward. While these are many seemingly independent requirements, we believe there is a simple scenario which explains most, if not all, of these properties.

First, let us assume that the magnetic field in the penumbra is essentially parallel to the penumbra. While this may be at odds with some recent attempts to determine the inclination of the penumbral magnetic field from analysis of line profiles (Lites, 1988), it is hard to avoid the visual impression one gets from just looking at pictures and movies of sunspots that penumbral filaments outline magnetic field lines. Furthermore, let us assume that there is a fairly sharp boundary between the magnetic and non-magnetic regions of a penumbra. In fact, let us assume that there is essentially a magnetic region 'floating' on a non-magnetic region. The jump in magnetic field strength across the interface must of course be matched by a corresponding jump in gas pressure. The pressure at the optical depth unity surface in the non-magnetic region must be quite similar to that of the surrounding photosphere; i.e., increasing from some $10^4 \, Nm^{-2}$ at the edge of the penumbra, to some $10^5 \, Nm^{-2}$ at the edge of the umbra (since this is about 600 km further down). Such a gas pressure can support a magnetic field strength jump of some $1 \, kG$ at the edge of the penumbra, and some $3 \, kG$ at the edge of the penumbra, provided that the interior of the magnetic field has a much smaller gas pressure. With the assumed topology of the field this must indeed be the case, since the interior of the magnetic field is connected to the umbra, and there must be approximate pressure equilibrium along magnetic field lines.

The actual field strength in the penumbra is a factor of somewhat less than two lower; more like several hundred Gauss at the edge of the penumbra, to a couple of kG at the edge of the umbra. This implies that the field is floating some distance *above* the optical depth unity surface of the penumbra, at a height of somewhat less that a pressure scale height (a factor of 2 in pressure, say).

This topology is essential for the ensuing discussion of the energy balance of the penumbral surface layers. Without an essentially field-free surface layer it is difficult to explain the relatively large surface brightness of the penumbra, being just a few tens of percent lower than that of the normal photosphere.

From numerical simulations of granular convection (Nordlund 1982, 1985; Nordlund & Stein 1988a, 1988b) we know the qualitative properties of the surface layers where convective energy transport gives way to radiation: the ascending flow is essentially adiabatic right up to the surface, where radiation is released in a very thin layer. The reason for the layer being so thin is the enormous temperature sensitivity of the continuum opacity ($\kappa \sim T^n$, $n \approx 10$); once the ascending flow becomes transparent, it cools, becomes more transparent, and thus rapidly emits a large fraction of its stored internal (ionization) energy (which is several eV per atom). At the surface, the flow is deflected horizontally by the large pressure fluctuations associated with the large temperature fluctuations, and rapidly flows to intergranular lanes, where it descends, having an entropy several Boltzman units lower than the ascending flow. The descending entropy deficient flow mixes with overturning fluid that never made it to the surface, and thus the temperature fluctuations are gradually reduced.

In the surface layers of the penumbra, a similar cycle must exist: An ascending flow of at least 2 kms^{-1} must reach the surface of optical depth unity to offset the radiative losses there, and the flow must satisfy conservation of mass. Namely, there must exist a substantial horizontal flow in the surface layers, and the horizontal flow must be supported by pressure fluctuations which are the consequence of large temperature differences between the (practically adiabatic) ascending flow and the descending flow.

Alternative scenarios, where convection occurs in rolls parallel to the field (Danielson 1961), or in sloshing motions where the direction of motion is along the field lines (Schmidt et al. 1986, Spruit 1987) probably would not work, since such topologies make it very hard to achieve the mixing of entropy deficient material (which has been cooled at the surface) and hence close the cycle.

The presence of the 'floating' magnetic interface above the penumbra surface influences the flow in several ways: (1) Such an interface can easily be distorted in the direction perpendicular to the field lines. Perturbations parallel to the field lines are much harder to induce. Undulating distortions are probably induced by the pushing of the ascending flows against the interface. These distortions in turn 'lock' the flow pattern into a much more long lived pattern than the granular flow in the photosphere (where the competition between the expansion of individual granules is a major factor shortening the lifetimes of granules). (2) The inclination of the magnetic and $\tau = 1$ interfaces implies that the isotherms near the surface are inclined almost paralell to the these interfaces. This creates a systematic horizontal pressure gradient towards the umbra in the surface layers, which deflects the ascending flows. (3) The ascending flows are also deflected by the pressure difference between the hot fluid in the bright filament and the cool fluid in the dark filament. Thus, a 'corrugated' surface is at the same time induced by and strongly influencing the ascending flows, as they are deflected into horizontal flows at the surface.

A quick look at the list of properties at the beginning of this subsection shows that such a scenario can easily explain properties 1-5: The energy flux in the penumbra is larger than in the umbra since convection reaches all the way to the surface (1), but is smaller than in the photosphere because the horizontal flow is constrained by the corrugated field pattern. The umbral/penumbral boundary marks the place where the interface descends through the surface of optical depth unity (3), and the penumbral boundary is where the inclined $\tau = 1$ surface reaches photospheric levels (4). The predominantly inward motion of bright penumbral features is presumably a result of the inclination of the penumbra; the strong cooling at an inclined surface produces systematic horizontal pressure gradients away from the surrounding photosphere and towards the interface; i.e. directed radially inwards (5).

The opposite directions of the Evershed flow (6) and the chromospheric inflow (7) are probably related to systematic pressure differences (leading to siphon flows of opposite sign) along the field lines that are very close to the interface and those that are far from it. Such systematic pressure differences may be a natural consequence of the random way spot field lines connect to surrounding plage flux elements. Any given umbral or penumbral fieldline is more likely to connect to the *interior* of a magnetic flux concentration in the surrounding plage, since there are 'many more' fieldlines in the interior than close to the boundaries of magnetic flux concentrations. If we compare the two ends of a magnetic field line that connects the penumbra to a plage flux element, the total pressure should be similar at a level just below the surrounding photosphere, but the magnetic field strength at the penumbral end is less than that in the plage flux element, so the gas pressure is greater at the penumbral end of the field line, and drives an outward siphon flow. Thus, tentatively, we would place the Evershed flow just above the magnetic interface, which is also at just the right height above the continuum layers (a few hundred kilometers) for it to be observed in typical photospheric lines. On the other hand, field lines that connect the interior of the spot umbra to the surrounding plage (and thus reach high into the chromosphere), have much lower gas pressures at the spot ends, and a siphon flow *from* the plage side towards the spot would be expected.

5 Numerical Experiments

In this section we discuss experiments that may be performed with the mhd code, to test the scenarios discussed in the previous section. For two of these scenarios (the pore and the umbra), we also have some preliminary results.

5.1 Network / Plages

The numerical experiments to be performed to test the network/plage scenario are relatively straightforward, although somewhat expensive. We would essentially like to run mhd simulations with a large enough model to have some meso-scale convection present (cf. Nordlund & Stein 1988ab). A number of simulations with varying average flux density should be run, to study how effective the quenching of the larger scale convective motion by the magnetic field is, and to what extent the large scale motion is able to push the magnetic field around. The reasons why such experiments are relatively expensive are the following: (1) Even pure hydrodynamic experiments with volumes large enough to develop meso-scale convection are expensive, at least if one wants to model convection on the granular scale accurately as well. (2) The mhd experiments *must* be performed with high spatial resolution, especially for the cases with relatively small average flux densities, where the magnetic filling factor is small, and consequently the sizes of the magnetic flux concentrations are small. (3) The (mass) densities in the evacuated interiors of the magnetic flux concentrations are small, while the magnetic flux densities are large; hence the Alfvén speed becomes large, which reduces the maximum allowed time step.

We estimate that each such experiment would require of the order of a thousand Cray-2 CPU hours, which means that a series of experiments would only be practical on a machine which is considerably faster than a Cray-2 (e.g., a Connection Machine).

5.2 Pores

A pore experiment is similar to the network/plage experiment discussed above, but for a couple of reasons it is considerably cheaper: (1) The purpose of such an experiment is to investigate how a small pore is formed when the average magnetic flux density is high enough. Thus only one or a few experiments are needed. (2) The numerical resolution

462

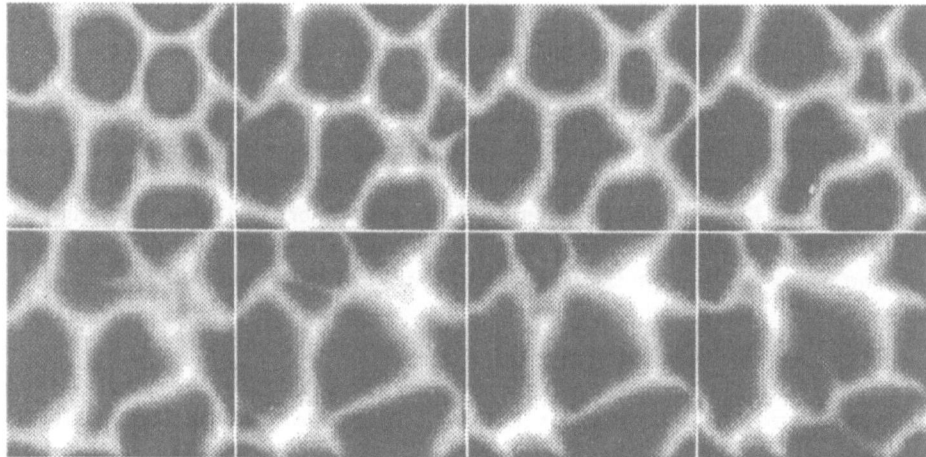

Figure 1: The magnetic flux density at the surface as a function of time. White corresponds to a field strength above 1500 G. The time interval between each panel is 4 solar minutes. The horizontal size is $3Mm$.

does not have to be as large as for the plage experiment, since the magnetic filling factor is larger and it is therefore easier to numerically resolve the magnetic flux concentrations.

In fact, we have already begun a such a numerical experiment. This is a case with an average flux density of 500 G, a horizontal box size of 3 Mm, a vertical extent of 1.5 Mm, and a numerical resolution of $63 \times 63 \times 63$.

So far, we have accumulated about 40 minutes of solar time, starting from a homogeneous vertical magnetic field superimposed on a snapshot from a simulation of granular convection without magnetic fields.

The time evolution of the magnetic field at the surface is illustrated in Fig. 1, which shows the magnetic flux density at the surface as a function of time. Initially (before the first panel in the figure), the magnetic field is rapidly swept to the boundaries of the granulation cells, where thin sheets are formed. These sheets isolate the flows in the granulation cells from each other, and stabilizes the flow in each cell. The cells become more regular (rounder) than normal granulation cells, and their lifetimes are increased considerably.

Over a time scale of several tens of minutes, the magnetic flux slowly creeps along the cell boundaries to concentrate further at a few locations. At the same time, some of the convection cells shrink into obliteration, thus increasing the size of the remaining granulation cells and contributing to the separation of magnetic flux and convective flow.

When the remaining granulation cells have become large enough, the familiar tendency for splitting of large granulation cells becomes operational. When a cell splits, the magnetic field "creeps" into the newly formed intergranular lane, since this is a region

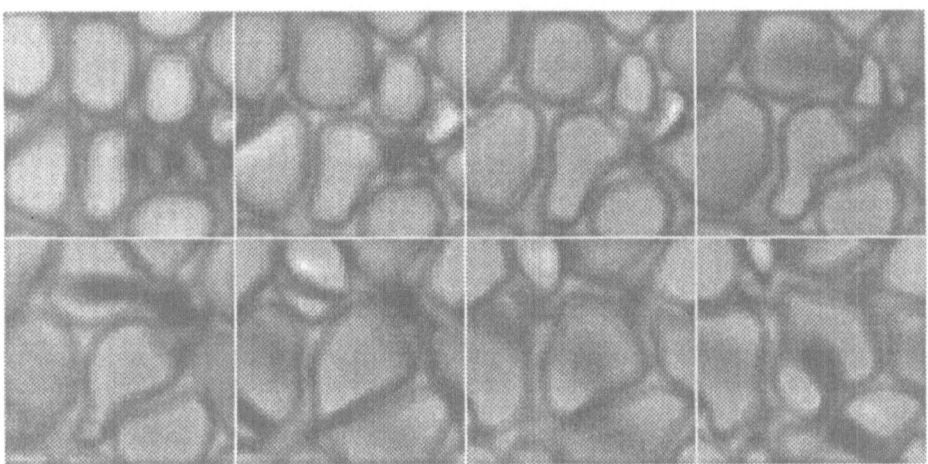

Figure 2: The total radiation intensity at the surface as a function of time, for the same time interval as in Fig. 1.

with relatively low gas pressure.

The locations of the magnetic field are clearly visible in a continuum intensity picture (Fig. 2), as intensity enhancements within the intergranular lanes. Also, the shapes of the granules are more regular than in a corresponding sequence of granulation without magnetic fields (Fig. 3).

The experiment has not reached a steady state yet; magnetic flux is still increasing in two rather well defined locations. However, the experiment has already provided support for the scenario discussed in the previous section. The tendency of the magnetic flux to concentrate in a few locations is clear, and has marginally led to the formation of dark features in the continuum pictures.

The concentration of the magnetic flux into pores is energetically favorable in that the cooling of the gas in the interior of large flux concentrations leads to a lowering of surfaces of constant mass column density, and hence to a lowering of the gravitational energy.

5.3 Umbra

We have also started a simple umbra experiment. This is an experiment with a model size of $3 \times 3 \times 3$ Mm, with a numerical resolution of $31 \times 31 \times 63$, and with an average vertical magnetic flux density of 2 kG. The initial condition was again a snapshot from a granulation simulation, with a superimposed homogeneous 2 kG vertical magnetic field. We expect to see a fairly smooth magnetic field (at least until the model starts to approach a steady state), and we are therefore satisfied with a lower horizontal resolution

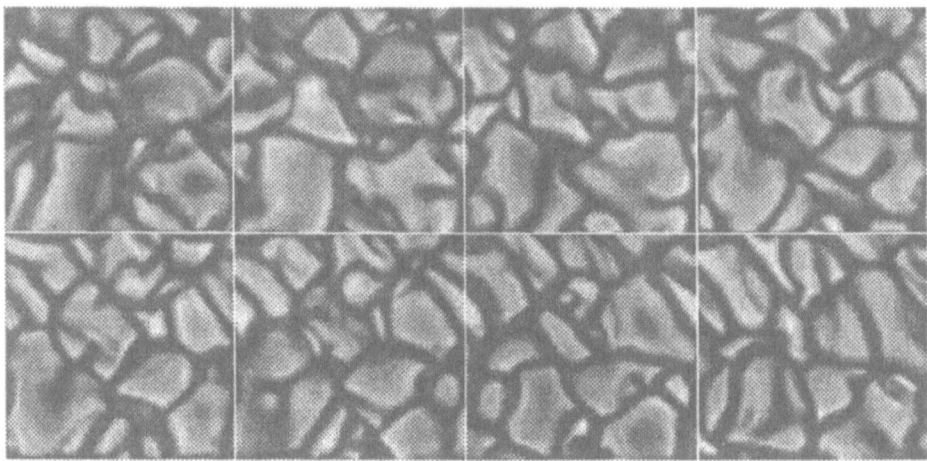

Figure 3: The total radiation intensity at the surface as a function of time, for normal granulation. Note that the horizontal size is $6Mm$.

than in the pore experiment. We prefer to keep a large number of vertical grid points to allow an accurate solution of the radiative transfer equation, and to allow room for the surface to descend.

The initial evolution of this experiment was – as expected – a strong suppression of the convection, followed by a rapid cooling of the surface. Fig. 4 shows the evolution of the average surface intensity with time. Within 10 minutes, the surface intensity has dropped to less than half the normal photospheric surface intensity. For reasons to be discussed below, the rate of decrease of the surface intensity falls off with time, and we expect the time needed to reach a steady state to be much longer than the time covered so far in the experiment.

Due to the cooling of the surface layers, the surface of optical depth unity recedes. Also, since the lower temperature of the surface layers implies a smaller pressure scale height, the pressure in the surface layers drops. Since the pressure is approximately equal to the mass column density times the acceleration of gravity (approximate pressure equilibrium), surfaces of constant mass column density must be receding as well. Fig. 5 shows the evolution of the average temperature as a function of time. The initial atmosphere was oscillating somewhat and the remaining vertical velocity is large enough to cause a slight 'bounce' at $t \approx 10\ min$, where the descent is temporarily halted.

The convective flux in the subsurface layers is very small during this entire period. As a consequence, the surface radiation varies only a few percent horizontally. The energy loss by radiation at the surface comes essentially from the internal energy of the cooling umbral gas. Since the density increases exponentially with depth, the cooling rate decreases rapidly with time. Also, since the surface intensity drops with time, the

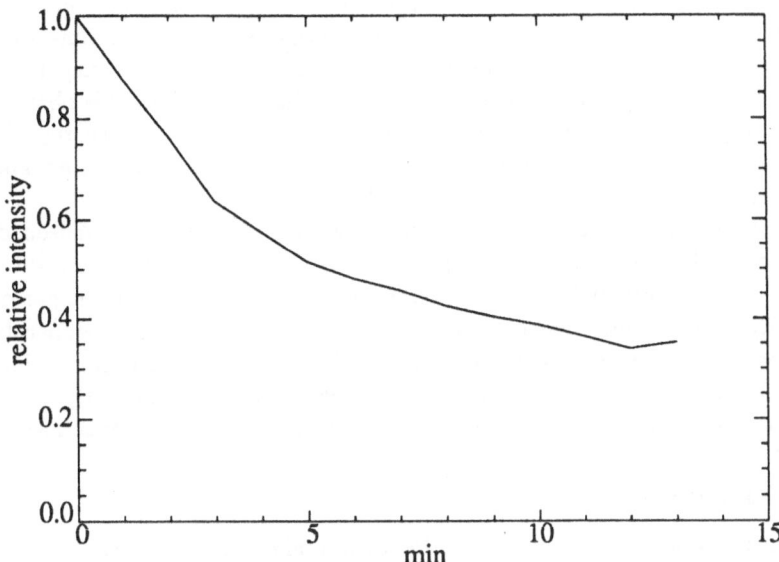

Figure 4: The average surface intensity as a function of time, relative to the average photospheric value.

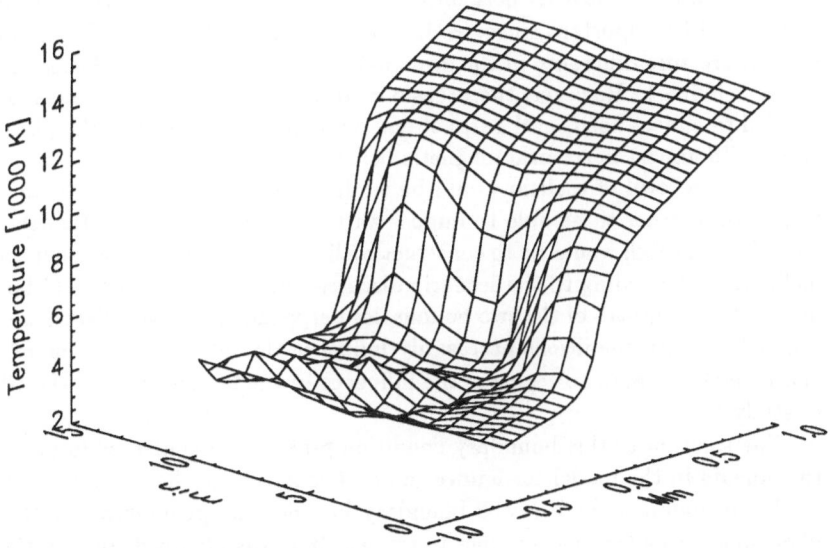

Figure 5: The average temperature as a function of time and depth. The total depth range shown in the figure is from $-0.5\ Mm$ to $+1.0\ Mm$, and the total time interval is 13 minutes.

radiative losses at the surface drop with time, which also decreases the rate of cooling.

Thus, a sunspot can sustain its surface brightness for some time with no energy transport to the surface. In this phase of the evolution of the sunspot, no subsurface convective energy transport is necessary. To estimate how long this phase may last, we adopt a 'canonical' Wilson depression of 600 km, estimate the energy content in the normal photosphere above that level, and divide by a quarter of the nominal solar flux. We get a time of the order of one to two hours.

Actually, we suspect that it is the other way around: The Wilson depression is determined by how far down a sunspot can sink before convection is turned back on again. The umbra eventually becomes convectively unstable again, when the gas pressure (and density) has become sufficiently large, and when the subsurface temperature gradient has become sufficiently steep, as discussed in the umbra scenario subsection above.

We are eagerly waiting to see if convection develops in sufficiently deep layers to go undetected as, for instance, convective blue-shifts of umbral spectral lines.

5.4 Penumbra

The penumbra scenario might be tested at several levels of approximation. The simplest test would be to impose a "floating" horizontal magnetic field on top of a model of the normal photosphere, and to study the behavior of the overturning convection under the horizontal interface. We would like to confirm the idea that the presence of a horizontal magnetic field at the top of the granulation layer changes the granules into structures elongated along the magnetic field.

One could argue that such an experiment might fail, noting that the magnetic field in the penumbra is not exactly horizontal and that the inclination might be significant. This may indeed be important since, as discussed in section 4.4, the inclination causes a systematic pressure difference in the direction of the inclination (and hence in the direction of the fieldlines). However, the occasional occurrence of elongated granules apparently parallel to almost horizontal parts of emerging flux indicates that the phenomenon may occur with almost horizontal magnetic fields.

The next level of realism would be to drop the periodic boundary condition in one direction, in order to be able to impose an inclined floating magnetic field. A difficulty with this approach would be to construct realistic boundary conditions in the direction of inclination. According to the scenario discussed previously, there would be a systematic inflow at the "upper" of the two boundaries perpendicular to the direction of inclination. Inflow boundary conditions necessarily influence the interior of the model significantly, and it might be hard to separate the influence of the boundary from the effects we wish to study.

The solution to this boundary condition problem, of course, is to include enough of the domain in the model to remove the ad hoc nature of the boundary condition. This implies including enough of the boundary between the penumbra and the surrounding photosphere to allow the fieldlines to ascend above the photosphere, and then come back down into a small fluxtube outside the penumbra. Such a model would be expensive, since the size along the field direction ideally should be several Mm. However, the size perpendicular to the magnetic field might not have to be that large, and thus the total number of grid points in the model might be acceptable. In this model, the Evershed flow aspect of the penumbra scenario could also be tested.

A further improvement on the model would be to change the shape of the model from a rectangular box to a "pie" shaped wedge. This could be done fairly simply, by changing the coordinate system to a cylindrical one, with periodic boundary conditions in the azimuthal direction. In the code, only the spatial derivative routines would have to be changed.

6 Concluding Remarks

In this paper, we have discussed a number of scenarios for the behavior of magnetic fields in the surface layers of the Sun. We have also discussed how numerical three-dimensional mhd modeling may be used to test these scenarios. We have not discussed how observations might be used for a similar purpose. The new and exciting developments of high resolution observations (Scharmer, 1987) and data analysis (Title et al. 1988, Simon et al. 1988, November & Simon 1988) indeed promise to provide new and important observational constraints, and to inspire greatly the development of conceptual scenarios for the interaction of magnetic fields and convection in the solar photosphere as well.

Acknowledgements

Å.N. gratefully acknowledges the hospitality of the faculty and staff at JILA during a one year Visiting Fellowship period. He also acknowledges support from the Danish Natural Science Research Council and the Carlsberg Foundation. R.F.S. thanks the National Science Foundation for partial support for this work under grant AST-83-16231. They both would like to acknowledge stimulating discussion with Fausto Cattaneo, Neal Hurlburt, Bruce Lites, Göran Scharmer, Jan-Olof Stenflo, Alan Title, and Juri Toomre. The penumbral scenario grew out of intense discussion between Å.N. and Henk Spruit, who continues to provide inspiring input every time they meet.

References

Beckers, J.M. 1977, *Ap. J.* **213**, 900

Beckers, J.M. 1980, in S.D. Jordan, ed., *The Sun as a Star*, p. 11

Danielson, R.R.: 1961, *Ap. J.* **134**, 275

Dravins, D., Nordlund, Å. 1988, (in preparation for *Astron. Astrophys.*)

Moreno-Insertis, F., Spruit, H.C. 1988, (submitted to Ap.J.)

Lites, B.W.: 1988, private communication

Nordlund, Å.: 1986, in "Small Scale Magnetic Flux Concentrations in the Solar Photosphere", eds. W. Deinzer, M. Knölker, H. Voigt, *Abh. der Akad. der Wissensch. in Göttingen*, **38**, 83

Nordlund, Å., Dravins, D.: 1988, (in preparation for *Astron. Astrophys.*)

Nordlund, Å., Stein, R.F.: 1988a, *Bull. Am. Astron. Soc.*, **Vol. 20**, No. 2, 702

Nordlund, Å., Stein, R.F.: 1988b, (in preparation for *Ap. J. Letters*),

November, L.J., Simon, G.W.: 1988, (submitted to *Ap. J.*)

468

Parker, E.N.: 1979a, *Ap. J.* **230**, 905

Parker, E.N.: 1979b, *Ap. J.* **234**, 333

Scharmer, G.B.: 1987, in *The Role of Fine-Scale Magnetic Fields on the Structure of the Solar Atmosphere*, ed. E. Scröter, M. Vazquez, and A. Wyller (Cambridge: Cambridge Univ. Press), p. 349

Schrijver, C.J.: 1987, *Astron. Astrophys.*, **180**, 241

Schrijver, C.J.: 1988, (to be submitted to *Solar Phys.*)

Schmidt, H.U., Spruit, H.C., Weiss, N.O.: 1986, *Astron. Astrophys.* **158**, 351

Simon, G.W., Title, A.M., Topka, K.P., Tarbell, T.D., Shine, R.A., Ferguson, S.H., Zirin, H., and the SOUP team: 1988, *Ap. J.*, **327**, 964

Spruit, H.C.: 1987, in *The Role of Fine-Scale Magnetic Fields on the Structure of the Solar Atmosphere*, ed. E. Scröter, M. Vazquez, and A. Wyller (Cambridge: Cambridge Univ. Press).

Stein, R.F., Nordlund, Å., and Kuhn, J., (this volume)

Title, A.M., Tarbell, T.D., Topka, K.P., Ferguson, S.H., Shine, R.A., and the SOUP team: 1988, "Statistical properties of the solar granulation derived from the SOUP instrument on Spacelab 2" (submitted to *Ap. J.*)

Wang, H.: 1988, BBSO #0286, submitted to *Solar Phys.*

Zwaan, C.: 1981, *Space Sci. Rev.*, **28**, 385

Discussion

TITLE — How do you account for the radial outflow of granules within the first 2-5 arcsec around spots that was observed with SOUP?

NORDLUND — This is presumably caused by the slight temperature excess under the penumbra and umbra, which itself is caused by the reduced surface cooling there.

FOX — Given the size and the time scales of actual sunspot umbrae and the reliance on granular convection suppression to block the energy, what differences in the evolution do you expect in this model case?
Does your discussion reaffirm the question: why is the umbra so bright, as distinct from why is it so dark?

NORDLUND — It does indeed, see above. The heat storage is significant in the spot evolution, but it is necessary to have convective heat transport up to a height just below the surface, in order to support the heat flux of longlived sunspot umbrae.

WEISS — Your scenarios describe a general consensus. I would differ from your assumption that the magnetic field floats above $\tau = 1$ in the penumbra; our model[1] required the magnetic boundary to lie about 100 km below the $\tau = 1$ level in dark filaments. The real difficulty in modelling penumbral structure lies at the inner boundary, were currents flow and the Lorentz force is significant: it is difficult to choose an appropiate boundary condition there. Finally, we should encourage the observers to use their new telescopes to measure magnetic fields in the penumbra.

NORDLUND — As I mentioned in my talk, the difficulty with having the magnetic field boundary below the surface is that it then becomes very hard topologically to arrange a sufficiently efficient energy sunspot to the surface.

KOUTCHMY — Why did you not discuss a scenario or a model for a small magnetic element (flux tube, etc.) imbedded in the granulation?

NORDLUND — That is a topic large enough for a talk in itself!

SIMON — My impression of flow observations (granules, magnetic elements) around sunspots is that they (almost) always move radially outward, not inward, during most or all of the spot's lifetime. The models should be consistent with these observations.

NORDLUND — This is the flow observed *just outside* the penumbra, which is probably part of the large scale circulation driven by the reduced surface cooling in spots which I mentioned above when answering Title's question, whereas the systematic pressure gradient caused by the inclined penumbral surface deflects the *small-scale* penumbral convective flow towards the umbra.

MÜLLER — One of the most intriguing properties of sunspots is the sharp penumbra-umbra boundary: within one or two arcsec, there is a transition from normal granules to elongated penumbral grains. If the magnetic field in the penumbra is

---

[1]H.U. Schmidt, H.C. Spruit and N.O. Weiss, 1986, *Astron. Astrophys.* **158**, 351.

slightly inclined and extends well away from the boundary, we should observe a larger and larger penumbra in the upper photospheric layers. This is not the case, however, so that the geometrical shape of the penumbra appears to be a vertical cilinder.

NORDLUND — In the scenario we suggest, normal granulation suddenly becomes possible at the point where the "floating" inclined magnetic interface reaches the level of the surrounding photosphere, or to be more precise, where the $\tau = 1$ surface (which is just below the magnetic interface) reaches the normal photospheric level. This is a precise boundary, given by the geometry of the magnetic field in the spot.

VAN BALLEGOOIJEN — To understand umbral and penumbral structure, it would be of interest to study the effect of the inclination of the magnetic field on the convection, not just the limiting cases of horizontal and vertical field.

NORDLUND — Yes, indeed, and as we discuss in the text, this may be done at various levels of refinement (and cost).

DAMÉ — In your simulations one can follow individual granules for more than 10 minutes. However, in the SOUP observations one cannot follow a single granule for more than 5 minutes. Is there an explanation for this difference?

NORDLUND — Some articifial increase of the lifetime may be caused by our finite (numerical) viscosity. However, this difference may also be a question of different definitions of the concept "lifetime".

MARTIČ — What would be the source of the umbral oscillations in your scenario? A resonant mode of fast magneto-atmospheric waves, trapped in the photosphere and subphotosphere and excited by overstable convection in the subphotosphere— or some other magneto-acoustic wave?

NORDLUND — I have not thought about umbral oscillations or penumbral waves, but I would guess that umbral oscillations are basically similar to photospheric p-modes, i.e. waves that are reflected at the surface because of the rapid drop of the density with height. The waves and the magnetic field are both vertical. These would be slow-mode type waves.

BECKERS — Do you have any idea what the lightbridges observed in sunspot umbrae are?

NORDLUND — Weak but systematic large-scale stresses that arise from large-scale flows or from large-scale magnetic field connectivity may pull a spot apart and allow "almost normal" granular convection on the fracture.

SIMON — The SOUP movies suggest that granules flow toward pores, but that they do not fall into them. Instead, the granules seem to bounce off an invisible (presumably magnetic) wall at the pore boundary, so that they are deflected aside rather than penetrate into the pores. Does your model confirm this observation?

NORDLUND — The pore is indeed an "obstacle", in the sense that flows do not penetrate into the magnetic field of the pore. Thus, conservation of mass constrains the flow patterns around the pore in the way you describe.

TIME DEPENDENT COMPRESSIBLE MAGNETOCONVECTION

N.O.Weiss
D.A.M.T.P
University of Cambridge
U.K.

ABSTRACT. Numerical experiments on two-dimensional compressible convection in the presence of a strong imposed vertical magnetic field reveal several different forms of time-dependent behaviour. As well as the familiar periodic oscillations (standing waves) there are travelling waves, vacillation following secondary instabilities, and streaming instabilities with reversal of the flow. These patterns of behaviour are relevant to sunspot umbrae and to intergranular magnetic fields.

1. Introduction

The availability of more powerful computers allows a greater range of investigations into astrophysical convection. On the one hand, it is possible to carry out ambitious simulations of three-dimensional, fully compressible convection at high Rayleigh numbers. On the other, it remains important to construct idealized model problems, designed to isolate different physical effects in the nonlinear regime, and to study them in detail by systematically varying the appropriate parameters. In this paper I shall describe some numerical experiments on two-dimensional fully compressible magnetoconvection, carried out by Derek Brownjohn, Neal Hurlburt, Michael Proctor and myself. We have found several unfamiliar patterns of time-dependent behaviour in strong magnetic fields. These results have applications to convection in sunspot umbrae and to magnetic fields in the lanes between granules or mesogranules. Full accounts will be submitted to the Journal of Fluid Mechanics and to Monthly Notices.

2. Travelling Waves and Standing Waves

In the Boussinesq limit convection in a plane layer of unit depth is characterized by the Rayleigh number R (proportional to the superadiabatic gradient) the Chandrasekhar number Q (proportional to the square of the imposed magnetic field strength) and by the ratio ς of the magnetic to the diffusivity. For $\varsigma < 1$ and Q sufficiently large convection sets in at an oscillatory bifurcation, giving rise to a branch of periodic solutions that typically terminates in a heteroclinic bifurcation on the unstable portion of the steady branch (Proctor & Weiss 1982; Hughe & Proctor 1988). With periodic lateral boundary conditions two solution branches, corresponding to travelling waves and to standing waves (periodic oscillations) emerge

R. J. Rutten and G. Severino (eds.), Solar and Stellar Granulation, 471–480.
© *1989 by Kluwer Academic Publishers.*

Figure 1. Travelling waves with $\beta = 6$, $R = 56000$. Streaklines (parallel to the instantaneous velocity) and magnetic field lines at equally spaced intervals in time.

from the oscillatory bifurcation. At the marginal state these solutions are composed of slow magnetoacoustic waves, travelling along the magnetic field with the Alfvén speed v_A. It can easily be shown that plane waves travelling at angles θ, $(\pi - \theta)$ to the vertical combine to give travelling waves with a wavelength $\lambda = 2 \cot \theta$, travelling with a velocity $v = \frac{1}{2} \lambda v_A$. Two waves travelling in opposite directions can then be combined to give a standing wave with period $\tau = 2/v_A$. In order to discover which of the two solutions is stable it is necessary to enter the nonlinear domain. Dangelmayr & Knobloch (1986) have shown that standing waves are preferred in Boussinesq magnetoconvection with λ of order unity.

We have carried out numerical experiments on two-dimensional convection in a fully compressible layer, using the code described by Hurlburt & Toomre (1988). The dimensionless temperature has values $T = 1/6$, $7/6$ at the top and bottom of the layer and in the absence of convection the atmosphere is polytropic with index $m = 1/4$ and $\gamma = 5/3$, so that the density contrast $\chi = 1.63$. A configuration is defined by the values of R, ς and the ratio of the gas pressure to the magnetic pressure (the 'plasma beta') $\beta \propto R/\varsigma Q$, all measured at the middle of the layer. For our calculations we take $\varsigma = 1/10$, which allows convection for $\beta \geq 5$.

For fixed $\beta < 200$ convection with $\lambda = 1$ or $\lambda = 2$ sets in at an oscillatory bifurcation when $R = R^{(o)}$ (cf. Cattaneo 1984). Immediately beyond this bifurcation standing wave solutions are preferred but as R is increased these give way to travelling waves if β is sufficiently small. Figure 1 shows streaklines (representing the velocity) and magnetic field lines for a travelling wave solution in a box with $\lambda = 2$ for $\beta = 6$ and $R = 16R^{(o)}$. The four rolls have a triangular form and travel to the right with a constant speed $v \approx 0.5v_A$. This is precisely the speed expected for travelling waves with $\lambda = 1$ in the Boussinesq approximation; yet these waves only appear at low β and it is clear that they are an essentially compressible phenomenon. When the field is sufficiently strong any significant convective motion concentrates magnetic flux at the upper boundary to give magnetic pressures that may exceed the ambient gas pressure. Even when the flux sheets are almost entirely evacuated the magnetic pressure gradients can only be balanced by inertial terms in the equation of motion. Thus there is a mechanism for destabilizing oscillations and propelling travelling waves - which nevertheless travel at the Alfvén speed.

3. Stratification

In the previous section we saw that standing waves and travelling waves appeared as the magnetic pressure was increased in a weakly stratified layer. Next we explore some effects of stratification in a moderately strong magnetic field. In our model the thermal conductivity is uniform, so the thermal diffusivity is inversely proportional to the density ρ and the ratio $\varsigma \propto \rho$. We have studied behaviour as R is increased for fixed Q in an atmosphere with a density contrast $\chi = 11$ for different choices of $\hat{\varsigma}$, the value of ς at the middle of the layer. For $\hat{\varsigma} = 6$ ($\varsigma \geq 1$) only steady solutions were obtained and for $\hat{\varsigma} = 0.6$ ($0.1 \leq \varsigma \leq 1.1$) we found periodic standing waves solutions for $R^{(o)} < R \leq 2R^{(o)}$. ($\varsigma$ was too large for travelling waves to be preferred.) With $\hat{\varsigma} = 1.2$ a more interesting possibility arose. In the lower part of the layer (with $\varsigma \leq 2.2$) steady convection would be expected to occur but in the upper (with $\varsigma \geq 0.2$) oscillations should be favoured. In fact convection set in at a stationary (pitchfork) bifurcation when $R = R^{(e)}$. For $R^{(e)} < R \leq 1.35R^{(e)}$ only steady solutions were found but these underwent a secondary oscillatory bifurcation, leading to vacillation about the steadily convecting state. Figure 2 shows streaklines and fieldlines for a periodic

Figure 2. Oscillatory convection in a stratified layer with $0.2 \leq \varsigma \leq 2.2$. Streaklines and field lines at equally spaced intervals during one period of the oscillation.

solution at $R \approx 1.6 R^{(e)}$ for $\lambda = 4/3$. At the base of the layer there are four rolls whose senses of rotation remain unchanged. In the course of one period first the outer pair and then the inner pair assume more prominence, so that motion reverses in the upper part of the layer. Roughly speaking, the fluid succeeds in combining oscillatory behaviour at the top with steady convection at the bottom of the layer, and motion is dominated by symmetrical upward spurts alternately at the centre and at the edges of the box.

4. Streaming Instabilities

In the weakly stratified layer time-dependent behaviour may persist even in the parameter range where convection sets in at a stationary bifurcation. Steady solutions with $\lambda = 1$ (which are stable in the Boussinesq limit) are unstable to spatially asymmetric modes which lead to tilted rolls and large-scale streaming motions. As a result the magnetic field is carried sideways and distorted until the Lorentz forces act back on the motion and reverse the flow. Figure 3 illustrates a periodic solution for $\beta = 1024$ and $R = 8000$, with a single roll within the spatially periodic domain. Note the strong tilted fields that are formed by the sheared horizontal velocities. This particular solution is a good example of the activity that should be expected when convective motions occur in a moderately strong magnetic field.

5. Applications to the Sun

The obvious application of these model problems is to sunspot umbrae, where there is a strong vertical magnetic field. At first sight one might expect to find travelling waves, since $\beta \approx 1$ and $\varsigma << 1$ at the photosphere. The surface plasma is, however, coupled to deeper layers where ς rapidly increases. The opacity rises owing to ionization, so the radiative conductivity falls and $\varsigma > 1$ at depths greater than 2000 km below the surface. This is the situation modelled in Figure 2, which shows that rapidly reversing motion in superficial layers may be coupled to slowly changing convection at greater depths. In a three-dimensional model subphotospheric convection cells might erupt aperiodically to produce an irregular pattern of strong upward motions at the surface, corresponding to umbral dots.

Travelling waves are likely to be generated in horizontal fields, so providing yet another mechanism for producing running penumbral waves. Another possibility is that travelling waves may be excited at the edge of the sunspot near the umbral-penumbral boundary, where the field is inclined. Electric currents flowing within the spot in this region are needed to provide a static equilibrium and the boundary may be unstable to interchanges as well as to convective modes. Waves travelling outwards might be responsible for altering the inclination of magnetic flux tubes and so controlling the evolution of penumbral filaments.

Finally, we consider strong magnetic fields in lanes between granules and mesogranules, where we expect to find flux sheets with $\beta \approx 1$. The examples presented above suggest that such flux sheets will be dynamically active. We do not pretend to have given a complete list of instabilities leading to time-dependent behaviour but it seems likely that with sufficiently high resolution it will be possible to observe a great variety of activity in intergranular magnetic fields.

Figure 3. Periodic oscillation with large-scale streaming motion. Streaklines and field lines during one period of the oscillation with $\beta = 1024$, $R = 8000$.

Acknowledgement

This research has been supported by grants from SERC.

References

Cattaneo, F. 1984. In *The Hydromagnetics of the Sun* ed. T.D.Guyenne, p.47, ESA SP-220.
Dangelmayr, G. & Knobloch, E. 1986. *Phys. Lett.* **117A**, 394.
Hughes, D.W. & Proctor, M.R.E. 1988. *Ann. Rev. Fluid Mech.* **20**, 187.
Hurlburt, N.E. & Toomre, J. 1988. *Astrophys. J.* **327**, 920.
Proctor, M.R.E. & Weiss, N.O. 1982. *Rep. Prog. Phys.* **45**, 317.

Discussion

MARTIČ — 1). What is the cell size in your periodic solutions?

2). How do you explain the disparity between granules and supergranulation in your results; unstable thermal boundary layer or shear instabilities?

WEISS — Ad 1: the preferred roll width in the absence of a magnetic field is slightly greater than the layer depth in these idealized calculations. Magnetic fields produce vertically elongated cells. I have shown an example here for which the preferred aspect ratio for rolls is one-third, and it might be much less than that in the real Sun. Thus, umbral dots may correspond to cells with diameters of about 300 km but with depths of 3000 km or more.

Ad 2: I agree that granules are most likely caused by some local instability in a boundary layer at the top of the convective zone. There are no indications of significant shear in supergranular or mesogranular flows near the photosphere—the granules are transported as though they were on a conveyor belt—so its seems more likely that they are produced by thermal instabilities. I am not convinced by the suggestion that granules are simply the products of a turbulent cascade driven by supergranular convection.

VAN BALLEGOOIJEN — What limits you to $\beta \geq 8$?

WEISS — We are limited by computational considerations: in order to obtain convection at lower β, we would have to reduce the magnetic and viscous diffusivities, and then we would be unable to resolve the relevant boundary layers.

ZHUGZHDA — I think that the penumbral travelling waves can be explained by the interaction of 5-minute oscillations and the inclined magnetic field of the penumbra. In this case, the 5-minute oscillations can transduce into the chromosphere only along the field. Consequently, outgoing penumbral waves arise[1].

WEISS — I agree that penumbral waves are most likely to be some form of global oscillation in the sunspot, but one should not overlook the possibility of convective excitation.

LOU — What are the boundary conditions, both on top and at the bottom of the layer?

WEISS — At the upper and lower boundaries the temperature is fixed, while the vertical velocity, the horizontal field and the tangential component of the viscous stress all vanish. These boundary conditions were chosen to match previous studies of Boussinesq magnetoconvection, which were conditioned by a preference for convenient eigenfunctions in the linear problem.

KOUTCHMY — You presented several examples of simulations showing inclined flows (or tilted streams?); are these realistic features which may be observable using, for example, high resolution spectroscopy?

WEISS — I think it likely that inclined velocities could, in principle, be observed in intergranular lanes, as a result of dynamic convective processes in a magnetic field.

---

[1] Y.D. Zhugzhda and N.S. Dzhalilov, 1984, *Astron. Astrophys.* **133**, 333.

LOU — What are the values of the ordinary and magnetic Prandtl numbers in your calculations?

WEISS — For the travelling wave solutions, we set $\sigma = \hat{\zeta} = 0.1$. We also experimented with other values,and found that travelling waves are still preferred with $\sigma = 1$ but are no longer stable when $\hat{\zeta} = 0.3$. In general, one expects more activity as the Prandtl number is decreased.

STEFFEN — What is the basic reason that up- and downflows have essentially the same amplitude in most of the simulations you showed?

WEISS — Most of the runs were for a density contrast of 1.63 across the layer, in order to separate the effects of increasing the magnetic pressure from the effects of stratification. As a result, the flow was almost Boussinesq, with only a small (but significant) difference between upward and downward flows.

FOX — In the phase diagram with Rayleigh number (R) and Chandrasekhar number (Q) which defines regions of stability, instability and overstability etc., have you investigated r and Q while changing the magnetic diffusivity?

Calculations which I have done for solar parameters indicate that the region where overstability sets in is much narrower than for the case of incompressible flow. This means that a narrow range of both R and Q is required to give convective overstability.

WEISS — For the results described we have set $\sigma = \hat{\zeta} = 0.1$, and the stability boundaries were principally determined by these values, which were limited by the resolution of our non-linear computations.

The extent of the 'overstable' range does indeed depend on the values of these diffusivity ratios.

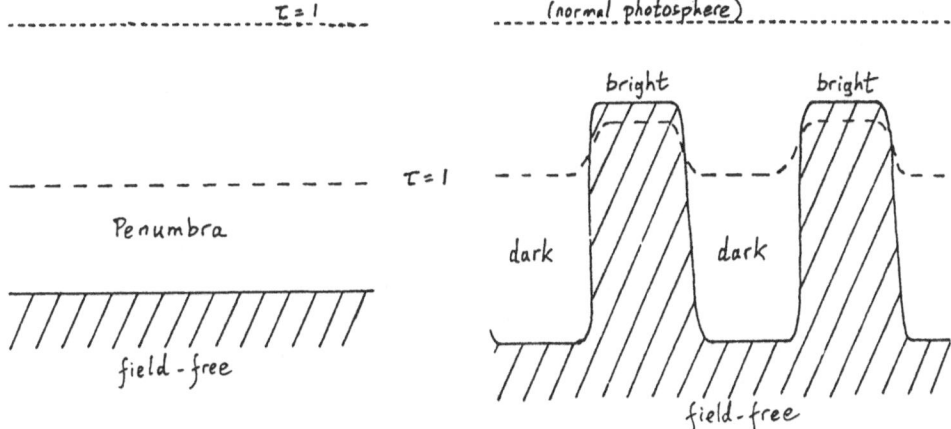

Figure 1: Sketch of penumbra models

NORDLUND — In your model of the penumbra, how do you get rid of the cool fluid in the dark penumbral filaments (which contain magnetic field), while at the same time allowing the ascending hot material to cool and to develop a significant entropy deficiency (which it can only do if at stays in the region $r \leq 1$)? Please draw a figure!

WEISS — In the model proposed by Schmidt, Spruit and myself[2] we supposed that energy is transferred across the magnetic discontinuity almost entirely by radiation. This is an inefficient process; hence the penumbra is cooled and the discontinuity is only a short distance below the level where $r = 1$. (A lower limit to the penumbral thickness is set by infrared observations.) We found that the mean penumbra is only 60 km thick, though the level where $r = 1$ is about 90 km lower than in the normal photosphere, as sketched above on the left. When we constructed a two-component model, the dark filaments were about 80 km thick but the magnetopause was actually above $r = 1$ in the bright filaments, as sketched above on the right. So, nonmagnetic plasma can lose energy by radiation.

A careful study of dark filaments at the outer boundary of the penumbra might allow us to decide between this model and the one that you outlined.

[2]H.U. Schmidt, H.C. Spruit and N.O. Weiss, 1986, *Astron. Astrophys.* **156**, 351.

MODELS OF MAGNETIC FLUX SHEETS [1]

U. Grossmann-Doerth[1], M. Knölker[2], M. Schüssler[1],
and E. Weisshaar[3]

[1]*Kiepenheuer-Institut für Sonnenphysik, Freiburg, FRG*
[2]*Universitäts-Sternwarte, Göttingen, FRG*
[3]*Institut für Theoretische Physik, Universität Bayreuth, FRG*

ABSTRACT. We present results of numerical model calculations of magnetic flux concentrations in the solar photosphere. Using a 2D slab geometry we solve the nonlinear MHD equations for a compressible medium incorporating partial ionization effects and a full radiative transfer (grey, LTE). An adaptive Moving Finite Element code is employed. We are able to analyze our results by computing the profiles of the four Stokes parameters (of various spectral lines) emerging from the flux concentrations and comparing them with observations. We find good agreement between synthetic and observed Stokes V–profiles in the visible and infrared. A major new result is the appearance of a temperature enhancement in the upper layers in the upper layers of the flux concentration which are heated by illumination from the hot bottom. We obtain a large positive continuum intensity contrast of the flux sheet.

1. Introduction

Most of the magnetic flux through the solar photosphere (in both quiet and active regions, outside sunspots) occurs in the form of concentrated, filamentary structures embedded in a virtually field-free medium. In continuum radiation the magnetic flux concentrations typically appear as points (Muller, 1983) and sheet-like, elongated bright structures (von der Lühe, 1987) located in intergranular lanes.

The formation of these concentrated structures is probably due to advection by flows, cooling by suppression of convective energy transport and magneto-convective instability. It is supposed that after formation the flux concentration reaches a static or stationary oscillating state. The energy budget is then dominated by the balance of vertical radiative loss and *lateral* inflow of radiation into the partially evacuated magnetic region.

The interaction of flux concentrations with the surrounding velocity fields (granules, downflow "fingers", vortices) is likely to excite MHD waves which are modified by the filamentary structure of the magnetic field (sausage, kink and torsional waves for a cylindrical structure). Such waves may contribute to heating stellar chromospheres and

---

[1]Mitteilungen aus dem Kiepenheuer-Institut Nr. 297

R. J. Rutten and G. Severino (eds.), Solar and Stellar Granulation, 481–492.

coronae. Another source could be the accumulation of magnetic stress and energy by magnetic flux concentrations that are carried around by photospheric flows.

A detailed knowledge of the MHD and thermodynamic structure of concentrated photospheric magnetic fields is essential not only for understanding the interaction of convection and magnetic fields in general, particularly in view of the transport of mechanical energy by excitation of oscillations and waves, but also for any reliable *quantitative* determination of solar and stellar magnetic fluxes and flux densities (*cf.* Grossmann-Doerth et al., 1987).

The nonlinearity of the equations, the importance of non-local radiative transport for the energy budget and the complicated spatial structure render numerical simulations of the MHD equations combined with a radiative transport code to an indispensable tool for understanding the basic formation processes, the equilibrium state (if any) and the interaction with the surrounding medium. The calculation of synthetic spectral line and Stokes profiles together with their contribution functions opens the possibility to use such simulations as a diagnostic tool for the interpretation of measurements.

2. Model description and previous results

The major features of elongated magnetic flux concentrations can be reasonably described by a flux sheet in a two-dimensional slab geometry in which all quantities are independent of the horizontal coordinate along the slab. We integrate numerically the time-dependent MHD equations for a compressible medium and incorporate partial ionization effects self-consistently. We employ the Moving Finite Element (MFE) method which is particularly useful for the resolution of the (generally moving) narrow transition layer between flux concentration and surrounding medium. More details about the model and the numerical method can be found in Deinzer et al. (1984a). Our previous investigations dealt with the stationary structure of small flux concentrations (*magnetic elements*) revealing the formation of a thermal circulation cell in the cool environment of the sheet (Deinzer et al., 1984b) and concentrated further on the influence of the transition layer width, temperature and continuum intensity structure, and the interchange (fluting) instability (Knölker et al., 1988). Models of larger structures with diameters between 500 km and 1000 km have recently been presented (Knölker and Schüssler, 1988).

3. Recent improvements and results

We recently improved our code significantly by

- replacement of the diffusion approximation by a full (grey, LTE) radiative transfer in the energy equation with Feautrier integration along rays of various angles,

- implementation of an *implicit* version of the MFE code with Newton-Raphson iteration using a Jacobian matrix which was determined analytically,

- addition of a diagnostic package which calculates synthetic Stokes parameter profiles of spectral lines using a modified version of the matrix algorithm proposed

by van Ballegooijen (1985) together with line depression contribution functions
(Grossmann-Doerth et al., 1988).

Since the new features described above have been successfully combined and tested only
recently, we have carried out only a limited number of calculations and not yet fully used
the potential of the new code and diagnostic package. Therefore, we restrict ourselves
to presenting preliminary results of a magnetic element model ($\approx$ 150 km diameter at
optical depth unity) which approaches but has not quite reached a stationary state. Fig.1
shows some properties of this model. The geometric height scale has its zero level at
the location of $\tau_c = 1$ (continuum, 5000 Å) of the 'undisturbed' solar atmosphere at the
right-hand-side boundary. Within the flux concentration, the level $\tau_c = 1$ is situated 180
km deeper as indicated in the frame on the upper right. The maximum velocity is about
1 km·s^{-1}. The most striking feature in Fig. 1 is the temperature enhancement in the
upper layers of the flux sheet which is caused by a radiative *illumination effect*: Material
in these layers "sees" the hot ($\approx$ 7200 K) bottom of the flux concentration at $\tau_c = 1$ while
material at the same height of the non-magnetic atmosphere is illuminated by a layer of
only $\approx$ 6400 K temperature. This leads to a local temperature enhancement of up to
400 K at the same geometrical level and more than 1000 K at the same *optical* depth.
Although these values may change somewhat after incorporation of a non-grey version
of the radiative transfer code, there is no doubt that the illumination effect leads to a
significant heating of the upper layers of small flux concentrations and that it contributes
to or may even represent the major cause of the well-known weakening of photospheric
spectral lines associated with concentrated magnetic fields.

Fig. 2 shows the appearance of the flux sheet in the continuum radiation for different
inclinations of the line of sight. For vertical incidence ($\mu = \cos\theta = 1$) the interior of the
flux concentration is brighter than the mean photosphere by a factor 1.6; this value is
compatible with the lower limit of 1.4 determined by Schüssler and Solanki (1988) on
the basis of the observed weakening of the FeI $\lambda5250$ line in plages.

To illustrate the possibilities for Stokes diagnostics of our models we show synthetic
profiles for the two FeI lines $\lambda5250$ and $\lambda15648$, both with Landé factor $g = 3$. Fig.
3 shows the I- and V-profiles together with the line depression contribution functions
(at the wavelength of the V-profile maximum) of the infrared line $\lambda15648$ for one ray
of vertical incidence ($\mu = 1$) in the plane of symmetry (center, $x = 0$) of the flux sheet.
The line is fully split and significantly weakened compared to the profile originating in
the non-magnetic atmosphere. The contribution functions are quite narrow (a similar
result is obtained for the $\lambda5250$ line) and illustrate the potential of a detailed probing of
the deep layers of the flux concentration using this spectral line.

For a comparison with observations, line profiles of single rays are not very useful
since observed profiles inevitably are spatial averages. In the case of the I-profile this
leads to a mixing of weakened and magnetically split line profiles from the flux concentrations and contributions from the non-magnetic atmosphere. Hence, the resulting average
profile critically depends on the averaging area, i.e. the spatial resolution attained. A
comparison with synthetic profiles is only possible if the filling factor of magnetic structures in the resolution element would be known which is generally not the case. For the
Stokes V-profiles, however, the situation is much more convenient since these originate
solely in the magnetic part of the atmosphere. Spatial averaging with non-magnetic

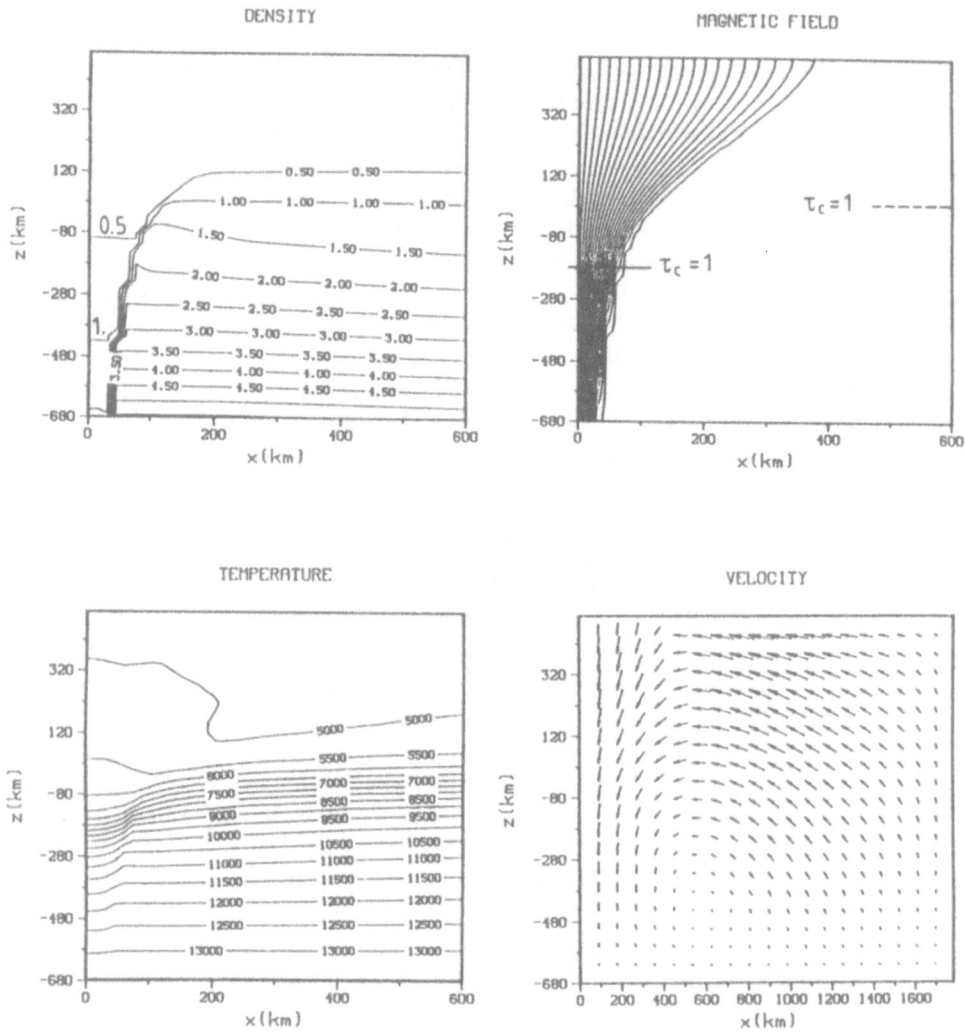

Figure 1: Properties of a model for solar *magnetic elements*. Only half of the symmetric flux sheet is shown and the horizontal coordinate has been stretched for better visibility of the structure (except for the velocity, bottom right). Densities are in units of $3 \cdot 10^{-7}$ $g \cdot cm^{-3}$, temperatures in Kelvin.

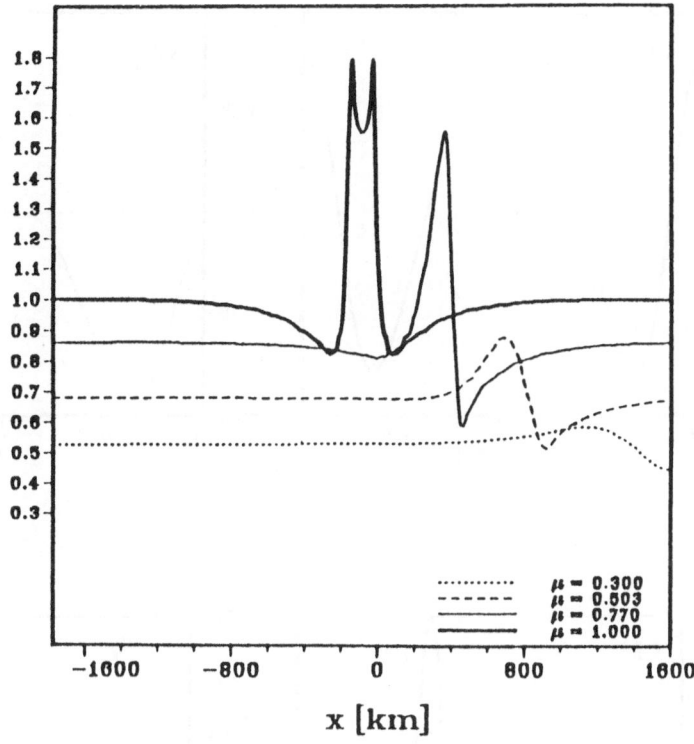

Figure 2: Horizontal profiles of emergent intensity (continuum, 5000 Å, in units of the mean solar intensity) for different inclinations $\mu = \cos\theta$ of the line of sight.

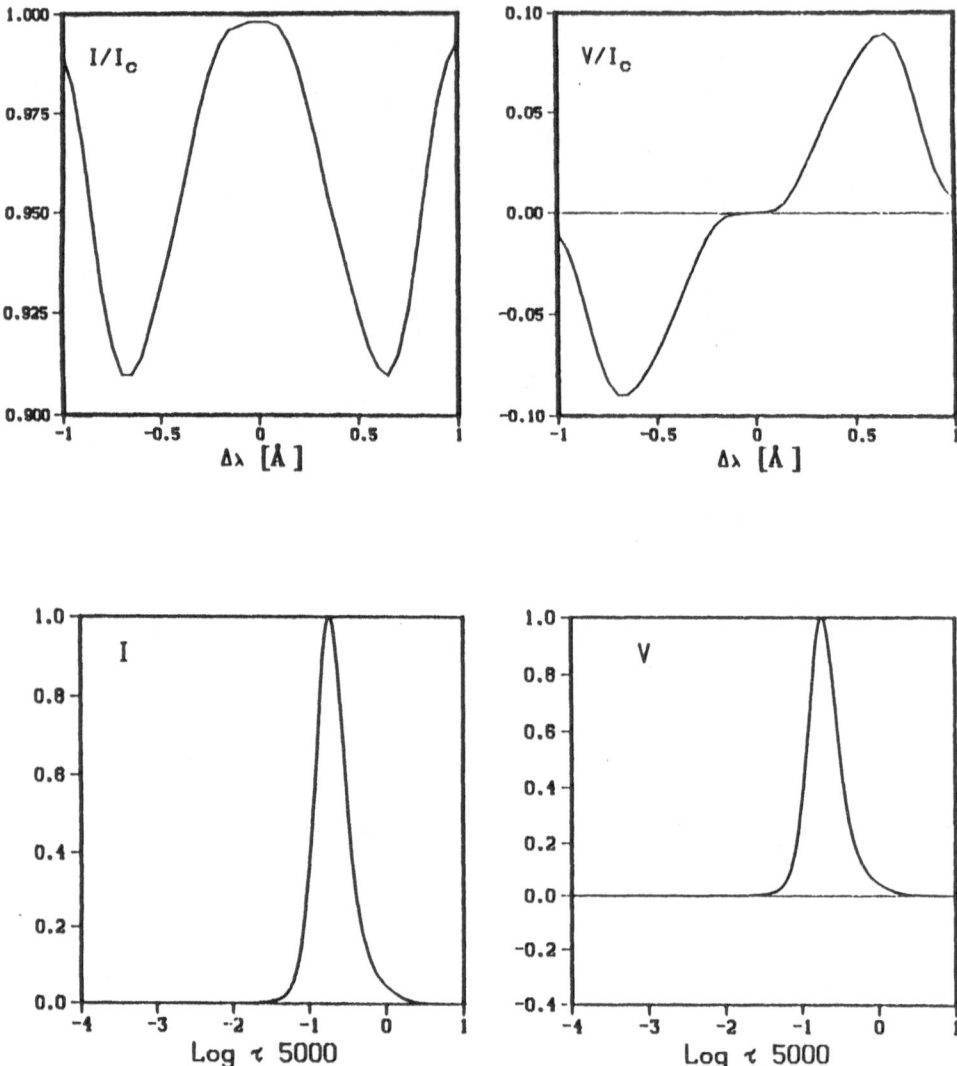

Figure 3: Stokes I- and V-profiles for the infrared line FeI $\lambda15648.54$ Å and Stokes line depression contribution functions at the wavelength of the V-profile maximum

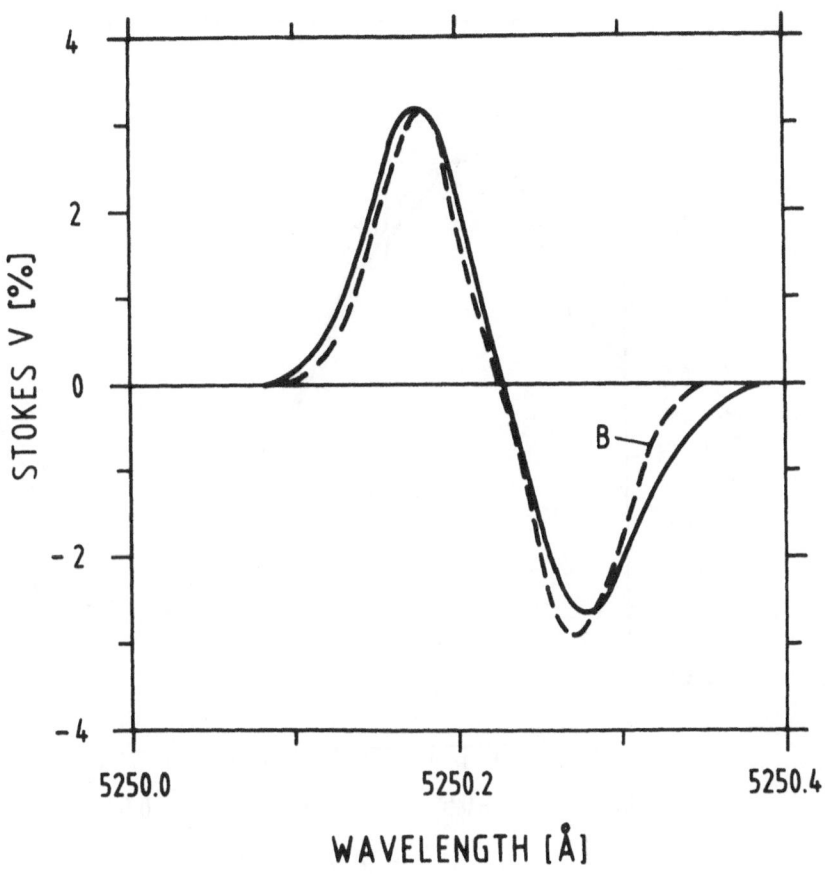

Figure 4: Observed (full) and averaged synthetic (dashed) V–profiles of FeI $\lambda5250.22$ Å for vertical incidence of the line of sight.

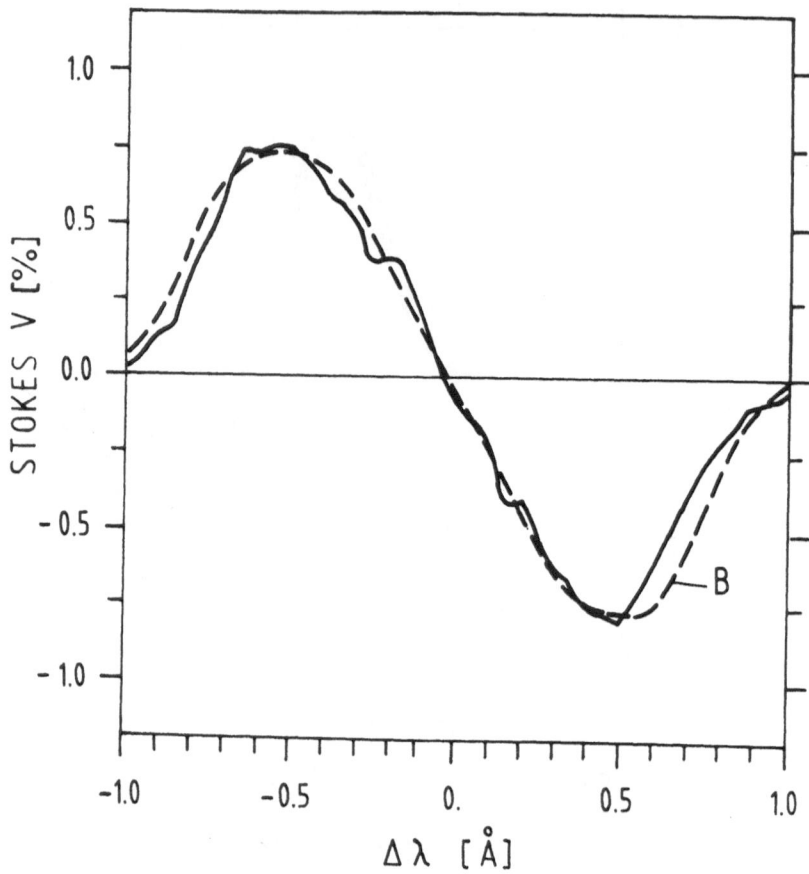

Figure 5: Same as Fig. 4, but for the infrared line FeI $\lambda15648.54$ Å

components amounts to multiplication of the whole profile with a constant factor, which essentially is the filling factor times the continuum intensity contrast. Fig. 4 shows a comparison between the observed V–profile of the $\lambda5250$ line in a plage region (Stenflo et al., 1984) shown by the full line and a synthetic V–profile averaged over the flux sheet model shown in Fig. 1 (model B, dashed line) and normalized to the maximum of the observed profile. Except for the fact that the synthetic profile is slightly too narrow (no micro- or macro-turbulence has been used) and does not exhibit the asymmetry of the observed profile, the agreement is satisfactory. Since we have used a large value of the viscosity in order to damp the initial dynamics of the approach to a stationary state we expect much higher velocity amplitudes of the external circulation which might well lead to a significant asymmetry of the synthetic profiles.

Fig. 5 shows a similar comparison between observed (Stenflo et al., 1987) and average synthetic V–profiles (model B) for the infrared line $\lambda15648$. Due to the deep origin of this line, rays at the periphery of the flux concentration traverse magnetic and non-magnetic parts of the atmosphere. They give contributions to the average profile with small amounts of Zeeman splitting and cause the large width of the synthetic average profile which compares very well with the observed profile.

4. Conclusions

We have presented first results of a new generation of our model calculations for solar magnetic flux concentrations. The models have been significantly improved by a new implicit, adaptive MFE code, inclusion of a full grey radiative transport and the implementation of a Stokes diagnostics package.

The more adequate treatment of radiative transfer revealed the *illumination effect* which leads to a significant radiative heating of the upper layers of flux concentrations and may well be the major cause of the observed weakening of photospheric lines. With a non-grey version of the radiative transport code we should be able to quantitatively test this possibility.

With aid of the new Stokes diagnostics we have been able to confirm the narrow contribution functions already indicated in the results of van Ballegooijen (1985). We are now in a position to confront our model results with observational data. Although the results presented here are of a somewhat preliminary nature since the stationary state has not been reached and the viscosity is rather large, we can conclude that the comparison with observed properties is rather promising. Continuum intensity, field strength and the V–profiles of the $\lambda5250$ and $\lambda15648$ lines agree well with observed quantities.

References

Deinzer, W., Hensler, G., Schüssler, M., Weisshaar, E.: 1984a,
    Astron. Astrophys. **139**, 426
Deinzer, W., Hensler, G., Schüssler, M., Weisshaar, E.: 1984b,
    Astron. Astrophys. **139**, 435
Grossmann-Doerth, U., Pahlke, K.-D., Schüssler, M.: 1987,

    Astron. Astrophys. **176**, 139

Grossmann-Doerth, U., Larsson, B., Solanki, S.K.: 1988,
    Astron. Astrophys. , in press

Knölker, M., Schüssler, M., Weisshaar, E.: 1988, *Astron. Astrophys.* **194**, 257

Knölker, M., Schüssler, M.: 1988, *Astron. Astrophys.*, in press

Müller, R.: 1985, *Solar Phys.* **85**, 113

Schüssler, M., Solanki, S.K.: 1988, *Astron. Astrophys.* **192**, 338

Stenflo, J.O., Harvey, J.W., Brault, J.W., Solanki, S.K.: 1984,
    Astron. Astrophys. **131**, 333

Stenflo, J.O., Solanki, S.K., Harvey, J.W.: 1987, *Astron. Astrophys.* **173**, 167

van Ballegooijen, A.A.: 1985, in *Measurements of Solar Vector Magnetic Fields*,
    ed. M.J.Hagyard, NASA Conf. Publ. 2374, p. 322

von der Lühe, O.: 1987, in *The Role of Fine-Scale Magnetic Fields
    on the Structure of the Solar Atmosphere*, eds. E.H. Schröter et al.,
    Cambridge University Press, p. 156

Discussion

VAN BALLEGOOIJEN — Can you reproduce the V-profile asymmetries that have been observed?

SCHÜSSLER — The asymmetries are there, but they are not large enough because the velocities are too small.

NORDLUND — I think that there is an asymmetry in the Stokes V, but that it is just too small. I believe that if you let a flow really develop (carry the whole convective flux), you would find that van Ballegooijen's[1] explanation for the asymmetry is correct.

SCHÜSSLER — Indeed we find a small asymmetry of the V-profiles. Since the viscosity in the model presented is large (to damp the vigorous dynamics of the approach to a stationary state), the velocities are still unrealistically small. The further evolution of the model will be calculated with a significantly smaller viscosity. This should increase the V-profile asymmetries. We have recently shown[2] that the spatial *separation* of the velocity and the magnetic field along the line of sight as proposed first by van Ballegooijen does not only produce asymmetries of the V-profiles, but that it yields *unshifted* zero crosses like the ones that are observed (e.g. Solanki[3]) as well.

KOUTCHMY — Can you reproduce the large center-to-limb brightening of magnetic elements (flux sheets) as they are observed in the continuum?

SCHÜSSLER — We cannot reproduce these with small (≈ 200 km) magnetic elements, while larger structures (500–1000 km) show a brightening similar to the observations[4]. The rapid decrease of the visibility of bright points when you observe outside disk center found by Müller[5], however, points to another explanation of limb brightening: the overlapping of hot clouds which are individually optically thin (cf. Schüssler[6]).

KNEER — Your calculation shows that the presence of the flux tube has large consequences for the structure of the upper atmosphere by radiative effects, i.e. illumination from below. This is fascinating because possible heating mechanisms of the chromosphere and the CO radiative instability near the temperature minimum region depend critically on this temperature structure (e.g. Kneer[7]). Would you be able to give an 'average' temperature enhancement over say 1 arcsec instead of just the center of the structure?

[1] A.A. van Ballegooijen, 1985, in: H.U. Schmidt (ed.), *Theoretical Problems in High Resolution Solar Physics*, MPA/LPARL Workshop, Max-Planck-Institut für Astrophysik, München, p. 177.

[2] U. Grossmann-Doerth et al., *Solar Phys.*, submitted.

[3] S.K. Solanki, 1986, *Astron. Astrophys.* **168**, 311.

[4] M. Knölker and M. Schüssler 1988, *Astron. Astrophys.*, in press.

[5] R. Müller and Th. Roudier, 1984, *Solar Phys.* **94**, 33.

[6] M. Schüssler, 1987, in E.H. Schroter et al., *On the role of small scale magnetic fields*, Cambridge Univ. Press, 1987.

[7] F. Kneer, 1983, *Astron. Astrophys.* **128**, 311.

SCHÜSSLER — The temperature enhancement in the magnetic structure is about 400 K. However, the precise value will change (probably increase) when we proceed to a non-grey radiative transfer. I think we should wait until such results are available before discussing more details.

RUTTEN — This radiative heating of a cloud above the flux concentration is much stronger in the non-grey case. In that case, the 2 million[+] lines of Fe II between 2000 and 4000 Å are very important. The $\tau = 1$ surface for the continuum is located quite deep geometrically, and so the ultraviolet radiation field from the walls is quite hot. The Fe II lines only start building up opacity where iron ionizes again, and are then quite thick in the cloud. They will absorb a lot of energy, more than in the grey case.

SCHÜSSLER — We are presently working on the non-grey case of the radiative transfer code.

NORDLUND — There will be a couple of changes when you go to more detailed descriptions.
1). The radiative equilibrium temperatures (both inside and outside) will decrease when you include line blanketing.
2). The temperature outside will decrease (due to adiabatic expansion effects) when you include the full velocity field.

LOU — Have you included a viscous dissipation term in your energy equation? It seems to me inconsistent to include viscosity in the momentum equation and not in the energy equation.

SCHÜSSLER — It hasn't been included in the previous calculations since viscosity was used there only to preserve numerical stability. For consistency, we will include the viscous dissipation term in the new code. However, viscous dissipation in the energy equation is entirely negligible compared to radiation and advection. The results do not depend on this term.

STELLAR GRANULATION: MODELING OF STELLAR SURFACES AND PHOTOSPHERIC LINE ASYMMETRIES

DAINIS DRAVINS
Lund Observatory
Box 43
S–22100 Lund
Sweden

ABSTRACT.   Numerical simulations of the three–dimensional structure and time evolution of stellar surface convection have been carried out for different *solar–type stars* with effective temperatures between 5200 and 6600 K.   Using the output from these simulations as a set of spatially and temporally varying model atmospheres, synthetic granulation images, and spectral line profiles have been computed.   The disk–integrated line profiles agree well with observed lineshapes and bisector patterns in *Procyon* (F5 IV–V), α *Cen A* (G2 V) and α *Cen B* (K1 V), and also permit the stellar rotation to be determined.

To infer the surface structure also in *early–type stars*, simple four–component models are used to reproduce observed bisectors.   It is found that rapidly rising 'granules' cover only a small fraction of the surface, a situation opposite to that in solar–type stars, and one likely to affect magnetic fields and stellar activity.   In *solar–type stars*, the granulation dynamics concentrate magnetic flux near the intense downdrafts that carry much of the convective energy, a mechanism that appears to be missing in *early–type stars*.

For the future, observations of convective *wavelength shifts* (in addition to line profiles and asymmetries) would be most useful for testing different models.   Greatly increased computing power is needed for detailed modeling of granulation throughout the Hertzsprung–Russell diagram, and may require custom–designed computers with novel computer architecture.

1. Introduction

Stellar granulation can be observed through various subtle signatures in high resolution spectra of integrated starlight, through the time variability of stellar irradiance, and by a few other means. The most promising observational parameter appears to be that of photospheric absorption line asymmetries, which (analogous to the solar case) may arise from correlated velocity and brightness patterns on stellar surfaces. Several authors have now reported observations of such asymmetries.

A number of detailed models of *solar* granulation have been worked out during recent years.   Their general success in describing observed properties of solar granulation patterns and spectral line shapes and shifts strongly suggest that their general description of the physical situation is now largely correct, at least in non–magnetic regions. Based on this experience, the time seems ripe to venture further out among other stars in the Hertzsprung–Russell diagram.

R. J. Rutten and G. Severino (eds.), Solar and Stellar Granulation, 493–519.
© *1989 by Kluwer Academic Publishers.*

2. Models of Stellar Granulation

The astrophysical aims of stellar granulation modeling are somewhat different from the solar equivalent, in particular because of the more fundamental questions that can be asked about surface structure and line formation in widely different types of stars. On the Sun, the surface inhomogeneities have been studied in great detail and with considerable spatial resolution while the corresponding properties in other stars are still largely unknown: the usual description of stellar atmospheres has in the past often been by homogeneous models only. Consequently, already rather primitive models could bring interesting new information. On the other hand, the more limited observational data that can be obtained for stars, seriously limits the number of ways in which any models may be tested. Primary among observable parameters are the disk–averaged line profiles, together with their asymmetries and wavelength shifts (Dravins, 1982; 1987a), and thus theoretical models should be able to confront observations by predicting (at least) these parameters.

2.1. THE ROLE OF NUMERICAL SIMULATIONS

An important role in stellar models can be played by numerical simulations. The limited observational data available make it awkward or impossible to obtain unique "inversions" of observations into a numerical description of stellar atmospheric inhomogeneities: the degrees of freedom in parametrized models of the three–dimensional atmospheric structure is potentially very large, and different combinations of parameters might result in e.g. similar line profiles. This circumstance appears to have discouraged some workers who wanted to apply traditional modeling techniques (i.e. to set up a model and adjust the parameters until the resulting output fits the observed data).

Such limitations can (and should) be avoided by instead using *computational experiments* to numerically simulate the three–dimensional and time–dependent stellar surface convection. Although the solar (and, by implication, stellar) surface structure at first sight may appear to be quite complex, the number of *fundamental* physical processes that determine the structure at large, is probably very limited (at least in non–magnetic regions). This makes the problem well suited to numerical simulations: the finite number of physical processes can be mathematically described by physical laws which are known and well understood (e.g. those of hydrodynamics and radiative transfer). The availability of supercomputers then makes it possible to perform numerical simulations with considerable detail and realism. In particular, it becomes possible to model nonlinear and complexly intercoupled phenomena, whose correct interpretation from observations and parametrized models only could be next to impossible. For a general introduction to this new role of supercomputer numerical laboratories in physics, see e.g. Winkler at al. (1987).

2.2. THREE–DIMENSIONAL MODELS OF SOLAR GRANULATION

In recent years it has become possible to solve numerically the sets of hydrodynamic equations that realistically describe the three–dimensional time evolution of solar granular convection (Nordlund, 1982; 1985). The equations of motion (which describe the time evolution of the velocity field) and the energy equation (which describes the time evolution of the temperature field) are used to step a three–dimensional numerical representation of the velocity and temperature fields forward in time. The use of realistic background physics (equation of state, absorption coefficients, etc.) taken from standard stellar atmosphere code, a detailed treatment of the radiative transfer (non–grey, three–dimensional), and the use of relevant boundary conditions (in particular, a constant heat input at the lower boundary), leads to a realistic simulation of the granular convective motions.

Using the output from such simulations as a set of time– and space–dependent model atmospheres, line profiles can be obtained as spatial and temporal averages over the simulation sequence, resulting in considerable similarities with observed solar line profiles, including their asymmetries and wavelength shifts (Dravins et al., 1981; 1986). The inclusion of the non–LTE iron ionization equilibrium (taking into account the radiation intensities at the most important ionization edges of Fe), leads to a detailed agreement with observed solar Fe I line parameters (Nordlund, 1985).

These numerical simulations contain no arbitrary *physical* parameters: the results depend only on the effective temperature, surface gravity and the chemical abundance. (These determine the heat flux into the simulation volume, the gravity forces stratifying the atmosphere, and the opacity of the gases, respectively.) Thus, these models can in principle be applied also to other stars.

The models naturally contain a number of approximations, e.g. the continuity equation is used in the anelastic approximation, thus eliminating sound waves (this is believed to be the most important limitation remaining in these models). Elsewhere at this conference, results from removing that restriction are discussed (Stein and Nordlund, 1989). Other items that may need improvement include the detailed description of the viscosity, also discussed at this conference. However, it should be stressed (especially for solar astronomers) that since the primary purpose of *stellar* granulation modeling is to obtain a first idea of the inhomogeneous structure of the still largely unknown stellar surfaces, many such details that still may of legitimate concern for solar studies will probably not be particularly significant in defining the gross stellar properties.

2.3. THREE–DIMENSIONAL MODELS OF STELLAR GRANULATION

These computer codes, initially developed for solar modeling, were suitably modified and improved, and such numerical simulations have now been completed for models of four different stars (Nordlund and Dravins, 1989). The temperature dependence of granulation along the main sequence is explored through models hotter and cooler than the Sun. A $T_{eff} = 6600$ K model has parameters close to those of *Procyon* (F5 IV–V), while a cooler one with $T_{eff} = 5200$ K corresponds to the K1 dwarf α *Centauri B*. The luminosity dependence is examined by two models at solar temperature ($T_{eff} = 5800$ K): one at one *half* solar surface gravity corresponds to the slightly evolved main–sequence star α *Centauri A* (G2 V), while another at one *quarter* solar surface gravity has parameters similar to the subgiant β *Hydri* (G2 IV). The simulations were made with a 32 × 32 × 32 spatial grid: 32 × 32 Fourier components horizontally, and 32 knot points defining cubic splines for the vertical dependence of different parameters. Each simulation was run for approximately 2 hours of stellar time, with a timestep of typically 2 seconds, requiring some tens of CPU–hours on a CRAY–1 computer.

This small grid of models thus allows the study of the temperature and luminosity dependence of granulation in the vicinity of the Sun in the Hertzsprung–Russell diagram. Samples from this work (Nordlund and Dravins, 1989) are presented in Section 3 below. The output from these model simulations has been used as a set of spatially and temporally variable model atmospheres, from which stellar surface images and spectral line profiles have been computed for a series of different parameters. This permits a study of the nature of the continuum and spectral line formation in realistic stellar photospheres and, after integrating the time–averaged spectral line flux from the full stellar disk, also permits a direct comparison with the line profiles and asymmetries in the four stars *Procyon*, *α Cen A*, *α Cen B*, and *β Hyi*, as observed with very high spectral resolution by Dravins (1987b). Samples from that work (Dravins and Nordlund, 1989) are shown in Sections 4–7 below.

3. The Structure and Evolution of Stellar Granulation

The output from the simulations shows the time evolution of the three–dimensional structure of temperature, pressure, velocities and other physical parameters in stellar photospheres.. In the surface regions of all models, hotter and generally rising features are seen, analogous to solar granules. The normal evolution of stellar granules is characterized by a gradual increase in size until they disintegrate due to various causes.

Relatively undisturbed granules often grow to larger sizes, until they "collapse under their own weight". In order to carry the stellar convective heat flux outward, there must be an upflow of hot gases at a few km/s from beneath the photosphere. Upon reaching optically thin layers, these gases cool off and eventually turn around to descend in the cooler intergranular areas. The necessary *horizontal acceleration* of the gases towards the sinking areas is accomplished by local overpressures that develop over granules (and also over intergranular lanes, in order to *decelerate* the incoming horizontal flow there). The larger the granule, the larger the overpressure required to accelerate away the material over the greater horizontal distances. Ultimately, the overpressure on the surface becomes so great that it impedes the arrival of additional hot gases from below. Having been cut off from its energy supply, the granular center begins to cool off, a dark center develops, and the now ring–shaped granule disintegrates, a process apparently similar to solar "exploding granules". This is a mechanism that limits the characteristic scale of granules on different stars: the granules cannot grow any larger because the required overpressure would block further convective energy supplies from beneath, thus strangling the granule through a lack of input heat. Characteristic scales range between $\simeq 10^3$ km in the cool *α Cen B* model to $\simeq 10^4$ km on the subgiant *β Hyi*.

Not all granules evolve in this manner. Some granules that are disturbed by velocity flows in neighboring granules may cleave and break apart into smaller fragments, which in turn grow and merge into new features. Some periods may be characterized by especially vigorous granulation with large and well correlated temperature and velocity amplitudes, while other periods (especially following the breakup of some especially large granule) may be characterized as "abnormal" granulation with apparently chaotic temperature and velocity fields, as the photosphere is gradually returning back to a more stable situation.

3.1. THE SURFACE STRUCTURE OF VELOCITY AND TEMPERATURE

As an example of stellar surface structure, Figure 1 shows a "snapshot" of the *Procyon* model at a representative instant, i.e. at a time when its spatially averaged properties are roughly similar to the average for the entire simulation sequence. Temperature, pressure and velocities are shown at a constant geometrical depth at

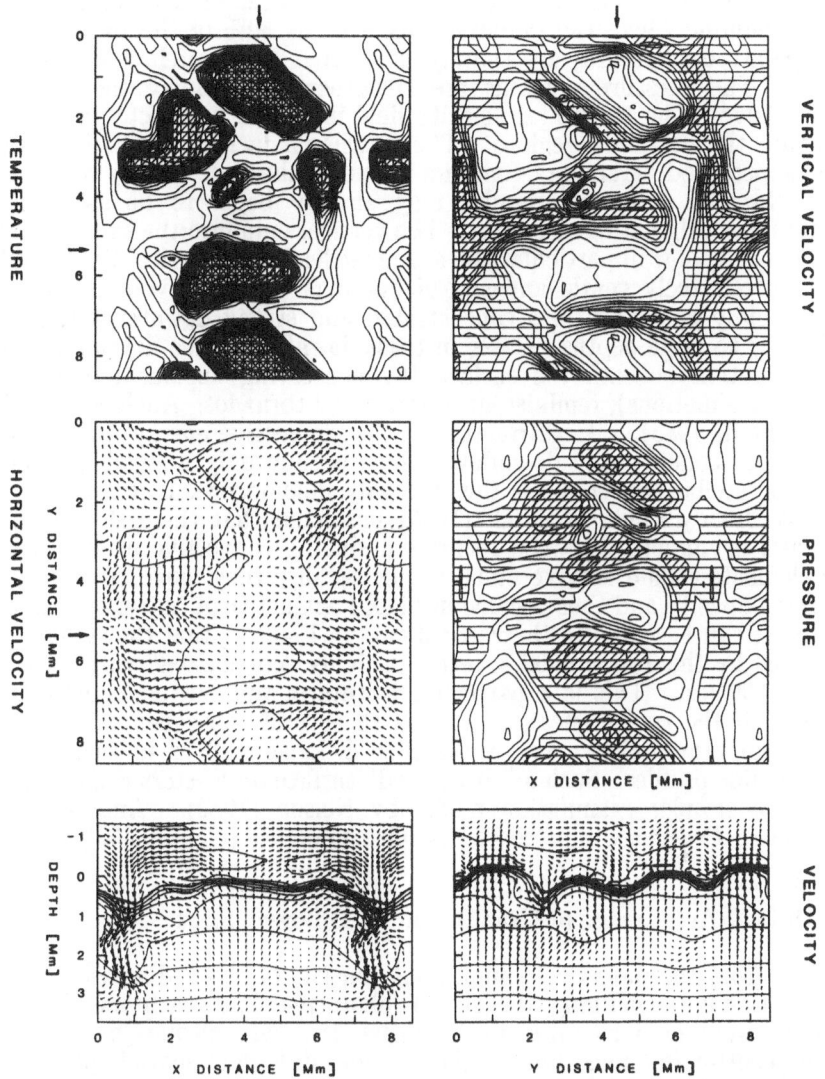

Figure 1. *Overall appearance* of the *Procyon* simulation at a representative time. The top four panels show temperature, pressure and velocities at a fixed geometrical height at the stellar "surface". Temperatures are shaded above the T_{eff} value of 6600 K, with contours every 500 K and shading increasing every 1500 K. Downward velocities are shaded with contours every 1.0 km/s and shading in steps of 3 km/s. Horizontal velocity vectors correspond to 40 seconds of gas motion. The superposed T = 6600 K contours delineate the granules. The logarithmic pressure plot is shaded above $10^{3.6}$ Pa, has contours every 0.05 dex, and increased shading every 0.1 dex. The Fourier representation makes the images periodic in X and Y, with a cycle of $\simeq$ 6840 km (slightly more than $^5/_4$ of one cycle is shown). The two bottom panels show the velocity components in vertical planes. The left cut into the simulation volume (at $Y \simeq 5.3$ Mm, as marked by arrows) goes across a large growing granule and its associated downflows, while the right one ($X \simeq 4.3$ Mm) goes across different smaller granules. The bold temperature contour marks $T_{eff} = 6600$ K, with other contours at 1500 K intervals. (Nordlund and Dravins, 1989)

the stellar "surface". This is here defined as the deepest horizontal level in the simulation volume where the radiative flux exceeds 50% of the total energy flux. The temperature structures are sharply delineated and correlate well with rising velocities. The pressure patterns are smoother and display the overpressures discussed above. The horizontal velocity field is obviously directed away from the upflows into the downflow regions. The occasional lack of a detailed correlation with the superposed "granular" contours is due to the time history of the different features: old and decaying granules may already begin to be swept away by horizontal flows originating elsewhere. The vertical cuts into the simulation volume (bottom panels) show in particular the asymmetries in the vertical flows. While the upflows are relatively gentle and spread out over extended volumes, the sinking material rapidly converges into concentrated and strong downflows that preserve their identity to large depths. Gases in these downflows often develop a significant rotational motion beneath the surface (by conserving angular momentum from random surface motions), reminiscent of terrestrial tornados. Analogous to the solar case (Nordlund, 1985), a large fraction of the convective energy flux is carried in these cool and concentrated downflows which occupy only a small fraction of the surface area: stellar convection is thus a highly inhomogeneous process. It may be noted that already Faulkner et al. (1965) predicted that there should be such a fundamental asymmetry between rising and sinking convective elements: cool and sinking elements should survive much longer because they move down into regions of increasing opacity and thus become insulated against their hotter surroundings.

Finally, it can be noted in the vertical cuts in Figure 1 that the stellar "surface" (now defined by the $T_{eff} = 6600$ K level) is highly "corrugated" with an amplitude of perhaps 500 km. This is a particular feature of this hotter model (rather less pronounced in the cooler ones), and leads to several consequences in the optical appearance of granulation across the stellar disk, as well as in asymmetries and shifts of the line profiles. Such a "corrugated" surface on F–stars could be predicted already in a simpler granulation model by Nelson (1980). Another model for granulation on *Procyon* was presented at this conference (Gadun, 1989).

4. The Optical Appearance of Granulation on Different Stars

The *output* from the numerical simulations consists of three–dimensional arrays of different parameters defining the stellar photosphere at each moment in time. These data were used as *input* for radiative transfer calculations to obtain the emerging continuum and line radiation for different spatial points, angles, wavelengths, and spectral line parameters. In Figure 2, synthetic continuum images are shown for the *Procyon* model, at the same representative instant of time as in Figure 1. As expected, the geometrical shapes of granules at stellar disk center closely correspond to the temperature features. However, the intensity *contrast* is much lower than could have been naively expected from the temperature contrast at a constant geometrical depth: the temperature–dependent opacities of overlying layers hide the larger temperature contrasts beneath. Nevertheless, by solar standards, the intensity contrast is rather high, and furthermore *increases* towards the stellar limb. This increase originates from the "corrugated" nature of the *Procyon* surface which, at larger inclination angles, occasionally makes it possible to glimpse hot and bright elements in deeper layers. This particular property is *not* shared by the cooler models, and the granulation contrast in e.g. the $T_{eff} = 5200$ K model is considerably lower and shows no tendency to increase towards the limb.

CONTINUUM BRIGHTNESS

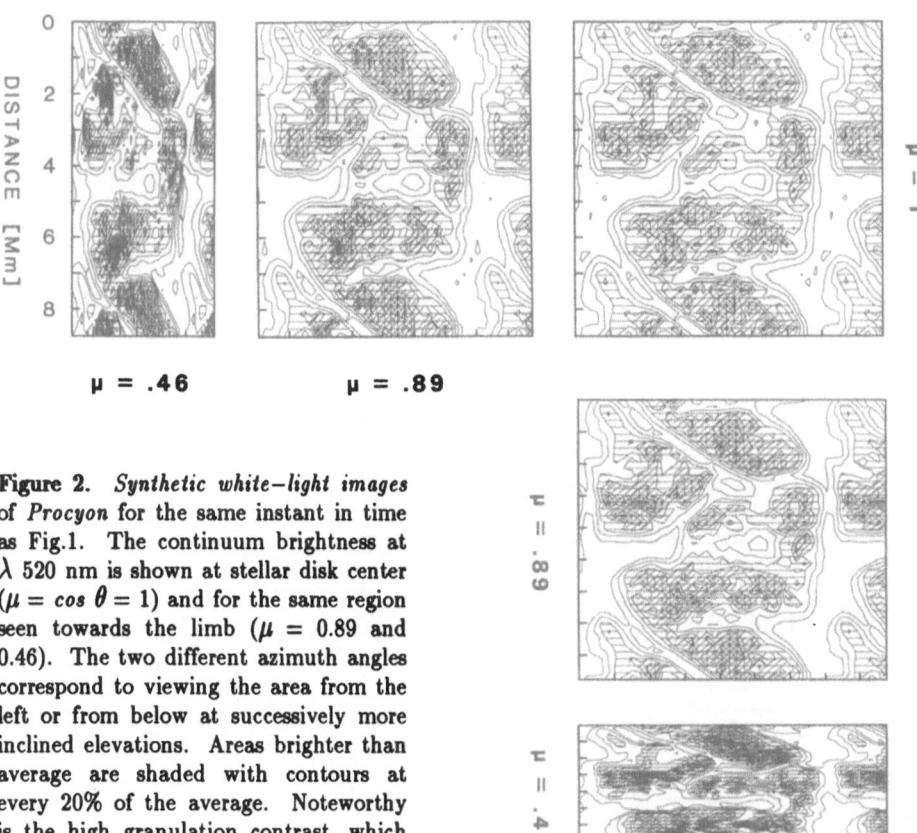

Figure 2. *Synthetic white—light images* of *Procyon* for the same instant in time as Fig.1. The continuum brightness at λ 520 nm is shown at stellar disk center ($\mu = cos\ \theta = 1$) and for the same region seen towards the limb ($\mu = 0.89$ and 0.46). The two different azimuth angles correspond to viewing the area from the left or from below at successively more inclined elevations. Areas brighter than average are shaded with contours at every 20% of the average. Noteworthy is the high granulation contrast, which *increases* from disk center towards the limb. (Dravins and Nordlund, 1989)

5. Line Formation in Stellar Photospheres

5.1. SPATIALLY RESOLVED LINE PROFILES

The simulation data were in particular used as an input to compute photospheric line profiles. The subsequent analysis of such profiles can give considerable insight in the physics of line formation in stellar photospheres and can predict observable quantities which can verify (or falsify) different models. Further, it may indicate the validity (if any) of classical approximations such as the concepts of *"micro-"* or *"macro-"* turbulence in stellar atmospheres.

The radiative transfer was solved for the inhomogeneous atmospheric structure along different rays throughout the simulation volume, for wavelengths corresponding to every 1 km/s within ± 10 km/s of the line's rest wavelength. For each spectral line, and each geometric angle, such computations were made for

500

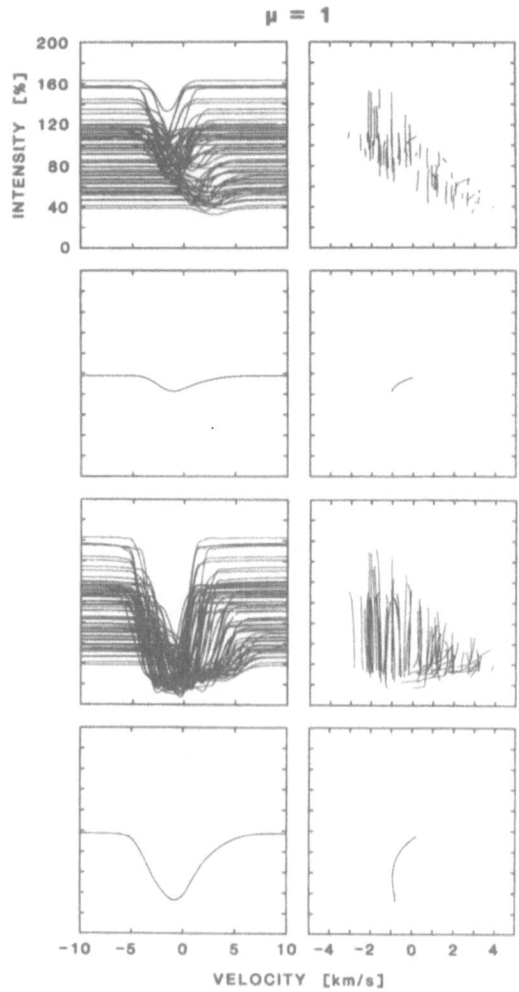

Figure 3. Spatially resolved line profiles at *Procyon* disk center, and their spatial averages at the same time as Figs.1–2. Line profiles and bisectors are shown for a grid of 8 × 8 = 64 spatial points out of the 32 × 32 = 1024 actually computed at each step in time. The top half of the figure shows data for a weak Fe I line, and the bottom for a strong one, both with $\chi = 3$ eV at λ 520 nm. 100% intensity corresponds to the average for the entire simulation sequence. The spatially averaged profile and its bisector at this instant in time are also shown. The average profile is *not* representative for spatially resolved points on the star: its width and asymmetry rather reflect the statistical distribution of spatial inhomogeneities. (Dravins and Nordlund, 1989)

$32 \times 32 = 1024$ horizontal spatial elements at each step in time, and for typically some 100 timesteps in each stellar simulation. The computations shown here were made with the assumption of local thermodynamic equilibrium.

Figure 3 shows a sample of such spatially resolved line profiles (and their spatial averages) at the *Procyon* disk center, at the same representative time as Figures 1–2. An important conclusion is that the spatially averaged profile is *not at all* typical for individual points on the stellar surface. The shape, asymmetry and shift of the average profile instead reflect the *statistical distribution functions* and *the ensemble average* of different profiles from different spatial points. The pronounced asymmetry of the spatially averaged profile is not frequent: it only occurs where there happen to be strong velocity gradients in the line–forming layers. Such intrinsic asymmetries are more common for strong lines with extended heights of formation, in particular in intergranular lanes. There the downflow velocities rapidly increase with depth, and this depth gradient can be manifest as asymmetries also in spatially resolved lines. This means that attempts to deduce stellar photospheric structure by interpreting observed line asymmetries as arising in horizontally homogeneous models, will likely lead to fortuitous depth–dependent velocity fields that are not at all present in any real stellar atmosphere.

The correlation between brightness and lineshift is particularly well visible for weaker lines (top panel in Figure 3). This of course indicates that hot elements generally are rising, and the correlation decreases near the stellar limb, where one mainly sees the effects of horizontal velocities. Closer to the limb, there is a noticeable increase in the velocity spread of the individual lines, reflecting that the horizontal velocities are generally of somewhat larger amplitude than the vertical ones.

5.2. SPATIALLY AVERAGED LINE PROFILES ACROSS STELLAR DISKS

Figure 3 also showed the *spatially averaged* line profiles at that particular point in time. New sets of line profiles were computed sufficiently often to follow the different phases of evolution in individual granular features over the full simulation sequence. Their spatial averages cover several granular features, and their time variability reflects not that of individual granules, but of larger convective patterns. The radiative flux *leaving* the simulation volume varies by some ± 10 % during the course of the simulation, although the convective flux *entering* the the bottom of the simulation volume remains constant. This time–variable accumulation and release of energy is manifest on the surface as occasional groups of especially bright and vigorous granules, perhaps followed by periods with fainter and apparently more disorganized granular features.

The effects of such processes is illustrated in Figure 4, which shows the longer time evolution of spatially averaged line profiles. The different fluxes at different times are visible as varying continuum levels. Just as in the spatially resolved case, there is a clear correlation between increased brightness and increased blueshift, at least near the stellar disk center. The scatter of the different line profiles around their average indicates the need to extend the model simulations for a sufficiently long time, in order to reach a stable average. It is noticeable that this spread increases towards the stellar limb. This phenomenon is connected with the "corrugated" nature of the stellar surface: near the limb, some parts of the surface are more or less hidden from view, leading to a more random character of the spatial averaging. The effect is known observationally for the Sun, where it forms one additional observational constraint on realistic models of granulation. For stellar models, it has the effect that, in order to obtain statistically stable line profiles for disk–integrated starlight (and thus for all values of $\mu = cos\ \theta$), the simulations must be extended over longer periods than would have been necessary for the disk center profiles only.

6. Line Profiles in Integrated Starlight

The steps of spatial and temporal averaging, leading to globally averaged line profiles were outlined in Figures 3–4. By suitable summation of such globally averaged data for different center–to–limb positions, disk–integrated line profiles are obtained. These profiles can be compared to observations, and constitute the

502

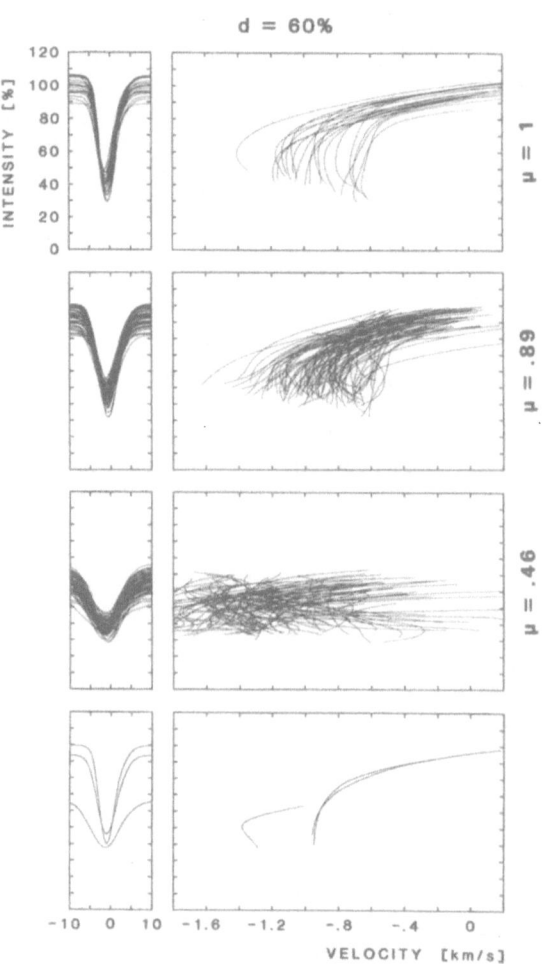

Figure 4. *Long–time evolution* over $1\frac{1}{2}$ hours of spatially averaged line profiles and bisectors in the *Procyon* model, leading to *global averages* (spatial and temporal) over the full simulation sequence. Each curve is a spatial average, time–averaged over 270 *s*. The sequence illustrates the amplitude of convective patterns after largely averaging away evolutionary effects in individual granules. Periods of more vigorous granulation may be followed by quieter ones, after which new granular patterns gradually develop. For $\mu = .89$ and .46, data for four different azimuth angles are shown. At bottom are the global averages at each μ for this $\chi = 3$ eV, λ 520 nm Fe I line of averaged disk–center depth $d = 60\%$ of the averaged continuum. The intensity scale is that of the average disk–center, and illustrates the limb darkening at $\mu = .89$ and .46. Just as on shorter timescales, the profile and bisector amplitudes around their averages increase from stellar disk center towards the limb. (Dravins and Nordlund, 1989)

most important diagnostic tool for stellar granulation studies.

6.1. EFFECTS OF STELLAR ROTATION

At this point it is necessary (for the first time in this modeling) to introduce a free parameter which can not be determined from outside the model: the projected stellar rotational velocity, *V sin i*. Stars obviously do rotate, and their rotational axes may have different inclinations relative to the Earth. Although, in principle, these quantities could be independently determined – e.g. the rotational period might be visible in the modulation of some chromospheric emission intensity – that is not trivial for ordinary stars of small rotational velocity. The main observable effect of stellar rotation is then the rotational broadening of spectral lines. Since both line asymmetry, rotational broadening, and limb darkening change with disk position, the disk–integrated profile reflects these properties in a complex manner.

This intercoupling might be turned to an advantage, possibly allowing one to determine not only the stellar rotational velocity, but also the line profile variations across stellar disks.

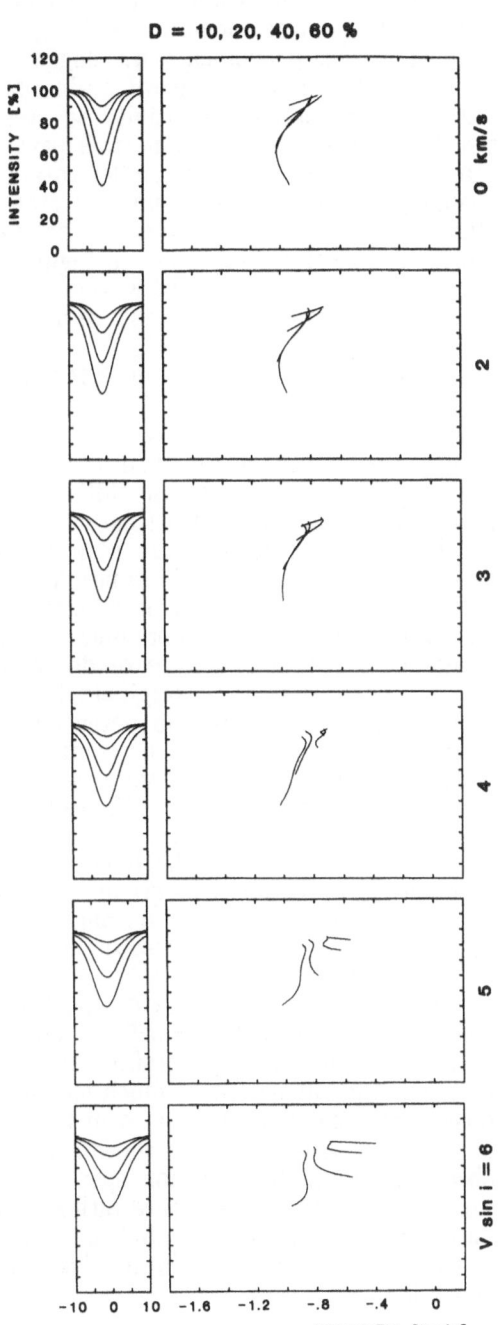

Figure 5. Line-depth dependence for line profiles and bisectors in the *disk-integrated flux* of the globally averaged *Procyon* model. Fe I lines ($\chi = 3$ eV; $\lambda = 520$ nm) with depths D in the *zero-rotation* flux spectrum of 10, 20, 40 and 60% of the flux continuum are shown for different values of the stellar rotation. This figure shows how the bisector patterns rapidly change when the rotational broadening reaches a significant fraction of the line-width. (Dravins and Nordlund, 1989)

Figure 5 shows disk–integrated line profiles and bisectors for different values of *V sin i* in the *Procyon* model. One might perhaps have expected that increased stellar rotation would merely smear out the line profiles and smoothen the bisectors. This also happens for modest rotational velocities. For more rapid rotation, however, when the rotational broadening reaches a significant fraction of the line–width, the bisector patterns change significantly, and the line asymmetries are enhanced. This phenomenon was suggested by Gray and Toner (1985; Gray, 1986), and studied in more detail for a simulated rapidly rotating Sun by Smith et al. (1987). The bisector sensitivity to stellar rotation indicates a potentially very sensitive method to determine this parameter. Since models predict spectral line properties not only for zero stellar rotation, but also their functional changes with changing rotation, they should be tested against observations of *groups* of similar stars with successively more rapid rotation. Such observations (e.g. a sequence of *Procyon*– like stars, analogous to Figure 5) may actually disentangle the changes of line profiles, asymmetries and wavelength shifts from center to limb, and thus give a first opportunity of realizing high resolution spectroscopy across stellar disks.

6.2. EXCITATION POTENTIAL, IONIZATION LEVEL, AND WAVELENGTH REGION

For the four models computed, the most pronounced differences in resulting line shapes are between those in different stars. However, for any given star, line asymmetries and wavelength shifts depend systematically on the line's atomic parameters, analogous to the solar case (Dravins et al., 1981; 1986).

The largest differences are seen among lines of different strength, reflecting different average heights of formation. The photospheric structure changes rapidly with height, and already slight differences in line formation conditions may lead to observable differences. For lines of a given strength, potentially observable differences exist among lines of different excitation potentials and/or ionization levels. These differences originate from the temperature sensitivity of the lines: those of a certain excitation potential will predominantly form in regions whose temperatures are sufficient to strongly populate the relevant atomic energy level, yet not so high as to ionize the species. Generally, Fe I lines of higher excitation potential show a larger blueshift, reflecting the more rapidly rising motion of the hotter elements, where these lines predominantly are formed. Smaller differences exist among otherwise similar lines in different wavelength regions. Such differences occur primarily because the granulation contrast changes with wavelength, and also because of changing continuum opacities. It is through such arrays of differential spectral line behavior that also detailed models of stellar granulation structure can be tested against observations.

Figure 6 illustrates some of these points. These synthetic line profiles and bisectors for the cool T_{eff} = 5200 K model should be compared to those for the 6600 K one in Figure 5. The profiles have rather different widths, differently shaped bisectors and (above all) a much smaller convective blueshift (only some 200 m/s rather than the 1000 m/s or so in the *Procyon* model). Lines of lower excitation potentials 1 and 3 eV are shown: with some effort, their differences might be observable in high–resolution spectra. These lineshapes encode the information from a stellar surface structure rather different from the *Procyon* one. The surface intensity contrast and the amplitudes of velocity and brightness are much smaller, and the stellar "surface" is far less "corrugated", leading to smaller line asymmetries and shifts.

Figure 6. Dependence on *excitation potential* for Fe I lines in the disk-integrated flux from the globally averaged model for the $T_{eff} = 5200$ K star α *Cen B* (K1 V). Lines with $\chi = 1$ and 3 eV are shown, each with zero-rotation depths $D = 30$ and 80% of the flux continuum at λ 520 nm. Although the line-depths are identical for zero rotation, their different center-to-limb behavior makes the $\chi = 3$ eV ones noticeably deeper already for $V \sin i = 3$ km/s. (Dravins and Nordlund, 1989)

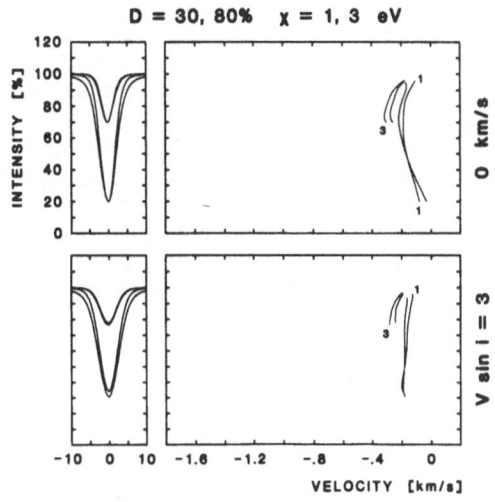

6.3. EFFECTS OF DEPARTURES FROM LTE

Effects of departures from local thermodynamic equilibrium were examined for some spectral lines by repeating the whole series of calculations, but now including an explicit treatment of the non–LTE iron ionization equilibrium, taking into account the radiation fluxes at the most important ionization edges of Fe I. This radiative ionization of iron above especially hot granules is believed to be the most important non–LTE effect for Fe I lines in the stars studied. Such hot granules are normally rapidly rising, and an LTE calculation generates strongly blueshifted absorption lines. These lines contribute relatively many photons to the average due to the high continuum brightness of the underlying granule. For such situations, a non–LTE calculation shows the neutral iron in the higher and cooler absorbing layers to be largely ionized by the strong ultraviolet flux from below, and a weaker absorption line results. In the global averaging, this means a smaller contribution from strongly blueshifted components, and thus a slightly smaller convective blueshift in the global average. Although this effect is clearly seen in different models, the amount of the effect is rather small, normally only a small fraction of the convective blueshift. Other non–LTE effects have not been studied, and nor have effects of multidimensional radiative transfer. It is not believed, however, that any of those effects will be even comparable in magnitude to the gross modifications of spectral line shapes and shifts caused by the dynamics in stellar surface convection.

6.4. FINITE SPECTROMETER RESOLUTION AND STELLAR OSCILLATIONS

Before detailed comparisons to observations can be made, the effects of the finite spectrometer resolution and possible asymmetries in its instrumental profile must be accounted for. Even if the instrumental profile is perfectly symmetric, its convolution with an intrinsically asymmetric stellar profile produces a line of different width, asymmetry and shift. Since different classes of stellar lines have different intrinsic asymmetries, they will be systematically affected, which demands particular care if differences between different classes of lines are searched for. The effects of finite spectral resolution are well visible in numerically simulated observations at different resolution (Dravins 1987a; Livingston and Huang, 1986).

Another effect that may have be to accounted for concerns *stellar oscillations* analogous to the global ("5–minute") solar ones. After stellar rotation, this is the second physical parameter in the modeling that, as yet, can not be independently determined. Such global oscillations are expected to occur over areas much larger than individual granules, so the expected effect in disk–integrated light is a slight "macroturbulent" broadening of the spectral line. On the Sun, and presumably on other solar–type stars, the oscillation amplitudes are much smaller than both granular velocities and normal spectrometer resolutions, and then the effect might simply be accounted for as a spectrum broadening with an "instrumental" profile with a width slightly larger than that due to the spectrometer alone.

7. Comparisons to Observations

The highest–resolution data available on stellar photospheric line profiles and their asymmetries have been obtained with the double–pass scanner of the coudé echelle spectrometer at European Southern Observatory (Dravins, 1987b). In this special scanner mode, the light passes an intermediate slit inside the spectrometer to effectively remove scattered light before making a second pass through the instrument. A very high spectral resolution ($\lambda/\Delta\lambda \simeq 200,000$) is obtained with a well–defined and very clean instrumental profile, but at a high cost in photon efficiency. Only a limited number of spectral lines in very bright stars could be observed in this manner.

The synthetic line profiles for disk–integrated starlight were convolved with profiles corresponding to the measured instrumental ones, but slightly broadened to include effects of plausible amplitudes of stellar surface oscillations. For each value of the stellar rotational velocity, this produces a grid of line profiles and bisectors for differently strong spectral lines. Observed line profiles and bisectors were then independently compared against these grids, and "best–fit" values of $V \sin i$ deduced.

The by far best–observed stars are *Procyon* and *α Cen A*, and Figures 7–10 show the resulting comparisons between synthetic and observed data. The synthetic data do not represent attempts to make the best fits to precisely the lines observed: the synthetic lines are computed for a representative excitation potential and wavelength region, which is not identical to that of each observed line.

7.1. PROCYON (F5 IV–V)

For *Procyon*, a fit of profiles for differently strong lines yields $V \sin i \simeq 3\ km/s$, while an independent fit to the bisector pattern yields $\simeq 2.7\ km/s$. As seen in Figure 7, the synthetic line *cores* agree well with observed ones, but there is inadequate absorption in the line wings outside about $\pm 5\ km/s$. The bisector patterns agree very well. Thus, while the *asymmetric* part of the line profiles is well modeled, there is still some *symmetric* broadening mechanism missing.

It is strongly believed that this is an artifact of the present hydrodynamically anelastic model, which does not handle sound waves. The granulation in *Procyon* is the most vigorous among our models, and horizontal velocities reach large amplitudes in the upper photosphere. Due to the inherent limitations in the anelastic approximation, the horizontal velocities had to be numerically constrained before becoming transonic. The removal of the anelastic approximation will allow

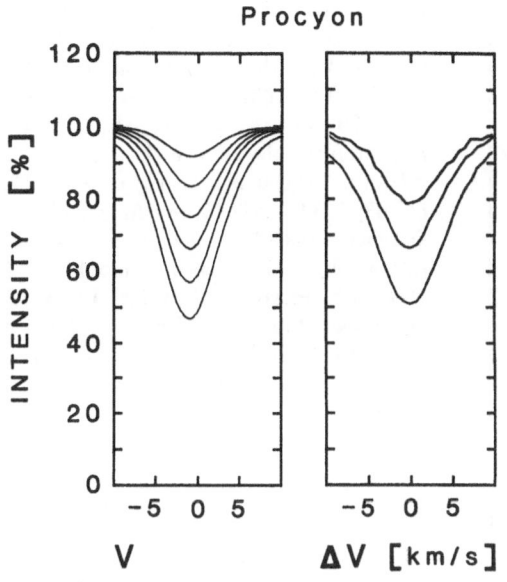

Procyon

INTENSITY [%]

V   ΔV [km/s]

Figure 7. *Synthetic* and *observed line profiles* for *Procyon*. Synthetic profiles ($\chi = 3$ eV, λ 520 nm) from the globally averaged simulation are shown at left. At right is a group of Fe I lines, observed with a spectral resolution $\lambda/\Delta\lambda \simeq$ 200,000 (λ 463.59, 536.54, 536.49 nm; for details see Dravins, 1987b). From a grid of rotationally broadened profiles, a "best–fit" rotation $V \sin i \simeq 3 \pm .5$ km/s was obtained. The synthetic profiles have in addition been convolved with a Gaussian of FWHM 2.1 km/s to approximate the spectrometer broadening and that from plausible stellar surface oscillations. The line *cores* agree well, but there is a somewhat inadequate absorption in the far line–wings. It is believed that this is an artifact of the present hydrodynamically anelastic model, where the horizontal velocities had to be numerically constrained before becoming transonic. (Dravins and Nordlund, 1989)

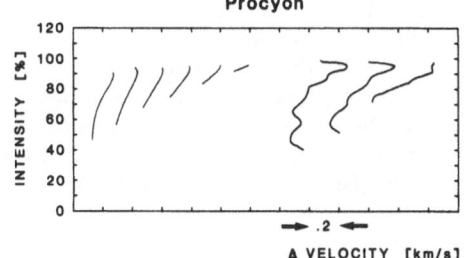

Procyon

INTENSITY [%]

Δ VELOCITY [km/s]

Figure 8. Comparison between *observed* and *"best–fit"* synthetic *bisectors* for *Procyon*, analogous to Fig.7. A bisector grid gives a best–fit stellar rotation value $V \sin i \simeq 2.7 \pm .5$ km/s, independently from profile fits. The strongest synthetic lines do not fully extend to the continuum because the line wings were only computed out to $\pm$ 10 km/s. The synthetic bisectors are placed without any absolute wavelength scale, similar to the observed ones at right. The observed data are averages for Fe I lines (Dravins, 1987b). The detailed agreement between synthetic and observed patterns shows that the *asymmetric* part of the *Procyon* lines is well modeled. (Dravins and Nordlund, 1989)

508

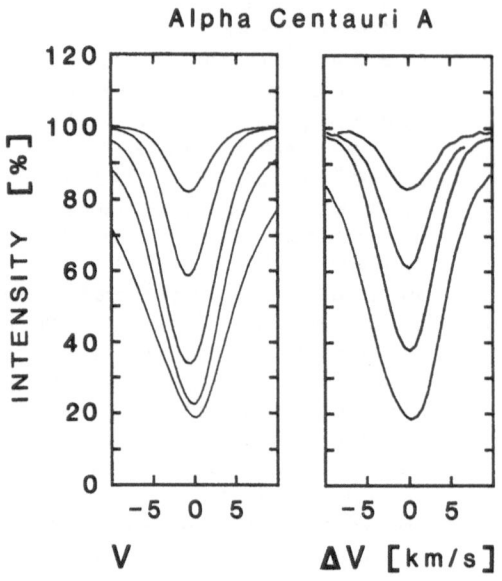

Alpha Centauri A

Figure 9. Comparison between *"best-fit"* synthetic and *observed line profiles* for the G2 V star α *Cen A*. Synthetic Fe I lines ($\chi = 3$ eV) with zero—rotation depths 20, 45, 69, 80 and 83% of the flux continuum at λ 520 nm are shown at left, and Fe I line profiles, observed with a resolution $\lambda/\Delta\lambda \simeq 200,000$ are at right (λ 685.57; 543.63; 633.53; 543.45 nm; for details see Dravins, 1987b). From a grid of rotationally broadened profiles, a "best-fit" rotation $V\,sin\,i \simeq 1.8 \pm .3$ km/s was obtained. There is a detailed agreement between observed and synthetic profiles. (Dravins and Nordlund, 1989)

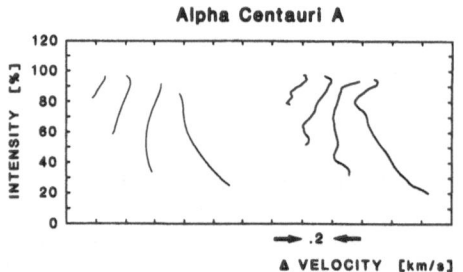

Alpha Centauri A

Figure 10. Comparison between *observed* and *"best-fit" synthetic bisectors* for α *Cen A*, analogous to Fig. 9. A bisector grid gives a best-fit stellar rotation $V\,sin\,i \simeq 1.7 \pm .3$ km/s, independently of profile fits. The observed bisectors are averages for Fe I lines from Dravins (1987b). The detailed agreement shows that also the asymmetric part of the α *Cen A* lines is well modeled. This model differs from the solar one only by slightly lower surface gravity. This leads to a slightly more vigorous convective overshoot, causing the characteristic signature of a more pronounced asymmetry in the cores of the strongest lines. (Dravins and Nordlund, 1989)

sound and shock waves to develop and contribute symmetric line broadening from horizontal features moving at sonic speed. Such modeling has now been carried out for the Sun (Stein and Nordlund, 1989) and the application of such more realistic models is a high priority item for better understanding the photosphere of *Procyon*.

7.2. ALPHA CENTAURI A (G2 V)

A certain effort had been put into the observations of α *Cen A* since, as a star with parameters very similar to solar ones, it was to be used as a stepping stone out into the Hertzsprung–Russell diagram. α *Cen A* is of the same spectral type as the *Sun* (G2 V) but, being slightly evolved, is slightly more luminous and occupies a place in the upper portion of the main–sequence band. The model parameters were $T_{eff} =$ 5800 K, at one half the solar surface gravity. As seen in Figures 9–10, the fit between observations and theory is excellent. The profile fit for differently strong lines yields $V \sin i \simeq 1.8$ *km/s*, in full agreement with the independent bisector fit giving $\simeq 1.7$ *km/s*. To the extent that this bisector fit can be considered an independent determination of stellar rotation, the stellar profiles of Figure 9 are predicted without any adjustable physical parameters: no "mixing–length", no "micro–" nor "macro–turbulence".

Despite the similarity to the Sun, the strongest lines in α *Cen A* are clearly more asymmetric than corresponding solar ones (Figure 10; compare with solar bisectors plotted on the same format in Dravins, 1987b). This enhanced asymmetry is very well reproduced by the model. This difference to the Sun can be traced back to be an effect of the lower surface gravity in α *Cen A*, which permits a more vigorous convective overshoot and a slightly larger vertical velocity amplitude in the high photosphere, where the cores of the strong lines often are formed. These velocities give contributions to the absorption not only near the spatially averaged line core, but also further out in the line flanks, in a manner to produce this distinct bisector signature. It is encouraging that current observations and models permit the study of such subtle differences between the velocity fields in two G2 V stars, revealing the effects on solar granulation of some 3 billion years of stellar evolution.

7.3. BETA HYDRI (G2 IV) AND ALPHA CENTAURI B (K1 V)

These stars are fainter, and the observations that could be made with the high spectral quality offered by the ESO double–pass scanner were more limited in both quantity and spectral resolution. Although this precludes equally detailed comparisons as those for *Procyon* or α *Cen A*, there is certainly a general agreement between theory and observations. Somewhat inadequate line absorption in the line wings of the subgiant β *Hyi* suggests (as in the *Procyon* case) that the missing broadening is an artifact of the anelastic approximation, where the largest horizontal velocities on β *Hyi* had to be numerically constrained before going transonic. There is no such disagreement in the narrower line profiles originating in the less violent photosphere of α *Cen B*, and its '\ '–shaped bisectors in the stronger lines are also well reproduced.

8. Granulation Modeling in Non Solar–Type Stars

Powerful as full numerical simulations may be, at the present time they still have limitations when it comes to modeling granulation in situations very different from the solar one. The problems arise mainly from the extensive computing power required, and the multidimensional nature of the problem. This makes it computationally very expensive to greatly increase the number of spatial resolution points, and thus the range of geometrical scales to be computed must be carefully optimized. This can be done for the Sun, where the sizes of granular features are rather well known, and intelligent guesses can also be made for those in other solar–type stars. However, to reliably estimate the relevant scales in giant stars, say, is much more difficult, and it has even been suggested that photospheric convection cells in red giants may subtend a significant fraction of the stellar surface (Schwarzschild, 1975). To be certain that features of all relevant sizes are modeled, one would need to greatly increase the number of resolution points in the horizontal coordinates, in the vertical one, as well as the time resolution. The timestep has to be sufficiently small to numerically resolve the motions not only at the stellar surface, but also those that may develop beneath it.

Further, any numerical simulation shows how a physical system evolves from the initial state. If these initial conditions (in terms of the structure of temperature inhomogeneities, say) are significantly different from the later statistical average, the simulation will show the gradual evolution from this initial state towards the statistically stable situation. This means that if the initial conditions for e.g. the temperature structure are atypical with, perhaps, cooler structures near the surface, the simulation will show how such cooler structures gradually decay and evolve into normal granulation. While this might conceivably be of interest to show the decay of some hypothetical starspot, it will not give representative data on normal stellar granulation until after the simulation has run for a time corresponding to some thermal timescale of the simulation volume. Unfortunately, such a timescale might compare unfavorably with the timescale of the possible computer time allocation. For our models discussed above, initial conditions were prepared starting from scaled solar models, which were verified by trial and error to actually be able to transport one stellar luminosity outward (rather than accumulate heat at the bottom, or to cool off the surface layers more rapidly than the energy replenishment from below).

While these circumstances should not deter future modeling of different stars, they do make it awkward to produce, in the immediate future, a *large* number of full simulations throughout the Hertzsprung–Russell diagram. Therefore, simpler and parametrized models must initially be applied. It is not suggested that such simple models could in any way replace the ultimate need for detailed modeling: their purpose is only to permit a first exploration of granulation properties for non–solar type stars.

8.1. FOUR–COMPONENT MODELS FOR STELLAR GRANULATION

From both solar observations and numerical simulations for stars, it is clear that spatial inhomogeneities is a main cause of the disk–averaged line asymmetries. In a simpler model, one should consequently somehow parametrize such inhomogeneities. After trying several alternatives, a *four–component* description of stellar photospheres was adopted. This includes hot and rising *"granules"* and *"granular centers"*, average–temperature and stationary *"neutral areas"*, and cooler and sinking *"intergranular lanes"*. Although each of these components may have different surface areas, velocities and brightnesses, all of these are not free parameters. Mass conservation requires the upflows in granular centers and granules to be balanced by a downflow in the intergranular lanes, and the spatially averaged brightness is equal to that of the "neutral areas". This means that, e.g., the intergranular lane properties are determined from the areas, brightnesses and velocities of the granules and granular centers, together with the area taken up by the neutral areas.

To synthesize a line profile, a summation of "spatially resolved" line profiles from the different components is made, appropriately weighted for their respective area coverage, brightness and Doppler shift due to the vertical velocity. For such line profiles of different strength, Voigt profiles were used, with Lorentzian damping wings in the strongest lines. The temperature (brightness) dependence of the line absorption was set to approximate that actually observed in stars of the relevant temperature range. No effort is made to model center–to–limb changes across the stellar disk (this would introduce too many free parameters), and the description is thus only that for a plane parallel stellar surface viewed head–on.

8.2. "INVERSE" LINE ASYMMETRIES IN EARLY–TYPE STARS

A particular challenge is to understand the origins of the "inverse" photospheric line asymmetries observed in early–type stars. While solar bisectors are often similar to the letter 'c', bisectors in F–giants have a forcefully *inverted* 'ɔ' shape. First observed for *Canopus* (F0 II) by Dravins and Lind (1984), this seems to be a general feature for supergiants earlier than about G0 (Gray and Toner, 1986). The generally more rapid rotation in early main–sequence stars makes their spectra more difficult to study, but there are indications of similarly inverted asymmetries in the relatively sharp–lined spectra of the dwarfs *Sirius* (A1 V; Dravins, 1987b) and τ *Sco* (B0 V; Smith and Karp, 1978; 1979).

While one does not expect "ordinary" surface convection driven by hydrogen ionization in the atmospheres of such hotter stars, there is no reason to believe their atmospheres should be static. Organized motions could well arise due to the He II ionization zone just below the photosphere, or through some other mechanism. Detailed models for the A0 V star *Vega* by Gigas (1988) indeed show its atmosphere to be hydrodynamically unstable, and developing an oscillatory behavior.

8.3. A FOUR–COMPONENT MODEL FOR GRANULATION ON CANOPUS

Observed photospheric line profiles in the bright giant *Canopus* (F0 II) show a depressed intensity in the far shortward wing of the line, leading to a characteristic blueward bend of the bisector near the continuum (Dravins, 1987b). One could wonder whether this might indicate an expanding atmosphere, with blueshifted absorption arising from an accelerating stellar wind. The argument against this is that the same type of asymmetry is seen in *all* photospheric lines, not predominantly in the strongest ones formed at greater heights, as would have been expected for a stellar wind. Thus these line asymmetries seem to be of photospheric origin. To explore a granular interpretation , the four–component model was applied.

The resulting "best–fit" to line asymmetries in *Canopus* is shown in Figure 11. Although there are slight differences between the fits for differently strong lines, a rising component at a Doppler shift of some 10 *km/s* is required to reproduce the observed bisectors. The strength of this component need not be large: it is sufficient to cover only some 10 % of the star with such rapidly rising granules, whose gases descend over a somewhat larger area (20 %, say), while the rest of the surface remains quiescent. These model fits are not especially sensitive to variations in continuum intensity, and for this reason the "best–fit" models were chosen as the simplest possible ones, without any intensity fluctuations at all.

Noticeable is how the modeled bisector changes shape between weak and strong lines: the blueward bend begins quite near the continuum in the faint line, but already halfway down the strong one (in agreement with observations). This originates from the extended Lorentzian wings in the spatially resolved strong line,

512

Figure 11. Four–component model of granulation on Canopus (F0 II), reproducing the line asymmetries for differently strong lines observed by Dravins (1987b). The upper curve in each panel shows the resulting absorption line after adding the contributions from the four different spatial components, seen at bottom. With successively more positive Doppler shifts, these are those from granular centers, granules, neutral areas, and intergranular lanes. Each contribution reflects that component's area coverage, velocity and photometric brightness. The asymmetry of the resulting line profile is shown by its dashed bisector, plotted on the tenfold expanded upper scale. The strong inversely 'C'–shaped bisectors are reproduced for velocity fields with rapidly rising motions covering only a very small fraction of the stellar surface. The pronounced damping wings for the strongest line are a primary cause for a different bisector curvature for these lines (bottom panel).

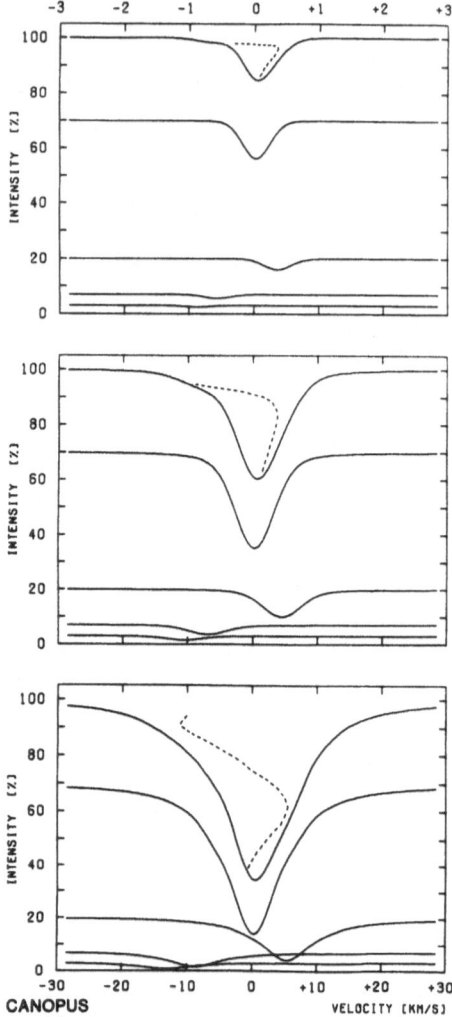

which actually makes it difficult to produce granulation models with a blueward bend very near the continuum in the strongest lines. Gray and Toner (1986) also applied some simpler modeling of line asymmetries in F supergiants, and arrived at similar conclusions: rapidly rising velocities of up to 25 km/s.

One might speculate what physical mechanism that possibly could constrain the upflow to such narrow channels. Possibly, the steep temperature dependence of the negative hydrogen ion opacity could be responsible. For solar surface temperatures, the opacity rapidly rises with increasing temperature, an effect that aids in constraining intergranular downflows by insulating their cool gases with hotter high–opacity material. Gases rising to the surface of a hotter star are initially too hot for the H^- opacity to be significant (> 10,000 K), but when approaching the surface, the temperature drops, and the increased H^- opacity that results might insulate the centers of the hot streams from the surrounding cooler material.

A visualization of the surface of *Canopus*, featuring rapidly upwelling currents in small ascending "geysers", whose output is balanced by a more sluggish downdraft over more extended areas, is certainly quite different from the familiar image of solar granulation. Since the photosphere is responsible for supplying the energy for heating stellar chromospheres and coronae, such differences in the granulation structure are certain to manifest themselves also in other aspects of stellar activity.

8.4. APPLYING FOUR-COMPONENT MODELS FOR SOLAR-TYPE STARS

This simple four-component modeling was applied to fit the observed bisector patterns also in other stars. Line profiles were computed for a large number of different four-component models, comprising virtually all reasonable combinations of areas, velocities and brightnesses for the four photospheric components. Besides for the modeling of non-solar type stars, the purpose of this grid was to understand how various hypothetical stellar surface patterns could manifest themselves in the resulting line profiles, and to get an indication of how well such descriptions could approximate the detailed hydrodynamic models now available for some stars.

Figure 12 shows a graphic summary of such model fits to two early-type stars, and to those for the *Sun* and *α Cen A*. For the latter two, detailed hydrodynamic models exist, and these fits can be compared to those. It is reassuring that already the simple 4-component description identifies the most characteristic surface properties of solar-type granulation: relatively large areas of bright upflows at a few *km/s*, balanced by more rapid and concentrated downflows of darker material over smaller areas. For *α Cen A* the same qualitative appearance is seen, and the somewhat larger velocity amplitudes identified. Encouraged by such a reasonable reproduction of solar-type granulation properties, one could then accept the deduced properties of earlier-type stars at face value, and examine what the consequences might be on photospheric magnetic fields and stellar activity.

8.5. A DIVIDING LINE FOR STELLAR ACTIVITY IN THE HR-DIAGRAM

The energy supply to stellar chromospheres and coronae must somehow come from the photosphere below. Although the details of that energy transfer are not at all completely understood, it is clear that a key role is played by the magnetic field. Irrespective of how that field has been generated, the field has to emerge through the photosphere, and will be modified by the granulation patterns there. The high electric conductivity of the photospheric material makes the magnetic fieldlines more or less "frozen-in", and they will be affected by the gas flows. Numerical simulations including also the three-dimensional effects of magnetic fields have been made for solar granulation by Nordlund (1986). Such simulations (as well as a general discussion of the situation) show that the interiors of granules are rapidly emptied from magnetic fields, which are instead advected to the intergranular lanes, forming magnetic flux concentrations. These flux concentrations are in quasi-static equilibrium with their non-stationary granulation surroundings, and through the granular evolution they are continually distorted and displaced.

As discussed in Section 3, one of the important topological effects seen in simulations of granulation in solar-type stars, is the development of spatially concentrated downdrafts, carrying a considerable fraction of the stellar convective energy flux. The conservation of angular momentum in these downdrafts leads to rapidly rotating motions below the visible surface. It is near these downdrafts in the intergranular lanes that the magnetic flux concentrations in solar-type stars do cluster. Even if the exact mechanisms are not known, one could argue that:

514

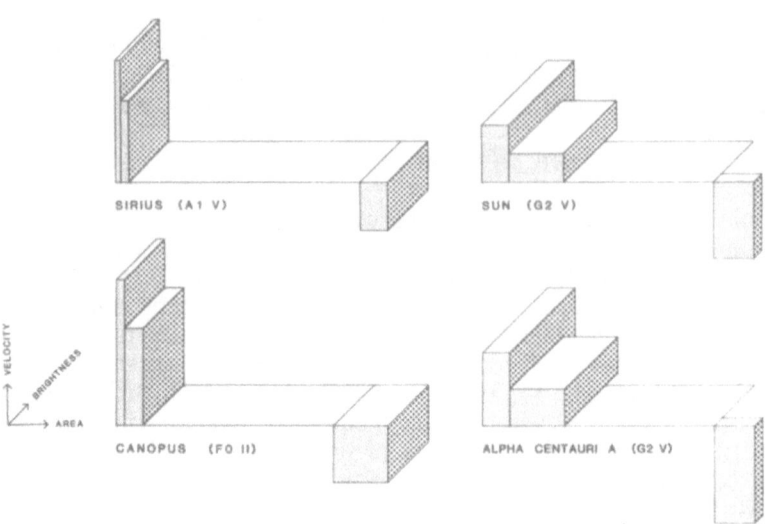

Figure 12. Graphic summary of the best–fit 4–component granulation models for different stars. The horizontal axis represents stellar surface area. From left to right, the horizontal extent shows the area coverages of granular centers, granules, neutral areas and intergranular lanes, respectively. The amplitudes of rising velocities are indicated upwards, and sinking ones downwards beneath the horizontal plane of the stationary neutral areas. Enhanced brightness is shown by an extent deeper than the average–brightness slab representing neutral areas. This summarizes the differences between earlier–type stars with their small concentrated upflows, and solar–type stars with larger granular areas, more concentrated downflows, and greater brightness fluctuations.

(1) In *solar–type stars*, the quiet–region magnetic fields may subtend only a small fraction of the stellar area. However, this area largely overlaps the narrow intergranular lanes where the magnetic fields must coexist with vigorously downstreaming and rotating motions in gases that carry a considerable fraction of the stellar convective energy flux. Even if the mechanisms for the coupling of mechanical energy to magnetic fieldlines are not well understood, it seems obvious that these velocities could be an important source for hydromagnetic wave disturbances, transporting energy up into the chromosphere and corona.

(2) For *early–type stars*, the magnetic fields can be similarly expected to be expelled from the concentrated granules through their horizontally outward velocities. Having been expelled from the granules, the magnetic flux is then dispersed over the more extended downdraft areas, lacking any obvious dynamic mechanism for flux concentration. Not only will the flux not be able to concentrate, but it will also not have spatial access to most of the dynamic energy of the photosphere, which is concentrated in the rapidly rising granules. Thus, the mechanism for tapping mechanical energy from the photosphere up into the chromosphere and corona may be reduced by orders of magnitude, and in such earlier–type stars only rather feeble chromospheric heating through magnetic effects could be expected.

515

9. Conclusions and Outlook For the Future

Perhaps the most important conclusion that can be drawn from the above discussion is that it *is* indeed possible to construct detailed models for stellar granulation, and that such models *can* be verified against existing observations. Foremost among new observations that are needed to further constrain or verify existing models are measurements of *differential wavelength shifts* among different classes of lines in the same star, and between different stars. Although high accuracies are required, such observations should be possible already with existing types of instrumentation. A confrontation between different models and wavelength shift observations could greatly strengthen the theoretical base for analyzing different classes of stars.

There clearly is a need to extend detailed granulation modeling across the Hertzsprung–Russell diagram. As discussed in Section 8.5, there is a tantalizing possibility that the presence or absence of chromospheric activity in different stars is directly connected to the different granulation topologies in their photospheres, but this suggestion needs detailed simulations for verification. In a shorter time perspective, it would be most important to perform simulations that include wave propagation and shock formation also for non–solar conditions, in particular for *Procyon*, where the "missing" spectral line absorption suggests material moving horizontally at near–sonic velocities.

In a longer time perspective, there is a need for *greatly* increased computing power to handle more realistic astrophysical situations. Even if computer performance may increase by perhaps *one* order of magnitude from one generation of supercomputers to the next, this is by far not sufficient to gain an order–of–magnitude better physical understanding of the situation on stellar surfaces. Fortunately, workers in computer science are currently putting considerable effort into the development of various "non–classical" concepts, e.g. employing novel computer architecture with many processors working in parallel. (For a general introduction to such concepts, see e.g. Bowler et al., 1987.) The greatly improved possibilities for computer–aided design and manufacture of unique electronic circuits at not unreasonable cost promises to make it possible to build a computer with circuits custom–designed for solving the problem of stellar surface convection. The combination of such circuits with novel computer architecture might well improve computing power by *several* orders of magnitude over today's "supercomputers", and then make it possible to realistically simulate the hydrodynamics of an entire stellar atmosphere, following the interplay between convective motions on different scales, global oscillations, and predicting not only spectral line profiles and stellar surface appearance, but also the spectrum of global oscillations, the variations in stellar irradiance, and perhaps also the development and decay of magnetically active regions. At last, then, will stellar granulation modeling have come of age !

ACKNOWLEDGEMENT

This work is supported by the Swedish Natural Science Research Council

References

Bowler, K.C., Bruce, A.D., Kenway. R.D., Pawley, G.S., Wallace, D.J.: 1987, *Physics Today* **40**, No. 10, p. 40
Dravins, D.: 1982, *Ann.Rev.Astron.Astrophys.* **20**, 61

516

Dravins, D.: 1987a, *Astron.Astrophys.* **172**, 200
Dravins, D.: 1987b, *Astron.Astrophys.* **172**, 211
Dravins, D., Larsson, B., Nordlund, Å.: 1986, *Astron.Astrophys.* **158**, 83
Dravins, D., Lind, J.: 1984, in S.L.Keil, ed. *Small−Scale Dynamical Processes in Quiet Stellar Atmospheres*, National Solar Observatory, Sacramento Peak, p. 414
Dravins, D., Lindegren, L., Nordlund, Å.: 1981, *Astron.Astrophys.* **96**, 345
Dravins, D., Nordlund, Å.: 1989, *Astron.Astrophys.*, to be submitted
Faulkner, J., Griffiths, K., Hoyle, F.: 1965, *Monthly Notices Roy.Astron.Soc.* **129**, 363
Gadun, A.S.: 1989, in R.F.Rutten, G.Severino, eds. *Solar and Stellar Granulation*, in press
Gigas, D.: 1989, in R.F.Rutten, G.Severino, eds. *Solar and Stellar Granulation*, in press
Gray, D.F.: 1986, *Publ.Astron.Soc.Pacific* **98**, 319
Gray, D.F., Toner, C.G.: 1985, *Publ.Astron.Soc.Pacific* **97**, 543
Gray, D.F., Toner, C.G.: 1986, *Publ.Astron.Soc.Pacific* **98**, 499
Livingston, W., Huang, Y.R.: 1986, in M.S.Giampapa, ed. *The SHIRSHOG Workshop*, National Solar Observatory, Tucson, p. 1
Nelson, G.D.: 1980, *Astrophys.J.* **238**, 659
Nordlund, Å.: 1982, *Astron.Astrophys.* **107**, 1
Nordlund, Å.: 1985, *Solar Phys.* **100**, 209
Nordlund, Å.: 1986, in W.Deinzer, M.Knölker, H.H.Voigt, eds. *Small Scale Magnetic Flux Concentrations in the Solar Photosphere*, Göttingen, p. 83
Nordlund, Å., Dravins, D.: 1989, *Astron.Astrophys.*, to be submitted
Schwarzschild, M.: 1975, *Astrophys.J.* **195**, 137
Smith, M.A., Huang, Y.R., Livingston, W.: 1987, *Publ.Astron.Soc.Pacific* **99**, 297
Smith, M.A., Karp, A.H.: 1978, *Astrophys.J.* **219**, 522
Smith, M.A., Karp, A.H.: 1979, *Astrophys.J.* **230**, 156
Stein, R.F., Nordlund, Å.: 1989, in R.J.Rutten, G.Severino, eds. *Solar and Stellar Granulation*, in press
Winkler, K.H.A., Chalmers, J.W., Hodson, S.W., Woodward, P.R., Zabusky, N.J.: 1987, *Physics Today* **40**, No. 10, p. 28

Discussion

FOX — Given the degree of comparison of your stellar models, can we now attribute the energy-carrying scales in these stars to what we know as granulation, or is it possible that there is another energy-carrying scale?

DRAVINS — The hydrodynamical models have been computed so far for solar-type stars only. Here, similar to the solar case, it appears that the energy transport to the surface can be well described on granular scales. However, the corresponding problem for non-solar type stars (e.g. red giants) is much more unclear.

WÖHL — You pointed out the importance of stellar rotation on the bisectors. What about the influence of differential rotation on the bisectors? Could you improve your fits or determine differential rotation?

DRAVINS — The accuracy of the synthetic line profiles is currently limited by approximations in the physical description of convection, and by the limited number of center-to-limb positions computed. Although they do permit comparisons with observed profiles for stars of different rotational velocities, any certain identification of higher-order effects is likely to require slightly more detailed models.

However, the level of modeling already used for the Sun, as presented by Nordlund and Stein at this meeting, should probably be sufficient. Thus, it seems likely that stellar models may already in the near future be sufficiently detailed to allow searches also for differential stellar rotation.

TITLE — How do you include stellar oscillations?

DRAVINS — We don't.

BECKERS — 1). Can you synthetize the convective blue shift of strong lines like the Mg-b lines or the Na-D lines? For the Sun, the convective blue shifts for these lines disappear, thus giving a zero point reference. Does that hold for these stars?
2). How do you determine radial velocities with Hipparcos?

DRAVINS — Ad 1): the current models are designed to permit accurate line profile calculations in the middle and lower photosphere only. They are *not* expected to accurately model all the physical processes around the temperature minimum or in the chromosphere. Consequently, they are not suited for accurate calculations of the cores of very strong lines such as Mg-b or Na-D.

Ad 2): astrometric radial velocities can realistically be obtained for stars in moving clusters, such as the Hyades. This is to be done by inverting the classical problem of distance determination from radial velocities and proper motions towards the cluster apex. Once the distance is accurately measured from trigonometric parallaxes, the varying proper motions across the cluster can be used to determine the heliocentric radial velocities, only assuming a constant velocity vector for all stars in the cluster.

VAN BALLEGOOIJEN — Is it possible to use binaries consisting of a hot and a cool star to measure the difference between the convective blueshifts of the two stars?

DRAVINS — That is indeed one of the suggested methods to search for different convective shifts in different stars. The orbit-averaged system velocity should be equal for both components, so that any remaining differences can be attributed to other causes, primarily gravitational and convective shifts.

STEFFEN — In real stellar atmospheres there must be substantial turbulent velocities on scales smaller than your grid size. Could this explain some of the missing line broadening?

DRAVINS — These simulations are not intended to realistically reproduce all the physical phenomena in upper photospheres, where the cores of spectral lines are formed. Unexplained line broadening is only one of them.

VINCE — How many free parameters do you use in your stellar spectral line bisector calculations?

DRAVINS — The number of free physical parameters is only three. They are:
1). Effective temperature, which determines the energy flux into the model.
2). Stellar surface gravity, which determines the vertical stratification in the model.
3). The chemical composition, which determines the opacities.
To be able to carry out the calculations it is of course necessary to choose suitable timesteps, spatial scales, physical approximations etc., but none of these parameters is treated as a variable in the models.

FOX — I assume that for the initial stellar models, you assume solar-type parameters (composition). Indications from stellar seismology on Procyon and α Cen A point to non-solar parameters. Could this have an effect on the comparisons?

DRAVINS — The chemical composition influences the models through the ensuing opacities. However, in order to have any significant effect on the model, drastic changes in the chemical composition are required. It is quite plausible that granulation in an extremely metal-poor star would be different from an otherwise similar solar-composition one, but the moderate differences discussed from stellar seismology results will not give such effects.

RUTTEN — There is a problem with these large-scale simulations for those researchers who wish to do smaller-scale experiments on the simulations without repeating the full effort. It is quite important to find a way of providing a sort of compressed computer file, with results intermediate between the full simulation and the final papers.

DRAVINS — I certainly agree, at least in principle. The practical problem is that *each* of our models is described by approximately 1000 Mbyte of data. We intend to publish some statistical data on e.g. the spatial scales, the temperatures and the radiative fluxes at different depths, in order to enable comparisons with other models and the construction of simpler, parametrized ones.

STEIN — The effect you find of convection being hidden by an overlying haze in your cool star is also seen when comparing Saturn and Jupiter. Saturn is cooler, and the atmospheric motions that are so apparent in pictures of Jupiter are nearly hidden by an overlying haze on Saturn.

DRAVINS — Very interesting. This illustrates that convection is a very important process in many different areas of astrophysics.

THREE-DIMENSIONAL SIMULATION OF CONVECTIVE MOTIONS IN THE PROCYON ENVELOPE

I.N.Atroshchenko, A.S.Gadun, R.I.Kostik
The Main Astronomical Observatory of the Academy of
Sciences of the Ukr.SSR
252127 Kiev Goloseevo, USSR

ABSTRACT. Scheme of three-dimensional numerical simulation (in the first approach) of convective motions in the Procyon photosphere, superadiabatic region and the results of calculations are given. Their short analysis is given. We notice positive correlation between tem - perature and vertical velicity fluctuations in the modeled region.

INTRODUCTION

Direct numerical simulation of dynamical effects in the solar and stellar envelopes has two main aims: to verify the physical processes treatment and their influence upon the solution; to obtain the sequence of three-dimensional nonhomogeneous models.

The verification of three-dimensional nonhomogeneous models must be run for the wide spectrum of empirical data (Nordlund,1984; Wohl, Nordlund,1985) such as: energy characteristics of models (the value of radiative flux, centre-to-limb darkening, energy distribution in the spectrum), statistical features of mechanical motions, reproduction of observed spectral line profiles. We can make conclusions about the quality of our models from the correspondence between the theoretical and empirical data.

The attempt of three-dimensional simulation of convective motions in the Procyon envelope is made in the present work. As the information about Procyon granulation layer is absent, so we can run verification of the models only through energy characteristics and spectral line profiles calculated on the basis of these models. We simplify the prob- lem more. Our requirements to the calculated models are: the models must reproduce observed integrated radiative flux and asymmetry of weak and moderate absorption lines.

521

R. J. Rutten and G. Severino (eds.), Solar and Stellar Granulation, 521–531.
© 1989 by Kluwer Academic Publishers.

1. IDEOLOGY OF NUMERICAL SIMULATION

Simplified task reduces considerably the requirements to the ideology of the "numerical experiment" and allows to apply the approach developed by Gadun(1986) for the construction of three-dimensional nonhomogeneous radiative gasodynamical models of solar envelope. This approach is based on the fact that large-scale motions at the large Reynolds number have sufficient "autonomy" i.e. they depend little upon the details of the dissipation mechanism. And because large-scale turbulence structures determine mainly the dynamics of the motion, we can use the numerical method of the first order of accuracy to obtain the solution of nonlinear averaged Reynolds equations for stellar envelopes at the short time of simulation and to apply approximative description of transfer phenomena on the sub-grid level.

We use simplified treatment of radiative energy transport in the LTE approach assuming that the absorption coefficient does not depend on the frequency and neglecting the radiative transport in the horizontal direction.

2. THE BASIC SYSTEM OF AVERAGED EQUATIONS

We shall consider the nonrotating star without magnetig and electric fields neglecting spherical effects. We shall describe the matter motion in the modeled region by the equations for compressible, gravitationally-stratified medium in the presence of radiation field. Applying Reynolds averaging procedure we receive:

$$\begin{cases} \dfrac{\partial \rho}{\partial t} + \dfrac{\partial \rho v_j}{\partial x_j} = 0 , \\[2mm] \dfrac{\partial \rho v_i}{\partial t} + \dfrac{\partial \rho v_i v_j}{\partial x_j} = -\dfrac{\partial p}{\partial x_i} - \dfrac{\partial}{\partial x_j} R_{ij} - \rho g \delta_{i3} , \\[2mm] \dfrac{\partial \rho E}{\partial t} + \dfrac{\partial \rho E v_j}{\partial x_j} = -\dfrac{\partial p v_j}{\partial x_j} - \dfrac{\partial}{\partial x_j} R_{ij} v_i + R_{ij} \dfrac{\partial v_i}{\partial x_j} + \rho q_D - Q_r - \rho g v_3 . \end{cases}$$

$$(1)$$

We neglect the mass and energy transfer effects on the sub-grid level as compared with convective velocities and molecular viscosity as compared with turbulent one. $R_{ij} = \rho \overline{v_i' v_j'}$ is the turbulent Reynolds tensor, q_D is the dissipation energy of sub-grid turbulence, $Q_r = \dfrac{\partial q_{r3}}{\partial x_3}$ is the divergence of radiative energy flux, E is the total energy (the sum of internal energy (e) and kinetic energy (V^2) of averaged flow). R_{ij} and q_D were calculated from the averaged velocity field:

$$R_{ij} = \frac{2}{3} \rho (q + 6 e_{\kappa\kappa}) \delta_{ij} - 2 6 e_{ij} , \qquad (2)$$

where 6 is the kinematic coefficient of turbulent viscosity in the Smagorinsky approach (Smagorinsky,1963) :

$$\sigma = \frac{2\,(c_\sigma \Delta)^2}{\sqrt{2}}\,(e_{ij}\,e_{ij})^{1/2}, \tag{3}$$

where $e_{ij} = \frac{1}{2}\left(\frac{\partial v_i}{\partial x_j} + \frac{\partial v_j}{\partial x_i}\right)$ is velocity deformation tensor. Kinetic energy of sub-grid turbulence:

$$q = \frac{1}{2}\left(\frac{\sigma}{c_\sigma \Delta}\right)^2, \tag{4}$$

then:

$$q_b = c_E\,\frac{q^{3/2}}{\Delta}. \tag{5}$$

c_σ and c_E are constants. We choose $c_\sigma = 0.176$ and $c_E = 1.2$ in our calculations. Δ is the step of the difference grid.

As the equation of state we use the equation of perfect gas state taking into account the radiative pressure and the possibility of electron concentration variation due to ionization of H and dowble ionization of He and 16 elements. The contribution of H_2, H_2^+ and H^- was considered (Gadun,1986). The equation of state was not solved in the calculation process, but linear logarithmic interpolation using the previously calculated tables of $P=P(e, \rho)$ and $T=T(e, \rho)$ was run.

3. RADIATIVE ENERGY TRANSPORT TREATMENT

For photospheric layers we solved equation of radiative energy transport in the LTE and pure absorption approach, neglecting the dependence of absorption coefficient on the frequency and the horizontal energy transfer:

$$\mu\,\frac{\partial I}{\partial z} = \varkappa_R \rho\,(B - I), \tag{6}$$

where B is Planck function, I is intensity of radiation fields, is Rosseland mean absorption coefficient (Alexander,1975; Cox,Tabor,1976). Equation (6) was solved for vertical columns of modeled region as for vertically stratified, horizontally homogeneous, plane-parallel atmosphere by Feautrier technique. Finally:

$$Q_r = 4\pi\rho\,\varkappa_R\,(B - J), \tag{7}$$

where:

$$J = \frac{1}{2}\int_{-1}^{1} I(\mu)\,d\mu = \int_0^1 u(\mu)\,d\mu, \tag{8}$$

$$u(\mu) = \frac{1}{2}\,(\,I(\mu) + I(-\mu)\,).$$

In the deep subphotospheric layers we used diffuse approximation.

4. NUMERICAL METHOD

System (1) was solved by the method "large particles". This method was developed on the basis of classic Harlow method of "particles in cells". It allows to run integration of nonlinear inhomogeneous differential equations in the particular derivatives in three stages. At the first stage convective members are exclude from the consideration, they are taken into account at the next two stages. The method has good stability due to the large scheme viscosity σ_{sc} :

$$\sigma_{sc} \sim \frac{1}{2} \Delta x \, |u| \, , \tag{10}$$

where Δx is the steep of the difference grid, u is the local velocity. The large scheme viscosity puts limitations on the time of simu - lation and on the character of small-scale turbulent structure dynamics.

5. BOUNDARY CONDITIONS

Lateral boundary conditions are based on the assumption that at the lateral bounds parts of the photosphere having analogous character of development of dynamic processes are placed. Thus we can assume absence of matter departure through the lateral sides.

Opened boundary conditions are set at the upper and lower borders. Velocities at the opened borders were calculated from the mass flux ba- lance condition and for the development of smooth shape of RMS fluctua- tion values at the boundary layers. Average values of thermodynamic para- meters are fixed by the initial conditions, but possibility of their fluctuations is provided in the correspondence with the profiles of thermodynamic values in the contiguous layers:

$$\rho^{*}(i,j) = \rho_{0} \left[1 + \left(\frac{\rho(i,j)}{\langle \rho \rangle} - 1 \right) f_{\rho} \right] , \tag{11}$$

$$e^{*}(i,j) = e_{0} \left[1 + \left(\frac{e(i,j)}{\langle e \rangle} - 1 \right) f_{e} \right] , \tag{12}$$

where $\rho^{*}(i,j)$ and $e^{*}(i,j)$ are density and internal energy of the cell (i,j) at the boundary, $\rho(i,j)$ and $e(i,j)$ are the same values for contiguous layer, ρ_{0} and e_{0} are thermodynamic values of initial model, $\langle \rho \rangle$ and $\langle e \rangle$ are horizontal averaged values in the contiguous layer, f_{ρ} and f_{e} are correction factors, which provide smoth shape of ρ and e fluctuations near the boundary.

6. INITIAL CONDITIONS AND CHOOSE OF MODELED REGION

Initial homogeneous distribution of thermodynamic values was determined by the Procyon atmosphere model (Kurucz,1979) with T_{eff}=6500 K and log(g)=4.00, which was combined with envelope model calculated in the Deupree modification of mixing-length theory (Atroshchenko,Gadun,1986). The solar chemical composition for the Procyon model was assumed.

Vertical size of modelling region was chosen so that we could calculated spectral line profiles using models obtained. The upper boundary
is 1000 km over the photosphere ($\log \tau_s \sim -3$) and its lower boundary is
2000 km under the photosphere. Horizontal size (4000 km) was determined
so that some typical granules could be placed in the modeled region.
The cells are cubes with $\Delta x = \Delta y = \Delta z = 100$ km. Total number of cells in the
modeled region is 40X40X30.

Initial velocity fields was chosen so that div(V)=0 for every large
particule and in the assumption of symmetry of the velocity field:

$$
\begin{cases}
u(i,j,k) = 10^5 \exp\left(-\frac{4\pi k}{NK}\right) \sin\left(\frac{2\pi(i-1)}{NI-1}\right) \cos\left(\frac{2\pi(j-1)}{NJ-1}\right), \\[2mm]
v(i,j,k) = 10^5 \exp\left(-\frac{4\pi k}{NK}\right) \cos\left(\frac{2\pi(i-1)}{NI-1}\right) \sin\left(\frac{2\pi(j-1)}{NJ-1}\right), \\[2mm]
w(i,j,k) = 10^5 \exp\left(-\frac{4\pi k}{NK}\right) \cos\left(\frac{2\pi(i-1)}{NI-1}\right) \cos\left(\frac{2\pi(j-1)}{NJ-1}\right). \quad (13)
\end{cases}
$$

where NK is the number of cells on the K-coordinate (K=1 corresponds to
the lower boundary of the region), NI,NJ are the numbers of cells on
the i- and j-coordinates. The profile of vertical velocity in the initial model is shown in Fig. 1.

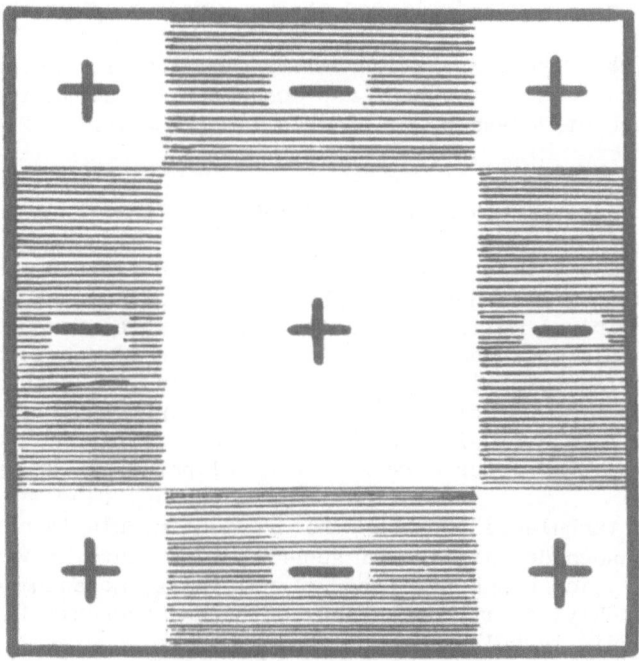

Figure 1. Profile of the vertical component (W) of the velocity vector
in the initial model.

7. RESULTS OF CALCULATIONS

Continuous duration of calculations is 28^m20^s (with time step 1^s) that approximately corresponds to the time of matter convolution in the modeled region. The development of the granular contrast is shown in Fig.2.

We notice that our initial conditions stimulate the development of convection at the initial stage. Next formation and development of structures takes place, their dynamics is due to buoyancy forces. But their further evolution is distorted by scheme viscosity (time 28^m20^s). Thus we have the sequence of models (15^m50^s - 26^m40^s) useful for analysis of convective motion dynamics.

The averaged by time and by horizontal layers distribution of thermodynamic values, RMS fluctuations of vertical and horizontal velocities, logarithm of RMS temperature fluctuations are shown in Fig. 3,4 for our models.

We can see that in the modeled region the effect of rising cool matter is absent in the photospheric layers as opposed to the solar photosphere (Gadun,1986). This may be due to the fact that the radiative energy transfer in the upper part of the Procyon photosphere plays more significant role than in the solar ones. Convective motion occurs with larger deviation from the adiabatic motion. Because the convective velocities can penetrate into steady layers at the larger heights than in the solar envelope.

This effect (positive correlation between the temperature and vertical velocity fluctuations in the weak and moderate line formation region) explains the absence of "C"-shape of Procyon line bisectors (Fig. 6).

8. DISCUSSION AND POSSIBLE DIRECTIONS OF FUTURE STUDIES

Three-dimensional gasodynamic models of the solar and Procyon photosphere and superadiabatic region describe physical conditions and dynamics better than semiempirical multiflux models. But our models are fit only for explanation of the most characteristic properties of convective mo - tions determined by large-scale structure dynamics and are not useful for explanation of fine effects such as radiative turbulent effects. Thus we suggest in future studies to consider possibilites of constructing more detailed second order models. The development of our approach may be as follows.

1. The treatment of turbulence based on the gradient models is not satisfactory because such description of sub-grid turbulence is very approximate, especially for rough real computional grids; momentum equations, for example Fredman-Keyller type equations of second approximation must be used for sub-grid turbulence treatment.

2. Finite difference methods with improved dissipative characteristics of higher order of accuracy may be applied for correct solution of such system nonlinear gasodynamic equations for compressible, nonhomogeneous, gravity-stratified medium in the presence of radiation field (Save - ljev,Tolstykh,1987).

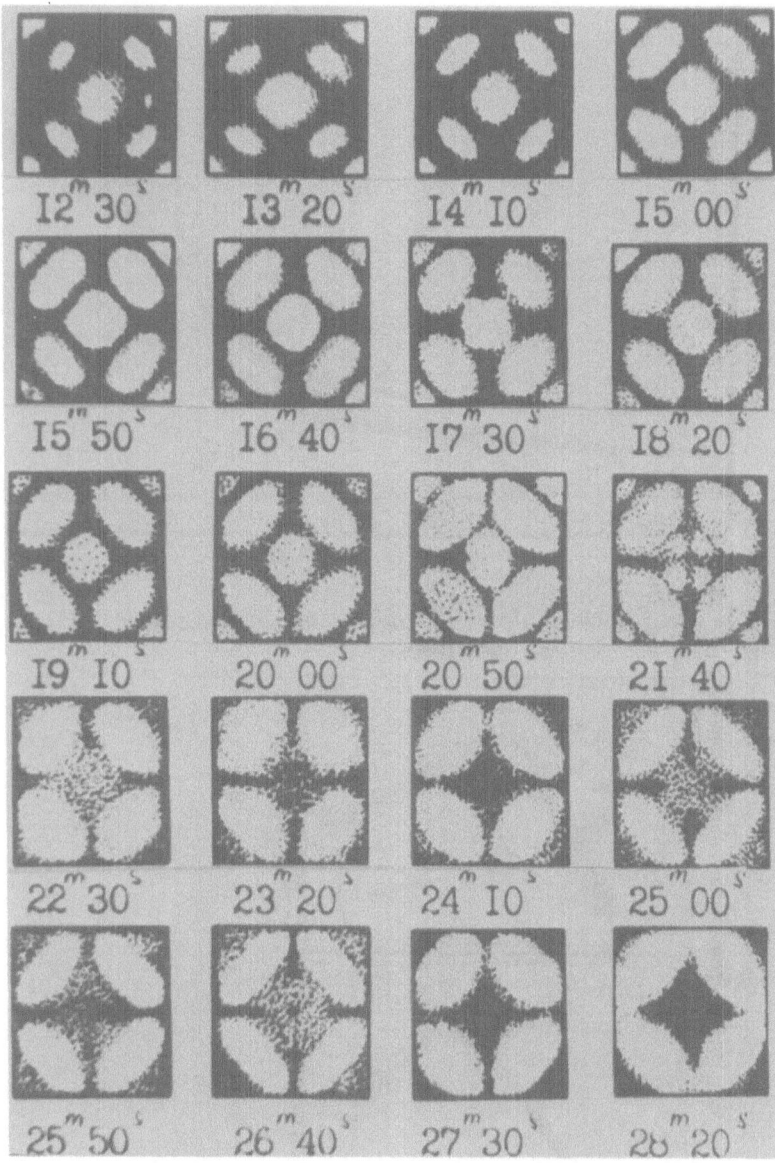

Figure 2. Evolution of photospheric granular contrast in the dynamic models.

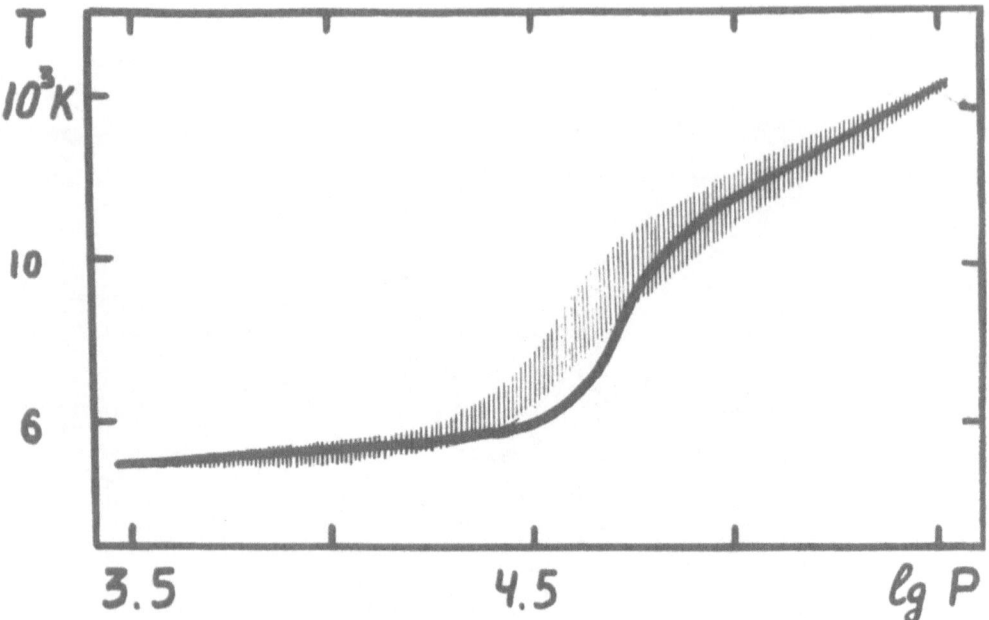

Figure 3. Dependence of T on log(P) for initial model (▬▬)
and for dynamic models (▓▓▓▓▓▓).

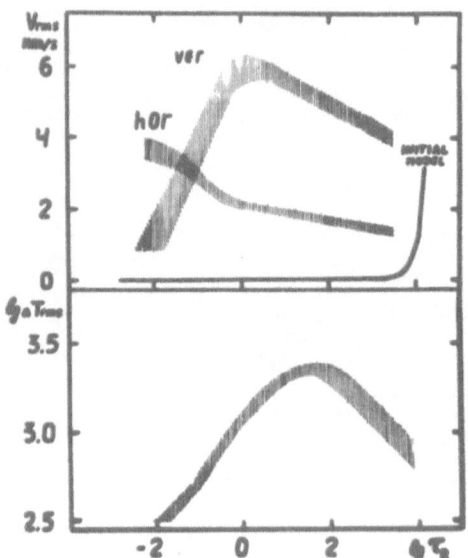

Figure 4. RMS distribution of vertical and horizontal velocities, logarithm of RMS temperature fluctuations.

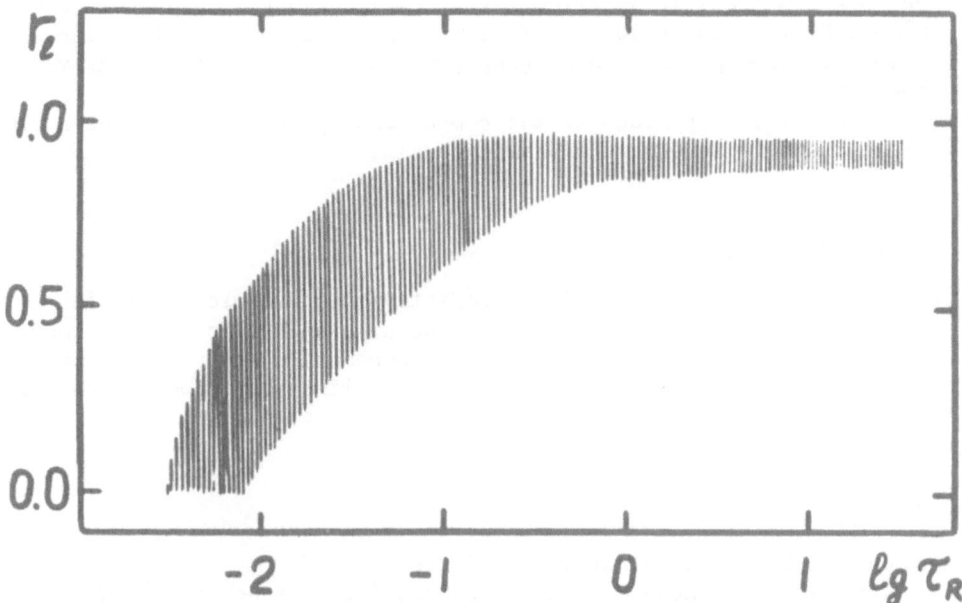

Figure 5. The correlation coefficient between the temperature and vertical velocity fluctuations.

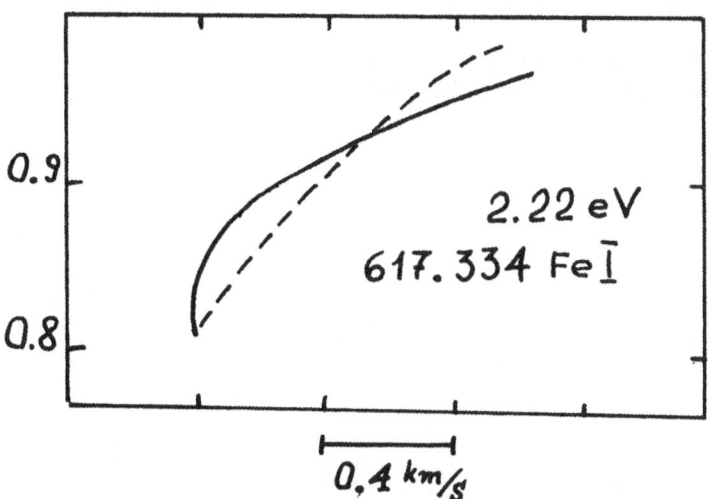

Figure 6. The Procyon Fe1 line bisectors (——— – calculated on the theoretical nonhomogeneous models, – – – – Griffin's atlas).

3.The improved treatment of radiative transport may be used which includes besides traditional effects (three-dimensional energy transfer,
monochromatic absorption coefficient for continuum and line absorption) consideration of averaging procedure for radiative flux vector components.
4.The more strict approach to the computional procedure of three-dimensional nonhomogeneous gasodynamic models.

CONCLUSION

The sequence of three-dimensional radiative gasodynamic models is
obtained for the Procyon photosphere and superadiabatic region. They
show positive correlation between the temperature and vertical velocity
fluctuations field in the weak and moderate line formation region which
is the cause of absence of "C"-shaped line bisectors.

REFERENCES

Alexander D.R. 1975, Astroph.J.Suppl.Ser.,29,363
Atroshchenko I.N.,Gadun A.S. 1986, Kinem. i Fizika Nebesn.Tel,2,21
Cox A.,Tabor J. 1976, Astroph.J.Suppl.Ser.,31,271
Gadun A.S. 1986, Preprint ITP of Ac.Sc.Ukr.SSR,ITP-86-106R
Kurucz R.L. 1979,Astroph.J.Suppl.Ser.,40,1
Nordlund A. 1984,Small-Scale Dynamical Processe in Quiet Stellar
   Atmospheres, Sunspot: Sacramento Peak Observatory, 181
Saveljev A.D., Tolstykh A.I. 1987, Soviet J.Comp.Mat. and Comp. Phys.,
   27,1709
Smagorinsky J. 1963, Mon.Weather Review,91,99
Wohl H.,Nordlund A. 1985, Solar Phys.,97,213

Discussion

NORDLUND — Do you experience any particular difficulties because of the stratification (large change in ρ)?

GADUN — The stratification is not a particular problem for the 'large particle' method; it does conserve mass. However, a strong stratification requires a better description of the turbulent effects in the equations.

DEUBNER — Since special emphasis was given to the fact that the coherence between vertical velocity and intensity does not drop to negative values with height, I should like to ask whether this is so because of the physics involved, or just because the model does not include the overshoot layers?

GADUN — The radiative energy transfer in the upper part of the Procyon photosphere plays a more significant role than in the solar case. The convective motions occur with larger deviations from adiabatic motions.

FOX — I would like to point out that this method is a 'particle-in-cell' (PIC) method, and not a true particle method. It does not have very good conservation properties, leading to difficulties with diffusion.

GADUN — In the Large Particle Method (MLP), the region in which the movement is studied is divided into a grid of fixed cells, with one 'large' particle in every Euler cell. In the classical Harlow 'particles-in-cells' method, there are several particles of fixed masses in every cell. This is the principal distinction between the two methods. The results of our analysis show that in the MLP the difference schemes are divergent-conservative and steady-dissipative.

The MLP cannot be used for problems with an inner structure determined by molecular viscous diffusion, because in the difference equations the molecular effects are small in comparison with the effects of the approximated viscosity.

WÖHL — You showed a comparison of the computed bisector with the observed one taken from Griffin's atlas. How many points on the disk did you use to take the center-to-limb variation into account?

GADUN — We used five points at $\mu = 1.0, 0.8, 0.6, 0.35$ and 0.15 respectively

NORDLUND — If I understood your horizontal boundary conditions correctly, they are somewhat impermeable. Do you see any particular effect of this?

GADUN — We use symmetrical lateral boundary conditions which perhaps affect the development of turbulent viscosity.

STEFFEN — Is there a principal upper limit to the duration in time that you can simulate with your numerical method?

GADUN — Yes, there is one. The numerical Large Particle Method typically has large artificial viscosity. This essential shortcoming affects especially the turbulent structures on small scales, and so puts an upper limit on the simulation time.

NUMERICAL SIMULATIONS OF GAS FLOWS IN AN A-TYPE STELLAR ATMOSPHERE

Detlef Gigas
Institut für Theoretische Physik
und Sternwarte der Universität Kiel
Olshausenstrasse 40
D-2300 Kiel 1
Federal Republic of Germany

ABSTRACT. First results of numerical simulations of gas flows in the atmosphere of an A0 V star (T_{eff} = 9500 K, log g = 3.90) are presented. The computations comprise $\sim$ 1 hour of stellar simulation time and were carried out for a 2-dimensional model assuming cylindrical symmetry. Effects of compressibility were fully retained. The resulting flow patterns are characterized by mainly vertical flows which exhibit an oscillatory behaviour in time. Typical time scales are of the order of 15 minutes. Maximum flow velocities increase with atmospheric height. Predicted line profiles were calculated for three Fe II spectral lines of different strengths. The time-averaged profiles are characterized by an additional non-thermal line broadening, which would have to be interpreted as a microturbulence of ξ = 1.6 km/sec in a classical model atmosphere analysis. Furthermore, line asymmetries occur, which are characterized by a slight depression of the blue line wing with a corresponding blueward bisector slope close to the continuum.

1. INTRODUCTION

Understanding hydrodynamical phenomena in the atmospheres of the Sun and other stars is one of the key topics of present-day astrophysics. In many respects our knowledge is currently incomplete: highly schematic approaches like mixing-length theory (MLT) can be handled easily, but lack precision if more than a coarse picture of convective energy transport is required. It is only through self-consistent modelling of stellar atmospheres that deeper insights into the physics of solar and stellar hydrodynamical phenomena can be obtained.

Due to the wealth of observational material available, first attempts have concentrated on the Sun. The first model calculation were successfully performed by Nordlund (1982, 1984). In fact, impressive agreement with solar granule lifetime and morphology (Wöhl and Nordlund, 1985) and solar line profiles and bisectors (Dravins et al., 1981; Nordlund, 1984) could be achieved. Similar attempts have recently been undertaken for cooler stars (e.g. Dravins, 1988). Little has been

R. J. Rutten and G. Severino (eds.), Solar and Stellar Granulation, 533–546.
© 1989 by Kluwer Academic Publishers.

published for early F and hotter objects.

According to the predictions of MLT, convective energy transport
in superficial atmospheric layers becomes unimportant in A-type stars.
It is for this reason that most A-type main-sequence model atmospheres
are computed considering only radiative energy transport (cf. Kurucz,
1979). However, the application of MLT may turn out to be inappropriate. In fact, studies of convective flows in deep atmospheric layers of
A-type stars have revealed significant deviations from the predictions
of MLT (cf. Toomre et al, 1976; Latour et al., 1981; Sofia and Chan,
1984).

Other striking problems remain: what is the meaning of 'microturbulence' derived from analyses of A-type stellar spectra (cf. reviews
by Gray, 1978, and Gehren, 1980)? Typical microturbulent velocities for
early-type A stars amount to $\sim$ 2 km/sec (Adelman, 1986) and are still
needed if deviations from LTE are taken into account (Gigas, 1986;
Freire Ferrero et al., 1983). Maximum convective velocities of only a
few hundred meters per second (as predicted by MLT) seem much too small
to account for these values. Why do hotter stars like Canopus (FO II)
and Sirius (A1 V) exhibit spectral line bisectors which are distinctly
different from those observed in solar-type stars (Dravins, 1987)?
These stars obviously do not simply possess a 'scaled' solar convection
pattern; a closer look at these objects thus seems worthwhile.

2. COMPUTATIONAL PROCEDURE

All computations were carried out with a hydrodynamics code derived
from that devised by Stefanik et al. (1984). Being based on the method
of bicharacteristics, this code solves the time-dependent nonlinear
equations of hydrodynamics and radiative transfer in two dimensions,
thereby assuming cylindrical symmetry. The influence of magnetic fields
or stellar rotation onto the equations of motion is not considered. No
other simplifying assumptions (such as linearizations or the anelastic
approximation) are made.

The effects of hydrogen ionization on the equation of state are
explicitly taken into account. Currently the solution of the radiative
transfer equation ist carried out in the grey approximation, thereby
employing a Rosseland opacity law. For each grid point the radiative
exchange with its environment is computed along a large number of rays
with different inclination angles.

A total of 36 (horizontal) × 30 (vertical) grid points was employed. Other 'technical' details are published elsewhere (Steffen and
Muchmore, 1988; Steffen et al., 1988).

3. THE ATMOSPHERIC MODEL

All results presented below were derived for a model atmosphere with an
effective temperature of T_{eff} = 9500 K and a surface gravity of log g =
3.90. Although no attempts were made to obtain a close match to a particular star, these parameters can be considered as being representa-

tive for Vega (A0 V), one of the most frequently investigated A-type main-sequence stars. The initial pressure and temperature stratifications were computed in the grey approximation.

Special pains should be taken to ensure realistic but still numerically tractable boundary conditions. According to the predictions of MLT, early A-type stars are characterized by two convection zones which are separated by a quiet intermediate layer. As we are interested in studying the dynamics of gas motions in spectroscopically accessible layers, the lower boundary of the computational domain was located in this intermediate layer below the upper convection zone. Vertical velocities were set equal to zero at this boundary; the radiative flux entering into the cylinder from below was computed from the star's effective temperature. The upper boundary, located at much smaller optical depths (log $\tau_{5000} \sim -2.4$), permits the transmission of sound waves.

The grid spacing was designed to allow a sufficient spatial resolution of gas flow phenomena. Vertically, it was varied between ~ 90 and 350 km, the smaller grid spacing being used in the convectively unstable domain close to the hydrogen ionization zone (total depth ~ 1500 km). Horizontally, a grid spacing of 150 km was employed. We thus arrive at a diameter of the (cylindrical) computational domain of 10500 km, which may be considered as being equal to a 'typical' solar granule (diameter ~ 1000 km) scaled with the ratio of the pressure scale heights of both stars. This, of course, cannot be regarded as being more than a reasonable first guess.

In order to start the numerical simulation, a velocity pattern resembling a two-stream convective model was introduced. The velocities were taken from the predictions of an ATLAS6 MLT model of an A0-type star. These velocities turned out to be much smaller than the gas flow velocities encountered later on; for an effective temperature of $T_{eff} = 9500$ K maximum convective velocities predicted by MLT amount to $v_{max} \sim 0.45$ km/sec; they decrease to ~ 0.25 km/sec for $T_{eff} = 9650$ K.

4. RESULTS OF SIMULATIONS

Starting from the initial model described above, the numerical simulation was advanced in time up to a total of about 1 hour of 'stellar' time. Numerically, these computations turned out to be more demanding than first expected. Several hours of CPU time on a CRAY-X/MP were needed to obtain the results presented here. The time step was ~ 0.1 sec.

Two 'snapshots' of the flow patterns (after 31 and 60 min simulation time) are plotted in Figs. 1 and 2. The characteristic flow patterns in an A-type stellar atmosphere turn out to be remarkably different from those occuring in simulations of solar granular flows using the same numerical code. After a short time, most horizontal differences disappear, and preferentially vertical flows persist. In general, these are characterized by an oscillatory behaviour in time with amplitudes dependent on geometrical height. Typical time scales are of the order of (5-15) minutes - a period remarkably close to those observed

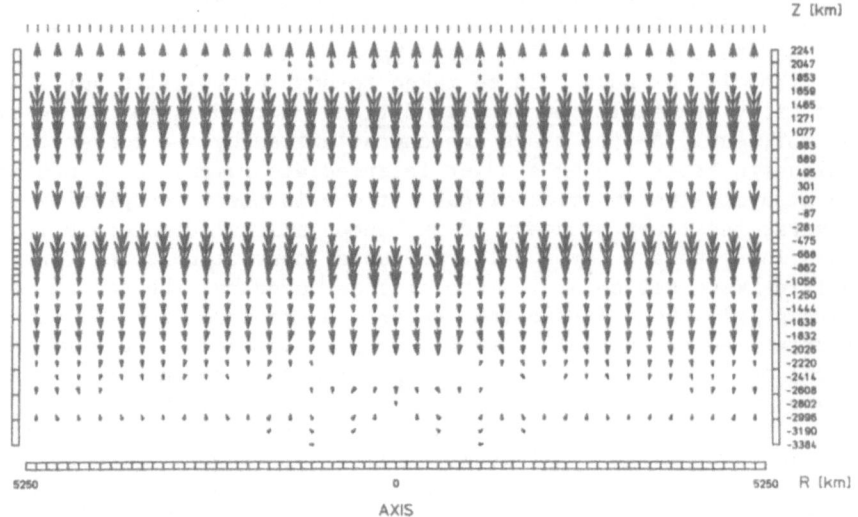

Fig. 1: Flow pattern after 31 min of simulation time. Flow velocities
are proportional to arrow size, maximum flow velocity is 3.20
km/sec. Optical depth $\tau = 1$ is close to $z = 0$

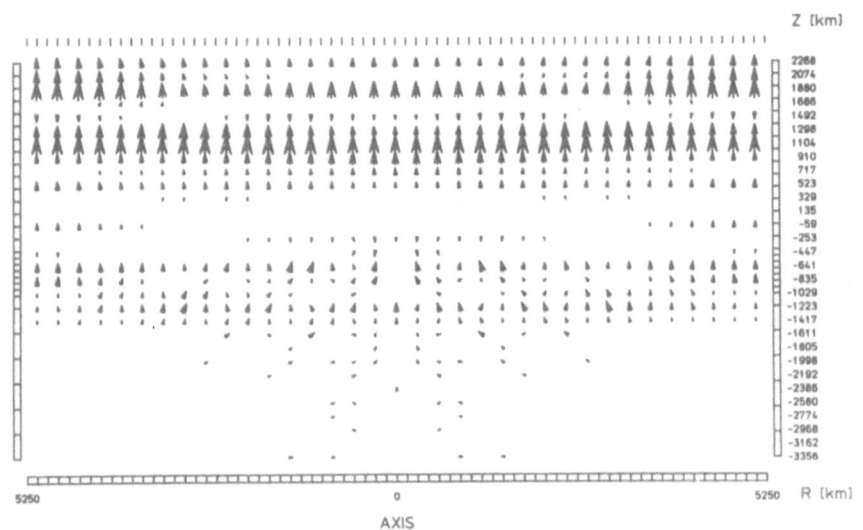

Fig. 2: Flow pattern after 60 min of simulation. Maximum flow
velocity is 2.44 km/sec. The zero point of the geometrical
depth scale has been adjusted

in some of the 'rapidly oscillating Ap stars' (cf. the review by Kurtz, 1986).

For a first coarse analysis, the computational domain may be divided into three parts. In the lower third (in geometrical height) hydrogen in almost completely ($\sim$ 99 %) ionized, which makes most gas constants close to those computed for an ideal gas mixture. Radiative equilibrium is largely fulfilled, temperature fluctuations are small compared with the average temperature, and vertical velocities are considerably smaller than in other parts of the computational domain. Averaged in time, the flow velocities amount to only a few tens of meters per second. With respect to these time-averaged properties, this region may be considered as a quiet, radiatively dominated region as usually predicted by mixing-length theory.

Conditions are much different in the middle third of the model atmosphere. It is there where considerable changes in most thermodynamic quantities take place. For instance, the hydrogen ionization fraction drops steeply from a percentage of $\sim$ 99 % in the lower part of this domain to $\sim$ 15 % in its upper part. The mass absorption coefficient reaches its peak value, thus creating a steep radiative temperature gradient and a large superadiabatic gradient. Coming from lower layers, the heat capacity C_p rises steeply, furthermore, a local density inversion occurs. This zone, located at optical depths of about log τ_{Ross} $\sim$ (+0.75 ... -0.75) and having a geometrical depth of only $\sim$ 2000 km ($\sim$ 1.5 pressure scale heights) should thus be regarded as a region with distinctly different properties. In fact, MLT predicts the onset of convective instability just in this layer, although with maximum convective velocities of only $\sim$ 0.4 km/sec.

The different physical conditions show up in the numerical results: typical vertical velocities in this domain are of the order of $\sim$ 1.5 km/sec with maximum values of $\sim$ 2.5 km/sec. Associated with this, temperature fluctuations occur which may amount to some hundred degrees ($\sim$ 5 % of the average temperature).

The upper third of the computational domain is characterized by a rather steep drop in mean pressure and density with a temperature approaching the boundary value of a grey atmosphere. Hydrogen can mainly be found in its neutral state (ionization fraction $\sim$ 15 %). Not considering the influence of the convectively unstable layer below, it should be stable against the onset of convection according to the Schwarzschild criterion. It is, however, considerably influenced by the velocity fields present in the thin convection zone underneath. Entering from below, these gas flows are able to induce wave-like velocity fields in the upper layer itself. Due to the decreasing density, the velocity amplitudes rise up to $\sim$ 5.0 km/sec in the uppermost layers. Temperature fluctuations are smaller than in the convection zone, amounting only to $\sim$ 50 K in the top layer.

To summarize one can say that the atmosphere is nearly static if merely time-averaged velocities are considered - upflows at one time step are followed by downflows with the same magnitude some time later. Viewed in more detail, a picture considerably different from the predictions of MLT emerges. A more detailed comparison of time-averaged *square* velocities $\sqrt{<v^2>}$ with the predictions of MLT (as computed with

the ATLAS6 code) is plotted in Fig. 3 versus Rosseland optical depth.
The results of a determination of the microturbulence $\xi(\tau)$ for the AO V
standard star Vega (as derived by Gigas, 1986, from lines of Fe I,
Fe II, and Ti II) are given as comparison. Of course it has to be

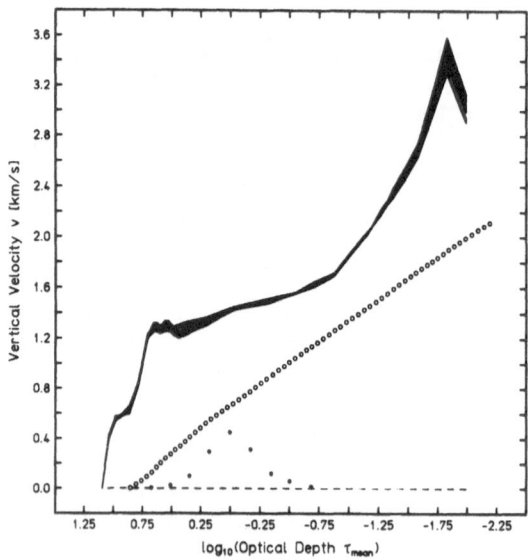

Fig. 3: Comparison of mean square velocities from hydrodynamic model (solid lines) with microturbulence for Vega as derived by Gigas, 1986 (circles). Dots: predicted convective velocities for Vega (ATLAS6 mixing length model). Abscissa: Rosseland optical depth

pointed out that mean square velocities are not necessarily identical
with microturbulence. Nevertheless, the derived flow velocities seem
large enough to account for what is classically interpreted as 'microturbulence' in Vega and similar A-type stars (see also below).

5. PREDICTED STELLAR LINE PROFILES

Before going into more detail, one fact should be expressed clearly:
the results presented here are no more than a preliminary study of gas
flows in an early-type star. Any comparison with observed stellar line
profiles should only be performed with great care. In particular, the
grey approximation becomes problematic in the outermost layers, which
presently precludes an accurate modelling of cores of strong spectral
lines due to the flat temperature gradient in the outer parts of the
atmosphere. Furthermore, the spectral synthesis is currently restricted
to the center of the stellar disc. We also do not consider the influence of rotational line broadening.

The spectral synthesis was performed under the assumption of LTE
by inserting temperatures, gas pressures, and velocity fields from the
hydrodynamical model into the Kiel line synthesis codes LINFOR and
ATMOS. Such procedure can be performed for every radial point of the
cylindrical computational domain. The line profile synthesis thus is
carried out in considerably more detail than the simple grey radiation
transport employed for the hydrodynamics calculations.

In order to minimize the influence of non-LTE effects on spectral
line formation (which are present for Fe I in A-type stars; see Gigas,
1986), line profiles were computed for <u>synthetic Fe II lines</u> situated
at 4500 Å. Line strengths were varied by adopting three different oscillator strengths; for other line data (e.g. damping constants) values
typical for visual Fe II spectral lines were adopted. A total of 136
model atmospheres, spaced in time by an average of ∿ 27 sec, was employed for the line synthesis.

<u>Time-averaged line profiles</u> for 'strong', 'medium', and 'weak'

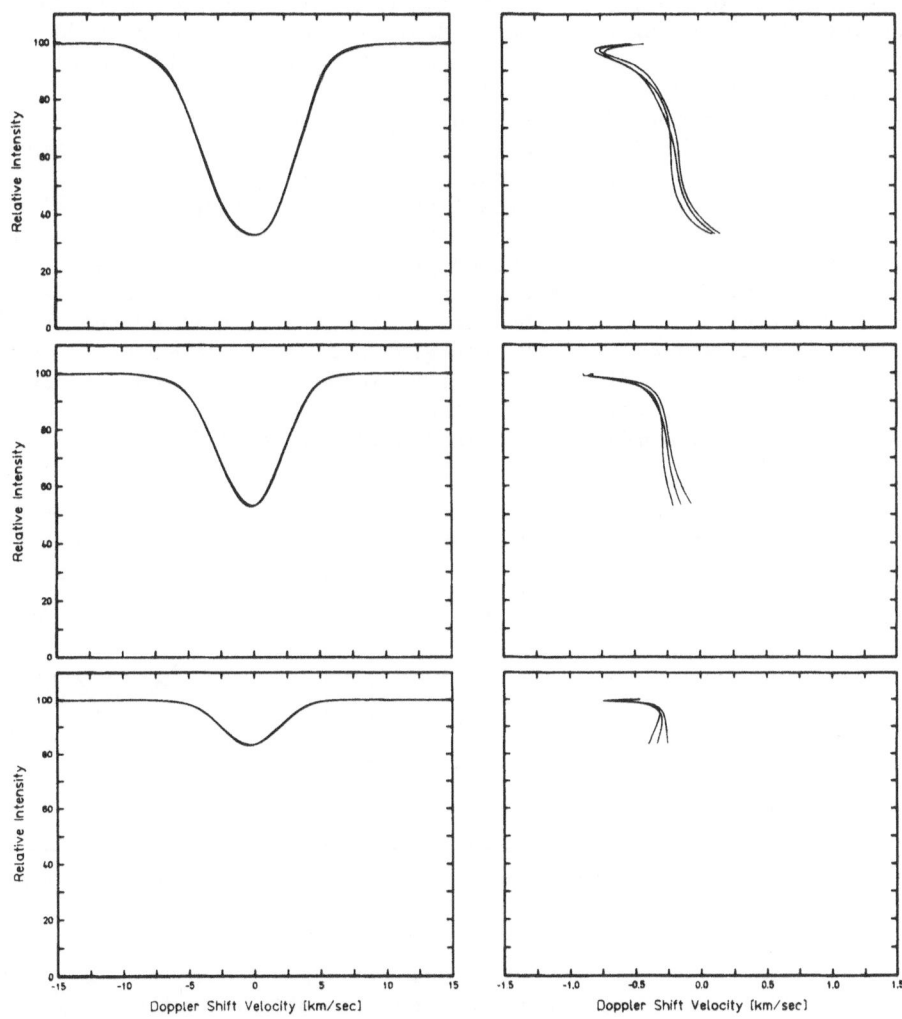

Fig. 4: Time-averaged synthetic Fe II spectral line profiles
('strong', 'medium', and 'weak' lines) and corresponding line bisectors for three radial atmospheres

Fe II lines are plotted in Fig. 4. On the right-hand side, the corresponding line bisectors are plotted for three selected radial atmospheres on a tenfold expanded abscissa. It should be pointed out that these profiles were obtained by considering depth- and time-dependent line shifts caused by hydrodynamical velocity fields as the only source of non-thermal line broadening; except for thermal, collisional, and natural broadening, no other broadening mechanisms (such as micro- or macroturbulence) were employed.

What is characteristic of all profiles is a slight <u>asymmetry</u> with a depressed blue wing. This also shows up in the corresponding bisectors: while showing no large variations at small intensities, the bisectors are bent sharply towards the blue close to the continuum. Bisectors computed for different radial atmospheres generally do not differ much due to the absence of large horizontal differences in temperature and velocity.

Can line broadening due to hydrodynamical velocity fields provide an explanation for the 'micro-' and 'macroturbulence' fit parameters employed in the analysis of stellar spectra? To this end, we have performed a line synthesis for the 'medium' strength Fe II line with different assumptions concerning the velocity fields (Fig. 5). On the left-hand side, the time-averaged profile as derived from our parameter-free hydrodynamical model is plotted (the velocity scale is different from that of Fig. 4). The same line profile, derived from a *single* grey atmosphere (obtained by averaging all model atmospheres) and by suppressing *all* non-thermal velocity fields, is shown on the right-hand side. It is deeper and narrower than its hydrodynamic counterpart; moreover, the equivalent width is smaller by about 15 %.

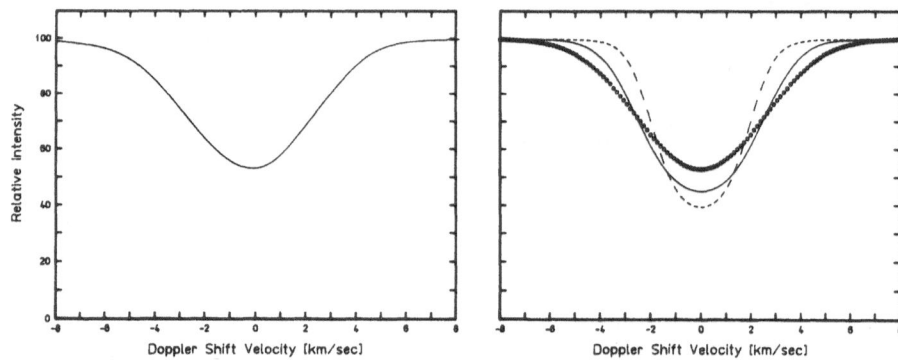

Fig. 5: Time-averaged 'medium' Fe II line from hydrodynamical model (left) compared with line profiles obtained without hydrodynamical velocity fields (right). Dashed line: without turbulent velocities, solid line: with microturbulence of ξ_{mic} = 1.61 km/sec, dotted line: after convolution with additional macroturbulence of ξ_{mac} = 2.03 km/sec

To obtain agreement with respect to equivalent width, a depth-independent microturbulence of ξ_{mic} = 1.61 km/sec had to be introduced; finally, the 'correct' central intensity was reproduced by convolving the profile with a Gaussian macroturbulence of ξ_{mac} = 2.03 km/sec. In terms of the classical 'turbulence' model, the resulting line profile thus would have to be interpreted by assuming these non-thermal velocities. Of course, no explanation can be provided for the line asymmetry in this classical scheme.

The explanation of the resulting bisector slope does not seem to be a trivial matter. As can be seen from Fig. 6, continuum intensities and line shapes are subject to considerable changes during the time evolution. Temperature and velocity fields, however, do not vary independently. It is by such correlation effects that asymmetries in spectral lines arise (cf. Cram, 1981, for the solar case).

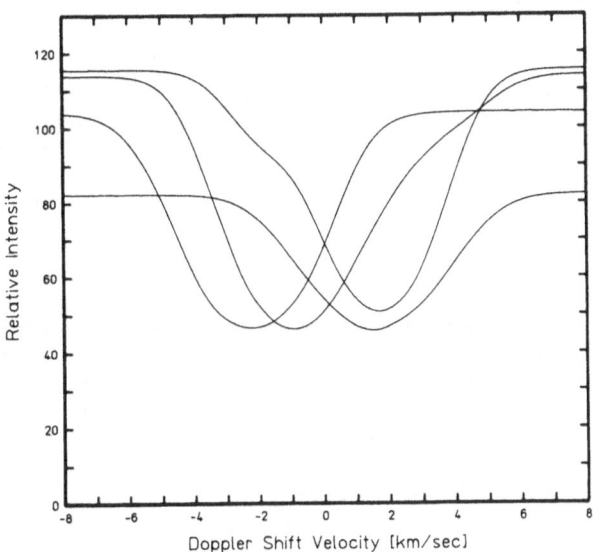

Fig. 6: Profiles of 'medium' Fe II spectral line at different time steps

6. COMPARISON WITH OBSERVED LINE PROFILES OF A-TYPE STARS

How do these first results compare with stellar observations? To begin with, there seems to be encouraging agreement of non-thermal broadening computed from the time evolution of a model atmosphere with 'typical' values found in the analysis of early A-type stars. In a non-LTE analysis of Vega, Gigas (1986) derived a microturbulence of ξ = 1.5 km/sec from visual Fe II lines; other LTE abundance determinations suggest values close to ξ = 2.0 km/sec (Sadakane et al., 1986). Moreover, such values seem typical for most 'normal' A-type stars in the same temperature range (Adelman, 1986). There are also indications that the microturbulence in Vega may increase with height (Gigas, 1986; Freire Ferrero et al., 1983).

Observations of <u>stellar bisectors</u> are considerably more demanding; they have been carried out only for a few of the brightest stars. Furthermore, rapid stellar rotation may complicate the interpretation of stellar line profiles. Nevertheless, the data obtained so far permit some conclusions to be drawn.

When going from cooler to hotter spectral types, the most remarkable change occurs with respect to <u>bisector slope</u>. Gray and Toner (1986) have shown that bisectors in K2 - F5 supergiants change from a redward bend close to the continuum to a blueward bend at spectral type G0 Ib. Basically the same results were obtained by Dravins (1987) in a study of prominent stars of different effective temperatures: the familiar 'C-shape' present in solar-type stars changes into an 'anti-C-shape' for early F-type stars. In particular, Canopus (F0 II) and Sirius (A1 V) proved to possess bisectors of this type. A slight depression of the blue line wing seems to persist even to earlier spectral types, as demonstrated by Smith and Karp (1978, 1979) for the slowly rotating B0 V star τSco.

Of course, our investigation is not intended to be more than a first attempt to provide a parameter-free description of how these bisectors may be formed. It is nevertheless encouraging to see that basic properties of early-type star line profiles can be reproduced from the self-consistent models of stellar atmospheres.

7. FUTURE OUTLOOK AND PROBLEMS

Although some properties of observed line profiles show good agreement with the results obtained so far, quite a lot of work remains to be done in this field. The following points seem most important:

7.1. Computations for other initial parameters

How do the flow patterns depend on effective temperature, surface gravity, and chemical composition? How does the transition from a solar-type convection pattern to the one present in A-type stars take place? This will be investigated in a series of future computations. In addition, the influence of several other parameters (such as grid spacing, size of computational domain, boundary conditions,...) will be studied.

7.2. Non-grey radiative transfer

One of the most serious limitations of our computations seems to be the restriction of the <u>radiative transfer</u> to one frequency point ('<u>grey approximation</u>'). Consequently, the temperature stratification is essentially given by those of a grey atmosphere in the outermost layers.

In the field of spectral line synthesis, this must result in a bad representation of the <u>cores of strong spectral lines</u>, which are formed in the outermost layers. Grey model atmospheres will not be able to reliably reproduce the dark central parts of those lines, but will yield brighter, 'box-shaped' cores instead. Moreover, this approximation will most likely have repercussions on <u>hydrodynamics</u> itself: a steeper temperature gradient in the outer layers may favour the penetration and overshoot of convective motions into the stable zone above.

A non-grey radiative transfer can be incorporated into the numerical code without problems by considering the radiative transfer in a number of representative wavelength intervals; this will be done in near future. It will, however, increase the time needed for the computation of hydrodynamical model atmospheres.

7.3. Helium ionization

Currently only hydrogen ionization is implemented into the equation of state. Other elements are either regarded as unionized or ionized by a constant degree. This approach needs further improvement. In the domain of hot main-sequence stars, helium ionization will play an important role in determining atmospheric dynamics. It may cause convective instabilities in deeper parts of the star's envelope or may modify the results obtained for the superficial hydrogen ionization zone. We plan to incorporate this effect into our numerical code.

7.4. Spectral line synthesis

As stated above, the spectral synthesis was performed for the center of the stellar disc. Judging from the solar case, one might expect disc-integrated line asymmetries to be at least qualitatively similar to those obtained for disc center. A self-consistent spectrum analysis should nevertheless include the contributions of all disc elements. Schematic assumptions, such as standard limb-darkening laws, should, if possible, be avoided. This requires the computation of the radiative transfer along a large number of oblique rays of various inclinations. Work in this field is in preparation.

7.5. Stellar rotation

In connection with stellar line synthesis, another problem is worth mentioning: the inclusion of stellar rotational line broadening, which is envisaged for future computations. Compared with late spectral types, early-type stars rotate more rapidly, projected rotational velocities of v sini $\sim$ (100–150) km/sec being normal for early A-type stars. Stars like Vega or Sirius with v sini $\sim$ 15 km/sec are slow rotators for A-type standards, but show a considerably larger rotational line broadening than solar-type stars. Apart from introducing additional uncertainties into the bisector determination, this also increases the probability of a line being disturbed by nearby lines.

Observations of extremely sharp-lined early-type stars (cf. Holweger et al., 1986a,b) could provide important data for checking the predictions of self-consistent numerical simulations for these objects.

7.6. Shocks

A further problem is the possible occurence of shocks. Steep velocity or density gradients can be found in the numerical simulations from time to time, which may result in the creation of shock fronts with an increased amount of dissipation. This may cause changes in the atmosphere's thermal stratification especially in the uppermost layers.

It also raises another question: can velocity fields which steepen up in the outer atmospheric layers contribute to a non-radiative heating of chromospheres and coronae in early-type stars? Acoustic fluxes computed from MLT models are subject to very large uncertainties in this temperature range (Fontaine et al., 1981); future investigations may prove rewarding.

REFERENCES

Adelman, S.J.: 1986, *Astron. Astrophys. Suppl. Ser.* **64**, 173
Cram, L.E.: 1981, *Astrophys. J.* **247**, 239
Dravins, D.: 1987, *Astron. Astrophys.* **172**, 211
Dravins, D.: 1988, in: *The Impact of Very High S/N Spectroscopy on Stellar Physics*, eds. G. Cayrel de Strobel and M. Spite, Kluwer, Dordrecht - Boston - London, p. 239
Dravins, D., Lindegren, L., Nordlund, Å.: 1981, *Astron. Astrophys.* **96**, 345
Fontaine, G., Villeneuve, B., Wilson, J.: 1981, *Astrophys. J.* **243**, 550
Freire Ferrero, R., Gouttebroze, P., Kondo, Y.: 1983, *Astron. Astrophys.* **121**, 59
Gehren, T.: 1980, in: *Stellar Turbulence*, ed. D.F. Gray and J.L. Linsky, Springer Lecture Notes in Physics **114**, 103
Gigas, D.: 1986, *Astron. Astrophys.* **165**, 170
Gray, D.F.: 1978, *Solar Phys.* **59**, 193
Gray, D.F., Toner, C.G.: 1986, *Publ. Astron. Soc. Pacific* **98**, 499
Holweger, H., Gigas, D., Steffen, M.: 1986a, *Astron. Astrophys.* **155**, 58
Holweger, H., Steffen, M., Gigas, D.: 1986b, *Astron. Astrophys.* **163**, 333
Kurtz, D.W.: 1986, in: *Seismology of the Sun and the Distant Stars*, ed. D.O. Gough, Reidel, Dordrecht - Boston, p. 417
Kurucz, R.L.: 1979, *Astrophys. J. Suppl.* **40**, 1
Latour, J., Toomre, J., Zahn, J.-P.: 1981, *Astrophys. J.* **248**, 1081
Nordlund, Å.: 1982, *Astron. Astrophys.* **107**, 1
Nordlund, Å.: 1984, in: *Small Scale Dynamical Processes in Quiet Stellar Atmospheres*, ed. S.L. Keil, p. 181
Sadakane, K., Nishimura, M., Hirata, R.: 1986, *Publ. Astron. Soc. Japan* **38**, 215
Smith, M.A., Karp, A.H.: 1978, *Astrophys. J.* **219**, 522
Smith, M.A., Karp, A.H.: 1979, *Astrophys. J.* **230**, 156
Sofia, S., Chan, K.L.: 1984, *Astrophys. J.* **282**, 550
Stefanik, R.P., Ulmschneider, P., Hammer, R., Durrant, C.J.: 1984, *Astron. Astrophys.* **134**, 77
Steffen, M., Muchmore, D.: 1988, *Astron. Astrophys.* **193**, 281
Steffen, M., Ludwig, H.-G., Krüß, A.: 1988 (in preparation)
Toomre, J., Zahn, J.-P., Latour, J., Spiegel, E.A.: 1976, *Astrophys. J.* **207**, 545
Wöhl, H., Nordlund, Å.: 1985, *Solar Phys.* **97**, 213

Discussion

VAN BALLEGOOIJEN — When you include helium ionization, there will be a second convection zone at larger depth. Does the neglect of this layer affect your calculations?

GIGAS — The inclusion of helium ionization will result in a superficial helium ionization zone close to the hydrogen ionization zone (first helium ionization), and another one located at much larger depth (second helium ionization). We are mainly interested in studying atmospheric dynamics in spectroscopically accessible layers, which may be influenced by the first helium ionization. Compared to this, the influence of the second helium ionization in deep layers is expected to be much smaller.

DRAVINS — It would be of considerable interest to make such models also for giants. The easier measurement of line asymmetries in the more sharp-lined and slowly rotating giants could make the testing of the models more straightforward.

GIGAS — Certainly. We indeed envisage to extend the model simulations to stellar atmospheres away from the main sequence. However, before proceeding that far, I feel that one should first try to obtain a more detailed understanding of convective phenomena in stars close to the main sequence.

NORDLUND — How would you characterize these motions? As overstable oscillations?

GIGAS — We have not yet arrived at a final understanding of the physical nature of these oscillations. Currently, it seems most likely to me that the oscillatory motions are due to a coupling of convective flows in the thin convection zone with subsequent wave propagation in the region above.

KNEER — The oscillations which you obtain from your calculations could be of significance to heat the outer atmospheric layers. A small amount of energy flux, e.g. 10^{-3} of the total flux, could be sufficient. Could you give any number on this possibility? One would need to know the correlation between the temperature and velocity fluctuations.

GIGAS — Currently, I do not have data concerning the amount of acoustic flux released at the upper boundary. The determination of this quantity (which will be done in future) should be of considerable interest because of the large uncertainties in the acoustic fluxes computed from mixing-length theory in this temperature range.

Our preliminary data suggest that there is a correlation between the temperature (and intensity) variations and the velocity variations. It is these correlation effects that contribute to the observed line asymmetries.

FOX — From what we understand in the more idealized models, we know that overstable oscillations are caused by external effects such as magnetic fields or rotation, and that the effect here is more like a wave effect (cf. Stein and Nordlund), which is better than overstability in the sense that the waves can carry much energy while overstable oscillations do not. Do you really expect sonic velocities to be important, given the low velocities predicted from mixing-length theory and the somewhat higher hydrodynamic velocities?

Have you looked at the modifications of the initial model adiabat, to see the effects of the hydrodynamics?

GIGAS — The flow velocities encountered in the lower part of the computational domain remain subsonic. They may, however, be close to the speed of sound from time to time in the uppermost atmospheric layers, where the densities are low.

There are only moderate modifications in the temperature versus optical depth relation of the time-averaged model relative to the initial model. A closer look at the results reveals typical deviations of about $\pm$ 150 K, not exceeding $\pm$ 300 K (corresponding to $\approx$ 3% of the average temperature only).

GRAY — Although you have made the point that your calculations are preliminary and that we should not rush to compare them with the observations, I would like to make a basic point in this direction. Since most of your model velocities are vertical, doing a full stellar-disk integration will reduce the velocity span of the bisector, compared to the disk-center values you have shown us. But the observed left-leaning bisectors have large velocity spans, ranging from over 1000 m/s near the main sequence to perhaps 3000–4000 m/s for the higher luminosity stars. So there is still considerable work to be done.

RUTTEN — If you fit an observed spectral line with the simulation and with classical micro/macroturbulence, will you get the same abundance value?

GIGAS — Almost the same line profiles (except for the line asymmetry) are obtained when comparing the time-averaged line profile from the hydrodynamic simulations with the line profile derived from one time-averaged model atmosphere without hydrodynamic velocity fields but with micro/macroturbulence and the same iron abundance.

There will be, however, a difference in the derived iron abundance if a grey model atmosphere, computed without regard to the hydrodynamic simulation, is choosen as a comparison. This is due to the fact that the thermal stratification of the time-averaged model atmosphere will be somewhat different from that of a radiative grey model. Test calculations suggest that the iron abundance derived from the hydro-dynamic simulations will be larger by about 0.08 dex for the 'medium'-strength Fe II line.

STEFFEN — These somewhat surprising results imply a warning. They demonstrate how easily one may go wrong by simply constructing empirical models. To explain the line asymmetries in A-type stars, empirical models would probably produce a granulation pattern with extremely dark or very narrow intergranular lanes, a picture which is not physically reasonable in view of the domination by the oscillation phenomena found from Gigas' numerical simulations.

Inhomogeneous chromospheric structures in the atmosphere of Arcturus generated by acoustic wave propagation

M. Cuntz[1] and D. Muchmore[1,2]
[1]Institut für Theoretische Astrophysik
Im Neuenheimer Feld 561, D-6900 Heidelberg, F.R.G.
[2]Departement de Physique, Universite de Montreal
C.P. 6128, succ. A, Montreal, P.Q. H3C-3J7, Canada

ABSTRACT. We investigate the propagation of a strong shock obtained in a time-dependent acoustic wave model of the outer atmosphere of Arcturus. Radiation damping due to CO and SiO molecules is considered. The development of the shock leads to a transition between a cool (molecule dominated) atmosphere and a hot chromospheric feature with a temperature similar to the semiempirical model of Ayres and Linsky. This behaviour suggests a possible explanation for the inhomogeneous chromospheric structures of late-type stars which are proposed by some observers.

1. INTRODUCTION

A typical example for a late-type giant star beyond the Linsky-Haisch division line is the well-studied star α Boo. The temperature and the density structure in the outer atmosphere of α Boo have been investigated by Ayres and Linsky (1975). For the temperature they find an increase, beginning from 3200 K at the temperature minimum layer and rising up to 20000 K. Additional results considering the effects of stellar wind flows are described by Drake (1985). The inner part of the Ayres-Linsky model is essentially confirmed by Judge (1986). Another characteristic as the influence of chromospheric heating is the importance of molecular cooling. This leads to reduced temperatures in static atmospheres (cf. Tsuji, 1964) and to thermal instabilities (Ayres, 1981; Kneer, 1983; Muchmore and Ulmschneider, 1985; Muchmore, 1986).

For α Boo a theoretical hydrodynamic model has been obtained by Ulmschneider et al. (1979) based on propagating acoustic shock waves. In this work for the radiation transport a simplified (grey, LTE) treatment of radiation losses employing the opacity table of Kurucz (1979) has been

547

R. J. Rutten and G. Severino (eds.), Solar and Stellar Granulation, 547–553.
© 1989 by Kluwer Academic Publishers.

used. Therefore, a wave-supported atmosphere with an increasing temperature beyond the temperature minimum region has been obtained. There are two reasons to include molecular cooling in such a model:

i) Observations of CO lines as discussed by Heasley et al. (1978) indicate cool atmospheric structures which are dominated by molecular cooling. The temperatures of these layers are not in aggreement with the Ayres-Linsky chromosphere. A similar result has recently been obtained by Ayres (1986).

ii) As found by Muchmore and Ulmschneider (1985) the temperature at the temperature minimum region of α Boo falls into a region of parameter space which is unstable to molecular cooling. Additionally Muchmore et al. (1987) have shown that in the atmospheres of late-type giant stars molecular cooling due to SiO occurs leading to a thermal instability with temperatures as low as 1000 K.

In the present work we investigate the dynamical effects of a strong shock in the outer atmosphere of α Boo considering radiation damping due to the CO and SiO molecules. The method which we have used is described in Section 2. Results are presented in Section 3. In Section 4 observations referring to the outer atmosphere of α Boo are discussed. Section 5 gives our conclusions.

2. METHOD

Our time-dependent model has been computed considering the propagation of acoustic waves which are introduced into a spherically symmetric atmosphere with height dependent gravity. We solve the time-dependent continuity, Euler and energy equation in a lagragian frame using the modified characteristics method as described by Ulmschneider et al. (1977, 1978). The inner boundary of the atmospheric layer is represented by a sinusoidally moving piston. As outer boundary condition we use a transmitting boundary which has been discussed by Muchmore and Ulmschneider (1985). This boundary condition which allows the transmission of waves with sufficient efficiency is needed and leads to acceptable results.

Radiative transfer is calculated with the method of Kalkofen and Ulmschneider (1977). For the computation of the net radiative cooling rate we use three frequency bands. One band represents the continuum with a Rosseland mean opacity from Kurucz (1979). This table extends only down to 3000 K, whereas the gas temperatures encountered in the models drop as low as 750 K. Where necessary, these opacities are extrapolated to lower temperatures - a procedure which should produce roughly acceptable results (cf. Alexander et al., 1983). The remaining two bands are used for the fundamental vibration-rotation bands of CO and SiO, which are two of the most abundant molecules to form in a cooling cosmic gas in

LTE, due both to the abundance of their constituent elements and to their high dissociation energies of 11.09 and 8.26 eV, respectively. The opacity used for each molecule is a restricted Planck mean opacity similar to that of Muchmore and Ulmschneider (1985). The bandpass over which the CO opacity is weighted has been changed to more accurately reflect the actual band. The SiO opacity is that of Muchmore et al. (1987).

To ensure energy conservation in the frequency quadrature while using very few frequency points, we use temperature dependent quadrature weights. The integration intervals are from 4.29 to 6.2 μm and from 7.5 to 10.0 μm for CO and SiO, respectively. Thereby the source function, summed over the three frequency bands, is the Planck function, so that the frequency quadrature does not introduce any spurious sources or sinks of energy.

3. RESULTS AND DISCUSSION

For our time-dependent wave calculation we take the stellar parameters of α Boo. As an initial energy flux we use $5 \cdot 10^7$ erg cm^{-2} s^{-1} as given by Bohn (1984). For the wave period we use $1.4 \cdot 10^4$ s appropriate to the model of Ulmschneider et al. (1979). For the abundances of C, O and Si we take solar values.

A typical model of the formation and the development of a hot feature is obtained by the switch-on phase of the atmosphere which continues over many wave periods. Monochromatic wave models which avoid the generation of hot features have to be constructed very artifically by slowly increasing the initial wave amplitude. The results for these models are described by Cuntz and Muchmore (1988). In our case the behaviour of the atmosphere is dominated by large scale oscillations and merging of shocks. Therefore we obtain a nonperiodic wave model which predicts in a very natural way the coexistence of hot and cool atmospheric layers. Figure 1 shows a series of snapshots which illustrate the development of a hot atmospheric feature over a time span of $8.5 \cdot 10^3$ s. The first frame is at a time $7.25 \cdot 10^5$ s, or 52 wave periods, after the piston at the base of the atmosphere was turned on. Only the outermost section of the model is shown, plotted in lagrangian coordinates. The geometrical heights are about $2.4 \cdot 10^{10}$ and $3.7 \cdot 10^{10}$ cm for the inner and outer boundary, respectively.

When one strong shock propagates through the atmosphere, it lifts the gas above its equilibrium position. The outermost parts of the atmosphere fall back, forming an almost stationary shock, akin to an accretion shock. Outgoing shock waves coalesce with the stationary shock, increasing its strength until it moves outward through the downflowing medium.

Fig. 1a shows such a shock about to be amplified by an outgoing wave. A strong shock with a strength of 4.60 is formed which leads to a hot layer with a temperature of more than 5700 K in the compression region (Figs. 1b and 1c). The elevated post-shock temperature destroys the CO and SiO molecules; the cooling due to the molecules vanishes. Now the continuum radiation becomes an important cooling mechanism. The radiation damping function is strongly peaked behind the shock. The rise of the temperature is stopped at 6150 K. Originally at the inner edge of the feature is a stationary discontinuity. When the shock becomes too strong to remain stationary, an episodic outflow occurs.

Fig. 1d shows that the hot atmospheric layer is moving outward with nearly constant velocity. At its base, the steep gradient of the flow speed implies an expansion, which reduces the temperature. The radiation damping function has now a second maximum in the transition layer to the molecule dominated region below the hot feature. This behaviour indicates the start of the regeneration of CO and SiO molecules.

Comparing Figs. 1d and 1e we see that the hot feature becomes broader, until cooling by the molecules reduces the temperature from the inner side faster than the shock moves outward. At time $8.5 \ 10^3$ s (Fig. 1f) we find that the hot atmospheric layer is now much smaller. The shock strength is again 4.50, corresponding to a temperature of 5110 K, and the shock has nearly stalled. After $1.2 \ 10^4$ s, the temperature drops down to 4700 K, but then another wave from deeper in the atmosphere merges with the shock, leading to a new enhancement and an increase of the temperature to 5600 K, similar to Fig. 1c. Finally, at $1.6 \ 10^4$ s, the hot feature propagates through the upper boundary of the atmosphere. The shock strength is now 4.37 and the maximum temperature 5150 K. Particle densities in the hot feature range between $6.4 \ 10^{11}$ cm^{-3} for the state shown in Fig. 1b and $1.5 \ 10^{10}$ cm^{-3} when the shock is transmitted.

We have followed this model for further 60 periods, during which 10 strong shock were formed. We have found that occasionally hot features are generated with temperatures as in the feature which we have described before.

4. COMPARISON WITH OBSERVATIONS

We have obtained that the time-dependent development of a strong shock leads to a nonperiodic wave model which predicts in a very natural way the coexistence of hot and cool atmospheric layers. Hot features with a temperature between 4500 K and 6000 K and particle densities in the order of 10^{10} and 10^{11} cm^{-3} are formed. These data suggest that an essential contribution of the observed chromospheric radiative emission can be given.

The high temperature values as reached during the dynamical development of a hot feature are in the same range as the temperature of the lower chromosphere of Arcturus obtained in the semiempirical model of Ayres and Linsky

(1975). These authors have constructed a semiempirical model for the photosphere and the lower chromosphere of Arcturus based on the H, K, and IR triplet lines of Ca II and the h and k lines of Mg II. An improved semiempirical chromosphere model which considers the influence of stellar wind flows is described by Drake (1985). Judge (1986) has analyzed a number of UV lines in IUE spectra in α Boo. For T < 8000 K, he confirmed the temperature-density stratification of the Ayres-Linsky model. He also found a larger emission measure for somewhat higher temperatures (T ~ 10000 K) and was able to conclude that there is no or little gas at T > 20000 K.

Heasley et al. (1978) have obtained high-spectral-resolution observations of the fundamental vibration-rotation bands of CO in the Arcturus spectrum and have compared them with synthetic spectra for a representative set of existing model atmospheres of this star. They have found that the Ayres and Linsky model of the lower chromo-sphere-upper photosphere cannot reproduce both the Ca II K line wings and the CO fundamental lines with a single-component model - not even after considering various uncertainties. They conclude that the discrepancies regarding the temperature structure might be reconciled if the atmosphere contains both a hot component (the Ayres-Linsky model), which is required to account for the K line observations, and a cold component to explain the CO lines. This result is a strong indication for an inhomogeneous chromosphere of Arcturus. The existence of cool layers in the atmosphere of Arcturus is meanwhile confirmed by Ayres (1986).

In the case of the Sun theoretical evidences for chromospheric inhomogeneities have been obtained by Ayres (1981). He first noted that the cooling influence of the CO molecules could lead to a thermal bifurcation of the outer atmosphere of a late-type star.

5. CONCLUSIONS

We have studied the development of a strong shock in the outer atmosphere of α Boo. We have obtained that hot features with temperatures between 4500 K and 6000 K are formed during the switch-on phase of the atmosphere. The features show a dynamical development controlled by an interplay between shock heating, formation and destruction of molecules and radiative cooling. Switch-on behaviour is typical before the atmosphere reaches a dynamical steady state. A dynamical steady state of a wave-supported atmo-sphere is caused by the periodicity of heating and momentum transfer of monochromatic waves. Cuntz (1987) has pointed out that shock waves with stochastically changing periods cannot lead to a dynamical steady state. Due to shock merging, shocks with different strength can be produced at the same atmospheric height. Especially shocks with large strengths are formed. This behaviour suggests that strong shocks can be generated in cool (e.g. molecule dominated) regions in the lower chromosphere of late-type stars which lead to a possible explanation for inhomogeneous chromospheric structures as proposed by Heasley et al. (1978) and Ayres (1986).

Acknowledgements. This work has been supported by the Sonderforschungsbereich 132 and the Deutsche Forschungsgemeinschaft Project Ul 57/11-1. We are grateful to P. Ulmschneider for helpful discussions.

REFERENCES

Alexander,D.R.,Johnson,H.R.,Rypma,R.L.: 1983, Astrophys.J. 272,773
Ayres,T.R.: 1981, Astrophys.J. 244,1046
Ayres,T.R.: 1986, Astrophys.J. 308,246
Ayres,T.R.,Linsky,J.L.: 1975, Astrophys.J. 200,660
Bohn,H.U.: 1984, Astron.Astrophys. 136,338
Cuntz,M.: 1987, Astron.Astrophys. 188,L5
Cuntz,M.,Muchmore,D.: 1988 (submitted)
Drake,S.A.: 1985, in: Progress in Stellar Spectral Line Formation Theory, eds. J.E.Beckman and L.Crivellari, Reidel, Dordrecht, p.351
Heasley,J.N.,Ridgway,S.T.,Carbon,D.F.,Milkey,R.W.,Hall, D.N.B.: 1978, Astrophys.J. 219,970
Judge,P.G.: 1986, Monthly Notices Roy. Astron. Soc. 221,119
Kalkofen,W.,Ulmschneider,P.: 1977, Astron.Astrophys. 57,193
Kneer,F.: 1983, Astron.Astrophys. 128,311
Kurucz,R.: 1979, Astrophys.J.Suppl. 40,1
Mäckle,R.,Holweger,H.,Griffin,R.,Griffin,R.: 1975, Astron.Astrophys. 38,238
Muchmore,D.: 1986, Astron.Astrophys. 155,172
Muchmore,D.,Ulmschneider,P.: 1985, Astron.Astrophys. 142,393
Muchmore,D.,Nuth III.,J.A.,Stencel,R.E.: 1987, Astrophys.J.Letters 315,L141
Tsuji,T.: 1964, Ann.Tokyo Astron.Obs. 2nd Ser., 9,1
Ulmschneider,P.,Nowak,T.,Bohn,H.U.: 1977, Astron.Astrophys. 54,61
Ulmschneider,P.,Schmitz,F.,Hammer,R.: 1979, Astron.Astrophys. 74,229
Ulmschneider,P.,Schmitz,F.,Kalkofen,W.,Bohn,H.U.: 1978, Astron.Astrophys. 70,487

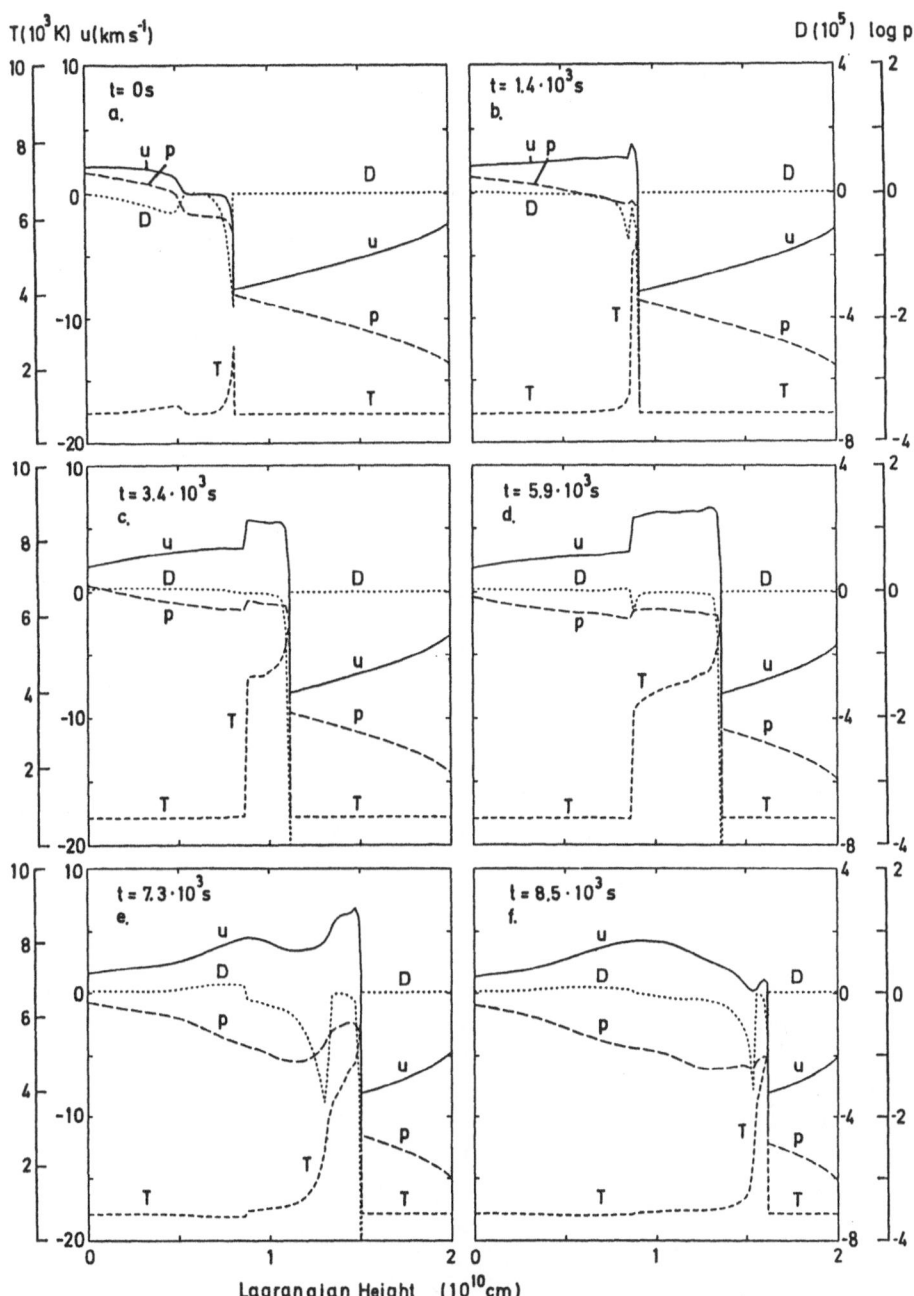

Fig. 1   A series of snapshots of a hot atmospheric feature generated by a strong accretion shock (M_S = 4.60 in Fig. 1b). The flow speed u, the temperature T, the pressure p (dyn cm^{-2}) and the radiation damping function D (10^5 erg cm^{-2} s^{-1} K^{-1}) are shown.

INTERACTION OF SOLAR GRANULATION WITH WEAK AND STRONG MAGNETIC FIELDS

Peter A. Fox

Center for Solar and Space Research, Yale University

Abstract. The modal expansion technique is used to investigate the complex non–linear interaction of granular scale convection with weak and strong mean magnetic fields on the Sun. Integration of the linearized equations also provides extra information on the role of the magnetic field strength and magnetic diffusion in determining the conditions for convective instability. In the weak mean field case (when significant field manipulation occurs) limits may be obtained on the turbulent magnetic resistivity η and in the strong mean field case (when overstable oscillations occur) the strength of the field influences the period of oscillation while η controls the rate of growth or decay of the oscillations. Some comparisons are made with the relevant solar observations in order to estimate the minimum magnetic field strength to produce convective overstability (oscillations).

1 Introduction

Magnetic fields play an important part in many of the observed solar surface features. Until recently, studies of these phenomena have been restricted to linearized equations, especially those involving magnetoacoustic waves, and very few workers have entered the non-linear domain without some very restrictive assumptions (e.g. Weiss 1981a,b).

The full magnetohydrodynamic equations are very complex, but so is the observed interaction between convection and magnetic fields, and a detailed non- linear theory is vital to attempt an understanding of even the most basic features.

Most of the fields on the Sun (what shall be termed background fields) are of the order of 2 to 10G (sometimes 50 to 100G). However, there are many regions of intense magnetic fields, 1000 to 3000G, distributed over most of the Sun. In fact, 90% of the magnetic flux is concentrated at supergranular boundaries and junctions (some 5% of the surface area).

It is therefore of interest to investigate the influence of weak and strong mean magnetic fields on convective motions. A linear instability analysis including magnetic field indicates that granular convection is not influenced by background magnetic fields but that it is inhibited by larger fields. It cannot, however, determine the extent of the influence of the convection on the magnetic field or demonstrate the observed concentration of magnetic flux into slender flux tubes or ropes at granular boundaries. These features must be studied by means of non-linear calculations.

R. J. Rutten and G. Severino (eds.), Solar and Stellar Granulation, 555–564.

Magnetic fields are likely to have varying effects depending on the spatial scales of convection they interact with. Considering the extent of the known surface magnetic fields and the somewhat unknown subsurface fields, this poses a difficult problem for theoretical treatment.

2 Observations

2.1 Granules

Granules are also observed to be unaffected by small scale mean magnetic fields of 2 to 10G in the quiet Sun (see Stenflo 1976) and they often manipulate the fields into concentrated regions at their edges. Granules are observable up to the edge of active regions (large magnetic field), and even within sunspot umbrae (Spruit 1981).

2.2 Periodic phenomena

In the presence of strong magnetic fields (active regions) many complex and diverse features are observed (Moore 1981a,b), and it is likely that convective processes play a role in the general behaviour of these features.

Many of the periodic phenomena in active regions are driven from sunspots by velocity oscillations in the umbral photosphere. The photospheric oscillations are probably excited by subphotospheric oscillatory convection in the umbra. These oscillations are virtually always present in sunspots and last for a few cycles of 2 to 3 minute periods with high characteristic velocities of ± 1 to ± 3 km sec^{-1} and typical cell sizes of 2" to 3" (1500 to 2500 km) staying in phase for up to 5" to 10" (Giovanelli 1972, Moore 1981a,b, Thomas 1981).

2.3 Aperiodic phenomena

Aperiodic phenomena in sunspots include umbral dots, penumbral grains and umbral granulation. The dots and grains appear to be manifestations of the same process but with vertical magnetic field in the umbra and near horizontal field in the penumbra, and they are similar to normal granules. Dots/grains are common to all spots and occur in all parts of the umbra/penumbra.

Umbral dots of a few hundred kilometers in diameter (Danielson 1964, Moore 1981b, Knobloch and Weiss 1984), last for 20 to 30 minutes (Danielson 1964, Moore 1981b), are less bright on average than the photosphere and have upward velocities of about 0.5 km sec^{-1} (Kneer 1973). Some controversy over the possible reduction of magnetic field strength in umbral dots remains, however, it appears that little or no reduction may occur(Spruit 1981). It is likely that non-oscillatory convection also contributes to umbral turbulence in sunspots with RMS vertical velocities of 1.3 ± 0.2 km sec^{-1}, and that RMS horizontal velocities of 1.5 ± 0.2 km sec^{-1} are still to be accounted for.

3   Theoretical treatment

An approach that has had considerable success in convection theory is the modal expansion technique. It treats the horizontal surfaces of various fields which are more or less homogeneous in some finite statistical sense, but retains the physical dimension in the vertical direction,where most changes in the structure occur.

The governing equations are generated by averaging in the horizontal plane using some variational or galerkin technique. Although the equations are complicated and of high order they are much easier to solve than the fully dimensional equations. The motivation for this type of representation arises from the seemingly cellular appearance (be they rolls, squares, hexagons or irregular polygons) of convective motions from laboratory, atmospheric and astronomical observations. Thus, the relevant observed pattern is included in the formulation of the equations.

This technique has the advantage that the effects of rotation, pulsation or magnetic fields, study overshooting or penetration into surrounding layers and model differing spatial scales of motion by including more modes in the expansion are easily incorporated. Apart from the suppression of horizontal variations of the field, the consequences of truncating the expansion to one or two terms (modes) are largely uncertain (see Gough, Spiegel and Toomre 1975, Toomre, Gough and Spiegel 1977, 1982 for laboratory comparisons and Marcus 1980).

4   Equations and Assumptions

The details of the equations will not be given here (refer to Fox 1985). This investigation will be carried out using a polytropic model of the thermodynamics. This model has been extremely successful in preliminary investigations of granulation and has the advantage of using constant coefficients of viscosity and conductivity and fixed layer depth to avoid the unnecessary complications that arise when more realistic solar models are studied (for more details see Fox 1985). Nevertheless, simple polytropic modeling should give an insight into the complex interaction of magnetic fields and convection.

The parameter values for the polytropic model (see Fox 1985) are for the polytropic index, $s = 1.409$, and for the stratification parameter $\phi = 0.597$. The ratios of the temperature, density and pressure at the upper (u) and lower (ℓ) boundaries are then $T_\ell/T_u = (1 + 1/\phi_0) = 2.68$, $\rho_\ell/\rho_u = (1 + 1/\phi_0)^s = 4.00$ and $p_\ell/p_u = 10.70$.

The domain is of granular size with a horizontal cell size (average diameter) of 2000 km and a non-dimensional wave number $a = \pi/\sqrt{2}$, corresponding to a depth of 530 km. The ratio of specific heats, γ, is taken as 1.25 which corresponds to a value of $\nabla - \nabla_{ad} = 0.21511$ occurring at a depth of 150 km according to mixing length models of the solar convection zone. The arbitrary selection of a constant buoyancy across the layer is an unfortunate feature of the polytropic model.

The average Prandtl number, σ, is to some extent a free parameter, however, it must be close to previously indicated values (0.2 in this case). The surface thermal diffusivity is $\kappa_0 = F_\odot (\gamma - 1)(s + 1)/\rho_u g = 4.644 \times 10^{12}$ cm^2 sec^{-1}.

The initial magnetic field, H_0, will be vertical and uniform, and perturbations from this initial field are important when studying the influence of convection on the field.

The permeability of the medium will be taken as unity and the magnetic resistivity η, which is of primary interest, will not vary across the layer.

In this study a hexagonal planform function $f(x,y)$ will be used.

5 Linear stability analysis

Table 1 demonstrates the effect of increasing magnetic field and decreasing magnetic resistivity upon the onset of convective instability for a granular size cell. The value H_ℓ (which is a minimum over a range of cell depths) indicates the point of maximum linear instability. The preferred depth is then the depth corresponding to that minimum H_ℓ. Table 1 indicates that the onset of granular convection is not influenced by background

H_0 (Gauss)	η ($cm^2 sec^{-1}$)	H_ℓ	Preferred depth (km) top at 500km
0	-	6.6390	850
2	3.0×10^{11}	6.6471	850
2	3.0×10^{10}	6.7170	850
2	3.0×10^{9}	7.4146	850
10	3.0×10^{11}	6.8334	850
10	3.0×10^{10}	8.5592	875
10	3.0×10^{9}	21.0000	900
50	3.0×10^{11}	10.9780	900
50	3.0×10^{10}	35.0410	1000
50	3.0×10^{9}	263.4280	1100
100	3.0×10^{10}	99.2200	1125
200	3.0×10^{11}	48.1680	1050

Table 1: The effect of increasing magnetic field and decreasing magnetic resistivity upon the onset of convective instability for a granular size cell of horizontal extent 2000 km.

fields below 50G, but is inhibited by fields larger than this value.

6 Weak Mean Magnetic Fields

In weak mean solar magnetic fields, the granular pattern does not become oscillatory, and it is often present, undisturbed, at the edge of active regions. This means that the turbulent magnetic resistivity η_t is likely to be large (certainly larger than the molecular value $\eta_m \approx 4 \times 10^6$ cm^2 sec^{-1} at the surface, as this value would severely inhibit convective motions), although its precise value is unknown.

It is possible to do numerical experiments in the weak mean field case to determine the influence of magnetic resistivity on granular scale motions. A lower limit on its value will be obtained by noting at what point granular characteristics are adversely affected. An upper limit on η may be deduced to ensure the existence of overstable motions when strong fields are present, i.e. $\eta < \kappa$ (Chandrasekhar 1961).

Table 2 gives the characteristics of granulation for various values of the initial field strength H_0 and magnetic resistivity η. For comparison, this table contains results obtained for the case without magnetic field as well.

After an integration time of 1135 secs (which is the duration in which the intensity fluctuation reaches $\approx$ 15 %), the maximum vertical velocities (for the cases considered in Table 2 except for $H_0 = 2G, \eta = 3 \times 10^7$ cm^2 sec^{-1} and $H_0 = 10G, \eta = 3 \times 10^8$ cm^2 sec^{-1}), reach a value of about 1.2 km sec^{-1}, while the RMS horizontal velocities are about 0.94 km sec^{-1}. Indeed, for $H_0 = 2G$, $\eta > 3 \times 10^8$ cm^2 sec^{-1} and $H_0 = 10G$, $\eta > 3 \times 10^9$ cm^2 sec^{-1}, the granular characteristics appear to be identical to those obtained in the absence of magnetic field (certainly to within observational accuracy).

The values of η that indicate a departure from the expected behaviour seem to depend on the field strength. However, a lower boundary of 3×10^8 cm^2 sec^{-1} still implies that there is a large interval in which η may lie, and still ensures the existence of overstable oscillations for stronger fields. Fortunately, this choice will be narrowed further in the next section due to consideration of oscillatory growth rates.

A small, initially uniform, magnetic field will be manipulated by convective motions to expel field from the central regions of the cell and concentrate it at the boundaries. This has been shown and discussed in more restrictive cases by Weiss (1981a,b) and Galloway and Weiss (1981) for example. For the weak fields that are being dealt with here the concentration of flux will be limited by diffusion.

On granular scale it is of interest to consider the limit on the field strength attained by such flux concentrations. Note, however, that the steady state effects of convection will not be considered, and that, although the short granular time scales do not prevent concentration, they may limit flux expulsion from the centre of cells.

H_0 (Gauss)	η (cm^2 sec^{-1})	Max Vert Vel (km sec^{-1})	RMS Horiz Vel (km sec^{-1})	Intensity %	Energy prop. %
0	-	1.2080	0.9416	15.427	1.5264
2	3.0×10^7	0.6925*	0.5310*	8.573*	0.5002*
2	3.0×10^8	1.2115	0.9441	15.481	1.5336
2	3.0×10^9	1.2095	0.9424	15.477	1.5249
2	3.0×10^{10}	1.2081	0.9412	15.420	1.5253
2	4.0×10^{11}	1.2081	0.9413	15.422	1.5256
10	3.0×10^8	0.8880**	0.6800**	12.347**	0.7872**
10	3.0×10^9	1.2020	0.9354	15.345	1.5067
10	3.0×10^{10}	1.1980	0.9319	15.218	1.5068
10	4.0×10^{11}	1.2000	0.9345	15.316	1.5068

Table 2: The main characteristics of normal granular flow for two weak mean magnetic field strengths and for various values of η, at $t = 1135$ sec. The values marked with asterisks are for $t = 1063$ sec.(*) and $t = 1095$ sec.(**) respectively. These times are the moments after which the motion is severely affected by the magnetic field in the two cases.

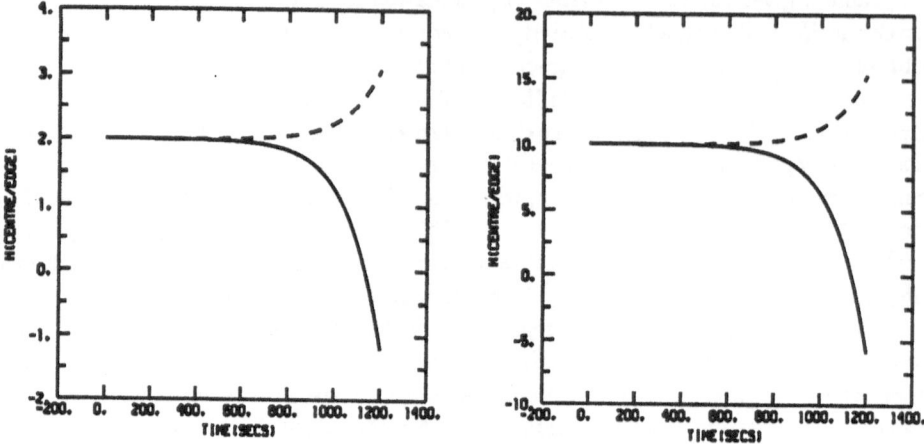

Figure 1: The surface values of the vertical (dominant) component of magnetic field at the centre and edge of the cell as a function of time for $H_0 = 2G$, $\eta = 3 \times 10^{10}$ cm^2 sec^{-1} (left) and $H_0 = 10G$, $\eta = 3 \times 10^{10}$ cm^2 sec^{-1} (right).

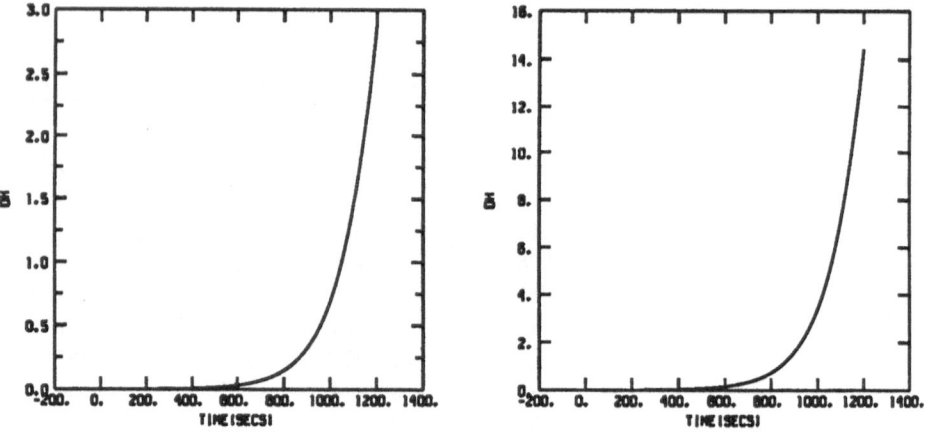

Figure 2: The RMS horizontal component of the surface magnetic field against time for the cases in Figure 1.

The surface values of the vertical (dominant) component of magnetic field at the centre and edge of the cell as a function of time give an idea of the migration of the field from centre to edge. Figure 1 shows these values against time for $H_0 = 2G$, $\eta = 3 \times 10^{10}$ cm^2 sec^{-1}, and $H_0 = 10G$, $\eta = 3 \times 10^{10}$ cm^2 sec^{-1} respectively, and it is clear that the process of expulsion begins at the first sign of convective motions. The RMS horizontal component of the magnetic field against time is shown in Figure 2 for both cases from Figure 3. It increases considerably and becomes comparable to the vertical component, demonstrating that the field is being pushed to the edges of the cell but does not remain predominantly vertical.

In the current formulation which uses the planform function $f(x, y)$ for hexagons, it appears that the magnitude of the field concentration is limited by the initial field strength. The choice of horizontal planform limits the amount of concentration as the magnetic field must satisfy $\nabla \cdot \mathbf{B} = 0$ within every hexagonal cell. The consequence is to generate negative magnetic fields (i.e. downward directed) at the centre of the cell after some time. These artificial fields are physically acceptable, but in reality the central field strength simply becomes negligible. As the magnetic resistivity is increased the process of magnetic field expulsion is slowed and thus the concentration at the boundaries is reduced.

7 Strong Magnetic Fields

Observations of oscillatory phenomena in active regions/sunspots are of a different nature than the ordinary non-oscillatory features. Velocity measurements are extracted as an oscillatory component from the general velocity field, usually with small but varying periods (about 3 min). The period of oscillation is an important measurable quantity, as are the lifetime of a particular pattern (number of periods) and the distance over which the patterns stay in phase. Measurements of intensity (temperature) fluctuations are rarely mentioned and are therefore probably negligible (N.B. in strong contrast to normal granular motions).

Indications from the small mean magnetic field case as presented in the previous section lead to limits on the resistivity η which can be used when strong fields are present. It is unclear what effect the magnetic field strength may have on the resistivity itself in addition to the induced velocity motions. It is true that the value of η influences the rate of growth (or decay) of the motion considerably once it is established. The growth (or decay) of the oscillatory motions depends on the properties of the medium like buoyancy, density stratification, viscosity etc., and on the magnetic resistivity and the initial strength of the magnetic field.

In the present model, the minimum magnetic field strength which can produce oscillations seems to be about 500G although this value is subject to some of the assumptions in this model.

The characteristics of overstable oscillations differ from the usual convective motions. The fluctuations in intensity and the amount of energy carried tend to be lower, but the velocities can become quite high, oscillating with small periods. Small time steps are often required to resolve these rapid changes, and when η is near the expected lower limit these oscillations are very difficult to resolve accurately. Further complications

562

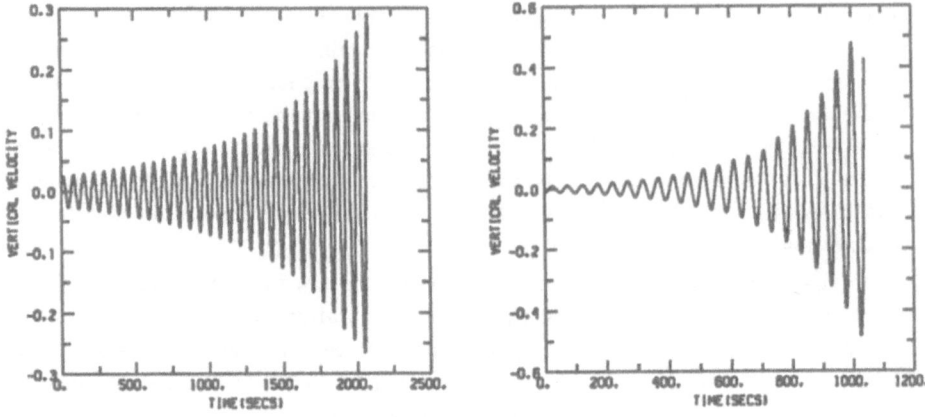

Figure 3: The time evolution of RMS vertical velocities near the surface for the strong mean field cases ($H_0 = 2000$G, $\eta = 3.0 \times 10^{10}$ cm^2 sec^{-1} [left] and $H_0 = 3000$G, $\eta = 4.0 \times 10^{11}$ cm^2 sec^{-1} [right]).

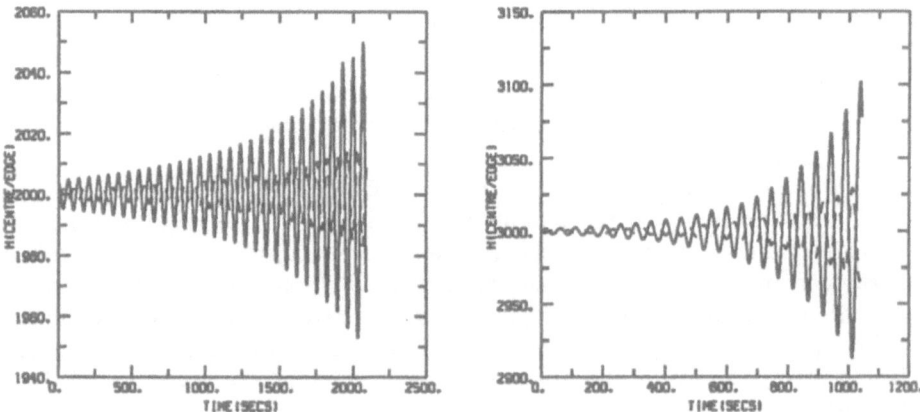

Figure 4: The surface values of the vertical (dominant) component of magnetic field at the centre and edge of the cell as a function of time for the cases in Figure 4.

arise from the depth dependence of the velocities where the fluid "turns over" or reverses its direction. The *strength* of the magnetic field determines the period of oscillation and does *not* seem to be influenced by the value of η.

H_0 (Gauss)	η (cm^2 sec^{-1})	Average Period (sec)	Vertical/Horizontal velocities (km sec^{-1})
500	4.0×10^{11}	≈ 560.9	growing*
1000	1.0×10^8	250.5	decaying
1000	1.0×10^9	250.5	decaying
1500	1.0×10^8	≈ 104.3	decaying
1500	1.0×10^9	≈ 104.3	decaying**
2000	3.0×10^{10}	68.0	$\pm 0.5/ \pm 0.4$
2000	3.0×10^{11}	68.0	decaying**
2500	3.0×10^9	56.8	$\pm 0.2/ \pm 0.1$
3000	3.0×10^{11}	48.0	$\pm 0.3/ \pm 0.25$

Table 3: The main characteristics of granular flow (overstable oscillations) for strong mean magnetic field strengths and for various values of η. Note * - growing but *very* slowly and ** - decaying but *very* slowly.

Table 3 gives the main characteristics of the flow for the range of field strengths and for various values of η. It should be noted that the velocity amplitudes are in good general agreement with observed oscillatory velocities, while the periods of the oscillation for observed field strengths are somewhat smaller than observed—probably due to the approximations inherent to the polytropic model.

The periods of oscillation tend to increase slightly as the velocity increases. An unfortunate feature of the modal (one mode) expansion technique is the inability to cause the oscillations to display time dependent limitations, therefore we cannot obtain an estimate of the lifetime. The present calculations were terminated when insufficient numerical accuracy existed to resolve the rapid changes that occur when the motions are fully developed.

The time evolution of vertical and horizontal velocities is given in Figure 3 for two cases from Table 3. The general magnetic field oscillates with very small amplitude around its mean initial value and the magnitude starts to increase slightly only when the velocity becomes significant (> 0.1 km sec^{-1}) see Figure 4. In short, no significant manipulation of the magnetic field by the convection is evident.

8 Summary

In this paper the fully compressible equations describing convective motions under the influence of magnetic field have been integrated using the modal expansion technique. The onset of normal granular convection seems to be inhibited by magnetic fields larger than about 50G. The non-linear interactions have enabled comparison with general observations of granular size motions in the presence of weak and strong mean magnetic fields. The upper and lower limits for the magnetic resistivity in the case of an initially weak uniform mean field are for $H_0 = 2G$, $\eta_{upper} = \kappa_0$, $\eta_{lower} \approx 10^8$ cm^2 sec^{-1} and for $H_0 = 10G$, $\eta_{upper} = \kappa_0$, $\eta_{lower} \approx 10^9$ cm^2 sec^{-1}, where H_0 is the magnetic field strength. This is an indication for the range over which convective motions are not appreciably affected by the magnetic field. The initially uniform weak fields are concentrated at granular boundaries and junctions and the field is approximately doubled, depending on the magnetic resistivity. The higher the resistivity, the weaker the concentration.

The process of concentration seems to be limited somewhat by granular lifetimes ($\approx$ 10 min), but mainly by the initial field strength and the choice of hexagonal planform function $f(x, y)$, which defines the horizontal dependence.

The existence of overstable motions for field strengths of 2000 to 3000G shows how the interaction of magnetic field and convection in this polytropic model can describe the generally observed solar characteristics to a fair degree. Velocity amplitudes are of the right order, but the periods of oscillation are somewhat small. The magnetic resistivity has a significant influence on the rate of growth or decay of overstable motions.

Again, the obvious restriction of the polytropic model may be influencing the finer points of the predictions.

The author wishes to acknowledge the support and encouragement of R. F. E. Van der Borght during the course of this work and S. Sofia for carefully reading the manuscript.

9 References

Chandrasekhar, S. 1961, Hydrodynamic and Hydromagnetic Stability, London, Oxford University Press.

Danielson, R. E. 1964, *Ap. J.*, **139**, 45.

Fox, P. A. 1985, *Ph. D. thesis*, Monash University, Australia.

Galloway, D. J. and Weiss, N. O. 1981, *Ap. J.*, **243**, 945.

Giovanelli, R. G. 1972, *Solar Phys.*, **27**, 71.

Gough, D. O., Spiegel, E. A. and Toomre, J. 1975, *J. Fluid Mech.*, **68**, 695.

Kneer, F. 1973, *Solar Phys.*, **28**, 361.

Knobloch, E. and Weiss, N. O. 1984, *Mon. Not. R. Astron. Soc.*, **207**, 203.

Marcus, P. S. 1980, *J. Fluid Mech.*, **103**, 241.

Moore, R. L. 1981a, in *Physics of Sunspots*, Ed. L. Cram and J. Thomas, Sacramento Peak Observatory, p 259.

Moore, R. L. 1981b, *Space Sci. Rev.*, **28**, 387.

Spruit, H. C. 1981, in *Physics of Sunspots*, Ed. L. Cram and J. Thomas, Sacramento Peak Observatory, p 359.

Stenflo, J. O. 1976, in *Basic mechanisms of Solar Activity*, Ed. V. Bumba and J. Kleczek, IAU, p 69.

Thomas, J. H. 1981, in *Physics of Sunspots*, Ed. L. Cram and J. Thomas, Sacramento Peak Observatory, p 345.

Toomre, J., Gough, D. O., and Spiegel, E. A. 1977, *J. Fluid Mech.*, **79**, 1.

Toomre, J., Gough, D. O., and Spiegel, E. A. 1982, *J. Fluid Mech.*, **125**, 99.

Weiss, N. O. 1981a,b, *J. Fluid Mech.*, **108**, 247, 283.

GRANULATION AND THE NLTE FORMATION OF K I 769.9

M.T. Gomez[1], G. Severino[1], R.J. Rutten[2]

[1]Osservatorio di Capodimonte, Napoli, Italy
[2]Sterrekundig Instituut Utrecht, The Netherlands

Abstract. The diagnostic use of solar spectral lines often requires NLTE analysis of their formation. Such NLTE modeling is usually done for static model atmospheres. When dynamical structures such as the granulation or the 5-minute oscillation are studied, it is convenient to assume that the velocities and the temperature and density perturbations do not affect the size of the departures from LTE, so that the atomic level populations may be obtained directly by applying the static departure coefficients to the dynamic LTE populations. This approach has been used by Keil and Canfield (1978) for selected Fe I lines, and by Severino *et al.* (1986) for the K I 769.9 nm resonance line. We discuss this assumption here for the latter line. It is of interest because the K I resonance line attracts much attention as a helioseismological diagnostic.

We have used an atomic model for K I similar to the one specified by Severino *et al.* (1986) together with the Carlsson (1986) radiative transfer code to study the effects of granular perturbations on the K I departure coefficients.

Previous discussions of the granular influence on NLTE departures have typically addressed metal lines such as Fe I lines. For Fe I it is clear that overionization due to strong ultraviolet radiation fields is the main NLTE mechanism to be taken into account (Lites 1972, see also Rutten 1988). Nordlund (1984) has shown that the horizontal fluctuations related to granulation alter the NLTE iron ionization equilibrium drastically, and that the non-local nature of the ionization equilibrium has to be taken into account for realistic modeling of photospheric iron lines.

We find that ultraviolet overionization is also important in the case of potassium but less than for iron. The ionization cross sections from the K I s-states are unusually small due to spin-orbit interaction (see Aymar et al. 1976, Sandner et al. 1981), and there is an effective compensation of the ultraviolet overionization due to radiative recombination to the 3D level. Recombination exceeds ionization for such low-energy bound-free transitions because they occur in the red part of the spectrum, for which the Planck function exceeds the mean intensity in the upper photosphere. In the case of K I, the ultraviolet overionization is canceled by this recombination, so that the ground state population is very close to LTE throughout the photosphere, and actually overpopulated in the deep photosphere. The populations of the higher levels are lower than the LTE values, however, due to photon losses in the lines. This is just the reverse of the behaviour of the optical Fe I lines, of which the source functions obey LTE while their opacities are below the LTE values.

565

R. J. Rutten and G. Severino (eds.), Solar and Stellar Granulation, 565–566.
© *1989 by Kluwer Academic Publishers.*

An additional NLTE phenomenon in the K I spectrum is the overpopulation of the 5P level, due to optical pumping from the ground state by violet radiation. A cascade down from this level to the ground state compensates the photon losses partially.

The K I population sensitivity to temperature fluctuations turns out to be also just the reverse of the Fe I population sensitivity. A granule, being much hotter than the mean photosphere in deep layers, produces a hotter-than-average radiation field in the upper photosphere where the K I and typical Fe I line cores are formed, even if at that height the granular temperature variation itself is negligible. This enhanced radiation produces increased photoionization and correspondingly larger departures from LTE in the case of Fe I. For K I there is also a similar increase in ultraviolet ionisation, but in addition the amount of overrecombination is reduced so that the overall K I populations are lowered. The ground state becomes underpopulated above a hot granule. In addition, the 4S–5P pumping loop is enhanced, and the source function of the 769.9 nm line is closer to LTE.

Our computations show that indeed the granular temperature perturbations in the deep photosphere affect the departure coefficients higher up. We find that the change in the departure coefficients by the hot radiation field from a granule, compared to the static atmosphere, is typically 30 % at the height where the 769.9 nm line is formed.

The changes in the K I departure coefficients that are due to granular density variations are smaller, typically less than 20 %. The density variations counteract the temperature variations, so that the combined changes of the departure coefficients are about 10%.

We conclude that full dynamical NLTE modeling is required only for detailed interpretation of observations employing the K I resonance line. However, one should be specifically aware of the effects of the bound-free transitions that 'feel' local differences in the radiation from the deep photosphere.

The paper will be submitted to Astronomy and Astrophysics.

References

Aymar, M., Luc-Koenig, E., Combet Farnoux, F.: 1976, J. Phys. B: Atom. Molec. Phys., **9**, 1279

Carlsson, M., 1986, Uppsala Astronomical Observatory Report No. 33

Keil, S.L. and Canfield, R.C.: 1978, Astron. Astrophys. **70**, 169

Lites, B.W.: 1972, *Observations and Analysis of the Solar Neutral Iron Spectrum*, NCAR Cooperative Thesis no. 28, University of Colorado, Boulder

Nordlund, Å.: 1984, in S.L. Keil (Ed.), *Small-Scale Dynamical Processes in Quiet Stellar Atmospheres*, National Solar Observatory Summer Conference, Sacramento Peak Observatory, Sunspot, p. 181

Rutten, R.J.: 1988, in R, Viotti, A. Vittone, M. Friedjung (Eds.), *Physics of Formation of Fe II Lines Outside LTE*, IAU Colloquium 94, p. 185

Sandner, W., Gallagher, T.F., Safinya, K.A., Gounand, F.: 1981, Physical Review A, **23**, 2732

Severino, G., Roberti, G., Marmolino, C., Gomez, M.T.: 1986, Solar Phys., **104**, 259

THE EFFECTS OF INHOMOGENEITIES IN 3-D MODELS ON ABUNDANCE DETERMINATIONS

G. Hammarbäck

Astronomiska Observatoriet, Box 515, S - 751 20 UPPSALA, Sweden

Abstract.
A project to estimate the errors made when using one-dimensional models to derive abundances for stars is presented. Comparison is made between the lines generated from a 3-D simulation and the lines generated from a 1-D model with the same basic parameters. The 3-D simulations have been made, and the analysis is in progress.

1 Introduction

Accurate abundance determinations using stellar spectra is an important tool in several branches of astrophysics. Using high resolution spectra and programs to synthesize the spectra observed, an accuracy of 0.2 or even 0.1 dex is often claimed. This may be a measure of the deviations in abundances obtained for different lines of the same species, but there may, however, be important systematic errors in the analysis due to the methods used. The programs used to generate the synthetic spectra usually assume the LTE approximation when calculating the lines, and they use either the plane-parallel or the spherically-symmetric approximation, i.e. the model atmospheres are one-dimensional, and take no account of the velocity fields (other than using the so called micro turbulence parameter) or the horizontal temperature variations known to exist in for example the sun.

In recent high spectral resolution surveys of solar type stars the relative abundance ratios for different chemical elements are studied as a function of overall metal abundance and age (cf., e.g., Nissen et al. (1985), Andersen et al. (1987)). The patterns found from these studies are important for the understanding of the evolution of the Galaxy. However, one may fear that the effects of inhomogeneities may be systematically different for stars of different metal abundance.

The purpose of this project is to try to estimate the systematic errors on abundances due to inhomogeneities, i.e. to see if there are systematic variations in the abundances derived from three-dimensional models with different overall metal abundances compared to one-dimensional models.

567

R. J. Rutten and G. Severino (eds.), Solar and Stellar Granulation, 567–569.
© *1989 by Kluwer Academic Publishers.*

2   Method

At first, Nordlund's anelastic 3-D code (cf. Nordlund (1982)) is used to generate models of solar type stars. Later, the new fully compressible code will be used both for solar like and for hotter stars. Spectral lines from different chemical elements are then calculated in the LTE approximation from these models using another of Nordlund's programs. Several models of the same star from different times are needed in order to simulate the light coming from different regions of the star with convection patterns that are independent of each other.

The question is how to make the comparison with the one-dimensional model. One way is to generate a 1-D model with the same basic parameters as the 3-D one, and make a best fit to the equivalent widths of the 3-D model by varying the abundance and the micro turbulence in the 1-D model.

If one generates the lines in the 3-D and 1-D model with the same code, then one may hopefully determine the absolute errors that arise when using the 1-D approximation. In practice this can be accomplished by expanding the 1-D model into a 3-D one and use this as input to Nordlund's line calculating code. Then one can alternatively introduce a micro turbulence parameter into the code, or take the velocity field from a 3-D simulation and use it for the expanded 1-D model.

Another way to get a 1-D model that is in some sense consistent with the 3-D simulation is to make a semi-empirical 1-D model using the continuum flux at different wavelengths and disk positions from the 3-D simulations. This mimics the procedure which is generally used to derive abundances for the sun.

3   Results

We have made 3-D simulations for a model with approximately solar abundance, and for a model with a tenth of solar abundance. Snapshots have been saved for every 5 minutes and the total simulation time obtained so far is of the order of 35 minutes.

Two 1-D models have been generated using the MARCS code developed in Uppsala (cf. Gustafsson et al. (1975)) with the same basic parameters as the 3-D simulations, and a micro turbulence with approximately the same value as that used for the sun. Equivalent widths for weak iron lines in the wavelength region 5000 – 9000 Å, with different excitation potential and log gf have been calculated both for the 1-D and the 3-D models. If one looks at the ratio of the equivalent width from the 1-D and the 3-D models one finds a large scatter between the different lines.

4   Conclusion

Nothing conclusive can be said from the results obtained so far. A more careful analysis along the lines outlined above is needed. One should also remember that the results that will be obtained are not based on a comparison between a real star and the corresponding 1-D model, but on a 3-D simulation of the star. As better codes are developed this analysis may have to be remade.

References

Andersen, J., Edvardsson, B., Gustafsson, B., Nissen, P.E., 1987, in *The impact of very high S/N spectroscopy on stellar physics*, Eds. G. Cayrel de Strobel and M. Spite, IAU Symposium 132, p. 441

Gustafsson, B., Bell, R.A., Eriksson, K., Nordlund, Å., 1975, *Astron. Astrophys.*, **42**, 407

Nissen, P.E., Edvardsson, B., Gustafsson, B., 1985, in *Production and distribution of CNO elements*, Eds. I.J. Danziger, F. Matteucci, K. Kjär, ESO workshop, p. 131

Nordlund, Å., 1982, *Astron. Astrophys.*, **107**, 1

References

2D FLUX TUBE IN RADIATIVE EQUILIBRIUM

W. Kalkofen[1], G. Bodo[2], S. Massaglia[3], and P. Rossi[3]

[1]Harvard-Smithsonian Center for Astrophysics, Cambridge, Massachusetts, USA
[2]Osservatorio Astronomico di Torino, Pino Torinese, Italy
[3]Istituto di Fisica Generale, Università di Torino, Torino, Italy

ABSTRACT: We solve the transfer equation for a magnetic flux tube in an atmosphere in radiative equilibrium with two-dimensional (2D) geometry. The tube is represented as a slab with reduced opacity. We describe an integral and a differential method for solving such transfer problems in a Cartesian grid system.

Thin flux tubes with a diameter of less than one photon mean free path have negligible influence on the structure of the surrounding atmosphere. The effect of thicker tubes depends on optical depth: Well below the outer surface of an atmosphere, their temperature and that of the surrounding medium are depressed, but closer to the surface the temperature is elevated. Transfer solutions in the diffusion approximation are qualitatively different, showing only depressed temperatures; they thus resemble the solutions of the heat flux equation of Spruit (1976). These results underline the importance of treating the transfer of radiation accurately, in particular, as far as discontinuities and boundaries are concerned.

1   INTRODUCTION

The main problem in the study of magnetic flux tubes in the solar atmosphere is the determination of their equilibrium state, namely, the temperature and density distributions within the tubes themselves and in the surrounding medium. For this the correct treatment of the energy transport is crucial. The approach generally followed is to use the diffusion approximation for the radiative transfer (*cf.* Spruit 1976, 1977; Deinzer *et al.* 1984a,b), which allows a detailed analysis of other characteristics of the problem. However, the diffusion approximation fails on scale sizes of the order of the photon mean free path and, in particular, close to boundaries. In flux tube models, where the opacity changes discontinuously at the tube wall, it is therefore necessary to solve the full radiative transfer equation instead of the diffusion equation. In doing this the computational difficulties grow considerably, and it is therefore prudent to increase the complexity of the problem in steps, especiallyas far as the geometry is concerned.

The equilibrium structure of an axially symmetric flux tube was analyzed by Ferrari *et al.* (1985) and Kalkofen *et al.* (1986) who solved the transfer equation under the assumptions that the temperature in the tube has plane-parallel structure and that the

R. J. Rutten and G. Severino (eds.), Solar and Stellar Granulation, 571–581.

tube does not affect the external medium. These approximations are justified if the radius of the flux tube is sufficiently small. However, since the general effect of a flux tube is to drain energy from the surrounding atmosphere we expect that, as the tube size increases, these approximations become eventually invalid and the task of a more general multi-dimensional treatment must be addressed. Here we present a first step in this direction. We treat a two-dimensional flux tube, assume grey opacity and a given shape of the flux tube, and ignore changes in the density and opacity brought about by changes in temperature; thus the opacity is assumed to be prescribed. The numerical solution of the full transfer equation is then compared with the solution of the diffusion equation in order to check the validity of the diffusion approximation.

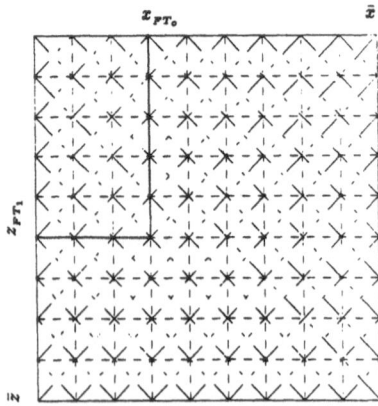

Fig. 1: Flux tube geometry with ray system for solving transfer equation.

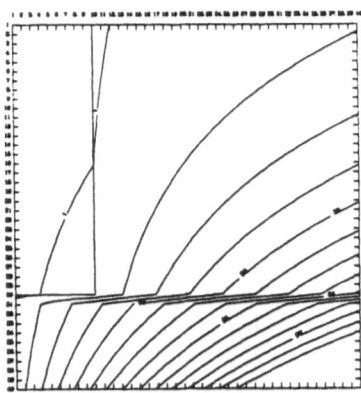

Fig. 2: Horizontal optical distance τ_x from symmetry line $x = 0$ for depth-dependent opacity. Labels are for $10 \times \tau_x$.

2 FLUX TUBE GEOMETRY

We consider a flux tube idealized as a slab of rectangular cross section residing inside an atmosphere that is semi-infinite in the vertical direction (z) and infinite in the horizontal directions $(x$ and $y)$. The slab representing the flux tube is located symmetrically about the midplane $x = 0$, extending horizontally from $x = -x_{FT}$ to $x = x_{FT}$, and vertically from $z = z_{FT_0}$ to the depth $z = z_{FT_1}$. In the y-direction there is no variation of parameters.

Since the atmosphere is symmetrical about the midplane $x = 0$ it is sufficient to treat only the half-space bounded by the mid-plane $x = 0$, the right boundary plane $x = \bar{x}$, the upper boundary of the atmosphere $z = 0$, and the lower boundary $z = \bar{z}$. In this computational space we solve the transfer equation. Transfer in the y-direction is ignored, although one may regard the third dimension as having been included by means of a rescaling of all lengths (with the factor of $\sqrt{3}$ of the Eddington approximation, for example).

The flux tube problem requires that the right boundary of the computational space at $\bar{x}$ be sufficiently far from the tube boundary at x_{FT} in order that the influence of the tube on the temperature structure of the atmosphere at $x = \bar{x}$ be negligible. The assumption is satisfied if the isotherms are nearly horizontal at the right boundary and if the horizontal component of the radiative flux is small compared to the vertical component.

The opacity (or the density) of the medium outside the flux tube is prescribed; it may have the z-dependence of a gravitating atmosphere or it may be constant; a dependence on the horizontal coordinate x is excluded for convenience. Inside the flux tube the opacity is reduced below that of the surrounding atmosphere at the same geometrical depth z by the factor $\bar{\beta}$ to account for the effect of the pressure exerted by the magnetic field.

On the space bounded in x by the lines (or planes) $x = 0$ and $x = \bar{x}$, and in z by the lines $z = 0$ and $z = \bar{z}$, we impose two grid systems, one equidistant in x and with the constant *geometrical* step size δx, and the other equidistant in z with the step size δz (*cf.* Figure 1). The *optical* grid distances are controlled by the opacity. If the opacity is constant, the optical distances in the vertical direction are given by a constant outside the flux tube, and by $1/\bar{\beta}$ times the same constant inside the tube. In the horizontal direction the optical distance between adjacent grid points is similarly given by a constant (the previous constant, scaled by the factor $\delta x/\delta z$), again with a jump at the tube wall, $x = x_{FT}$. Of course, when the opacity varies with depth z the optical distances between neighboring grid points depend on depth z. Such a case is shown in Figure 2 for the "thin" magnetic flux tube discussed below, with lines of constant horizontal optical distance τ_x from the symmetry line $x = 0$; the opacity increases exponentially, $\kappa_j = \kappa_0 \times \exp(j/15)$, where j is the vertical grid point index, with $\kappa_0 = 0.1$ outside the tube and $\kappa_0 = 0.01$ inside. Note the effect of the opacity reduction in the tube for the curve for $\tau_x = 0.1$, labeled 1; the other curves are for $\tau_x = 0.9, 1.7, 2.5, 3.3$ (labeled 33), *etc.*

3 RADIATIVE TRANSFER

We solve the transfer problem, consisting of the transfer equation and the equation of radiative equilibrium, by means of two methods, based either on integral or on differential equations. We describe the integral method and sketch the differential method. We also solve the same problem by means of the diffusion approximation.

3.1 THE INTEGRAL APPROACH

In the integral equation approach, the mean intensity $J(x_i, z_j)$ at the point (x_i, z_j) due to emission by internal sources at $(x_{i'}, z_{j'})$ and due to the transmission of radiation incident at the boundaries is expressed by an equation of the form

$$J(x_i, z_j) = \Lambda(x_i, z_j, x_{i'}, z_{j'})B(x_{i'}, z_{j'}) + G(x_i, z_j) , \qquad (1)$$

where B is the Planck function, Λ the operator describing the transfer of diffuse radiation from the source point $(x_{i'}, z_{j'})$ to the field point (x_i, z_j), and G is the boundary term; summation over the primed quantities is implied. Energy conservation for grey opacity is expressed by

$$J(x_i, z_j) = B(x_i, z_j) , \qquad (2)$$

stating the equality of absorbed and emitted energy.

The two equations are combined into the single integral equation

$$\mathcal{L}(x_i, z_j, x_{i'}, z_{j'}) J(x_{i'}, z_{j'}) = G(x_i, z_j) , \tag{3}$$

with the matrix $\mathcal{L}$ given by

$$\mathcal{L}(x_i, z_j, x_{i'}, z_{j'}) = \delta(x_i, x_{i'})\delta(z_j, z_{j'}) - \Lambda(x_i, z_j, x_{i'}, z_{j'}) , \tag{4}$$

where $\delta(x, x')$ [=1 if $x = x'$, and =0 otherwise] is the Kronecker delta. The solution of the transfer problem is then obtained by the solution of the system (3) of linear equations, i.e.,

$$J = \mathcal{L}^{-1}G . \tag{5}$$

It may be worth noting that if this grey transfer problem included true emission as well as isotropic scattering, it would lead to exactly the same definition of the integral operator (4) and the solution (5). The only difference would be the relation between geometrical and optical distances which would then contain both the true absorption coefficient and the scattering coefficient.

The construction of the integral operator $\mathcal{L}$ and of the boundary term G parallels the solution method by means of differential equations, in which the transfer equation is solved along rays. We now consider the transfer along rays.

We describe the specific intensity of the radiation field in terms of its even and odd parts,

$$J_\mu = \tfrac{1}{2}(I_\mu^+ + I_\mu^-) , \tag{6}$$

and

$$H_\mu = \tfrac{1}{2}(I_\mu^+ - I_\mu^-) , \tag{7}$$

i.e., by the Feautrier variables (cf. Mihalas 1978), which are the intensity mean and the net flux along a ray specified by the index μ; I_μ^+ and I_μ^- are the specific intensities in the two directions along the ray. The mean intensity is given by the sum

$$J = \sum_\mu \omega_\mu J_\mu , \tag{8}$$

which extends over the four values of μ corresponding to the four types of rays shown in Figure 1.

The transfer equation along a ray specified by μ is

$$(1 - d^2/d\tau_\mu^2)J_\mu = B , \tag{9}$$

subject to boundary conditions. These prescribe either the incident specific intensity $I_{bc}^\pm$, with the boundary condition given by the equation

$$(1 \pm d/d\tau_\mu)J_\mu = I_{bc}^\pm , \tag{10}$$

or the boundary flux H_μ, for which the boundary equation takes the form

$$\frac{d}{d\tau_\mu}J_\mu = H_\mu . \tag{11}$$

The optical depth τ_μ is measured along the ray μ.

We treat the differential equations of transfer passing through each point in the two-dimensional grid along rays in four directions, namely, the vertical, the horizontal, and along the two diagonals that arise in each basic rectangular cell $(\delta x, \delta z)$. The notation for the angle index μ is μ_1 for horizontal rays, μ_2 for diagonal rays inclined towards the left of the outward normal, μ_3 for vertical rays, and μ_4 for diagonal rays inclined towards the right of the outward normal — Note that because of the asymmetry introduced into the medium by the flux tube the two diagonal rays are not equivalent, unlike the case of a plane-parallel atmosphere. — For convenience we assume that the number of grid points in x is the same as in z and is equal to N.

Thus we consider five kinds of rays (cf. Figure 1): Vertical rays enter the two-dimensional medium at the lower boundary $(z = \bar{z})$, where the incident specific intensity is $I_{bc,\mu_3}^+(\bar{z})$, and leave at the upper boundary $(z = 0)$, where the incident intensity is $I_{bc,\mu_3}^-(0)$; horizontal rays enter at the right $(x = \bar{x})$, where the incident intensity is given by the Planck function $B(z)$, and leave at the left $(x = 0)$, where the net flux is zero $(H_{\mu_1} = 0)$; inclined rays entering at the right $(x = \bar{x})$ $(I_{incident} = I_{bc,\mu_2}^+(z))$ leave at the top $(z = 0)$ $(I_{incident} = I_{bc,\mu_2}^-(0) = 0)$; inclined rays entering at the bottom $(z = \bar{z})$ and going to the right $(I_{incident} = I_{bc,\mu_4}^+(\bar{z}))$ leave at the right $(x = \bar{x})$ $(I_{incident} = I_{bc,\mu_4}^-(z))$; and inclined rays entering at the bottom $(z = \bar{z})$ and going to the left $(I_{incident} = I_{bc,\mu_2}^+(\bar{z}))$ are reflected at $x = 0$, then go to the right and leave at the top $(I_{incident} = I_{bc,\mu_2}^-(0))$.

Along each ray passing through the grid points in the two-dimensional grid, the transfer equation is described by the finite-differenced form of the transfer equation (9) (plus eq. 10 or 11), i.e., the Feautrier equation, which we write as

$$-a_{\mu,l}J_{\mu,l-1} + b_{\mu,l}J_{\mu,l} - c_{\mu,l}J_{\mu,l+1} = q_{\mu,l} \quad , \tag{12}$$

where l is the running index denoting grid points along the particular ray; the source term q is given by the Planck function at interior points, and by the incident intensity or the net flux at boundary points. At the upper (or the left) boundary the coefficient $a_{\mu,1}$ is zero, and at the lower (or the right) boundary, $c_{\mu,N} = 0$ (for the values of the coefficients, see, for example, Mihalas 1978, or Auer 1984). Thus the transfer along rays is described by tridiagonal operators acting on the mean specific intensity J_μ. The inverse of the differential operator furnishes us with the Λ_μ matrix along the same ray, as well as with the damping terms for the incident intensities at the two boundaries. Thus,

$$(1 - d^2/d\tau_\mu^2)^{-1} \sim \Lambda_\mu , \tag{13}$$

(cf. Cannon 1985). The integrated Λ matrix of equation (1) is then built up by summing over all specific Λ_μ operators along all rays,

$$\Lambda(x_i, z_j, x_{i'}, z_{j'}) = \sum_\mu \omega_\mu \Lambda_\mu(x_i, z_j, x_{i'}, z_{j'}) , \tag{14}$$

and a similar sum gives the boundary term G of the transfer equation (1).

3.2 THE DIFFERENTIAL APPROACH

For the differential equation statement of the transfer problem (3) we note that the vertical and the two inclined rays passing through a point with the x and z indices (i,j) connect on the $(j-1)^{th}$ row to points with column indices i or $i \pm 1$, respectively. The horizontal ray passing through (i,j) connects to the points $(i-1,j)$ and $(i+1,j)$. Thus the equation for the whole transfer problem in terms of the Feautrier equation must collect all intensity components at all points in row z_j into a single vector. This Feautrier equation then connects rows $j-1$, j, and $j+1$ in a tridiagonal matrix equation. In order to take also the energy conservation equation (2) into account we define a vector $\tilde{J}_j$ that consists of the intensity components along the four rays passing through each one of the N points at the depth z_j and of the N components of the Planck function at z_j. Thus each intensity component $\tilde{J}_j$ has itself $5N$ components, and the whole vector $\tilde{J}$ (i.e., $\tilde{J}_j$, $j = 1, N$) contains all intensity values J_{μ_i} and Planck function values B at all grid points in the two-dimensional $N \times N$ grid. The transfer problem in the 2D grid subject to the condition of radiative equilibrium is then given by an equation of the form

$$-A_j \tilde{J}_{j-1} + B_j \tilde{J}_j - C_j \tilde{J}_{j+1} = Q_j , \qquad (15)$$

in which the inhomogeneous term Q_j has non-zero values only at boundaries. Note that the radiative equilibrium condition contributes only to the diagonal term (i.e., to B_j) in this equation.

It is clear that the transfer problem could just as easily be described in terms of an intensity and a Planck function vector in which the intensity components $J_{\mu_k}(x_i, z_j)$ and Planck function components $B(x_i, z_j)$ are ordered according to columns in the 2D grid. In the present case of N grid points in both x and z there is no advantage to either definition. But in a rectangular space it would be more efficient numerically to define the vector $\tilde{J}$ in terms of the components ordered along the narrower side.

3.3 THE DIFFUSION APPROXIMATION

In the diffusion approximation to the radiative transfer equation, the flux can be expressed as

$$\vec{H} = \frac{4}{3\kappa} \nabla B , \qquad (16)$$

where κ is the absorption coefficient. The equation governing flux conservation now reads

$$\nabla \cdot \vec{H} = 0 \quad . \qquad (17)$$

We solve the equation in the geometry described above using a finite-difference method, with the boundary condition

$$\frac{\partial B}{\partial x} = 0 \qquad (18)$$

on the left and right boundaries. At the top and bottom boundaries, respectively, we use a boundary condition analogous to equation (10) for the specific mean intensity, viz.

$$(1 \mp \frac{d}{d\tau})B = I_{bc}^{\mp} , \qquad (19)$$

where $I_{bc}^- = 0$ and I_{bc}^+ is obtained from the two-stream solution of the transfer problem without a flux tube; τ is optical depth measured normal to the respective surface. Alternatively, we have used the vertical component of equation (16) at the bottom boundary, with the vertical flux component prescribed. At the flux tube boundaries we demand continuity of the flux component normal to the interface.

4 Results and Discussion

The first question we need to address concerns the lower boundary condition for the atmosphere with a flux tube. There are two simple limiting cases: either the net radiative flux or the incident intensity are prescribed, neither of them being modified to account for the reduced opacity inside the flux tube.

Even though the convective flux in the tube is strongly inhibited by the magnetic field, the net radiative flux is not changed by the field *per se*. However, the reduction of the gas pressure and, hence, of the opacity in the tube increases the temperature gradient relative to optical depth, which drives a larger radiative flux than flows through the field-free atmosphere. Thus, the unmodified net radiative flux should present a *lower* limit on a realistic value of the radiative flux entering the magnetic flux tube, at least in the case of fairly thin flux tubes. On the other hand, the unmodified incident intensity must be an *upper* limit on the boundary condition in our case. This limit is appropriate for sufficiently thin flux tubes that do not influence the structure of the surrounding atmosphere. Then the isotherms of the medium are horizontal throughout and extend without change into the flux tube. Thus boundary conditions can be stated in terms of the temperature and its gradient. Since our flux tubes end above the lower boundary of the computational space, this condition can be replaced by a condition on the intensity. — It is obvious that of the two boundary conditions chosen, the one giving the incident intensity is apt to be more fitting since it must be realistic for sufficiently thin flux tubes. If the flux tube extended through the bottom boundary, this condition should be replaced by a condition on the temperature, which would then lead to a higher intensity because of the increased temperature derivative relative to optical depth inside the tube.

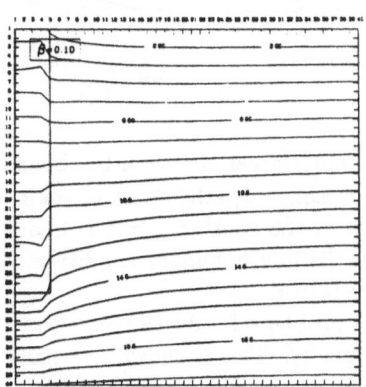

Fig. 3: Isotherms for thin flux tube; constant opacity; intensity at bottom boundary prescribed.

We first solve the transfer problem for a relatively thin flux tube in an atmosphere with constant opacity and for the boundary conditions fixed in terms of the incident intensity (*cf.* Figure 3). The horizontal as well as the vertical thickness of the computational space is equal to $\delta\tau = 10$. The flux tube extends to a depth of $\tau_{ext} = 7.5$ in the external medium ($\tau_{int} = 0.75$ in the tube). The tube "diameter", $2\delta\tau_R$, *i.e.*, twice the optical thickness of the slab from $i = 1$ to $i = 5$, is equal to $2\delta\tau_R = 2.5$ (measured in units of the external photon mean free path). The isotherms are labeled by the Planck function B ($\propto T^4$); at the surface $\tau_z = 0$, B is approximately equal to unity.

The numerical results show that there is a small influence of the flux tube on the surrounding atmosphere and that the temperature inside the tube is slightly different from the temperature outside. Near the surface of the atmosphere, the temperature inside as well as near the tube is higher than in the undisturbed medium, in accordance with the expectations for thin flux tubes of Kalkofen *et al.*(1986) but at variance with the results of Spruit (1976). At the external depth of $\tau_{ext} = 5$ ($\tau_{int} = 0.5$), the temperature in the tube is depressed by about 2%.

Figure 4 shows a tube with twice the "diameter" of the previous tube ($2\delta\tau_R = 5$); the temperature depression on the tube axis is much larger, amounting to more than 5% at $\tau_{ext} = 5$, and the layers with depressed temperature reach closer to the surface, although the top layers inside and near the tube remain hotter than the outside atmosphere.

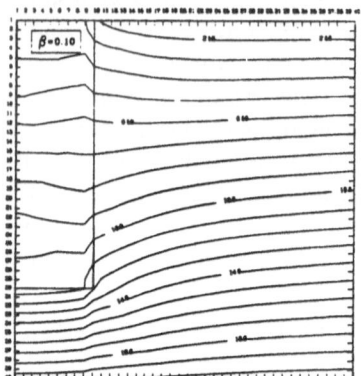

Fig. 4: Isotherms for thick flux tube; constant opacity; intensity at bottom boundary prescribed.

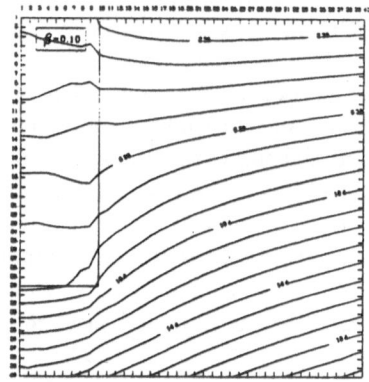

Fig. 5: Isotherms for thick flux tube; constant opacity; net flux at bottom boundary prescribed.

For a flux tube with the same, larger, diameter but boundary conditions fixed in terms of the net radiative flux (*cf.* Figure 5), the effect of the tube is still more pronounced. The temperature depression at the depth $\tau_{ext} = 5$ is about 12%. Recall, however, that the value of the net flux in the flux boundary condition without the effect of the tube may be too small by a large factor.

The solutions of transfer problems for depth-dependent opacity are qualitatively similar to those for constant opacity. This is shown in Figure 6 for the stratified medium described in Figure 2, in which the opacity grows as $\kappa_j \propto e^{j/15}$.

The changes in the net radiative flux brought about by the magnetic flux tube are shown in Figures 7a-7c for the tube discussed in Figure 5: There is a strong increase of the vertical flux inside the tube (Fig. 7a), consistent with the increase in the temperature gradient relative to optical depth. Outside the tube, the vertical flux is largely unchanged. But now there is also a finite horizontal flux (Fig. 7b), which was zero for the atmosphere without the tube. As a consequence of these changes, the net flux vector (Fig. 7c) shows a horizontal flow into the flux tube and an increased net radiative flux towards the surface of the atmosphere. It is this energy flux that drains energy from the outside medium and causes the depressed temperature in and near the flux tube.

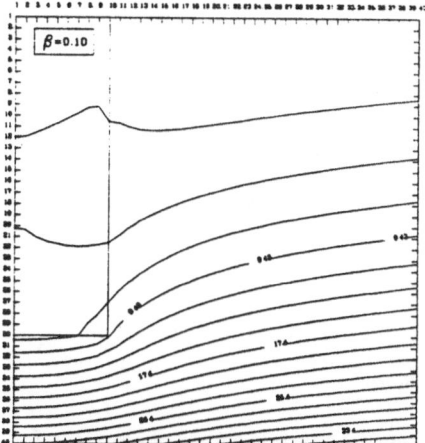

Fig. 6: Isotherms for thick flux tube; depth-dependent opacity; net flux at bottom boundary prescribed.

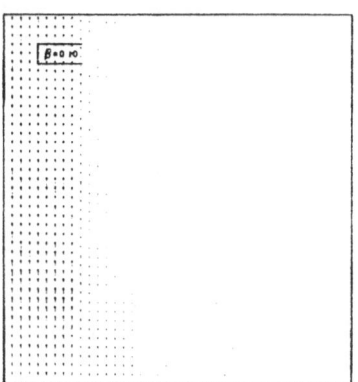

Fig. 7a: Changed net vertical flux for the thick flux tube of Fig. 3.

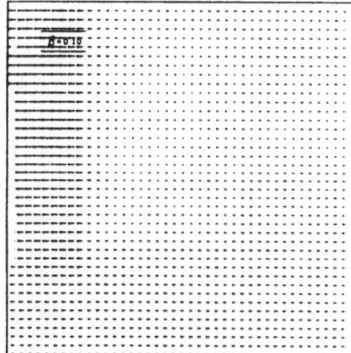

Fig. 7b: Net horizontal flux for the thick flux tube.

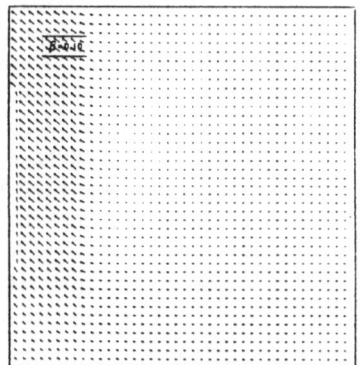

Fig. 7c: Changed net flux for the thick flux tube.

It is interesting to note that all tubes show regions where the temperature inside the tube is higher than outside; the larger the diameter of a tube, the more this phenomenon is confined to the surface layers. For sufficiently thin tubes we expect the inside temperature to be (mildly) elevated in *all* layers of a radiative equilibrium atmosphere. This is not the case for solutions of the diffusion equation (*cf.* Figure 8), which differ qualitatively from solutions of the normal transfer equation. Even the thin flux tube of Figure 3 shows only depressed temperatures, inside as well as outside the tube. In this respect, the diffusion equation solutions are reminiscent of the flux tube models of Spruit (1976), who solved the diffusion equation modelling radiative as well as convective energy transport.

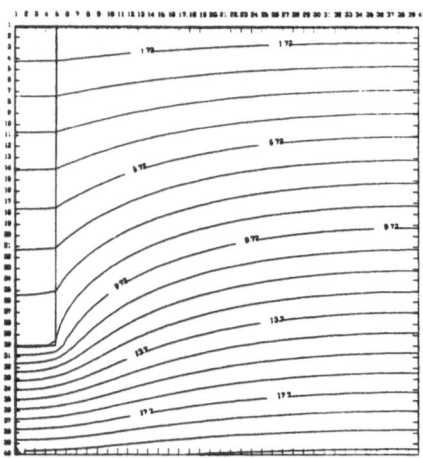

Fig. 8: Isotherms for thin flux tube; constant opacity; diffusion approximation solution.

The diffusion equation solutions are not very sensitive to our choice of equation (19) as a condition to be satisfied at the top and bottom boundaries. If equation (19) is replaced by a condition on the temperature, as used by Deinzer *et al.*(1984a,b) or by Spruit (1976, 1977), the results are qualitatively the same.

The differences in the temperature structure between the solution of the radiative transfer problem in the diffusion approximation (17) and the solution of the radiative transfer equation (1) emphasizes the importance of treating the boundaries accurately. Thus, while the diffusion approximation to the radiative transfer may give acceptable results far from the flux tube and far from the outer surface of the atmosphere, thus allowing an assessment of the influence of the flux tube on distant regions of the surrounding medium, the region inside the flux tube and close to it as well as the layers near the outer surface of the atmosphere should be treated by solving the radiative transfer equation accurately, *i.e.*,by satisfying the correct boundary conditions at boundaries and interfaces. This is particularly important for studies of the spectroscopic signature of flux tubes since their effect on excitation and ionization conditions may be magnified by the temperature differences.

REFERENCES

Auer, L. H. 1984, *Methods in Radiative Transfer*, W. Kalkofen ed., Cambridge University Press, Cambridge, 79.

Cannon, C. J. 1985, *The Transfer of Spectral Line Radiation*, Cambridge University Press, Cambridge.

Deinzer, W., Hensler, G., Schüssler, M., Weisshaar E. 1984a, *Astron. Astroph.*, **139**, 426.

———————— 1984b, *Astron. Astroph.*, **139**, 435.

Ferrari, A., Massaglia, S., Kalkofen, W., Rosner, R., Bodo, G. 1985, *Ap. J.*, **298**, 181.

Kalkofen, W., Rosner, R., Ferrari, A., Massaglia, S. 1986, *Ap. J.*, **304**, 519.

Mihalas, D. 1978, *Stellar Atmospheres*, Second Edition, W. H. Freeman & Co., San Francisco.

Spruit, H., C. 1976, *Sol. Phys.*, **50**, 269.

———————— 1977, *Sol. Phys.*, **55**, 3.

ISOLATED MAGNETOHYDRODYNAMIC THERMAL EDDIES IN A THERMALLY STRATIFIED FLUID

YU-QING LOU
Advanced Study Program and
High Altitude Observatory,
National Center for Atmospheric Research,[1]
P.O. Box 3000,
Boulder, Colorado 80307,
U.S.A.

1. Introduction

Finite-amplitude Alfvén waves in a thermally stratified fluid were discussed by Parker (1984) in the Boussinesq approximation. We note that the convective force (due to the uniform vertical temperature gradient) plays an analogous role, as the "β force" (due to the variation of Coriolis parameter) does in the motion of a "Rossby modon" in the β-plane approximation (Rossby 1939, Larichev and Reznik 1976). By a similar mathematical treatment, we obtain a two-dimensional nonlinear magnetohydrodynamic (MHD) solution for an isolated thermal vortex moving horizontally along the direction of a background magnetic field in a thermally stratified fluid. This kind of isolated MHD thermal vortex is driven primarily by convective force modified by the presence of magnetic field, and can be regarded as a form of convection. In the absence of magnetic field, the solution reduces to an isolated hydrodynamic thermal vortex which moves horizontally. These kinds of eddy phenomena may well exist in the Earth's core and in a convective overshoot layer in the solar atmosphere (with magnetic field), or in the troposphere and mesosphere of the Earth (without magnetic field).

---

[1] The National Center for Atmospheric Research is sponsored by the National Science Foundation

R. J. Rutten and G. Severino (eds.), Solar and Stellar Granulation, 583–588.
© 1989 by Kluwer Academic Publishers.

2. A General Formulation

The background equilibrium is characterized by a uniform, negative temperature gradient in the vertical $\hat{y}$-direction, a uniform gravitational acceleration in the downward direction, and a uniform magnetic field in the horizontal $\hat{x}$-direction. In the presence of perturbations, diffusive time scales are assumed to be much longer than a typical time scale of the isolated vortex motion so that ideal MHD equations in the Boussinesq approximation are adopted. These equations are:

$$\frac{\partial \vec{v}}{\partial t} + (\vec{v} \cdot \nabla)\vec{v} = -\frac{\nabla p}{\rho_0} + \frac{[\nabla \times (\vec{B} + B_0\hat{x})] \times (\vec{B} + B_0\hat{x})}{4\pi\rho_0} + \alpha g T \hat{y}, \qquad (2.1)$$

$$\frac{\partial \vec{B}}{\partial t} = \nabla \times [\vec{v} \times (\vec{B} + B_0\hat{x})], \qquad \frac{\partial T}{\partial t} + \vec{v} \cdot \nabla(T + T_0) = 0, \qquad (2.2, 2.3)$$

$$\nabla \cdot \vec{v} = 0, \qquad \nabla \cdot \vec{B} = 0, \qquad (2.4, 2.5)$$

where $\vec{v}$, $\vec{B}$, $B_0\hat{x}$, T, T_0, ρ_0, p, and g are the velocity perturbation, the magnetic field perturbation, the background magnetic field, the temperature perturbation, the background temperature, the background fluid mass density, the fluid pressure perturbation, and the uniform gravitational acceleration, respectively; α is the coefficient of linear thermal expansion of density perturbation ρ.

In a Cartesian coordinate system, we consider two-dimensional motions in the (x, y)-plane with zero $\hat{z}$-component of velocity v_z, but with nonzero $\hat{z}$-component of magnetic field $B_z(x, y, t)$. To satisfy equations (2.4) and (2.5), a stream function $\psi(x, y, t)$ and a magnetic flux function $A(x, y, t)$ are introduced so that

$$v_x = -\frac{\partial \psi}{\partial y}, \qquad v_y = \frac{\partial \psi}{\partial x}, \qquad (2.6)$$

$$B_x = -\frac{\partial A}{\partial y}, \qquad B_y = \frac{\partial A}{\partial x}. \qquad (2.7)$$

We seek solutions that have the functional form $\psi = \psi(x', y)$, where $x' \equiv x - Ut$, with U a constant speed. It follows from equations (2.2), (2.6), and (2.7), and from equations (2.3) and (2.6), that

$$A + A_0 = G(\Phi), \quad B_z = Z(\Phi), \quad \text{and} \quad T + T_0 = W(\Phi), \qquad (2.8, 2.9, 2.10)$$

respectively, where $G(\Phi)$, $Z(\Phi)$, and $W(\Phi)$ are three free functions, and

$$A_0 \equiv -B_0 y, \qquad \Phi \equiv \psi + Uy. \qquad (2.11, 2.12)$$

From the momentum equation (2.1) together with equations (2.6) - (2.8), and (2.10), it is straightforward to obtain

$$\nabla^2 \psi - \frac{1}{4\pi\rho_0}\frac{dG}{d\Phi}\nabla^2 A - \alpha g\frac{dW}{d\Phi}y = F(\Phi), \qquad (2.13)$$

where $F(\Phi)$ is the fourth free function and ∇^2 is the two-dimensional Laplacian operator. When the four free functions $G(\Phi)$, $Z(\Phi)$, $W(\Phi)$, and $F(\Phi)$ are prescribed, the stream function ψ is solved from equations (2.12) and (2.13); v_x and v_y are obtained from equation (2.6); B_x and B_y are obtained from equations (2.7) and (2.8); B_z and T are obtained from (2.9) and (2.10) respectively; finally, ρ and p are obtained from linear thermal expansion and the momentum equation respectively.

3. Isolated MHD Thermal Vortex

In order to construct an isolated MHD thermal vortex, we choose the free functions in the following way,

$$\frac{dG}{d\Phi} = Q, \qquad \frac{dW}{d\Phi} = R, \qquad (3.1)$$

where Q and R are two constants, and $F(\Phi)$ is continuous but of piece-wise linear form, viz.,

$$F(\Phi) = -m^2(\psi_{in} + Uy) \qquad \text{for } r < a, \qquad (3.2a)$$

$$F(\Phi) = l^2(\psi_{ex} + Uy) \qquad \text{for } r > a, \qquad (3.2b)$$

where m^2 and l^2 are two constants, $r \equiv [(x'^2 + y^2]^{1/2}$ is the radial distance from the center of the vortex and a is the radius of the vortex. We then solve equations (2.12) and (2.13) for ψ. Conditions for a quiet exterior at infinity are: zero flow, uniform horizontal magnetic field B_0, and uniform vertical temperature gradient $-H_0$. In order to match the interior ($r < a$) and the exterior ($r > a$) solutions, we require the stream function ψ to have continuous derivatives up to second order across the circle $r = a$. Thus, in a polar coordinate system (r, ϕ) with $x' = r \cos \phi$ and $y = r \sin \phi$, the interior solution is

$$\psi_{in} = \left[\tilde{A}J_1(\mu r) + \left(\frac{\alpha gR}{m^2} - U\right)r\right]\sin \phi \qquad \text{for } (r < a), \qquad (3.3a)$$

586

and the exterior solution is

$$\psi_{ex} = \widetilde{B} K_1(\sigma r) \sin \phi \qquad \text{for } (r > a), \tag{3.3b}$$

where $J_1(u)$ and $K_1(u)$ are Bessel and McDonald functions respectively;

$$\mu^2 \equiv \frac{m^2}{1 - Q^2/4\pi\rho_0} > 0, \qquad \sigma^2 \equiv \frac{l^2}{1 - Q^2/4\pi\rho_0} > 0; \tag{3.4a, 3.4b}$$

from matching conditions at $r = a$, we have

$$\widetilde{A} = -\frac{\alpha g R a}{m^2 J_1(\mu a)}, \qquad \widetilde{B} = \frac{\alpha g R a}{l^2 K_1(\sigma a)}, \tag{3.5a, 3.5b}$$

and a typical "modon relation"

$$-\frac{J_2(\mu a)}{\mu a J_1(\mu a)} = \frac{K_2(\sigma a)}{\sigma a K_1(\sigma a)}, \tag{3.6}$$

which relates the translation speed U (through σ; see equation 3.7), the size a, and the circulation strength (through μ) of the vortex (Flierl 1987). From conditions at infinity, the translation speed U of the MHD thermal vortex is

$$U = \pm \left(C_A^2 + \frac{\alpha g H_0}{\sigma^2} \right)^{1/2}, \tag{3.7}$$

where C_A^2 is the square of Alfvén speed. Since $K_1(\sigma r)$ is exponentially decaying, this solution describes an isolated MHD thermal vortex moving in a thermally stratified fluid layer. The generation process and the stability property of the isolated MHD thermal eddy need further investigation.

4. Applications

The background conditions for which an isolated MHD thermal vortex may exist can be found in various physcial systems, at least in a certain local sense. We shall briefly describe a few examples in the following.

4.1. THE EARTH'S MOLTEN CORE

The molten core of the Earth has inner and outer radii at 1200 km and 3500 km respectively (Melchoir 1986). Within the molten core, the temperature decreases from 4200 K° to 3200 K° with the increase of radial distance. A toroidal magnetic field strength (estimated from theoretical considerations of geomagnetic dynamo process) varies from 50 to 300 gauss (Braginsky 1964, Hide 1966). The estimate for the coefficient of thermal expansion is $\alpha \sim 2 \times 10^{-5} K^{\circ -1}$. Poloidal magnetic fields advected by an isolated MHD thermal vortex may have observable effects over the Earth's surface.

4.2. THE SOLAR CONVECTIVE OVERSHOOT LAYER

In order to account for the total magnetic flux emerging during the solar cycle (Golub *et al.* 1981), the estimated toroidal magnetic field strength in the convective overshoot layer (Spiegel and Weiss 1980; Galloway and Weiss 1981) is $\sim 10^4$ gauss for a layer thickness $\sim 2 \times 10^4$ km (Schmitt and Rosner 1983). Since a negative temperature gradient must exist across this layer, as it does throughout the solar interior, isolated MHD thermal vortices may well exist in such a layer. Since the fluid is compressible and stratified, to justify the usage of the Boussinesq approximation, the radius of a MHD thermal vortex should be smaller than local scale heights.

4.3. THE EARTH'S TROPOSPHERE AND MESOSPHERE

Within both troposphere and mesosphere of the Earth (without magnetic field), temperatures decrease with heights. Isolated hydrodynamic thermal vortices should be observable as long as their sizes are smaller than local scale heights.

ACKNOWLEDGMENTS: I thank T. Bogdan, B. C. Low, and R. Rosner for useful discussions.

References

Braginsky, S. I., "Magnetohydrodynamics of the Earth's core," *Geomagnetism Aeronomy* **4**, 698-712 (1964).

Flierl, G. R., "Isolated eddy models in geophysics," *Ann. Rev. Fluid Mech.* **19**, 493-530 (1987).

Galloway, D. J. and Weiss, N. O., "Convection and magnetic fields in stars," *Ap. J.* **243**, 945-953 (1981).

Golub, L., Rosner, R., Vaiana, G. S. and Weiss, N. O., "Solar magnetic fields: the generation of emerging flux," *Ap. J.* **243**, 309-316 (1981).

Hide, R., "Free hydromagnetic oscillations of the earth's core and the theory of the geomagnetic secular variations," *Phil. Trans. Roy. Soc. London, A* **259**, 615-647 (1966).

Larichev, V. and Reznik, G., "Two dimensional Rossby soliton: An exact solution," *POLYMODE News* No. **19**, 1-6 (1976).

Melchior, P. J., *The Physics of the Earth's Core*, Pergamon Press, Oxford, pp. 1-256 (1986).

Parker, E. N., "Alfvén waves in a thermally stratified fluid," *Geophys. Astrophys. Fluid Dynamics* **29**, 1-12 (1984).

Rossby, C.-G. and Collaborators, "Relation between variations in the intensity of the zonal circulation of the atmosphere and the displacements of the semi-permanent centers of action," *J. Marine Res.* **2**, 38-55 (1939).

Schmitt, J. H. M. M. and Rosner, R., "Doubly diffusive magnetic buoyancy instability in the solar interior," *Ap. J.* **265**, 901-924 (1983).

Spiegel, E. A. and Weiss, N. O., "Magnetic activity and variations in solar luminosity," *Nature* **287**, 616-617 (1980).

WHAT DO THE MG II LINES TELL US ABOUT WAVES AND MAGNETIC FIELDS?

W. RAMMACHER AND P. ULMSCHNEIDER
Institut für Theoretische Astrophysik,
Universität Heidelberg,
Im Neuenheimer Feld 561
D-6900 Heidelberg, Federal Republic of Germany

ABSTRACT. For different flux tube distributions over a solar surface element and for various energies of MHD-waves propagating along the tubes, the Mg II k and Ca II K lines integrated over the area are simulated. We discuss the influence of various physical parameters on the line shape.

1. Introduction

It is very difficult to observe magnetohydrodynamic waves propagating along solar magnetic flux tubes because the tube diameters as a rule are much below the resolution limit of about one arc sec. Spectral line observations therefore average over a whole ensemble of different surface features which in addition to non-magnetic regions usually include many flux tubes. Due to the spreading of the flux tubes with height this averaging process is even more complicated. However, not all of the information on the waves and the filling factor or geometry of the magnetic field is lost. In the present work we show some first results of our investigation of how the spatially integrated Mg II k and Ca II K line shapes change, if the physical parameters of a solar surface element and of the waves propagating along the tubes are varied.

2. Method and Results

We assume a solar surface element to be uniformly covered with vertically directed magnetic flux tubes. Fig. 1 shows a vertical slice through the flux tube field. The tube diameter at the bottom is $d_{bot} = 100\ km$. The tubes are assumed to spread with height reaching eventually a maximum diameter d_{max} where the field fills out the entire available space. The distance of neighbouring tubes is d_a. For the non-magnetic atmosphere outside the tube we have computed a nongrey NLTE H$^-$ and Mg II line radiative equilibrium atmosphere. Using a considerably modified code after Herbold et al. (1985) we have calculated a series of longitudinal MHD waves along the tubes. Various wave energies and a wave period of 30 s were chosen. It should be noted that different choices for the flux tube geometry (e.g. d_a and d_{max}) will strongly influence the physics of the wave propagation, the radiation damping and the shock formation in the tube as has been discussed by Ulmschneider et al. (1987). The radiation damping in the tube is due to nongrey NLTE H$^-$ and Mg II line emission. The tube is assumed optically thin in H$^-$ with the mean intensity given by the outside atmosphere. For the outside atmosphere H$^-$ has been computed by solving the transfer

R. J. Rutten and G. Severino (eds.), Solar and Stellar Granulation, 589–593.

equation. For the Mg II line it is assumed that the transfer is mainly in vertical direction (Fig. 1). Here and for the Ca II K line complete redistribution has been assumed. After completion of the wave computation the solar surface element is sampled every 40 km at equidistantly spaced points (Fig. 2, boxes) and the transfer along the different rays, which for a given aspect angle α contribute to the integrated line profile, is evaluated (Fig. 1, dashed). For preliminary computations we have used linear sampling where only rays from the bottom row of boxes in Fig. 2 was used. This overestimates the flux tube contribution compared to the non-magnetic contribution.

Fig. 3 shows the temperature profile of three different phases of the waves for an *active region* flux tube distribution ($d_{max} = d_a = 220\ km$). Here a longitudinal MHD tube wave with an energy flux of $1.0 \cdot 10^8 erg/cm^2 s$ and a period $P = 30s$ was assumed. Phase A is at time $t = 6155\ s$, B at $t = 6200\ s$ and C at $t = 6270\ s$. The radiative equilibrium temperature distribution of the atmosphere outside the flux tube is shown dotted. Phase A and C show the moments where shocks are transmitted at the top boundary. It is seen that at great heights the temperature rises steeply due to unbalanced shock heating when the main emitter gets destroyed due to the ionization of Mg II to Mg III. This behaviour is different if Lyman α emission were included but a similar effect would eventually happen at greater heights when hydrogen ionizes, as has been discussed by Ulmschneider et al. (1987). Fig. 5 gives the corresponding theoretical, not instrumentally degraded, integrated, linearly sampled Mg II k line shape of the three different phases of Fig. 3. Note that due to our two level treatment the line is a singlet. It is seen that the transition-layer-like rapid temperature rise at great heights does not influence the line core and that the core is formed near 1000 km height.

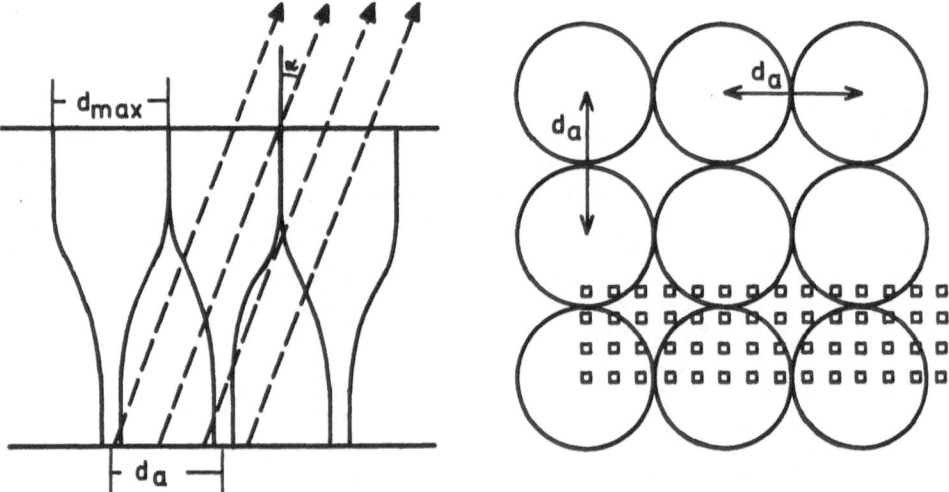

Fig. 1 Flux tube geometry, side view   Fig. 2 Surface area sampling

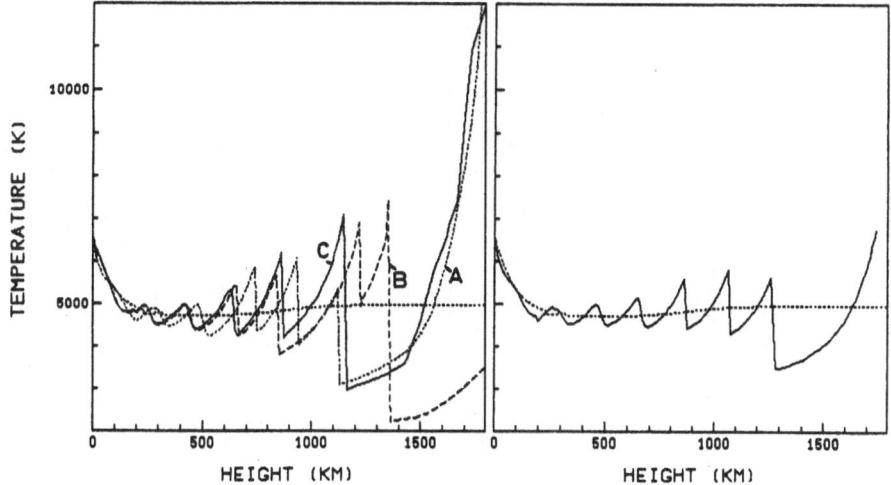

Fig. 3 *Wave phases in an active region*   Fig. 4 *Wave in a quiet region*

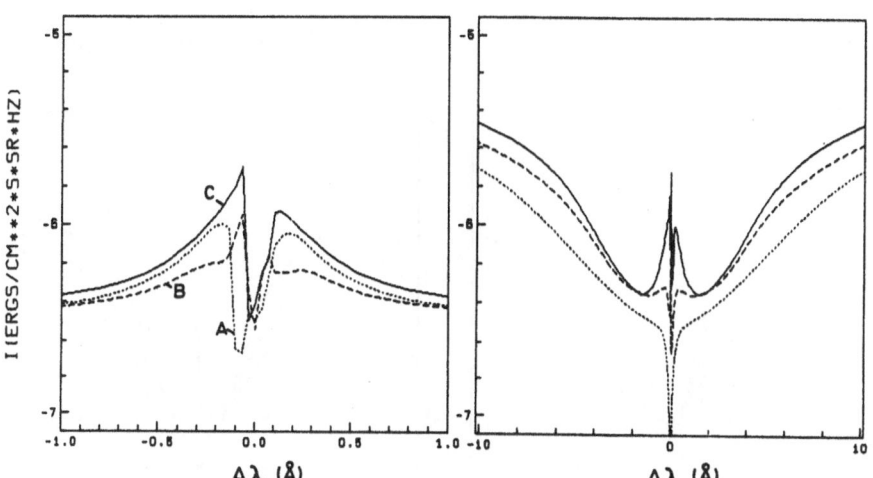

Fig. 5 *Mg II profiles for different phases*   Fig. 6 *Change of the filling factor*

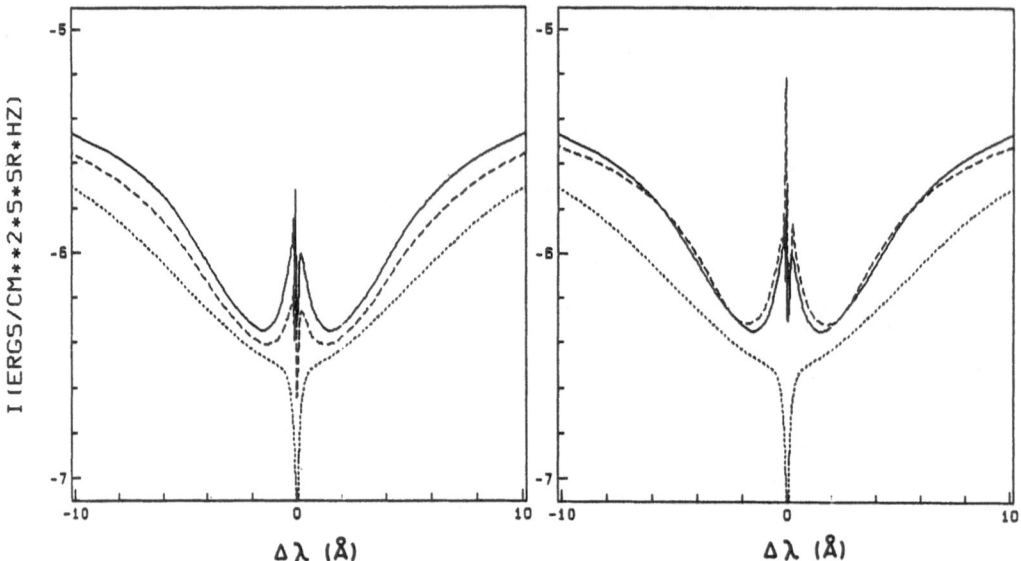

Fig. 7 Change of the geometry

Fig. 8 Increase of the wave flux

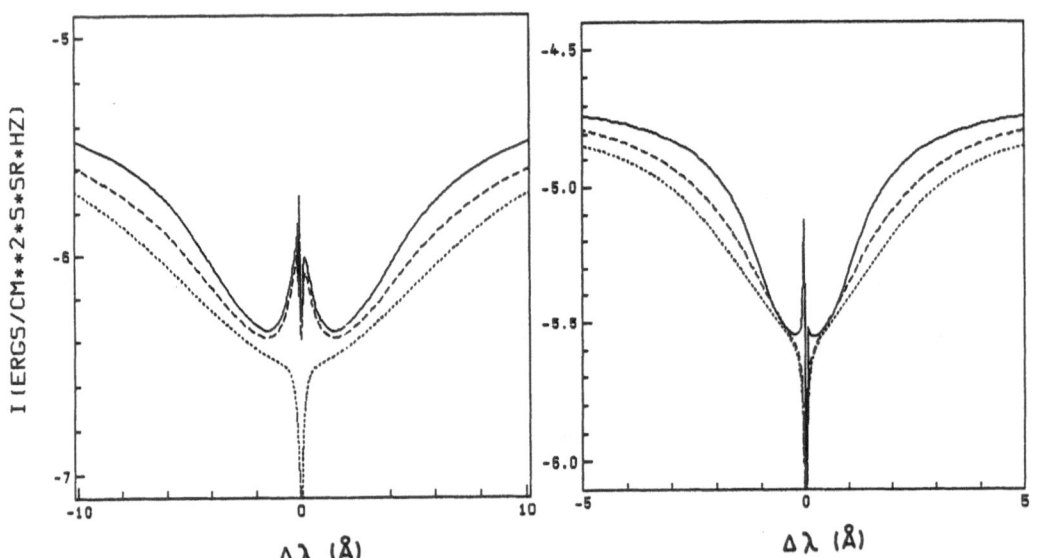

Fig. 9 Linear and 2D sampling

Fig. 10 Ca II line similar as Fig. 6

Fig. 4 displays the temperature profile for a wave of the same energy and period as in Fig. 3 but propagating in a flux tube relevant for a *quiet region* ($d_{max} = d_a = 600\ km$). It is seen that this wave, due to the much greater spreading of the tube, has a much smaller amplitude growth. Fig. 6 shows the comparison of the integrated line shapes for the active region (phase C, solid) and the quiet region (dashed). The radiative equilibrium line profile (Fig. 6, dotted) shows that the absence of waves leads to a pure absorption line. It is seen that for the chosen wave type the main effect of increasing the filling factor of magnetic fields is an enhancement of the wings and the central core of the line. This is due to the fact that in a narrower tube, shock formation occurs at lower height, which increases the mean temperature at these heights. The smaller tube diameter at large height generates greater wave amplitudes which leads to stronger line core emission. In addition the larger filling factor implies more wave energy per surface area.

To see the purely geometrical effect of increased crowding of flux tubes we have computed the active region case C of $d_{max} = 220\ km$, making the unphysical assumption $d_a = 440\ km$, where the non-magnetic space outside the tube is increased by a factor of four. This decrease of the filling factor displaces the line profile (Fig. 7, dashed) towards the radiative equilibrium profile. In Fig. 7 and the subsequent Figs. the active region case of $d_{max} = 220\ km$ is shown solid and the radiative equilibrium profile dotted.

Fig. 8 shows the effect of increasing the wave energy. For the case of the active region tube we have increased the wave flux to $3.0 \cdot 10^8 erg/cm^2 s$. It is seen that the more energetic wave (Fig. 8, dashed) leads to increased emission both in the near wings and the line core compared to the less energetic wave. The results shown so far have been obtained by sampling the solar surface along a linear slab (c.f. Fig. 2) which goes through the tube centers. Fig. 9 shows the comparison of linear sampling (solid) with two-dimensional sampling (dashed) which covers the entire solar surface element. The 2D sampling leads to an increased contribution of the non-magnetic areas.

Fig. 10 shows a computation similar to Fig. 6 except that now the Ca II K line profile is simulated. As in Fig. 6 the active region profile is shown solid, the quiet region profile dashed and the radiative equilibrium line profile (from the non-magnetic area) dotted.

3. Conclusions

We have simulated Mg II k and Ca II K line profiles from a solar surface element which is filled by flux tubes of various size and by non-magnetic areas. Longitudinal MHD-waves are assumed to propagate along the spreading magnetic flux tubes. The variation of the magnetic field filling factor, the tube geometry and the wave energy result in characteristic changes of the integrated line profiles. This shows that, despite of a severe loss of horizontal resolution, the integrated line shapes of the chromospheric lines of stars contain significant information about the magnetic field, its geometry and filling factor and about the waves which heat the chromosphere.

Acknowledgement The authors want to thank the Deutsche Forschungsgemeinschaft for support for Project Ul 57/11-1.

References

Herbold, G., Ulmschneider, P., Spruit, H.C., Rosner, R.: 1985, Astron. Astrophys. **145**, 157
Ulmschneider, P., Muchmore, D., Kalkofen, W.: 1987, Astron. Astrophys. **177**, 292

A SIMPLE MODEL OF MESOGRANULAR AND SUPERGRANULAR FLOWS

G.W.Simon   and   N.O.Weiss
Air Force Geophysics Laboratory   D.A.M.T.P
National Solar Observatory   University of Cambridge
N.O.A.O.*, U.S.A.   U.K.

*Operated by AURA Inc. under contract with NSF. Partial support for NSO is provided by the USAF under a memorandum of understanding with NSF.

ABSTRACT. We represent a convection plume by an axisymmetric source with a radial outflow described by a simple analytic function. By superposing a number of such sources we can produce regular or irregular patterns of cellular convection. This procedure has been used to simulate the large-scale horizontal flow patterns observed by the SOUP instrument on Spacelab 2. Passive test particles moving with this velocity accumulate in a network that coincides with the magnetic network observed at Big Bear Solar Observatory.

1. The Model

The radial outflow from an isolated source is represented by a velocity

$$v = V \left(\frac{r}{R} \right) \exp \left[- \left(\frac{r}{R} \right)^2 \right], \tag{1}$$

where V and R are the strength and radius of the source. Then continuity implies that the vertical velocity w is proportional to the divergence of the horizontal flow,

$$\Delta = \frac{1}{r} \frac{d}{dr}(rv) = \frac{2V}{R} \left[1 - \left(\frac{r}{R} \right)^2 \right] \exp \left[- \left(\frac{r}{R} \right)^2 \right]. \tag{2}$$

Figure 1 shows the variation with radius (r/R) of the normalised radial and vertical components of the velocity in an isolated plume. In order to model an observed velocity pattern we locate all maxima of the divergence Δ and fit outflows with the correct maximum value $\Delta_{\max}$ to all significant sources, setting R equal to the mean radius at which $\Delta = 0$. In Figure 2 we show radial scans (dots) from the centre of the divergence peak with co-ordinates (37,41) arc sec in Figure 2 of the preceding paper in these proceedings (Paper I). The averages of these radial scans are shown as circles and can be compared with the model (2) which yields values of Δ indicated by + signs. The two curves are in excellent agreement to a distance of 7 arc sec, beyond which large asymmetries are created by neighbouring sources and sinks.

R. J. Rutten and G. Severino (eds.), Solar and Stellar Granulation, 595–599.
© 1989 by Kluwer Academic Publishers.

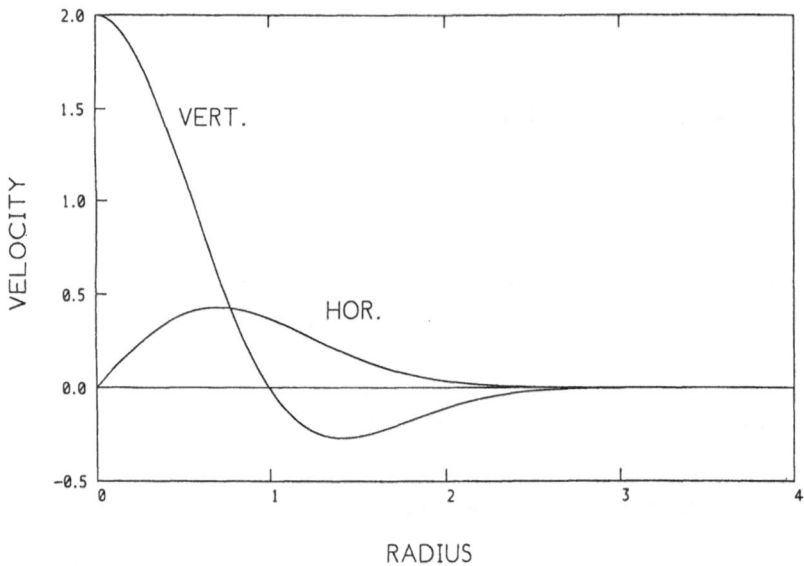

Figure 1. Radial dependence of the normalised velocity components.

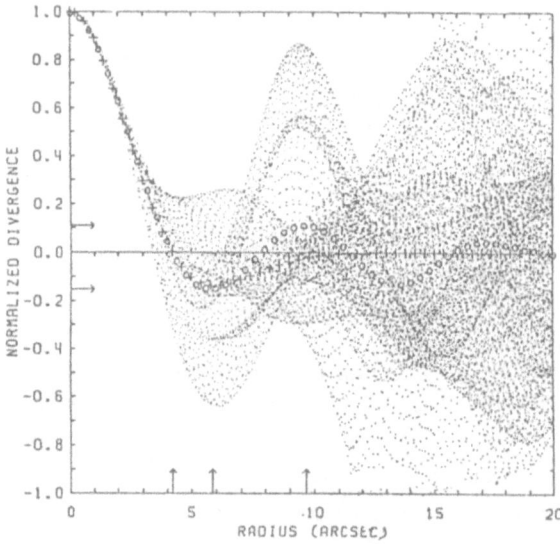

Figure 2. Comparison between the model and an observed mesogranular source.

2. Results

We consider a magnetically quiet area within the region observed by the SOUP instrument; this 35×35 arc sec area has co-ordinates $95 \leq x \leq 130$, $8 \leq y \leq 43$ arc sec in Figure 2 of Paper I. In Figure 3(a) we show the measured velocity u_H and its divergence Δ, which can be compared with the velocity u'_H and its divergence Δ', in Figure 3(b), obtained by fitting all sources in the area and its immediate neighbourhood with $\Delta_{max} \geq 0.009$.

(b)

(a)

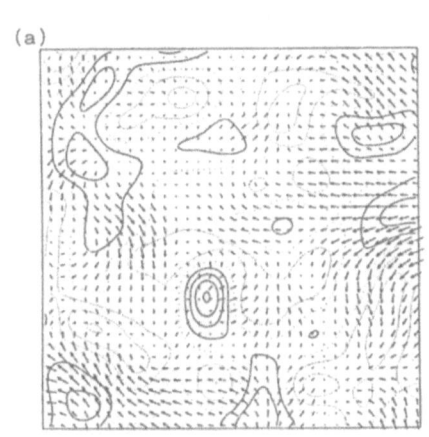 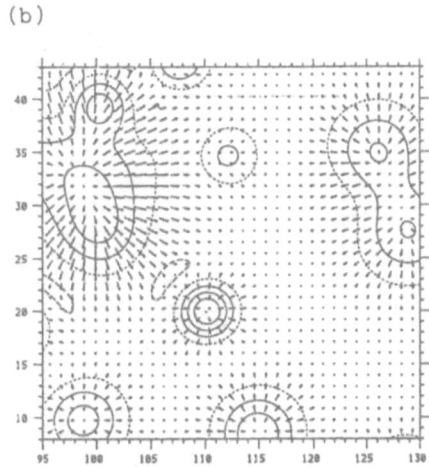

Figure 3. Horizontal velocity (indicated by arrows) and vertical velocity contours of Δ (proportional to vertical velocity); full lines indicate positive values and broken lines negative values of Δ. (a) Velocities measured from SOUP data in a 35×35 arc sec area. (b) Velocities obtained by fitting the data with 16 axisymmetric sources.

In order to model the behaviour of magnetic fields we have computed the positions of passive test particles (corks) moving with the velocity u'_H. The corks start on a uniform square grid with approximately 1 arc sec spacing at time $t = 0$. Figure 4 shows the positions of the corks after 1hr, 2hr, 4hr, 8hr, 1d and 2d have elapsed. In a few hours the corks move away from the centres of mesogranules and supergranules within the region. After about eight hours they form a network outlining the mesogranular structure. The most prominent features are a supergranule at the right of the region and a mesogranule just below the centre. At the end of a day the network is narrower and over the next day the particles slowly migrate along the network to accumulate at junctions where Δ is a minimum.

3. Discussion

Our results show that mesogranular and supergranular flows in the solar photosphere can be modelled by matching outflows from sources (where Δ is a maximum). The model flow u'_H does not describe the streaming motions e.g. at the bottom of Figure 3(a); moreover, the actual flows have sinks that are not represented in Figure 3(b) and the horizontal velocity u_H is not irrotational. Nevertheless, these

598

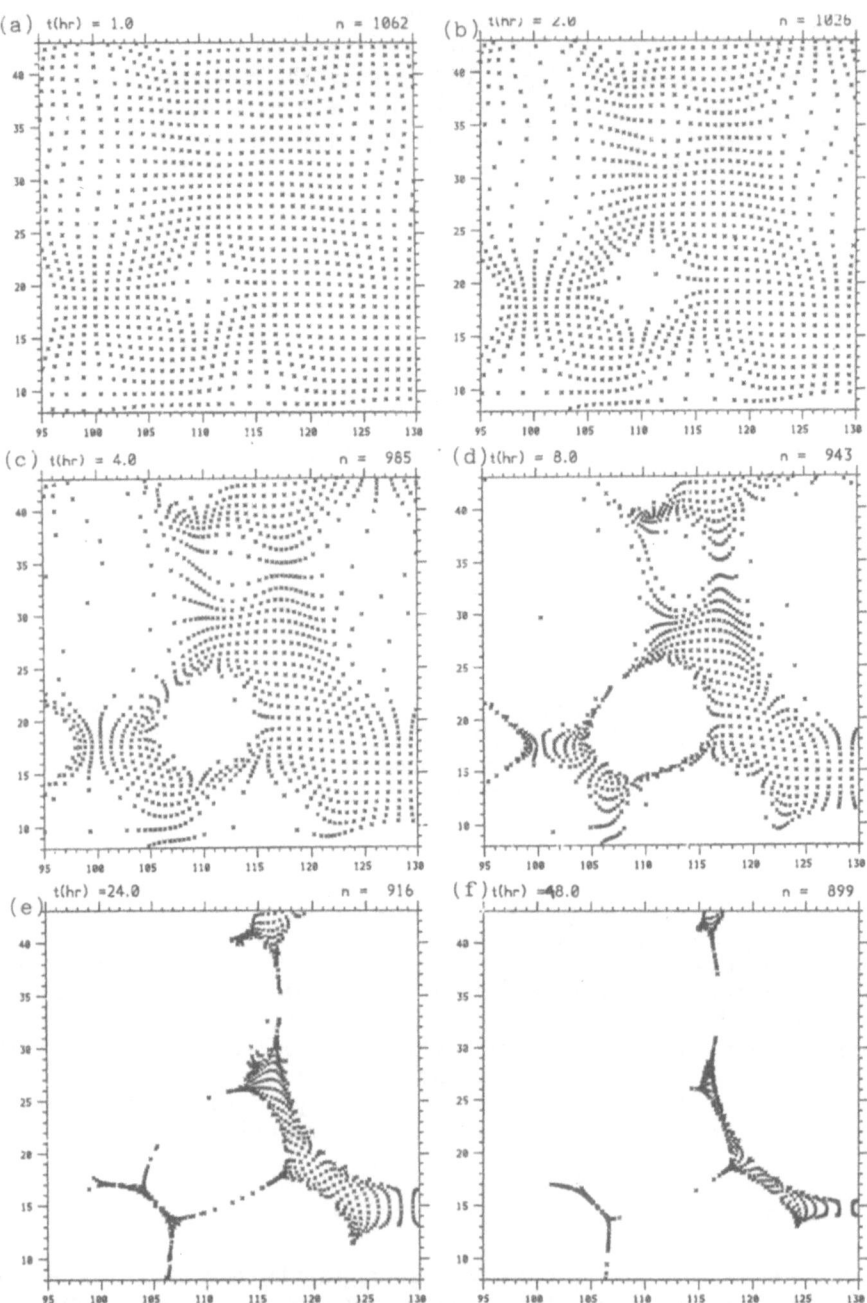

Figure 4. Positions of test particles (corks) moving with the velocity in Figure 3(b) after (a) 1 hr, (b) 2hr, (c) 4hr, (d) 8hr, (e) 24hr and (f) 48hr have elapsed.

differences do not significantly affect the cork motions. The network in Fiugre 4 resembles that obtained from the observed velocity and corresponds closely with the location of the magnetic fields observed at Big Bear Solar Observatory at the same time as the SOUP observations were made (Simon et al. 1988). A full account of this work will be submitted to the Astrophysical Journal.

Acknowledgements

We thank Alan Title and Ted Tarbell for permission to use SOUP results and Larry November for computing the resulting velocities.

Reference

Simon, G.W. et al.   1988   *Astrophys. J.* **327**, 964.

PROBING OF SUNSPOT UMBRAL STRUCTURE BY OSCILLATIONS

Y.D. Zhugzhda (a), J. Staude (b) and V. Locans (c)
(a) IZMIRAN, USSR Academy of Sciences, SU-142092
    Troitsk
(b) Central Institute of Astrophysics, Solar Obser-
    vatory 'Einsteinturm', DDR-1560 Potsdam
(c) Radioastrophys. Observatory, Latvian Academy
    of Sciences, SU-226524 Riga

EXTENDED ABSTRACT. Recently the present authors explained
oscillations observed in the umbral chromosphere and transition region (TR) by the action of a chromospheric resonator, that is by slow, quasi-longitudinal magneto-atmospheric
waves which are partly trapped between temperature minimum
(T_{min}) and TR. Subsequently the model has been extended to
- a photospheric-chromospheric resonator (a) by considering additional reflections from the lower photosphere;
- two subphotospheric resonators for the fast (quasi-longitudinal, acoustic) wave (b) and for the slow (quasi-trans-
  verse) wave (c) and their coupling with the atmospheric
  resonator (a) have been investigated as well.
    In our detailed modelling of the longitudinal waves we
calculated the resonant transmission or filtering of a
broad-band flux of wave energy incident from deeper convective layers. Our code provides information on the partial
transmissions and reflections from each intermediate height
level, moreover, on the resonance frequencies and on the
height dependence of amplitudes and phases of oscillations
of both velocity and thermodynamic quantities. The theoretical predictions turned out to be in good agreement with
the basic features of observed umbral oscillations.
    Details of the oscillations critically depend on the
structure of the assumed umbral model. In this way we obtain a useful method for probing not only the subphotosphere, but also the atmospheric umbral layers by seismology,
thus completing customary spectroscopic diagnostics. E.g.,
- the extent between T_{min} and TR determines the spectrum of
  resonance frequencies, thus providing a method to determine temperature gradients in the umbral chromosphere;
- the values of T_{min} in umbral fine structures can be estimated by investigating the quality of their resonators;
- observations of UV lines provide information on the umbral TR, where the observed oscillations are probably
  concentrated in cold fine structure elements.

R. J. Rutten and G. Severino (eds.), Solar and Stellar Granulation, 601–605.

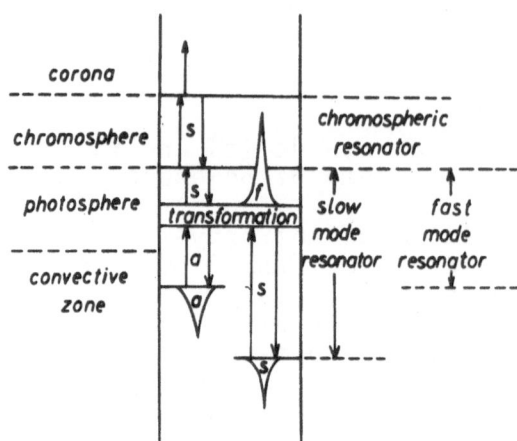

Fig. 1   Simplified scheme of resonators in the sunspot
umbra. Arrows show the direction of wave propagation,
tips mean evanescent waves. The letters s, f, and a
indicate slow, fast, and quasi-acoustic waves, respectively.

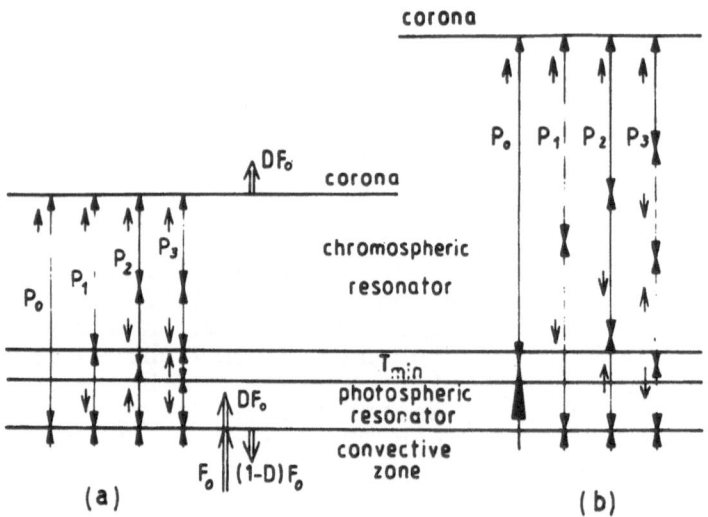

Fig. 2   Scheme of the photospheric-chromospheric resonator in
two different models of the umbral atmosphere: (a) 'thin'
(shallow) chromosphere, (b) 'thick' (extended) chromosphere. Short arrows indicate the direction of velocity
in a given phase for each mode with resonance period P_i
encounters of triangles represent nodes, double arrows
describe the wave energy flux.

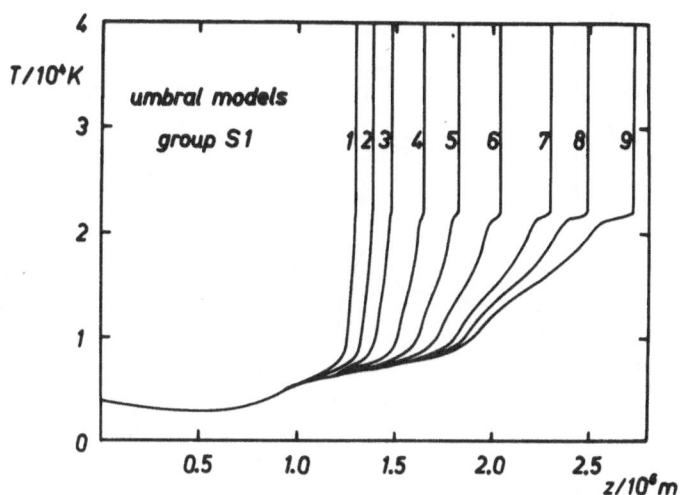

Fig. 3   Example of a grid of umbral models (temperature T
          versus geometrical height z) with different extents
          of the chromosphere.

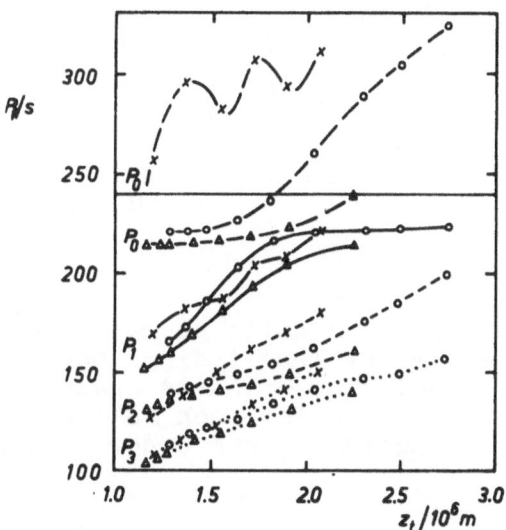

Fig. 4   Resonance periods P_i versus chromospheric extent z_t
          (height of TR) for three different grids of umbral
          models. The horizontal line shows the acoustic cutoff
          frequency for $T_{min} \approx 3000$ K. Circles relate to grid S1
          as shown in the figure above.

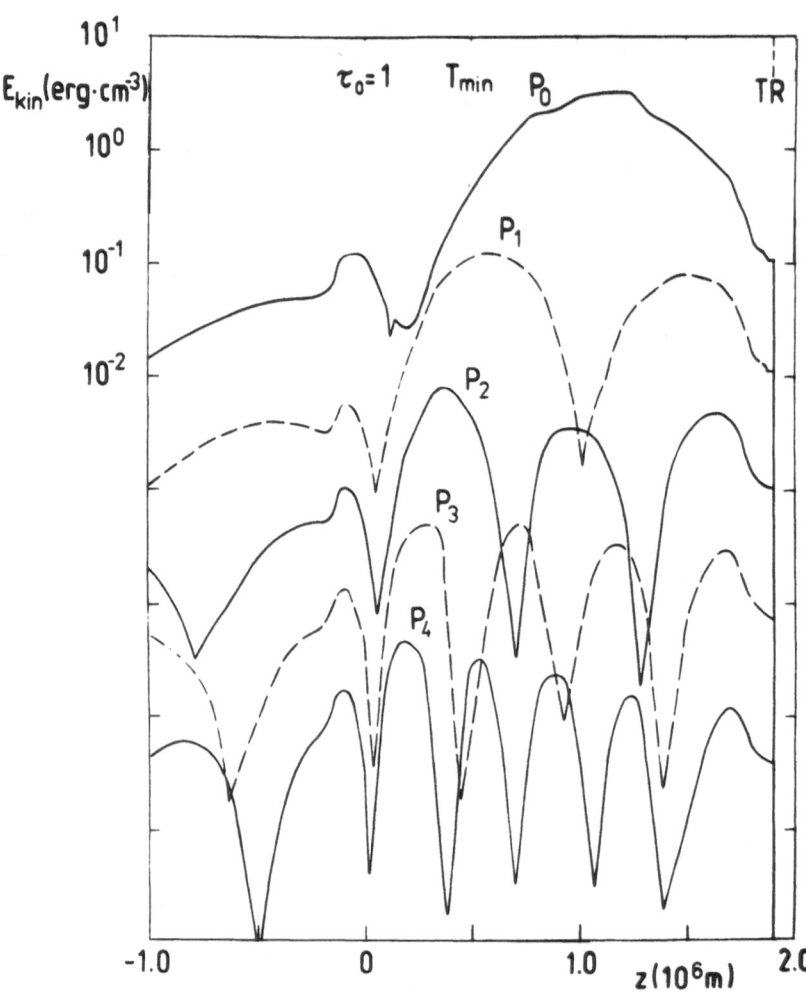

Fig. 5   Height dependence of kinetic energy density of different modes P_i of longitudinal oscillations in an umbral model. The ordinate scale relates to P_0, while the curve for each following P_i has been shifted downwards by one order of magnitude of E_{kin}.

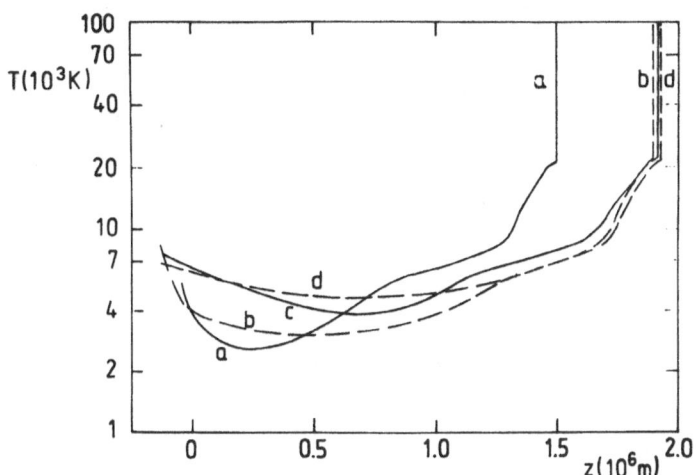

Fig. 6   Temperature T(z) in 4 umbral models with different
values of T_{min}.

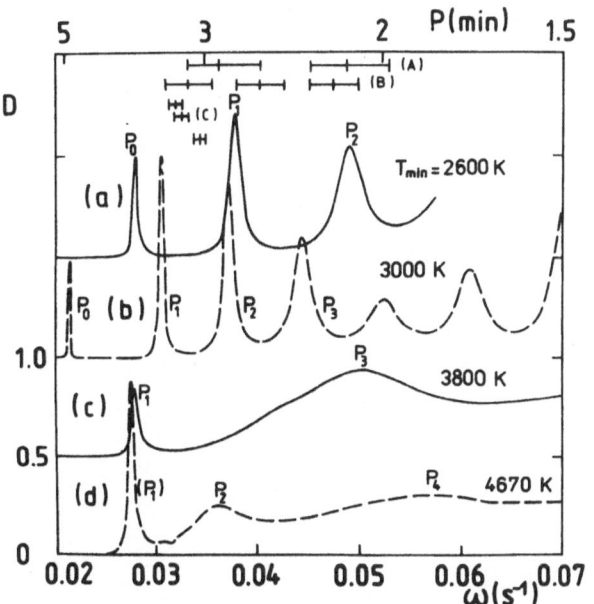

Fig. 7   Transmission (D versus frequency ω) of the energy of
waves passing through each of the umbral models shown
in the figure above. The ordinate refers to model (d),
while the other curves are shifted from each other by
$\Delta D = 0.5$. For comparison some resonance frequencies
recently observed in the umbral chromosphere and TR
are drawn above, together with the error limits (frequency resolution) given by the authors.

WORKSHOP IMPRESSIONS

Jacques Maurice Beckers*

Advanced Development Program, National Optical Astronomy Observatories†,

Tucson, AZ 85726 USA

1   Introduction

Rather than giving a summary of the workshop as the organizers requested, I will give you my personal impressions of the contributions presented here in the last 5 days. I will not attempt to be encyclopedic and include references to everything presented. Instead, this will be a description of the progress in this exciting field of research as seen through the eyes of someone who used to be active in the field but who left it almost a decade ago.

In those ten years there have been very significant developments in the studies of solar and stellar granulation, in its observational and computational tools, in its observations and computer modelling, and as a result in our understanding. There are many new words in the field like 'Mesogranulation', 'C-shapes', 'Cork-Movies', 'r-zero', 'SOUP', 'THEMIS', and 'LEST', but the researcher is still motivated by the same old desires. These are lofty ones, like the desire to study perhaps the only example of observable astrophysical convection, or the interaction of the convective motions with magnetic fields and the resulting modification of the atmospheric energy transport and the heating of the chromosphere and corona. Or they are more mundane and human like the beauty of the subject under study, the observational challenge, the satisfaction derived from the construction of functioning, complex computer codes, and the ability to manipulate an enormous data set in N different ways. Whatever the motivations may be, we have learned a lot as is clear from this workshop, and much more is to come in the next decade.

2   Tools for the Study of Granulation

We heard about the many new tools that have become available or that are likely to become available in the near future, to enable the solar and stellar astronomer to study granulation and other velocity and magnetic fields. They are of many different kinds, including:

---

*Presently at the European Southern Observatory, D-8046 Garching bei München, West Germany

†Operated by the Association of Universities for Research in Astronomy Inc., under contract with the National Science Foundation

R. J. Rutten and G. Severino (eds.), Solar and Stellar Granulation, 607–613.

608

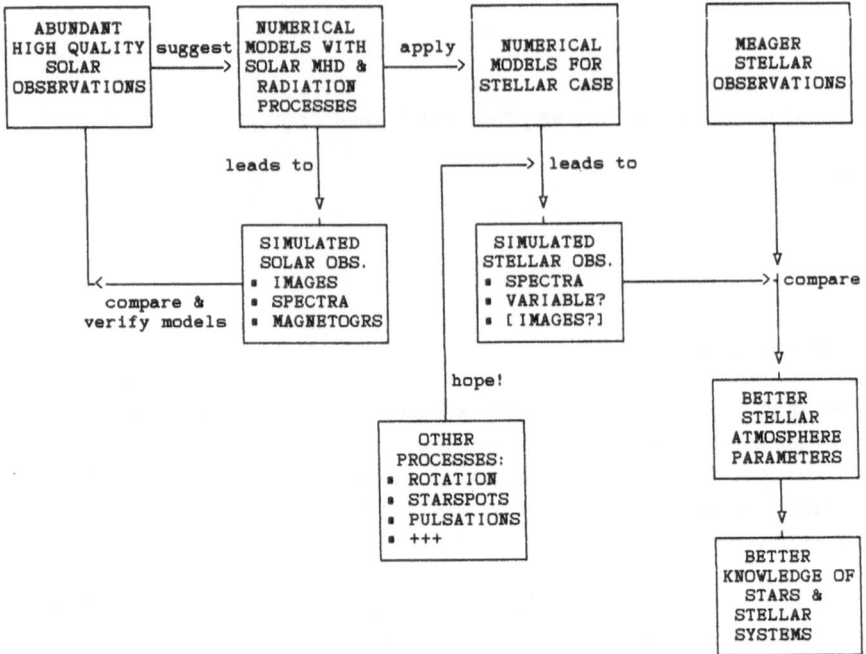

Figure 1: Interaction between tools

2.1 Telescopes

Optical telescopes in space (SOUP, and in the future the successor to SOT: the Orbiting Solar Laboratory) and on the ground (the Swedish and German telescopes in the Canary Islands, THEMIS, and in the future LEST) are posing serious challenges to the existing high-resolution facilities at Pic du Midi, Sacramento Peak, and Big Bear. Angular resolutions near 1/8 arc second appear within grasp, both at visible wavelengths from space and at infrared wavelengths from the ground.

2.2 Image Analysis Techniques

The detailed analysis of the SOUP observations, and the application of the resulting algorithms to the observations of the Swedish telescope at La Palma, are providing important tools for the objective reduction and analysis of the data.

2.3 Image Reconstruction Techniques

Image selection has always been a powerful technique for improving image quality before detection. It has now been put on a quantitative basis, and it appears that further improvements may be possible in the future. Factors of 3 to 4 improvement over average image quality appear possible.

A number of adaptive optics systems for solar and stellar observations are now being put into operation. We may expect them to be a common component of solar ground-based telescopes within the next decade. Multiconjugate adaptive optics provides a way to increase the area on the sun corrected by adaptive optics.

I am fascinated by the technique developed by Högbom to recover diffraction-limited images from two uncorrected, or partially corrected, images taken simultaneously at two focal positions.

2.4 Computer Modelling

Many of the most dramatic results presented at this workshop came from the rapid advances that are being made in the computer modelling of hydrodynamics, radiation transfer, and magneto hydrodynamic processes in the outer solar layers. As computers become more powerful, this type of numerical experimentation promises to develop rapidly.

2.5 Stellar Capabilities

To make the stellar connection it is essential that high spectral resolution ($>\sim 100,000$) and accurate (0.1 to 1%) stellar observations be gathered. The next generation of giant 8 to 16 meter telescopes coupled efficiently to high resolution spectrum analyzers (a real challenge!) should make this possible.

2.6 Interactions Between These Tools

In figure 1 I have tried to illustrate the interactions between the applications of these different tools.

Solar observations and numerical modelling of physical processes acting in the outer solar layers are highly interactive activities. They cover the major part of this workshop. Abundant, accurate, and high-resolution observations of solar granulation, oscillations, and other dynamic and magnetic phenomena will place important constraints on the numerical models, which as a result will acquire the sophistication needed to make accurate predictions of the much more limited stellar observations. With the uncertainties of the modelling minimized, it will then be possible to extract the maximum amount of information from the stellar data. The main uncertainty in this process is probably the existence of other non-modellable phenomena on stellar surfaces like stellar rotation, star patches, etc. Nevertheless, the process shown in figure 1 must be the one leading to the best understanding of stellar atmospheres, and of the spectra of stars and stellar systems (like galaxies and globular clusters).

3 Solar Granulation

We used to talk about "abnormal granulation" as distinct from "normal granulation". After seeing the modern observations here I almost have to conclude that all granulation is abnormal in the sense that none of it fits the old picture of Bénard-type cells resulting from a nice stationary convection pattern. I was often reminded of the comments by V. Krat that granulation is not convection but rather a wave-type phenomenon.

I will comment below on a number of items that particularly struck me during this workshop:

3.1 Smallest Sizes and Numbers per 100 Arc Seconds

There remains a disagreement on the total number of granules on the solar surface, with Schröter arguing that this is a rather well-defined quantity and Title on the other hand pointing out that this must critically depend on the angular resolution of the observations and on the often unclear definition of the granule itself.

There was some discussion about the smallest-scale structure that one might hope to observe on the sun. Some of the attendants are of the opinion that the photon mean free path sets the lower limit to the size of the observable phenomena on the sun. I remain unconvinced of that. It is certainly not the case in the solar corona. It has been argued that in an optically thick medium like the photosphere it is the case, but then we are looking to a large extent at a region above the photosphere which is optically not thick.

3.2 Distribution of Granule Sizes

The question is whether the break in the size distribution seen in the Pic-du-Midi observations at about 1.5 arc seconds is real and whether the size distribution below this break is "Kolmogoroffian". The break is seen in the area-periphery distribution, in the velocity power spectrum, and in the number- vs-size distribution. The observations appear to conflict, so the issue remains to be settled in the future.

Even if there is such a break, and even when the distribution is Kolmogoroffian, one has to be careful in concluding that we are dealing on the small scale with turbulent eddy decay and that the "upper scale of turbulence" in the solar atmosphere is near

1000 km. The shape of the distribution, and maybe the break, are necessary but not sufficient conditions for this case.

3.3   Exploding Granules

We learned that "exploding granules" are abundant in regions of so-called positive divergence (the centers of the so-called mesogranulation) whereas they are hard to find in the regions of negative divergence. They remind me of the so-called vortex rings studied by Musman in the 1970's.

3.4   Mesogranulation

Although originally a sceptic on the subject of the existence of mesogranulation, I have become convinced of the existence of a distinct structure at the mesogranular scale ($\approx$ 7000 km) as a result of the so-called "cork movies" and the low-altitude horizontal velocity power spectra shown at this workshop.

It is not clear to me, however, that the postulate made by many that the apparent (proper) motions in the "cork-movies" correspond to real physical motions is correct. If it were correct I would expect the Doppler shifts of the 80-1000 m/s horizontal motions of the mesogranulation to dominate those of the $\approx$ 300 m/s supergranules in full disc solar dopplergrams. The same should be the case for the vertical motions.

3.5   C-Shapes and Bisectors

We have seen some fine examples of calculated line asymmetries from the computer models as well as observations, at scales both smaller than and similar to the granule scale. Both show local line asymmetries much larger in displacement magnitude than the displacements associated with the so-called C-shapes of the average solar spectral lines. The latter is the small residual of the average of many lines with much larger bisector shifts. The question is therefore to what extent the C-shape (the only thing observable on stars) is a useful tool for the study of stellar atmospheric motions. Are the differential lineshifts between weak and strong lines, or between lines of low and high excitation potential, perhaps a better tool? The latter has the additional advantage of requiring less spectral resolution.

3.6   Absolute Wavelengths?

It is not clear to me that a tabulation of the absolute wavelengths of solar spectral lines is needed, as was suggested during this workshop. What is needed is a good calibration of solar spectrograms with (pressure controlled?) hollow cathode sources or with intermediate spectral line sources like iodine or fluoride absorption tubes. The same is even more the case for stellar spectroscopy since the absolute stellar Doppler shift with respect to the observer is not known from independent sources.

3.7   Pressure Shifts of Spectral Lines

Pressure shifts of some solar spectral lines appear to be more of a curiosity and a nuisance rather than a useful tool for the study of the sun. There are plenty of lines with small enough pressure shift to be usable for solar velocity measurements.

3.8   Limb Shift

I was surprised to hear almost nothing about solar limb shifts. The solar limb effect is a very powerful tool for the study of small scale (unresolved) solar velocities, temperatures, and pressure variations.

3.9   Limb Observations

Because the horizontal scale of granular phenomena starts approaching the vertical pressure and optical depth scale in the solar atmosphere, limb observations will become a very powerful tool in the study of solar granulation. Earlier observations have already shown clear geometrical foreshortening effects, perspective effects, and projection effects. We heard little about this at this meeting. As in the case of the limb effect, I expect to see more work in the near future.

3.10   Observations in 1.6 μm Atmospheric Window

At 1.6 μm one looks about 25–30 km deeper in the sun than at optical wavelength. Although this does not sound like much, one should remember that this increase in depth places one below the convective overshoot region into the actual solar convection zone. One might therefore expect significant differences from visible light observations.

An added advantage at 1.6 μm is the larger magnitude of the Zeeman shifts at that wavelength. It should be possible to fully resolve the Zeeman splittings in the magnetic elements so that vector magnetic field measurements will become more reliable.

4   Computer Modelling

The progress made in computer modelling of solar and stellar granulation is truly stunning. This work holds the real clue to the extrapolation of the solar work to stellar observations. I especially appreciated the clarity of presentation by the researchers in this field and their efforts to place their results into the observational context.

This interaction between theory, modelling, and observations is one of the best astrophysics efforts today. In the near future we expect to see more work and understanding in the fields of granule–wave interaction, magnetic field modulation and its effect on chromosphere/corona structuring and heating, sunspot umbra and penumbra processes, and the coupling of interior solar convection and energy transfer to the surface structure. What is needed is access to bigger and faster computers.

5 Stellar Granulation

We have heard about the first applications of the computer modelling to stellar conditions and observations. As things improve in both observations and modelling we expect to see a wealth of information forthcoming.

6 Conclusion

I leave this workshop with the strong conviction that this field of astrophysical research is in the middle of a strong revival with its climax still to come, as high-resolution ground and space solar observatories come into being, as computer modelling becomes more powerful, and as the new generation of powerful ground-based telescopes is being built. The story of this exciting research should be made very visible so that our other astronomy colleagues can enjoy it as well.

I finally want to express my thanks to all who through their hard work have made this workshop possible. The outstanding success of the workshop is the best thank-you gift to you, it truly made your efforts worthwhile.

Subject Index

1-D modelling – 567
2-D modelling – 401*ff*, 415*ff*, 425*ff*, 471*ff*, 481*ff*, 533*ff*, 571*ff*, 583*ff*
3-D Fourier analysis – 61*ff*, 165, 217, 225*ff*
3-D modelling – 381*ff*, 415*ff*, 425, 453*ff*, 493*ff*, 521*ff*, 555, 567

A-type stars – 91, 533*ff*
abnormal granulation – 114, 265, 365
acoustic waves – 30, 195*ff*, 207*ff*, 225*ff*, 325, 333*ff*, 381*ff*, 401*ff*, 421, 508, 545,
   547*ff*, 589*ff*
activity, solar – 113, 217, 253*ff*, 283*ff*, 289*ff*, 313*ff*, 327*ff*, 359*ff*
activity cycle – 101*ff*, 118, 456
Adams-Bashforth algorithm – 384, 455
adaptive optics – 29, 43*ff*, 61*ff*, 77*ff*, 608
agile optics – 333
Air Force Geophysical Laboratory (AFGL) – 39, 333
Airy disk – 46, 62
Alfvén waves – 583
ANA software – 164
anelastic approximation – 406, 426, 493*ff*, 506, 568
anisoplanatism – 77, 88
Ap stars – 537
Arcturus – 547
astrometry – 157
asymmetry, line – 119, 135*ff*, 145*ff*, 153*ff*, 187*ff*, 273*ff*, 425*ff*, 493*ff*, 540
atmosphere, terrestrial – 49, 125*ff*, 349, 402
atmospheric seeing – 43*ff*, 55*ff*, 61*ff*, 77*ff*, 161*ff*, 349
atomic collisions – 91, 93*ff*
autocollimation – 18
autocorrelation – 62, 225*ff*

Big Bear Solar Observatory – 228, 372, 595, 599, 608
bisectors – 71, 93*ff*, 110, 125*ff*, 135*ff*, 145*ff*, 187*ff*, 273*ff*, 295, 425*ff*, 493*ff*,
   521*ff*, 533*ff*, 542, 611
bogus – 422
α Boo – 127, 547
ξ Boo A – 73, 145

Bibliography on
Solar and Stellar Granulation

As an accompaniment to these proceedings for the Workshop on Solar and Stellar Granulation, Robert Rutten asked me to compile a bibliography on granulation. The sources for this bibliography are an extensive list of papers from Frank Hill, Physics Abstracts (extracted by John Cornett), Astronomy and Astrophysics Abstracts (1969 to 1987), Astronomischer Jahresbericht (1899 to 1968), and the second edition of Solar Granulation by R. Bray and R. Loughhead. All the papers appearing in the proceedings of this workshop also are included. This is by no means a complete list of references on the subject of granulation, but hopefully it will prove useful to the reader.

I wish to express my special thanks to Frank Hill and John Cornett for their assistance in preparing this bibliography and to the National Optical Astronomy Observatories and the National Solar Observatory for making the computer and library services available for this project.

Karen L. Harvey

Acton, D.S., *see* Peri, M.L. *et al.* 1989

Acton, L.W., *see* Foing, B. *et al.* 1984

Acton, L., *see* Title, A.M. *et al.* 1987

Adam, J.A.: 1977, 'Hydrodynamic Instability of Convectively Unstable Atmospheres in Shear Flow', *Astrophys. Space Sci.* **5**, 493

Adjabshirzadeh, A.: 1980, 'Solar Granulation and Bright Dots in Sunspot Umbra: Comparison of Power Spectra', *Compt. Rend. Acad. Sci. Ser. B* **290**, 541

Adjabshirzadeh, A. and Koutchmy, S.: 1980, 'Photometric Analysis of the Sunspot Umbral Dots. I: Dynamical and Structural Behaviour', *Astron. Astrophys.* **89**, 88

Adjabshirzadeh, A. and Koutchmy, S.: 1982 , 'About the Foreshortening Effect on Sunspot Umbral Dots', *Solar Phys.* **75**, 71

Aime, C.: 1973 , 'Statistical Analysis of a Solar Granulation Plate', *Solar Phys.* **30**, 15

Aime, C.: 1976 , 'Solar Seeing and the Statistical Properties of the Photospheric Solar Granulation. I: Noise in Michelson and Speckle Interferometry', *Astron. Astrophys.* **47**, 5

Aime, C.: 1978 , 'A Morphological Interpretation of the Spatial Power Spectrum of the Solar Granulation', *Astron. Astrophys.* **67**, 1

Aime, C.: 1978, 'Some Solar Granulation Observations Using Speckle Interferometry Techniques', *Oss. e Mem. Oss. Arcetri* **106**, 110

Aime, C.: 1979, 'Speckle Interferometric Techniques Applied to the Study of the Solar Granulation', *J. Opt.* **10**, 318

Aime, C.: 1979, 'Interferometric Techniques Applied to High Resolution Observation of the Solar Granulation', in J. Davis and W.J. Tango (eds.), *High Angular Resolution Stellar Interferometry, IAU Colloq. 50*, Chatterson Astronomy Department, University of Sydney, Sydney, p. 30-1

Aime, C. and Ricort, G.: 1975, 'Sur l'Analyse Statistique de la Granulation Solaire', *Astron. Astrophys.* **39**, 319

Aime, C. and Ricort, G.: 1980, 'Speckle Interferometric Techniques Applied to the Observation of the Solar Photosphere', *Proc. Soc. Photo-Opt. Instrum. Eng.* **243**, 58

Aime, C., Borgnino, J., Druesne, P., Harvey, J.W., Martin, F., and Ricort, G.: 1985 , 'Speckle Interferometry Technique Applied to the Study of Granular Velocities', in R. Muller (ed.), *High Resolution in Solar Physics*, Proc. 8th IAU European Regional Meeting, *Lecture Notes in Physics* **233**, Springer-Verlag, Berlin, p. 103

Aime, C., Borgnino, F., Druesne, F., Martin, F., and Ricort, G.: 1989, 'Solar Granulation Speckle Interferometry Using Cross-Spectrum Techniques', *In These Proceedings*, p. 75

Aime, C., Borgnino, J., Martin, F., and Ricort, G.: 1977, 'Comments on the Low Wavenumber Power of Granulation Brightness Fluctuations', *Solar Phys.* **53**, 189

Aime, C., Kadiri, S., Martin, F., and Ricort, G.: 1981, 'Temporal Autocorrelation Functions of Solar Speckle Patterns', *Opt. Commun.* **39**, 287

Aime, C., Martin, F., Grec, G., and Roddier, F.: 1979, 'Statistical Determination of a Morphological Parameter in Solar Granulation: Spatial Distribution of Granules', *Astron. Astrophys.* **79**, 1

Aime, C., Ricort, G., and Grec, G.: 1975, 'Study of the Solar Photospheric Granulation by Michelson Interferometry-First Results', *Astron. Astrophys.* **43**, 313

Aime, C., Ricort, G., and Grec, G.: 1977, 'Solar Seeing and the Statistical Properties of the Photospheric Solar Granulation. II: Power Spectrum Calibration via Michelson Stellar Interferometry', *Astron. Astrophys.* **54**, 505

Aime, C., Ricort, G., and Harvey, J.: 1978, 'One-Dimensional Speckle Interferometry of the Solar Granulation', *Astrophys. J.* **221**, 362

Aime, C., Ricort, G., Roddier, C., and Lago, G.: 1978, 'Changes in the Atmospheric-Lens Modulation Transfer Function Used for Calibration in Solar Speckle Interferometry', *J. Opt. Soc. Am.* **68**, 1063

Aime, C., *see* Martin, F. 1979

Aime, C., *see* Ricort, G. 1979

Aime, C., *see* Ricort, G. *et al.* 1974

Aime, C., *see* Ricort, G. *et al.* 1981

Aime, C., *see* Ricort, G. *et al.* 1981

Aime, C., *see* Ricort, G. *et al.* 1981

Alamanni, N., *see* Cavallini, G. *et al.* 1987

Albregtsen, F. and Hansen, T.L.: 1977, 'The Wavelength Dependence of Granulation (0.38-2.4 μm)', *Solar Phys.* **54**, 31

Alfvén, H.: 1947, 'Granulation, Magnetohydrodynamic Waves and the Heating of the Solar Corona', *Monthly Notices Roy. Astron. Soc.* **107**, 211

Alissandrakis, C.E., Dialetis, D., and Tsiropoula, G.: 1987, 'Determination of the Mean Lifetime of Solar Features from Photographic Observations', *Astron. Astrophys.* **174**, 275

Alissandrakis, C.E., Macris, C.J., and Zachariadis, T.G.: 1982, 'Measurements of the Granule-Intergranular Lane Contrast at 5200 Å and 6300 Å', *Solar Phys.* **76**, 129

Allen, M.S. and Musman, S.: 1973, 'The Location of Exploding Granules', *Solar Phys.* **32**, 311

Altrock, R.C.: 1976, 'The Horizontal Variation of Temperature in the Low Solar Photosphere', *Solar Phys.* **47**, 517

Altrock, R.C. and Musman, S.: 1976, 'Physical Conditions in Granulation', *Astrophys. J.* **203**, 533

Altrock, R.C., Musman, S., and Cook, M.C.: 1983, 'The Evolution of an Average Solar Granule', in S.L. Keil (ed.), *Small Scale Dynamical Processes in Quiet Stellar Atmospheres*, National Solar Observatory/Sacramento Peak, Sunspot, p. 130

Andersen, B.N.: 1984, 'A New Method For Measurement of Granular Velocities', *Inst. Theor. Astrophys. Blinderen Oslo Rept.* **61**, p. 1

Andersen, B.N.: 1984, 'Limb Effect of Solar Absorption Lines', *Solar Phys.* **94**, 49

Andersen, B.N.: 1987, 'Solar Limb Effect and Meridional Flows. Results for Fe I 512.4, 543.4, and 709.0 nm', *Solar Phys.* **114**, 207

Andersen, B.N., Barth, S., Hansteen, V., Leifsen, T., Lilje, P.B., and Vikanes, F.: 1985, 'The Limb Effect of the K I Resonance Line, 769.9 nm', *Solar Phys.* **99**, 17

Andreassen, O., Engvold, O., and Muller, R.: 1985, 'Simulated Correlation Tracking on Solar Granulation', in R. Muller (ed.), *High Resolution in Solar Physics*, Proc. 8th IAU European Regional Meeting, *Lecture Notes in Physics* **233**, Springer-Verlag, Berlin, p. 91

Andreiko, A.V., Karpinsky, V.N., and Kotljar, L.M.: 1972, 'Method and Some Results of Detemrining the Spatial Spectrum of Photometric Tracings of Solar Granulation', *Soln. Dannye 1972 Byull.*, No. 10, p. 93

Andreiko, A.V., *see* Pravdyuk, L.M. *et al.* 1974

Antia, H.M., Chitre, S.M., and Narasimha, D.: 1982, 'Overstability of Acoustic Modes and the Solar Five-Minute Oscillations', *Solar Phys.* **77**, 303

Antia, H.M., Chitre, S.M., and Narasimha, D.: 1983, 'Influence of Turbulent Pressure on Solar Convective Modes', *Monthly Notices Roy. Astron. Soc.* **204**, 865

Antia, H.M., Chitre, S.M., and Pandey, S.K.: 1981, 'Granulation and Supergranulation as Convective Modes in the Solar Envelope', *Solar Phys.* **70**, 67

Athay, R.G., *see* Bruner, E.C. *et al.* 1976

Atroshchenko, I.N., Gadun, A.S., and Kostik, R.I.: 1989, 'Asymmetry of Absorption Lines in the Solar and Procyon Spectra', *In These Proceedings*, p. 135

Atroshchenko, I.N., Gadun, A.S., and Kostik, R.I.: 1989, 'Three-Dimensional Simulation of Convective Motions in the Procyon Envelope', *In These Proceedings*, p. 521

Bahng, J. and Schwarzschild, M.: 1961, 'Lifetime of Solar Granules', *Astrophys. J.* **134**, 312

Bahng, J. and Schwarzschild, M.: 1961, 'The Temperature Fluctuations in the Solar Granulation', *Astrophys. J.* **134**, 337

Balthasar, H.: 1985, 'On the Contribution of Horizontal Granular Motions to observed Limb Effect Curves', *Solar Phys.* **99**, 31

Banos, G., *see* Fang, C. *et al.*1984

Banos, G.J., *see* Macris, C.J.*et al.*1961

Barth, S., *see* Andersen, B.N. *et al.* 1985

Bartholomew, C.F.: 1976, 'The Discovery of the Solar Granulation', *Quart. J. Roy. Astron. Soc.* **17**, 263

Bässgen, M. and Deubner, F.-L.: 1982, 'On the Magnitude and the Height Dependence of the Granular Vertical Flow Velocity', *Astron. Astrophys.* **111**, L1

Beckers, J.M.: 1964, 'On the Relation between Solar Granules and Spicules', *Astrophys. J.* **140**, 1339

Beckers, J.M.: 1968, 'High-Resolution Measurements of Photosphere and Sunspot Velocity and Magnetic Fields Using a Narrow-Band Birefringent Filter', *Solar Phys.* **3**, 258

Beckers, J.M.: 1980, 'Examples of Non-Thermal Motions as Seen on the Sun', in D.F. Gray and J.L. Linsky (eds.), *Stellar Turbulence, IAU Colloq. 51*, Lecture Notes in Physics **114**, Springer-Verlag, Berlin, p. 85

Beckers, J.M.: 1981, 'Dynamics of the Solar Photosphere', in S. Jordan (ed.), *The Sun as a Star*, NASA SP-450, p. 11

Beckers, J.M.: 1989, 'Solar Image Restoration by Adaptive Optics', *In These Proceedings*, p. 43

Beckers, J.M.: 1989, 'Improving Solar Image Quality by Image Selection', *In These Proceedings*, p. 55

Beckers, J.M. and Canfield, R.C.: 1976, 'Spatially Resolved Motions in the Solar Atmosphere', in R.C. Cayrel and M. Steinberg (eds.), *Physique des Mouvements dans les Atmosphères Stellaires*, CNRS, Paris, p. 207

Beckers, J.M. and Morrison, R.A.: 1970, 'The Interpretation of Velocity Filtergrams. III: Velocities inside Solar Granules', *Solar Phys.* **14**, 280

Beckers, J.M. and Nelson, G.D.: 1978, 'Some Comments on the Limb Shift of Solar Lines. II: The Effect of Granular Motions', *Solar Phys.* **58**, 243

Beckers, J.M. and Parnell, R.L.: 1969, 'The Interpretation of Velocity Filtergrams. II: The Velocity and Intensity Field of the Central Solar Disk', *Solar Phys.* **9**, 39

Bédard, J., *see* Nadeau, D. *et al.* 1989

Bernière, G., Michard, R., and Rigal, G.: 1962, Étude Statistique des Fluctuations Locales de Brilliance et de Vitesse dans la Photosphére', *Ann. Astrophys.* **25**, 279

Bhatnagar, A., *see* Howard, R.F. 1969

Bhatnagar, A., *see* Sheeley, N.R., Jr. 1971

Bhatnagar, A., *see* Sheeley, N.R., Jr. 1971

Biermann, L.: 1948, 'Über die Ursache der Chromosphärischen Turbulenz und des UV-Exzesses der Sonnenstrahlung', *Z. Astrophys.* **25**, 161

Biermann, L., Kippenhahn, Lüst, R., and Temesváry, S.: 1959, 'Beiträge zur Theorie der Sonnengranulation', *Z. Astrophys.* **48**, 172

Birkle, K.: 1967, 'Über das Verhalten der Photosphärischen Granulation im Fleckenzyklus', *Z. Astrophys.* **66**, 252

Blackwell, D.E., Dewhirst, D.W., and Dollfus, A.: 1957, 'Solar Granulation and Its Observation from a Free Ballon', *Nature* **180**, 211

Blackwell, D.E., Dewhirst, D.W., and Dollfus, A.: 1959, 'The Observation of Solar Granulation from a Manned Balloon. I: Observational Data and Measurement of Contrast', *Monthly Notices Roy. Astron. Soc.* **119**, 98

Blamont, J.E. and Carpentier, G.: 1967, 'Mise en Evidence de la Granulation Solaire à 2000 Å', *Solar Phys.* **1**, 180

Blamont, J.E. and Carpentier, G.: 1968, 'Photographies de Soleil a Haute Résolution à 2000 Å', *Ann. Astrophys.* **31**, 333

Bodo, G., *see* Kalkofen, W. *et al.* 1989

Böhm-Vitense, E.: 1961, 'Considerations on Localized Velocity Fields in Stellar Atmospheres: Prototype - The Solar Atmosphere. Convection and Granulation', in R.N. Thomas (ed.), *Aerodynamic Phenomena in Stellar Atmospheres, IAU Symp. 12, Nuovo Cim. Suppl. Ser. 10* **22**, p. 330

Böhm, K.H.: 1962, 'Die Temperaturschwankungen in der Sonnengranulation', *Z. Astrophys.* **54**, 217

Bonet, J.A., Márquez, I., and Vázquez, M.: 1989, 'The Limb Effect of the K I 769.9 nm Line in Quiet Regions', *In These Proceedings*, p. 299

Bonet, J.A., Márquez, I., Vázquez, M., and Wöhl, H.: 1988, 'Temporal and Center-To-Limb Variations of the K I 769.9 nm Line Profiles in Quiet and Active Solar Regions', *Astron. Astrophys.* **198**, 322

Bonnet, R.M., *see* Foing, B.H. 1984

Bonnet, R.M., *see* Foing, B.H. 1984

Bonnet, R.M., *see* Foing, B. *et al.* 1984

Bonnet, R.M., *see* Foing, B. *et al.* 1986

Van der Borght, R.: 1974, 'Nonlinear Thermal Convection in a Layer with Imposed Energy Flux (Solar Granulation)', *Australian J. Phys.* **27**, 481

Van der Borght, R.: 1975, 'Finite-Amplitude Convection in a Compressible Medium and Its Application to Solar Granulation', *Monthly Notices Roy. Astron. Soc.* **173**, 85

Van der Borght, R. and Fox, P.: 1983, 'A Convective Model of Solar Granulation', *Proc. Astron. Soc. Australia* **5**, 166

Van der Borght, R. and Fox, P.: 1983, 'Granulation and Supergranulation as a Diagnostic Test of Solar Structure', *Proc. Astron. Soc. Australia* **5**, 168

Van der Borght, R. and Fox, P.: 1983, 'Accuracy of the Anelastic Approximation in the Theory of Compressible convection', *Proc. Astron. Soc. Australia* **5**, 170

Van der Borght, R. and Fox, P.: 1984, 'Solar Granulation', in B. Hidayat and M.W. Feast (eds.), *Proc. Second Asian-Pacific Regional Meeting on Astronomy*, Tira Pustaka Publ. House, Jakarta, p. 483

Van der Borght, R. and Fox, P.: 1984, 'Convective Motions as an Indicator of Solar Structure', in A. Maeder and A. Renzini (eds.), *Observational Tests of the Stellar Evolution Theory, IAU Symp. 105*, D. Reidel Publ. Co., Dordrecht, p. 71

Van der Borght, R. and Fox, P.: 1984, 'A Multimode Investigation of Granular and Supergranular Motions. I: Boussinesq Model', in A. Berger (ed.), *The Big Bang and Georges Lemaitre*, D. Reidel Publ. Co., Dordrecht, p. 269

Van der Borght, R. and Murphy, J.O.: 1973, 'The Combined Effect of Rotation and Magnetic Field on Finite-Amplitude Thermal Convection', *Australian J. Phys.* **26**, 617

Van der Borght, R., *see* Fox, P. 1985

Borgnino, J., *see* Aime, C. *et al.* 1977

Borgnino, J., *see* Aime, C. *et al.* 1985

Borgnino, F., *see* Aime, C. *et al.* 1989

Borgnino, J., *see* Ricort, G. *et al.* 1981

Borgnino, J., *see* Ricort, G. *et al.* 1981

Brandt, P.N.: 1986, 'Andertsich die Struktur der Granulation mit dem Solaren Aktivitätszydlus?', *Sterne Weltraum, 25. Jahrg.*, No. 3, p. 128

Brandt, P.N., Scharmer, G.B., Ferguson, S.H., Shine, R.A., Tarbell, T.D., and Title, A.M.: 1989, 'Vortex Motions of the Solar Granulation', *In These Proceedings*, p. 305

Bray, R.J.: 1982, 'The Wavelength Variation of the Granule/Intergranule Contrast', *Solar Phys.* **77**, 299

Bray, R.J. and Loughhead, R.E.: 1958, 'Observations of Changes in the Photospheric Granules', *Astralian J. Phys.* **11**, 507

Bray, R.J. and Loughhead, R.E.: 1959, 'High Resolution Observations of the Granular Structure of Sunspot Umbrae', *Australian J. Phys.* **12**, 320

Bray, R.J. and Loughhead, R.E.: 1961, 'Facular Granule Lifetimes Determined with a Seeing-Monitored Photoheliograph', *Australian J. Phys.* **14**, 14

Bray, R.J. and Loughhead, R.E.: 1974, *The Solar Chromosphere*, Chapman and Hall, London, p. 322

Bray, R.J. and Loughhead, R.E.: 1977, 'A New Determination of the Granule/Intergranule Contrast', *Solar Phys.* **54**, 319

Bray, R.J. and Loughhead, R.E.: 1978, 'A Convective Correction to Measured Solar Wavelengths', *Astrophys. J.* **224**, 276

Bray, R.J. and Loughhead, R.E.: 1984, 'Isophotes of a Large Granule', *Astron. Astrophys.* **133**, 409

Bray, R.J., Loughhead, R.E., and Durrant, C.J.: 1984, *Solar Granulation (second edition)*, Cambridge Press, Cambridge

Bray, R.J., Loughhead, R.E., and Tappere, E.J.: 1974, 'High-Resolution Photography of the Solar Chromosphere. XV: Preliminary Observations in Fe I λ6569.2', *Solar Phys.* **39**, 323

Bray, R.J., Loughhead, R.E., and Tappere, E.J.: 1976, 'Convective Velocities Derived from Granule Contrast Profiles in Fe I λ 6569.2', *Solar Phys.* **49**, 3

Bray, R.J., *see* Loughhead, R.E. 1959

Bray, R.J., *see* Loughhead, R.E. 1960

Bray, R.J., *see* Loughhead, R.E. 1975

Brown, J.D., *see* Goldberg, L. *et al.* 1957

Brown, J., *see* Goldberg, L. *et al.* 1960

Brown, W.A., *see* Foing, B. *et al.* 1984

Bruggencate, P.T. and Grotrian, W.: 1936, 'Die Bestimmung der Mittleren Lebensdauer der Granulation', *Z. Astrophys.* **12**, 323

Bruggencate, P.T. and Müller, H.: 1942, 'Untersuchungen der Granulation der Sonne', *Z. Astrophys.* **21**, 198

Bruls, J.H.M.J., Uitenbroek, H., and Rutten, R.: 1989, 'The Granulation Sensitivity of Neutral Metal Lines', *In These Proceedings*, p. 311

Bruner, E.C., Jr., Chipman, E.G., Lites, B.W., Rottman, G.J., Shine, R.A., Athay, R.G., and White, O.R.: 1976, 'Preliminary Rèsults from the Orbiting Solar Observatory 8: Transition Zone Dynamics over a Sunspot', *Astrophys. J. Letters* **210**, L97

Bruner, M., *see* Foing, B. *et al.* 1984

Bruner, M., *see* Foing, B. *et al.* 1986

Bruning, D.H and Saar, S.H.: 1989, 'Line Asymmetries in Late-Type Dwarf Photospheres', *In These Proceedings*, p. 145

Bumba, V., Hejna, L., and Suda, J.: 1975, 'Granular-Like Pattern in Sunspot Umbrae', *Bull. Astron. Inst. Czech.* **26**, 315

Caccin, B., Falciani, R., Gomez, M.T., Marmolino, C., Roberti, G., Severino, G., and Smaldone, L.A.: 1981, 'Magnetic Fine Structures and Granular Velocities', *Space Sci. Rev.* **29**, 373

Cambell, E., *see* Keil, S.L. *et al.* 1989

Canfield, R.C.: 1976, 'The Height Variation of Granular and Oscillatory Velocities', *Solar Phys.* **50**, 239

Canfield, R.C. and Mehltretter, J.P.: 1973, 'Fluctuations of Brightness and Vertical Velocity at Various Heights in the Photosphere', *Solar Phys.* **33**, 33

Canfield, R.C., *see* Beckers, J.M. 1976

Canfield, R.C., *see* Keil, S.L. 1978

Cannon, C.J.: 1969, 'Power Spectral Analysis of Chromospheric Inhomogeneities', *Proc. Astron. Soc. Australia* **1**, 197

Cannon, C.J. and Wilson, P.R.: 1970, 'Center-Limb Observations of Inhomogeneities in the Solar Atmosphere. I: The Mg b Lines', *Solar Phys.* **14**, 29

Cannon, C.J. and Wilson, P.R.: 1971, 'Center Limb Observations of Inhomogeneities in the Solar Atmosphere. II: The Na D and Na 5688 Doublets and the Mg I 4571 Line', *Solar Phys.* **17**, 288

Carlier, A., Chauveau, F., Hugon, M., and Rösch, J.: 1968, 'Cinématographie à Haute Résolution Spatiale de la Granulation Photosphérique', *Compt. Rend. Acad. Sci. Paris B* **266**, 199

Carpentier, G., *see* Blamont, J.E. 1967

Carpentier, G., *see* Blamont, J.E. 1968

Cattaneo, F., Hurlburt, N.E., and Toomre, J.: 1989, 'Two- and Three-Dimensional Simulations of Compressible Convection', *In These Proceedings*, p. 415

Cavallini, F. and Ceppatelli, G.: 1989, 'Line Bisectors in and out of Magnetic Regions', *In These Proceedings*, p. 283

Cavallini, F., Ceppatelli, G., and Righini, A.: 1986, 'Solar Limb Effect and Meridional Flow: Results on the Fe I Lines at 5569.6 Å and 5576.1 Å', *Astron. Astrophys.* **163**, 219

Cavallini, F., Ceppatelli, G., and Righini, A.: 1987, 'Interpretation of Shifts and Asymmetries of Fe I Lines in Solar Facular Areas', *Astron. Astrophys.* **173**, 155

Cavallini, F., Ceppatelli, G., Righini, A., and Tsiropoula, G.: 1987, '5-Min Oscillations in the Wings and Bisectors of Solar Photospheric Fe I Lines', *Astron. Astrophys.* **173**, 161

Ceppatelli, G., *see* Cavallini, F. 1989

Ceppatelli, G., *see* Cavallini, F. *et al.* 1986

Ceppatelli, G., *see* Cavallini, G. *et al.* 1987

Ceppatelli, G., *see* Cavallini, G. *et al.* 1987

Chan, K.L., Sofia, S.; and Wolff, C.L.: 1982, 'Turbulent Compressible Convection in a Deep Atmosphere. I: Preliminary Two-Dimensional Results', *Astrophys. J.* **263**, 935

Chapman, G.A., *see* Lynch, D.K. 1975

Chauveau, F., *see* Carlier, A. *et al.* 1968

Chen, P.: 1960, 'Theory of Solar Granulation', *Acta. Astron. Sinica* **8**, 43

Chipman, E.G., *see* Bruner, E.C. *et al.* 1976

Chitre, S.M., *see* Antia, H.M. *et al.* 1981

630

Chitre, S.M., *see* Antia, H.M. *et al.* 1982

Chitre, S.M., *see* Antia, H.M. *et al.* 1983

Chitre, S.M., *see* Narasimha, D. *et al.* 1980

Chevalier, S.: 1908, 'Contribution to the Study of the Photosphere', *Astrophs. J.* **27**, 329

Chevalier, S.: 1914, Étude Photographique de la Photosphère Solaire', *Ann. l'Obs. Astron. de Zô-Sè* **8**, C1

Cloutman, L.D.: 1976, 'Dynamics of the Density Inversion Zone of a One Solar Mass Star', in A.N. Cox and R.G. Deupree (eds.), *Solar and Stellar Pulsation Conference Proceedings*, Los Alamos Scientific Laboratory, Los Alamos, p. 36

Cloutman, L.D.: 1979, 'A Physical Model of the Solar Granulation', *Astrophys. J.* **227**, 614

Cloutman, L.D.: 1980, 'Numerical Simulation of the Solar Granulation', *Space Sci. Rev.* **27**, 293

Collados, M. and Vázquez, M.: 1987, 'A New Determination of the Solar Granulation Contrast', *Astron. Astrophys.* **180**, 223

Collados, M., Marco, E., Del Toro, J.C., and Vázquez, M.: 1986, 'Granulation Deformation near and in Sunspot Regions', *Solar Phys.* **105**, 17

Cook, M.C., *see* Altrock, R.C. *et al.* 1983

Coupinot, G., *see* Roudier, T. *et al.* 1985

Cuntz, M. and Muchmore, D.: 1989, 'Inhomogeneous Chromospheric Structures in the Atmosphere of Arcturus Generated by Acoustic Wave Propagation', *In These Proceedings*, p. 547

D'Angelo, N.: 1969, 'Heating of the Solar Corona', *Solar Phys.* **7**, 321

Damé, L.: 1985, 'Meso-Scale Structures: An Oscillatory Phenomenon?', in H.U. Schmidt (ed.), *Theoretical Problems in High Resolution Solar Physics*, Max Planck Institut Publ. MPA **212**, p. 244

Damé, L. and Martic, M.: 1987, 'Observation and Oscillatory Properties of Mesostructures in the Solar Chromosphere', *Astrophys. J. Letters* **314**, L15

Damé, L. and Martic, M.: 1988, 'Oscillatory Properties of Meso-Scale Intensity Structures at Chromospheric Level', in J. Christensen-Dalsgaard, and S. Frandsen (eds.) *Advances in Helio- and Asteroseismology, IAU Symp. 123*, D. Reidel Publ. Co., Dordrecht, p. 433

Damé, L., *see* Martic, M. 1989

Darvann, T.A. and Kusoffsky, U.: 1989, 'Granule Lifetimes and Dimensions as Function of Distance from a Solar Pore Region', *In These Proceedings*, p. 313

Dawes, W.R.: 1864, 'Results of Some Recent Observations of the Solar Surface, with Remarks', *Monthly Notices Roy. Astron. Soc.* **24**, 161

Decaudin, M., *see* Foing, B. *et al.* 1984

Del Toro, J.C., *see* Collados, M. *et al.* 1986

Deubner, F.-L.: 1974, 'On the Energy Distribution in Wavenumber Spectra of the Granular Velocity Field', *Solar Phys.* **36**, 299

Deubner, F.-L.: 1974, 'Some Properties of Velocity Fields in the Solar Photosphere. V: Spatio-Temporal Analysis of High Resolution Spectra', *Solar Phys.* **39**, 31

Deubner, F.-L.: 1976, 'Photospheric Magnetic Flux Concentrations and the Granular Velocity Field', *Astron. Astrophys.* **47**, 475

Deubner, F.-L.: 1977, 'Velocity Fields in the Solar Atmosphere (Observations)', *Mem. Soc. Astron. Ital.* **48**, 499

Deubner, F.-L.: 1988, 'Has Turbulent Granular Decay Been Observed?', *Astron. Astrophys.* **204**, 301

Deubner, F.-L.: 1989, 'Granulation and Waves?', *In These Proceedings*, p. 195

Deubner, F.-L. and Fleck, B.: 1989, 'Spatio-Temporal Analysis of Waves in the Solar Atmosphere', *In These Proceedings*, p. 325

Deubner, F.-L. and Mattig, W.: 1975, 'New Observations of the Granular Intensity Fluctuations', *Astron. Astrophys.* **45**, 167

Deubner, F.-L., *see* Bässgen, M. 1982

Deubner, F.-L., *see* Ricort, G. *et al.* 1981

Deubner, F.-L., *see* Schmidt, W. *et al.* 1979

Dewhirst, D.W., *see* Blackwell, D.E. *et al.* 1959

Dialetis, D., Macris, C., Muller, R., and Prokakis, T.: 1988, 'A Possible Relation between Lifetime and Location of Solar Granules', *Astron. Astrophys.* **204**, 275

Dialetis, D., Macris, C., Muller, R., and Prokakis, T.: 1989, 'On the Granule Lifetime Near and Far Away from Sunspots', *In These Proceedings*, p. 327

Dialetis, D., Macris, C., and Prokakis, T.: 1985, 'The Variability of Photospheric Granulation and Total Radio Flux of Quiet Sun in the Centimetric Band during a Solar Cycle', in R. Muller (ed.), *High Resolution in Solar Physics*, Proc. 8th IAU European Regional Meeting, *Lecture Notes in Physics* **233**, Springer-Verlag, Berlin, p. 289

Dialetis, D., Macris, C., Prokakis, T., and Sarris, E.: 1985, 'Fine Structure and Evolution of Solar Granulation', in R. Muller (ed.), *High Resolution in Solar Physics*, Proc. 8th IAU European Regional Meeting, *Lecture Notes in Physics* **233**, Springer-Verlag, Berlin, p. 164

Dialetis, D., Macris, C., Prokakis, T., and Sarris, E.: 1986, 'The Lifetime and Evolution of Solar Granules', *Astron. Astrophys.* **168**, 330

Dialetis, D., *see* Alissandrakis, C.E. *et al.* 1987

Dialetis, D., *see* Macris, C. *et al.* 1989

Diemel, W.E., *see* Namba, O. 1969

Dimitrijević, M.S., *see* Vince, I. 1989

Dollfus, A., *see* Blackwell, D.E. *et al.* 1959

Dravins, D.: 1976, 'Observation of Convection in Stellar Photospheres', in R. Cayrel and M. Steinberg (eds), *Physique des Mouvements dans les Atmosphère Stellaires*, CNRS, Paris, No. 250, p. 459

Dravins, D.: 1987, 'Stellar Granulation. I: The Observability of Stellar Photospheric Convection', *Astron. Astrophys.* **172**, 200

Dravins, D.: 1987, 'Stellar Granulation. II: Stellar Photospheric Line Asymmetries', *Astron. Astrophys.* **172**, 211

Dravins, D.: 1988, 'Stellar Granulation and Photospheric Line Asymmetries', in G. Cayrel de Strobel and M. Spite (eds.), *The Impact of Very High S/N Spectroscopy on Stellar Physics*, *IAU Symp. 132*, Kluwer Academic Publ., Dordrecht, p. 239

Dravins, D.: 1989, 'Challenges and Opportunities in Stellar Granulation Observations', *In These Proceedings*, p. 153

Dravins, D.: 1989, 'Stellar Granulation: Modeling of Stellar Surfaces and Photospheric Line Asymmetries', *In These Proceedings*, p. 493

Dravins, D. and Lind, J.: 1984, 'Observing Stellar Granulation', in S.L. Keil (ed.), *Small Scale Dynamical Processes in Quiet Stellar Atmospheres*, National Solar Observatory/Sacramento Peak, Sunspot, p. 414

Dravins, D., Larsson, B., and Nordlund, Å.: 1986, 'Solar Fe II Line Asymmetries and Wavelength Shifts', *Astron. Astrophys.* **158**, 83

Dravins, D., Lindegren, L., and Nordlund, Å.: 1981, 'Solar Granulation: Influence of Convection on Spectral Line Asymmetries and Wavelength Shifts', *Astron. Astrophys.* **96**, 345

Druesne, P., *see* Aime, C. *et al.* 1985

Druesne, F., *see* Aime, C. *et al.* 1989

Dumont, S., *see* Fang, C. *et al.*1984

Duncan, D., *see* Title, A.M. *et al.* 1987

Dunn, R.B. and November, L.J.: 1985, 'Collages of Granulation Pictures', in R. Muller (ed.), *High Resolution in Solar Physics*, Proc. 8th IAU European Regional Meeting, *Lecture Notes in Physics* 233, Springer-Verlag, Berlin, p. 85

Dunn, R.B., and November, L.: 1985, 'Solar Granulation Movie', in H.U. Schmidt (ed.), *Theoretical Problems in High Resolution Solar Physics*, Max Planck Institut Publ. MPA 212, p. 27

Dunn, R.B., and Spence, G.E.: 1984, 'Granulation Data Set', in S.L. Keil (ed.), *Small Scale Dynamical Processes in Quiet Stellar Atmospheres*, National Solar Observatory/Sacramento Peak, Sunspot, p. 88

Dunn, R.B. and Zirker, J.B.: 1973, 'The Solar Filigree', *Solar Phys.* 33, 281

Dunn, R.B., *see* von der Lühe, O. 1987

Durrant, C.J. and Nesis, A.: 1981, 'Vertical Structure of the Solar Photosphere', *Astron. Astrophys.* 95, 221

Durrant, C.J. and Nesis, A.: 1982, 'Vertical Structure of the Solar Photosphere. II: The Small-Scale Velocity Field', *Astron. Astrophys.* 111, 272

Durrant, C.J., Kneer, F.J., and Maluck, G.: 1981, 'The Analysis of Solar limb Observations. II: Geometrical Smearing', *Astron. Astrophys.* 104, 211

Durrant, C.J., Mattig, W., Nesis, A., Reiss, G., and Schmidt, W.: 1979, 'Studies of Granular Velocities. VIII: The Height Dependence of the Vertical Granular Velocity Component', *Solar Phys.* 61, 251

Durrant, C.J., Mattig, W., Nesis, A., and Schmidt, W.: 1983, 'Balloon-Borne Imagery of the Solar Granulation. IV: The Centre-to-Limb Variation of the Intensity Fluctuations', *Astron. Astrophys.* 123, 319

Durrant, C.J., *see* Bray, R.J. *et al.* 1984

Durrant, C.J., *see* Kaisig, M. 1984

Durrant, C.J., *see* Kaisig, M. 1984

Durrant, C.J., *see* Wiesmeier, A. 1981

Durrant, C.J., *see* Nesis, A. *et al.* 1984

Durrant, C.J., *see* Nesis, A. *et al.* 1988

Edmonds, F.N., Jr.: 1960, 'On Solar Granulation', *Astrophys. J.* 131, 57

Edmonds, F.N., Jr.: 1962, 'A Statistical Photometric Analysis of Granulation across the Solar Disk', *Astrophys. J. Suppl.* 6, 357

Edmonds, F.N., Jr.: 1962, 'A Coherence Analysis of Fraunhofer Line Fine Structure and Continuum Brightness Fluctuations near the Center of the Solar Disk', *Astrophys. J.* 136, 507

Edmonds, F.N., Jr.: 1966, 'A Coherence Analysis of Fraunhofer Line Fine Structure and Continuum Brightness Fluctuations near the Center of the Solar Disk. II', *Astrophys. J.* 144, 733

Edmonds, F.N., Jr.: 1967, 'Amplitude Distributions of Solar Photospheric Fluctuations', *Solar Phys.* 1, 5

Edmonds, F.N., Jr.: 1972, 'Spectral Analyses of Solar Photospheric Fluctuations. II: Profile Fluctuations in the Wings of the λ 5183.6 Mg I b_1 Line', *Solar Phys.* 23, 47

Edmonds, F.N., Jr.: 1975, 'Spectral Analyses of Solar Photospheric Fluctuations. IV: The Low-Wavenumber Power of Granulation Brightness Fluctuations', *Solar Phys.* 44, 293

Edmonds, F.N., Jr. and Hinkle, K.H.: 1977, 'Spectral Analyses of Solar Photospheric Fluctuations. V: A Two-Dimensional Analysis of Granulation at the Center of the Disk', *Solar Phys.* **51**, 273

Edmonds, F.N., Jr. and Hsu, J.-H.: 1983, 'A Statistical Analysis of Na I D_1 Profile Fluctuations at the Center of the Solar Disk. I: Data Reduction and Resolvable Velocities', *Solar Phys.* **83**, 217

Edmonds, F.N., Jr. and McCullough, J.R.: 1966, 'The Evidence for an Oscillatory Component in Solar Granulation Brightness Fluctuations', *Astrophys. J.* **144**, 754

Elias, D., *see* Macris, C.J. 1955

Elste, G.: 1985, 'Temperature Gradients in the Solar Granulation', in R. Muller (ed.), *High Resolution in Solar Physics*, Proc. 8th IAU European Regional Meeting, *Lecture Notes in Physics* **233**, Springer-Verlag, Berlin, p. 169

Elste, G.: 1987, 'Manganese and Carbon Lines as Temperature Indicators', *Solar Phys.* **107**, 47

Elste, G.H., *see* Teske, R.G. 1979

Engvold, O., *see* Andreassen, O. *et al.* 1985

Evans, J.W.: 1964, 'Inclined Inhomogeneities in the Solar Photosphere', *Astrophys. Norv.* **9**, 33

Evans, J.W.: 1968, 'Color in Solar Granulation', *Solar Phys.* **3**, 344

Falciani, R., *see* Caccin, B. *et al.* 1981

Fang, C., Mouradian, Z., Banos, G., Dumont, S., and Pecker, J.C.: 1984, 'Structure and Physics of Solar Faculae. IV: Chromospheric Granular Structure', *Solar Phys.* **91**, 61

Fellgett, P.: 1959, 'On the Interpretation of Solar Granulation', *Monthly Notices Roy. Astron. Soc.* **119**, 475

Fellgett, P.: 1976, 'A Note on the 'Discovery of Solar Granulation'', *Quart. J. Roy. Astron. Soc.* **18**, 159

Ferguson, S.H., *see* Brandt, P.N. *et al.* 1989

Ferguson, S.H., *see* November, L.J. *et al.* 1987

Ferguson, S.H., *see* Simon, G.W. *et al.* 1988

Ferguson, S.H., *see* Simon, G.W. *et al.* 1989

Ferguson, S., *see* Title, A.M. *et al.* 1987

Ferguson, S.H., *see* Title, A.M. *et al.* 1989

Ferguson, S., *see* Title, A.M. *et al.* 1989

Finch, M., *see* Title, A.M. *et al.* 1987

Finlay-Freundlich, E. and Forbes, E.G.: 1956, 'On the Red Shift of the Solar Lines', *Ann. Astrophys.* **19**, 183

Finlay-Freundlich, E. and Forbes, E.G.: 1956, 'On the Red Shift of the Solar Lines. II', *Ann. Astrophys.* **19**, 215

Fleck, B., *see* Deubner, F.-L. 1988

Fleig, K.H., *see* Nesis, A. *et al.* 1987

Fleig, K.H., *see* Nesis, A. *et al.* 1989

Foing, B.H. and Bonnet, R.M.: 1984, 'Results from the Transition Region Camera', *Adv. Space Res.* **4** (8), 43

Foing, B.H. and Bonnet, R.M.: 1984, 'Characteristic Structures of the Solar Disc Observed on Rocket UV Filtergrams', *Astron. Astrophys.* **136**, 133

Foing, B.H., Bonnet, R.M., Acton, L.W., Brown, W.A., Bruner, M., and Decaudin, M.: 1984, 'The Transition Region Camera Experiment: High Resolution Ultraviolet Filtergrams of the Sun', in S.L. Keil (ed.), *Small Scale Dynamical Processes in Quiet Stellar Atmospheres*, National Solar Observatory/Sacramento Peak, Sunspot, p. 99

Foing, B., Bonnet, R.M., and Bruner, M.: 1986, 'New High-Resolution Rocket Ultraviolet Filtergrams of the Solar Disc', *Astron. Astrophys.* **162**, 292

Forbes, E.G., *see* Finlay-Freundlich, E. 1956

Forbes, E.G., *see* Finlay-Freundlich, E. 1956

Fossat, E., *see* Ricort, G. *et al.* 1974

Foukal, P.V. and Fowler, L.: 1984, 'A Photometric Study of Heat Flow at the Solar Photosphere', *Astrophys. J.* **281**, 442

Foukal, P.V., *see* Reeves, E.M. *et al.* 1974

Fowler, L., *see* Foukal, P.V. 1984

Fox, P.A.: 1989, 'How Much Can Theoretical Models of Compressible Convection Tell Us about Solar Granulation?', *In These Proceedings*, p. 401

Fox, P.A.: 1989, 'Interaction of Solar Granulation with Weak and Strong Magnetic Fields', *In These Proceedings*, p. 555

Fox, P. and van der Borght, R.: 1985, 'Solar Granulation: The Influence of Viscosity Laws on Theoretical Models', *Proc. Astron. Soc. Australia* **6**, 60

Fox, P., *see* van der Borght, R. 1983

Fox, P., *see* van der Borght, R. 1983

Fox, P., *see* Van der Borght, R. 1983

Fox, P., *see* van der Borght, R. 1984

Fox, P., *see* Van der Borght, R. 1984

Fox, P., *see* Van der Borght, R. 1984

Frank, Z.A., *see* Peri, M.L. *et al.* 1989

Frank, Z., *see* Title, A.M. *et al.* 1987

Frazier, E.N.: 1976, 'The Photosphere - Magnetic and Dynamic State', *Phil. Trans. Roy. Soc. London, Ser. A* **281**, 295

Frenkiel, F.N.: 1954, 'On the Spectrum of Turbulence in the Photosphere of the Sun', in M.H. Wrubel (ed.), *Proc. National Science Foundation Conference on Stellar Atmospheres*, Indiana University, Bloomington, p. 23

Frenkiel, F.N. and Schwarzschild, M.: 1952, 'Preliminary Analysis of the Turbulence Spectrum of the Solar Photosphere at Long Wave Lengths', *Astrophys. J.* **116**, 422

Frenkiel, F.N. and Schwarzschild, M.: 1955, 'Additional Data for Turbulence Spectrum of Solar Photosphere at Long Wave Lengths', *Astrophys. J.* **121**, 216

Gadun, A.S., *see* Atroshchenko, I.N. *et al.* 1989

Gadun, A.S., *see* Atroshchenko, I.N. *et al.* 1989

Galloway, D.J. and Moore, D.R.: 1979, 'Axisymmetric Convection in the Presence of a Magnetic Field', *Geophys. Astrophys. Fluid Dyn.* **12**, 75

Gaustad, J. and Schwarzschild, M.: 1960, 'Note on the Brightness Fluctuation in the Solar Granulation', *Monthly Notices Roy. Astron. Soc.* **121**, 260

Gebbie, K.B., *see* Hill, F. *et al.* 1984

Gebbie, K.B., *see* November, L.J. *et al.* 1981

Gebbie, K.B., *see* November, L.J. *et al.* 1982

Gehlsen, M., *see* Holweger, H. *et al.* 1978

Gerbil'skij, M.G. and Sobolev, V.M.: 1978, 'On the Influence of the Granulation Structure on a Weak Absorption Line', *Soln. Dannye 1977 Byull.*, No. 11, p. 79

Gigas, D.: 1989, 'Numberical Simulations of Gas Flows in an A-Type Stellar Atmosphere', *In These Proceedings*, p. 533

Gigas, D., *see* Steffen, M. 1985

Giovanelli, R.G.: 1961, 'On the Center-Limb Variation of Granule Contrast', *Monthly Notices Roy. Astron. Soc.* **122**, 523

Giovanelli, R.G.: 1970, 'Solar Magnetic Fields and Velocities', *Proc. Astron. Soc. Australia* **1**, 363

Giovanelli, R.G.: 1974, 'Chromospheric Granulation', *Solar Phys.* **37**, 301

Giovanelli, R.: 1975, 'Wave Systems in the Chromosphere', *Solar Phys.* **44**, 299

Goldberg, L., Mohler, O.C., and Brown, J.D.: 1957, 'A Connexion between the Granulation and the Structure of the Low Chromosphere', *Nature* **179**, 369

Goldberg, L., Mohler, O.C., Unno, W., and Brown, J.: 1960, 'The Measurement of the Local Doppler Shift of Fraunhofer Lines', *Astrophys. J.* **132**, 184

Goldberg-Rogosinskaja, N.M, *see* Krat, V.A. 1956

Gomez, M.T., Marmolino, C., Roberti, G., and Severino, G.: 1987, 'Temporal Variations of Solar Spectral Line Profiles Induced by the 5-minute Photospheric Oscillation', *Astron. Astrophys.* **188**, 169

Gomez, M.T., Servino, G., and Rutten, R.J.: 1989, 'Granulation and the NLTE Formation of K I 769.9', *In These Proceedings*, p. 565

Gomez, M.T., *see* Caccin, B. *et al.* 1981

Graves, J.E. and Pierce, A.K.: 1986, 'The Morphology of Solar Granulations and Dark Networks', *Solar Phys.* **106**, 249

Gray, D.F.: 1980, 'Measurements of Spectral Line Asymmetries for Arcturus and the Sun', *Astrophys. J.* **235**, 508

Gray, D.F.: 1984, 'Stellar Granulation', in T.D. Guyenne and J.J. Hunt (eds.), *The Hydromagnetics of the Sun*, Proc. Fourth European Meeting on Solar Physics, ESA SP-220, p. 211

Gray, D.F.: 1986, 'The Rotation Effect - A Mechanism for Measuring Granulation Velocities in Stars', *Publ. Astron. Soc. Pacific* **98**, 319

Gray, D.F.: 1989, 'Granulation in the Photospheres of Stars', *In These Proceedings*, p. 71

Gray, D.F. and Toner, C.G.: 1985, 'Inferred Properties of Stellar Granulation', *Publ. Astron. Soc. Pacific* **97**, 543

Grec, G., *see* Aime, C. *et al.* 1975

Grec, G., *see* Aime, C. *et al.* 1977

Grec, G., *see* Aime, C. *et al.* 1979

Grevesse, N.: 1981, 'La Photosphére', *Ciel Terre* **97**, 111

Grossmann-Doerth, U., Knölker, M., Schüssler, and Weisshaar, E.: 1989, 'Models of Magnetic Flux Sheets', *In These Proceedings*, p. 481

Grossmann-Doerth, U., *see* Schmidt, W. *et al.* 1979

Grotrian, W., *see* Bruggencate, P.T. 1936

Hadjebi, B.: 1981, 'On the Temporal Variation of the Solar Continuous Brightness Fluctuations. Time Dependence of the Spatial Power Spectra', *Solar Phys.* **73**, 25

Hafkenscheid, G.A.M., *see* Namba, O. *et al.* 1983

Hammarbäck, G.: 1989, 'The Effects of Inhomogeneities in 3-D Models on Abundance Determinations', *In These Proceedings*, p. 567

Hanslmeier, A., *see* Mattig, W. *et al.* 198

Hanslmeier, A., *see* Münzer, H. *et al.* 1989

Hansen, T.L. *see* Albregtsen, F. 1977

Hansky, A.: 1908,'Mouvement des Granules sur la Surface du Soleil', *Mitt. Astron. Hauptobs. Pulkovo* **3**, 1

Hansteen, V., *see* Andersen, B.N. *et al.* 1985

Harvey, J.W.: 1985, 'Optical Interferometry of Solar Fine Structure', in H.U. Schmidt (ed.), *Theoretical Problems in High Resolution Solar Physics*, Max Planck Institut Publ. MPA **212**, p. 183

Harvey, J.W. and Ramsey, H.E.: 1963, 'Photospheric Granulation in the Near Ultraviolet', *Publ. Astron. Soc. Pacific* **75**, 283

Harvey, J.W. and Schwarzschild, M.: 1975, 'Photoelectric Speckle Interferometry of the Solar Granulation', *Astrophys. J.* **196**, 221

Harvey, J., *see* Aime, C. *et al.* 1978

Harvey, J.W. *see* Aime, C. *et al.* 1985

Harvey, J.W., *see* Title, A.M. *et al.* 1987

Hasan, S.S., *see* Venkatakrishnan, P. 1981

Hecquet, J., *see* Roudier, T. *et al.* 1985

Hejna, L.: 1977, 'Elementary Statistical Analysis of Some Parameters of the Photospheric and Umbral Granulation', *Bull. Astron. Inst. Czech.* **28**, 126

Hejna, L.: 1980, 'A Comment on the Character of the Photospheric Granular Net', *Bull. Astron. Inst. Czech.* **31**, 362

Hejna, L., *see* Bumba, V. *et al.* 1975

Hersè, W.: 1979, 'High Resolution Photographs of the Sun near 200nm', *Solar Phys.* **63**, 35

Hiei, E., *see* Suemoto, Z. *et al.* 1987

Higgs, L.A.: 1960, 'The Solar Red Shift', *Monthly Notices Roy. Astron. Soc.* **121**, 421

Hill, F., Toomre, J., November, L.J., and Gebbie, K.B.: 1984, 'On the Determination of the Lifetime of Vertical Velocity Patterns in Mesogranulation and Supergranulation', in S.L. Keil (ed.), *Small Scale Dynamical Processes in Quiet Stellar Atmospheres*, National Solar Observatory/Sacramento Peak, Sunspot, p. 160

Hinkle, K.H., *see* Edmonds, F.N., Jr. 1977

Högbom, J.A.: 1989, 'Reconstruction from Focal Volume Information', *In These Proceedings*, p. 61

Holweger, H.: 1987, 'Atmospheric Structure and the Activity Cycle', in E.H. Schröter, M. Vázquez, and A.A. Wyller (eds.), *The Role of Fine-Scale Magnetic Fields on the Structure of the Solar Atmosphere*, Cambridge University Press, Cambridge, p. 1

Holweger, H. and Kneer, F.: 1989, 'Spatially Resolved Spectra of Solar Granules', *In These Proceedings*, 173

Holweger, H., Gehlsen, M., and Ruland, F.: 1978, 'Spatially Averaged Properties of the Photospheric Velocity Field', *Astron. Astrophys.* **70**, 537

Howard, R.F. and Bhatnagar, A.: 1969, 'On the Spectrum of Granular and Intergranular Regions', *Solar Phys.* **10**, 245

Hsu, J.-H., *see* Edmonds, F.N., Jr. 1983

Huber, M.C.E., *see* Reeves, E.M. *et al.* 1974

Hudson, H.S., *see* Lindsey, C.A. 1976

Huggins;, W.: 1866, 'Results of Some Observations on the Bright Granules of the Solar Surface, with Remarks on the Nature of These Bodies', *Monthly Notices Roy. Astron. Soc.* **26**, 260

Hugon, M., *see* Carlier, A. *et al.* 1968

Hugon, M., *see* Rösch, J. 1959

Hurlburt, N.E., *see* Cattaneo, F. *et al.* 1989

Ilias, D.P.: 1956, 'Die Veränderlichkeit der Sonnengranulation und ihre Auswirkung', *Bull. Milit. Geogr. Amt* 1956, p. 49

de Jager, C.: 1974, 'The Influence of a Photospheric Spectrum of Turbulence on the Profiles of Weak Fraunhofer Lines', *Solar Phys.* **34**, 91

de Jager, C. and Pecker, J.-C.: 1951, 'Interprétation des Mesures de Vitesses Radiales dans les Granules Solaires', *Compt. Rend. Acad. Sci. Paris* **232**, 1645

de Jager, C., *see* Vermue, J. 1979

Janssen, J.: 1896, 'Mémoire sur la Photographie Solaire', *Ann. l'Obs. Astron. Phys. Paris sis Parc de Meudon* **1**, 91

Jones, C.A. and Moore, D.R.: 1979, 'The Stability of Axisymmetric Convection', *Geophys. Astrophys. Fluid Dyn.* **11**, 245

Judina, I.W., *see* Krat, V.A. 1961

Jurač, S.: 1982, 'On Solar Granulation', *Vasiona, Année 30*, No. 4, p. 81

Kadiri, S., *see* Aime, C. *et al.* 1981

Kaisig, M. and Durrant, C.J.: 1984, 'The Analysis of Solar Limb Observations. III: Geometrical Effects in Weighting Functions', *Astron. Astrophys.* **130**, 171

Kaisig, M. and Durrant, C.J.: 1984, 'Geometrical Effects in Weighting Functions', in S.L. Keil (ed.), *Small Scale Dynamical Processes in Quiet Stellar Atmospheres*, National Solar Observatory/Sacramento Peak, Sunspot, p. 406

Kalkofen, W., Bodo, G., Massaglia, S., and Rossi, P.: 1989, '2-D Flux Tube in Radiative Equilibrium', *In These Proceedings*, p. 571

Kandel, R.S. and Keil, S.L.: 1973, 'Solar Granulation, Limb Flux, and Oblateness', *Solar Phys.* **33**, 3

Karpinsky, V.N.: 1979, 'Brightness and Radial Velocity Are Uncorrelated in the Fine Structure of the Lower Solar Photosphere', *Soviet Astron. Letters* **5**, 295

Karpinsky, V.N.: 1980, 'Morphological Elements and Characteristics of the Fine Structure of the Photospheric Brightness Field near the Solar Disc Center', *Soln. Dannye 1980 Byull.*, No. 7, p. 94

Karpinsky, V.N.: 1981, 'Photometric Profile of a Granule and Horizontal Derivatives of the Brightness Temperature in Granulation', *Soln. Dannye 1981 Byull.*, No. 1, p. 88

Karpinsky, V.N. and Kostjukevich, V.I.: 1971, 'On High-Resolution Photographs of the Solar Photosphere', *Soln. Dannye 1971 Byull.*, No. 3, p. 88

Karpinsky, V.N. and Mekhanikov, V.V.: 1977, 'Two-Dimensional Spatial Spectrum of the Photospheric Brightness Field near to the Solar Disc Center', *Solar Phys.* **54**, 25

Karpinsky, V.N. and Pravdjuk, L.M.: 1973, 'Variations of Solar Granulation with Wavelength (from λ 3900Å to λ 6600Å)', *Soln. Dannye 1973 Byull.*, No. 10, p. 79

Karpinsky, V.N., *see* Andreiko, A.V. *et al.* 1972

Karpinsky, V.N., *see* Pravdyuk, L.M. *et al.* 1974

Kavetsky, A. and O'Mara, B.J.: 1984, 'The Solar O I λ 7773 Triplet. I: Spatially Resolved Profiles', *Solar Phys.* **92**, 47

Kawaguchi, I.: 1980, 'Morphological Study of the Solar Granulation. II: The Fragmentation of Granules', *Solar Phys.* **65**, 207

Kawaguchi, I., *see* Kitai, R. 1979

Keenan, P.C.: 1938, 'Dimensions of the Solar Granules', *Astrophys. J.* **88**, 360

Keil, S.L.: 1974, 'Models of Solar Granular Structure and the Interpretation of Photospheric Observations', *Thesis*

Keil, S.L.: 1977, 'A New Measurement of the Center-To-Limb Variation of the rms Granular Contrast', *Solar Phys.* **53**, 359

Keil, S.L.: 1980, 'The Interpretation of Solar Line Shift Observations', *Astron. Astrophys.* **82**, 144

Keil, S.L.: 1980, 'The Structure of Solar Granulation. I: Observations of the Spatial and Temporal Behavior of Vertical Motions', *Astrophys. J.* **237**, 1024

Keil, S.L.: 1980, 'The Structure of Solar Granulation. II: Models of Vertical Motion', *Astrophys. J.* **237**, 1035

Keil, S.L.: 1984, 'Line Asymmetries of Partially Resolved Granular Profiles', in S.L. Keil (ed.), *Small Scale Dynamical Processes in Quiet Stellar Atmospheres*, National Solar Observatory/Sacramento Peak, Sunspot, p. 148

Keil, S.L. and Canfield, R.C.: 1978, 'The Height Variation of Velocity and Temperature Fluctuations in the Solar Photosphere', *Astron. Astrophys.* **70**, 169

Keil, S. and Mossman, A.: 1989, 'Observations of High Frequency Waves in the Solar Atmosphere', *In These Proceedings*, p. 333

Keil, S.L. and Yackovich, F.H.: 1981, 'Photospheric Line Asymmetry and Granular Velocity Models', *Solar Phys.* **69**, 213

Keil, S.L., Roudier, Th., Cambell, E., Koo, B.C., and Marmolino, C.: 1989, 'Observations and Interpretation of Photospheric Line Asymmetry Changes near Active Regions', *In These Proceedings*, p. 273

Keil, S.L., *see* Kandel, R.S. 1973

Keil, S.L., *see* Muller, R. 1983

Kelly, G., *see* Title, A.M. *et al.* 1987

Kerimbekov, M.B.: 1961, 'Untersuchung der Sonnengranulation mittels Kinematographischer Aufnahmem', *Sonnendaten 1960*, No. 9, p. 66

Kerimbekov, M.B.: 1966, 'Über die Beobachtungen der Sonnengranulation am Astrophysikalischen Observatorium Schemacha. I', *Sonnendaten 1966*, No. 2, p. 59

Kerimbekov, M.B.: 1968, 'Observations of the Solar Granulation at Shemakha Astrophysical Observatory. II', *Izv. AN Azerb. SSR. Ser. fiz. techn. i matem.*, No. 3, p. 3

Kerimbekov, M.B.: 1970, 'Observations of the Solar Granulation at Shemakha Astrophysical Observatory. III', *Soln. Dannye 1969 Byull.*, No. 12, p. 88

Kerimbekov, M.B.: 1971, 'On Observations of the Solar Granulation', *Soln. Dannye 1971 Byull.*, No. 4, p. 96

Kerimbekov, M.B.: 1980, 'About the Channels of the Waves on the Surface of the Sun', *Izv. Akad. Nauk Az. SSR Ser. Fiz.-Tekh. and Mat. Nauk*, No. 2, p. 75

Kerimbekov, M.B., Magerramov, V.A., and Ovchinnokova, M.I.: 1973, 'On the Investigation of the Solar Surface Structure. I', *Soln. Dannye 1973 Byull.*, No. 5, p. 84

Kerimbekov, M.B., Magerramov, V.A., and Ovchinnokova, M.I.: 1973, 'On the Investigation of the Solar Surface Structure. III', *Soln. Dannye 1973 Byull.*, No. 9, p. 71

Kinahan, B.F.: 1976, 'On the Power Spectrum of the Solar Granulation at High Wavenumbers', *Astrophys. J.* **209**, 282

Kippenhahn, R., *see* Biermann, L. *et al.* 1959

Kirk, J.G. and Livingston, W.C.: 1968, 'A Solar Granulation Spectrogram', *Solar Phys.* **3**, 510

Kitai, R. and Kawaguchi, I.: 1979, 'Morphological Study of the Solar Granulation. I: Dark Dot Formation in the Cell', *Solar Phys.* **64**, 3

Klinkmann, W., *see* Schlosser, W. 1974

von Klüber, H.: 1956, 'Granulation of the Solar Surface', *Observatory* **76**, 68

Kneer, F.J.: 1984, 'Interpretation of High Resolution Measurements, Selected Problems', in S.L. Keil (ed.), *Small Scale Dynamical Processes in Quiet Stellar Atmospheres*, National Solar Observatory/Sacramento Peak, Sunspot, p. 110

Kneer, F.J., and Wiehr, E.: 1989, 'The Gregory-Coudé-Telescope at the Observatorio Del Teide, Tenerife', *In These Proceedings*, p. 13

Kneer, F.J., Mattig, W., Nesis, A., and Werner, W.: 1980, 'Coherence Analysis of Granular Intensity', *Solar Phys.* **68**, 31

Kneer, F., *see* Wiehr, E. 1988

Kneer, F., *see* Holweger, H. 1989

Kneer, F.L., *see* Trujillo-Bueno, J. 1989

Kneer, F.J., *see* Durrant, C.J. *et al.* 1981

Knölker, M., *see* Schmidt, W. *et al.* 1979

Knölker, M., *see* Grossmann-Doerth, U. *et al.* 1989

Kokhan, E.K. and Krat, V.A.: 1981, 'On the Asymmetry in Fraunhofer Line Profiles', *Soviet Astron. Letters* **7**, 274

Kondo, M.-A. and Unno, W.: 1983, 'Convection and Gravity Waves in Two Layer Models. II: Overstable Convection of Large Horizontal Scales', *Geophys. Astrophys. Fluid Dyn.* **27**, 229

Kononovich, Eh.V.: 1973, 'Possible Interpretation of Solar Granulation Contrast', *Astron. Tsirk.*, No. 786, p. 8

Kononovich, Eh.V.: 1979, 'A Unified Interpretation of Photospheric and Facular Granules', *Soviet Astron. Letters* **5**, 50

Kononovich, Eh.V.: 1986, 'Solar Granulation Pattern as Magnetic Tube System', *Astron. Tsirk.*, No. 1419, p. 3

Kononovich, Eh.V. and Kupryakov, Yu.A.: 1978, 'Model of an Exploding Photospheric Granule', *Astron. Tsirk.*, No. 1026, p. 2

Koo, B.C., *see* Keil, S.L. *et al.* 1989

Kopecký, M. and Obridko, V.: 1968, 'On the Energy Release by Magnetic Field Dissipation in the Solar Atmosphere', *Solar Phys.* **5**, 354

Korduba, B.M., Musievskaya, M.E., Olijnyk, P.A., and Sobolev, V.M.: 1976, 'Determining Differences of Excitation Temperatures of a Granule-Intergranular Region', *Soln. Dannye 1976 Byull.*, No. 10, p. 79

Korduba, B.M., Musievskaya, M.E., Olijnyk, P.A., and Sobolev, V.M.: 1978, 'On Interpretation of the Equivalent Widths of Fraunhofer Lines of Solar Granulation Elements', *Tsirk. L'vov. Astron. Obs.*, No. 53, p. 21

Korn, J.: 1937, 'Zur Bestimmung der Mittleren Lebensdauer der Granulation', *Astron. Nachr.* **263**, 397

Kostik, R.I.: 1982, 'Sizes of Turbulence Elements in the Solar Atmosphere', *Soviet Astron.* **26**, 703

Kostik, R.I., *see* Atroshchenko, I.N. *et al.* 1989

Kostik, R.I., *see* Atroshchenko, I.N. *et al.* 1989

Kostjukevich, V.I., *see* Karpinsky, V.N. 1973

Kotljar, L.M., *see* Andreiko, A.V. *et al.* 1972

Koutchmy, S.: 1989, 'Granulation in and out of Magnetic Regions', *In These Proceedings*, p. 253

Koutchmy, S.: 1989, 'Photoelectric Analysis of the Solar Granulation in the Infrared', *In These Proceedings*, p. 347

Koutchmy, S. and Lebecq, C.: 1986, 'The Solar Granulation. II: Photographic and Photoelectric Analysis of Photospheric Intensity Fluctuations at the Mesogranulation Scale', *Astron. Astrophys.* **169**, 323

Koutchmy, S. and Legait, A.: 1980, 'The Solar Granulation. I: Two Dimensional Power Spectrum Analysis Using Optical Data Processing Methods', *Astron. Astrophys.* **88**, 345

Koutchmy, S., *see* Adjabshirzadeh, A. 1980

Koutchmy, S., *see* Adjabshirzadeh, A. 1982

Koval'chuk, M.M. and Olijnyk, P.A.: 1985, 'Application of a Two-Dimensional Model to the Interpretation of the Equivalent Widths of Absorption Lines of Solar Granulation', *Vestn. L'vov. univ. Ser. Astron.*, No. 59, p. 11

Kovshov, V.I.: 1977, 'Cellular Model of Solar Granulation', *Astron. Tsirk.*, No. 965, p. 3

Kovshov, V.I.: 1979, 'Equations of Stellar Convection. I: Ensemble Averaging', *Astrophysics* 15, 207

Koyama, S., *see* Namba, O. *et al.* 1983

Krsljanin, V.: 1989, 'On Pressure Shifts of Fe I Lines in Stellar Atmospheres', *In These Proceedings*, p. 91

Krat, V.A.: 1954, 'Untersuchung der Granulation der Sonnenphotosphäre. I', *Mitt. Astron. Hauptobs. Pulkovo* 19, No. 5, p. 1

Krat, V.A.: 1961, 'Motion of Solar Granules', *Mitt. Astr. Hauptobs. Pulkovo* 22, No. 4, p. 2

Krat, V.A.: 1961, 'Die Helligkeitsfluktuationen auf der Sonnenscheibe und die Helligkeit der Granulen' *Sonnendaten 1960*, No. 10, p. 69

Krat, V.A.: 1961, 'Über die Radialgeshcwindigkeiten der Photosphärischen Granulen' *Sonnendaten 1961*, No. 2, p. 5

Krat, V.A.: 1962, 'Granulation of the Solar Photosphere and Motion in the Solar Atmosphere', *Mitt. Astr. Hauptobs. Pulkovo* 22, No. 5, p. 2

Krat, V.A.: 1973, 'On the Structure of the Solar Photosphere', *Solar Phys.* 32, 307

Krat, V.A.: 1982, 'Granulation, Supergranulation and Atmospheric Waves', in W. Fricke and G. Teleki (eds.), *Sun and Planetary System*, D. Reidel Publ. Co., Dordrecht, p. 81

Krat, V.A. and Goldberg-Rogosinskaja, N.M.: 1956, 'Untersuchung der Granulation der Sonnenphotosphäre. II' *Mitt. Astron. Hauptobs. Pulkovo* 20, No. 2, p. 17

Krat, V.A. and Judina, I.W.: 1961, 'Photoelektrische Photometrie Photosphäre Granulen' *Sonnendaten 1960*, No. 9, p. 63

Krat, V.A., Makarov, V.I., and Tavastsherna, K.S.: 1980, 'On the Large Scale Brightness Fluctuations in the Solar Atmosphere', *Solar Phys.* 68, 237

Krat, V.A. and Pravdjuk, L.M.: 1960, 'Besonderheiten von auf der Erdoberfläche aufgenommenen Photographien der Sonnengranulation', *Sonnendaten 1959*, No. 11, p. 74

Krat, V.A. and Shpitalnaya, A.A.: 1974, 'On the Motion of Solar Plasma in Photospheric Granules', *Soln. Dannye 1974 Byull.*, No. 2, p. 63

Krat, V.A., *see* Kokhan, E.K. 1981

Kuhn, J.R., *see* Stein, R.F. *et al.* 1989

Kulčár, L.: 1983, 'Granulation and Supergranulation on the Sun as Observed from the OSO-8 Satellite', *Kozmos* 14, 82

Kuklin, G.V., *see* Vainstein, S.I. *et al.* 1977

Kupryakov, Yu.A., *see* Kononovich, Eh.V. 1978

Kusoffsky, U., *see* Darvann, T.A. 1989

LaBonte, B.J.: 1984, 'Recent Ground-Based Observations of the Global Properties of the Sun', in B.J. LaBonte, G.A. Chapman, H.S. Hudson, and R.C. Willson (eds.), *Solar Irradiance Variations on Active Region Time Scales*, NASA CP 2310, p. 151

Lago, G., *see* Aime, C. *et al.* 1978

Lambert, D.L. and Mallia, E.A.: 1968, 'Absolute Wavelengths of Fraunhofer Lines: Convective Motions in the Solar Photosphere and the Gravitational Red Shift', *Solar Phys.* 3, 499

Landman, D.A., *see* Lindsey, C.A. 1980

Langley, S.P.: 1874, 'On the Minute Structure of the Solar Photosphere', *Am. J. of Sci., Series* 3, 7, 87

Langley, S.P.: 1874, 'On the Structure of the Solar Photosphere', *Monthly Notices Roy. Astron. Soc.* **34**, 255

Larsson, B., *see* Dravins, D. *et al.* 1986

Laufer, J.: 1981, 'Räumliche Lokalisation Kurzperiodischer Schallwellenfelder in der Photosphäre der Sonne, Würzburg', *Thesis*

Lawrence, J.K.: 1983, 'The Spatial Distribution of Umbral Dots and Granules', *Solar Phys.* **87**, 1

Lebecq, C., *see* Koutchmy, S. 1986

Lebedinsky, A.I.: 1940, 'Über Beobachtungen der Granulation', *Ann. Leningrand State Univ., Astron. Series,* No. 53, p. 59

Ledoux, P., Schwarzschild, M., and Spiegel, E.A.: 1961, 'On the Spectrum of Turbulent Convection', *Astrophys. J.* **133**, 184

Legait, A.: 1986, 'Radiative Convection in a Stratified Atmosphere', *Astron. Astrophys.* **168**, 173

Legait, A., *see* Koutchmy, S. 1980

Leibacher, J.W., *see* Title, A.M. *et al.* 1987

Leifsen, T., *see* Andersen, B.N. *et al.* 1985

Leighton, R.B.: 1957, 'Some Observations of Solar Granulation', *Publ. Astron. Soc. Pacific* **69**, 497

Leighton, R.B.: 1963, 'The Solar Granulation', *Ann. Rev. Astron. Astrophys.* **1**, 19

Léna, P.J., Livingston, W.C., and Slaughter, C.: 1968, 'Wavelength Dependence of Solar Granulation - A Preliminary Report', Astrophys. J. Suppl. **73**, S67

Léna, P., *see* Turon, P.J. 1973

Lequeux, J.: 1958, 'L'Observation de la Granulation Solaire en France', *La Nature* **86**, 502

Lévy, M.: 1967, 'Analyse Photométrique Statistique de la Granulation', *Ann. Astrophys.* **30**, 887

Lévy, M.: 1971, 'Analyse Photométrique Statistique de la Granulation Corrigée de l'Influence de l'Atmosphère Terrestre et de l'Instrument', *Astron. Astrophys.* **14**, 15

Lévy, M.: 1974, 'Temperature Fluctuations in Solar Granulation', *Astron. Astrophys.* **31**, 451

Lilje, P.B., *see* Andersen, B.N. *et al.* 1985

Lind, J., *see* Dravins, D. 1984

Lindegren, L., *see* Dravins, D. *et al.* 1981

Lindgren, R., *see* Title, A.M. *et al.* 1987

Lindsey, C.A. and Hudson, H.S.: 1976, 'Solar Limb Brightening in Submillimeter Wavelengths', *Astrophys. J.* **203**, 753

Lindsey, C.A. and Landman, D.A.: 1980, 'Effects of Granular Convection in the Response of C I 5380Å to Solar Luminosity Variations', *Astrophys. J.* **237**, 999

Lites, B.W.: 1983, 'An Estimation of the Fluctuations in the Extreme Limb of the Sun', *Solar Phys.* **85**, 193

Lites, B.W., Nordlund, Å., and Scharmer, G.B.: 1989, 'Constraints Imposed by Very High Resolution Spectra and Images on Theoretical Simulations of Granular Convection', *In These Proceedings*, p. 349

Lites, B.W., *see* Bruner, E.C. *et al.* 1976

Livingston, W.C.: 1983, 'Magnetic Fields and Convection: New Observations', in J.O. Stenflo (ed.), *Solar and Magnetic Fields: Origins and Coronal Effects, IAU Symp. 102*, D. Reidel Publ. Co., Dordrecht, p. 149

Livingston, W.C.: 1984, 'Line Asymmetry and Magnetic Fields', *Kodaikanal Obs. Bull.* **4**, 7

Livingston, W.C.: 1984, 'Secular Change of Full Disk Line Asymmetry', in S.L. Keil (ed.), *Small Scale Dynamical Processes in Quiet Stellar Atmospheres*, National Solar Observatory/Sacramento Peak, Sunspot, p. 330

Livingston, W.C.: 1987, 'Line Asymmetry and the Activity Cycle', in E.H. Schröter, M. Vázquez, and A.A. Wyller (eds.), *The Role of Fine-Scale Magnetic Fields on the Structure of the Solar Atmosphere*, Cambridge University Press, Cambridge, p. 14

Livingston, W.C., *see* Kirk, J.G. 1968

Livingston, W.C., *see* Lena, P.J. *et al.* 1968

Livingston, W.C., *see* Title, A.M. *et al.* 1987

Locans, V., *see* Zhugzhda, Y.D. *et al.* 1989

LoPresto, J.C. and Pierce, A.K.: 1985, 'The Center to Limb Shift of a Number of Fraunhofer Lines', *Solar Phys.* **102**, 21

Lou, Y.Q.: 1989, 'Isolated Magnetohydrodynamic Thermal Eddies in a Thermally Stratified Fluid', *In These Proceedings*, p. 583

Loughhead, R.E. and Bray, R.J.: 1959, '"Turbulence" and the Photospheric Granulation' *Nature* **183**, 240

Loughhead, R.E. and Bray, R.J.: 1960, 'Granulation near the Extreme Solar Limb', *Australian J. Phys.* **13**, 738

Loughhead, R.E. and Bray, R.J.: 1975, 'Visibility of the Photospheric Granulation in Fe I λ 6569.2', *Solar Phys.* **45**, 35

Loughhead, R.E., *see* Bray, R.J. 1958

Loughhead, R.E., *see* Bray, R.J. 1959

Loughhead, R.E., *see* Bray, R.J. 1961

Loughhead, R.E., *see* Bray, R.J. 1974

Loughhead, R.E., *see* Bray, R.J. 1977

Loughhead, R.E., *see* Bray, R.J. 1978

Loughhead, R.E., *see* Bray, R.J. 1984

Loughhead, R.E., *see* Bray, R.J. *et al.* 1974

Loughhead, R.E., *see* Bray, R.J. *et al.* 1976

Loughhead, R.E., *see* Bray, R.J. *et al.* 1984

von der Lühe, O.: 1981, 'A Comparison of Optical and Digital Fourier Transformation of Solar Granulation', *Astron. Astrophys.* **101**, 277

von der Lühe, O.: 1983, 'A Study of a Correlation Tracking Method to Improve Imaging Quality of Ground Based Solar Telescopes', *Astron. Astrophys.* **119**, 85

von der Lühe, O.: 1984, 'Estimating Fried's Parameter from a Time Series of an Arbitrary Resolved Object Imaged through Atmospheric Turbulence', *J. Opt. Soc. Am. A.***1**, 510

von der Lühe, O.: 1985, 'High Resolution Speckle Imaging of Solar Small-Scale Structure: The Influence of Anisoplanatism', in R. Muller (ed.), *High Resolution in Solar Physics*, Proc. 8th IAU European Regional Meeting, *Lecture Notes in Physics* **233**, Springer-Verlag, Berlin, p. 96

von der Lühe, O.: 1987, 'Calibration Problems in Solar Speckle Interferometry', in J.W. Goad (ed.), *Interferometric Imaging in Astronomy*, European Southern Observatory, Garching, and National Optical Astronomy Observatories, Tucson, p. 9

von der Lühe, O. and Dunn, R.B.: 1987, 'Solar Granulation Power Spectra from Speckle Interferometry', *Astron. Astrophys.***177**, 265

Lüst, R., *see* Biermann, L. *et al.* 1959

Lynch, D.K. and Chapman, G.A.: 1975, 'Solar Granulation and Oscillations as Spatially Random Processes', *Astrophys. J.* **197**, 241

Macris, C.J.: 1949, 'Sur les Dimensions des Granules Photosphériques', *Compt. Rend. Acad. Sci. Paris* **228**, 1792

Macris, C.J.: 1949, 'Sur les Dimensions des Granules Photosphériques au Voisinage des Taches Solaires', *Compt. Rend. Acad. Sci. Paris* **228**, 2010

Macris, C.J.: 1949, 'Sur les Déplacements Latéraux des Granules Photosphériques', *Compt. Rend. Acad. Sci. Paris* **229**, 112

Macris, C.J.: 1952, 'Sur la Vie Moyenne des Granules Faculaires et des Plages Lumineuses dans la Pénombre des Taches', *Compt. Rend. Acad. Sci. Paris* **235**, 868

Macris, C.J.: 1953, 'Recherches sur la Granulation Photosphérique', *Ann. Astrophys.* **16**, 19

Macris, C.: 1955, 'The Possible Variation of Photospheric Granules', *Observatory* **75**, 122

Macris, C.: 1959, 'The Dimensions of the Photospheric Granules in Various Spectral Regions', *Observatory* **79**, 22

Macris, C.J.: 1978, 'Sur la Variation des Dimensions des Granules Photosphériques au Voisinage des Taches Solaires', *Compt. Rend. Acad. Sci. Paris Ser. B* **286**, 315

Macris, C.J.: 1979, 'The Variation of the Mean Diameters of the Photospheric Granules near the Sunspots', *Astron. Astrophys.* **78**, 186

Macris, C.J.: 1988, 'The Influence of the Variability of Photospheric Granulation on the Radio Flux of the Quiet Sun in the Centimetre Wave Band', *Compt. Rend. Acad. Sci. II, Mec. Phys. Chem. Sci.* **306**, 707

Macris, C.J. and Banos, G.J.: 1961, 'Mean Distance between Photospheric Granules and Its Change with the Solar Activity', *Mem. Natl. Obs. Athens, Series 1*, No. 8, p. 1

Macris, C.J. and Elias, D.: 1955, 'Sur une Variation du Nombre des Granules Photosphériques en Fonction de l'Activité Solaire', *Ann. Astrophys.* **18**, 143

Macris, C.J. and Prokakis, T.J.: 1962, 'Sur une Différence des Dimensions des Granules Photosphériques au Voisinage et Loin de la Pénombre des Taches Solaires', *Compt. Rend. Acad. Sci. Paris* **255**, 1862

Macris, C.J. and Prokakis, T.J.: 1963, 'New Results on the Lifetime of the Solar Granules', *Mem. Natl. Obs. Athens, Series 1*, No. 10, p. 1

Macris, C.J. and Rösch, J.: 1983, 'Variations des Mailles du Réseau de la Granulation Photosphérique en Fonction de l'Activité Solaire', *Compt. Rend. Acad. Sci. Paris Ser. II* **296**, 265

Macris, C.J., Muller, R., Rösch, J., and Roudier, T.: 1984, 'Variation of the Mesh of the Granular Network along the Solar Cycle', in S.L. Keil (ed.), *Small Scale Dynamical Processes in Quiet Stellar Atmospheres*, National Solar Observatory/Sacramento Peak, Sunspot, p. 265

Macris, C., Prokakis, T., Dialetis, D., and Muller, R.: 1989, 'The Variation of the Mean Size of the Photospheric Granules Near and Far from a Sunspot', *In These Proceedings*, p. 359

Macris, C., *see* Dialetis, D. *et al.* 1985

Macris, C., *see* Dialetis, D. *et al.* 1985

Macris, C., *see* Dialetis, D. *et al.* 1986

Macris, C., *see* Dialetis, D. *et al.* 1989

Macris, C., *see* Dialetis, D. *et al.* 1988

Macris, C.J., *see* Alissandrakis, C.E. *et al.* 1982

Magerramov, V.A.: 1982, 'Investigation of the Gradient of Solar Granulation Fields', *Soobshch. Shemakh. Astrofiz. Obs.* **8**, 178

Magerramov, V.A., *see* Kerimbekov, M.B. *et al.* 1973

Magerramov, V.A., *see* Kerimbekov, M.B. *et al.* 1973

Maillard, J.P., *see* Nadeau, D. *et al.* 1989

Makarov, V.I., *see* Krat, V.A. *et al.* 1980

Maksimov, V.P., *see* Vainstein, S.I. *et al.* 1977

Mallia, E.A., *see* Lambert, D.L. 1968

Maluck, G., *see* Durrant, C.J. *et al.* 1981

Marco, E., *see* Collados, M. *et al.* 1986

Margrave, T.E., Jr.: 1968, 'Review of Visual Observations of Solar Granulation', *J. Washington Acad. Sci.* **58**, 26

Margrave, T.E., Jr. and Swihart, T.L.: 1969, 'Inhomogeneities in the Solar Photosphere', *Solar Phys.* **6**, 12

Margrave, T.E., Jr. and Swihart, T.L.: 1969, 'More on Granulation Models', *Solar Phys.* **9**, 315

Marmolino, C. and Severino, G.: 1981, 'The Third Central Moment of Photospheric Lines as a Measure of Velocity Gradients and Line Shifts', *Astron. Astrophys.* **100**, 191

Marmolino, C., Roberti, G., and Severino, G.: 1985, 'Line Asymmetries and Shifts in Presence of Granulation and Oscillations: The CLV of the K I 7699 Resonance Line', in H.U. Schmidt (ed.), *Theoretical Problems in High Resolution Solar Physics*, Max Planck Institut Publ. MPA **212**, p. 89

Marmolino, C., Roberti, G., and Severino, G.: 1987, 'Line Asymmetries and Shifts in the Presence of Granulation and Oscillations: The CLV of the K I 7699 Resonance Line', *Solar Phys.* **108**, 21

Marmolino, C., Roberti, G., and Severino, G.: 1988, 'Fe II Lines in the Presence of Photospheric Oscillations', in R. Viotti, A. Vittone, and M. Friedjung (eds.), *Physics of Formation of Fe II Lines Outside LTE, IAU Colloq. 94*, D. Reidel Publ. Co., Dordrecht, p. 217

Marmolino, C., *see* Caccin, B. *et al.* 1981

Marmolino, C., *see* Gomez, M.T. *et al.* 1987

Marmolino, C., *see* Keil, S.L. *et al.* 1989

Márquez, I., *see* Bonet, J.A. *et al.* 1988

Márquez, I., *see* Bonet, J.A. *et al.* 1988

Martic, M. and Damé, L.: 1989, 'High Resolution Diagnostic of the Mesocells in the Solar Temperature Minimum Region', *In These Proceedings*, p. 207

Martic, M., *see* Damé, L. 1987

Martic, M., *see* Damé, L. 1988

Martin, F. and Aime, C.: 1979, 'Some Spectral Properties of Random Arrays of Grains (Solar Granulations)', *J. Opt. Soc. Am.* **69**, 1315

Martin, F., *see* Aime, C. *et al.* 1977

Martin, F., *see* Aime, C. *et al.* 1979

Martin, F., *see* Aime, C. *et al.* 1981

Martin, F., *see* Aime, C. *et al.* 1985

Martin, F., *see* Aime, C. *et al.* 1989

Massaglia, S. *see* Kalkofen, W. *et al.* 1989

Mattig, W.: 1980, 'The Mean Vertical Velocity of Solar Granulation', *Astron. Astrophys.* **83**, 129

Mattig, W.: 1985, 'The High Resolution Structure of the Sun', in R. Muller (ed.), *High Resolution in Solar Physics*, Proc. 8th IAU European Regional Meeting, *Lecture Notes in Physics* **233**, Springer-Verlag, Berlin, p. 141

Mattig, W. and Nesis, A.: 1974, 'Studies of Granular Velocities. IV: Statistical Analysis of Granular Doppler Shifts', *Solar Phys.* **36**, 3

Mattig, W. and Nesis, A.: 1974, 'Studies of Granular Velocities. VI: Changes in the Granular Velocity Field around Sunspots', *Solar Phys.* **38**, 337

Mattig, W. and Nesis, A.: 1976, 'Studies of Granular Velocities. VII: Granular Velocities around Sunspots', *Solar Phys.* **50**, 255

Mattig, W. and Schlebbe, H.: 1974, 'Studies of Granular Velocities. V: The Height Dependence of the Granular Doppler Shifts', *Solar Phys.* **34**, 299

Mattig, W., Hanslmeier, A., and Nesis, A.: 1989, 'Granulation Line Asymmetries', *In These Proceedings*, p. 187

Mattig, W., Mehltretter, J.P., and Nesis, A.: 1969, 'Studies of Granular Velocities. I: Granular Doppler Shifts and Convective Motion', *Solar Phys.* **10**, 254

Mattig, W., Mehltretter, J.P., and Nesis, A.: 1981, 'Granular Size Horizontal Velocities in the Solar Atmosphere', *Astron. Astrophys.* **96**, 96

Mattig, W., Nesis, A., Reiss, G.: 1977, 'Korrektur von Gemessenen Granularen Geschwindig Keiten aus Sonnenfinsternisbeobachtungen', *Mitt. Astron. Ger.*, No. 42, p. 114

Mattig, W., *see* Deubner, F.-L. 1983

Mattig, W., *see* Durrant, C.J. *et al.* 1979

Mattig, W., *see* Durrant, C.J. *et al.* 1983

Mattig, W., *see* Kneer, F.J. *et al.* 1980

Mattig, W., *see* Nesis, A. *et al.* 1984

Mattig, W., *see* Nesis, A. *et al.* 1987

Mattig, W., *see* Nesis, A. *et al.* 1988

Mattig, W., *see* Nesis, A. *et al.* 1989

Mattig, W., *see* Ricort, G. *et al.* 1981

Mattig, W., *see* Schmidt, W. *et al.* 1979

McCullough, J.R., *see* Edmonds, F.N., Jr. 1986

McNally, D.: 1964, 'The Sun - Granulation', *Sci. Progr.* **52**, 84

McMath, R.R., Mohler, O.C., and Pierce, A.K.: 1955, 'Doppler Shifts in Solar Granules', *Astrophys. J.* **122**, 565

Mehltretter, J.P.: 1971, 'Studies of Granular Velocities. II: Statistical Analysis of Two High Resolution Spectrograms', *Solar Phys.* **16**, 253

Mehltretter, J.P.: 1971, 'On the rms Intensity Fluctuation of Solar Granulation', *Solar Phys.* **19**, 32

Mehltretter, J.P.: 1973, 'Studies of Granular Velocities. III: The Influence of Finite Spatial and Spectral Resolution upon the Measurement of Granular Doppler Shifts', *Solar Phys.* **30**, 19

Mehltretter, J.P.: 1976, 'Die Zeitliche Entwicklung der Granulation', *Mitt. Astron. Ges.*, No. 40, p. 162

Mehltretter, J.P.: 1977, 'Über die Lebensdaurer der Photosphärischen Granulation', *Mitt. Astron. Ges.*, No. 42, p. 113

Mehltretter, J.P.: 1978, 'Was wissen wir von der Granulation?', *Sterne Weltraum, Jahrg.* **17**, 207

Mehltretter, J.P.: 1978, 'Balloon-Borne Imagery of the Solar Granulation. II: The Lifetime of Solar Granulation', *Astron. Astrophys.* **62**, 311

Mehltretter, J.P., *see* Canfield, R.C. 1973

Mehltretter, J.P., *see* Wittmann, A. 1977

Mehltretter, J.P., *see* Wittmann, A. 1977

Mehltretter, J.P., *see* Mattig, W. *et al.* 1969

Mehltretter, J.P., *see* Mattig, W. *et al.* 1981

Mehltretter, J.P., *see* Schmidt, W. *et al.* 1979

Mein, P.: 1968, Étude Spatio-Temporelle de la Granulation Solaire. Amélioration des Images Déformées par l'Agitation Atmosphérique', *Ann. Astrophys.* **31**, 115

Mein, P.: 1976, 'The Photosphere', *Trans. IAU* **16A**, 55

Mekhanikov, V.V., *see* Karpinsky, V.N. 1977

Mena, B., *see* Muller, R. 1987

Meyer, F. and Schmidt, H.U.: 1967, 'Die Erzeugung von Schwingungen in der Sonnenatmosphäre durch einzelne Granula', *Z. Astrophys.* **65**, 274

Michard, R., *see* Bernière, G. *et al.* 1962

Mitchell, W.E., Jr.: 1982, 'Large Scale Brightness Inhomogeneities in the Solar Atmosphere', *Solar Phys.* **80**, 3

Mohler, O.C., *see* Goldberg, L. *et al.* 1957

Mohler, O.C., *see* Goldberg, L. *et al.* 1960

Mohler, O.C., *see* McMath, R.R. *et al.* 1955

Moore, D.R.: 1979, 'Can Granulation be Understood as Convective Overshoot?', in F.-L. Deubner, C.J. Durrant, M. Stix, and E.H. Schröter (eds.), *Small Scale Motions on the Sun*, Keipenheuer Inst. No. 179, Freiburg, p. 55

Moore, D.R.: 1981, 'Penetrative Convection', in S. Jordan (ed.), *The Sun as a Star*, NASA SP-450, p. 253

Moore, D.R., *see* Galloway, D.J. 1979

Moore, D.R., *see* Jones, C.A. 1979

Morrill, M., *see* Title, A.M. *et al.* 1987

Morrison, R.A., *see* Beckers, J.M. 1970

Mossman, A., *see* Keil, S. 1989

Mouradian, Z., *see* Fang, C. *et al.*1984

Muchmore, D., *see* Cuntz, M. 1989

Muchmore, D., *see* Steffen, M. 1988

Müller, H., *see* Bruggencate, P.T. 1942

Muller, R.: 1977, 'Morphological Properties and Origin of the Photospheric Facular Granules', *Solar Phys.* **52**, 249

Muller, R.: 1983, 'Variation of the Distribution of Magnetic Flux Tubes in the Quiet Photosphere along the Solar Cycle', in J.-C. Pecker and Y. Uchida (eds.), *Proc. Japan-France Seminar on Active Phenomena in the Outer Atmosphere of the Sun and Stars*, CNRS, Paris, p. 382

Muller, R.: 1985, 'The Fine Structure of the Quiet Sun', *Solar Phys.* **100**, 237

Muller, R.: 1985, 'Solar Granulation', *Trans. IAU* **19A**, 103

Muller, R.: 1989, 'Observations of the Solar Granulation at the Pic du Midi Observatory and with THEMIS', *In These Proceedings*, p. 9

Muller, R.: 1989, 'Solar Granulation: Overview', *In These Proceedings*, p. 101

Muller, R.: 1989, 'Abnormal Granulation', *In These Proceedings*, p. 365

Muller, R. and Keil, S.L.: 1983, 'The Characteristic Size and Brightness of Facular Points in the Quiet Photosphere', *Solar Phys.* **87**, 243

Muller, R. and Mena, B.: 1987, 'Motions around a Decaying Sunspot', *Solar Phys.* **112**, 295

Muller, R. and Roudier, T.: 1984, 'Variability of the Quiet Photospheric Network', *Solar Phys.* **94**, 33

Muller, R. and Roudier, T.: 1984, 'Variability of the Quiet Photospheric Network', in T.D. Guyenne and J.J. Hunt (eds.), *The Hydromagnetics of the Sun*, Proc. Fourth European Meeting on Solar Physics, ESA SP-220, p. 239

Muller, R. and Roudier, T.: 1984, 'Variability of the Structure of the Granulation Over the Solar Activity Cycle', in T.D. Guyenne and J.J. Hunt (eds.), *The Hydromagnetics of the Sun*, Proc. Fourth European Meeting on Solar Physics, ESA SP-220, p. 51

Muller, R. and Roudier, T.: 1985, 'The Structure of the Solar Granulation', in R. Muller (ed.), *High Resolution in Solar Physics*, Proc. 8th IAU European Regional Meeting, *Lecture Notes in Physics* 233, Springer-Verlag, Berlin, p. 242

Muller, R. and Roudier, T.: 1985, 'Variability of the Quiet Photospheric Network', in R. Muller (ed.), *High Resolution in Solar Physics*, Proc. 8th IAU European Regional Meeting, *Lecture Notes in Physics* 233, Springer-Verlag, Berlin, p. 277

Muller, R., Roudier, T., and Vigneau, J.: 1989, 'Supergranular Pattern Formed by the Large Granules', *In These Proceedings*, p. 367

Muller, R., *see* Roudier, T. 1986

Muller, R., *see* Andreassen, O. *et al.* 1985

Muller, R., *see* Dialetis, D. *et al.* 198ᐢ

Muller, R., *see* Dialetis, D. *et al.* 1988

Muller, R., *see* Macris, C.J. *et al.* 1984

Muller, R., *see* Macris, C. *et al.* 1989

Muller, R., *see* Roudier, T. *et al.* 1985

Münzer, H., Hanslmeier, A., Schröter, E.H., and Wöhl, H.: 1989, 'Pole-Equator Difference of the Size of the Chromospheric CaII-K Network in the Quiet and Active Solar Regions', *In These Proceedings*, p. 217

Murphy, J.O., *see* van der Borght, R. 1973

Musievskaya, M.E., *see* Korduba, B.M. *et al.* 1976

Musievskaya, M.E., *see* Korduba, B.M. *et al.* 1978

Musievskaya, M.E., *see* Olijnyk, P.A. *et al.* 1979

Musman, S.: 1969, 'The Effect of Finite Resolution on Solar Granulation', *Solar Phys.* 7, 178

Musman, S.: 1972, 'A Mechanism for the Exploding Granule Phenomenon', *Solar Phys.* 26, 290

Musman, S.: 1974, 'The Origin of the Solar Five-Minute Oscillation', *Solar Phys.* 36, 313

Musman, S. and Nelson, G.D.: 1976, 'The Energy Balance of Granulation', *Astrophys. J.* 207, 981

Musman, S., *see* Allen, M.S. 1973

Musman, S., *see* Altrock, R.C. 1976

Musman, S., *see* Nelson, G.D. 1977

Musman, S., *see* Nelson, G.D. 1977

Musman, S., *see* Parvey, M.I. 1971

Musman, S., *see* Altrock, R.C. *et al.* 1983

Nadeau, D., Bédard, J., and Maillard, J.P.: 1989, 'Line Shifts in the Infrared Spectra of Late-Type Stars', *In These Proceedings*, p. 125

Nakagawa, Y. and Priest, E.R.: 1972, 'A Possible New Interpretation of Power Spectra of Solar Granulation Brightness Fluctuations', *Astrophys. J.* 178, 251

Nakagomi, Y., *see* Suemoto, Z. *et al.* 1987

Namba, O.: 1986, 'Evolution of "Exploding Granules"', *Astron. Astrophys.* 161, 31

Namba, O. and Diemel, W.E.: 1969, 'A Morphological Study of the Solar Granulation', *Solar Phys.* 7, 167

Namba, O. and van Rijsbergen, R.: 1977, 'Evolution Pattern of the Exploding Granules', in E.A. Spiegel and J.P. Zahn (eds.), *Problems of Stellar Convection, IAU Colloq. 38*, Springer-Verlag, Berlin, p. 119

Namba, O., Hafkenscheid, G.A.M., and Koyama, S.: 1983, 'Profiles and Shifts of the C I 5052 Å Line in the Granulation Spectrum', *Astron. Astrophys.* **117**, 277

Narasimha, D., Pandey, S.K., and Chitre, S.M.: 1980, 'Convective Instability in the Solar Envelope', *J. Astrophys. Astron.* **1**, 165

Narasimha, D., *see* Antia, H.M. *et al.* 1982

Narasimha, D., *see* Antia, H.M. *et al.* 1983

Nasmyth, J.: 1862, 'On the Structure of the Luminous Envelope of the Sun', *Mem. Lit. Phil. Soc. of Manchester, Series 3*, **1**, 407

Nelson, G.D.: 1978, 'A Two-Dimensional Solar Model', *Solar Phys.* **60**, 5

Nelson, G.D.: 1980, 'Granulation in a Main Sequence F-Type Star', *Astrophys. J.* **238**, 659

Nelson, G.D. and Musman, S.: 1977, 'A Dynamical Model of Solar Granulation', *Astrophys. J.* **214**, 912

Nelson, G.D. and Musman, S.: 1978, 'The Scale of Solar Granulation', *Astrophys. J. Letters* **222**, L69

Nelson, G.D., *see* Beckers, J.M. 1978

Nelson, G.D., *see* Musman, S. 1976

Nesis, A.: 1977, 'Granulation Observations', in E.A. Spiegel and J.P. Zahn (eds.), *Problems of Stellar Convection, IAU Colloq. 38*, Springer-Verlag, Berlin, p. 126

Nesis, A.: 1979, 'Convection and the Solar Granulation', in U. Muller, K.G. Roesner, and B. Schmidt (eds.), *Recent Developments in Theoretical and Experimental Fluid Mechanics. Compressible and Incompressible Flows*, Springer-Verlag, Berlin, p. 38

Nesis, A.: 1984, 'A Model for the Dynamics of the Deep Photosphere', in T.D. Guyenne and J.J. Hunt (eds.), *The Hydromagnetics of the Sun*, Proc. Fourth European Meeting on Solar Physics, ESA SP-220, p. 203

Nesis, A.: 1985, 'A Model for the Run of the Horizontal and Vertical Velocities in the Deep Photosphere', in R. Muller (ed.), *High Resolution in Solar Physics*, Proc. 8th IAU European Regional Meeting, *Lecture Notes in Physics* **233**, Springer-Verlag, Berlin, p. 249

Nesis, A.: 1987, 'Overshoot of the Solar Granulation', in E.H. Schröter, M. Vázquez, and A.A. Wyller (eds.), *The Role of Fine-Scale Magnetic Fields on the Structure of the Solar Atmosphere*, Cambridge University Press, Cambridge, p. 322

Nesis, A.: 1988, 'Convective Overshoot and Upper Boundary Conditions', in J. Christensen-Dalsgaard, and S. Frandsen (eds.) *Advances in Helio- and Asteroseismology, IAU Symp. 123*, D. Reidel Publ. Co., Dordrecht, p. 443

Nesis, A. and Severino, G.: 1987, 'Velocity Variations of Small Scale Solar Structures, and Physical Problems Related to the Overshoot Layer', in J.L. Linsky and R.E. Stencel (eds.), *Cool Stars, Stellar Systems and the Sun*, Springer-Verlag, Berlin, p. 154

Nesis, A., Durrant, C.J., and Mattig, W.: 1984, 'Studies of Overshoot', in S.L. Keil (ed.), *Small Scale Dynamical Processes in Quiet Stellar Atmospheres*, National Solar Observatory/Sacramento Peak, Sunspot, p. 243

Nesis, A., Durrant, C.J., and Mattig, W.: 1988, 'Overshoot of Horizontal and Vertical Velocities in the Deep Solar Photosphere', *Astron. Astrophys.* **201**, 153

Nesis, A., Fleig, K.H., and Mattig, W.: 1989, 'RMS Velocities in Solar Active Regions', *In These Proceedings*, p. 289

Nesis, A., Mattig, W., Fleig, K.H., and Wiehr, E.: 1987, 'The Gradient of the Small Scale Velocity Fluctuation in the Solar Atmosphere', *Astron. Astrophys.* **182**, L5

Nesis, A., *see* Durrant, C.J. 1981

Nesis, A., *see* Durrant, C.J. 1982

Nesis, A., *see* Mattig, W. 1974

Nesis, A., *see* Mattig, W. 1974

Nesis, A., *see* Mattig, W. 1976

Nesis, A., *see* Durrant, C.J. *et al.* 1979

Nesis, A., *see* Durrant, C.J. *et al.* 1983

Nesis, A., *see* Kneer, F.J. *et al.* 1980

Nesis, A., *see* Mattig, W. *et al.* 1969

Nesis, A., *see* Mattig, W. *et al.* 1977

Nesis, A., *see* Mattig, W. *et al.* 1981

Nesis, A., *see* Mattig, W. *et al.* 1989

Nordlund, Å.: 1976, 'A Two-Component Representation of Stellar Atmospheres with Convection', *Astron. Astrophys.* **50**, 23

Nordlund, Å.: 1977, 'Convective Overshooting in the Solar Photosphere: A Model Granular Velocity Field', in E.A. Spiegel and J.P. Zahn (eds.), *Problems of Stellar Convection, IAU Colloq. 38*, Springer-Verlag, Berlin, p. 237

Nordlund, Å.: 1978, 'Solar Granulation and the Nature of "Microturbulence"', in A. Reiz and T. Andersen (eds.), *Astronomical Papers Dedicated to Bengt Strömgren*, Copenhagen University Observatory, Copenhagen, p. 95

Nordlund, Å.: 1980, 'Numerical Simulation of Granular Convection: Effects on Photospheric Spectral Line Profiles', in D.F. Gray and J.L. Linsky (eds.), *Stellar Turbulence, IAU Colloq. 51*, Springer-Verlag, Berlin, p. 213

Nordlund, Å.: 1982, 'Numerical Simulations of the Solar Granulation. I: Basic Equations and Methods', *Astron. Astrophys.* **107**, 1

Nordlund, Å.: 1983, 'Numerical 3-D Simulations of the Collapse of Photospheric Flux Tubes', in J.O. Stenflo (ed.), *Solar and Magnetic Fields: Origins and Coronal Effects, IAU Symp. 102*, D. Reidel Publ. Co., Dordrecht, p. 79

Nordlund, Å.: 1984, 'A Reevaluation of the Granular ΔI_{rms}', in S.L. Keil (ed.), *Small Scale Dynamical Processes in Quiet Stellar Atmospheres*, National Solar Observatory/Sacramento Peak, Sunspot, p. 174

Nordlund, Å.: 1984, 'The 3-D Structure of the Magnetic Field and Its Interaction with Granulation', in S.L. Keil (ed.), *Small Scale Dynamical Processes in Quiet Stellar Atmospheres*, National Solar Observatory/Sacramento Peak, Sunspot, p. 181

Nordlund, Å.: 1985, 'The Dynamics of Granulation and Its Interaction with the Radiation Field', in H.U. Schmidt (ed.), *Theoretical Problems in High Resolution Solar Physics*, Max Planck Institut Publ. MPA **212**, p. 1

Nordlund, Å.: 1985, 'Modelling of Small-Scale Dynamical Processes: Convection and Wave Generation', in H.U. Schmidt (ed.), *Theoretical Problems in High Resolution Solar Physics*, Max Planck Institut Publ. MPA **212**, p. 101

Nordlund, Å: 1985 , 'Solar Convection', *Solar Phys.* **100**, 209

Nordlund, Å.: 1986, '3-D Model Calculations', in W. Deinzer, M. Knölker, and H.H. Voigt (eds.), *Small Scale Magnetic Flux Concentrations in the Solar Photosphere*, Vanderhoeck and Ruprecht, Göttingen, p. 83

Nordlund, Å. and Stein, R.F.: 1989, 'Simulating Magnetoconvection', *In These Proceedings*, p. 453

Nordlund, Å., *see* Wöhl, H. 1985

Nordlund, Å., *see* Dravins, D. *et al.* 1981

Nordlund, Å., *see* Dravins, D. *et al.* 1986

Nordlund, Å., *see* Lites, B.W. *et al.* 1989

Nordlund, Å., *see* Stein, R.F. *et al.* 1989

650

November, L.J.: 1979, 'Mesogranulation and Supergranulation in the Sun', *Thesis*

November, L.J. and Simon, G.W.: 1988, 'Precise Proper-Motion Measurement of Solar Granulation', *Astrophys. J.* **333**, 427

November, L.J., Simon, G.W., Tarbell, T.D., Title, A.M., and Ferguson, S.H.: 1987, 'Large-Scale Horizontal Flows from SOUP Observations of Solar Granulation', in R.G. Athay and D.S. Spicer (eds.), *Theoretical Problems in High Resolution Solar Physics II*, NASA CP2483, p. 121

November, L.J., Toomre, J., Gebbie, K.B., and Simon, G.W.: 1981, 'The Detection of Mesogranulation on the Sun', *Astrophys. J. Letters* **245**, L123

November, L.J., Toomre, J., Gebbie, K.B., and Simon, G.W.: 1982, 'Vertical Flows of Supergranular and Mesogranular Scale Observed on the Sun with OSO 8', *Astrophys. J.* **258**, 846

November, L., *see* Dunn, R.B. 1985

November, L.J., *see* Dunn, R.B. 1985

November, L.J., *see* Hill, F. *et al.* 1984

November, L.J., *see* Simon, G.W. *et al.* 1989

November, L.J., *see* Title, A.M. *et al.* 1987

Noyes, R.W.: 1967, 'Observational Studies of Velocity Fields in the Solar Photosphere and Chromosphere', in R.N. Thomas (ed.), *Aerodynamic Phenomena in Stellar Atmospheres*, *IAU Symp. 28*, Academic Press, New York, p. 293

Noyes, R.W., *see* Reeves, E.M. *et al.* 1974

Obridko, V., *see* Kopecký, M. 1968

Oda, N.: 1984, 'Morphological Study of the Solar Granulation. III: The Mesogranulation', *Solar Phys.* **93**, 243

Olijnyk, P.A., Musievskaya, M.E., and Sobolev, V.M.: 1979, 'Study of Absorption Line Equivalent Widths of Solar Granulation Elements', in S.K. Vsekhsvyatskij (ed.), *Problems of Cosmic Physics. Vypusk 14*, Respublikanskij Mezhvedomstevennyj Nauchnyj Sbornik. Izdatel'stvo pri Kievskom Gosudarstvennom Universitete Izdatel'skogo Obedineniy "Vishcha Shkolo", Kiev, p. 113

Olijnyk, P.A., *see* Korduba, B.M. *et al.* 1976

Olijnyk, P.A., *see* Korduba, B.M. *et al.* 1978

Olijnyk, P.A., *see* Koval'chuk, M.M. 1985

Olson, E.C.: 1962, 'Observations of Solar Line Asymmetries', *Astrophys. J.* **136**, 946

O'Mara, B.J., *see* Kavetsky, A. 1984

Osherovich, V.A.: 1985, 'Quasi-Potential Magnetic Fields in Stellar Atmospheres. I: Static Model of Magnetic Granulation', *Astrophys. J.* **298**, 235

Ovchinnokova, M.I., *see* Kerimbekov, M.B. *et al.* 1973

Ovchinnokova, M.I., *see* Kerimbekov, M.B. *et al.* 1973

Pandey, S.K., *see* Antia, H.M. *et al.* 1981

Pandey, S.K., *see* Narasimha, D. *et al.* 1980

Parfinenko, L.D: 1981, 'On the Phenomenon of Chains in the Photospheric Granulation', *Soln. Dannye 1981 Byull.*, No. 10, p. 101

Parfinenko, L.D: 1985, 'On Some Peculiarities of the Fine Structure of Radial Velocities of Solar Granulation', *Soln. Dannye 1985 Byull.*, No. 8, p. 68

Parker, E.N.: 1963, 'Kinematical Hydromagnetic Theory and Its Application to the Low Solar Photosphere', *Astrophys. J.* **138**, 552

Parker, E.N.: 1965, 'Some Remarks on the Influence of the Granulation on the Magnetic Field', in R. Lüst (ed.), *Stellar and Solar Magnetic Fields, IAU Symp. 22*, N. Holland, Amsterdam, p. 232

Parker, E.N.: 1974, 'Hydraulic Concentration of Magnetic Fields in the Solar Photosphere. I: Turbulent Pumping', *Astrophys. J.* **189**, 563

Parker, E.N.: 1976, 'Hydraulic Concentration of Magnetic Fields in the Solar Photosphere. III: Fields of One or Two Kilogauss', *Astrophys. J.* **204**, 259

Parnell, R.L., *see* Beckers, J.M. 1969

Parvey, M.I. and Musman, S.: 1971, 'Bright-Dark Asymmetry in Solar Granulation', *Solar Phys.* **18**, 385

Pecker, J.C., *see* Fang, C. *et al.*1984

Peckover, R.S. and Weiss, N.O.: 1978, 'On the Dynamic Interaction between Magnetic Fields and Convection', *Monthly Notices Roy. Astron. Soc.* **182**, 189

Peri, M.L., Smithson, R.C., Acton, D.S., Frank, Z.A., and Title, A.M.: 1989, 'Active Optics, Anoisoplanatism and the Correction of Astronomical Images', *In These Proceedings*, p. 77

Pierce, A.K., *see* Graves, J.E. 1986

Pierce, A.K., *see* LoPresto, J.C. 1985

Pierce, A.K., *see* McMath, R.R. *et al.* 1955

Plaskett, H.H.: 1936, 'Solar Granulation', *Monthly Notices Roy. Astron. Soc.* **96**, 402

Plaskett, H.H.: 1954, 'Motions in the Sun at the Photospheric Level. V: Velocities of Granules and of Other Localized Regions', *Monthly Notices Roy. Astron. Soc.* **114**, 251

Plaskett, H.H.: 1955, 'Physical Conditions in the Solar Photosphere', in A. Beer (ed.), *Vistas in Astronomy* 1, Pergamon Press, London, p. 637

Pope, T., *see* Title, A.M. *et al.* 1987

Poulakos, C., *see* Xanthakis, J. 1985

Pravdyuk, L.M., Karpinsky, V.N., and Andreiko, A.V.: 1974, 'Distribution of the Brightness Amplitude in Photospheric Granulation', *Soln. Dannye 1974 Byull.*, No. 2, p. 70

Pravdjuk, L.M., *see* Karpinsky, V.N. 1973

Pravdjuk, L.M., *see* Krat, W.A. 1960

Priest, E.R., *see* Nakagawa, Y. 1972

Prokakis, T.J., *see* Macris, C.J. 1962

Prokakis, T.J., *see* Macris, C.J. 1963

Prokakis, T., *see* Dialetis, D. *et al.* 1985

Prokakis, T., *see* Dialetis, D. *et al.* 1985

Prokakis, T., *see* Dialetis, D. *et al.* 1986

Prokakis, T., *see* Dialetis, D. *et al.* 1988

Prokakis, T., *see* Dialetis, D. *et al.* 1989

Prokakis, T., *see* Macris, C. *et al.* 1989

Punetha, L.M.: 1974, 'Fine Structure in Ca II K Line', *Bull. Astron. Inst. Czech.* **25**, 207

Rammacher, W. and Ulmschneider, P.: 1989, 'What Do the Mg II Lines Tell Us about Waves and Magnetic Fields', *In These Proceedings*, p. 589

Ramsey, H.E., Schoolman, S.A., and Title, A.M.: 1977, 'On the Size, Structure and Strength of the Small Scale Solar Magnetic Field', *Astrophys. J. Letters* **215**, L41

Ramsey, H.E., *see* Harvey, J.W. 1963

Ramsey, H.E., *see* Schoolman, S.A. 1976

Read, P.A.: 1958, 'Photographing the Solar Granulation. Two New and Unusual Methods', *Southern Stars* **17**, 116

Reeves, E.M., Foukal, P.V., Huber, M.C.E., Noyes, R.W., Schmahl, E.J., Timothy, J.G., Vernazza, J.E., and Withbroe, G.L.: 1974, 'Observations of the Chromospheric Network: Initial Results from the Apollo Telescope Mount', *Astrophys. J. Letters* **188**, L27

Reeves, R., *see* Title, A.M. *et al.* 1987

Rehse, R., *see* Title, A.M. *et al.* 1987

Reiling, H.: 1971, 'A Power Spectrum Analysis of Granular Intensity Fluctuations and Velocities', *Solar Phys.* **19**, 297

Reiss, G., *see* Durrant, C.J. *et al.* 1979

Reiss, G., *see* Mattig, W. *et al.* 1977

Richardson, R.S. and Schwarzschild, M.: 1950, 'On the Turbulent Velocities of Solar Granules', *Astrophys. J.* **111**, 351

Ricort, G. and Aime, C.: 1979, 'Solar Seeing and the Statistical Properties of the Photospheric Solar Granulation. III: Solar Speckle Interferometry', *Astron. Astrophys.* **76**, 324

Ricort, G., Aime, C., Deubner, F.-L., and Mattig, W.: 1981, 'Solar Granulation Study in Partial Eclipse Conditions Using Speckle Interferometric Techniques', *Astron. Astrophys.* **97**, 114

Ricort, G., Aime, C., and Fossat, E.: 1974, 'Statistical Analysis of a Plate Corrected for Photographic Noise', *Astron. Astrophys.* **37**, 105

Ricort, G., Aime, C., Roddier, C., and Borgnino, J.: 1981, 'Determination of Fried's Parameter r_0 Prediction of the Observed rms Contrast in Solar Granulation', *Solar Phys.* **69**, 223

Ricort, G., Borgnino, J., and Aime, C.: 1982, 'A Comparison between Estimations of Fried's Parameter r_0 Simultaneously Obtained by Measurements of Solar Granulation Contrast and of the Variance of Angle-of-Arrival Fluctuations', *Solar Phys.* **75**, 377

Ricort, G. 1975, *see* Aime, C. 1975

Ricort, G. 1980, *see* Aime, C. 1980

Ricort, G., *see* Aime, C. *et al.* 1975

Ricort, G., *see* Aime, C. *et al.* 1977

Ricort, G., *see* Aime, C. *et al.* 1977

Ricort, G., *see* Aime, C. *et al.* 1978

Ricort, G., *see* Aime, C. *et al.* 1978

Ricort, G., *see* Aime, C. *et al.* 1981

Ricort, G., *see* Aime, C. *et al.* 1985

Ricort, G., *see* Aime, C. *et al.* 1989

Rigal, G., *see* Bernière, G. *et al.* 1962

Righini, A. and Schröter, E.H.: 1989 'Report on the Position and Discussion Session on Solar Data', *In These Proceedings*, p. 295

Righini, A., *see* Cavallini, F. *et al.* 1986

Righini, A., *see* Cavallini, G. *et al.* 1987

Righini, A., *see* Cavallini, G. *et al.* 1987

van Rijsbergen, R., *see* Namba, O. 1977

Roberti, G., *see* Caccin, B. *et al.* 1981

Roberti, G., *see* Gomez, M.T. *et al.* 1987

Roberti, G., *see* Marmolino, C. *et al.* 1985

Roberti, G., *see* Marmolino, C. *et al.* 1987

Roberti, G., *see* Marmolino, C. *et al.* 1988

Roberts, B.: 1979, 'Spicules: The Resonant Response to Granular Buffeting?', *Solar Phys.* **61**, 23

Roddier, C., *see* Aime, C. *et al.* 1978

Roddier, C., *see* Ricort, G. *et al.* 1981

Roddier, F.: 1965, Étude a Haute Résolution de Quelques Raies de Fraunhofer par Observation de la Résonance Optique d'un Jet Atomique. II: Résultat des Observations Solaires', *Ann. Astrophys.* **28**, 478

Roddier, F., see Aime, C. et al. 1979

Rösch, J.: 1955, 'Données Morphologiques Nouvelles sur la Granulation Photosphérique Solaire', *Compt. Rend. Acad. Sci. Paris* **240**, 1630

Rösch, J.: 1956, 'Photographie à Cadence Rapide del la Photosphère et des Taches Solaires', *Compt. Rend. Acad. Sci. Paris* **243**, 478

Rösch, J.: 1959, 'Observations sur la Photosphère Solaire. I: Technique des Observations Photographiques (objectif de 23 cm)', *Ann. Astrophys.* **22**, 571

Rösch, J.: 1959, 'Observations sur la Photosphère Solaire. II: Numération et Photométrie Photographique des Granules dans le Domaine Spectral 5900-6000 Å', *Ann. Astrophys.* **22**, 584

Rösch, J.: 1961, 'Preview on Granulation - Observational Studies. Observations from the Pic du Midi', in R.N. Thomas (ed.), *Aerodynamic Phenomena in Stellar Atmospheres, IAU Symp. 12, Nuovo Cim. Suppl. Ser. 10* **22**, p. 313

Rösch, J. and Hugon, M.: 1959, 'Sur l'Evolution dans le Temps de la Granulation Photosphérique', *Compt. Rend. Acad. Sci. Paris* **249**, 625

Rösch, J., see Macris, C.J. 1983

Rösch, J., see Carlier, A. et al. 1968

Rösch, J., see Macris, C.J. et al. 1984

Rossi, P., see Kalkofen, W. et al. 1989

Rottman, G.J., see Bruner, E.C. et al. 1976

Roudier, T. and Muller, R.: 1986, 'Structure of the Solar Granulation', *Solar Phys.* **107**, 11

Roudier, T., Coupinot, G., Hecquet, J., and Muller, R.: 1985, 'Filtrage Numérique des Clichés à Faible Contraste avec Possibilité d'Analyse Structurale pa Sementation en Domaines Connexee', *J. Opt.* **16**, 107

Roudier, T., see Muller, R. 1984

Roudier, T., see Muller, R. 1984

Roudier, T., see Muller, R. 1984

Roudier, T., see Muller, R. 1985

Roudier, T., see Muller, R. 1985

Roudier, T., see Keil, S.L. et al. 1989

Roudier, T., see Macris, C.J. et al. 1984

Roudier, T., see Muller, R. et al. 1989

Ruland, F., see Holweger, H. et al. 1978

Rush, J.H., see Stuart, F.E. 1954

Rutten, R., see Bruls, J.H.M.J. et al. 1989

Rutten, R.J., see Gomez, M.T. et al. 1989

Saar, S.H., see Bruning, D.H 1989

Sarris, E., see Dialetis, D. et al. 1985

Sarris, E., see Dialetis, D. et al. 1986

Scharmer, G.B.: 1988, 'High Resolution Observations from La Palma: Techniques and First Results', *In These Proceedings*, p. 161

Scharmer, G.B., see Brandt, P.N. et al. 1989

Scharmer, G.B., see Lites, B.W. et al. 1989

Schatzman, E.: 1954, 'Sur une Nouvell Théorie de la Granulation Solaire. II: Propagation dans une Atmosphere Isotherme', *Bull. Cl. Sci. Acad. Roy. Belgique* **40**, 139

Schlebbe, H., *see* Mattig, W. 1974

Schlosser, W. and Klinkmann, W.: 1974; 'Laseroptische Analyze der Sonnengranulation', *Astron. Astrophys.* **32**, 29

Schmahl, E.J., *see* Reeves, E.M. *et al.* 1974

Schmidt, H.U., *see* Meyer, F. 1967

Schmidt, H.U., Simon, G.W., and Weiss, N.O.: 1985, 'Buoyant Magnetic Flux Tubes. II: Three Dimensional Behaviour in Granules and Supergranules', *Astron. Astrophys.* **148**, 191

Schmidt, W., Deubner, F.-L., Mattig, W., and Mehltretter, J.P.: 1979, 'On the Center to Limb Variation of the Granular Brightness Fluctuations', *Astron. Astrophys.* **75**, 223

Schmidt, W., Grossmann-Doerth, U., and Schröter, E.H.: 1988, 'The Solar Granulation in the Vicinity of Sunspots', *Astron. Astrophys.* **197**, 306

Schmidt, W., Knölker, M., and Schröter, E.H.: 1981, 'RMS Value and Power Spectrum of the Photospheric Intensity Fluctuations', *Solar Phys.* **73**, 217

Schmidt, W., *see* Durrant, C.J. *et al.* 1979

Schmidt, W., *see* Durrant, C.J. *et al.* 1983

Schneider, G., *see* Strebel, H. 1935

Schoolman, S.A. and Ramsey, H.E.: 1976, 'The Dark Component of the Photospheric Network', *Solar Phys.* **50**, 25

Schoolman, S.A., *see* Ramsey, H.E. *et al.* 1977

Schröter, E.H.: 1962, 'Einige Beobachtungen und Messungen an StratoskopI-Negativen', *Z. Astrophys.* **56**, 183

Schröter, E.H.: 1963, 'Die Sonnengranulation. I', *Sterne Weltraum* **2**, 100

Schröter, E.H.: 1963, 'Die Sonnengranulation. II', *Sterne Weltraum* **2**, 127

Schröter, E.H., *see* Righini, A. 1989

Schröter, E.H., *see* Münzer, H. *et al.* 1989

Schröter, E.H., *see* Schmidt, W. *et al.* 1979

Schröter, E.H., *see* Schmidt, W. *et al.* 1979

Schüssler, M.: 1984, 'The Interchange Instability of Small Flux Tubes', *Astron. Astrophys.* **140**, 453

Schüssler, M., *see* Grossmann-Doerth, U. *et al.* 1989

Schwarzschild, M.: 1948, 'On Noise Arising from the Solar Granulation', *Astrophys. J.* **107**, 1

Schwarzschild, M.: 1959, 'Photographs of the Solar Granulation Taken from the Stratosphere', *Astrophys. J.* **130**, 345

Schwarzschild, M., *see* Bahng, J. 1961

Schwarzschild, M., *see* Bahng, J. 1961

Schwarzschild, M., *see* Frenkiel, F.N. 1952

Schwarzschild, M., *see* Frenkiel, F.N. 1955

Schwarzschild, M., *see* Gaustad, J. 1960

Schwarzschild, M., *see* Harvey, J.W. 1975

Schwarzschild, M., *see* Richardson, R.S. 1950

Schwarzschild, M., *see* Ledoux, P. *et al.* 1961

Sellers, F.J.: 1943, 'Surface Granulation on the Sun', *Observatory* **65**, 77

Semel, M: 1962, 'Sur la Détermination du Champ Magnétique de la Granulation Solaire', *Compt. Rend. Hebdomadaires Séances Acad. Sci.* **254**, 3978

Secchi, A.: 1875, *Le Soleil, second edition*, Vol. 1, Gauthier-Villars, Paris

Servajean, R.: 1961, 'Contribution à lÉtude de la Cinématique de la Matière dans les Taches et la Granulation Solaires', *Ann. Astrophys.* **24**, 1

Severino, G., *see* Marmolino, C. 1981

Severino, G., *see* Nesis, A. 1987

Severino, G., *see* Caccin, B. *et al.* 1981

Severino, G., *see* Gomez, M.T. *et al.* 1987

Severino, G., *see* Gomez, M.T. *et al.* 1988

Severino, G., *see* Marmolino, C. *et al.* 1985

Severino, G., *see* Marmolino, C. *et al.* 1987

Severino, G., *see* Marmolino, C. *et al.* 1988

Severny, A.B.: 1965, 'The Nature of Solar Magnetic Fields: The Fine Structure of the Field', *Soviet Astron.* **9**, 171

Seykora, E.J.: 1985, 'Observations of Very Low Contrast White Light Solar Structures Utilizing Differential Photometry', *Solar Phys.* **99**, 39

Sheeley, N.R., Jr.: 1969, 'The Evolution of the Photospheric Network', *Solar Phys.* **9**, 347

Sheeley, N.R., Jr. and Bhatnagar, A.: 1971, 'The Reduction of the Solar Velocity Field into Its Oscillatory and Slowly Varying Components', *Solar Phys.* **18**, 195

Sheeley, N.R., Jr. and Bhatnagar, A.: 1971, 'Measurements of the Oscillatory and Slowly Varying Components of the Solar Velocity Field', *Solar Phys.* **18**, 379

Shine, R.A., *see* Brandt, P.N. *et al.* 1989

Shine, R.A., *see* Bruner, E.C. *et al.* 1976

Shine, R.A., *see* Simon, G.W. *et al.* 198

Shine, R., *see* Title, A.M. *et al.* 1987

Shine, R.A., *see* Title, A.M. *et al.* 1987

Shine, R.A., *see* Title, A.M. *et al.* 198

Shine, R.A., *see* Title, A.M. *et al.* 1988

Shine, R.A., *see* Title, A.M. *et al.* 1989

Shpitalnaya, A.A., *see* Krat, V.A. 1974

Siedentoph, H: 1933, 'Konvektion in Sternatmosphären. I', *Astron. Nachr.* **247**, 297

Siedentoph, H: 1941, 'Sonnengranulation und Zeellulare Konvektion', *Vierteljahressehriff* **76**, 185

Simon, G.W.: 1966, 'On the Correlation between Granule and Supergranule Intensity Fields', *Astrophys. J.* **145**, 411

Simon, G.W.: 1967, 'Observations of Horizontal Motions in Solar Granulation: Their Relation to Supergranulation', *Z. Astrophys.* **65**, 345

Simon, G.W., Title, A.M., Topka, K.P., Tarbell, T.D., Shine, R.A., Ferguson, S.H., Zirin, H., and the SOUP Team: 1988, 'On the Relation between Photospheric Flow Fields and the Magnetic Field Distribution on the Solar Surface', *Astrophys. J.* **327**, 964

Simon, G.W., November, L.J., Ferguson, S.H., Shine, R.A., Tarbell, T.D., Title, A.M., Topka, K.P., and Zirin, H.: 1989, 'Details of Large Scale Solar Motions Revealed by Granulation Test Particles', *In These Proceedings*, p. 371

Simon, G.W. and Weiss, N.O.: 1989, 'A Simple Model of Mesogranular and Supergranular Flows', *In These Proceedings*, p. 595

Simon, G.W., *see* November, L.J. 1988

Simon, G.W., *see* November, L.J. *et al.* 1981

Simon, G.W., *see* November, L.J. *et al.* 1982

Simon, G.W., *see* November, L.J. *et al.* 1987

Simon, G.W., *see* Schmidt, H.U. *et al.* 1985

Simon, G.W., *see* Title, A.M. *et al.* 1986

Simon, G.W., *see* Title, A.M. *et al.* 1987

Simon, G.W., *see* Title, A.M. *et al.* 1987

Sivaraman, K.R. and Venkitachalam, P.P.: 1977, 'Intensity Fluctuations in the Solar Chromosphere', *Kodaikanal Obs. Bull. Ser. A* **2**, 34

Skumanich, A.: 1955, 'On Bright-Dark Symmetry of Solar Granulation', *Astrophys. J.* **121**, 404

Slaughter, C. and Wilson, A.M.: 1972, 'Space and Time Variations of the Solar Na D Line Profiles', *Solar Phys.* **24**, 43

Slaughter, C., *see* Lena, P.J. *et al.* 1968

Smaldone, L.A., *see* Caccin, B. *et al.* 1981

Smithson, R.C., *see* Peri, M.L. *et al.* 1989

Snider, J.L.: 1970, 'Atomic Beam Study of the Solar 7699 Å Potassium Line and the Solar Gravitational Red Shift', *Solar Phys.* **12**, 352

Sobolev, V.M.: 1972, 'On the Equivalent Widths of the Spectral Lines of Solar Granulation Elements', *Soln. Dannye 1972 Byull.*, No. 11, p. 95

Sobolev, V.M.: 1974, 'On the Equivalent Widths of the Spectral Lines of Solar Granulation Elements, II', *Soln. Dannye 1974 Byull.*, No. 2, p. 55

Sobolev, V.M.: 1975, 'On the Profile Characteristics of the Fraunhofer Lines in the Granules and Intergranular Regions', *Soln. Dannye 1975 Byull.*, No. 6, p. 68

Sobolev, V.M., *see* Gerbil'skij, M.G. 1978

Sobolev, V.M., *see* Korduba, B.M. *et al.* 1976

Sobolev, V.M., *see* Korduba, B.M. *et al.* 1978

Sobolev, V.M., *see* Olijnyk, P.A. *et al.* 1979

Sofia, S., *see* Chan, K.L. *et al.* 1982

Soltau, D.: 1989, 'The Status of the Latest German Solar Facility on Tenerife', *In These Proceedings*, p. 17

Souffrin, P.: 1963, 'Remarques sur les Mouvements Transitoires des Systèmes Instables. Application a un Modèle de Granulation Solaire', *Ann. Astrophys.* **26**, 170

Spence, G.E., *see* Dunn, R.B. 1984

Spiegel, E.A.: 1961, 'Preview on Granulation - Observational Studies. The Princeton Balloon Observations', in R.N. Thomas (ed.), *Aerodynamic Phenomena in Stellar Atmospheres, IAU Symp. 12, Il Nuovo Cimento Sup. Ser. 10* **22**, p. 319

Spiegel, E.A., *see* Ledoux, P. *et al.* 1961

Spruit, H.C.: 1984, 'Interaction of Flux Tubes with Convection', in S.L. Keil (ed.), *Small Scale Dynamical Processes in Quiet Stellar Atmospheres*, National Solar Observatory/Sacramento Peak, Sunspot, p. 249

Staude, J.: 1978, 'Models of Heat Flux in the Subphotospheric Layers of Sunspots and the Interpretation of Umbral Granulation', *Bull. Astron. Inst. Czech.* **29**, 71

Staude, J., *see* Zhugzhda, Y.D. *et al.* 1989

Steffen, M.: 1987, 'A 2D Study of Compressible Granular Flow and Predicted Spectroscopic Properties', in E.H. Schröter, M. Vázquez, and A.A. Wyller (eds.), *The Role of Fine-Scale Magnetic Fields on the Structure of the Solar Atmosphere*, Cambridge University Press, Cambridge, p. 47

Steffen, M.: 1988, 'Interaction of Convection and Oscillations in the Solar Atmosphere', in J. Christensen-Dalsgaard, and S. Frandsen (eds.) *Advances in Helio- and Asteroseismology, IAU Symp. 123*, D. Reidel Publ. Co., Dordrecht, p. 379

Steffen, M.: 1989, 'Spectrocopic Properties of Solar Granulation Obtained from 2-D Numerical Simulations', *In These Proceedings*, p. 425

Steffen, M. and Gigas, D.: 1985, 'Solar Granulation: Numerical Simulation and Resulting Disc-Center Line Profiles', in H.U. Schmidt (ed.), *Theoretical Problems in High Resolution Solar Physics*, Max Planck Institut Publ. MPA **212**, p. 95

657

Steffen, M. and Muchmore, D.: 1988, 'Can Granular Fluctuations in the Solar Photosphere Produce Temperature Inhomogeneities at the Height of the Temperature Minimum?', *Astron. Astrophys.* **193**, 281

Stein, R.F., Nordlund, Å., and Kuhn, J.R.: 1989, 'Convection and Waves', *In These Proceedings*, p. 381

Stein, R.F., *see* Nordlund, Å. 1989

Steshenko, N.V.: 1960, 'On the Determination of Magnetic Fields of Solar Granulation', *Izv. Krym. Astrofiz. Obs.* **22**, 49

Strebel, H.: 1933, 'Beitrag zum Problem der Sonnengranulation', *Z. Astrophys.* **6**, 313

Strebel, H. and Schneider, G.: 1935, 'Neue Strhlungsmessungen auf der Sonnenphotosphäre zur Exakten Definition und Bestimmung der Strhlung der Granula, Fackeln, des Hintergrundes sowie der Umbra', *Astron. Nachr.* **254**, 169

Stuart, F.E. and Rush, J.H.: 1954, 'Correlation Analyses of Turbulent Velocities and Brightness of the Photospheric Granulation', *Astrophys. J.* **120**, 245

Suda, J.: 1976, 'Granulation in Sunspot Umbrae', *Fiz. Soln. Pyaten. Moskva, Nauka*, p. 42

Suda, J., *see* Bumba, V. *et al.* 1975

Suemoto, Z., Hiei, E., and Nakagomi, Y.: 1987, 'Bright Threads in the Inner Wing of Solar Ca II K Line', *Solar Phys.* **112**, 59

Swihart, T.L., *see* Margrave, T.E., Jr. 1969

Swihart, T.L., *see* Margrave, T.E., Jr. 1969

Tappere, E.J., *see* Bray, R.J. *et al.* 1974

Tappere, E.J., *see* Bray, R.J. *et al.* 1976

Tarbell, T.D., *see* Brandt, P.N. *et al.* 1989

Tarbell, T.D., *see* November, L.J. *et al.* 1987

Tarbell, T.D., *see* Simon, G.W. *et al.* 1988

Tarbell, T.D., *see* Simon, G.W. *et al.* 1989

Tarbell, T.D., *see* Title, A.M. *et al.* 1986

Tarbell, T., *see* Title, A.M. *et al.* 1987

Tarbell, T.D., *see* Title, A.M. *et al.* 1987

Tarbell, T.D., *see* Title, A.M. *et al.* 1987

Tarbell, T.D., *see* Title, A.M. *et al.* 1989

Tarbell, T.D., *see* Title, A.M. *et al.* 1989

Tarbell, T.D., *see* Title, A.M. *et al.* 1989

Tavastsherna, K.S., *see* Krat, V.A. *et al.* 1980

Temesváry, S.: *see* Biermann, L. *et al.* 1959

Teske, R.G.: 1974, 'Two-Dimensional Spatial Power Spectra of Photospheric Velocity Fluctuations', *Solar Phys.* **39**, 363

Teske, R.G. and Elste, G.H.: 1979, 'Dependence of the Correlation of Small Scale Photospheric Structures upon Resolution', *Solar Phys.* **62**, 241

Thiessen, G.: 1955, 'Zur Sonnengranulation', *Z. Astrophys.* **35**, 237

Thüring, B.: 1937, 'Photometrische Untersunchung eines UV-Bildes der Sonnengranulation', *Astron. Nachr.* **264**, 117

Timothy, J.G., *see* Reeves, E.M. *et al.* 1974

Title, A.: 1985, 'High Resolution Solar Observations', in R. Muller (ed.), *High Resolution in Solar Physics*, Proc. 8th IAU European Regional Meeting, *Lecture Notes in Physics* **233**, Springer-Verlag, Berlin, p. 51

Title, A.: 1989, 'An Overview of the Orbiting Solar Laboratory', *In These Proceedings*, p. 29

Turon, P.: 1975, 'Inhomogeneous Model of the Photosphere', *Solar Phys.* **41**, 271

Turon, P.J. and Léna, P.: 1973, 'First Observations of the Granulation at 1.65 μ, Center to Limb Variation of the Contrast', *Solar Phys.* **30**, 3

Uberoi, M.S.: 1955, 'Investigation of Turbulence in the Solar Atmosphere', *Astrophys. J.* **121**, 400

Uberoi, M.S.: 1955, 'On the Solar Granules', *Astrophys. J.* **122**, 466

Uitenbroek, H., *see* Bruls, J.H.M.J. *et al.* 198

Ulmschneider, P., *see* Rammacher, W. 1989

Unno, W., *see* Goldberg, L. *et al.* 1960

Unno, W., *see* Kondo, M.-A. 1983

Unso:old, A.: 1930, 'Konvektion in der Sonnenatmosphα:are', *Z. Astrophys.* **1**, 138

Vainstein. S.I.: 1979, 'A Possible Explanation for the Fine Structure of Magnetic Fields on the Sun', *Soviet Astron.* **23**, 734

Vainstein, S.I., Kuklin, G.V., and Maksimov, V.P.: 1977, 'On a Possible Mechanism of Solar Faculae Heating', *Solar Phys.* **53**, 15

Vandakurov, Yu.V.: 1975, 'On Convection in the Sun', *Solar Phys.* **40**, 3

Vassiljeva, G.Y.: 1967, 'Photoelectric Photometry of Solar Granulation in Several Regions of the Continuum', *Solar Phys.* **1**, 16

Vassiljeva, G.: 1968, 'On the Difference between the Photometric Inhomogeneities on the Solar Surface in Two Colors', *Solar Phys.* **4**, 300

Vázquez, M., *see* Collados, M. 1987

Vázquez, M., *see* Bonet, J.A. *et al.* 198

Vázquez, M., *see* Bonet, J.A. *et al.* 1988

Vázquez, M., *see* Collados, M. *et al.* 1986

Venkatakrishnan, P. and Hasan, S.S.: 1981, 'Time-Dependent Interaction of Granules with Magnetic Flux Tubes', *J. Astrophys. Astron.* **2**, 133

Venkitachalam, P.P., *see* Sivaraman, K.R. 1977

Vermue, J. and de Jager, C.: 1979, 'Macro- and Micro-Turbulent Filter Functions for Weak Lines in Stellar Atmospheres', *Astrophys. Space Sci.* **61**, 129

Vernazza, J.E., *see* Reeves, E.M. *et al.* 1974

Vigneau, J., *see* Muller, R. *et al.* 1989

Vikanes, F., *see* Andersen, B.N. *et al.* 1985

Vince, I. and Dimitrijevič, M.S.: 1989, 'Pressure Broadening and Solar Spectral Line Bisectors', *In These Proceedings*, p. 93

Vyal'shin, G.F.: 1985, 'On the Magnetic Field of Solar Granulation', *Astron. Tsirk.*, No. 1405, p. 7

Weiss, N.O.: 1978, 'Small Scale Magnetic Fields and Convection in the Solar Photosphere', *Monthly Notices Roy. Astron. Soc.* **183**, 63P

Weiss, N.O.: 1984, 'Problems of Flux Tube Formation', in S.L. Keil (ed.), *Small Scale Dynamical Processes in Quiet Stellar Atmospheres*, National Solar Observatory/Sacramento Peak, Sunspot, p. 287

Weiss, N.O.: 1985, 'Theoretical Interpretation of Small-Scale Solar Features', in R. Muller (ed.), *High Resolution in Solar Physics*, Proc. 8th IAU European Regional Meeting, *Lecture Notes in Physics* **233**, Springer-Verlag, Berlin, p. 217

Weiss, No.O.: 1989, 'Time-Dependent Compressible Magnetoconvection', *In These Proceedings*, p. 471

Weiss, N.O., *see* Peckover, R.S. 1978

Weiss, N.O., *see* Schmidt, H.U. *et al.* 1985

Weiss, N.O., *see* Simon, G.W. 1989

Weisshaar, E., *see* Grossmann-Doerth, U. *et al.* 1989

Werner, W., *see* Kneer, F.J. *et al.* 1980

White, O.R., *see* Bruner, E.C. *et al.* 1976

Whitney, C.A.: 1958, 'Granulation and Oscillations of the Solar Atmosphere', *Smithsonian Contrib. Astrophys.* **2**, 365

Wiehr, E. and Kneer, F.: 1988, 'Spectroscopy of the Solar Photosphere with High Spatial Resolution', *Astron. Astrophys.* **195**, 310

Wiehr, E., *see* Nesis, A. *et al.* 1987

Wiehr, E., *see* Kneer, F.J. 1989

Wiesmeier, A. and Durrant, C.J.: 1981, 'The Analysis of Solar Limb Observations. I: Restoration of Data in a Tilted Reference Frame', *Astron. Astrophys.* **104**, 207

Wilson, A.M., *see* Slaughter, C. 1972

Wilson, P.R.: 1962, 'The Application of the Equation of Transfer to the Interpretation of Solar Granulation', *Monthly Notices Roy. Astron. Soc.* **123**, 287

Wilson, P.R.: 1963, 'An Interpretation of Edmonds' Granulation Data', *Astrophys. J.* **137**, 606

Wilson, P.R.: 1964, 'Photospheric Structure and RMS Fluctuation Data', *Astrophys. J.* **140**, 1148

Wilson, P.R.: 1969, 'Temperature Fluctuations and Convective Modes in the Solar Photosphere', *Proc. Astron. Soc. Australia* **1**, 195

Wilson, P.R.: 1969, 'Temperature Fluctuations in the Solar Photosphere', *Solar Phys.* **6**, 364

Wilson, P.R.: 1969, 'On Granulation Models', *Solar Phys.* **8**, 20

Wilson, P.R., *see* Cannon, C.J. 1970

Wilson, P.R., *see* Cannon, C.J. 1971

Withbroe, G.L., *see* Reeves, E.M. *et al.* 1974

Wittmann, A.: 1979, 'Observations of Solar Granulation - A Review', in F.-L. Deubner, C.J. Durrant, M. Stix, and E.H. Schröter (eds.), *Small Scale Motions on the Sun*, Keipenheuer Inst., No. 179, Freiburg, p. 29

Wittmann, A.: 1981, 'Balloon-Borne Imagery of the Solar Granulation. III: Digital Analysis of a White Light Time Series', *Astron. Astrophys.* **99**, 90

Wittmann, A.: 1981, 'Digitale Bildverarbeitung und Photometrische Analyse einer Zeitserie von Ballonaufnahmen der Sonnengranulation', *Mitt. Astron. Ger,* No. 52, p. 165

Wittmann, A. and Mehltretter, J.P.: 1977, 'Balloon-Borne Imagery of the Solar Granulation. I: Digital Image Enhancement and Photometric Properties', *Astron. Astrophys.* **61**, 75

Wittmann, A. and Mehltretter, J.P.: 1977, 'Computer-Processed Granulation Pictures of Project "Spectro-Stratoscope"', *Mitt. Astron. Ges.,* No. 42, p. 114

Wlérick, G.: 1957, 'Studies of Solar Granulation. I: The Statistical Interpretation of Granule Structure from One Dimensional Microphotometer Tracings', *Smithsonian Contrib. Astrophys.* **2**, 25

Wöhl, H. and Nordlund, Å.: 1985, 'A Comparison of Artifical Solar Granules with Real Solar Granules', *Solar Phys.* **97**, 213

Wöhl, H., *see* Tönjes, K. 1982

Wöhl, H., *see* Bonet, J.A. *et al.* 1988

Wöhl, H., *see* Münzer, H. *et al.* 1988

Wolff, C.L.: 1973, 'What is the Horizontal Scale of the 5-min Oscillations?', *Solar Phys.* **32**, 31

Wolff, C.L., *see* Chan, K.L. *et al.* 1982

Wolfson, C.J, *see* Title, A.M. *et al.* 1989

Wooley, R.v.d.R: 1943, 'Note on Convection in the Sun', *Monthly Notices Roy. Astron. Soc.* **103**, 191

Xanthakis, J. and Poulakos, C.: 1985, 'Long and Short Term Variation of the 10.7 cm Solar Flux. The Photospheric Granules and the Zürich Numbers', *Astrophys. Space Sci.* **111**, 179

Yackovich, F.H., *see* Keil, S.L. 1981

Yudina, I.V.: 1974, 'The Contrast of Granulation in Different Regions of the Continuum', *Soln. Dannye 1974 Byull.*, No. 11, p. 71

Yudina, I.V.: 1974, 'Photoelectric Photometry of the Solar Granulation in Two Regions of the Continuum', *Soln. Dannye 1974 Byull.*, No. 2, p. 107

Zachariadis, T.G., *see* Alissandrakis, C.E. *et al.* 1982

Zhugzhda, Y.D.: 1983, 'Nonadiabatic Oscillations in an Isothermal Atmosphere', *Soviet Astron. Letters* **9**, 329

Zhugzhda, Y.D.: 1989, 'Temperature Waves and Solar Granulation', *In These Proceedings*, p. 219

Zhugzhda, Y.D., Staude, J., and Locans, V.: 1989, 'Probing of Sunspot Umbral Structure by Oscillations', *In These Proceedings*, p. 601

Zirin, H., *see* Simon, G.W. *et al.* 1989

Zirin, H. *see* Simon, G.W. *et al.* 1988

Zirin, H., *see* Title, A.M. *et al.* 1987

Zirker, J.B., *see* Dunn, R.B. 1973

Zirker, J.B., *see* Title, A.M. *et al.* 1987

Zwaan, C.: 1985, 'Atmospheric Fine Structure as a Probe for the Solar Interior', in R. Muller (ed.), *High Resolution in Solar Physics*, Proc. 8th IAU European Regional Meeting, *Lecture Notes in Physics* **233**, Springer-Verlag, Berlin, p. 263